Lecture Notes in Computer Science 9200

Commenced Publication in 1973
Founding and Former Series Editors:
Gerhard Goos, Juris Hartmanis, and Jan van Leeuwen

More information about this series at http://www.springer.com/series/7407

Narciso Martí-Oliet · Peter Csaba Ölveczky
Carolyn Talcott (Eds.)

Logic, Rewriting, and Concurrency

Essays Dedicated to José Meseguer
on the Occasion of His 65th Birthday

 Springer

Editors
Narciso Martí-Oliet
Universidad Complutense de Madrid
Madrid
Spain

Carolyn Talcott
SRI International
Menlo Park, CA
USA

Peter Csaba Ölveczky
University of Oslo
Oslo
Norway

Cover illustration: It shows an *enso* (a circle drawn by a single brush motion) drawn by a renowned calligrapher in a Buddhist monastry near Kyoto, Japan.

Photograph on p. V: The photograph of the honoree was provided by José Meseguer himself. Used with permission.

ISSN 0302-9743   ISSN 1611-3349 (electronic)
Lecture Notes in Computer Science
ISBN 978-3-319-23164-8   ISBN 978-3-319-23165-5 (eBook)
DOI 10.1007/978-3-319-23165-5

Library of Congress Control Number: 2015947411

LNCS Sublibrary: SL1 – Theoretical Computer Science and General Issues

Springer Cham Heidelberg New York Dordrecht London

Printed on acid-free paper

Springer International Publishing AG Switzerland is part of Springer Science+Business Media
(www.springer.com)

José Meseguer

Preface

This Festschrift volume contains 28 refereed papers—including personal memories, essays, and regular research papers—by close collaborators and friends of José Meseguer to honor him on the occasion of his 65th birthday. These papers were presented at a symposium at the University of Illinois at Urbana-Champaign during September 23–25, 2015. The symposium also featured invited talks by Claude and Hélène Kirchner and by Patrick Lincoln. The foreword of this volume adds a brief overview of some of José's many scientific achievements followed by a bibliography of papers written by José.

We are grateful for having the opportunity to express our gratitude and admiration to José Meseguer by editing this volume and organizing the Festschrift symposium. Two of us were early PhD students of José who were inspired by the elegance and practicality of José's ideas, and one of has had the privilege of collaborating with José for many years. We have continued to be inspired by the wealth of ideas coming from José, and have to a certain extent lived on "crumbs from the great table of José," to paraphrase Aeschylus. Our scientific inspiration and admiration have since grown into a deep appreciation for José and into treasured lifelong friendships.

Each of the 35 submissions that we received were reviewed by between two and four expert reviewers. We would like to thank the authors for their contributions and the reviewers for their timely and very helpful reviews, which together have contributed to make this Festschrift worthy of José. We thank Claude and Hélène Kirchner and Patrick Lincoln for accepting our invitations to give invited talks reflecting on José's work from different perspectives. We thank Grigore Roşu for being responsible for the local arrangements, Donna Coleman for her invaluable help with administrative and a range of other matters, the Department of Computer Science at the University of Illinois at Urbana-Champaign for providing the workshop facilities, and Alberto Verdejo for preparing and organizing both José's bibliography included in this volume and the list of José's PhD students in the foreword. We are grateful to Alfred Hofmann for enthusiastically agreeing to publish this Festschrift in Springer's LNCS series.

Finally, we would like to congratulate José on his 65th birthday. We look forward to many more years of friendship and inspiring scientific leadership!

June 2015

Narciso Martí-Oliet
Peter Csaba Ölveczky
Carolyn Talcott

Organization

Program Committee

Narciso Martí-Oliet   Universidad Complutense de Madrid, Spain
Peter Csaba Ölveczky   University of Oslo, Norway
Carolyn Talcott   SRI International, USA

Local Organization Chair

Grigore Roşu   University of Illinois at Urbana-Champaign, USA

Additional Reviewers

Agha, Gul
Aguirre, Luis
Alpuente, María
Anastasio, Thomas J.
Asavoae, Mihail
Bae, Kyungmin
Braga, Christiano
Bruni, Roberto
Chen, Shuo
Dalla Preda, Mila
Durán, Francisco
Eker, Steven
Escobar, Santiago
Frutos Escrig, David
Futatsugi, Kokichi
Gonzalez Burgueño,
   Antonio

Gunter, Carl A.
Hanus, Michael
Hendrix, Joe
Hennicker, Rolf
Kapur, Deepak
Knapp, Alexander
Lluch Lafuente, Alberto
Lucanu, Dorel
Lucas, Salvador
Lynch, Christopher
López-Fraguas, Francisco
   Javier
Meadows, Catherine
Misra, Jayadev
Montanari, Ugo
Moreau, Pierre-Etienne
Mossakowski, Till

Mosses, Peter D.
Orejas, Fernando
Owre, Sam
Palomino, Miguel
Peña, Ricardo
Pita, Isabel
Rabe, Florian
Ravindran, Balaraman
Riesco, Adrián
Rosa-Velardo, Fernando
Roşu, Grigore
Rusu, Vlad
Santiago, Sonia
Sasse, Ralf
Stehr, Mark-Oliver
Verdejo, Alberto
Wirsing, Martin

Contents

José Meseguer:
Scientist and Friend Extraordinaire

Narciso Martí-Oliet[1], Peter Csaba Ölveczky[2]([⊠]), and Carolyn Talcott[3]

[1] Facultad de Informática, Universidad Complutense de Madrid, Madrid, Spain
[2] Department of Informatics, University of Oslo, Oslo, Norway
peterol@ifi.uio.no
[3] Computer Science Laboratory, SRI International, Menlo Park, CA, USA

1 José's Origins and Positions

José was born in Murcia, Spain, in 1950, and obtained his PhD in Mathematics in 1975 from the University of Zaragoza with a dissertation on "Primitive Recursion in Monoidal Categories" advised by Michael Pfender, with José Luis Viviente and Roberto Moreno-Díaz as co-advisors. In February 1974, at the age of 23, José met Joseph Goguen at the First International Symposium on Category Theory Applied to Computation and Control in San Francisco. This meeting led to many years of close and successful collaboration, as well as a life-long friendship between the two, with their first joint paper [24] published in 1977.

After a year as an assistant professor in Zaragoza and a year as a researcher in Santiago de Compostela, José left Spain to become a postdoctoral researcher at UC Berkeley from 1977 to 1980. During these years José also had appointments at UCLA, where Goguen worked, and with the "ADJ" group at IBM Yorktown Heights that pioneered the initial algebra approach to abstract data types.

Joseph Goguen moved to the Computer Science Laboratory at SRI International in 1979, and in 1980 José became a computer scientist at SRI, occupying the office next door to another influential computer scientist, Leslie Lamport.

The SRI years were very productive and allowed José to focus on research in a world-class environment. However, except for advising those of us who were lucky enough to have our universities allow us to study under José, it was not possible for José to pursue his passion for teaching at SRI. He therefore left SRI in 2001 to become a professor at the University of Illinois at Urbana-Champaign, where he has been thriving ever since.

2 José's Research

José is a leading researcher with seminal contributions in many fields of theoretical computer science and beyond; from general logics to vision algorithms for robots. His work is characterized by innovation, conceptual elegance and rigor combined with practical applicability. Instead of attempting the futile exercise of trying to mention the various topics in which José has made profound contributions, we focus here on the core vision and a few cornerstones in José's research,

© Springer International Publishing Switzerland 2015
N. Martí-Oliet et al. (Eds.): Meseguer Festschrift, LNCS 9200, pp. 1–47, 2015.
DOI: 10.1007/978-3-319-23165-5_1

and refer to the bibliography included in this volume for a full list of his research papers and to José's own survey papers [41,42] for an overview of his research program and the applications of rewriting logic, respectively.

It is useful to start by separating *system specifications*, specifying how a systems works, and *property specifications*, specifying the properties that the system should satisfy. To formally verify whether a system satisfies its specification, the system specification must also be a mathematical object. Furthermore, system specifications should be executable, so that we can view specification and programming as mathematical modeling. Which leads us to José's main vision: *programming/modeling language design is logic design*. In many ways, José's work has been a quest for logics/languages that satisfy the seemingly mutually contradictory goals of being:

- mathematical;
- amenable to formal reasoning and verification;
- executable;
- expressive enough to deal with complex state-of-the-art systems; and
- simple and intuitive.

If language design is logic design, then what is a logic? A key contribution here is José's work on *general logics* [34], which gives a theory of what a logic and a declarative language are. General logics therefore provide a design space for logics, and hence for declarative languages. In addition, general logics can be used to define and reason about different concepts in a logic/language-independent way. For example, in joint work with Salvador Lucas, José has worked on defining and analyzing termination for *any declarative language* using general logics [32]; the notion of termination is nontrivial for, say, conditional rewrite systems, where the evaluation of the condition of a rewrite rule may loop forever even though the rewrite relation is Noetherian. But we digress.

Joseph Goguen was a pioneer in the development of *algebraic specifications*. However, many-sorted equational logics were not very good at dealing with partiality. A key innovation in Joseph's and José's quest for an expressive yet simple and elegant (or "lean and mean," as José writes in [41]) computational logic was the development in the early eighties of *order-sorted equational logic* [26] and *order-sorted rewriting* [21,28], supported by the algebraic specification languages OBJ2 [20] and OBJ3 [22], that increased the expressiveness of algebraic specifications and allowed them to deal with partiality.[1] Another important novel feature of OBJ2 and OBJ3 was the possibility of defining parameterized modules with *semantic* requirements, which makes possible a powerful discipline of parameterized programming.

In the late eighties and early nineties, José led the *rewrite rule machine* project, which sought to "produce a highly parallel computer for enormously efficient term rewriting" [2,23]. This project developed compilers and hardware prototypes which could execute rewriting more than a thousand times faster

---

[1] Coincidentally, the first research paper the second author ever read, as a M.Sc. student in Oslo, was the paper [26] on order-sorted equational logic.

than a single processor. In a sign of what was to come, the 1986 paper [23] realized that term rewriting is naturally concurrent and proposed "concurrent term rewriting as a model of computation."

A significant limitation of the OBJ languages was that they could not specify concurrent, and therefore nondeterministic, systems in an elegant way, since they were based on equational logic. A new computational logic was needed.

Inspired by his work on the rewrite rule machine and on the categorical semantics of Petri nets (with Ugo Montanari) [44] and linear logic (with Narciso Martí-Oliet) [33], José realized that the intuitive notion of *rewriting* is not just the operational counterpart of equational logic, but can be seen as a very simple and powerful computational logic in which a large number of concurrent systems, including concurrent object-based systems, can naturally be represented. Thus, José's most significant contribution, *rewriting logic*, was born, and first presented in 1990 [36]. To quote [41]: "The whole point of rewriting logic and Maude was to unify within a single logic and associated declarative language: (i) equational/functional programming; (ii) object-oriented programming; and (iii) concurrent/distributed programming." A crucial prerequisite for this new form of nonterminating and nonconfluent rewriting was the development in [35,37] of a model theory, using category theory, with initial models and soundness and completeness results for the new logic.

Although order-sorted equational logic is more powerful than the many-sorted version, it is nevertheless somewhat limited since it cannot define *semantic* (sub)sorts; this implied that meaningful expressions such as $\frac{4}{(3-1)}$ were not well-defined (since it is impossible to determine by syntactic means that $3-1$ belongs to a sort of *non-zero* numbers). A key innovation which significantly increased both the expressiveness and elegance of rewriting logic was the development of *membership equational logic* [5,40] as its equational sublogic. This logic supports the definition of semantic sorts using *sort constraints*, and has a very nice solution to problems with partiality that had bedeviled the algebraic specification community for so long: Expressions like $\frac{4}{0}$ and $\frac{4}{(3-1)}$ are "legal" terms of a given *kind*, and can therefore be reduced, giving these potentially meaningful expressions "the benefit of doubt," but only terms such as $\frac{4}{2}$ have sorts.

Specification in rewriting logic with membership equational logic as its equational sublogic, and the formal analysis of such rewriting logic specifications, have been supported by the language and tool *Maude* [8], developed by José in joint work with the "Maude team" and expertly implemented by Steven Eker, since the mid-nineties. We refer to the paper "Two Decades of Maude" in this volume for a brief history of the Maude language, and to the survey [42] for an overview of rewriting logic and Maude as well as their applications.

An early question was how to execute a possibly nondeterministic and non-terminating rewriting logic specification. To guide the execution of such models, one could of course have a fixed set of extra-logical *strategy constructs* whose interpreter would have to be hard-coded.

A main breakthrough and novelty in Maude was based on the work of José and Manuel Clavel which showed that rewriting logic is a *reflective logic*:

a rewriting logic theory T could be represented as a *term* $\overline{T}$ in another rewriting logic theory U which, furthermore, could simulate the rewrites in T [7,9,10]. Reflection made it possible to define (and reason about) different execution strategies for rewriting logic specifications *in rewriting logic itself*, as opposed to using extra-logical strategies. Equally importantly, reflection allows us to define and reason about any computable module operation using rewriting logic, instead of having to resort to a fixed repertoire of extra-logical structuring mechanisms for developing large specifications in a modular fashion. For example, if the term $\overline{T}$ has sort Module in U, then a module operation f, such as module union, can be defined in rewriting logic as a function f : Module $\times$ Module $\rightarrow$ Module. This allowed making the algebra of module operations user-definable, extensible, and specifiable within rewriting logic [11].

Reflection is supported in Maude with built-in data types for meta-representing terms and theories and with a number of descent functions providing efficient implementation of key metalevel functionality. Maude's reflective features have been absolutely crucial to the Maude project, since they allow anyone to define and implement extensions of the Maude language *in Maude* itself—in a much clearer way and with much less effort than if one had to implement it in, say, C++. Tools like Full Maude [8]—extending Maude with convenient syntax for object-oriented specification and powerful module operations—, the Real-Time Maude tool [50] extending Maude to real-time systems, the Maude-NPA state-of-the-art cryptanalysis tool [15], the JavaFAN analyzer for Java and JVM code [19], theorem proving and other analysis tools for Maude such as its coherence and Church-Rosser checkers [12], and Pathway Logic—a framework and tool for representing and analyzing (executable) models of cellular processes [14,30]—are all implemented in Maude using Maude's reflective features.

The success of rewriting logic and Maude, and the ease with which Maude can be extended and provide a modeling and analysis environment for other languages have inspired José and his colleagues to further extend the applicability of rewriting logic and Maude. We have already mentioned the extension to real-time systems [51,52]. Other extensions include probabilistic rewrite theories that extend rewriting logic to probabilistic systems [1,29], and the extension to stochastic hybrid systems [48] and to hybrid systems whose components influence each other's continuous behaviors [17].

Rewriting logic and Maude have shown their mettle on a large class of applications, including as a semantic framework and formal analysis tool for a wide array of models of concurrency and modeling and programming languages [39,46,47], as well as for formally modeling and analyzing large and complex state-of-the-art systems, which are typically beyond the scope of less expressive formal tools, such as: large network protocols of different kinds, cloud computing systems [13,27,31], biological systems [30,53], cyber-physical systems [3,18], web browser security [6] (which uncovered a number of subtle previously unknown security flaws in Internet Explorer), and so on.

On the property specification and verification side, Maude supports simulation, reachability analysis, and LTL model checking, and its extensions support verifying inductive theorems and checking critical properties such as being Church-Rosser and coherent [12]. Together with Miguel Palomino, Narciso Martí-Oliet, Kyungmin Bae, and others, José has also developed abstraction techniques for rewriting logic [4,45]. Together with Bae, Ölveczky, Mu Sun, Lui Sha, and others, José has developed a number of formal design patterns that reduce the modeling and verification problems for certain complex systems to much simpler ones [43]; the point is that the cost needed to verify the correctness of the pattern can be amortized over the many instances of the pattern.

More recently, José has focused on developing symbolic analysis techniques. One key idea has been to use narrowing modulo axioms to solve *existential reachability queries* $\exists \overline{x} \, . \, t(\overline{x}) \longrightarrow^* u(\overline{x})$, already envisioned as MaudeLog in 1992 [38]. Narrowing for reachability analysis in rewriting logic took off in work with Prasanna Thati [49]. Narrowing requires efficient unification algorithms, and the work with Santiago Escobar and Ralf Sasse [16] brought a breakthrough on finitary unification algorithms for Maude, which enables efficient narrowing modulo theories for large classes of specifications. Maude-NPA [15], the latest generation of the NRL Protocol Analyzer, uses such variant narrowing to perform backwards reachability analysis to check whether some initial states can be reached (backwards) from a set of compromised states given as state patterns containing variables.

In this brief foreword we cannot even mention the vast majority of the many truly innovative ideas that have come out of José's brilliant mind, and that are reflected in the more than 350 research papers he has written or coauthored. Let us nevertheless point out that Joseph Goguen and José's 1982 paper [25] on noninterference—which formalizes security properties restricting information flow, such as, for example, that commands of high-level users cannot be observed by lower-level users—remains one of the most cited papers in computer security.

3 José the Person and Scientist

We end this foreword with a few words about José the person and scientist.

José tries to keep the highest ethical and scientific standards—in his research and in his dealings with colleagues, students, and other people around him; new students are treated with the same courtesy as leading researchers. This is reflected in the fact that many former students continue to collaborate with José when they become more senior researchers. His scientific approach is best described by his dictum *"beauty is our business."* José takes you seriously and gives you a lot; he therefore also expects a serious effort from you in order for you to reach your maximal potential.

José is not afraid of venturing into domains far away from what one would think are his comfort zones; *au contraire*, he relishes the opportunity to clarify and impose rigor on other fields and to encounter new challenges that inspire (and necessitate) new theoretical developments. José is remarkably generous in sharing and discussing the many ideas emanating from his fertile mind, and is genuinely curious and appreciative of other opinions. This collaborative spirit is also witnessed by him having 132 coauthors listed on DBLP.

Despite being one of our generation's leading computer scientists, many of José's collaborators first and foremost consider José to be a remarkably loyal and true friend. José takes great interest in literature, theology, history, and philosophy. It might be less well known that José also is an excellent "fast food" cook. It is always a pleasure to have a dinner *chez Meseguer*, which may include a healthy treatise on some *culturilla* or advanced philosophy which—possibly aided by a glass or two of Spanish wine—easily leaves your brain spinning.

José's contagious intellectual enthusiasm is complemented by his enthusiasm for other aspects of life. We fondly remember him encouraging us to: hike with him over Half Dome in Yosemite, camp in the High Sierras, enjoy a special hot chocolate in a Barcelona café, taste for the first time sophisticated Japanese food, and organize a scientific workshop among the polar bears in Spitsbergen.

We are convinced that we speak on behalf of a large number of colleagues, friends, and others when we write: Dear José, for your valued friendship, for your research leadership and mentoring, for showing us how beautiful theoretical computer science can be, and for your remarkable scientific contributions, we congratulate you on your first 65 years and wish you a long, happy, and scientifically productive life!

4 José's PhD Students

This section lists every PhD student whose PhD thesis was supervised by José. As already mentioned, José had fairly limited possibility to advise graduate students until 2001. The majority of José's students have themselves become professors and have supervised PhD theses, which are also listed in the following "PhD tree" of José, which, for each student, gives the title of the dissertation, the department where the PhD was granted, and the month of the PhD defense.

➤ **Narciso Martí-Oliet**
   About Two Categorical Logics: Linear Logic and Order-Sorted Algebra
   Universidad Complutense de Madrid, Spain
   June 1991
   → **Isabel Pita Andreu**
      Rewriting-Logic-Based Techniques for the Formal Specification of Object-Oriented Systems
      Universidad Complutense de Madrid, Spain
      March 2003
   → **Alberto Verdejo**
      Maude as an Executable Semantic Framework
      Universidad Complutense de Madrid, Spain
      March 2003
   → **Adrián Riesco Rodríguez**
      Declarative Debugging and Heterogeneous Verification in Maude
      (with Alberto Verdejo)
      Universidad Complutense de Madrid, Spain
      June 2011

➤ **Manuel G. Clavel**
Reflection in Rewriting Logic: Metalogical Foundations and Metaprogramming Applications
Universidad de Navarra, Spain
February 1998

↪ **Marina Egea**
An Executable Formal Semantics for OCL with Applications to Model Analysis and Validation
Universidad Complutense de Madrid, Spain
November 2008

↪ **Miguel A. García de Dios**
Model-Driven Development of Secure Data-Management Applications
Universidad Complutense de Madrid, Spain
April 2015

➤ **Francisco Durán**
A Reflective Module Algebra with Applications to the Maude Language
Universidad de Málaga, Spain
April 1999

↪ **José Eduardo Rivera Cabaleiro**
On the Semantics of Real-Time Domain Specific Modelling Languages
(with Antonio Vallecillo)
Universidad de Málaga, Spain
October 2010

↪ **Manuel Roldán Castro**
Strategies for Guiding and Monitoring the Execution of Systems in Maude
Universidad de Málaga, Spain
September 2011

➤ **Peter Csaba Ölveczky**
Specifying and Analyzing Real-Time and Hybrid Systems in Rewriting Logic
University of Bergen, Norway
December 2000

↪ **Jon Grov**
Transactional Data Management for Multi-Site Systems: New Approaches and Formal Analysis
University of Oslo, Norway
June 2014

↪ **Muhammad Fadlisyah**
A Rewriting-Logic-Based Approach for the Formal Modeling and Analysis of Interacting Hybrid Systems
(with Erika Ábrahám)
University of Oslo, Norway
September 2014

↪ **Daniela Lepri**
Timed Temporal Logic Model Checking of Real-Time Systems: A Rewriting-Logic-Based Approach

(with Erika Ábrahám)
University of Oslo, Norway
April 2015
→ **Lucian Bentea**
Formal Modeling and Analysis of Probabilistic Real-Time Systems in Rewriting Logic: A Probabilistic Strategy Language Approach
(with Olaf Owe)
University of Oslo, Norway
June 2015
➤ **Christiano Braga**
Rewriting Logic as a Semantic Framework for Modular Structural Operational Semantics
(with Edward Hermann Haeusler and Peter D. Mosses)
Pontifícia Universidade Católica do Rio de Janeiro, Brazil
September 2001
➤ **Mark-Oliver Stehr**
Programming, Specification, and Interactive Theorem Proving: Towards a Unified Language based on Equational Logic, Rewriting Logic, and Type Theory
(with Rüdiger Valk)
Universität Hamburg, Germany
September 2002
➤ **Miguel Palomino**
Reflection, Abstraction, and Simulation in Rewriting Logic
(with Narciso Martí-Oliet)
Universidad Complutense de Madrid, Spain
March 2005
→ **Ignacio Fábregas Alfaro**
Coalgebraic and Categorical Techniques to Study Process Semantics
(with David de Frutos Escrig)
Universidad Complutense de Madrid, Spain
March 2012
➤ **Azadeh Farzan**
Static and Dynamic Formal Analysis of Concurrent Systems and Languages: A Semantics-Based Approach
University of Illinois at Urbana-Champaign, US
May 2007
→ **Niloofar Razavi**
Effective Heuristic-Based Test Generation Techniques for Concurrent Software
University of Toronto, Canada
October 2013
→ **Zachary Kincaid**
Parallel Proofs for Parallel Programs
University of Toronto, Canada
December 2015

➤ **Artur Boronat**
MOMENT: A Formal Framework for Model Management
(with Isidro Ramos and José Á. Carsí)
Universitat Politècnica de València, Spain
December 2007

↳ **Nissreen A. S. El-Saber**
  CMMI-CM Compliance Checking of Formal BPMN Models using Maude
  (with Reiko Heckel)
  University of Leicester, UK
  January 2015

➤ **Joseph Hendrix**
Decision Procedures for Equationally Based Reasoning
University of Illinois at Urbana-Champaign, US
September 2008

➤ **Musab A. AlTurki**
Rewriting-Based Formal Modeling, Analysis and Implementation of Real-Time Distributed Services
University of Illinois at Urbana-Champaign, US
May 2011

➤ **Michael Katelman**
A Meta-Language for Functional Verification
University of Illinois at Urbana-Champaign, US
July 2011

➤ **Ralf Sasse**
Security Models in Rewriting Logic for Cryptographic Protocols and Browsers
University of Illinois at Urbana-Champaign, US
July 2012

➤ **Camilo Rocha**
Symbolic Reachability Analysis for Rewrite Theories
University of Illinois at Urbana-Champaign, US
October 2012

➤ **Mu Sun**
Formal Patterns for Medical Device Safety
(with Lui Sha)
University of Illinois at Urbana-Champaign, US
December 2013

➤ **Kyungmin Bae**
Rewriting-Based Model Checking Methods
University of Illinois at Urbana-Champaign, US
June 2014

References

1. Agha, G.A., Meseguer, J., Sen, K.: PMaude: rewrite-based specification language for probabilistic object systems. Electron. Notes Theoret. Comput. Sci. **153**(2), 213–239 (2006)
2. Aida, H., Goguen, J.A., Meseguer, J.: Compiling concurrent rewriting onto the rewrite rule machine. In: Okada, M., Kaplan, S. (eds.) CTRS 1990. LNCS, vol. 516, pp. 320–332. Springer, Heidelberg (1991)
3. Bae, K., Krisiloff, J., Meseguer, J., Ölveczky, P.C.: Designing and verifying distributed cyber-physical systems using Multirate PALS: an airplane turning control system case study. Sci. Comput. Program. **103**, 13–50 (2015)
4. Bae, K., Meseguer, J.: Predicate abstraction of rewrite theories. In: Dowek, G. (ed.) RTA-TLCA 2014. LNCS, vol. 8560, pp. 61–76. Springer, Heidelberg (2014)
5. Bouhoula, A., Jouannaud, J., Meseguer, J.: Specification and proof in membership equational logic. Theoret. Comput. Sci. **236**(1–2), 35–132 (2000)
6. Chen, S., Meseguer, J., Sasse, R., Wang, H.J., Wang, Y.: A systematic approach to uncover security flaws in GUI logic. In: Proceedings of the IEEE Symposium on Security and Privacy, pp. 71–85. IEEE Computer Society (2007)
7. Clavel, M.: Reflection in Rewriting Logic: Metalogical Foundations and Metaprogramming Applications. CSLI Publications, Stanford (2000)
8. Clavel, M., Durán, F., Eker, S., Lincoln, P., Martí-Oliet, N., Meseguer, J., Talcott, C. (eds.): All About Maude. LNCS, vol. 4350. Springer, Heidelberg (2007)
9. Clavel, M., Meseguer, J.: Reflection and strategies in rewriting logic. Electron. Notes Theoret. Comput. Sci. **4**, 126–148 (1996)
10. Clavel, M., Meseguer, J.: Reflection in conditional rewriting logic. Theoret. Comput. Sci. **285**(2), 245–288 (2002)
11. Durán, F., Meseguer, J.: Maude's module algebra. Sci. Comput. Program. **66**(2), 125–153 (2007)
12. Durán, F., Meseguer, J.: On the Church-Rosser and coherence properties of conditional order-sorted rewrite theories. J. Logic Algebraic Program. **81**(7–8), 816–850 (2012)
13. Eckhardt, J., Mühlbauer, T., AlTurki, M., Meseguer, J., Wirsing, M.: Stable availability under denial of service attacks through formal patterns. In: de Lara, J., Zisman, A. (eds.) FASE 2012. LNCS, vol. 7212, pp. 78–93. Springer, Heidelberg (2012)
14. Eker, S., Knapp, M., Laderoute, K., Lincoln, P., Meseguer, J., Sonmez, K.: Pathway logic: symbolic analysis of biological signaling. In: Proceedings of the Pacific Symposium on Biocomputing, pp. 400–412, January 2002. http://psb.stanford.edu/psb-online/proceedings/psb02/eker.pdf
15. Escobar, S., Meadows, C., Meseguer, J.: Maude-NPA: cryptographic protocol analysis modulo equational properties. In: Aldini, A., Barthe, G., Gorrieri, R. (eds.) FOSAD 2007/2008/2009. LNCS, vol. 5705, pp. 1–50. Springer, Heidelberg (2009)
16. Escobar, S., Sasse, R., Meseguer, J.: Folding variant narrowing and optimal variant termination. J. Logic Algebraic Programm. **81**(7–8), 898–928 (2012)
17. Fadlisyah, M., Ölveczky, P.C., Ábrahám, E.: Object-oriented formal modeling and analysis of interacting hybrid systems in HI-Maude. In: Barthe, G., Pardo, A., Schneider, G. (eds.) SEFM 2011. LNCS, vol. 7041, pp. 415–430. Springer, Heidelberg (2011)
18. Fadlisyah, M., Ölveczky, P.C., Ábrahám, E.: Formal modeling and analysis of interacting hybrid systems in HI-Maude: what happened at the 2010 Sauna World Championships? Sci. Comput. Program. **99**, 95–127 (2015)

19. Farzan, A., Chen, F., Meseguer, J., Roşu, G.: Formal analysis of Java programs in JavaFAN. In: Alur, R., Peled, D.A. (eds.) CAV 2004. LNCS, vol. 3114, pp. 501–505. Springer, Heidelberg (2004)
20. Futatsugi, K., Goguen, J.A., Jouannaud, J., Meseguer, J.: Principles of OBJ2. In: Proceedings of the ACM Symposium on Principles of Programming Languages, pp. 52–66. ACM Press (1985)
21. Goguen, J.A., Jouannaud, J., Meseguer, J.: Operational semantics for order-sorted algebra. In: Brauer, W. (ed.) ICALP 1985. LNCS, vol. 194, pp. 221–231. Springer, Heidelberg (1985)
22. Goguen, J.A., Kirchner, C., Kirchner, H., Mégrelis, A., Meseguer, J., Winkler, T.C.: An introduction to OBJ3. In: Kaplan, S., Jouannaud, J.-P. (eds.) CTRS 1987. LNCS, vol. 308, pp. 258–263. Springer, Heidelberg (1988)
23. Goguen, J.A., Kirchner, C., Meseguer, J.: Concurrent term rewriting as a model of computation. In: Fasel, J.H., Keller, R.M. (eds.) Graph Reduction 1986. LNCS, vol. 279, pp. 53–93. Springer, Heidelberg (1987)
24. Goguen, J.A., Meseguer, J.: Correctness of recursive flow diagram programs. In: Gruska, J. (ed.) MFCS 1977. LNCS, vol. 53, pp. 580–595. Springer, Heidelberg (1977)
25. Goguen, J.A., Meseguer, J.: Security policies and security models. In: Proceedings of the IEEE Symposium on Security and Privacy, pp. 11–20. IEEE Computer Society (1982)
26. Goguen, J.A., Meseguer, J.: Order-sorted algebra I: equational deduction for multiple inheritance, overloading, exceptions and partial operations. Theoret. Comput. Sci. **105**(2), 217–273 (1992)
27. Grov, J., Ölveczky, P.C.: Increasing consistency in multi-site data stores: Megastore-CGC and its formal analysis. In: Giannakopoulou, D., Salaün, G. (eds.) SEFM 2014. LNCS, vol. 8702, pp. 159–174. Springer, Heidelberg (2014)
28. Kirchner, C., Kirchner, H., Meseguer, J.: Operational semantics of OBJ3 (extended abstract). In: Lepistö, T., Salomaa, A. (eds.) ICALP 1988. LNCS, vol. 317, pp. 287–301. Springer, Heidelberg (1988)
29. Kumar, N., Sen, K., Meseguer, J., Agha, G.: A rewriting based model for probabilistic distributed object systems. In: Najm, E., Nestmann, U., Stevens, P. (eds.) FMOODS 2003. LNCS, vol. 2884, pp. 32–46. Springer, Heidelberg (2003)
30. Lincoln, P.D., Talcott, C.: Symbolic systems biology and Pathway Logic. In: Symbolic Systems Biology, pp. 1–29. Jones and Bartlett (2010)
31. Liu, S., Rahman, M.R., Skeirik, S., Gupta, I., Meseguer, J.: Formal modeling and analysis of Cassandra in Maude. In: Merz, S., Pang, J. (eds.) ICFEM 2014. LNCS, vol. 8829, pp. 332–347. Springer, Heidelberg (2014)
32. Lucas, S., Meseguer, J.: Localized operational termination in general logics. In: De Nicola, R., Hennicker, R. (eds.) Software, Services, and Systems. LNCS, vol. 8950, pp. 91–114. Springer, Heidelberg (2015)
33. Martí-Oliet, N., Meseguer, J.: From Petri nets to linear logic. In: Dybjer, P., Pitts, A.M., Pitt, D.H., Poigné, A., Rydeheard, D.E. (eds.) CTCS 1989. LNCS, vol. 389, pp. 313–340. Springer, Heidelberg (1989)
34. Meseguer, J.: General logics. In: Proceedings of the Logic Colloquium 1987, pp. 275–329. North-Holland (1989)

35. Meseguer, J.: Conditional rewriting logic: deduction, models and concurrency. In: Okada, M., Kaplan, S. (eds.) CTRS 1990. LNCS, vol. 516, pp. 64–91. Springer, Heidelberg (1991)
36. Meseguer, J.: Rewriting as a unified model of concurrency. In: Baeten, J.C.M., Klop, J.W. (eds.) CONCUR 1990. LNCS, vol. 458, pp. 384–400. Springer, Heidelberg (1990)
37. Meseguer, J.: Conditional rewriting logic as a unified model of concurrency. Theoret. Comput. Sci. **96**(1), 73–155 (1992)
38. Meseguer, J.: Multiparadigm logic programming. In: Kirchner, H., Levi, G. (eds.) ALP 1992. LNCS, vol. 632, pp. 158–200. Springer, Heidelberg (1992)
39. Meseguer, J.: Rewriting logic as a semantic framework for concurrency: a progress report. In: Sassone, V., Montanari, U. (eds.) CONCUR 1996. LNCS, vol. 1119, pp. 331–372. Springer, Heidelberg (1996)
40. Meseguer, J.: Membership algebra as a logical framework for equational specification. In: Parisi-Presicce, F. (ed.) WADT 1997. LNCS, vol. 1376, pp. 18–61. Springer, Heidelberg (1998)
41. Meseguer, J.: From OBJ to Maude and beyond. In: Futatsugi, K., Jouannaud, J.-P., Meseguer, J. (eds.) Algebra, Meaning, and Computation. LNCS, vol. 4060, pp. 252–280. Springer, Heidelberg (2006)
42. Meseguer, J.: Twenty years of rewriting logic. J. Logic Algebraic Program. **81**(7–8), 721–781 (2012)
43. Meseguer, J.: Taming distributed system complexity through formal patterns. Sci. Comput. Program. **83**, 3–34 (2014)
44. Meseguer, J., Montanari, U.: Petri nets are monoids. Inf. Comput. **88**(2), 105–155 (1990)
45. Meseguer, J., Palomino, M., Martí-Oliet, N.: Algebraic simulations. J. Logic Algebraic Program. **79**(2), 103–143 (2010)
46. Meseguer, J., Roşu, G.: The rewriting logic semantics project. Theoret. Comput. Sci. **373**(3), 213–237 (2007)
47. Meseguer, J., Roşu, G.: The rewriting logic semantics project: a progress report. In: Owe, O., Steffen, M., Telle, J.A. (eds.) FCT 2011. LNCS, vol. 6914, pp. 1–37. Springer, Heidelberg (2011)
48. Meseguer, J., Sharykin, R.: Specification and analysis of distributed object-based stochastic hybrid systems. In: Hespanha, J.P., Tiwari, A. (eds.) HSCC 2006. LNCS, vol. 3927, pp. 460–475. Springer, Heidelberg (2006)
49. Meseguer, J., Thati, P.: Symbolic reachability analysis using narrowing and its application to verification of cryptographic protocols. Higher-Order Symb. Comput. **20**(1–2), 123–160 (2007)
50. Ölveczky, P.C.: Real-Time Maude and its applications. In: Escobar, S. (ed.) WRLA 2014. LNCS, vol. 8663, pp. 42–79. Springer, Heidelberg (2014)
51. Ölveczky, P.C., Meseguer, J.: Specification of real-time and hybrid systems in rewriting logic. Theoret. Comput. Sci. **285**(2), 359–405 (2002)
52. Ölveczky, P.C., Meseguer, J.: Semantics and pragmatics of Real-Time Maude. Higher-Order Symb. Comput. **20**(1–2), 161–196 (2007)
53. Talcott, C., Dill, D.L.: Multiple representations of biological processes. In: Priami, C., Plotkin, G. (eds.) Transactions on Computational Systems Biology VI. LNCS (LNBI), vol. 4220, pp. 221–245. Springer, Heidelberg (2006)

Bibliography of José Meseguer
from 1973 to 2015

Books Authored or Edited, and Special Issues

1. Meseguer, J. (ed.): Rewriting Logic and Its Applications, Electronic Notes in Theoretical Computer Science, vol. 4. Elsevier (1996)
2. Futatsugi, K., Goguen, J.A., Meseguer, J. (eds.): OBJ/Cafe OBJ/Maude at Formal Methods'99. Formal Specification, Proof and Applications. Theta Foundation (1999)
3. Martí-Oliet, N., Meseguer, J. (eds.): Rewriting Logic and Its Applications, Special issue of Theoretical Computer Science, vol. 285(2). Elsevier (2002)
4. Futatsugi, K., Jouannaud, J., Meseguer, J. (eds.): Algebra, Meaning, and Computation - Essays Dedicated to Joseph A. Goguen on the Occasion of His 65th Birthday, LNCS, vol. 4060. Springer (2006)
5. Clavel, M., Durán, F., Eker, S., Lincoln, P., Martí-Oliet, N., Meseguer, J., Talcott, C.L.: All About Maude - A High-Performance Logical Framework, How to Specify, Program and Verify Systems in Rewriting Logic, LNCS, vol. 4350. Springer (2007)
6. Degano, P., De Nicola, R., Meseguer, J. (eds.): Concurrency, Graphs and Models - Essays Dedicated to Ugo Montanari on the Occasion of His 65th Birthday, LNCS, vol. 5065. Springer (2008)
7. Meseguer, J., Roşu, G. (eds.): Algebraic Methodology and Software Technology, 12th International Conference, AMAST 2008, Urbana, IL, USA, July 28–31, 2008, Proceedings, LNCS, vol. 5140. Springer (2008)
8. Agha, G., Danvy, O., Meseguer, J. (eds.): Formal Modeling: Actors, Open Systems, Biological Systems - Essays Dedicated to Carolyn Talcott on the Occasion of Her 70th Birthday, LNCS, vol. 7000. Springer (2011)
9. Iida, S., Meseguer, J., Ogata, K. (eds.): Specification, Algebra, and Software - Essays Dedicated to Kokichi Futatsugi, LNCS, vol. 8373. Springer (2014)

Book Chapters

1. Goguen, J.A., Meseguer, J., Plaisted, D.: Programming with parameterized abstract objects in OBJ. In: Ferrari, D., Goguen, J.A. (eds.) Theory and Practice of Software Technology. North Holand (1983)
2. Meseguer, J., Goguen, J.A.: Initiality, induction, and computability. In: Reynolds, J., Nivat, M. (eds.) Algebraic Methods in Semantics, pp. 459–541. Cambridge University Press (1985)
3. Goguen, J.A., Meseguer, J.: EQLOG: equality, types, and generic modules for logic programming. In: DeGroot, D., Lindstrom, G. (eds.) Logic Programming: Functions, Relations, and Equations, pp. 295–363. Prentice Hall (1986)
4. Goguen, J.A., Meseguer, J.: Unifying functional, object-oriented and relational programming with logical semantics. In: Shriver, B., Wegner, P. (eds.) Research Directions in Object-Oriented Programming, pp. 417–478. The MIT Press (1987)

5. Futatsugi, K., Goguen, J.A., Meseguer, J., Okada, K.: Parameterized programming and its application to rapid prototyping in OBJ2. In: Matsumoto, Y., Ohmo, Y. (eds.) Japanese Perspectives in Software Engineering, pp. 77–102. Addison Wesley (1989)
6. Goguen, J.A., Nutt, W., Meseguer, J., Smolka, G.: Order-sorted equational computation. In: Nivat, M., Aït-Kaci, H. (eds.) Resolution of Equations in Algebraic Structures, pp. 297–367. Academic Press (1989)
7. Meseguer, J., Goguen, J.A., Smolka, G.: Order-sorted unification. In: Kirchner, C. (ed.) Unification, pp. 457–487. Academic Press (1990)
8. Aida, H., Leinwand, S., Meseguer, J.: Architectural design of the rewrite rule machine ensemble. In: Delgado-Frias, J.G., Moore, W.R. (eds.) VLSI for Artificial Intelligence and Neural Networks. Springer (1991)
9. Goguen, J.A., Meseguer, J.: Modeli i raventsvo v logicheskom programmirovanii (in Russian). In: Zakhazjaschev, M.V., Yanov, Y.I. (eds.) Matematiskaya logika v Programmirovanii, pp. 274–310. Mir (1991)
10. Meseguer, J., Martí-Oliet, N.: An algebraic axiomatization of linear logic models. In: Reed, G.M., Rocoe, A.W., Wachter, R.F. (eds.) Topology and Category Theory in Computer Science, pp. 335–355. Oxford University Press (1991)
11. Meseguer, J.: A logical theory of concurrent objects and its realization in the Maude language. In: Agha, G., Wegner, P., Yonezawa, A. (eds.) Research Directions in Object-Based Concurrency, pp. 314–390. MIT Press (1993)
12. Martí-Oliet, N., Meseguer, J.: General logics and logical frameworks. In: Gabbay, D. (ed.) What is a Logical System?, pp. 355–392. Oxford University Press (1994)
13. Meseguer, J., Montanari, U., Sassone, V.: Representation theorems for Petri nets. In: Freksa, C., Jantzen, M., Valk, R. (eds.) Foundations of Computer Science: Potential - Theory - Cognition, to Wilfried Brauer on the occasion of his sixtieth birthday. LNCS, vol. 1337, pp. 239–249. Springer (1997)
14. Martí-Oliet, N., Meseguer, J.: Action and change in rewriting logic. In: Pareschi, R., Fronhöfer, B. (eds.) Dynamic Worlds: From the Frame Problem to Knowledge Management, pp. 1–53. Kluwer Academic Publisher (1999)
15. Meseguer, J.: Research directions in rewriting logic. In: Berger, U., Schwichtenberg, H. (eds.) Computational Logic. NATO Advanced Study Institute Series F, vol. 165, pp. 345–398. Springer (1999)
16. Clavel, M., Durán, F., Eker, S., Meseguer, J.: Building equational proving tools by reflection in rewriting logic. In: Futatsugi, K., Nakagawa, A.T., Tamai, T. (eds.) CAFE: An Industrial-Strength Algebraic Formal Method, pp. 1–31. Elsevier (2000)
17. Goguen, J.A., Winkler, T., Meseguer, J., Futatsugi, K., Jouannaud, J.: Introducing OBJ. In: Goguen, J.A., Malcolm, G. (eds.) Software Engineering with OBJ: Algebraic Specification in Action, pp. 3–167. Kluwer (2000)
18. Stehr, M., Meseguer, J., Ölveczky, P.C.: Rewriting logic as a unifying framework for Petri nets. In: Ehrig, H., Juhás, G., Padberg, J., Rozenberg, G. (eds.) Unifying Petri Nets, Advances in Petri Nets. LNCS, vol. 2128, pp. 250–303. Springer (2001)

19. Martí-Oliet, N., Meseguer, J.: Rewriting logic as a logical and semantic framework. In: Gabbay, D., Guenthner, F. (eds.) Handbook of Philosophical Logic, 2nd Edition, pp. 1–87. Kluwer Academic Publisher (2002)

20. Meseguer, J.: Software specification and verification in rewriting logic. In: Broy, M., Pizka, M. (eds.) Models, Algebras, and Logic of Engineering Software. NATO Advanced Study Institute, pp. 133–193. IOS Press (2003)

21. Stehr, M., Meseguer, J.: Pure type systems in rewriting logic: Specifying typed higher-order languages in a first-order logical framework. In: Owe, O., Krogdahl, S., Lyche, T. (eds.) From Object-Orientation to Formal Methods, Essays in Memory of Ole-Johan Dahl. LNCS, vol. 2635, pp. 334–375. Springer (2004)

22. Meseguer, J.: Functorial semantics of rewrite theories. In: Kreowski, H., Montanari, U., Orejas, F., Rozenberg, G., Taentzer, G. (eds.) Formal Methods in Software and Systems Modeling, Essays Dedicated to Hartmut Ehrig, on the Occasion of His 60th Birthday. LNCS, vol. 3393, pp. 220–235. Springer (2005)

23. Meseguer, J.: From OBJ to Maude and beyond. In: Futatsugi, K., Jouannaud, J., Meseguer, J. (eds.) Algebra, Meaning, and Computation, Essays Dedicated to Joseph A. Goguen on the Occasion of His 65th Birthday. LNCS, vol. 4060, pp. 252–280. Springer (2006)

24. De Nicola, R., Degano, P., Meseguer, J.: Ugo Montanari in a nutshell. In: Degano, P., De Nicola, R., Meseguer, J. (eds.) Concurrency, Graphs and Models, Essays Dedicated to Ugo Montanari on the Occasion of His 65th Birthday. LNCS, vol. 5065, pp. 1–8. Springer (2008)

25. Meseguer, J.: The temporal logic of rewriting: A gentle introduction. In: Degano, P., De Nicola, R., Meseguer, J. (eds.) Concurrency, Graphs and Models - Essays Dedicated to Ugo Montanari on the Occasion of His 65th Birthday. LNCS, vol. 5065, pp. 354–382. Springer (2008)

26. Sha, L., Meseguer, J.: Design of complex cyber physical systems with formalized architectural patterns. In: Wirsing, M., Banâtre, J., Hölzl, M.M., Rauschmayer, A. (eds.) Software-Intensive Systems and New Computing Paradigms - Challenges and Visions, LNCS, vol. 5380, pp. 92–100. Springer (2008)

27. Escobar, S., Meadows, C., Meseguer, J.: Maude-NPA: Cryptographic protocol analysis modulo equational properties. In: Aldini, A., Barthe, G., Gorrieri, R. (eds.) Foundations of Security Analysis and Design V, FOSAD 2007/2008/2009 Tutorial Lectures. LNCS, vol. 5705, pp. 1–50. Springer (2009)

28. Meseguer, J.: Order-sorted parameterization and induction. In: Palsberg, J. (ed.) Semantics and Algebraic Specification - Essays Dedicated to Peter D. Mosses on the Occasion of His 60th Birthday. LNCS, vol. 5700, pp. 43–80. Springer (2009)

29. Meseguer, J.: Maude. In: Padua, D.A. (ed.) Encyclopedia of Parallel Computing, pp. 1095–1102. Springer (2011)

30. Rocha, C., Meseguer, J.: Mechanical analysis of reliable communication in the alternating bit protocol using the Maude invariant analyzer tool. In: Iida, S., Meseguer, J., Ogata, K. (eds.) Specification, Algebra, and Software - Essays Dedicated to Kokichi Futatsugi. LNCS, vol. 8373, pp. 603–629. Springer (2014)
31. Liu, S., Ölveczky, P.C., Meseguer, J.: Formal analysis of leader election in MANETs using Real-Time Maude. In: De Nicola, R., Hennicker, R. (eds.) Software, Services, and Systems - Essays Dedicated to Martin Wirsing on the Occasion of His Retirement from the Chair of Programming and Software Engineering. LNCS, vol. 8950, pp. 231–252. Springer (2015)
32. Lucas, S., Meseguer, J.: Localized operational termination in general logics. In: De Nicola, R., Hennicker, R. (eds.) Software, Services, and Systems - Essays Dedicated to Martin Wirsing on the Occasion of His Retirement from the Chair of Programming and Software Engineering. LNCS, vol. 8950, pp. 91–114. Springer (2015)

Journal Papers

1. Meseguer, J., Sols, I.: On a categorical tensor calculus for automata. Bulletin de l'Académie Polonaise des Sciences. Série des Sciences Mathématiques, Astronomiques et Physiques 23(11), 1161–1166 (1975)
2. Kühnel, W., Meseguer, J., Pfender, M., Sols, I.: Primitive recursive algebraic theories and program schemas. Bulletin of the Australian Mathematical Society 17, 207–233 (1977)
3. Meseguer, J.: Varieties of chain-complete algebras. Journal of Pure and Applied Algebra 19, 347–383 (1980)
4. Kühnel, W., Meseguer, J., Pfender, M., Sols, I.: Algebras with actions and automata. International Journal of Mathematics and Mathematical Sciences 5(1), 61–85 (1982)
5. Goguen, J.A., Meseguer, J.: Correctness of recursive parallel nondeterministic flow programs. Journal of Computer and System Sciences 27(2), 268–290 (1983)
6. Meseguer, J.: Order completion monads. Algebra Universalis 16, 63–82 (1983)
7. Goguen, J.A., Meseguer, J.: Equality, types, modules, and (why not?) generics for logic programming. The Journal of Logic Programming 1(2), 179–210 (1984)
8. Meseguer, J., Sols, I.: Fascieaux semi-stables de rang 2 sur $\mathbf{P}^3$. Comptes Rendus de l'Académie des Sciences - Series I - Mathematics 298(20), 525–528 (1984)
9. Goguen, J.A., Meseguer, J.: Completeness of many-sorted equational logic. Houston Journal of Mathematics 11(3), 307–334 (1985)
10. Guessarian, I., Meseguer, J.: On the axiomatization of if-then-else. SIAM Journal on Computing 16(2), 332–357 (1987)
11. Meseguer, J., Goguen, J.A.: Order-sorted unification. Journal of Symbolic Computation 8(4), 383–413 (1989)

12. Meseguer, J., Montanari, U.: Petri nets are monoids. Information and Computation 88(2), 105–155 (1990)

13. Casley, R., Crew, R.F., Meseguer, J., Pratt, V.R.: Temporal structures. Mathematical Structures in Computer Science 1(2), 179–213 (1991)

14. Martí-Oliet, N., Meseguer, J.: From Petri nets to linear logic. Mathematical Structures in Computer Science 1(1), 69–101 (1991)

15. Martí-Oliet, N., Meseguer, J.: From Petri nets to linear logic through categories: A survey. International Journal of Foundations of Computer Science 2(4), 297–399 (1991)

16. Goguen, J.A., Meseguer, J.: Order-sorted algebra I: equational deduction for multiple inheritance, overloading, exceptions and partial operations. Theoretical Computer Science 105(2), 217–273 (1992)

17. Meseguer, J.: Conditional rewriting logic as a unified model of concurrency. Theoretical Computer Science 96(1), 73–155 (1992)

18. Moss, L.S., Meseguer, J., Goguen, J.A.: Final algebras, cosemicomputable algebras and degrees of unsolvability. Theoretical Computer Science 100(2), 267–302 (1992)

19. Meseguer, J., Goguen, J.A.: Order-sorted algebra solves the constructor-selector, multiple representation, and coercion problems. Information and Computation 103(1), 114–158 (1993)

20. Degano, P., Meseguer, J., Montanari, U.: Axiomatizing the algebra of net computations and processes. Acta Informatica 33(7), 641–667 (1996)

21. Goguen, J.A., Nguyen, D., Meseguer, J., Luqi, Zhang, D., Berzins, V.: Software component search. Journal of Systems Integration 6(1/2), 93–134 (1996)

22. Martí-Oliet, N., Meseguer, J.: Inclusions and subtypes I: first-order case. Journal of Logic and Computation 6(3), 409–438 (1996)

23. Martí-Oliet, N., Meseguer, J.: Inclusions and subtypes II: higher-order case. Journal of Logic and Computation 6(4), 541–572 (1996)

24. Meseguer, J., Montanari, U., Sassone, V.: Process versus unfolding semantics for place/transition Petri nets. Theoretical Computer Science 153(1&2), 171–210 (1996)

25. Cerioli, M., Meseguer, J.: May I borrow your logic? (transporting logical structures along maps). Theoretical Computer Science 173(2), 311–347 (1997)

26. Meseguer, J., Montanari, U., Sassone, V.: On the semantics of place/transition Petri nets. Mathematical Structures in Computer Science 7(4), 359–397 (1997)

27. Bouhoula, A., Jouannaud, J., Meseguer, J.: Specification and proof in membership equational logic. Theoretical Computer Science 236(1–2), 35–132 (2000)

28. Bruni, R., Meseguer, J., Montanari, U., Sassone, V.: Functorial models for Petri nets. Information and Computation 170(2), 207–236 (2001)

29. Bruni, R., Meseguer, J., Montanari, U.: Symmetric monoidal and cartesian double categories as a semantic framework for tile logic. Mathematical Structures in Computer Science 12(1), 53–90 (2002)

30. Clavel, M., Durán, F., Eker, S., Lincoln, P., Martí-Oliet, N., Meseguer, J., Quesada, J.F.: Maude: specification and programming in rewriting logic. Theoretical Computer Science 285(2), 187–243 (2002)
31. Clavel, M., Meseguer, J.: Reflection in conditional rewriting logic. Theoretical Computer Science 285(2), 245–288 (2002)
32. Martí-Oliet, N., Meseguer, J.: Rewriting logic: roadmap and bibliography. Theoretical Computer Science 285(2), 121–154 (2002)
33. Ölveczky, P.C., Meseguer, J.: Specification of real-time and hybrid systems in rewriting logic. Theoretical Computer Science 285(2), 359–405 (2002)
34. Durán, F., Meseguer, J.: Structured theories and institutions. Theoretical Computer Science 309(1–3), 357–380 (2003)
35. Basin, D.A., Clavel, M., Meseguer, J.: Reflective metalogical frameworks. ACM Transactions on Computational Logic 5(3), 528–576 (2004)
36. Lucas, S., Marché, C., Meseguer, J.: Operational termination of conditional term rewriting systems. Information Processing Letters 95(4), 446–453 (2005)
37. Martí-Oliet, N., Pita, I., Fiadeiro, J.L., Meseguer, J., Maibaum, T.S.E.: A verification logic for rewriting logic. Journal of Logic and Computation 15(3), 317–352 (2005)
38. Bruni, R., Meseguer, J.: Semantic foundations for generalized rewrite theories. Theoretical Computer Science 360(1–3), 386–414 (2006)
39. Escobar, S., Meadows, C., Meseguer, J.: A rewriting-based inference system for the NRL protocol analyzer and its meta-logical properties. Theoretical Computer Science 367(1–2), 162–202 (2006)
40. Ölveczky, P.C., Meseguer, J., Talcott, C.L.: Specification and analysis of the AER/NCA active network protocol suite in Real-Time Maude. Formal Methods in System Design 29(3), 253–293 (2006)
41. Thati, P., Meseguer, J.: Complete symbolic reachability analysis using back-and-forth narrowing. Theoretical Computer Science 366(1–2), 163–179 (2006)
42. Clavel, M., Meseguer, J., Palomino, M.: Reflection in membership equational logic, many-sorted equational logic, Horn logic with equality, and rewriting logic. Theoretical Computer Science 373(1–2), 70–91 (2007)
43. Durán, F., Meseguer, J.: Maude's module algebra. Science of Computer Programming 66(2), 125–153 (2007)
44. Meseguer, J., Roşu, G.: The rewriting logic semantics project. Theoretical Computer Science 373(3), 213–237 (2007)
45. Meseguer, J., Thati, P.: Symbolic reachability analysis using narrowing and its application to verification of cryptographic protocols. Higher-Order and Symbolic Computation 20(1–2), 123–160 (2007)
46. Ölveczky, P.C., Meseguer, J.: Semantics and pragmatics of Real-Time Maude. Higher-Order and Symbolic Computation 20(1–2), 161–196 (2007)
47. Rocha, C., Meseguer, J.: A rewriting decision procedure for Dijkstra-Scholten's syllogistic logic with complements. Revista Colombiana de Computación 8(2) (2007)

48. Durán, F., Lucas, S., Marché, C., Meseguer, J., Urbain, X.: Proving operational termination of membership equational programs. Higher-Order and Symbolic Computation 21(1–2), 59–88 (2008)
49. Lucas, S., Meseguer, J.: Termination of just/fair computations in term rewriting. Information and Computation 206(5), 652–675 (2008)
50. Meseguer, J., Palomino, M., Martí-Oliet, N.: Equational abstractions. Theoretical Computer Science 403(2–3), 239–264 (2008)
51. Şerbănuţă, T., Roşu, G., Meseguer, J.: A rewriting logic approach to operational semantics. Information and Computation 207(2), 305–340 (2009)
52. Boronat, A., Meseguer, J.: An algebraic semantics for MOF. Formal Aspects of Computing 22(3–4), 269–296 (2010)
53. Meseguer, J., Palomino, M., Martí-Oliet, N.: Algebraic simulations. The Journal of Logic and Algebraic Programming 79(2), 103–143 (2010)
54. Durán, F., Meseguer, J.: On the Church-Rosser and coherence properties of conditional order-sorted rewrite theories. The Journal of Logic and Algebraic Programming 81(7–8), 816–850 (2012)
55. Escobar, S., Sasse, R., Meseguer, J.: Folding variant narrowing and optimal variant termination. The Journal of Logic and Algebraic Programming 81(7–8), 898–928 (2012)
56. Katelman, M., Keller, S., Meseguer, J.: Rewriting semantics of production rule sets. The Journal of Logic and Algebraic Programming 81(7–8), 929–956 (2012)
57. Meseguer, J.: Twenty years of rewriting logic. The Journal of Logic and Algebraic Programming 81(7–8), 721–781 (2012)
58. Meseguer, J., Ölveczky, P.C.: Formalization and correctness of the PALS architectural pattern for distributed real-time systems. Theoretical Computer Science 451, 1–37 (2012)
59. Meseguer, J., Roşu, G.: The rewriting logic semantics project: A progress report. Information and Computation 231, 38–69 (2013)
60. Alpuente, M., Escobar, S., Espert, J., Meseguer, J.: A modular order-sorted equational generalization algorithm. Information and Computation 235, 98–136 (2014)
61. Bae, K., Meseguer, J., Ölveczky, P.C.: Formal patterns for multirate distributed real-time systems. Science of Computer Programming 91, 3–44 (2014)
62. Escobar, S., Meadows, C., Meseguer, J., Santiago, S.: State space reduction in the Maude-NRL protocol analyzer. Information and Computation 238, 157–186 (2014)
63. Meseguer, J.: Taming distributed system complexity through formal patterns. Science of Computer Programming 83, 3–34 (2014)
64. Bae, K., Krisiloff, J., Meseguer, J., Ölveczky, P.C.: Designing and verifying distributed cyber-physical systems using Multirate PALS: an airplane turning control system case study. Science of Computer Programming 103, 13–50 (2015)
65. Bae, K., Meseguer, J.: Model checking linear temporal logic of rewriting formulas under localized fairness. Science of Computer Programming 99, 193–234 (2015)

66. Eckhardt, J., Mühlbauer, T., Meseguer, J., Wirsing, M.: Semantics, distributed implementation, and formal analysis of KLAIM models in Maude. Science of Computer Programming 99, 24–74 (2015)
67. Gutiérrez, R., Meseguer, J., Rocha, C.: Order-sorted equality enrichments modulo axioms. Science of Computer Programming 99, 235–261 (2015)
68. Cholewa, A., Escobar, S., Meseguer, J.: Constrained narrowing for conditional equational theories modulo axioms. Science of Computer Programming (2015) To appear.
69. Liu, S., Ölveczky, P.C., Meseguer, J.: Modeling and analyzing mobile ad hoc networks in Real-Time Maude. Journal of Logical and Algebraic Methods in Programming (2015) To appear.
70. Lucas, S., Meseguer, J.: Normal forms and normal theories in conditional rewriting. Journal of Logical and Algebraic Methods in Programming (2015) To appear.

Invited Conference Papers

1. Meseguer, J.: A Birkhoff-like theorem for algebraic classes of interpretations of program schemes. In: Díaz, J., Ramos, I. (eds.) Formalization of Programming Concepts, International Colloquium, Peñíscola, Spain, April 19–25, 1981, Proceedings. LNCS, vol. 107, pp. 152–168. Springer (1981)
2. Goguen, J.A., Meseguer, J.: Models and equality for logical programming. In: Ehrig, H., Kowalski, R.A., Levi, G., Montanari, U. (eds.) TAPSOFT'87: Proceedings of the International Joint Conference on Theory and Practice of Software Development, Pisa, Italy, March 23–27, 1987, Volume 2: Advanced Seminar on Foundations of Innovative Software Development II and Colloquium on Functional and Logic Programming and Specifications (CFLP). LNCS, vol. 250, pp. 1–22. Springer (1987)
3. Meseguer, J.: General logics. In: Ebbinghaus, H.D., Fernández-Prida, J., Garrido, M., Lascar, D., Rodríguez-Artalejo, M. (eds.) Proceedings Logic Colloquium'87. pp. 275–329 (1989)
4. Meseguer, J.: Conditional rewriting logic: Deduction, models and concurrency. In: Kaplan, S., Okada, M. (eds.) Conditional and Typed Rewriting Systems, 2nd International CTRS Workshop, Montreal, Canada, June 11–14, 1990, Proceedings. LNCS, vol. 516, pp. 64–91. Springer (1990)
5. Meseguer, J.: Multiparadigm logic programming. In: Kirchner, H., Levi, G. (eds.) Algebraic and Logic Programming, Third International Conference, Volterra, Italy, September 2–4, 1992, Proceedings. LNCS, vol. 632, pp. 158–200. Springer (1992)
6. Meseguer, J., Winkler, T.C.: Parallel programming in Maude. In: Banâtre, J., Métayer, D.L. (eds.) Research Directions in High-Level Parallel Programming Languages, Mont Saint-Michel, France, June 17–19, 1991, Proceedings. LNCS, vol. 574, pp. 253–293. Springer (1992)
7. Lincoln, P., Martí-Oliet, N., Meseguer, J.: Specification, transformation, and programming of concurrent systems in rewriting logic. In: Blelloch, G., Chandy, K.M., Jagannathan, S. (eds.) Specification of Parallel Algorithms. American Mathematical Society DIMACS Series, vol. 18, pp. 309–339 (1994)

8. Meseguer, J., Montanari, U., Sassone, V.: On the model of computation of place/transition Petri nets. In: Valette, R. (ed.) Application and Theory of Petri Nets 1994, 15th International Conference, Zaragoza, Spain, June 20–24, 1994, Proceedings. LNCS, vol. 815, pp. 16–38. Springer (1994)

9. Meseguer, J., Martí-Oliet, N.: From abstract data types to logical frameworks. In: Astesiano, E., Reggio, G., Tarlecki, A. (eds.) Recent Trends in Data Type Specification, 10th Workshop on Specification of Abstract Data Types Joint with the 5th COMPASS Workshop, S. Margherita, Italy, May 30 - June 3, 1994, Selected Papers. LNCS, vol. 906, pp. 48–80. Springer (1995)

10. Martí-Oliet, N., Meseguer, J.: Rewriting logic as a logical and semantic framework. In: Meseguer, J. (ed.) Proceedings of the First International Workshop on Rewriting Logic and its Applications, WRLA'96, Asilomar, California, September 3–6, 1996. ENTCS, vol. 4, pp. 190–225. Elsevier (1996)

11. Meseguer, J.: Rewriting logic as a semantic framework for concurrency: a progress report. In: Montanari, U., Sassone, V. (eds.) CONCUR'96, Concurrency Theory, 7th International Conference, Pisa, Italy, August 26–29, 1996, Proceedings. LNCS, vol. 1119, pp. 331–372. Springer (1996)

12. Bouhoula, A., Jouannaud, J., Meseguer, J.: Specification and proof in membership equational logic. In: Bidoit, M., Dauchet, M. (eds.) TAPSOFT'97: Theory and Practice of Software Development, 7th International Joint Conference CAAP/FASE, Lille, France, April 14–18, 1997, Proceedings. LNCS, vol. 1214, pp. 67–92. Springer (1997)

13. Meseguer, J.: Membership algebra as a logical framework for equational specification. In: Parisi-Presicce, F. (ed.) Recent Trends in Algebraic Development Techniques, 12th International Workshop, WADT'97, Tarquinia, Italy, June 1997, Selected Papers. LNCS, vol. 1376, pp. 18–61. Springer (1998)

14. Meseguer, J., Montanari, U.: Mapping tile logic into rewriting logic. In: Parisi-Presicce, F. (ed.) Recent Trends in Algebraic Development Techniques, 12th International Workshop, WADT'97, Tarquinia, Italy, June 1997, Selected Papers. LNCS, vol. 1376, pp. 62–91. Springer (1998)

15. Meseguer, J.: A logical framework for distributed systems and communication protocols. In: Budkowski, S., Cavalli, A.R., Najm, E. (eds.) Formal Description Techniques and Protocol Specification, Testing and Verification, FORTE XI / PSTV XVIII'98, IFIP TC6 WG6.1 Joint International Conference on Formal Description Techniques for Distributed Systems and Communication Protocols (FORTE XI) and Protocol Specification, Testing and Verification (PSTV XVIII), 3–6 November, 1998, Paris, France. IFIP Conference Proceedings, vol. 135, pp. 327–333. Kluwer (1998)

16. Basin, D.A., Clavel, M., Meseguer, J.: Rewriting logic as a metalogical framework. In: Kapoor, S., Prasad, S. (eds.) Foundations of Software Technology and Theoretical Computer Science, 20th Conference, FST TCS 2000 New Delhi, India, December 13–15, 2000, Proceedings. LNCS, vol. 1974, pp. 55–80. Springer (2000)

17. Meseguer, J.: Rewriting logic and Maude: a wide-spectrum semantic framework for object-based distributed systems. In: Smith, S.F., Talcott, C.L. (eds.) Formal Methods for Open Object-Based Distributed Systems IV, IFIF TC6/WG6.1 Fourth International Conference on Formal Methods for Open Object-Based Distributed Systems (FMOODS 2000), September 6–8, 2000, Stanford, California, USA. IFIP Conference Proceedings, vol. 177, pp. 89–117. Kluwer (2000)
18. Meseguer, J.: Rewriting logic and Maude: Concepts and applications. In: Bachmair, L. (ed.) Rewriting Techniques and Applications, 11th International Conference, RTA 2000, Norwich, UK, July 10–12, 2000, Proceedings. LNCS, vol. 1833, pp. 1–26. Springer (2000)
19. Meseguer, J.: Specifying, analyzing and programming communication systems in Maude. In: Hommel, D. (ed.) Communication Based Systems. Kluwer (2000)
20. Meseguer, J., Roşu, G.: Towards behavioral Maude: Behavioral membership equational logic. In: Moss, L.S. (ed.) Proceedings of the Fifth Workshop on Coalgebraic Methods in Computer Science, CMCS 2002, Satellite Event of ETAPS 2002, Grenoble, France, April 6–7, 2002. ENTCS, vol. 65(1), pp. 197–253. Elsevier (2002)
21. Meseguer, J., Talcott, C.L.: Semantic models for distributed object reflection. In: Magnusson, B. (ed.) ECOOP 2002 - Object-Oriented Programming, 16th European Conference, Málaga, Spain, June 10–14, 2002, Proceedings. LNCS, vol. 2374, pp. 1–36. Springer (2002)
22. Meseguer, J.: Executable computational logics: Combining formal methods and programming language based system design. In: 1st ACM & IEEE International Conference on Formal Methods and Models for Co-Design (MEMOCODE 2003), 24–26 June 2003, Mont Saint-Michel, France, Proceedings. p. 3. IEEE Computer Society (2003)
23. Meseguer, J., Roşu, G.: Rewriting logic semantics: From language specifications to formal analysis tools. In: Basin, D.A., Rusinowitch, M. (eds.) Automated Reasoning - Second International Joint Conference, IJCAR 2004, Cork, Ireland, July 4–8, 2004, Proceedings. LNCS, vol. 3097, pp. 1–44. Springer (2004)
24. Meseguer, J.: A rewriting logic sampler. In: Hung, D.V., Wirsing, M. (eds.) Theoretical Aspects of Computing - ICTAC 2005, Second International Colloquium, Hanoi, Vietnam, October 17–21, 2005, Proceedings. LNCS, vol. 3722, pp. 1–28. Springer (2005)
25. Meseguer, J., Roşu, G.: The rewriting logic semantics project. In: Mosses, P., Ulidowski, I. (eds.) Proceedings of the Second Workshop on Structural Operational Semantics, SOS 2005, Lisbon, Portugal, July 10, 2005. ENTCS, vol. 156(1), pp. 27–56. Elsevier (2006)
26. Eker, S., Martí-Oliet, N., Meseguer, J., Verdejo, A.: Deduction, strategies, and rewriting. In: Archer, M., de la Tour, T.B., Muñoz, C. (eds.) Proceedings of the 6th International Workshop on Strategies in Automated Deduction, STRATEGIES 2006, Seattle, WA, USA, August 16, 2006. ENTCS, vol. 174(11), pp. 3–25. Elsevier (2007)

27. Escobar, S., Meseguer, J., Thati, P.: Narrowing and rewriting logic: from foundations to applications. In: López-Fraguas, F.J. (ed.) Proceedings of the 15th Workshop on Functional and (Constraint) Logic Programming, WFLP 2006, Madrid, Spain, November 16–17, 2006. ENTCS, vol. 177, pp. 5–33. Elsevier (2007)

28. Meseguer, J.: Twenty years of rewriting logic. In: Ölveczky, P.C. (ed.) Rewriting Logic and Its Applications - 8th International Workshop, WRLA 2010, Held as a Satellite Event of ETAPS 2010, Paphos, Cyprus, March 20–21, 2010, Revised Selected Papers. LNCS, vol. 6381, pp. 15–17. Springer (2010)

29. Meseguer, J., Roşu, G.: The rewriting logic semantics project: A progress report. In: Owe, O., Steffen, M., Telle, J.A. (eds.) Fundamentals of Computation Theory - 18th International Symposium, FCT 2011, Oslo, Norway, August 22–25, 2011. Proceedings. LNCS, vol. 6914, pp. 1–37. Springer (2011)

30. Meseguer, J.: Taming distributed system complexity through formal patterns. In: Arbab, F., Ölveczky, P.C. (eds.) Formal Aspects of Component Software - 8th International Symposium, FACS 2011, Oslo, Norway, September 14–16, 2011, Revised Selected Papers. LNCS, vol. 7253, pp. 1–2. Springer (2012)

31. Meseguer, J.: Lógica, represetanción del conocimiento y verificación formal. In: Oriol Salgado, M. (ed.) Proc. of the Madrid International Conference on Inteligencia y Filosofía, Madrid, Spain, pp. 25–54. Marova (2012)

32. Meseguer, J.: Symbolic formal methods: Combining the power of rewriting, narrowing, SMT solving and model checking. In: Arbab, F., Sirjani, M. (eds.) Fundamentals of Software Engineering - 5th International Conference, FSEN 2013, Tehran, Iran, April 24–26, 2013, Revised Selected Papers. LNCS, vol. 8161, p. XVI. Springer (2013)

33. Meseguer, J.: Why formal modeling language semantics matters. In: Dingel, J., Schulte, W., Ramos, I., Abrahão, S., Insfrán, E. (eds.) Model-Driven Engineering Languages and Systems - 17th International Conference, MODELS 2014, Valencia, Spain, September 28 - October 3, 2014. Proceedings. LNCS, vol. 8767, pp. XIX–XX. Springer (2014)

Contributed Conference Papers

1. Meseguer, J., Sols, I.: Categorical tensor representation of deterministic, relational, and probabilistic finite functions. In: Actas II Jornadas Matemáticas Hispano-Lusitanas. pp. 340–346. Madrid, Spain (1973)

2. Meseguer, J., Sols, I.: Automata in semimodule categories. In: Manes, E.G. (ed.) Category Theory Applied to Computation and Control, Proceedings of the First International Symposium, San Francisco, CA, USA, February 25–26, 1974, Proceedings. LNCS, vol. 25, pp. 193–198. Springer (1974)

3. Kühnel, W., Meseguer, J., Pfender, M., Sols, I.: Primitive recursive algebraic theories with applications to program schemes. In Proc. 2ᵉ Colloque sur l'Algebre des Categories, Amiens, France. Cahiers de Topologie et Géométrie Différentielle Catégoriques, 16(3), 271–273 (1975)

4. Goguen, J.A., Meseguer, J.: Correctness of recursive flow diagram programs. In: Gruska, J. (ed.) Mathematical Foundations of Computer Science 1977, 6th Symposium, Tatranska Lomnica, Czechoslovakia, September 5–9, 1977, Proceedings. LNCS, vol. 53, pp. 580–595. Springer (1977)
5. Meseguer, J.: On order-complete universal algebra and enriched functorial semantics. In: FCT. pp. 294–301 (1977)
6. Meseguer, J.: Completions, factorizations and colimits for ω-posets. In: Mathematical Logic in Computer Science, Salgotarjan, 1978. Colloquia Mathematica Societatis Janos Bolyai, vol. 26, pp. 509–545. North Holand (1981)
7. Meseguer, J., Sols, I., Strømme, S.A.: Compactification of a family of vector bundles on $\mathbf{P}^3$. In: Proceedings of the 18th Scandinavian Congress of Mathematics. Progress in Mathematics, vol. 11, pp. 474–494. Birkhauser Verlag (1981)
8. Dolev, D., Meseguer, J., Pease, M.C.: Finding safe paths in a faulty environment. In: Probert, R.L., Fischer, M.J., Santoro, N. (eds.) ACM SIGACT-SIGOPS Symposium on Principles of Distributed Computing, Ottawa, Canada, August 18–20, 1982. pp. 95–103. ACM (1982)
9. Goguen, J.A., Meseguer, J.: Rapid prototyping in the OBJ executable specification language. In: Proceedings of ACM SIGSOFT Rapid Prototyping Workshop. ACM Software Engineering Notes, vol. 7(5), pp. 75–84 (1982)
10. Goguen, J.A., Meseguer, J.: Security policies and security models. In: 1982 IEEE Symposium on Security and Privacy, Oakland, CA, USA, April 26–28, 1982. pp. 11–20. IEEE Computer Society (1982)
11. Goguen, J.A., Meseguer, J.: Universal realization, persistent interconnection and implementation of abstract modules. In: Nielsen, M., Schmidt, E.M. (eds.) Automata, Languages and Programming, 9th Colloquium, Aarhus, Denmark, July 12–16, 1982, Proceedings. LNCS, vol. 140, pp. 265–281. Springer (1982)
12. Kremers, J., Blahnik, C., Brain, A., Cain, R., DeCurtins, J., Meseguer, J., Peppers, N.: Development of a machine-vision based robotic arc-welding system. In: Proceedings 13th International Symposium on Industrial Robotics. pp. 14.19–14.33. Chicago, Illinois (1983)
13. Goguen, J.A., Meseguer, J.: Equality, types, modules and generics for logic programming. In: Tärnlund, S. (ed.) Proceedings of the Second International Logic Programming Conference, Uppsala University, Uppsala, Sweden, July 2–6, 1984. pp. 115–125. Uppsala University (1984)
14. Goguen, J.A., Meseguer, J.: Unwinding and inference control. In: Proceedings of the 1984 IEEE Symposium on Security and Privacy, Oakland, California, USA, April 29 - May 2, 1984. pp. 75–87. IEEE Computer Society (1984)
15. Futatsugi, K., Goguen, J.A., Jouannaud, J., Meseguer, J.: Principles of OBJ2. In: Deusen, M.S.V., Galil, Z., Reid, B.K. (eds.) Conference Record of the Twelfth Annual ACM Symposium on Principles of Programming Languages, New Orleans, Louisiana, USA, January 1985. pp. 52–66. ACM Press (1985)

16. Goguen, J.A., Jouannaud, J., Meseguer, J.: Operational semantics for order-sorted algebra. In: Brauer, W. (ed.) Automata, Languages and Programming, 12th Colloquium, Nafplion, Greece, July 15–19, 1985, Proceedings. LNCS, vol. 194, pp. 221–231. Springer (1985)

17. Goguen, J.A., Kirchner, C., Meseguer, J.: Concurrent term rewriting as a model of computation. In: Fasel, J.H., Keller, R.M. (eds.) Graph Reduction, Proceedings of a Workshop, Santa Fé, New Mexico, USA, September 29 - October 1, 1986. LNCS, vol. 279, pp. 53–93. Springer (1986)

18. Goguen, J.A., Meseguer, J.: Foundations and extensions of object-oriented programming. In: Proceedings Workshop on Object-Oriented Programming. SIGPLAN Notices, vol. 21(10), pp. 153–162 (1986)

19. Futatsugi, K., Goguen, J.A., Meseguer, J., Okada, K.: Parameterized programming in OBJ2. In: Riddle, W.E., Balzer, R.M., Kishida, K. (eds.) Proceedings, 9th International Conference on Software Engineering, Monterey, California, USA, March 30 - April 2, 1987. pp. 51–60. ACM Press (1987)

20. Goguen, J.A., Kirchner, C., Kirchner, H., Mégrelis, A., Meseguer, J., Winkler, T.C.: An introduction to OBJ3. In: Kaplan, S., Jouannaud, J. (eds.) Conditional Term Rewriting Systems, 1st International Workshop, Orsay, France, July 8–10, 1987, Proceedings. LNCS, vol. 308, pp. 258–263. Springer (1987)

21. Goguen, J.A., Kirchner, C., Meseguer, J., Winkler, T.: OBJ as a language for concurrent programming. In: Proceedings of Supercomputing'87. vol. I, pp. 196–198. International Supercomputing Institute (1987)

22. Goguen, J.A., Meseguer, J.: Order-sorted algebra solves the constructor-selector, multiple representation and coercion problems. In: Proceedings of the Symposium on Logic in Computer Science (LICS'87), Ithaca, New York, USA, June 22–25, 1987. pp. 18–29. IEEE Computer Society (1987)

23. Moss, L.S., Meseguer, J., Goguen, J.A.: Final algebras, cosemicomputable algebras, and degrees of unsolvability. In: Pitt, D.H., Poigné, A., Rydeheard, D.E. (eds.) Category Theory and Computer Science, Edinburgh, UK, September 7–9, 1987, Proceedings. LNCS, vol. 283, pp. 158–181. Springer (1987)

24. Goguen, J.A., Meseguer, J.: Software for the rewrite rule machine. In: Proceedings 1998 Fifth Generation Computer Systems Conference. pp. 628–637. Institute for New Generation Computer Technology (1988)

25. Kirchner, C., Kirchner, H., Meseguer, J.: Operational semantics of OBJ3 (extended abstract). In: Lepistö, T., Salomaa, A. (eds.) Automata, Languages and Programming, 15th International Colloquium, ICALP88, Tampere, Finland, July 11–15, 1988, Proceedings. LNCS, vol. 317, pp. 287–301. Springer (1988)

26. Meseguer, J., Montanari, U.: Petri nets are monoids: A new algebraic foundation for net theory. In: Proceedings of the Third Annual Symposium on Logic in Computer Science (LICS '88), Edinburgh, Scotland, UK, July 5–8, 1988. pp. 155–164. IEEE Computer Society (1988)

27. Casley, R., Crew, R.F., Meseguer, J., Pratt, V.R.: Temporal structures. In: Pitt, D.H., Rydeheard, D.E., Dybjer, P., Pitts, A.M., Poigné, A. (eds.) Category Theory and Computer Science, Manchester, UK, September 5–8, 1989, Proceedings. LNCS, vol. 389, pp. 21–51. Springer (1989)

28. Degano, P., Meseguer, J., Montanari, U.: Axiomatizing net computations and processes. In: Proceedings of the Fourth Annual Symposium on Logic in Computer Science (LICS '89), Pacific Grove, California, USA, June 5–8, 1989. pp. 175–185. IEEE Computer Society (1989)

29. Martí-Oliet, N., Meseguer, J.: From Petri nets to linear logic. In: Pitt, D.H., Rydeheard, D.E., Dybjer, P., Pitts, A.M., Poigné, A. (eds.) Category Theory and Computer Science, Manchester, UK, September 5–8, 1989, Proceedings. LNCS, vol. 389, pp. 313–340. Springer (1989)

30. Meseguer, J.: Relating models of polymorphism. In: Conference Record of the Sixteenth Annual ACM Symposium on Principles of Programming Languages, Austin, Texas, USA, January 11–13, 1989. pp. 228–241. ACM Press (1989)

31. Aida, H., Goguen, J.A., Meseguer, J.: Compiling concurrent rewriting onto the rewrite rule machine. In: Kaplan, S., Okada, M. (eds.) Conditional and Typed Rewriting Systems, 2nd International CTRS Workshop, Montreal, Canada, June 11–14, 1990, Proceedings. LNCS, vol. 516, pp. 320–332. Springer (1990)

32. Meseguer, J.: A logical theory of concurrent objects. In: Yonezawa, A. (ed.) Conference on Object-Oriented Programming Systems, Languages, and Applications / European Conference on Object-Oriented Programming (OOPSLA/ECOOP), Ottawa, Canada, October 21–25, 1990, Proceedings. pp. 101–115. ACM (1990)

33. Meseguer, J.: Rewriting as a unified model of concurrency. In: Baeten, J.C.M., Klop, J.W. (eds.) CONCUR '90, Theories of Concurrency: Unification and Extension, Amsterdam, The Netherlands, August 27–30, 1990, Proceedings. LNCS, vol. 458, pp. 384–400. Springer (1990)

34. Aida, H., Goguen, J.A., Lincoln, P., Meseguer, J., Taheri, B., Winkler, T.: Simulation and performance estimation for the rewrite rule machine. In: Proceedings of the 1992 Symposium on the Frontiers of Massively Parallel Computing. pp. 336–344. IEEE (1992)

35. Meseguer, J., Futatsugi, K., Winkler, T.: Using rewriting logic to specify, program, integrate, and reuse open concurrent systems of cooperating agents. In: Proceedings of the 1992 International Symposium on New Models for Software Architecture, Tokyo, Japan, November 1992. pp. 61–106. Research Institute of Software Engineering (1992)

36. Meseguer, J., Montanari, U., Sassone, V.: On the semantics of Petri nets. In: Cleaveland, R. (ed.) CONCUR '92, Third International Conference on Concurrency Theory, Stony Brook, NY, USA, August 24–27, 1992, Proceedings. LNCS, vol. 630, pp. 286–301. Springer (1992)

37. Cerioli, M., Meseguer, J.: May I borrow your logic? In: Borzyszkowski, A.M., Sokolowski, S. (eds.) Mathematical Foundations of Computer Science 1993, 18th International Symposium, MFCS'93, Gdansk, Poland, August 30 - September 3, 1993, Proceedings. LNCS, vol. 711, pp. 342–351. Springer (1993)

38. Meseguer, J.: Solving the inheritance anomaly in concurrent object-oriented programming. In: Nierstrasz, O. (ed.) ECOOP'93 - Object-Oriented Programming, 7th European Conference, Kaiserslautern, Germany, July 26–30, 1993, Proceedings. LNCS, vol. 707, pp. 220–246. Springer (1993)

39. Meseguer, J., Qian, X.: A logical semantics for object-oriented databases. In: Buneman, P., Jajodia, S. (eds.) Proceedings of the 1993 ACM SIGMOD International Conference on Management of Data, Washington, D.C., May 26–28, 1993. pp. 89–98. ACM Press (1993)

40. Lincoln, P., Martí-Oliet, N., Meseguer, J., Ricciulli, L.: Compiling rewriting onto SIMD and MIMD/SIMD machines. In: Halatsis, C., Maritsas, D.G., Philokyprou, G., Theodoridis, S. (eds.) PARLE '94: Parallel Architectures and Languages Europe, 6th International PARLE Conference, Athens, Greece, July 4–8, 1994, Proceedings. LNCS, vol. 817, pp. 37–48. Springer (1994)

41. Lincoln, P., Meseguer, J., Ricciulli, L.: The rewrite rule machine node architecture and its performance. In: Buchberger, B., Volkert, J. (eds.) Parallel Processing: CONPAR 94 - VAPP VI, Third Joint International Conference on Vector and Parallel Processing, Linz, Austria, September 6–8, 1994, Proceedings. LNCS, vol. 854, pp. 509–520. Springer (1994)

42. Clavel, M., Eker, S., Lincoln, P., Meseguer, J.: Principles of Maude. In: Meseguer, J. (ed.) Proceedings of the First International Workshop on Rewriting Logic and its Applications, WRLA'96, Asilomar, California, September 3–6, 1996. ENTCS, vol. 4, pp. 65–89. Elsevier (1996)

43. Clavel, M., Meseguer, J.: Reflection and strategies in rewriting logic. In: Meseguer, J. (ed.) Proceedings of the First International Workshop on Rewriting Logic and its Applications, WRLA'96, Asilomar, California, September 3–6, 1996. ENTCS, vol. 4, pp. 126–148. Elsevier (1996)

44. Ölveczky, P.C., Meseguer, J.: Specifying real-time systems in rewriting logic. In: Meseguer, J. (ed.) Proceedings of the First International Workshop on Rewriting Logic and its Applications, WRLA'96, Asilomar, California, September 3–6, 1996. ENTCS, vol. 4, pp. 284–309. Elsevier (1996)

45. Ricciulli, L., Lincoln, P., Meseguer, J.: Distributed simulation of parallel executions. In: Proceedings 29st Annual Simulation Symposium (SS'96), April 8–11, 1996, New Orleans, LA, USA. pp. 15–24. IEEE Computer Society (1996)

46. Clavel, M., Meseguer, J.: Reflection in rewriting logic and its applications in the Maude language. In: Proceedings of the 1997 International Symposium on New Models for Software Architecture. pp. 128–139. Information-Technology Promotion Agency, Japan (1997)

47. Ishikawa, H., Meseguer, J., Watanabe, T., Futatsugi, K., Nakashima, H.: On the semantics of GAEA — An object-oriented specification of a concurrent reflective language in rewriting logic. In: Proceedings of the 1997 International Symposium on New Models for Software Architecture. pp. 70–109. Information-Technology Promotion Agency, Japan (1997)
48. Bruni, R., Meseguer, J., Montanari, U., Sassone, V.: A comparison of Petri net semantics under the collective token philosophy. In: Hsiang, J., Ohori, A. (eds.) Advances in Computing Science - ASIAN'98, 4th Asian Computing Science Conference, Manila, The Philippines, December 8–10, 1998, Proceedings. LNCS, vol. 1538, pp. 225–244. Springer (1998)
49. Bruni, R., Montanari, U., Meseguer, J.: Internal strategies in a rewriting implementation of tile systems. In: Kirchner, C., Kirchner, H. (eds.) Proceedings of the Second International Workshop on Rewriting Logic and its Applications, WRLA'98, Pont-à-Mousson, France, September 1–4, 1998. ENTCS, vol. 15, pp. 263–284. Elsevier (1998)
50. Clavel, M., Durán, F., Eker, S., Lincoln, P., Martí-Oliet, N., Meseguer, J.: Metalevel computation in Maude. In: Kirchner, C., Kirchner, H. (eds.) Proceedings of the Second International Workshop on Rewriting Logic and its Applications, WRLA'98, Pont-à-Mousson, France, September 1–4, 1998. ENTCS, vol. 15, pp. 331–352. Elsevier (1998)
51. Clavel, M., Durán, F., Eker, S., Lincoln, P., Martí-Oliet, N., Meseguer, J., Quesada, J.F.: Maude as a metalanguage. In: Kirchner, C., Kirchner, H. (eds.) Proceedings of the Second International Workshop on Rewriting Logic and its Applications, WRLA'98, Pont-à-Mousson, France, September 1–4, 1998. ENTCS, vol. 15, pp. 147–160. Elsevier (1998)
52. Durán, F., Meseguer, J.: An extensible module algebra for Maude. In: Kirchner, C., Kirchner, H. (eds.) Proceedings of the Second International Workshop on Rewriting Logic and its Applications, WRLA'98, Pont-à-Mousson, France, September 1–4, 1998. ENTCS, vol. 15, pp. 174–195. Elsevier (1998)
53. Meseguer, J., Talcott, C.L.: Mapping OMRS to rewriting logic. In: Kirchner, C., Kirchner, H. (eds.) Proceedings of the Second International Workshop on Rewriting Logic and its Applications, WRLA'98, Pont-à-Mousson, France, September 1–4, 1998. ENTCS, vol. 15, pp. 33–54. Elsevier (1998)
54. Bruni, R., Meseguer, J., Montanari, U.: Executable tile specifications for process calculi. In: Finance, J. (ed.) Fundamental Approaches to Software Engineering, Second International Conference, FASE'99, Held as Part of the European Joint Conferences on the Theory and Practice of Software, ETAPS'99, Amsterdam, The Netherlands, March 22–28, 1999, Proceedings. LNCS, vol. 1577, pp. 60–76. Springer (1999)
55. Bruni, R., Meseguer, J., Montanari, U., Sassone, V.: Functorial semantics for Petri nets under the individual token philosophy. In: Hofmann, M., Rosolini, G., Pavlovic, D. (eds.) Proceedings of CTCS99, 8th conference on Category Theory and Computer Science. ENTCS, vol. 29, p. 21 (1999)

56. Clavel, M., Durán, F., Eker, S., Lincoln, P., Martí-Oliet, N., Meseguer, J., Quesada, J.F.: The Maude system. In: Narendran, P., Rusinowitch, M. (eds.) Rewriting Techniques and Applications, 10th International Conference, RTA-99, Trento, Italy, July 2–4, 1999, Proceedings. LNCS, vol. 1631, pp. 240–243. Springer (1999)

57. Clavel, M., Durán, F., Eker, S., Meseguer, J., Stehr, M.: Maude as a formal meta-tool. In: Wing, J.M., Woodcock, J., Davies, J. (eds.) FM'99 - Formal Methods, World Congress on Formal Methods in the Development of Computing Systems, Toulouse, France, September 20–24, 1999, Proceedings, Volume II. LNCS, vol. 1709, pp. 1684–1703. Springer (1999)

58. Denker, G., García-Luna-Aceves, J.J., Meseguer, J., Ölveczky, P.C., Raju, J., Smith, B., Talcott, C.L.: Specification and analysis of a reliable broadcasting protocol in Maude. In: Hajek, B., Sreenivas, R.S. (eds.) Proceedings of the 37th Allerton Conference on Communication, Control and Computation. pp. 738–747. University of Illinois at Urbana-Champaign (1999)

59. Durán, F., Meseguer, J.: Structured theories and institutions. In: Hofmann, M., Rosolini, G., Pavlovic, D. (eds.) Proceedings of CTCS99, 8th conference on Category Theory and Computer Science. ENTCS, vol. 29, pp. 23–41 (1999)

60. Meseguer, J., Talcott, C.L.: A partial order event model for concurrent objects. In: Baeten, J.C.M., Mauw, S. (eds.) CONCUR '99: Concurrency Theory, 10th International Conference, Eindhoven, The Netherlands, August 24–27, 1999, Proceedings. LNCS, vol. 1664, pp. 415–430. Springer (1999)

61. Braga, C., Haeusler, E.H., Meseguer, J., Mosses, P.D.: Maude action tool: Using reflection to map action semantics to rewriting logic. In: Rus, T. (ed.) Algebraic Methodology and Software Technology. 8th International Conference, AMAST 2000, Iowa City, Iowa, USA, May 20–27, 2000, Proceedings. LNCS, vol. 1816, pp. 407–421. Springer (2000)

62. Clavel, M., Durán, F., Eker, S., Lincoln, P., Martí-Oliet, N., Meseguer, J., Quesada, J.F.: Towards Maude 2.0. In: Futatsugi, K. (ed.) Proceedings of the Third International Workshop on Rewriting Logic and its Applications, WRLA 2000, Kanazawa, Japan, September 18–20, 2000. ENTCS, vol. 36, pp. 294–315. Elsevier (2000)

63. Clavel, M., Durán, F., Eker, S., Lincoln, P., Martí-Oliet, N., Meseguer, J., Quesada, J.F.: Using Maude. In: Maibaum, T.S.E. (ed.) Fundamental Approaches to Software Engineering, Third International Conference, FASE 2000, Held as Part of the European Joint Conferences on the Theory and Practice of Software, ETAPS 2000, Berlin, Germany, March 25 - April 2, 2000, Proceedings. LNCS, vol. 1783, pp. 371–374. Springer (2000)

64. Coglio, A., Giunchiglia, F., Meseguer, J., Talcott, C.L.: Composing and controlling search in reasoning theories using mappings. In: Kirchner, H., Ringeissen, C. (eds.) Frontiers of Combining Systems, Third International Workshop, FroCoS 2000, Nancy, France, March 22–24, 2000, Proceedings. LNCS, vol. 1794, pp. 200–216. Springer (2000)

65. Denker, G., Meseguer, J., Talcott, C.L.: Formal specification and analysis of active networks and communication protocols: The Maude experience. In: Koob, G., Maughan, D., Saydjari, S. (eds.) Proceedings of the DARPA Information Survivability Conference and Exposition, DISCEX 2000, Hilton Head Island, South Carolina, January 25–27, 2000. pp. 251–265. IEEE Computer Society Press (2000)

66. Denker, G., Meseguer, J., Talcott, C.L.: Rewriting semantics of meta-objects and composable distributed services. In: Futatsugi, K. (ed.) Proceedings of the Third International Workshop on Rewriting Logic and its Applications, WRLA 2000, Kanazawa, Japan, September 18–20, 2000. ENTCS, vol. 36, pp. 405–425. Elsevier (2000)

67. Durán, F., Eker, S., Lincoln, P., Meseguer, J.: Principles of Mobile Maude. In: Kotz, D., Mattern, F. (eds.) Agent Systems, Mobile Agents, and Applications, Second International Symposium on Agent Systems and Applications and Fourth International Symposium on Mobile Agents, ASA/MA 2000, Zürich, Switzerland, September 13–15, 2000, Proceedings. LNCS, vol. 1882, pp. 73–85. Springer (2000)

68. Durán, F., Meseguer, J.: Parameterized theories and views in Full Maude 2.0. In: Futatsugi, K. (ed.) Proceedings of the Third International Workshop on Rewriting Logic and its Applications, WRLA 2000, Kanazawa, Japan, September 18–20, 2000. ENTCS, vol. 36, pp. 316–338. Elsevier (2000)

69. Fiadeiro, J.L., Maibaum, T.S.E., Martí-Oliet, N., Meseguer, J., Pita, I.: Towards a verification logic for rewriting logic. In: Bert, D., Choppy, C., Mosses, P.D. (eds.) Recent Trends in Algebraic Development Techniques, 14th International Workshop, WADT'99, Château de Bonas, France, September 15–18, 1999, Selected Papers. LNCS, vol. 1827, pp. 438–458. Springer (2000)

70. Ölveczky, P.C., Meseguer, J.: Real-Time Maude: A tool for simulating and analyzing real-time and hybrid systems. In: Futatsugi, K. (ed.) Proceedings of the Third International Workshop on Rewriting Logic and its Applications, WRLA 2000, Kanazawa, Japan, September 18–20, 2000. ENTCS, vol. 36, pp. 361–382. Elsevier (2000)

71. Wang, B., Meseguer, J., Gunter, C.A.: Specification and formal analysis of a PLAN algorithm in Maude. In: Lai, T. (ed.) Proceedings of the 2000 ICDCS Workshops, April 10, 2000, Taipei, Taiwan, ROC. pp. E49–E56 (2000)

72. Clavel, M., Durán, F., Eker, S., Lincoln, P., Martí-Oliet, N., Meseguer, J.: Language prototyping in the Maude metalanguage. In: Orejas, F., Cuartero, F., Cazorla, D. (eds.) Actas de las Primeras Jornadas sobre Programación y Lenguajes, PROLE 2001, Almagro (Ciudad Real), España, Noviembre 23–24, 2001. pp. 93–110. Universidad de Castilla La Mancha (2001)

73. Naumov, P., Stehr, M., Meseguer, J.: The HOL/NuPRL proof translator (A practical approach to formal interoperability). In: Boulton, R.J., Jackson, P.B. (eds.) Theorem Proving in Higher Order Logics, 14th International Conference, TPHOLs 2001, Edinburgh, Scotland, UK, September 3–6, 2001, Proceedings. LNCS, vol. 2152, pp. 329–345. Springer (2001)

74. Ölveczky, P.C., Keaton, M., Meseguer, J., Talcott, C.L., Zabele, S.: Specification and analysis of the AER/NCA active network protocol suite in Real-Time Maude. In: Hußmann, H. (ed.) Fundamental Approaches to Software Engineering, 4th International Conference, FASE 2001 Held as Part of the Joint European Conferences on Theory and Practice of Software, ETAPS 2001 Genova, Italy, April 2–6, 2001, Proceedings. LNCS, vol. 2029, pp. 333–348. Springer (2001)

75. Stehr, M.O., Meseguer, J., Ölveczky, P.C.: Representation and execution of Petri nets using rewriting logic as a uniform framework. In: Ehrig, H., Ermel, C., Padberg, J. (eds.) Proceedings of the.Uniform Approaches to Graphical Process Specification Techniques Satellite Event of ETAPS 2001, UNIGRA 2001, Genova, Italy, March 31-April 1, 2001. ENTCS, vol. 44(4), pp. 140–162. Elsevier (2001)

76. Eker, S., Knapp, M., Laderoute, K., Lincoln, P., Meseguer, J., Sönmez, M.K.: Pathway logic: Symbolic analysis of biological signaling. In: Pacific Symposium on Biocomputing. pp. 400–412 (2002)

77. Meseguer, J., Ölveczky, P.C., Stehr, M., Talcott, C.L.: Maude as a wide-spectrum framework for formal modeling and analysis of active networks. In: 2002 DARPA Active Networks Conference and Exposition (DANCE 2002), 29–31 May 2002, San Francisco, CA, USA. pp. 494–510. IEEE Computer Society (2002)

78. Meseguer, J., Roşu, G.: A total approach to partial algebraic specification. In: Widmayer, P., Ruiz, F.T., Bueno, R.M., Hennessy, M., Eidenbenz, S., Conejo, R. (eds.) Automata, Languages and Programming, 29th International Colloquium, ICALP 2002, Málaga, Spain, July 8–13, 2002, Proceedings. LNCS, vol. 2380, pp. 572–584. Springer (2002)

79. Braga, C., Haeusler, E.H., Meseguer, J., Mosses, P.D.: Mapping modular SOS to rewriting logic. In: Leuschel, M. (ed.) Logic Based Program Synthesis and Transformation, 12th International Workshop, LOPSTR 2002, Madrid, Spain, September 17–20, 2002, Revised Selected Papers. LNCS, vol. 2664, pp. 262–277. Springer (2003)

80. Bruni, R., Meseguer, J.: Generalized rewrite theories. In: Baeten, J.C.M., Lenstra, J.K., Parrow, J., Woeginger, G.J. (eds.) Automata, Languages and Programming, 30th International Colloquium, ICALP 2003, Eindhoven, The Netherlands, June 30 - July 4, 2003. Proceedings. LNCS, vol. 2719, pp. 252–266. Springer (2003)

81. Bruni, R., Meseguer, J., Montanari, U., Sassone, V.: Algebraic theories for contextual pre-nets. In: Blundo, C., Laneve, C. (eds.) Theoretical Computer Science, 8th Italian Conference, ICTCS 2003, Bertinoro, Italy, October 13–15, 2003, Proceedings. LNCS, vol. 2841, pp. 256–270. Springer (2003)

82. Clavel, M., Durán, F., Eker, S., Lincoln, P., Martí-Oliet, N., Meseguer, J., Talcott, C.L.: The Maude 2.0 system. In: Nieuwenhuis, R. (ed.) Rewriting Techniques and Applications, 14th International Conference, RTA 2003, Valencia, Spain, June 9–11, 2003, Proceedings. LNCS, vol. 2706, pp. 76–87. Springer (2003)

83. Eker, S., Meseguer, J., Sridharanarayanan, A.: The Maude LTL model checker and its implementation. In: Ball, T., Rajamani, S.K. (eds.) Model Checking Software, 10th International SPIN Workshop. Portland, OR, USA, May 9–10, 2003, Proceedings. LNCS, vol. 2648, pp. 230–234. Springer (2003)

84. Kumar, N., Sen, K., Meseguer, J., Agha, G.: A rewriting based model for probabilistic distributed object systems. In: Najm, E., Nestmann, U., Stevens, P. (eds.) Formal Methods for Open Object-Based Distributed Systems, 6th IFIP WG 6.1 International Conference, FMOODS 2003, Paris, France, November 19.21, 2003, Proceedings. LNCS, vol. 2884, pp. 32–46. Springer (2003)

85. Meseguer, J., Palomino, M., Martí-Oliet, N.: Equational abstractions. In: Baader, F. (ed.) Automated Deduction - CADE-19, 19th International Conference on Automated Deduction Miami Beach, FL, USA, July 28 - August 2, 2003, Proceedings. LNCS, vol. 2741, pp. 2–16. Springer (2003)

86. Roşu, G., Eker, S., Lincoln, P., Meseguer, J.: Certifying and synthesizing membership equational proofs. In: Araki, K., Gnesi, S., Mandrioli, D. (eds.) FME 2003: Formal Methods, International Symposium of Formal Methods Europe, Pisa, Italy, September 8–14, 2003, Proceedings. LNCS, vol. 2805, pp. 359–380. Springer (2003)

87. Braga, C., Meseguer, J.: Modular rewriting semantics in practice. In: Martí-Oliet, N. (ed.) Proceedings of the Fifth International Workshop on Rewriting Logic and its Applications, WRLA 2004, Barcelona, Spain, March 27-April 4, 2004. ENTCS, vol. 117, pp. 393–416. Elsevier (2004)

88. Bruni, R., Meseguer, J., Montanari, U.: Tiling transactions in rewriting logic. In: Gadducci, F., Montanari, U. (eds.) Proceedings of the Fourth International Workshop on Rewriting Logic and its Applications, WRLA 2002, Pisa, Italy, September 19–21, 2002. ENTCS, vol. 71, pp. 90–109. Elsevier (2004)

89. Clavel, M., Meseguer, J., Palomino, M.: Reflection in membership equational logic, many-sorted equational logic, Horn logic with equality, and rewriting logic. In: Gadducci, F., Montanari, U. (eds.) Proceedings of the Fourth International Workshop on Rewriting Logic and its Applications, WRLA 2002, Pisa, Italy, September 19–21, 2002. ENTCS, vol. 71, pp. 110–116. Elsevier (2004)

90. Durán, F., Lucas, S., Meseguer, J., Marché, C., Urbain, X.: Proving termination of membership equational programs. In: Heintze, N., Sestoft, P. (eds.) Proceedings of the 2004 ACM SIGPLAN Workshop on Partial Evaluation and Semantics-based Program Manipulation, 2004, Verona, Italy, August 24–25, 2004. pp. 147–158. ACM (2004)

91. Eker, S., Meseguer, J., Sridharanarayanan, A.: The Maude LTL model checker. In: Gadducci, F., Montanari, U. (eds.) Proceedings of the Fourth International Workshop on Rewriting Logic and its Applications, WRLA 2002, Pisa, Italy, September 19–21, 2002. ENTCS, vol. 71, pp. 162–187. Elsevier (2004)

92. Escobar, S., Meseguer, J., Thati, P.: A general natural rewriting strategy. In: Lucas, S., Gallardo, M., Pimentel, E. (eds.) Actas de las IV Jornadas sobre Programación y Lenguajes, PROLE 2004, Málaga, España, Noviembre 11–12, 2004 (2004)

93. Farzan, A., Chen, F., Meseguer, J., Roşu, G.: Formal analysis of Java programs in JavaFAN. In: Alur, R., Peled, D. (eds.) Computer Aided Verification, 16th International Conference, CAV 2004, Boston, MA, USA, July 13–17, 2004, Proceedings. LNCS, vol. 3114, pp. 501–505. Springer (2004)

94. Farzan, A., Meseguer, J., Roşu, G.: Formal JVM code analysis in JavaFAN. In: Rattray, C., Maharaj, S., Shankland, C. (eds.) Algebraic Methodology and Software Technology, 10th International Conference, AMAST 2004, Stirling, Scotland, UK, July 12–16, 2004, Proceedings. LNCS, vol. 3116, pp. 132–147. Springer (2004)

95. Martí-Oliet, N., Meseguer, J., Palomino, M.: Theoroidal maps as algebraic simulations. In: Fiadeiro, J.L., Mosses, P.D., Orejas, F. (eds.) Recent Trends in Algebraic Development Techniques, 17th International Workshop, WADT 2004, Barcelona, Spain, March 27–29, 2004, Revised Selected Papers. LNCS, vol. 3423, pp. 126–143. Springer (2004)

96. Martí-Oliet, N., Meseguer, J., Verdejo, A.: Towards a strategy language for Maude. In: Martí-Oliet, N. (ed.) Proceedings of the Fifth International Workshop on Rewriting Logic and its Applications, WRLA 2004, Barcelona, Spain, March 27-April 4, 2004. ENTCS, vol. 117, pp. 417–441. Elsevier (2004)

97. Meseguer, J., Braga, C.: Modular rewriting semantics of programming languages. In: Rattray, C., Maharaj, S., Shankland, C. (eds.) Algebraic Methodology and Software Technology, 10th International Conference, AMAST 2004, Stirling, Scotland, UK, July 12–16, 2004, Proceedings. LNCS, vol. 3116, pp. 364–378. Springer (2004)

98. Meseguer, J., Thati, P.: Symbolic reachability analysis using narrowing and its application to verification of cryptographic protocols. In: Martí-Oliet, N. (ed.) Proceedings of the Fifth International Workshop on Rewriting Logic and its Applications, WRLA 2004, Barcelona, Spain, March 27-April 4, 2004. ENTCS, vol. 117, pp. 153–182. Elsevier (2004)

99. Ölveczky, P.C., Meseguer, J.: Real-Time Maude 2.1. In: Martí-Oliet, N. (ed.) Proceedings of the Fifth International Workshop on Rewriting Logic and its Applications, WRLA 2004, Barcelona, Spain, March 27-April 4, 2004. ENTCS, vol. 117, pp. 285–314. Elsevier (2004)

100. Ölveczky, P.C., Meseguer, J.: Specification and analysis of real-time systems using Real-Time Maude. In: Wermelinger, M., Margaria, T. (eds.) Fundamental Approaches to Software Engineering, 7th International Conference, FASE 2004, Held as Part of the Joint European Conferences on Theory and Practice of Software, ETAPS 2004 Barcelona, Spain, March 29 - April 2, 2004, Proceedings. LNCS, vol. 2984, pp. 354–358. Springer (2004)

101. Escobar, S., Meadows, C., Meseguer, J.: A rewriting-based inference system for the NRL protocol analyzer: grammar generation. In: Atluri, V., Samarati, P., Küsters, R., Mitchell, J.C. (eds.) Proceedings of the 2005 ACM workshop on Formal methods in security engineering, FMSE 2005, Fairfax, VA, USA, November 11, 2005. pp. 1–12. ACM (2005)

102. Escobar, S., Meseguer, J., Thati, P.: Natural rewriting for general term rewriting systems. In: Etalle, S. (ed.) Logic Based Program Synthesis and Transformation, 14th International Symposium, LOPSTR 2004, Verona, Italy, August 26–28, 2004, Revised Selected Papers. LNCS, vol. 3573, pp. 101–116. Springer (2005)

103. Escobar, S., Meseguer, J., Thati, P.: Natural narrowing for general term rewriting systems. In: Giesl, J. (ed.) Term Rewriting and Applications, 16th International Conference, RTA 2005, Nara, Japan, April 19–21, 2005, Proceedings. LNCS, vol. 3467, pp. 279–293. Springer (2005)

104. Hendrix, J., Clavel, M., Meseguer, J.: A sufficient completeness reasoning tool for partial specifications. In: Giesl, J. (ed.) Term Rewriting and Applications, 16th International Conference, RTA 2005, Nara, Japan, April 19–21, 2005, Proceedings. LNCS, vol. 3467, pp. 165–174. Springer (2005)

105. Lucas, S., Meseguer, J.: Termination of fair computations in term rewriting. In: Sutcliffe, G., Voronkov, A. (eds.) Logic for Programming, Artificial Intelligence, and Reasoning, 12th International Conference, LPAR 2005, Montego Bay, Jamaica, December 2–6, 2005, Proceedings. LNCS, vol. 3835, pp. 184–198. Springer (2005)

106. Meseguer, J.: Localized fairness: A rewriting semantics. In: Giesl, J. (ed.) Term Rewriting and Applications, 16th International Conference, RTA 2005, Nara, Japan, April 19–21, 2005, Proceedings. LNCS, vol. 3467, pp. 250–263. Springer (2005)

107. Palomino, M., Meseguer, J., Martí-Oliet, N.: A categorical approach to simulations. In: Fiadeiro, J.L., Harman, N., Roggenbach, M., Rutten, J.J.M.M. (eds.) Algebra and Coalgebra in Computer Science: First International Conference, CALCO 2005, Swansea, UK, September 3–6, 2005, Proceedings. LNCS, vol. 3629, pp. 313–330. Springer (2005)

108. Thati, P., Meseguer, J.: Complete symbolic reachability analysis using back-and-forth narrowing. In: Fiadeiro, J.L., Harman, N., Roggenbach, M., Rutten, J.J.M.M. (eds.) Algebra and Coalgebra in Computer Science: First International Conference, CALCO 2005, Swansea, UK, September 3–6, 2005, Proceedings. LNCS, vol. 3629, pp. 379–394. Springer (2005)

109. Agha, G.A., Meseguer, J., Sen, K.: PMaude: Rewrite-based specification language for probabilistic object systems. In: Cerone, A., Wiklicky, H. (eds.) Proceedings of the Third Workshop on Quantitative Aspects of Programming Languages, QAPL 2005, Edinburgh, UK, April 2–3, 2005. ENTCS, vol. 153(2), pp. 213–239. Elsevier (2006)
110. Farzan, A., Meseguer, J.: State space reduction of rewrite theories using invisible transitions. In: Johnson, M., Vene, V. (eds.) Algebraic Methodology and Software Technology, 11th International Conference, AMAST 2006, Kuressaare, Estonia, July 5–8, 2006, Proceedings. LNCS, vol. 4019, pp. 142–157. Springer (2006)
111. Garrido, A., Meseguer, J.: Formal specification and verification of Java refactorings. In: Proceedings of the Sixth IEEE International Workshop on Source Code Analysis and Manipulation, SCAM 2006, Philadelphia, Pennsylvania, September 27–29, 2006. pp. 165–174. IEEE (2006)
112. Hendrix, J., Meseguer, J., Ohsaki, H.: A sufficient completeness checker for linear order-sorted specifications modulo axioms. In: Furbach, U., Shankar, N. (eds.) Automated Reasoning, Third International Joint Conference, IJCAR 2006, Seattle, WA, USA, August 17–20, 2006, Proceedings. LNCS, vol. 4130, pp. 151–155. Springer (2006)
113. Meseguer, J., Sharykin, R.: Specification and analysis of distributed object-based stochastic hybrid systems. In: Hespanha, J.P., Tiwari, A. (eds.) Hybrid Systems: Computation and Control, 9th International Workshop, HSCC 2006, Santa Barbara, CA, USA, March 29–31, 2006, Proceedings. LNCS, vol. 3927, pp. 460–475. Springer (2006)
114. AlTurki, M., Meseguer, J.: Real-time rewriting semantics of Orc. In: Leuschel, M., Podelski, A. (eds.) Proceedings of the 9th International ACM SIGPLAN Conference on Principles and Practice of Declarative Programming, July 14–16, 2007, Wroclaw, Poland. pp. 131–142. ACM (2007)
115. Chen, S., Meseguer, J., Sasse, R., Wang, H.J., Wang, Y.: A systematic approach to uncover security flaws in GUI logic. In: 2007 IEEE Symposium on Security and Privacy (S&P 2007), 20–23 May 2007, Oakland, California, USA. pp. 71–85. IEEE Computer Society (2007)
116. Clavel, M., Durán, F., Hendrix, J., Lucas, S., Meseguer, J., Ölveczky, P.C.: The Maude formal tool environment. In: Mossakowski, T., Montanari, U., Haveraaen, M. (eds.) Algebra and Coalgebra in Computer Science, Second International Conference, CALCO 2007, Bergen, Norway, August 20–24, 2007, Proceedings. LNCS, vol. 4624, pp. 173–178. Springer (2007)
117. Escobar, S., Meadows, C., Meseguer, J.: Equational cryptographic reasoning in the Maude-NRL Protocol Analyzer. In: Fernández, M., Kirchner, C. (eds.) Proceedings of the First International Workshop on Security and Rewriting Techniques, SecReT 2006, Venice, Italy, July 15, 2006. ENTCS, vol. 171(4), pp. 23–36. Elsevier (2007)
118. Escobar, S., Meseguer, J.: Symbolic model checking of infinite-state systems using narrowing. In: Baader, F. (ed.) Term Rewriting and Applications, 18th International Conference, RTA 2007, Paris, France, June 26–28, 2007, Proceedings. LNCS, vol. 4533, pp. 153–168. Springer (2007)

119. Farzan, A., Meseguer, J.: Partial order reduction for rewriting semantics of programming languages. In: Denker, G., Talcott, C. (eds.) Proceedings of the Sixth International Workshop on Rewriting Logic and its Applications, WRLA 2006, Vienna, Austria, April 1–2, 2006. ENTCS, vol. 176(4), pp. 61–78. Elsevier (2007)

120. Hendrix, J., Meseguer, J.: On the completeness of context-sensitive order-sorted specifications. In: Baader, F. (ed.) Term Rewriting and Applications, 18th International Conference, RTA 2007, Paris, France, June 26–28, 2007, Proceedings. LNCS, vol. 4533, pp. 229–245. Springer (2007)

121. Katelman, M., Meseguer, J.: A rewriting semantics for ABEL with applications to hardware/software co-design and analysis. In: Denker, G., Talcott, C. (eds.) Proceedings of the Sixth International Workshop on Rewriting Logic and its Applications, WRLA 2006, Vienna, Austria, April 1–2, 2006. ENTCS, vol. 176(4), pp. 47–60. Elsevier (2007)

122. Ölveczky, P.C., Meseguer, J.: Abstraction and completeness for Real-Time Maude. In: Denker, G., Talcott, C. (eds.) Proceedings of the Sixth International Workshop on Rewriting Logic and its Applications, WRLA 2006, Vienna, Austria, April 1–2, 2006. ENTCS, vol. 176(4), pp. 5–27. Elsevier (2007)

123. Ölveczky, P.C., Meseguer, J.: Recent advances in Real-Time Maude. In: Fernández, M., Lämmel, R. (eds.) Proceedings of the 7th International Workshop on Rule Based Programming, RULE 2006, Seattle, WA, USA, August 11, 2006. ENTCS, vol. 174(1), pp. 65–81. Elsevier (2007)

124. Sasse, R., Meseguer, J.: Java+ITP: A verification tool based on Hoare logic and algebraic semantics. In: Denker, G., Talcott, C. (eds.) Proceedings of the Sixth International Workshop on Rewriting Logic and its Applications, WRLA 2006, Vienna, Austria, April 1–2, 2006. ENTCS, vol. 176(4), pp. 29–46. Elsevier (2007)

125. Şerbănuţă, T.F., Roşu, G., Meseguer, J.: A rewriting logic approach to operational semantics (extended abstract). In: van Glabbeek, R., Hennessy, M. (eds.) Proceedings of the Fourth Workshop on Structural Operational Semantics, SOS 2007, Wroclaw, Poland, July 9, 2007. ENTCS, vol. 192(1), pp. 125–141. Elsevier (2007)

126. Alpuente, M., Escobar, S., Meseguer, J., Ojeda, P.: A modular equational generalization algorithm. In: Hanus, M. (ed.) Logic-Based Program Synthesis and Transformation, 18th International Symposium, LOPSTR 2008, Valencia, Spain, July 17–18, 2008, Revised Selected Papers. LNCS, vol. 5438, pp. 24–39. Springer (2008)

127. AlTurki, M., Meseguer, J.: Reduction semantics and formal analysis of Orc programs. In: Ballis, D., Escobar, S., Marchiori, M. (eds.) Proceedings of the 3rd International Workshop on Automated Specification and Verification of Web Systems, WWV 2007, Venice, Italy, December 14, 2007. ENTCS, vol. 200(3), pp. 25–41. Elsevier (2008)

128. Boronat, A., Knapp, A., Meseguer, J., Wirsing, M.: What is a multi-modeling language? In: Corradini, A., Montanari, U. (eds.) Recent Trends in Algebraic Development Techniques, 19th International Workshop, WADT 2008, Pisa, Italy, June 13–16, 2008, Revised Selected Papers. LNCS, vol. 5486, pp. 71–87. Springer (2008)

129. Boronat, A., Meseguer, J.: An algebraic semantics for MOF. In: Fiadeiro, J.L., Inverardi, P. (eds.) Fundamental Approaches to Software Engineering, 11th International Conference, FASE 2008, Held as Part of the Joint European Conferences on Theory and Practice of Software, ETAPS 2008, Budapest, Hungary, March 29-April 6, 2008. Proceedings. LNCS, vol. 4961, pp. 377–391. Springer (2008)

130. Chadha, R., Gunter, C.A., Meseguer, J., Shankesi, R., Viswanathan, M.: Modular preservation of safety properties by cookie-based DoS-protection wrappers. In: Barthe, G., de Boer, F.S. (eds.) Formal Methods for Open Object-Based Distributed Systems, 10th IFIP WG 6.1 International Conference, FMOODS 2008, Oslo, Norway, June 4–6, 2008, Proceedings. LNCS, vol. 5051, pp. 39–58. Springer (2008)

131. Durán, F., Lucas, S., Meseguer, J.: MTT: the Maude termination tool (system description). In: Armando, A., Baumgartner, P., Dowek, G. (eds.) Automated Reasoning, 4th International Joint Conference, IJCAR 2008, Sydney, Australia, August 12–15, 2008, Proceedings. LNCS, vol. 5195, pp. 313–319. Springer (2008)

132. Escobar, S., Meadows, C., Meseguer, J.: State space reduction in the Maude-NRL protocol analyzer. In: Jajodia, S., López, J. (eds.) Computer Security - ESORICS 2008, 13th European Symposium on Research in Computer Security, Málaga, Spain, October 6–8, 2008. Proceedings. LNCS, vol. 5283, pp. 548–562. Springer (2008)

133. Escobar, S., Meseguer, J., Sasse, R.: Effectively checking the finite variant property. In: Voronkov, A. (ed.) Rewriting Techniques and Applications, 19th International Conference, RTA 2008, Hagenberg, Austria, July 15–17, 2008, Proceedings. LNCS, vol. 5117, pp. 79–93. Springer (2008)

134. Katelman, M., Meseguer, J., Escobar, S.: Directed-logical testing for functional verification of microprocessors. In: 6th ACM & IEEE International Conference on Formal Methods and Models for Co-Design (MEMOCODE 2008), June 5–7, 2008, Anaheim, CA, USA. pp. 89–100. IEEE (2008)

135. Katelman, M., Meseguer, J., Hou, J.C.: Redesign of the LMST wireless sensor protocol through formal modeling and statistical model checking. In: Barthe, G., de Boer, F.S. (eds.) Formal Methods for Open Object-Based Distributed Systems, 10th IFIP WG 6.1 International Conference, FMOODS 2008, Oslo, Norway, June 4–6, 2008, Proceedings. LNCS, vol. 5051, pp. 150–169. Springer (2008)

136. Lucas, S., Meseguer, J.: Order-sorted dependency pairs. In: Antoy, S., Albert, E. (eds.) Proceedings of the 10th International ACM SIGPLAN Conference on Principles and Practice of Declarative Programming, July 15–17, 2008, Valencia, Spain. pp. 108–119. ACM (2008)

137. Martí-Oliet, N., Meseguer, J., Palomino, M.: Algebraic stuttering simulations. In: Pimentel, E. (ed.) Proceedings of the Seventh Spanish Conference on Programming and Computer Languages, PROLE 2007, Zaragoza, Spain, September 12–14, 2007. ENTCS, vol. 206, pp. 91–110. Elsevier (2008)

138. Meseguer, J., Roşu, G.: Computational logical frameworks and generic program analysis technologies. In: Meyer, B., Woodcock, J. (eds.) Verified Software: Theories, Tools, Experiments, First IFIP TC 2/WG 2.3 Conference, VSTTE 2005, Zurich, Switzerland, October 10–13, 2005, Revised Selected Papers and Discussions. LNCS, vol. 4171, pp. 256–267. Springer (2008)

139. Ölveczky, P.C., Meseguer, J.: The Real-Time Maude tool. In: Ramakrishnan, C.R., Rehof, J. (eds.) Tools and Algorithms for the Construction and Analysis of Systems, 14th International Conference, TACAS 2008, Held as Part of the Joint European Conferences on Theory and Practice of Software, ETAPS 2008, Budapest, Hungary, March 29-April 6, 2008. Proceedings. LNCS, vol. 4963, pp. 332–336. Springer (2008)

140. Rocha, C., Meseguer, J.: Theorem proving modulo based on Boolean equational procedures. In: Berghammer, R., Möller, B., Struth, G. (eds.) Relations and Kleene Algebra in Computer Science, 10th International Conference on Relational Methods in Computer Science, and 5th International Conference on Applications of Kleene Algebra, RelMiCS/AKA 2008, Frauenwörth, Germany, April 7–11, 2008. Proceedings. LNCS, vol. 4988, pp. 337–351. Springer (2008)

141. Alpuente, M., Escobar, S., Meseguer, J., Ojeda, P.: Order-sorted generalization. In: Falaschi, M. (ed.) Proceedings of the 17th International Workshop on Functional and (Constraint) Logic Programming, WFLP 2008, Siena, Italy, July 3–4, 2008. ENTCS, vol. 246, pp. 27–38. Elsevier (2009)

142. AlTurki, M., Meseguer, J., Gunter, C.A.: Probabilistic modeling and analysis of DoS protection for the ASV protocol. In: Dougherty, D.J., Escobar, S. (eds.) Proceedings of the Third International Workshop on Security and Rewriting Techniques, SecReT 2008, Pittsburgh, PA, USA, June 22, 2008. ENTCS, vol. 234, pp. 3–18. Elsevier (2009)

143. Boronat, A., Heckel, R., Meseguer, J.: Rewriting logic semantics and verification of model transformations. In: Chechik, M., Wirsing, M. (eds.) Fundamental Approaches to Software Engineering, 12th International Conference, FASE 2009, Held as Part of the Joint European Conferences on Theory and Practice of Software, ETAPS 2009, York, UK, March 22–29, 2009. Proceedings. LNCS, vol. 5503, pp. 18–33. Springer (2009)

144. Boronat, A., Meseguer, J.: Algebraic semantics of OCL-constrained metamodel specifications. In: Oriol, M., Meyer, B. (eds.) Objects, Components, Models and Patterns, 47th International Conference, TOOLS EUROPE 2009, Zurich, Switzerland, June 29-July 3, 2009. Proceedings. Lecture Notes in Business Information Processing, vol. 33, pp. 96–115. Springer (2009)

145. Boronat, A., Meseguer, J.: MOMENT2: EMF model transformations in Maude. In: Vallecillo, A., Sagardui, G. (eds.) XIV Jornadas de Ingeniería del Software y Bases de Datos (JISBD 2009), San Sebastián, Spain, September 8–11, 2009. pp. 178–179 (2009)

146. Clavel, M., Durán, F., Eker, S., Escobar, S., Lincoln, P., Martí-Oliet, N., Meseguer, J., Talcott, C.L.: Unification and narrowing in Maude 2.4. In: Treinen, R. (ed.) Rewriting Techniques and Applications, 20th International Conference, RTA 2009, Brasília, Brazil, June 29 - July 1, 2009, Proceedings. LNCS, vol. 5595, pp. 380–390. Springer (2009)
147. Durán, F., Lucas, S., Meseguer, J.: Methods for proving termination of rewriting-based programming languages by transformation. In: Almendros-Jiménez, J.M. (ed.) Proceedings of the Eighth Spanish Conference on Programming and Computer Languages, PROLE 2008, Gijón, Spain, October 8–10, 2008. ENTCS, vol. 248, pp. 93–113. Elsevier (2009)
148. Durán, F., Lucas, S., Meseguer, J.: Termination modulo combinations of equational theories. In: Ghilardi, S., Sebastiani, R. (eds.) Frontiers of Combining Systems, 7th International Symposium, FroCoS 2009, Trento, Italy, September 16–18, 2009. Proceedings. LNCS, vol. 5749, pp. 246–262. Springer (2009)
149. Durán, F., Lucas, S., Meseguer, J., Gutiérrez, F.: Web services and interoperability for the Maude termination tool. In: Almendros-Jiménez, J.M. (ed.) Proceedings of the Eighth Spanish Conference on Programming and Computer Languages, PROLE 2008, Gijón, Spain, October 8–10, 2008. ENTCS, vol. 248, pp. 83–92. Elsevier (2009)
150. Escobar, S., Meseguer, J., Sasse, R.: Variant narrowing and equational unification. In: Roşu, G. (ed.) Proceedings of the Seventh International Workshop on Rewriting Logic and its Applications, WRLA 2008, Budapest, Hungary, March 29–30, 2008. ENTCS, vol. 238(3), pp. 103–119. Elsevier (2009)
151. Lucas, S., Meseguer, J.: Operational termination of membership equational programs: the order-sorted way. In: Roşu, G. (ed.) Proceedings of the Seventh International Workshop on Rewriting Logic and its Applications, WRLA 2008, Budapest, Hungary, March 29–30, 2008. ENTCS, vol. 238(3), pp. 207–225. Elsevier (2009)
152. Martí-Oliet, N., Meseguer, J., Verdejo, A.: A rewriting semantics for Maude strategies. In: Roşu, G. (ed.) Proceedings of the Seventh International Workshop on Rewriting Logic and its Applications, WRLA 2008, Budapest, Hungary, March 29–30, 2008. ENTCS, vol. 238(3), pp. 227–247. Elsevier (2009)
153. Miller, S.P., Cofer, D., Sha, L., Meseguer, J., Al-Nayeem, A.: Implementing logical synchrony in integrated modular avionics. In: Proc. 28th Digital Avionics Systems Conference. IEEE (2009)
154. Santiago, S., Talcott, C.L., Escobar, S., Meadows, C., Meseguer, J.: A graphical user interface for Maude-NPA. In: Lucio, P., Moreno, G., Peña, R. (eds.) Proceedings of the Ninth Spanish Conference on Programming and Languages, PROLE 2009, San Sebastián, Spain, September 9–11, 2009. ENTCS, vol. 258(1), pp. 3–20. Elsevier (2009)

155. Shankesi, R., AlTurki, M., Sasse, R., Gunter, C.A., Meseguer, J.: Model-checking DoS amplification for VoIP session initiation. In: Backes, M., Ning, P. (eds.) Computer Security - ESORICS 2009, 14th European Symposium on Research in Computer Security, Saint-Malo, France, September 21–23, 2009. Proceedings. LNCS, vol. 5789, pp. 390–405. Springer (2009)

156. Alarcón, B., Lucas, S., Meseguer, J.: A dependency pair framework for $A \vee C$-termination. In: Ölveczky, P.C. (ed.) Rewriting Logic and Its Applications - 8th International Workshop, WRLA 2010, Held as a Satellite Event of ETAPS 2010, Paphos, Cyprus, March 20–21, 2010, Revised Selected Papers. LNCS, vol. 6381, pp. 35–51. Springer (2010)

157. AlTurki, M., Meseguer, J.: Dist-Orc: A rewriting-based distributed implementation of Orc with formal analysis. In: Ölveczky, P.C. (ed.) Proceedings First International Workshop on Rewriting Techniques for Real-Time Systems, RTRTS 2010, Longyearbyen, Norway, April 6–9, 2010. EPTCS, vol. 36, pp. 26–45 (2010)

158. Bae, K., Meseguer, J.: The linear temporal logic of rewriting Maude model checker. In: Ölveczky, P.C. (ed.) Rewriting Logic and Its Applications - 8th International Workshop, WRLA 2010, Held as a Satellite Event of ETAPS 2010, Paphos, Cyprus, March 20–21, 2010, Revised Selected Papers. LNCS, vol. 6381, pp. 208–225. Springer (2010)

159. Durán, F., Meseguer, J.: A Church-Rosser checker tool for conditional order-sorted equational Maude specifications. In: Ölveczky, P.C. (ed.) Rewriting Logic and Its Applications - 8th International Workshop, WRLA 2010, Held as a Satellite Event of ETAPS 2010, Paphos, Cyprus, March 20–21, 2010, Revised Selected Papers. LNCS, vol. 6381, pp. 69–85. Springer (2010)

160. Durán, F., Meseguer, J.: A Maude coherence checker tool for conditional order-sorted rewrite theories. In: Ölveczky, P.C. (ed.) Rewriting Logic and Its Applications - 8th International Workshop, WRLA 2010, Held as a Satellite Event of ETAPS 2010, Paphos, Cyprus, March 20–21, 2010, Revised Selected Papers. LNCS, vol. 6381, pp. 86–103. Springer (2010)

161. Escobar, S., Meadows, C., Meseguer, J., Santiago, S.: Sequential protocol composition in Maude-NPA. In: Gritzalis, D., Preneel, B., Theoharidou, M. (eds.) Computer Security - ESORICS 2010, 15th European Symposium on Research in Computer Security, Athens, Greece, September 20–22, 2010. Proceedings. LNCS, vol. 6345, pp. 303–318. Springer (2010)

162. Escobar, S., Sasse, R., Meseguer, J.: Folding variant narrowing and optimal variant termination. In: Ölveczky, P.C. (ed.) Rewriting Logic and Its Applications - 8th International Workshop, WRLA 2010, Held as a Satellite Event of ETAPS 2010, Paphos, Cyprus, March 20–21, 2010, Revised Selected Papers. LNCS, vol. 6381, pp. 52–68. Springer (2010)

163. Hendrix, J., Kapur, D., Meseguer, J.: Coverset induction with partiality and subsorts: A powerlist case study. In: Kaufmann, M., Paulson, L.C. (eds.) Interactive Theorem Proving, First International Conference, ITP 2010, Edinburgh, UK, July 11–14, 2010. Proceedings. LNCS, vol. 6172, pp. 275–290. Springer (2010)

164. Katelman, M., Keller, S., Meseguer, J.: Concurrent rewriting semantics and analysis of asynchronous digital circuits. In: Ölveczky, P.C. (ed.) Rewriting Logic and Its Applications - 8th International Workshop, WRLA 2010, Held as a Satellite Event of ETAPS 2010, Paphos, Cyprus, March 20–21, 2010, Revised Selected Papers. LNCS, vol. 6381, pp. 140–156. Springer (2010)

165. Katelman, M., Meseguer, J.: vlogsl: A strategy language for simulation-based verification of hardware. In: Barner, S., Harris, I.G., Kroening, D., Raz, O. (eds.) Hardware and Software: Verification and Testing - 6th International Haifa Verification Conference, HVC 2010, Haifa, Israel, October 4–7, 2010. Revised Selected Papers. LNCS, vol. 6504, pp. 129–145. Springer (2010)

166. Katelman, M., Meseguer, J.: Using the PALS architecture to verify a distributed topology control protocol for wireless multi-hop networks in the presence of node failures. In: Ölveczky, P.C. (ed.) Proceedings First International Workshop on Rewriting Techniques for Real-Time Systems, RTRTS 2010, Longyearbyen, Norway, April 6–9, 2010. EPTCS, vol. 36, pp. 101–116 (2010)

167. Meredith, P.O., Katelman, M., Meseguer, J., Roşu, G.: A formal executable semantics of Verilog. In: 8th ACM/IEEE International Conference on Formal Methods and Models for Codesign (MEMOCODE 2010), Grenoble, France, 26–28 July 2010. pp. 179–188. IEEE (2010)

168. Meseguer, J., Ölveczky, P.C.: Formalization and correctness of the PALS architectural pattern for distributed real-time systems. In: Dong, J.S., Zhu, H. (eds.) Formal Methods and Software Engineering - 12th International Conference on Formal Engineering Methods, ICFEM 2010, Shanghai, China, November 17–19, 2010. Proceedings. LNCS, vol. 6447, pp. 303–320. Springer (2010)

169. Ölveczky, P.C., Boronat, A., Meseguer, J.: Formal semantics and analysis of behavioral AADL models in Real-Time Maude. In: Hatcliff, J., Zucca, E. (eds.) Formal Techniques for Distributed Systems, Joint 12th IFIP WG 6.1 International Conference, FMOODS 2010 and 30th IFIP WG 6.1 International Conference, FORTE 2010, Amsterdam, The Netherlands, June 7–9, 2010. Proceedings. LNCS, vol. 6117, pp. 47–62. Springer (2010)

170. Ölveczky, P.C., Meseguer, J.: Specification and verification of distributed embedded systems: A traffic intersection product family. In: Ölveczky, P.C. (ed.) Proceedings First International Workshop on Rewriting Techniques for Real-Time Systems, RTRTS 2010, Longyearbyen, Norway, April 6–9, 2010. EPTCS, vol. 36, pp. 137–157 (2010)

171. Rocha, C., Meseguer, J.: Constructors, sufficient completeness, and deadlock freedom of rewrite theories. In: Fermüller, C.G., Voronkov, A. (eds.) Logic for Programming, Artificial Intelligence, and Reasoning - 17th International Conference, LPAR-17, Yogyakarta, Indonesia, October 10–15, 2010. Proceedings. LNCS, vol. 6397, pp. 594–609. Springer (2010)

172. Sasse, R., Escobar, S., Meadows, C., Meseguer, J.: Protocol analysis modulo combination of theories: A case study in Maude-NPA. In: Cuéllar, J., Lopez, J., Barthe, G., Pretschner, A. (eds.) Security and Trust Management - 6th International Workshop, STM 2010, Athens, Greece, September 23–24, 2010, Revised Selected Papers. LNCS, vol. 6710, pp. 163–178. Springer (2010)

173. Sun, M., Meseguer, J.: Distributed real-time emulation of formally-defined patterns for safe medical device control. In: Ölveczky, P.C. (ed.) Proceedings First International Workshop on Rewriting Techniques for Real-Time Systems, RTRTS 2010, Longyearbyen, Norway, April 6–9, 2010. EPTCS, vol. 36, pp. 158–177 (2010)

174. Sun, M., Meseguer, J., Sha, L.: A formal pattern architecture for safe medical systems. In: Ölveczky, P.C. (ed.) Rewriting Logic and Its Applications - 8th International Workshop, WRLA 2010, Held as a Satellite Event of ETAPS 2010, Paphos, Cyprus, March 20–21, 2010, Revised Selected Papers. LNCS, vol. 6381, pp. 157–173. Springer (2010)

175. AlTurki, M., Meseguer, J.: PVeStA: A parallel statistical model checking and quantitative analysis tool. In: Corradini, A., Klin, B., Cîrstea, C. (eds.) Algebra and Coalgebra in Computer Science - 4th International Conference, CALCO 2011, Winchester, UK, August 30 - September 2, 2011. Proceedings. LNCS, vol. 6859, pp. 386–392. Springer (2011)

176. Bae, K., Meseguer, J.: State/event-based LTL model checking under parametric generalized fairness. In: Gopalakrishnan, G., Qadeer, S. (eds.) Computer Aided Verification - 23rd International Conference, CAV 2011, Snowbird, UT, USA, July 14–20, 2011. Proceedings. LNCS, vol. 6806, pp. 132–148. Springer (2011)

177. Bae, K., Ölveczky, P.C., Al-Nayeem, A., Meseguer, J.: Synchronous AADL and its formal analysis in Real-Time Maude. In: Qin, S., Qiu, Z. (eds.) Formal Methods and Software Engineering - 13th International Conference on Formal Engineering Methods, ICFEM 2011, Durham, UK, October 26–28, 2011. Proceedings. LNCS, vol. 6991, pp. 651–667. Springer (2011)

178. Boronat, A., Meseguer, J.: Automated model synchronization: A case study on UML with Maude. ECEASST 41 (2011)

179. Dave, N., Katelman, M., King, M., Arvind, Meseguer, J.: Verification of microarchitectural refinements in rule-based systems. In: Singh, S., Jobstmann, B., Kishinevsky, M., Brandt, J. (eds.) 9th IEEE/ACM International Conference on Formal Methods and Models for Codesign, MEMOCODE 2011, Cambridge, UK, 11–13 July, 2011. pp. 61–71. IEEE (2011)

180. Durán, F., Eker, S., Escobar, S., Meseguer, J., Talcott, C.L.: Variants, unification, narrowing, and symbolic reachability in Maude 2.6. In: Schmidt-Schauß, M. (ed.) Proceedings of the 22nd International Conference on Rewriting Techniques and Applications, RTA 2011, May 30 - June 1, 2011, Novi Sad, Serbia. LIPIcs, vol. 10, pp. 31–40. Schloss Dagstuhl - Leibniz-Zentrum fuer Informatik (2011)

181. Escobar, S., Kapur, D., Lynch, C., Meadows, C., Meseguer, J., Narendran, P., Sasse, R.: Protocol analysis in Maude-NPA using unification modulo homomorphic encryption. In: Schneider-Kamp, P., Hanus, M. (eds.) Proceedings of the 13th International ACM SIGPLAN Conference on Principles and Practice of Declarative Programming, July 20–22, 2011, Odense, Denmark. pp. 65–76. ACM (2011)

182. Rocha, C., Meseguer, J.: Proving safety properties of rewrite theories. In: Corradini, A., Klin, B., Cîrstea, C. (eds.) Algebra and Coalgebra in Computer Science - 4th International Conference, CALCO 2011, Winchester, UK, August 30 - September 2, 2011. Proceedings. LNCS, vol. 6859, pp. 314–328. Springer (2011)

183. Schernhammer, F., Meseguer, J.: Incremental checking of well-founded recursive specifications modulo axioms. In: Schneider-Kamp, P., Hanus, M. (eds.) Proceedings of the 13th International ACM SIGPLAN Conference on Principles and Practice of Declarative Programming, July 20–22, 2011, Odense, Denmark. pp. 5–16. ACM (2011)

184. Bae, K., Krisiloff, J., Meseguer, J., Ölveczky, P.C.: PALS-based analysis of an airplane multirate control system in Real-Time Maude. In: Ölveczky, P.C., Artho, C. (eds.) Proceedings First International Workshop on Formal Techniques for Safety-Critical Systems, FTSCS 2012, Kyoto, Japan, November 12, 2012. EPTCS, vol. 105, pp. 5–21 (2012)

185. Bae, K., Meseguer, J.: Model checking LTLR formulas under localized fairness. In: Durán, F. (ed.) Rewriting Logic and Its Applications - 9th International Workshop, WRLA 2012, Held as a Satellite Event of ETAPS, Tallinn, Estonia, March 24–25, 2012, Revised Selected Papers. LNCS, vol. 7571, pp. 99–117. Springer (2012)

186. Bae, K., Meseguer, J.: A rewriting-based model checker for the linear temporal logic of rewriting. In: Kniesel, G., Pinto, J.S. (eds.) Proceedings of the Ninth International Workshop on Rule-Based Programming, RULE 2008, Hagenberg Castle, Austria, June 18, 2008. ENTCS, vol. 290, pp. 19–36. Elsevier (2012)

187. Bae, K., Meseguer, J., Ölveczky, P.C.: Formal patterns for multi-rate distributed real-time systems. In: Pasareanu, C.S., Salaün, G. (eds.) Formal Aspects of Component Software, 9th International Symposium, FACS 2012, Mountain View, CA, USA, September 12–14, 2012. Revised Selected Papers. LNCS, vol. 7684, pp. 1–18. Springer (2012)

188. Bae, K., Ölveczky, P.C., Meseguer, J., Al-Nayeem, A.: The SynchAADL2Maude tool. In: de Lara, J., Zisman, A. (eds.) Fundamental Approaches to Software Engineering - 15th International Conference, FASE 2012, Held as Part of the European Joint Conferences on Theory and Practice of Software, ETAPS 2012, Tallinn, Estonia, March 24 - April 1, 2012. Proceedings. LNCS, vol. 7212, pp. 59–62. Springer (2012)

189. Eckhardt, J., Mühlbauer, T., AlTurki, M., Meseguer, J., Wirsing, M.: Stable availability under denial of service attacks through formal patterns. In: de Lara, J., Zisman, A. (eds.) Fundamental Approaches to Software Engineering - 15th International Conference, FASE 2012, Held as Part of the European Joint Conferences on Theory and Practice of Software, ETAPS 2012, Tallinn, Estonia, March 24 - April 1, 2012. Proceedings. LNCS, vol. 7212, pp. 78–93. Springer (2012)
190. Erbatur, S., Escobar, S., Kapur, D., Liu, Z., Lynch, C., Meadows, C., Meseguer, J., Narendran, P., Santiago, S., Sasse, R.: Effective symbolic protocol analysis via equational irreducibility conditions. In: Foresti, S., Yung, M., Martinelli, F. (eds.) Computer Security - ESORICS 2012 - 17th European Symposium on Research in Computer Security, Pisa, Italy, September 10–12, 2012. Proceedings. LNCS, vol. 7459, pp. 73–90. Springer (2012)
191. Gutiérrez, R., Meseguer, J., Rocha, C.: Order-sorted equality enrichments modulo axioms. In: Durán, F. (ed.) Rewriting Logic and Its Applications - 9th International Workshop, WRLA 2012, Held as a Satellite Event of ETAPS, Tallinn, Estonia, March 24–25, 2012, Revised Selected Papers. LNCS, vol. 7571, pp. 162–181. Springer (2012)
192. Hendrix, J., Meseguer, J.: Order-sorted equational unification revisited. In: Kniesel, G., Pinto, J.S. (eds.) Proceedings of the Ninth International Workshop on Rule-Based Programming, RULE 2008, Hagenberg Castle, Austria, June 18, 2008. ENTCS, vol. 290, pp. 19–36. Elsevier (2012)
193. Lluch-Lafuente, A., Meseguer, J., Vandin, A.: State space c-reductions of concurrent systems in rewriting logic. In: Aoki, T., Taguchi, K. (eds.) Formal Methods and Software Engineering - 14th International Conference on Formal Engineering Methods, ICFEM 2012, Kyoto, Japan, November 12–16, 2012. Proceedings. LNCS, vol. 7635, pp. 430–446. Springer (2012)
194. Sasse, R., King, S.T., Meseguer, J., Tang, S.: IBOS: A correct-by-construction modular browser. In: Pasareanu, C.S., Salaün, G. (eds.) Formal Aspects of Component Software, 9th International Symposium, FACS 2012, Mountain View, CA, USA, September 12–14, 2012. Revised Selected Papers. LNCS, vol. 7684, pp. 224–241. Springer (2012)
195. Wirsing, M., Eckhardt, J., Mühlbauer, T., Meseguer, J.: Design and analysis of cloud-based architectures with KLAIM and Maude. In: Durán, F. (ed.) Rewriting Logic and Its Applications - 9th International Workshop, WRLA 2012, Held as a Satellite Event of ETAPS, Tallinn, Estonia, March 24–25, 2012, Revised Selected Papers. LNCS, vol. 7571, pp. 54–82. Springer (2012)
196. Bae, K., Escobar, S., Meseguer, J.: Abstract logical model checking of infinite-state systems using narrowing. In: van Raamsdonk, F. (ed.) 24th International Conference on Rewriting Techniques and Applications, RTA 2013, June 24–26, 2013, Eindhoven, The Netherlands. LIPIcs, vol. 21, pp. 81–96. Schloss Dagstuhl - Leibniz-Zentrum fuer Informatik (2013)

197. Eckhardt, J., Mühlbauer, T., Meseguer, J., Wirsing, M.: Statistical model checking for composite actor systems. In: Martí-Oliet, N., Palomino, M. (eds.) Recent Trends in Algebraic Development Techniques, 21st International Workshop, WADT 2012, Salamanca, Spain, June 7–10, 2012, Revised Selected Papers. LNCS, vol. 7841, pp. 143–160. Springer (2013)
198. Erbatur, S., Escobar, S., Kapur, D., Liu, Z., Lynch, C., Meadows, C., Meseguer, J., Narendran, P., Santiago, S., Sasse, R.: Asymmetric unification: A new unification paradigm for cryptographic protocol analysis. In: Bonacina, M.P. (ed.) Automated Deduction - CADE-24 - 24th International Conference on Automated Deduction, Lake Placid, NY, USA, June 9–14, 2013. Proceedings. LNCS, vol. 7898, pp. 231–248. Springer (2013)
199. Skeirik, S., Bobba, R.B., Meseguer, J.: Formal analysis of fault-tolerant group key management using ZooKeeper. In: 13th IEEE/ACM International Symposium on Cluster, Cloud, and Grid Computing, CCGrid 2013, Delft, Netherlands, May 13–16, 2013. pp. 636–641. IEEE Computer Society (2013)
200. Alpuente, M., Escobar, S., Espert, J., Meseguer, J.: ACUOS: A system for modular ACU generalization with subtyping and inheritance. In: Fermé, E., Leite, J. (eds.) Logics in Artificial Intelligence - 14th European Conference, JELIA 2014, Funchal, Madeira, Portugal, September 24–26, 2014. Proceedings. LNCS, vol. 8761, pp. 573–581. Springer (2014)
201. Bae, K., Meseguer, J.: Infinite-state model checking of LTLR formulas using narrowing. In: Escobar, S. (ed.) Rewriting Logic and Its Applications - 10th International Workshop, WRLA 2014, Held as a Satellite Event of ETAPS, Grenoble, France, April 5–6, 2014, Revised Selected Papers. LNCS, vol. 8663, pp. 113–129. Springer (2014)
202. Bae, K., Meseguer, J.: Predicate abstraction of rewrite theories. In: Dowek, G. (ed.) Rewriting and Typed Lambda Calculi - Joint International Conference, RTA-TLCA 2014, Held as Part of the Vienna Summer of Logic, VSL 2014, Vienna, Austria, July 14–17, 2014. Proceedings. LNCS, vol. 8560, pp. 61–76. Springer (2014)
203. Bae, K., Ölveczky, P.C., Meseguer, J.: Definition, semantics, and analysis of Multirate Synchronous AADL. In: Jones, C.B., Pihlajasaari, P., Sun, J. (eds.) FM 2014: Formal Methods - 19th International Symposium, Singapore, May 12–16, 2014. Proceedings. LNCS, vol. 8442, pp. 94–109. Springer (2014)
204. Escobar, S., Meadows, C., Meseguer, J., Santiago, S.: A rewriting-based forwards semantics for Maude-NPA. In: Proceedings of the 2014 Symposium and Bootcamp on the Science of Security. pp. 3:1–3:12. HotSoS '14, ACM, New York, NY, USA (2014)
205. González-Burgueño, A., Santiago, S., Escobar, S., Meadows, C., Meseguer, J.: Analysis of the IBM CCA security API protocols in Maude-NPA. In: Chen, L., Mitchell, C. (eds.) Security Standardisation Research. LNCS, vol. 8893, pp. 111–130. Springer (2014)

206. Liu, S., Ölveczky, P.C., Meseguer, J.: A framework for mobile ad hoc networks in Real-Time Maude. In: Escobar, S. (ed.) Rewriting Logic and Its Applications - 10th International Workshop, WRLA 2014, Held as a Satellite Event of ETAPS, Grenoble, France, April 5–6, 2014, Revised Selected Papers. LNCS, vol. 8663, pp. 162–177. Springer (2014)

207. Liu, S., Rahman, M.R., Skeirik, S., Gupta, I., Meseguer, J.: Formal modeling and analysis of Cassandra in Maude. In: Merz, S., Pang, J. (eds.) Formal Methods and Software Engineering - 16th International Conference on Formal Engineering Methods, ICFEM 2014, Luxembourg, Luxembourg, November 3–5, 2014. Proceedings. LNCS, vol. 8829, pp. 332–347. Springer (2014)

208. Lucas, S., Meseguer, J.: 2D dependency pairs for proving operational termination of CTRSs. In: Escobar, S. (ed.) Rewriting Logic and Its Applications - 10th International Workshop, WRLA 2014, Held as a Satellite Event of ETAPS, Grenoble, France, April 5–6, 2014, Revised Selected Papers. LNCS, vol. 8663, pp. 195–212. Springer (2014)

209. Lucas, S., Meseguer, J.: Models for logics and conditional constraints in automated proofs of termination. In: Aranda-Corral, G.A., Calmet, J., Martín-Mateos, F.J. (eds.) Artificial Intelligence and Symbolic Computation - 12th International Conference, AISC 2014, Seville, Spain, December 11–13, 2014. Proceedings. LNCS, vol. 8884, pp. 9–20. Springer (2014)

210. Lucas, S., Meseguer, J.: Strong and weak operational termination of order-sorted rewrite theories. In: Escobar, S. (ed.) Rewriting Logic and Its Applications - 10th International Workshop, WRLA 2014, Held as a Satellite Event of ETAPS, Grenoble, France, April 5–6, 2014, Revised Selected Papers. LNCS, vol. 8663, pp. 178–194. Springer (2014)

211. Lucas, S., Meseguer, J.: Proving operational termination of declarative programs in general logics. In: Danvy, O. (ed.) Proceedings of the 16th International ACM SIGPLAN Conference on Principles and Practice of Declarative Programming, September 8–10, 2014, Canterbury, UK. pp. 111–122. ACM (2014)

212. Rocha, C., Meseguer, J., Muñoz, C.A.: Rewriting modulo SMT and open system analysis. In: Escobar, S. (ed.) Rewriting Logic and Its Applications - 10th International Workshop, WRLA 2014, Held as a Satellite Event of ETAPS, Grenoble, France, April 5–6, 2014, Revised Selected Papers. LNCS, vol. 8663, pp. 247–262. Springer (2014)

213. Santiago, S., Escobar, S., Meadows, C., Meseguer, J.: A formal definition of protocol indistinguishability and its verification using Maude-NPA. In: Mauw, S., Jensen, C.D. (eds.) Security and Trust Management - 10th International Workshop, STM 2014, Wroclaw, Poland, September 10–11, 2014. Proceedings. LNCS, vol. 8743, pp. 162–177. Springer (2014)

214. Sun, M., Meseguer, J.: Formal specification of button-related fault-tolerance micropatterns. In: Escobar, S. (ed.) Rewriting Logic and Its Applications - 10th International Workshop, WRLA 2014, Held as a Satellite Event of ETAPS, Grenoble, France, April 5–6, 2014, Revised Selected Papers. LNCS, vol. 8663, pp. 263–279. Springer (2014)

215. Yang, F., Escobar, S., Meadows, C., Meseguer, J., Narendran, P.: Theories of homomorphic encryption, unification, and the finite variant property. In: Danvy, O. (ed.) Proceedings of the 16th International ACM SIGPLAN Conference on Principles and Practice of Declarative Programming, September 8–10, 2014, Canterbury, UK. pp. 123–134. ACM (2014)
216. Liu, S., Nguyen, S., Ganhotra, J., Rahman, M., Gupta, I., Meseguer, J.: Quantitative analysis of consistency in NoSQL key-value stores. In: Campos, J., Haverkort, B. (eds.) 12th International Conference on Quantitative Evaluation of SysTems, QEST 2015, Madrid, Spain, September 1–3, 2015. Proceedings. LNCS, Springer (2015) To appear.
217. Meseguer, J., Skeirik, S.: Equational formulas and pattern operations in initial order-sorted algebras. In: Falaschi, M. (ed.) Logic-Based Program Synthesis and Transformation, 25th International Symposium, LOPSTR 2015, Siena, Italy, July 13–15, 2015, Revised Selected Papers. LNCS. Springer (2015) To appear.

Newsletter Articles

1. Goguen, J.A., Meseguer, J.: Completeness of many-sorted equational logic. SIGPLAN Notices 17(1), 9–17 (1982)
2. Goguen, J.A., Meseguer, J.: Remarks on remarks on many-sorted algebras with possibly empty carrier sets. Bulletin of the EATCS 30, 66–73 (1986)
3. Goguen, J.A., Kirchner, C., Leinwand, S., Meseguer, J.: Progress report on the rewrite rule machine. IEEE Technical Committee on Computer Architecture Newsletter, March 1996, 7–21 (1986)
4. Goguen, J.A., Meseguer, J.: Remarks on remarks on many-sorted equational logic. SIGPLAN Notices 22(4), 41–48 (1987)
5. Meseguer, J.: Rewriting as a unified model of concurrency. OOPS Messenger 2(2), 86–88 (1991)
6. Meseguer, J.: Why OOP needs new semantic foundations. ACM Comput. Surv. 28(4es), 159 (1996)
7. Meseguer, J., Talcott, C.L.: Maude. ACM SIGSOFT Software Engineering Notes 25(1), 104 (2000)
8. Meseguer, J., Talcott, C.L.: Semantic interoperation of open systems. ACM SIGSOFT Software Engineering Notes 25(1), 64–65 (2000)
9. Meseguer, J.: Report on ETAPS 2000. ACM SIGSOFT Software Engineering Notes 26(1), 39 (2001)
10. Futatsugi, K., Jouannaud, J., Meseguer, J.: Joseph Goguen (1941–2006). Bulletin of the EATCS 90, 199–201 (2006)

Sentence-Normalized Conditional Narrowing Modulo in Rewriting Logic and Maude

Luis Aguirre, Narciso Martí-Oliet$^{(\boxtimes)}$, Miguel Palomino, and Isabel Pita

Facultad de Informática, Universidad Complutense de Madrid, Madrid, Spain
{luisagui,narciso,miguelpt,ipandreu}@ucm.es

Abstract. This work studies the relationship between verifiable and computable answers for reachability problems in rewrite theories with an underlying membership equational logic. These problems have the form

$$(\exists \bar{x}) t(\bar{x}) \rightarrow^* t'(\bar{x})$$

with $\bar{x}$ some variables, or a conjunction of several of these subgoals. A calculus that solves this kind of problems working always with *normalized terms and substitutions* has been developed. Given a reachability problem in a rewrite theory, this calculus can compute any normalized answer that can be checked by rewriting, or one that can be instantiated to that answer. Special care has been taken in the calculus to keep membership information attached to each term, to make use of it whenever possible.

Keywords: Maude · Narrowing · Reachability · Rewriting logic · Unification · Membership equational logic

Dedicated to José Meseguer on occasion of his 65th birthday. This paper is based on work he has been developing during the last 30 years, from order-sorted algebra to narrowing, going through membership equational logic, rewriting logic, and much more.

1 Introduction

Rewriting logic is a computational logic that has been around for more than twenty years [20], whose semantics [6] has a precise mathematical meaning allowing mathematical reasoning for proving properties, providing a flexible framework for the specification of concurrent systems. It turned out that it can express both concurrent computation and logical deduction, allowing its application in many areas such as automated deduction, software and hardware specification and verification, security, etc. [19,22].

A deductive system is specified in rewriting logic as a rewrite theory $\mathcal{R} = (\Sigma, \mathcal{E}, R)$, with $(\Sigma, \mathcal{E})$ an underlying equational theory (in this paper we will

---

Partially supported by MINECO Spanish project StrongSoft (TIN2012–39391–C04–04) and Comunidad de Madrid program N-GREENS Software (S2013/ICE-2731).

N. Martí-Oliet et al. (Eds.): Meseguer Festschrift, LNCS 9200, pp. 48–71, 2015.
DOI: 10.1007/978-3-319-23165-5_2

consider *membership equational logic*), where terms are given an algebraic data type, allowing us to identify as semantically equal two syntactically different terms, and R a set of rules that specify how the deductive system can derive one term from another. *Order-sorted, many-sorted,* and *unsorted* theories can be formulated as special cases of membership equational logic (MEL) theories.

Reachability problems have the form

$$(\exists \bar{x}) t(\bar{x}) \rightarrow^* t'(\bar{x})$$

with t, t' terms with variables in $\bar{x}$, or a conjunction of several of these subgoals. They can be solved by model-checking methods for finite state spaces. When the initial term t has no variables, i.e. it is a ground term, and under certain admissibility conditions, rewriting can be used in a breadth-first way to traverse the state space, trying to find a suitable matching of $t'(\bar{x})$ in each traversed node. In the general case where $t(\bar{x})$ is not a ground term, a technique known as *narrowing* [14] that was first proposed as a method for solving equational goals (*unification*), has been extended to cover also reachability goals [24], leaving equational goals as a special case. The strength of narrowing can be found in that it enables us to manage complex concurrent and deductive systems that cannot be handled by faster, but more limited, specialized methods. Under the admissibility conditions for rewrite theories, which allow for conditional rules and equations with extra variables in the conditions under some requirements, and the assumption of the existence of an $\mathcal{E}$-unification algorithm, we can use narrowing modulo $\mathcal{E}$ to perform symbolic analysis of the possibly infinite set of initial states $t(\bar{x})$ in the state space and determine the actual values of $\bar{x}$ that allow us to derive $t'(\bar{x})$ from $t(\bar{x})$.

What is most striking is the fact that an $\mathcal{E}$-unification algorithm can itself make use of narrowing at another level for finding the solution to its equational goals. Specific $\mathcal{E}$-unification algorithms exist for a small number of equational theories, but if the equational theory $(\Sigma, \mathcal{E})$ can be decomposed as $E \cup A$, where A is a set of axioms having a unification algorithm, and the equations E can be turned into a set of rules $\overrightarrow{E}$, by orienting them, such that the rewrite theory $\overrightarrow{\mathcal{E}} = (\Sigma, A, \overrightarrow{E})$ is admissible in the above sense, then narrowing can be used on $\overrightarrow{\mathcal{E}}$ to solve the $\mathcal{E}$-unification goals generated by performing narrowing on $\mathcal{R}$. For these equational goals the idea of *variants of a term* has been applied in recent years to narrowing. A strategy known as *folding variant narrowing* [13], which computes a complete set of variants of any term, has been developed by Escobar, Sasse, and Meseguer, allowing unification modulo a set of unconditional equations and axioms. The strategy terminates on any input term on those systems enjoying the *finite variant property*, and it is optimally terminating. It is being used for cryptographic protocol analysis [24], with tools like Maude-NPA [12], termination algorithms modulo axioms [9], algorithms for checking confluence and coherence of rewrite theories modulo axioms [10], and infinite-state model checking [4]. Recent development in conditional narrowing have been made for order-sorted equational theories [7] and also for narrowing with constraint solvers [26].

Conditional narrowing without axioms for rewrite theories with an order-sorted type structure has been thoroughly studied for increasingly complex categories of term rewriting systems. A wide survey can be found in [25]. The literature that can be found is scarce when we allow for extra variables in conditions (e.g. [15,16]), conditional narrowing modulo axioms (e.g. [7]), or conditional narrowing modulo a set of equations(e.g. [5]). Conditional narrowing modulo axioms for MEL theories has not been addressed, to the best of our knowledge, one of the main reasons being the lack of fast and effective unification algorithms modulo axioms for MEL theories. Nonetheless, there are plenty of algebraic data types, including all types that imply some kind of order between subterms, that are better expressed inside a MEL theory, so there is a need to give an answer to these cases. In this paper we focus on conditional narrowing modulo axioms for rewrite theories with an underlying equational MEL theory.

Our main contribution in this work is the proposal of two narrowing calculi for computing normalized answers to unification and reachability problems in membership equational logic and membership conditional rewrite theories, respectively. These calculi normalize all terms before applying any unification or reachability rule, and only generate normalized instantiations of reachability terms and intermediate (*matching*) variables, greatly reducing the state space. They have been proved sound and weakly complete, i.e., complete with respect to idempotent normalized answers.

The work is structured as follows: in Sect. 2 all needed definitions and properties for rewriting and narrowing are introduced. In Sect. 3 we present a rewrite theory, that only generates normalized substitutions on matching variables, for checking normalized solutions to unification problems. Section 4 introduces the narrowing calculus for equational unification. Section 5 introduces the narrowing calculus for reachability. Section 6 shows the calculi at work. In Sect. 7, related work, conclusions, and future lines of investigation for this work are presented. This paper is a continuation of a previous one [1], where non-normalized terms were allowed by the calculus.

2 Preliminaries

We assume familiarity with term rewriting and rewriting logic [6]. Rewriting logic is always parameterized by an underlying equational logic. This work is focused in membership equational logic [21], an equational logic that generalizes both many-sorted and order-sorted equational theories and that can also handle partial functions. There are several language implementations of rewriting logic, one of them being Maude [8], a language whose underlying logic is membership equational logic.

2.1 Search Tree Example

A search tree implementation will be used as running example to explain the definitions in a less abstract way. We review the needed terms, enclosing shortcuts for the definitions between brackets. We have Keys (abbreviated to k) a,...,f lexicographically ordered, and Values (v) 1,...,9. A pair Key Value forms a Record (r). A SearchTree (st) can be an EmptyTree (et), which we call empty, or non-empty, NeSearchTree (nt), containing in this case a Record at its root and two sub-SearchTrees, left and right. All the Keys in the left sub-SearchTree must be smaller than the key in the root and all the Keys in the right sub-SearchTree must be greater than the key in the root. t is the Boolean (b) result of a valid comparison ($<$), max and min return the Record on a NeSearchTree with the highest and lowest Key respectively, and key returns the Key of a Record. We also have a set of records RecordSet (rs) with none as identity atom, which are intended to be nondeterministically inserted (ins) in the SearchTree, hence yielding a NeSearchTree, deleted (del) (we admit deletion attempts on EmptyTrees), or omitted. Finally, there is a list Path (p), with nil as identity atom, holding the history - inserted, deleted, or omitted (i, d, o) - of already processed records. A triple SearchTree, RecordSet, Path forms a State (s).

2.2 Membership Equational Logic

Let's take a partially ordered set $(S, \leq)$ of *sorts*, whose *connected components* are the equivalence classes corresponding to the least equivalence relation $\equiv_\leq$ containing $\leq$.

A *membership equational logic* (MEL) *signature* [6] is defined by a *kind-complete* triple $\Sigma = (K, \Omega, S)$ meaning that:

- K is a set of *kinds*.
- S is split into a K-kinded family of disjoint sets of sorts S_k, i.e. $S = \bigcup_{k \in K} S_k$, such that if $s_i \leq s_j$ and $s_i \in S_k$ then $s_j \in S_k$. We write $[s_i] = k$ and say that the kind of s_i is k, i.e., each connected component of $(S, \leq)$ has the same kind. $\leq$ is extended so that $s_i \leq k$ iff $s_i \in S_k$, i.e., k is the top sort of its connected component (we also define $[k] = k$ if $k \in K$ for simplicity of notation).
- $\Omega = \{\Sigma_{\bar{\kappa},\kappa}\}_{(\bar{\kappa},\kappa)\in(K\cup S)^*\times(K\cup S)}$ is an algebraic signature of *function symbols*, where for each symbol $f \in \Sigma_{\kappa_1...\kappa_n,\kappa}$ if $n \geq 1$ and at least one of the subindices is not a kind, then there is another function symbol $f \in \Sigma_{[\kappa_1]...[\kappa_n],[\kappa]}$.

When $f \in \Sigma_{\epsilon,\kappa}$ (ϵ is the empty word), we say that f is a *constant* with *type* (meaning sort or kind) κ. We write $f \in \Sigma_\kappa$ instead of $f \in \Sigma_{\epsilon,\kappa}$.

If $f \in \Sigma_{\kappa_1...\kappa_n,\kappa}$, then we display f as $f : \kappa_1 ... \kappa_n \rightarrow \kappa$, and say that f has *arity* n. We call this a *rank* declaration for symbol f. Constant symbols have only one rank declaration $f : \rightarrow \kappa$ (plus the mandatory $f : \rightarrow [\kappa]$ if κ is not a kind). We extend the order $\leq$ on $K \cup S$ to $(K \cup S)^*$, component-wise. Then Ω must also satisfy a *monotonicity condition*: $f \in \Sigma_{\kappa_1...\kappa_n,\kappa} \cap \Sigma_{\kappa_1'...\kappa_n',\kappa'}$ and $\kappa_1 ... \kappa_n \leq \kappa_1' ... \kappa_n'$ imply $\kappa \leq \kappa'$. If $f \in \Sigma_{\kappa_1...\kappa_n,\kappa}$ and $t_1, ..., t_n$ have type κ_1, ..., κ_n respectively, then the term $f(t_1,...,t_n)$ has type κ. If $\kappa \leq \kappa'$ and the term t has type κ, then t has also type κ'. This means that a term may have several types. In fact, as for every sort s we have that $s \leq [s]$, if a term has only one type then it must be a kind.

A MEL Σ-algebra $\mathcal{A}$ contains a set $\mathcal{A}_k$ for each kind $k \in K$, an n-ary function $\mathcal{A}_f : \mathcal{A}_{\kappa_1} \ldots \mathcal{A}_{\kappa_n} \to \mathcal{A}_\kappa$ for each function $f \in \Sigma_{\kappa_1 \ldots \kappa_n, \kappa}$, and a subset $\mathcal{A}_s \subseteq \mathcal{A}_k$ for each sort $s \in S_k$ such that if $s_i \leq s_j$ then $\mathcal{A}_{s_i} \subseteq \mathcal{A}_{s_j}$, and if $f \in \Sigma_{\kappa_1 \ldots \kappa_n, \kappa} \cap \Sigma_{\kappa_1' \ldots \kappa_n', \kappa'}$ and $\kappa_1 \ldots \kappa_n \leq \kappa_1' \ldots \kappa_n'$ then $\mathcal{A}_f : \mathcal{A}_{\kappa_1} \ldots \mathcal{A}_{\kappa_n} \to \mathcal{A}_\kappa$ equals $\mathcal{A}_f : \mathcal{A}_{\kappa_1'} \ldots \mathcal{A}_{\kappa_n'} \to \mathcal{A}_{\kappa'}$ on $\mathcal{A}_{\kappa_1} \ldots \mathcal{A}_{\kappa_n}$.

In membership equational logic the elements in a sort are well-defined, while the elements in a kind that don't belong to any sort are usually meant to refer to error or undefined elements. Kinds also provide a general way of dealing with partial functions in equational specifications. For instance, in the search tree example a `NeSearchTree` must have its `Keys` correctly ordered. Otherwise we have an error term with kind `[NeSearchTree]` (we haven't defined a sort `Tree`), so the constructor function for `NeSearchTrees` becomes total on `[NeSearchTree]`.

We allow *mix-fix* notation in Ω, where the symbol _ is used to identify the position of each $\kappa_i \in \bar{\kappa}$. For instance, $\_<\_ : Int\ Int \to Bool$ is a rank declaration stating that $5 < 4$ is a term with sort $Bool$. If omitted we assume the usual functional notation $f(\kappa_1, \ldots, \kappa_n)$, which is an alternative notation admitted for all functions. We call $\mathcal{X} = \bigcup_{\kappa \in (K \cup S)} \mathcal{X}_\kappa$, where $\mathcal{X}_\kappa = \{x_\kappa^i\}$ for $\kappa \in (K \cup S)$, is a family of pairwise disjoint infinitely countable sets of variables. If κ is a sort then x_κ^i has sort κ (and kind $[\kappa]$), otherwise x_κ^i has kind κ but no sort. The set of variables is potentially infinite, but any computation will only require a finite number of variables. A term that has no variables in it is said to be *ground*. A term where each variable occurs only once is said to be *linear* (ground terms are linear).

The sets $T_{\Sigma,\kappa}$, $T_\Sigma(\mathcal{X})_\kappa$ denote, respectively, the set of ground Σ-terms with sort or kind κ and the set of Σ-terms with sort or kind κ over $\mathcal{X}$. We ambiguously use the notation T_Σ to refer to the initial Σ-algebra and as a shortcut for $\bigcup_{\kappa \in (K \cup S)} T_{\Sigma,\kappa}$. We also ambiguously use the notation $T_\Sigma(\mathcal{X})$ to refer to the free Σ-algebra on $\mathcal{X}$ and as a shortcut for $\bigcup_{\kappa \in (K \cup S)} T_\Sigma(\mathcal{X})_\kappa$. $Var(t) \subseteq \mathcal{X}$ denotes the set of variables in $t \in T_\Sigma(\mathcal{X})$. Σ is assumed to be *sensible* meaning that if $f \in \Sigma_{\kappa_1 \ldots \kappa_n, \kappa}$, $f \in \Sigma_{\kappa_1' \ldots \kappa_n', \kappa'}$ and $[\kappa_i] = [\kappa_i']$ for $i = 1, \ldots, n$ then $[\kappa] = [\kappa']$. We also assume that Σ has non-empty sorts, i.e., $T_{\Sigma,s} \neq \emptyset$ for all $s \in S$.

In the search tree example we have, omitting the implied kinded definition for each function in Ω, that $\Sigma = (K, \Omega, S)$ is:

$K = \{[\mathsf{rs}], [\mathsf{st}], [\mathsf{k}], [\mathsf{v}], [\mathsf{b}], [\mathsf{p}], [\mathsf{s}]\}, S = \{\mathsf{r}, \mathsf{et}, \mathsf{st}, \mathsf{nt}, \mathsf{k}, \mathsf{v}, \mathsf{b}, \mathsf{rs}, \mathsf{p}, \mathsf{s}\},$
$S_{[\mathsf{rs}]} = \{\mathsf{r}, \mathsf{rs}\}, S_{[\mathsf{st}]} = \{\mathsf{et}, \mathsf{nt}, \mathsf{st}\}, S_{[\mathsf{k}]} = \{\mathsf{k}\}, S_{[\mathsf{v}]} = \{\mathsf{v}\}, S_{[\mathsf{b}]} = \{\mathsf{b}\},$
$S_{[\mathsf{p}]} = \{\mathsf{p}\}, S_{[\mathsf{s}]} = \{\mathsf{s}\},$
$\Omega = \{\{\_<\_\}_{\mathsf{k}\,\mathsf{k},\mathsf{b}}, \{\_\_\}_{\mathsf{k}\,\mathsf{v},\mathsf{r}}, \{\_;\_\}_{\mathsf{rs}\,\mathsf{rs},\mathsf{rs}}, \{\mathsf{key}\}_{\mathsf{r},\mathsf{k}}, \{\_\mathsf{i}\_, \_\mathsf{d}\_, \_\mathsf{o}\_\}_{\mathsf{p}\,\mathsf{r},\mathsf{p}},$
$\{\mathsf{ins}\}_{\mathsf{st}\,\mathsf{k}\,\mathsf{v},\mathsf{nt}}, \{\mathsf{del}\}_{\mathsf{st}\,\mathsf{k},\mathsf{st}}, \{\_|\_|\_\}_{\mathsf{st}\,\mathsf{rs}\,\mathsf{p},\mathsf{s}}, \{\_[\_]\_\}_{\mathsf{st}\,\mathsf{r}\,\mathsf{st},\mathsf{nt}},$
$\{\mathsf{min}, \mathsf{max}\}_{\mathsf{nt},\mathsf{r}}, \{\mathsf{a}, \ldots, \mathsf{f}\}_\mathsf{k}, \{1, \ldots, 9\}_\mathsf{v}, \{\mathsf{t}\}_\mathsf{b}, \{\mathsf{empty}\}_{\mathsf{et}}, \{\mathsf{none}\}_{\mathsf{rs}}, \{\mathsf{nil}\}_\mathsf{p}\}.$

We explain the notation used in Ω: $\{\_[\_]\_\}_{\mathsf{st}\,\mathsf{r}\,\mathsf{st},\mathsf{nt}}$ means that there is a mix-fix function symbol $\_[\_]\_$ such that if t_1, t_3 are terms with sort `SearchTree` and t_2 is a term with sort `Record` then $t_1[t_2]t_3$ is a term with sort `NeSearchTree`.

Positions in a term t: as previously said, a term t can be always expressed in functional notation as $f(t_1, \ldots, t_n)$. Then we can picture t as a tree with root

f and children $t_1, \ldots, t_n$. We refer to the root position of t as ϵ and to the other positions of t as strings of nonzero natural numbers, $i_1 \ldots i_m$, meaning the position $i_2 \ldots i_m$ of t_{i_1}. The set of positions of a term is written $Pos(t)$. The set of nonvariable positions of a term is written $Pos_\Sigma(t)$. $t|_p$ is the subtree of t below position p. $t[u]_p$ is the replacement in t of the subterm at position p with a term u. As an example, if t is $f(g(a,b),c)$, then $t|_1$ is $g(a,b)$, $t|_{12}$ is b, and $t[d]_{12}$ is $f(g(a,d),c)$.

A MEL signature Σ is said to be *preregular* iff for each n, for every function symbol f with arity n, and for every $\kappa_1 \ldots \kappa_n \in (K \cup S)^n$, if the set S_f containing all the sorts s' that appear in rank declarations in Σ of the form $f : \kappa'_1 \ldots \kappa'_n \to \kappa'$ such that $\kappa_i \leq \kappa'_i$, for $1 \leq i \leq n$, is not empty (so a term $f(t_1, \ldots, t_n)$ where t_i has type κ_i for $1 \leq i \leq n$ would be a Σ-term), then S_f has a least sort. Preregularity guarantees that every Σ-term t has a *least sort*, denoted $ls(t)$, among all the sorts that t has because of the different rank declarations that can be applied to t, which is the most accurate classification for t.

A *substitution* $\sigma : \mathcal{X} \to T_\Sigma(\mathcal{X})$ is a function that matches the identity function in all $\mathcal{X}$ except for a finite set of variables $\mathcal{Y} \subseteq \mathcal{X}$, verifying that for each variable $y_\kappa \in \mathcal{Y}$ we have that $ls(y_\kappa\sigma) \leq \kappa$. Substitutions are written as $\sigma = \{y^1_{\kappa_1} \mapsto t_1, \ldots, y^n_{\kappa_n} \mapsto t_n\}$ where $Dom(\sigma) = \{y^1_{\kappa_1}, \ldots, y^n_{\kappa_n}\}$ and $Ran(\sigma) = \bigcup_{i=1}^n Var(t_i)$. The identity substitution is displayed as id. Substitutions are homomorphically extended to terms in $T_\Sigma(\mathcal{X})$ (and also to the rest of syntactic structures introduced along this paper, such as equations, goals, etc.). The restriction $\sigma|_\mathcal{V}$ of σ to a set of variables $\mathcal{V}$ is defined as $x\sigma|_\mathcal{V} = x\sigma$ if $x \in \mathcal{V}$ and $x\sigma|_\mathcal{V} = x$ otherwise. Composition of two substitutions is denoted by $\sigma\sigma'$, with $x(\sigma\sigma') = (x\sigma)\sigma'$. If $\sigma\sigma = \sigma$ we say that σ is *idempotent*. For substitutions σ and σ' where $Dom(\sigma) \cap Dom(\sigma') = \emptyset$, we denote their union by $\sigma \cup \sigma'$.

A Σ-*equation* is an expression of the form $t = t'$. A Σ-*equation* $t = t'$ is said to be:

- *Regular* if $Var(t) = Var(t')$.
- *Sort-preserving* if for each substitution σ, we have $t\sigma \in T_\Sigma(\mathcal{X})_\kappa$ ($\kappa \in K \cup S$) implies $t'\sigma \in T_\Sigma(\mathcal{X})_\kappa$ and vice versa.
- *Left (or right) linear* if t (resp. t') is linear.
- *Linear* if it is both left and right linear.

A set of equations E is said to be regular, or sort-preserving, or (left or right) linear, if each equation in it is so.

A MEL theory [6] is a pair $(\Sigma, \mathcal{E})$, where Σ is a MEL signature and $\mathcal{E}$ is a finite set of (possibly labeled) MEL sentences, either conditional equations or conditional memberships of the forms:

$$t = t' \text{ if } A_1 \wedge \ldots \wedge A_n, \qquad t : s \text{ if } A_1 \wedge \ldots \wedge A_n,$$

where $t = t'$ is a Σ-equation, $t : s$, $s \in S$, is a unary membership predicate stating that t is a term with sort s, provided that the condition holds, and each A_i can be of the form $t = t'$, $t : s$ or $t := t'$ (a *matching* equation). Matching equations are treated as ordinary Σ-equations. They are a warning that new *extra* variables appear in t, imposing a limitation in the syntax of admissible MEL theories, as we will see. We also admit unconditional sentences in $\mathcal{E}$. $x_{s_1} : s_2$ is an unconditional membership expressing $s_1 \leq s_2$. For each variable $x_s \in \mathcal{X}_s$, where $s \in S$, we have that $x_s : s \in \mathcal{E}$. As an exception, there are two types of unconditional memberships over kinds, instead of sorts, that are implied by the MEL signature: if $f \in \Sigma_{\kappa_1 \ldots \kappa_n, k}$, $k \in K$ then $f(x_{\kappa_1}, \ldots, x_{\kappa_n}) : k \in \mathcal{E}$; also for each variable $x_\kappa \in \mathcal{X}_\kappa$ such that $[\kappa] = k$, $x_\kappa : k \in \mathcal{E}$.

Throughout this paper we will assume that all signatures are preregular and all their equations and memberships $t=t'$, $t:=t'$ and $t:s$, $s \in S$, satisfy the conditions $[ls(t)] = [ls(t')]$ and $[ls(t)] = [s]$, that is, they are well-formed.

Given a MEL sentence ϕ, we denote by $\mathcal{E} \vdash \phi$ the fact that ϕ can be deduced from $\mathcal{E}$ using the rules in Fig. 1 [3,6]; for an equation $t = t'$, $\mathcal{E} \vdash t = t'$ is also written $t =_{\mathcal{E}} t'$. These rules, where the symbol = stands for = or := indistinctly, specify a sound and complete calculus. A MEL theory $(\Sigma, \mathcal{E})$ has an *initial algebra* $(T_{\Sigma/\mathcal{E}})$, whose elements are equivalence classes $[t]_{\mathcal{E}} \subseteq T_\Sigma$ of ground terms identified by the equations in $\mathcal{E}$.

$$\frac{t \in T_\Sigma(\mathcal{X})}{t = t} \text{ Reflexivity} \qquad \frac{t = t'}{t' = t} \text{ Symmetry}$$

$$\frac{t':s \quad t = t'}{t:s} \text{ Membership} \qquad \frac{t_1 = t_2 \quad t_2 = t_3}{t_1 = t_3} \text{ Transitivity}$$

$$\frac{f \in \Sigma_{k_1 \ldots k_n, k} \quad t_i = t'_i \quad t_i, t'_i \in T_\Sigma(X)_{k_i}, 1 \leq i \leq n}{f(t_1, \ldots, t_n) = f(t'_1, \ldots, t'_n)} \text{ Congruence}$$

$$\frac{(A_0 \text{ if } \bigwedge_i A_i) \in E \quad \sigma: X \to T_\Sigma(Y) \quad A_1\sigma \ldots A_n\sigma}{A_0\sigma} \text{ Replacement}$$

Fig. 1. Deduction rules for membership equational logic.

The MEL theory for the search tree example consists of $\Sigma = (K, \Omega, S)$ and the following set $\mathcal{E}$ of MEL sentences, one of them labeled (i1), where the first line of MEL sentences represents the subsort ordering in S. We omit the implicit subsorts for each kind, and the implicit memberships for each variable and kinded function. For executability requirements of the theory, that will be later defined, associativity, commutativity, and identity axioms are defined over kinds:

$x_{\text{nt}} : \text{st}, x_{\text{et}} : \text{st}, x_{\text{r}} : \text{rs}$,

$(x_{[\text{rs}]}; y_{[\text{rs}]}); z_{[\text{rs}]} = x_{[\text{rs}]}; (y_{[\text{rs}]}; z_{[\text{rs}]})$ (associativity),

$x_{[\text{rs}]}; y_{[\text{rs}]} = y_{[\text{rs}]}; x_{[\text{rs}]}$ (commutativity), $x_{[\text{rs}]}; \text{none} = x_{[\text{rs}]}$ (identity),

$\text{empty}[x_{\text{r}}]\text{empty} : \text{nt}$, $\text{empty}[x_{\text{r}}]r_{\text{nt}} : \text{nt}$ *if* $\text{key}(x_{\text{r}}) < \text{key}(\min(r_{\text{nt}})) = \text{t}$,

$l_{\text{nt}}[x_{\text{r}}]\text{empty} : \text{nt}$ *if* $\text{key}(\max(l_{\text{nt}})) < \text{key}(x_{\text{r}}) = \text{t}$,

$l_{\text{nt}}[x_{\text{r}}]r_{\text{nt}} : \text{nt}$ *if* $\text{key}(\max(l_{\text{nt}})) < \text{key}(x_{\text{r}}) = \text{t} \wedge \text{key}(x_{\text{r}}) < \text{key}(\min(r_{\text{nt}})) = \text{t}$,

$\text{a} < \text{b} = \text{t} \ldots \text{a} < \text{f} = \text{t}, \text{b} < \text{c} = \text{t} \ldots \text{b} < \text{f} = \text{t} \ldots \text{e} < \text{f} = \text{t}, \text{key}(y_k z_v) = y_k$,

$\min(\text{empty}[x_{\text{r}}]r_{\text{st}}) = x_{\text{r}}$, $\min(l_{\text{nt}}[x_{\text{r}}]r_{\text{nt}}) = \min(l_{\text{nt}})$,

$\max(l_{\text{st}}[x_{\text{r}}]\text{empty}) = x_{\text{r}}$, $\max(l_{\text{nt}}[x_{\text{r}}]r_{\text{nt}}) = \max(r_{\text{nt}})$,

$\mathtt{ins}(\mathtt{empty}, y_\mathrm{k}, z_\mathrm{v}) = \mathtt{empty}[y_\mathrm{k}, z_\mathrm{v}]\mathtt{empty},$

$\mathtt{ins}(l_\mathrm{st}[y_\mathrm{k}z_\mathrm{v}']r_\mathrm{st}, y_\mathrm{k}, z_\mathrm{v}) = l_\mathrm{st}[y_\mathrm{k}z_\mathrm{v}]r_\mathrm{st},$

(i1) $\mathtt{ins}(l_\mathrm{st}[x_\mathrm{r}]r_\mathrm{st}, y_\mathrm{k}, z_\mathrm{v}) = \mathtt{ins}(l_\mathrm{st}, y_\mathrm{k}, z_\mathrm{v})[x_\mathrm{r}]r_\mathrm{st}$ if $y_\mathrm{k} < \mathtt{key}(x_\mathrm{r}) = \mathtt{t},$

$\mathtt{ins}(l_\mathrm{st}[x_\mathrm{r}]r_\mathrm{st}, y_\mathrm{k}, z_\mathrm{v}) = l_\mathrm{st}[x_\mathrm{r}]\mathtt{ins}(r_\mathrm{st}, y_\mathrm{k}, z_\mathrm{v})$ if $\mathtt{key}(x_\mathrm{r}) < y_\mathrm{k} = \mathtt{t},$

$\mathtt{del}(\mathtt{empty}, y_\mathrm{k}) = \mathtt{empty}, \mathtt{del}(\mathtt{empty}[y_\mathrm{k}z_\mathrm{v}]r_\mathrm{st}, y_\mathrm{k}) = r_\mathrm{st},$

$\mathtt{del}(l_\mathrm{st}[y_\mathrm{k}z_\mathrm{v}]\mathtt{empty}, y_\mathrm{k}) = l_\mathrm{st},$

$\mathtt{del}(l_\mathrm{nt}[y_\mathrm{k}z_\mathrm{v}]r_\mathrm{nt}, y_\mathrm{k}) = l_\mathrm{nt}[\mathtt{min}(r_\mathrm{nt})]\mathtt{del}(r_\mathrm{nt}, \mathtt{key}(\mathtt{min}(r_\mathrm{nt}))),$

$\mathtt{del}(l_\mathrm{st}[x_\mathrm{r}]r_\mathrm{st}, y_\mathrm{k}) = \mathtt{del}(l_\mathrm{st}, y_\mathrm{k})[x_\mathrm{r}]r_\mathrm{st}$ if $y_\mathrm{k} < \mathtt{key}(x_\mathrm{r}) = \mathtt{t},$

$\mathtt{del}(l_\mathrm{st}[x_\mathrm{r}]r_\mathrm{st}, y_\mathrm{k}) = l_\mathrm{st}[x_\mathrm{r}]\mathtt{del}(r_\mathrm{st}, y_\mathrm{k})$ if $\mathtt{key}(x_\mathrm{r}) < y_\mathrm{k} = \mathtt{t}.$

These axioms correspond to an algebraic specification of search trees similar to the one found in [8].

2.3   Unification

Given a MEL theory $(\Sigma, \mathcal{E})$, the $\mathcal{E}$-*subsumption* preorder $\ll_{\mathcal{E}}$ on $T_\Sigma(\mathcal{X})_k$ is defined by $t \ll_{\mathcal{E}} t'$ if there is a substitution σ such that $t =_{\mathcal{E}} t'\sigma$. For substitutions σ, ρ and a set of variables $\mathcal{V}$ we define $\sigma|_{\mathcal{V}} \ll_{\mathcal{E}} \rho|_{\mathcal{V}}$ if there is a substitution η such that $\sigma|_{\mathcal{V}} =_{\mathcal{E}} (\rho\eta)|_{\mathcal{V}}$. Then we say that ρ is more general than σ with respect to $\mathcal{V}$. When $\mathcal{V}$ is not specified, we assume that $Dom(\rho) \subseteq Dom(\sigma)$ and say that ρ is more general than σ.

A *system of equations* F is a conjunction of the form $t_1 = t_1' \wedge \ldots \wedge t_n = t_n'$ where for $1 \leq i \leq n$, $t_i = t_i'$ is a Σ-equation. We define $Var(F) = \bigcup_i (Var(t_i) \cup Var(t_i'))$. An $\mathcal{E}$-*unifier* for F is a substitution σ such that $t_i\sigma =_{\mathcal{E}} t_i'\sigma$ for $1 \leq i \leq n$. For $\mathcal{V} = Var(F) \subseteq \mathcal{W}$, a set of substitutions $CSU_{\mathcal{E}}^{\mathcal{W}}(F)$ is said to be a *complete set of unifiers modulo $\mathcal{E}$ of F away from $\mathcal{W}$* if

- each $\sigma \in CSU_{\mathcal{E}}^{\mathcal{W}}(F)$ is an $\mathcal{E}$-unifier of F;
- for any $\mathcal{E}$-unifier ρ of F there is a $\sigma \in CSU_{\mathcal{E}}^{\mathcal{W}}(F)$ such that $\rho|_{\mathcal{V}} \ll_{\mathcal{E}} \sigma|_{\mathcal{V}}$;
- for all $\sigma \in CSU_{\mathcal{E}}^{\mathcal{W}}(F)$, $Dom(\sigma) \subseteq \mathcal{V}$ and $Ran(\sigma) \cap \mathcal{W} = \emptyset$.

An $\mathcal{E}$-unification algorithm is *complete* if for any given system of equations it generates a complete set of $\mathcal{E}$-unifiers, which may not be finite. A unification algorithm is said to be *finite* and complete if it terminates after generating a finite and complete set of solutions.

2.4   Rewriting Logic

A rewrite theory $\mathcal{R} = (\Sigma, \mathcal{E}, R)$ consists of a MEL theory $(\Sigma, \mathcal{E})$ together with a finite set R of *conditional rewrite rules* each of which has the form

$$l \to r \ if \ \bigwedge_h p_h = q_h \wedge \bigwedge_i u_i := v_i \wedge \bigwedge_j w_j : s_j \wedge \bigwedge_k l_k \to r_k,$$

where l, r, and also each pair l_k, r_k, are Σ-terms of the same kind, and the rest of conditions fulfill the same requirements pointed out for MEL sentences. We will sometimes write $l \to r \ if \ c$ as a shortcut. Rewrite rules can also be unconditional. Equational and membership conditions are intended to be solved within the MEL theory $(\Sigma, \mathcal{E})$, i.e., no rewriting with rules from R is allowed on those conditions, whereas *reachability conditions* $l_k \to r_k$ are intended to be inferred using the deduction rules for rewrite theories below.

Such a rewrite rule specifies a *one-step transition* from a state $t[l\theta]_p$ to the state $t[r\theta]_p$, denoted by $t[l\theta]_p \to_R^1 t[r\theta]_p$, provided the condition holds. The subterm $t|_p$ is called a *redex*.

In the search tree example, R has as elements the nondeterministic (labeled) rewrite rules:

(I1) $x_{\text{st}} \mid y_k\, z_v; v_{\text{rs}} \mid w_p \;\to\; \text{ins}(x_{\text{st}}, y_k, z_v) \mid v_{\text{rs}} \mid w_p\, \text{i}\, y_k\, z_v.$
(D1) $x_{\text{st}} \mid y_k\, z_v; v_{\text{rs}} \mid w_p \;\to\; \text{del}(x_{\text{st}}, y_k) \mid v_{\text{rs}} \mid w_p\, \text{d}\, y_k\, z_v.$
(O1) $x_{\text{st}} \mid y_k\, z_v; v_{\text{rs}} \mid w_p \;\to\; x_{\text{st}} \mid v_{\text{rs}} \mid w_p\, \text{o}\, y_k\, z_v.$

All three rules have the same left term $x_{\text{st}} \mid y_k\, z_v; v_{\text{rs}} \mid w_p$, hence the nondeterminism of its application. As an example, rule (I1) states that from a State formed by any SearchTree (x_{st}), a non-empty RecordSet ($y_k\, z_v; v_{\text{rs}}$), and any Path (w_p), we can reach the State formed by the NeSearchTree obtained by inserting the Record $y_k\, z_v$ into x_{st}, the RecordSet v_{rs}, and the Path formed with the function symbol _i_ applied to the Path w_p and the Record $y_k\, z_v$. Recall that as RecordSets are commutative and associative, the Record is also nondeterministically chosen.

The inference rules for rewrite theories [3,6] in Fig. 2 specify a sound and complete calculus for the system specified by $\mathcal{R}$. We can reach a state v from a state u, written $\mathcal{R} \vdash u \to v$, if $u \to v$ can be inferred from $\mathcal{R}$ using these rules.

$$\frac{t \in T_\Sigma(\mathcal{X})}{t \to t}\ \textbf{Reflexivity} \qquad \frac{t_1 \to t_2, t_2 \to t_3}{t_1 \to t_3}\ \textbf{Transitivity}$$

$$\frac{f \in \Sigma_{k_1 \cdots k_n, k}\quad t_i \to t_i'\quad t_i, t_i' \in T_\Sigma(\mathcal{X})_{k_i}, 1 \le i \le n}{f(t_1, \ldots, t_n) \to f(t_1', \ldots, t_n')}\ \textbf{Congruence}$$

$$\frac{(l \to r \text{ if } \bigwedge_i p_i = q_i \wedge \bigwedge_j w_j : s_j \wedge \bigwedge_k l_k \to r_k) \in R \quad \theta : X \to T_\Sigma(Y)}{\begin{array}{c} \mathcal{E} \vdash p_i\theta = q_i\theta \text{ for all } i, \quad \mathcal{E} \vdash w_j\theta : s_j \text{ for all } j, \quad l_k\theta \to r_k\theta \text{ for all } k \\ \hline l\theta \to r\theta \end{array}}\ \textbf{Replacement}$$

Fig. 2. Deduction rules for rewrite theories.

The relation $\to_{R/\mathcal{E}}^1$ on $T_\Sigma(\mathcal{X})$ is $=_\mathcal{E} \circ \to_R^1 \circ =_\mathcal{E}$. $\to_{R/\mathcal{E}}^1$ on $T_\Sigma(\mathcal{X})$ induces a relation $\to_{R/\mathcal{E}}^1$ on $T_{\Sigma/\mathcal{E}}(\mathcal{X})$, the equivalence relation modulo $\mathcal{E}$, by $[t]_\mathcal{E} \to_{R/\mathcal{E}}^1 [t']_\mathcal{E}$ iff $t \to_{R/\mathcal{E}}^1 t'$. The transitive (resp. transitive and reflexive) closure of $\to_{R/\mathcal{E}}^1$ is denoted $\to_{R/\mathcal{E}}^+$ (resp. $\to_{R/\mathcal{E}}^*$).

A rewrite rule $l \to r$ if c, is *sort-decreasing* if for each substitution σ we have that $l\sigma \in T_\Sigma(\mathcal{X})_\kappa$ ($\kappa \in K \cup S$) and $c\sigma$ is verified implies $r\sigma \in T_\Sigma(\mathcal{X})_\kappa$.

For any relation $\to_R^1$ we say that a term t is $\to_R$-*irreducible* (or just R-irreducible) if there is no term t' such that $t \to_R^1 t'$ and we say that a substitution is R-*normalized* (or normalized if R can be deduced from the context) if $x\sigma$ is R-irreducible for all $x \in Dom(\sigma)$. We also say that a term t is *strongly* R-irreducible if for every R-normalized substitution σ the term $t\sigma$ is R-irreducible.

The relation $\rightarrow^1_R$ is *terminating* if there are no infinite rewriting sequences in $\rightarrow^1_R$. The relation $\rightarrow^1_R$ is *confluent* if whenever $t\rightarrow^*_R t'$ and $t\rightarrow^*_R t''$, there exists a term t''' such that $t'\rightarrow^*_R t'''$ and $t''\rightarrow^*_R t'''$. In a confluent, terminating, sort-decreasing, membership rewrite theory, for each term $t \in T_\Sigma(\mathcal{X})$, there is a unique (up to $\mathcal{E}$-equivalence) $R/\mathcal{E}$-irreducible term t' obtained by rewriting to *canonical* form, denoted by $t \rightarrow^!_{R/\mathcal{E}} t'$, or $t \downarrow_{R/\mathcal{E}}$ when t' is not relevant, which we call $can_{R/\mathcal{E}}(t)$.

2.5   Executable Rewrite Theories

For a rewrite theory $\mathcal{R} = (\Sigma, \mathcal{E}, R)$, whether a one step rewrite $t \rightarrow^1_{R/\mathcal{E}} t'$ holds is undecidable in general, because it involves searching a potentially infinite, and even non-computable, set $[t]_\mathcal{E}$ and checking if for any of its elements t_i we have that $t_i \rightarrow^1_R t''$ and $t'' =_\mathcal{E} t'$. The approach taken to solve this problem is to decompose $\mathcal{E}$ into a disjoint union $E \cup A$, with A a set of equational axioms (such as associativity, and/or commutativity, and/or identity) and define a new relation on $T_\Sigma(\mathcal{X})$ which, under certain assumptions on $\mathcal{R}$, will make $t \rightarrow^1_{R/\mathcal{E}} t'$ decidable.

Associated Rewrite Theory. Any MEL theory $(\Sigma, E\cup A)$ has a corresponding rewrite theory $\mathcal{R}_E = (\Sigma', A, R_E)$ associated to it [9], that allows us to check solutions for MEL conditions using rewriting instead of the deduction rules. Under certain restrictions this will be a finite process. The associated rewrite theory is constructed in the following way: we add a new connected component with sort *Truth* and a constant tt of this sort to Σ, and for each kind $k \in K$ a function symbol $eq_{k,k} : k\ k \rightarrow Truth$. We will use the symbols $\rightarrow^1$ and $\rightarrow$ to refer to single rewrite steps or full rewriting paths in this rewrite theory. There are rules $eq_{k,k}(x_k, x_k)\rightarrow tt$ in R_E for each kind $k \in K$. For each equation or membership in E

$$t = t' \text{ if } A_1 \wedge \ldots \wedge A_n \qquad t{:}s \text{ if } A_1 \wedge \ldots \wedge A_n,$$

R_E has a conditional rule or membership of the form

$$t\rightarrow t' \text{ if } A'_1 \wedge \ldots \wedge A'_n \qquad t{:}s \text{ if } A'_1 \wedge \ldots \wedge A'_n$$

where if $A_i \equiv t_i{:}s_i$ then A'_i is $t_i{:}s_i$, if $A_i \equiv t_i{:=}t'_i$ then A'_i is $t'_i\rightarrow t_i$, and if $A_i \equiv t_i{=}t'_i$ then A'_i is $eq_{k_i,k_i}(t_i, t'_i)\rightarrow tt$, where $k_i = [ls(t_i)]$.

The *inference rules for membership rewriting in* $\mathcal{R}_E$ are the ones in Fig. 3, adapted from [9, Fig. 4, p. 71], where the rules are defined for context-sensitive membership rewriting.

Definition 1 (E,A Rewriting). *The relation* $\rightarrow^1_{E,A}$ *is defined as:* $t \rightarrow^1_{E,A} t'$ *if there is a position* $p \in Pos(t)$, *a rule* $l \rightarrow r$ *if* $\bigwedge_{i\in I} A'_i$ *in* R_E, *and a substitution* σ *such that* $t|_p =_A l\sigma$ *(A-matching),* $t' = t[r\sigma]_p$, *and for all* $i \in I$:

$$\frac{t_1\to^1 t_2,\ t_2\to t_3}{t_1\to t_3}\ \text{Transitivity} \qquad \frac{t\to^1 t',\, t':s}{t:s}\ \text{Subject Reduction}$$

$$\frac{t=_A t'}{t\to t'}\ \text{Reflexivity} \qquad \frac{t_i\to^1 t_i'}{f(t_1,\ldots,t_i,\ldots,t_n)\to^1 f(t_1,\ldots,t_i',\ldots,t_n)}\ \text{Congruence}$$

$$\frac{l\to r\ \text{if}\ A_1'\wedge\ldots\wedge A_n'\in R_E\ \text{and}\ t=_A l\sigma}{A_1'\sigma\ldots A_n'\sigma}\ \text{Replacement}$$
$$t\to^1 r\sigma$$

$$\frac{l:s\ \text{if}\ A_1'\wedge\ldots\wedge A_n'\in R_E\ \text{and}\ t=_A l\sigma}{A_1'\sigma\ldots A_n'\sigma}\ \text{Membership}$$
$$t:s$$

Fig. 3. Inference rules for membership rewriting.

(i) *If A_i' is of the form $t_i \to t_i'$ then there is a term t_i'' such that $t_i\sigma \to^*_{E,A} t_i'' =_A t_i'\sigma$.*

(ii) *If A_i' is of the form $t_i : s_i$ then there is a term t_i'', a conditional membership $u : s_i$ if $\bigwedge_{j\in J} B_j'$ in R_E, and a substitution ρ such that $t_i\sigma \to^*_{E,A} t_i''$, $t_i'' =_A u\rho$, and $B_j'\rho$ satisfies one of these same two conditions for all $j \in J$.*

The relation should have been called $\to^1_{R_E,A}$, but we prefer $\to^1_{E,A}$ for simplicity. It is important to point out that after zero or more rewrite steps from $t_i\sigma$ in $\to^1_{E,A}$ we must check for A-equality the resulting term t_i'' against $t_i'\sigma$ in case *i*, and against $u\rho$ in case *ii*. This corresponds to an application of the reflexivity inference rule in R_E, which allows us to derive a rewrite step from an A-equality without applying any rewrite rule from R_E. The substitutions σ and ρ are difficult to find, but the conditions that we are going to impose on the rewrite theories, in order to make them executable, will make this task decidable.

Definition 2 (R,A Rewriting). *The relation $\to^1_{R,A}$ is defined as: $t \to^1_{R,A} t'$ if there is a position $p \in Pos(t)$, a rule $l \to r$ if $\bigwedge_{i\in I} A_i$ in R, and a substitution σ such that $t|_p =_A l\sigma$, $t' = t[r\sigma]_p$, and for all $i \in I$:*

- *if A_i is of the form $t_i \to t_i'$ then there is a term t_i'' such that $t_i\sigma \to^*_{R,A} t_i'' =_A t_i'\sigma$.*
- *if A_i is of the form $t_i : s_i$ then there is a term t_i'', a conditional membership $u : s_i$ if $\bigwedge_{j\in J} B_j'$ in R_E, and a substitution ρ such that $t_i\sigma \to^*_{E,A} t_i''$, $t_i'' =_A u\rho$, and $B_j'\rho$ satisfies this same condition or the following one for all $j \in J$.*
- *else we consider A_i', as in R_E, which is of the form $t_i \to t_i'$. Then there must be a term t_i'' such that $t_i\sigma \to^*_{E,A} t_i'' =_A t_i'\sigma$.*

We define: $\to_{E,A}$ as $\to^*_{E,A} \circ =_A$; $\to_{R,A}$ as $\to^*_{R,A} \circ =_A$; $\to^1_{R\cup E,A}$ as $\to^1_{R,A} \cup \to^1_{E,A}$; $\to_{R\cup E,A}$ as $\to^*_{R\cup E,A} \circ =_A$.

We have replaced searching in $\mathcal{E}$ with matching modulo and rewriting with $\to^1_{E,A}$. There are some problems with $\to^1_{E,A}$, and also with $\to^1_{R,A}$, that must be solved to make this approach executable. Consider a rewrite theory $\mathcal{R}$

with only one sort s, and whose only rule is $f(a, b) \rightarrow c$, where f is associative and commutative. The term $f(f(a, a), b)$ is a normal form in $\rightarrow^1_{R,A}$, but $f(f(a, a), b) \rightarrow^1_{R/A} f(a, c)$, because $f(f(a, a), b) =_A f(a, f(a, b))$, so the relations are different. This problem would not happen if $\mathcal{R}$ had another rule $f(x_s, f(a, b)) \rightarrow f(x_s, c)$ that could be applied on top of the term $f(f(a, a), b)$ with matching $x_s \mapsto a$, modulo associativity and commutativity, leading to $f(f(a, a), b) \rightarrow^1_{R,A} f(a, c)$. Rewrite theories that have these rules, avoiding such problems, are called *closed under A-extensions* [23].

A problem that can arise when trying to decide $t \rightarrow^*_R t'$ in a rewrite theory is that although $\rightarrow^1_R$ is terminating the proof of a condition may generate a recursive infinite check of conditions. This leads us to the notion of *operational termination*.

Definition 3 (Operational Termination of $\rightarrow^1_R$). *The relation $\rightarrow^1_R$ is operationally terminating if there are no infinite well-formed proof trees [18].*

This notion of operational termination was presented by Lucas, Marché and Meseguer [17] in an attempt to exclude those conditional term rewriting systems like the one consisting of the single conditional rule:

$$a \rightarrow b \ \ if \ f(a) \rightarrow b$$

The absence of unconditional rules makes the relation $\rightarrow$ trivially empty, hence terminating. Nevertheless, when trying to reduce the term a, most implementations will loop because of the following infinite derivation tree:

$$\frac{\dfrac{\dfrac{\cdots}{a \rightarrow b}}{f(a) \rightarrow b}}{a \rightarrow b}$$

The condition of operational termination states that such derivation trees don't exist. It may be argued that implementations can be enhanced to identify repeated terms in the derivation tree and block further derivations. This would not solve the problem. For instance, let's consider a specification of natural numbers in a MEL theory, which is easy to develop. In this specification we call the sort for natural numbers Nat, and the successor operation $s \in \Sigma_{\text{Nat,Nat}}$. Now we extend this specification with the declaration of a new sort Inf, with Inf $\leq$ Nat, and add the following equations:

$$x_{\text{Inf}} : \text{Nat} \ (i.e. \ \text{Inf} \leq \text{Nat})$$

$$s(x_{\text{Nat}}) : \text{Inf} \ if \ s(s(x_{\text{Nat}})) : \text{Inf}$$

If we try to derive $s(y_{\text{Nat}}) : \text{Inf}$, we get the following infinite derivation tree:

$$\frac{\dfrac{\dfrac{\cdots}{s(s(s(y_{\text{Nat}}))) : \text{Inf}}}{s(s(y_{\text{Nat}})) : \text{Inf}}}{s(y_{\text{Nat}}) : \text{Inf}}$$

Now, implementations usually get stuck here and there are no repeated terms that they can use to end the infinite loop. To avoid this problem, we will restrict ourselves to operationally terminating rewrite theories.

Another problem with rewrite theories when trying to apply a rule $l \rightarrow r$ if c is the value given to new *extra* variables in c, i.e., variables appearing in c and not in l. In order not to have to "guess" these values, which may make a rewrite theory untractable, rewrite theories must be *admissible*. We will first define *deterministic* and *strongly deterministic* rewrites theories, which is admissibility for rewrite theories with no Σ-equations at all, and later extend the definition to cover all rewrite theories.

Definition 4 (Deterministic Rewrite Theory). *Let* $\mathcal{R} = (\Sigma, A, R)$ *be a rewrite theory. We call* $\mathcal{R}$ *deterministic iff for each* $l \rightarrow r$ *if* $\bigwedge_{i=1}^{n} A_i \in R$, A_i *is of the form* $u_i \rightarrow v_i$ *or* $u_i : s$, *and for each* i, $1 \leq i \leq n$, *we have* $Var(u_i) \subseteq Var(l) \cup \bigcup_{j=1}^{i-1} Var(v_j)$.

In a deterministic rewrite theory, a condition $u_i \rightarrow v_i$ in a rule is satisfied if before attempting a rewriting step there exists a substitution σ such that $u_i =_A v_i\sigma$ (A-matching). For executability purposes of rewriting and efficiency of narrowing we limit this checking only to normal forms by using strongly deterministic rewrite theories.

Definition 5 (Strongly Deterministic Rewrite Theory). *A deterministic rewrite theory* $\mathcal{R} = (\Sigma, A, R)$ *is called* strongly deterministic *iff for each* $l \rightarrow r$ *if* $\bigwedge_{i=1}^{n} A_i \in R$, *and for each* i, $1 \leq i \leq n$ *where* A_i *is of the form* $u_i \rightarrow v_i$, v_i *is strongly* R, A-*irreducible.*

As previously said $R/\mathcal{E}$-rewriting may be non-computable. Under certain conditions we can replace it with $R \cup E$, A-rewriting and A-matching, and make a rewrite theory *executable*.

Definition 6 (Executable Rewrite Theory). *A rewrite theory* $\mathcal{R} = (\Sigma, E \cup A, R)$ *is executable, and also its underlying* MEL *theory* $(\Sigma, E \cup A)$, *if* Σ *is preregular modulo* A; E, A, *and* R *are finite, and the following conditions hold:*

1. *E and R are admissible [8] and closed under A-extensions, and $\rightarrow^1_{E/A}$ and $\rightarrow^1_{R/A}$ are operationally terminating. Admissibility is a translation of strong determinism to* MEL *theories forcing all matching variables in conditions to be instantiated by A-matching.*
2. *The axioms in A are regular, linear, and sort-preserving. Any variable appearing in an axiom must be a kinded variable. Furthermore, equality modulo A must be decidable and there must exist a finite matching algorithm modulo A producing a finite number of A-matching substitutions, $Match_A(t_1, t_2) = \{\sigma_i\}_{i=1}^{n}$ meaning that $t_1 =_A t_2\sigma_i$ for $i = 1, \ldots, n$, or failing otherwise.*
3. *The relation $\rightarrow^1_{E/A}$ is sort-decreasing, terminating, and confluent (where we again prefer to use $\rightarrow^1_{E/A}$ instead of $\rightarrow^1_{R_E/A}$).*

4. $\rightarrow^1_{R,A}$ is $\mathcal{E}$-consistent with $\rightarrow^1_{E,A}$, i.e., for all t_1, t_2, t_3 we have $t_1 \rightarrow^1_{R,A} t_2$ and $t_1 \rightarrow^*_{E,A} t_3$ implies that there exist t_4, t_5 such that $t_3 \rightarrow^*_{E,A} t_4$, $t_4 \rightarrow^1_{R,A} t_5$, and $t_2 =_{\mathcal{E}} t_5$. We represent this property by using a diagram with filled lines for universal quantification and dotted lines for existential quantification:

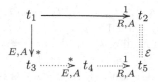

$$\mathcal{E}\text{-consistency of } \rightarrow^1_{R,A} \text{ with } \rightarrow^1_{E,A}$$

Under the above assumptions $\rightarrow^1_{E,A}$ is *strictly coherent*, i.e., for all t_1, t_2, t_3 if $t_1 \rightarrow^1_{E,A} t_2$ and $t_1 =_A t_3$ then there exists t_4 such that $t_3 \rightarrow^1_{E,A} t_4$ and $t_2 =_A t_4$ [23]. Also $\rightarrow^1_{R,A}$ is *strictly coherent*, i.e., for all t_1, t_2, t_3 we have $t_1 \rightarrow^1_{R,A} t_2$ and $t_1 =_A t_3$ implies that there exists t_4 such that $t_3 \rightarrow^1_{R,A} t_4$ and $t_2 =_A t_4$.

(a) strict coherence of $\rightarrow^1_{E,A}$ (b) strict coherence of $\rightarrow^1_{R,A}$

Technically, strict coherence means that the weaker relations $\rightarrow^1_{E,A}$ and $\rightarrow^1_{R,A}$ become semantically equivalent to the stronger relations $\rightarrow^1_{E/A}$ and $\rightarrow^1_{R/A}$. Under these conditions we can implement $\rightarrow^1_{R/\mathcal{E}}$ on terms using $\rightarrow^1_{R\cup E,A}$.

The following lemma links $\rightarrow^1_{R/\mathcal{E}}$ with $\rightarrow^1_{E,A}$ and $\rightarrow^1_{R,A}$.

Lemma 1. *Let $\mathcal{R} = (\Sigma, \mathcal{E}, R)$ be an executable rewrite theory, that is, it has all the properties above. Then $t_1 \rightarrow^1_{R/\mathcal{E}} t_2$ if and only if $t_1\!\downarrow \rightarrow^1_{R,A} t_3$ for some $t_3 =_{\mathcal{E}} t_2$, which can be verified by checking $t_3\!\downarrow =_A t_2\!\downarrow$.*

Lemma 2. *The rewrite theory $\mathcal{R}_E = (\Sigma', A, R_E)$ associated to any executable* MEL *theory $(\Sigma, E \cup A)$ is strongly deterministic.*

The rewrite theory for the search tree example is executable if we decompose $\mathcal{E}$ in the following way: the set A contains the associative, commutative, and identity equations in $\mathcal{E}$; the set E contains the rest of equations and all memberships in $\mathcal{E}$.

2.6 Reachability Goals

Given a rewrite theory $\mathcal{R} = (\Sigma, \mathcal{E}, R)$, a *reachability goal* G is a conjunction of the form $t_1 \rightarrow^* t_1' \wedge \ldots \wedge t_n \rightarrow^* t_n'$ where for $1 \le i \le n$, $t_i, t_i' \in T_\Sigma(\mathcal{X})_{\kappa_i}$ for appropriate κ_i. We define $Var(G) = \bigcup_i Var(t_i) \cup Var(t_i')$. A substitution σ is a *solution* of G if $t_i\sigma \rightarrow^*_{R/\mathcal{E}} t'_i\sigma$ for $1 \leq i \leq n$. If the substitution is idempotent we also say that the solution is *idempotent*. We define $E(G)$ to be the system of equations $t_1 = t'_1 \wedge \ldots \wedge t_n = t'_n$. We say σ is a *trivial solution* of G if it is an $\mathcal{E}$-unifier for $E(G)$. We say G is trivial if the identity substitution id is a trivial solution of G.

For goals $G : t_1 \rightarrow^* t_2 \wedge \ldots \wedge t_{2n-1} \rightarrow^* t_{2n}$ and $G' : t'_1 \rightarrow^* t'_2 \wedge \ldots \wedge t'_{2n-1} \rightarrow^* t'_{2n}$ we say $G =_{\mathcal{E}} G'$ if $t_i =_{\mathcal{E}} t'_i$ for $1 \leq i \leq 2n$. We say $G \rightarrow^1_R G'$ if there is an odd i such that $t_i \rightarrow_R t'_i$ and for all $j \neq i$ we have $t_j = t'_j$. That is, G and G' differ only in one subgoal ($t_i \rightarrow t_{i+1}$ vs $t'_i \rightarrow t_{i+1}$), but $t_i \rightarrow t'_i$, so when we rewrite t_i in G to t'_i we get G'. The relation $\rightarrow^1_{R/\mathcal{E}}$ over goals is defined as $=_{\mathcal{E}} \circ \rightarrow^1_R \circ =_{\mathcal{E}}$.

Systems of equations in $(\Sigma, E \cup A)$ with form $G \equiv \bigwedge^m_{i=1}(t_i = t'_i)$, with $[ls(t_i)] = k_i$, become reachability goals in $\mathcal{R}_E$ of the form $\bigwedge^m_{i=1}(eq_{k_i,k_i}(t_i, t'_i) \rightarrow tt)$. Using $\mathcal{R}_E$ we can verify whether a substitution σ is a solution of G by checking $eq_{k_i,k_i}(t_i\sigma, t'_i\sigma) \rightarrow tt$ for $i = 1, \ldots, m$. We will extend our notion of system of equations to include well-formed sentences $\bigwedge^n_{j=1} t_j : \kappa_j$, i.e. with $\kappa_j \in (S \cup K)$ and $[ls(t_j)] = [\kappa_j]$ for $j = 1, \ldots, n$, equivalent to an equation $t_j = x_{\kappa_j}$ with x_{κ_j} a fresh variable, which become reachability goals of the same form. Then σ is a solution of G if $t_j\sigma : \kappa_j$ is derivable in $\mathcal{R}_E$ for $j = 1, \ldots, n$, and $eq_{k_i,k_i}(t_i\sigma, t'_i\sigma) \rightarrow tt$ for $i = 1, \ldots, m$.

2.7 Narrowing

Let t be a Σ-term and $\mathcal{W}$ be a set of variables such that $Var(t) \subseteq \mathcal{W}$. The R, A-*narrowing* relation on $T_\Sigma(\mathcal{X})$ is defined as follows: $t \leadsto_{p,\sigma,R,A} t'$ if there is a non-variable position $p \in Pos_\Sigma(t)$, a rule $l \rightarrow r$ if $c \in R$, properly renamed, such that $Var(l) \cap \mathcal{W} = \emptyset$, and a unifier $\sigma' \in CSU^{\mathcal{W}'}_A(t|_p = l)$ for $\mathcal{W}' = \mathcal{W} \cup Var(l)$, such that $\sigma = \sigma'\sigma''$ for some σ'', $t' = (t[r]_p)\sigma$ and $c\sigma''$ holds. Similarly E, A-narrowing and $R \cup E, A$-narrowing relations are defined.

The substitution σ'' appears because the use of conditional rules usually forces the recursive resolution of $c\sigma'$ with narrowing, returning some σ'' as solution. Then $\sigma = \sigma'\sigma''$ is the desired substitution such that $t \leadsto_{p,\sigma,R,A} t'$.

Example 1. Consider a rewrite theory $\mathcal{R} = (\Sigma, E \cup A, R)$, where $S = \{s\}$, $\Omega = \{\{a, b, c\}_s, \{f, g\}_{ss,s}\}$, with $A = E = \emptyset$, and $R = \{g(b, c) \rightarrow c, f(a, z_s) \rightarrow b$ if $g(b, z_s) \rightarrow c\}$.

If we try to narrow the term $f(x_s, y_s)$ with rule $f(a, z_s) \rightarrow b$ if $g(b, z_s) \rightarrow c$ and unifier $\sigma' = \{x_s \mapsto a, y_s \mapsto w_s, z_s \mapsto w_s\}$ we have to prove the condition $g(b, w_s) \rightarrow c$, which can be narrowed with rule $g(b, c) \rightarrow c$ and substitution $\sigma'' = \{w_s \mapsto c\}$, so $g(b, z_s) \leadsto_{\sigma'',R,A} c$. Then, by composition of the substitutions σ' and σ'', we get $\sigma = \{x_s \mapsto a, y_s \mapsto c, z_s \mapsto c\}$ and we have $f(x_s, y_s) \leadsto_{\sigma,R,A} b$. As a consequence, that will be later proved, $f(x_s, y_s)\sigma \rightarrow^1_{R,A} b$.

3   Sentence-Normalized Rewriting

In this paper we are going to prove properties of our calculus with respect to $R \cup E, A$-normalized solutions. Executable rewrite and MEL theories force us to check whether there is a match between u_i and v_i or not before each single rewriting step for any condition $u_i \to_{E,A} v_i$ or $u_i \to_{R,A} v_i$. We are going to restrict ourselves to *FPP (Fresh Pattern Property)* executable rewrite and MEL theories [7], meaning that any variable appearing in the left side of a matching equation is new. These theories allow us to reduce every term to normal form before each narrowing step in a safe way.

When we check whether some substitution γ is a solution of a reachability goal G we can assume that $G\gamma$ is a ground term. If $Var(G\gamma) \neq \emptyset$ we can extend the signature Σ' in a way that all these variables become constants in the extended signature. By doing so, the only variables that will appear at any rewriting step will be matching variables.

We are going to incrementally construct the substitution used in the replacement and membership rules on FPP executable MEL theories. From now on we will write $\downarrow$ instead of $\downarrow_{E,A}$. If the ground term t matches l using σ_0 ($t =_A l\sigma_0$), then $l\sigma_0$ is a ground term and σ_0 is a ground substitution. If A_i has no matching variables then $\sigma_i = id$. If $A_i \equiv t_i' \to t_i$ has matching variables, then $Dom(\bigcup_{j=0}^{i-1} \sigma_j) \cap Var(t_i) = \emptyset$ and $eq_{k_i,k_i}(t_i \bigcup_{j=0}^{i-1} \sigma_j, t_i' \bigcup_{j=0}^{i-1} \sigma_j) = eq_{k_i,k_i}(t_i, t_i' \bigcup_{j=0}^{i-1} \sigma_j)$. Let σ_i be an A-matching of t_i with $(t_i' \bigcup_{j=0}^{i-1} \sigma_j)\downarrow$, that is $t_i\sigma_i =_A (t_i' \bigcup_{j=0}^{i-1} \sigma_j)\downarrow$, which is a ground term. As we are matching against an E, A-normalized ground term and t_i is strongly irreducible, σ_i must be ground E, A-normalized. The only exception is the first substitution σ_0 which is ground but may not be E, A-normalized. The *extended* substitution σ that we need to apply the replacement rule or the membership rule is $\sigma = \bigcup_{i=0}^{n} \sigma_i$, where the instantiation of all matching variables, $\bigcup_{i=1}^{n} \sigma_i$, is ground E, A-normalized.

Definition 7 (Sentence-Normalized Substitution). *Given an FPP executable* MEL *theory* $(\Sigma, E \cup A)$, *its associated rewrite theory* $\mathcal{R}_E = (\Sigma', A, R_E)$, *for any conditional sentence,* $c \equiv l \to r$ *if* $(\bigwedge_{i=1}^{n} A_i)$ *or* $c \equiv l : s$ *if* $(\bigwedge_{i=1}^{n} A_i)$, *and substitution* σ, *the* sentence-normalized *substitution* σ_c *is defined as* $\sigma_c = \sigma|_{Var(l)} \cup \sigma\downarrow|_{Mat(c)}$, *where* $Mat(c)$ *is the set of matching variables in* c.

Proposition 1. *Given an FPP executable* MEL *theory* $(\Sigma, E \cup A)$, *its associated rewrite theory* $\mathcal{R}_E = (\Sigma', A, R_E)$, *and a ground term* $t \in T_\Sigma$, *if* $t \to^1 r\sigma$ *or* $t : s$ *is derived with a conditional sentence* $c \equiv l \to r$ *if* $\bigwedge_{i=1}^{n} A_i'$ *or* $c \equiv l : s$ *if* $\bigwedge_{i=1}^{n} A_i'$ *and a substitution* σ, *then* $t \to^1 r\sigma_c$ *or* $t : s$ *is derivable using the same sentence* c *and* σ_c.

We are interested in derivations for reachability goals where we only rewrite with sentence-normalized substitutions, hence reducing the state space.

Definition 8 (Sentence-Normalized Rewriting). *When we want to imply that only sentence-normalized substitutions are applied in a rewrite step or a derivation with* $\to_{E,A}^1$, *will use the terms* sentence-normalized rewriting (SNR) *or* SNR-derivable, *and write* $t :_N s$, $t \to_N^1 t'$, *or* $t \to_N t'$.

Lemma 3 (Completeness of Sentence-Normalized Rewriting). *Given an FPP executable* MEL *theory* $(\Sigma, E \cup A)$ *and its associated rewrite theory* $\mathcal{R}_E =$

(Σ', A, R_E) if $eq_k(t, t') \rightarrow tt$, $t' \rightarrow t''$ or $t : s$, with t'' a normal form, are derivable in $\mathcal{R}_E$, then $eq_k(t, t') \rightarrow_N tt$, $t' \rightarrow_N t''$ or $t :_N s$ are SNR-derivable in $\mathcal{R}_E$ respectively.

As a direct consequence we get the following theorem telling us that with respect to reachability goals the solutions are the same using $\rightarrow^1_{E,A}$ or $\rightarrow^1_N$.

Theorem 1 (Equivalence of SNR for Reachability Goal Solutions).
Given an FPP executable MEL *theory* $(\Sigma, E \cup A)$ *and its associated rewrite theory* $\mathcal{R}_E = (\Sigma', A, R_E)$, σ *is a solution of* $\bigwedge_{i=1}^{m}(eq_{k_i,k_i}(t_i, t'_i) \rightarrow tt) \wedge \bigwedge_{j=1}^{n} t''_j : s_j$ *if and only if* $eq_{k_i,k_i}(t_i\sigma, t'_i\sigma) \rightarrow_N tt$, $i = 1, \ldots, m$, *and* $t''_j \sigma :_N s_j$, $j = 1, \ldots, n$, *are SNR-derivable.*

Now we prove that conditions and reduced conditions have the same solutions. This result is important because it will allow us to reduce the state space in our narrowing problems.

Proposition 2. *Given an FPP executable* MEL *theory* $(\Sigma, E \cup A)$ *and its associated rewrite theory* $\mathcal{R}_E = (\Sigma', A, R_E)$, *for any conditional* MEL *sentence* $c \equiv s$ *if* $\bigwedge_{i=1}^{n} A_i \in E$ *and corresponding rule or membership* $c' \equiv s'$ *if* $\bigwedge_{i=1}^{n} A'_i \in \mathcal{R}_E$ *if there is a substitution* σ *such that* $A'_1\downarrow\sigma_c, \ldots, A'_n\downarrow\sigma_c$ *are derivable in* $\mathcal{R}_E$ *then* $A'_1\sigma_c, \ldots, A'_n\sigma_c$ *are SNR-derivable in* $\mathcal{R}_E$.

Proposition 3. *Given an FPP executable* MEL *theory* $(\Sigma, E \cup A)$ *and its associated rewrite theory* $\mathcal{R}_E = (\Sigma', A, R_E)$, *for any conditional* MEL *sentence* $c \equiv s$ *if* $\bigwedge_{i=1}^{n} A_i \in E$ *and corresponding rule or membership* $c' \equiv s'$ *if* $\bigwedge_{i=1}^{n} A'_i \in \mathcal{R}_E$ *if there is a substitution* σ *such that* $A'_1\sigma_c, \ldots, A'_n\sigma_c$ *are derivable in* $\mathcal{R}_E$ *then* $A'_1\downarrow\sigma_c, \ldots, A'_n\downarrow\sigma_c$ *are SNR-derivable in* $\mathcal{R}_E$.

As SNR-derivations are derivations, we have proved the most strict properties in both directions. Again, as reachability goals are a special case of sentence conditions, we have as a direct consequence of the last two propositions the desired result.

Lemma 4 (Equivalence of Solutions for Reduced Reachability Goals).
Given an FPP executable MEL *theory* $(\Sigma, E \cup A)$ *and its associated rewrite theory* $\mathcal{R}_E = (\Sigma', A, R_E)$, σ *is a solution of* $\bigwedge_{i=1}^{m}(eq_{k_i,k_i}(t_i, t'_i) \rightarrow tt) \wedge \bigwedge_{j=1}^{n} t''_j : s_j$ *if and only if it is a solution of* $\bigwedge_{i=1}^{m}(eq_{k_i,k_i}(t_i, t'_i)\downarrow \rightarrow tt) \wedge \bigwedge_{j=1}^{n} t''_j\downarrow : s_j$.

4 Conditional Narrowing Modulo Unification

Narrowing allows us to assign values to variables in such a way that a reachability goal holds. We implement narrowing using a calculus with the following properties:

1. The calculus is *weakly complete*, i.e., for any idempotent E, A-normalized solution of a reachability goal G, the calculus can compute a more general answer for G.
2. The calculus is *sound*, i.e., if the calculus computes an answer σ for a reachability goal G, then σ is a solution of G.

We are going to split the calculus into two parts: the one that solves unification problems and the one that solves reachability problems.

4.1   Transformations and Calculus Rules for Unification

We assume that we have an A-unification algorithm that returns a *complete set of unifiers modulo A (CSU_A)* for any pair of terms. A unification equation has the form $t : \kappa = t' : \kappa'$, as a shorthand for $t = t' \wedge t : \kappa \wedge t' : \kappa'$. A unification goal is a conjunction of unification equations.

First we introduce an extension for the signature of the transformation described in Sect. 2.5 (Associated Rewrite Theory) in which we add syntax useful for the calculus, to force left to right solving of subgoals, and change the rules in R_E by turning conditional memberships into conditional rules, adding a new transformation for matching equations in conditions. Both changes will be useful in the context of the narrowing calculus.

Transformations of the Associated Rewrite Theory. We associate to any FPP executable MEL theory $(\Sigma, E \cup A)$ another rewrite theory $\tilde{\mathcal{R}}_E = (\tilde{\Sigma}, A, \tilde{R}_E)$, where $\tilde{\Sigma}$ is an extension of Σ' in $\mathcal{R}_E$. We add in $\tilde{\Sigma}$:

- A function symbol $\_ \wedge \_ : [\mathit{Truth}]\ [\mathit{Truth}] \rightarrow [\mathit{Truth}]$.
- A function symbol $eq_{s,s'} : [s]\ [s] \rightarrow \mathit{Truth}$ for each pair of sorts $s, s' \in S$ such that $[s] = [s']$
- A function symbol $\_ : \kappa : [\kappa] \rightarrow \mathit{Truth}$ for each $\kappa \in (S \cup K)$.
- For each conditional equation $(l = r\ if\ \bigwedge_{i=1}^n A_i)$ or conditional membership $(l : \kappa\ if\ \bigwedge_{i=1}^n A_i)$ in E, we add to $\tilde{R}_E$ a conditional rule $l \rightarrow r\ if\ A_1^\bullet \wedge \ldots \wedge A_n^\bullet$ or $l : \kappa \rightarrow tt\ if\ A_1^\bullet \wedge \ldots \wedge A_n^\bullet$, where each condition has an implicit $\rightarrow tt$, if $A_i \equiv t : s$ then $A_i^\bullet$ is $t : s$, and if $A_i \equiv t = t'$ or $A_i \equiv t := t'$ then $A_i^\bullet$ is $eq_{k,k}(t, t')$ with $k = [ls(t)]$.

A unification goal $\bigwedge_{i=1}^n (t_i : \kappa_i = t_i' : \kappa_i')$ has an associated narrowing problem $\bigwedge_{i=1}^n eq_{\kappa_i, \kappa_i'}(t_i, t_i') \downarrow \wedge\ tt$ (with an implicit $\rightsquigarrow tt$). The trailing $\wedge\ tt$ will allow us to work with a simpler set of calculus rules.

Calculus Rules for Unification. As we are computing $R \cup E, A$-normalized solutions, one general rule in our calculus is that we only apply a calculus rule if the composition of all computed substitutions remains E, A-normalized with respect to all matching variables and $R \cup E, A$-normalized with respect to the variables in the initial reachability problem. Our calculus is defined by the following inference rules, where t' has sort Truth and t'' has kind $[\mathit{Truth}]$:

- $[t]$ *transitivity*

$$\frac{t' \wedge t''}{t' \rightsquigarrow^1 x_{[\mathit{Truth}]}, x_{[\mathit{Truth}]} \wedge t''}$$

- $[n]$ *narrowing*

$$\frac{t \rightsquigarrow^1 x_k, t''}{(((c) \wedge t'')\rho\theta)\downarrow}$$

where $l \rightarrow r\ (if\ c) \in \tilde{R}_E, \theta \in CSU_A(t = l), \rho = \{x_k \mapsto r\}$

– [c] *congruence*

$$\frac{f(\bar{t}) \rightsquigarrow^1 x_k, t''}{t_i \rightsquigarrow^1 x'_{k_i}, t''\theta}$$

where $t_i \notin \mathcal{X}, \theta = \{x_k \mapsto f(t_1, \ldots, t_{i-1}, x'_{k_i}, t_{i+1}, \ldots, t_n)\}, k_i = [ls(t_i)]$

– [e] *elimination*

$$\frac{tt \wedge t''}{t''}$$

– [u] *unification*

$$\frac{eq_{\kappa,\kappa'}(t_1, t_2) \wedge t''}{((t_1 : \kappa'' \wedge t'')\theta)\downarrow}$$

where $\theta \in CSU_A(t_1 = t_2), \kappa''$ maximal such that $\kappa'' \leq \kappa, \kappa'' \leq \kappa'$

Definition 9 (Computed Answer). *Given a unification goal u and the corresponding narrowing problem G, if there is a narrowing path from G to tt, $G = G_0 \rightsquigarrow_{\sigma_1} G_1 \rightsquigarrow_{\sigma_2} \ldots \rightsquigarrow_{\sigma_n} tt$, where if in the step i we apply the transitivity rule or the elimination rule then $\sigma_i = id$, and $\sigma = \sigma_1\sigma_2 \ldots \sigma_n$, then we write $G \rightsquigarrow_\sigma^* tt$ and call $\sigma|_{Var(G)}$ a computed answer for u.*

Theorem 2. *The calculus for unification is sound and weakly complete, i.e., complete with respect to E, A-normalized solutions.*

5   Reachability by Conditional Narrowing

In this part of the calculus we define a new type of reachability goal $G \equiv \bigwedge_{i=1}^n t_i:\kappa_i \Rightarrow t'_i:\kappa'_i$, where the κ's are meant to carry sort information throughout the calculus in order to make use of it. Given an FPP executable rewrite theory with an underlying FPP MEL theory $\mathcal{R} = (\Sigma, E \cup A, R)$, we try to find an idempotent $R \cup E, A$-normalized solution σ (ground or not) such that $t_i\sigma \rightarrow_{R \cup E,A} t'_i\sigma$, $t_i\sigma$ has type κ_i and $t'_i\sigma$ has type κ'_i for all $i = 1, \ldots, n$. Since we need additional syntax to handle reachability goals, we extend $\tilde{R}_E$ to $\tilde{R}$ in the following way:

– For each pair $\kappa, \kappa' \in (S \cup K)$ such that $[\kappa] = [\kappa']$ we add the function symbol $\_:\kappa \Rightarrow \_:\kappa'$ to $\Sigma_{\kappa\kappa', Truth}$ to handle the new reachability goals. As rewrite theories are not sort-decreasing in general, we need a function symbol for each pair of types with same kind.
– For each kind $k \in K$ we add the function symbol $\_\Rightarrow^1\_$ to $\Sigma_{kk, Truth}$. This function symbol is *ad hoc* overloaded, i.e. the arguments of each overloaded function symbol have different kind, and stands for an actual narrowing step performed using the rules in R.

A conditional rewrite rule $l \rightarrow r$ if $\bigwedge_i A_i$ becomes $l \Rightarrow r$ if $\bigwedge_i A_i^\bullet$, where conditions of the form $l_i \rightarrow r_i$ become $l_i:k_i \Rightarrow r_i:k_i$, with $k_i = [ls(t_i)]$ and the rest of conditions are transformed as in the calculus for unification.

We turn the reachability goal G into a narrowing problem $G' \equiv \bigwedge_i t_i{\downarrow}{:}\kappa_i \Rightarrow t'_i{\downarrow}{:}\kappa'_i \wedge tt \rightsquigarrow tt$. Admissible goals, or simply goals, are now extended to be a conjunction of terms $t{:}\kappa \Rightarrow t'{:}\kappa'$, $t \Rightarrow^1 t'$, $t \rightsquigarrow tt$, $t \rightsquigarrow^1 x_k$ and tt, ending with tt. As in the calculus for unification we will use t as a shortcut for $t \rightsquigarrow tt$.

Any reachability subgoal in our calculus of the form $t{:}\kappa \Rightarrow t'{:}\kappa'$ is equivalent to $t \Rightarrow t' \wedge t{=}x_\kappa \wedge t'{=}y_{\kappa'}$ (or $t \Rightarrow t' \wedge t{:}\kappa \wedge t'{:}\kappa'$). We will solve normalized reachability goals $t_i{\downarrow}{:}\kappa \Rightarrow t'_i{\downarrow}{:}\kappa'$, which have the same solutions that the reachability goals $t_i{:}\kappa \Rightarrow t'_i{:}\kappa'$ have, because if $t_i\sigma \Rightarrow t'_i\sigma$, as $t_i =_\varepsilon t_i{\downarrow}$ and $t'_i =_\varepsilon t'_i{\downarrow}$ then $t_i{\downarrow}\sigma \Rightarrow t'_i{\downarrow}\sigma$ and vice versa. Also if $t_i{\downarrow}\sigma$ has type κ, as $t_i\sigma \rightarrow^* t_i{\downarrow}\sigma$, by subject reduction $t_i\sigma$ has type κ (and $t'_i\sigma$ has type κ'), and if $t_i\sigma$ has type κ, as $\rightarrow$ is sort-decreasing then $t_i{\downarrow}\sigma$ has type κ (and $t'_i{\downarrow}\sigma$ has type κ').

5.1   Calculus Rules for Reachability

Reachability by conditional narrowing is achieved using the calculus rules presented in Sect. 4, extended with the following calculus rules based on the deduction rules for rewrite theories in Fig. 2. It is very important to point out that we only apply replacement on $t|_p$ with substitution θ if *the whole term $t\theta$ is E, A-normalized*, achieving another reduction in the state space:

- $[X]$ *reflexivity*

$$\frac{t : \kappa \Rightarrow t' : \kappa' \wedge G'}{eq_{\kappa,\kappa'}(t,t') \wedge G'}$$

- $[N]$ *narrowing*

$$\frac{t : \kappa \Rightarrow t' : \kappa' \wedge G'}{t \rightsquigarrow^1 x_{[\kappa]}, x_{[\kappa]} : \kappa \Rightarrow t' : \kappa' \wedge G'}$$
$$\text{where } t \notin \mathcal{X}$$

- $[T]$ *transitivity*

$$\frac{t : \kappa \Rightarrow t' : \kappa' \wedge G'}{t \Rightarrow^1 x_{[\kappa]}, x_{[\kappa]} : [\kappa] \Rightarrow t' : \kappa' \wedge t : \kappa \wedge G'}$$
$$\text{where } t \notin \mathcal{X}$$

- $[I]$ *imitation*

$$\frac{t|_p \Rightarrow^1 x^p_{k_p}, t[x^p_{k_p}]_p : [\kappa] \Rightarrow t' : \kappa' \wedge t : \kappa \wedge G'}{t_j \Rightarrow^1 x^{pj}_{k_{pj}}, t[x^{pj}_{k_{pj}}]_{pj} : [\kappa] \Rightarrow t' : \kappa' \wedge t : \kappa \wedge G'}$$
$$\text{where } t|_p = f(t_1,\ldots,t_n), t_j \notin \mathcal{X}, k_{pj} = [ls(t_j)], 1 \le j \le n$$

- $[R]$ *replacement*

$$\frac{t|_p \Rightarrow^1 x^p_{k_p}, t[x^p_{k_p}]_p : [\kappa] \Rightarrow t' : \kappa' \wedge t : \kappa \wedge G'}{(((c) \wedge t[r]_p : [\kappa] \Rightarrow t' : \kappa' \wedge t : \kappa \wedge G')\theta){\downarrow}}$$
$$\text{where } l \Rightarrow r \ (if \ c) \in \tilde{R}, \theta \in CSU_A(t|_p = l), t\theta E, A\text{-normalized}$$

Theorem 3. *The calculus for reachability is sound and weakly complete, i.e., complete with respect to $R \cup E$, A-normalized solutions.*

6   Example

As an example of application of our calculus we use the specification of search trees in Sect. 2. We will abbreviate empty to ϵ, and consider the reachability goal $\epsilon\,[x_k^1\,y_v^1]\,\epsilon\mid f\,y_v^2\;;\;x_k^2\,2\mid nil:s \Rightarrow (\epsilon\,[x_k^3\,2]\,\epsilon)\,[c\,y_v^3]\,\epsilon\mid w_{rs}\mid z_p:s$, where from a State composed of a SearchTree with only one Record $x_k^1 y_v^1$ on the root, a RecordSet with two Records, $f\,y_v^2$ and $x_k^2\,2$ and the nil Path, we want to reach a State composed of a SearchTree with two Records, $x_k^3\,2$ on the left branch of the root and $c\,y_v^2$ on the root, some RecordSet, w_{rs}, and some Path, z_p. The reachability goal is already normalized, so we only have to add $\wedge\,tt$ to get the reachability problem (where we call $T_1 = (\epsilon\,[x_k^3\,2]\,\epsilon)\,[c\,y_v^3]\,\epsilon\mid w_{rs}\mid z_p$ and $T_2 = \epsilon\,[x_k^1\,y_v^1]\,\epsilon\mid f\,y_v^2;\;x_k^2\,2\mid nil)$:

1. $\epsilon\,[x_k^1\,y_v^1]\,\epsilon\mid f\,y_v^2\;;\;x_k^2\,2\mid nil:s \Rightarrow T_1:s\wedge tt \rightsquigarrow_{[T]}$
   The transitivity rule is always needed before a narrowing step.
2. $\epsilon\,[x_k^1\,y_v^1]\,\epsilon\mid f\,y_v^2\;;\;x_k^2\,2\mid nil:s \Rightarrow^1 x_{[s]}, x_{[s]}:[s] \Rightarrow T_1:s\wedge T_2:s\wedge tt \rightsquigarrow_{[R],I1}$
   Rule I1 is able to insert the record $x_k^2\,2$ thanks to the commutative axiom on RecordSets. We call $P_1 = nil\;i\;x_k^2\,2$.
3. $ins(\epsilon\,[x_k^1\,y_v^1]\,\epsilon, x_k^2, 2)\mid f\,y_v^2\mid P_1:[s] \Rightarrow T_1:s\wedge tt\wedge tt \rightsquigarrow_{[N]}$
   Now the calculus generates a narrowing for unification step.
4. $ins(\epsilon\,[x_k^1\,y_v^1]\,\epsilon, x_k^2, 2)\mid f\,y_v^2\mid P_1 \rightsquigarrow^1 x_{[s]}^4, x_{[s]}^4:[s] \Rightarrow T_1:s\wedge tt\wedge tt \rightsquigarrow_{[c]}$
   Congruence chooses the first subterm.
5. $ins(\epsilon\,[x_k^1\,y_v^1]\,\epsilon, x_k^2, 2) \rightsquigarrow^1 x_{[t]}^5, x_{[t]}^5\mid f\,y_v^2\mid P_1:[s] \Rightarrow T_1:s\wedge tt\wedge tt \rightsquigarrow_{[n],i1}$
   We apply narrowing for unification with conditional rule i1 :
   $ins(l_{st}[x_r]r_{st}, y_k, z_v) = ins(l_{st}, y_k, z_v)[x_r]r_{st}\;if\;y_k < key(x_r) = t.$
6. $eq_{[b],[b]}(x_k^2 < x_k^1, t)\wedge (\epsilon\,[x_k^2\,2]\,\epsilon)\,[x_k^1\,y_v^1]\,\epsilon\mid f\,y_v^2\mid P_1:[s] \Rightarrow T_1:s\wedge tt\wedge tt \rightsquigarrow_{[t]}$
   Transitivity generates another narrowing for unification step.
7. $eq_{[b],[b]}(x_k^2 < x_k^1, t) \rightsquigarrow^1 x_{[Truth]}^6, x_{[Truth]}^6\wedge (\epsilon\,[x_k^2\,2]\,\epsilon)\,[x_k^1\,y_v^1]\,\epsilon\mid f\,y_v^2\mid P_1:[s] \Rightarrow$
   $T_1:s\wedge tt\wedge tt \rightsquigarrow_{[c]}$
   Congruence chooses the first subterm.
8. $x_k^2 < x_k^1 \rightsquigarrow^1 x_{[b]}^7, eq_{[b],[b]}(x_{[b]}^7, t)\wedge (\epsilon\,[x_k^2\,2]\,\epsilon)\,[x_k^1\,y_v^1]\,\epsilon\mid f\,y_v^2\mid P_1:[s] \Rightarrow$
   $T_1:s\wedge tt\wedge tt \rightsquigarrow_{[n],b<c=t}, \{x_k^2 \mapsto b, x_k^1 \mapsto c\}$
   We apply narrowing for unification.
9. $tt\wedge (\epsilon\,[b\,2]\,\epsilon)\,[c\,y_v^1]\,\epsilon\mid f\,y_v^2\mid nil\;i\;b\,2:[s] \Rightarrow T_1:s\wedge tt\wedge tt \rightsquigarrow_{[e]}$
   Elimination removes leading tt's.
10. $(\epsilon\,[b\,2]\,\epsilon)\,[c\,y_v^1]\,\epsilon\mid f\,y_v^2\mid nil\;i\;b\,2:[s] \Rightarrow T_1:s\wedge tt\wedge tt \rightsquigarrow_{[X]}$
    The reflexivity rule turns the reachability problem into a unification one.
11. $eq_{[s],s}((\epsilon\,[b\,2]\,\epsilon)\,[c\,y_v^1]\,\epsilon\mid f\,y_v^2\mid nil\;i\;b\,2, (\epsilon\,[x_k^3\,2]\,\epsilon)\,[c\,y_v^3]\,\epsilon\mid w_{rs}\mid z_p)\wedge tt\wedge tt$
    $\rightsquigarrow_{[u],\{x_k^3\mapsto b, y_v^1\mapsto y_v^4, y_v^2\mapsto y_v^5, y_v^3\mapsto y_v^4, w_{rs}\mapsto f\,y_v^5, z_p\mapsto nil\;i\;b\,2\}}$
    The unification rule solves the problem.
12. $((\epsilon\,[b\,2]\,\epsilon)\,[c\,y_v^4]\,\epsilon\mid f\,y_v^5\mid nil\;i\;b\,2:s\wedge tt\wedge tt)\downarrow \equiv$
    $tt\wedge tt\wedge tt \rightsquigarrow_{[e]}$
13. $tt\wedge tt \rightsquigarrow_{[e]}$
14. tt

$\sigma = \{x_k^1 \mapsto \text{c}, x_k^2 \mapsto \text{b}, x_k^3 \mapsto \text{b}, y_v^1 \mapsto y_v^4, y_v^2 \mapsto y_v^5, y_v^3 \mapsto y_v^4, w_{\text{rs}} \mapsto \text{f } y_v^5, z_{\text{p}} \mapsto$ nil i b 2} is the computed answer (substitutions in steps 8 and 11). The calculus has found for the given reachability goal an instance $\epsilon \,[\text{c } y_v^4] \,\epsilon \,|\, \text{f } y_v^5, \text{b} \,2\,|\, \text{nil} : \text{s} \Rightarrow$ $(\epsilon \,[\text{b} \,2] \,\epsilon) \,[\text{c } y_v^4] \,\epsilon \,|\, \text{f } y_v^5 \,|\, \text{nil i b 2} : \text{s}$ where in the final State the SearchTree has two Records, b 2 and c y_v^4, there is still one Record, f y_v^5, left in the RecordSet to process, and the Path nil i b 2 tells us that the system has chosen to insert the Record b 2 in the SearchTree when building the computed answer.

7 Related Work, Conclusions and Future Work

A classic reference in equational conditional narrowing modulo is the work of Bockmayr [5]. The topic is addressed here for Church-Rosser equational conditional term rewriting systems with empty axioms, but non-terminating axioms (like ACU) are not allowed. Non-conditional narrowing modulo order-sorted equational logics is covered by Meseguer and Thati [24] and it is being used for cryptographic protocol analysis. Equivalence of $R/\mathcal{E}$ and $R \cup E, A$ rewriting was proved by Viry [27] for unsorted rewrite theories. Membership equational logic was defined by Meseguer [21]. A rewrite system for MEL theories that allows unification by rewriting is presented by Durán, Lucas et al. [9]. Strategies, which also play a main role in narrowing, have been studied by Antoy, Echahed, and Hanus [2]. Their needed narrowing strategy, for inductively sequential rewrite systems, generates only narrowing steps leading to a computed answer. Recently Escobar, Sasse, and Meseguer [13] have developed the concepts of variant and folding variant, a narrowing strategy for order-sorted unconditional rewrite theories that terminates on those theories having the *finite variant property*. Foundations for order-sorted conditional rewriting have been published by Meseguer [23]. Cholewa, Escobar, and Meseguer [7] have defined a new hierarchical method, called *layered constraint narrowing*, to solve narrowing problems in order-sorted conditional equational theories and given new theoretical results on that matter. Order-sorted conditional narrowing with constraint solvers has been addressed by Rocha and Meseguer [26].

In this work we have developed narrowing calculi for unification in membership equational logic and reachability in rewrite theories with an underlying membership equational logic. The main features in these calculi are that membership information is taken into account, all terms are normalized after each calculus step and only normalized instantiations are allowed for reachability terms and matching variables, greatly reducing the state space. The calculi have been proved sound and weakly complete.

Previous work with non-normalized terms and substitutions is available at http://maude.sip.ucm.es/cnarrowing/, where a strategy for applying the calculi was shown and implemented using Maude. This new version of the calculi has not been implemented yet, but we plan to make use of it in our current line of investigation, that concerns the extension of the calculi to handle constraints and their connection with external constraint solvers for domains such as finite domains, integers, Boolean values, etc., that could greatly improve the performance of any implementation.

Acknowledgments. We are very grateful to the anonymous referees for all their helpful comments.

References

1. Aguirre, L., Martí-Oliet, N., Palomino, M., Pita, I.: Conditional narrowing modulo in rewriting logic and Maude. In: Escobar [11], pp. 80–96
2. Antoy, S., Echahed, R., Hanus, M.: A needed narrowing strategy. In: Boehm, H., Lang, B., Yellin, D.M. (eds.) Conference Record of POPL 1994: 21st ACM SIGPLAN-SIGACT Symposium on Principles of Programming Languages, Portland, Oregon, USA, 17–21 January 1994, pp. 268–279. ACM Press (1994)
3. Bae, K., Meseguer, J.: Model checking LTLR formulas under localized fairness. In: Durán, F. (ed.) WRLA 2012. LNCS, vol. 7571, pp. 99–117. Springer, Heidelberg (2012)
4. Bae, K., Meseguer, J.: Infinite-state model checking of LTLR formulas using narrowing. In: Escobar [11], pp. 113–129
5. Bockmayr, A.: Conditional narrowing modulo a set of equations. Appl. Algebra Eng. Commun. Comput. **4**, 147–168 (1993)
6. Bruni, R., Meseguer, J.: Semantic foundations for generalized rewrite theories. Theor. Comput. Sci. **360**(1–3), 386–414 (2006)
7. Cholewa, A., Escobar, S., Meseguer, J.: Constrained narrowing for conditional equational theories modulo axioms. Technical report, C.S. Department, University of Illinois at Urbana-Champaign, August 2014. http://hdl.handle.net/2142/50289
8. Clavel, M., Durán, F., Eker, S., Lincoln, P., Martí-Oliet, N., Meseguer, J., Talcott, C.: All About Maude - A High-Performance Logical Framework: How to Specify, Program, and Verify Systems in Rewriting Logic. LNCS, vol. 4350. Springer, Heidelberg (2007)
9. Durán, F., Lucas, S., Marché, C., Meseguer, J., Urbain, X.: Proving operational termination of membership equational programs. High. Order Symb. Computat. **21**(1–2), 59–88 (2008)
10. Durán, F., Meseguer, J.: On the Church-Rosser and coherence properties of conditional order-sorted rewrite theories. J. Log. Algebr. Program. **81**(7–8), 816–850 (2012)
11. Escobar, S. (ed.): WRLA 2014. LNCS, vol. 8663. Springer, Heidelberg (2014)
12. Escobar, S., Meadows, C., Meseguer, J.: Maude-NPA: cryptographic protocol analysis modulo equational properties. In: Aldini, A., Barthe, G., Gorrieri, R. (eds.) FOSAD 2007/2008/2009. LNCS, vol. 5705, pp. 1–50. Springer, Heidelberg (2009)
13. Escobar, S., Sasse, R., Meseguer, J.: Folding variant narrowing and optimal variant termination. J. Log. Algebr. Program. **81**(7–8), 898–928 (2012)
14. Fay, M.: First-order Unification in an Equational Theory. University of California (1978)
15. Giovannetti, E., Moiso, C.: A completeness result for E-unification algorithms based on conditional narrowing. In: Boscarol, M., Aiello, L.C., Levi, G. (eds.) Foundations of Logic and Functional Programming. LNCS, vol. 306, pp. 157–167. Springer, Heidelberg (1986)
16. Hamada, M.: Strong completeness of a narrowing calculus for conditional rewrite systems with extra variables. Electr. Notes Theor. Comput. Sci. **31**, 89–103 (2000)
17. Lucas, S., Marché, C., Meseguer, J.: Operational termination of conditional term rewriting systems. Inf. Process. Lett. **95**(4), 446–453 (2005)

18. Lucas, S., Meseguer, J.: Operational termination of membership equational programs: the order-sorted way. Electr. Notes Theor. Comput. Sci. **238**(3), 207–225 (2009)
19. Martí-Oliet, N., Meseguer, J.: Rewriting logic: roadmap and bibliography. Theor. Comput. Sci. **285**(2), 121–154 (2002)
20. Meseguer, J.: Rewriting as a unified model of concurrency. In: Baeten, J., Klop, J. (eds.) CONCUR 1990 Theories of Concurrency: Unification and Extension. LNCS, vol. 458, pp. 384–400. Springer, Heidelberg (1990)
21. Meseguer, J.: Membership algebra as a logical framework for equational specification. In: Parisi-Presicce, F. (ed.) WADT 1997. LNCS, vol. 1376, pp. 18–61. Springer, Heidelberg (1998)
22. Meseguer, J.: Twenty years of rewriting logic. J. Log. Algebr. Program. **81**(7–8), 721–781 (2012)
23. Meseguer, J.: Strict coherence of conditional rewriting Modulo axioms. Technical report, C.S. Department, University of Illinois at Urbana-Champaign, August 2014. http://hdl.handle.net/2142/50288
24. Meseguer, J., Thati, P.: Symbolic reachability analysis using narrowing and its application to verification of cryptographic protocols. High. Order Symb. Comput. **20**(1–2), 123–160 (2007)
25. Middeldorp, A., Hamoen, E.: Completeness results for basic narrowing. Appl. Algebra Eng. Commun. Comput. **5**, 213–253 (1994)
26. Rocha, C.: Symbolic reachability analysis for rewrite theories. Ph.D. thesis, C.S. Department, University of Illinois at Urbana-Champaign, February 2013. http:// hdl.handle.net/2142/42200
27. Viry, P.: Rewriting: an effective model of concurrency. In: Halatsis, C., Philokyprou, G., Maritsas, D., Theodoridis, S. (eds.) PARLE 1994. LNCS, vol. 817, pp. 648–660. Springer, Heidelberg (1994)

Combining Runtime Checking and Slicing to Improve Maude Error Diagnosis

María Alpuente[1], Demis Ballis[2], Francisco Frechina[1], and Julia Sapiña[1($\boxtimes$)]

[1] DSIC-ELP, Universitat Politècnica de València,
Camino de Vera s/n, Apdo 22012, 46071 Valencia, Spain
{alpuente,ffrechina,jsapina}@dsic.upv.es
[2] DIMI, University of Udine, Via Delle Scienze, 206, 33100 Udine, Italy
demis.ballis@uniud.it

Abstract. This paper introduces the idea of using assertion checking for enhancing the dynamic slicing of Maude computation traces. Since trace slicing can greatly simplify the size and complexity of the analyzed traces, our methodology can be useful for improving the diagnosis of erroneous Maude programs. The proposed methodology is based on (i) a logical notation for specifying two types of user-defined assertions that are imposed on execution runs: functional assertions and system assertions; (ii) a runtime checking technique that dynamically tests the assertions and is provably safe in the sense that all errors flagged are definite violations of the specifications; and (iii) a mechanism based on equational least general generalization that automatically derives accurate criteria for slicing from falsified assertions.

1 Introduction

Back in the mid-80s, while the scientific research in Computer Science was just taking off in Spain, some magnetic manuscripts written by Joseph Goguen and José Meseguer came into our hands [17,18]. Although some predicted that no language as ambitious as that described in [17,18,25] could reach widespread or practical use, these ground-breaking documents were, for many of us, the starting point for pursuing the advancement of multi-paradigm declarative languages and their development environments. Towards this endeavor, the aim of this work is to contribute to further advancing the state-of-the-art of the leading-edge multi-paradigm language Maude.

Assertion checking is the problem of deciding whether a certain assertion holds at a given program (or execution) point. Although not universally used, assertion checking seems to have widely infiltrated common programming

This work has been partially supported by the EU (FEDER) and the Spanish MINECO project ref. TIN2013-45732-C4-01 (DAMAS), and by Generalitat Valenciana ref. PROMETEOII/2015/013 (SmartLogic). F. Frechina was supported by FPU-ME grant AP2010-5681, and J. Sapiña was supported by FPI-UPV grant SP2013-0083.

© Springer International Publishing Switzerland 2015
N. Martí-Oliet et al. (Eds.): Meseguer Festschrift, LNCS 9200, pp. 72–96, 2015.
DOI: 10.1007/978-3-319-23165-5_3

practice as witnessed by the growth of assertion capabilities in widely used programming languages such as C♯, C++, and Java [12]. Assertions may be used statically to support program analysis and also for secondary purposes, such as documentation and to provide information to an optimizer during code generation. A brief history of the research ideas that have contributed to the assertion capabilities of modern programming languages and development tools can be found in [12]. The most obvious way to dynamically use assertions is to test them at runtime and report any detected violations. By finding inconsistencies between asserted properties and the program code, runtime assertion checking can be used to reveal program faults and to obtain information about their locations. Since an assertion failure usually reports an error, the user can direct its attention to the location at which the logical inconsistency is detected and (hopefully) trace the errors back to their sources more easily.

Program slicing [20] automatically identifies a subset of program statements that either (i) contribute to the values of a set of variables at a given point, or (ii) are influenced by the values of a given set of variables. The first approach corresponds to forms of backward slicing, whereas the second corresponds to forward slicing. Automatic slicing plays an important role in program diagnosis and understanding since it allows one to focus on code fragments that are relevant to a given slicing criterion, that is, the relevant information we want to track (backwards or forwards) from a given execution point.

Maude [13] is a high-level language and high-performance system that supports both equational and rewriting logic computations. Maude modules correspond to specifications in rewriting logic [24], which is a logic that allows the representation of many models of concurrent and distributed systems. In [6,8], a rich and highly dynamic parameterized scheme for exploring rewriting logic computations is developed that can significantly reduce the size and complexity of the runs under examination by automatically slicing both programs and computation traces [4].

The aim of this work is to provide Maude with runtime assertion-checking capabilities by first introducing a simple assertion language that suffices for the purpose of improving error diagnosis and debugging of Maude programs, while remaining tractable. We follow the approach of modern specification and verification systems such as Spec♯ or the Java Modeling Language (JML) where the specification language is typically an extension of the underlying programming language and specifications are used as *contracts* that guarantee certain properties to hold at a number of execution states, e.g., before or after a given function call [22]. We believe that this choice of a language is of practical interest because it facilitates the job of programmers. Even if Maude is a highly declarative language that supports a programming style where no conceptual difference exists between programs and high-level specifications, a separate description given by the assertions may help developers identify essential program behaviors to be preserved when modifying code.

We distinguish two groups of assertions: (1) functional assertions, for specifying properties of functions defined by an equational theory; and (2) system

assertions, which allow one to express properties concerning the system's execution. The assertions we support allow us to express properties that are quite
general, including user–defined programs. However, for the functional assertions,
we require the user to ensure that the execution of any property terminates for
any possible initial state and that the resulting verdict is unique, in the same
spirit of Maude's (canonical) equational theories. In the proposed framework,
if an assertion evaluates to false at runtime, an assertion failure results, which
typically causes execution to abort while delivering a huge execution trace. By
automatically inferring deft slicing criteria from falsified assertions, we derive
a self-initiating, enhanced dynamic slicing technique that automatically starts
slicing the trace backwards at the time the assertion violation occurs, without
having to manually determine the slicing criterion in advance. As a by-product of
the trace slicing process, we also compute a dynamic program slice that preserves
the program behavior for the considered program inputs [20].

The Maude Formal Environment (MFE) is a recent effort to integrate and
interoperate most of the available Maude analysis and verification tools. These
include among others an inductive theorem prover, a declarative debugger, and
Maude's model checkers [23]. Maude supports strong typing and subtyping assertions via *membership axioms*, which are used to automatically 'narrow' the type
T of a value into a subtype of T. Nevertheless, to the best of our knowledge,
no general built-in support is provided in Maude or the MFE for the runtime
checking of user-defined assertions. Related to our work, generic strategies are
defined in [16,28] to guarantee that a set of invariants (that can be expressed
in different logics) are satisfied at every computed state. This is achieved by
avoiding the execution of actions that otherwise would conduct the system to
states that do not satisfy the constraints. This is in contrast to our approach
in two ways. On the one hand, our assertions are *external* and evaluated at
runtime, whereas driving the system's execution in such a way that every computation state complies with the constraints makes the assertions *internal* to the
programmed strategy. On the other hand, the strategy of [16,28] never results
in violated assertions, which is essential for automatic trace slicing to be fired
according to our approach. As another difference, we are able to check assertions
that regard: (1) the normalizations carried out by using the equational part of
the rewriting theory; and (2) system properties that are not necessarily global
invariants but can only hold in those states that match a given state template.

Following the discussion above, this work can be seen as the first framework that exploits the synergies we can find between runtime assertion checking
and automated (program and program trace) transformations for improving the
diagnosis of Maude programs.

Plan of the paper. The paper is organized as follows. Section 2 provides a brief
introduction to rewriting logic and Maude and introduces the running example
that we use throughout the paper: a conditional rewrite theory that models a
simple, distributed banking system. Section 3 introduces a very simple assertion
language and the notions of functional and system assertions whose violation
helps signal functional and system *error symptoms*. Section 4 recalls a trace slic-

ing methodology for simplifying rewriting logic computations. Section 5 enriches the slicing methodology with runtime assertion checking in order to improve the diagnosis of erroneous Maude programs, and describes its implementation in the ABETS tool. Section 6 concludes. More details and examples, and a thorough comparison with the related literature, can be found in an extended version of this article, which is available at [7].

2 Rewriting Logic and Maude

Let us recall some important notions that are relevant to this work. We assume some basic knowledge of term rewriting [29] and rewriting logic [24] (RWL). Some familiarity with the Maude language [13,14] is also required. Throughout the paper, Maude notation will be introduced "on the fly" as required.

2.1 Preliminaries

Let Σ be a *signature* that allows operators to be specified together with their type structure by means of suitable sets of sorts and kinds. By $\tau(\Sigma)$, we specify the term algebra that includes all the ground terms built over Σ, while $\tau(\Sigma, \mathcal{V})$ is the usual nonground term algebra built over Σ and the set of variables $\mathcal{V}$. Each operator in Σ is defined along with its sort and axiom declarations that may specify algebric laws such as associativity (`assoc`), commutativity (`comm`), and identity (`id`).

A *position* w in a term t is represented by a sequence of natural numbers that addresses a subterm of t (Λ denotes the empty sequence, i.e., the root position). Given a term t, we let $\mathcal{P}os(t)$ denote the set of positions of t. By $t_{|w}$, we denote the *subterm* of t at position w, and by $t[s]_w$, we denote the result of *replacing the subterm* $t_{|w}$ by the term s in t.

A *substitution* $\sigma \equiv \{x_1/t_1, x_2/t_2, \ldots, x_n/t_n\}$ is a mapping from the set of variables $\mathcal{V}$ to the set of terms $\tau(\Sigma, \mathcal{V})$, which is equal to the identity almost everywhere except over a set of variables $\{x_1, \ldots, x_n\}$. By $\{\}$, we denote the *identity* substitution. The application of a substitution σ to a term t, denoted $t\sigma$, is defined by induction on the structure of terms as usual [10]. Given two terms t and t', we say that t is *more general* than t' iff there exists a substitution σ such that $t\sigma = t'$. We also say that t' is an *instance* of t.

Given a syntactic expression e, by $\mathcal{V}ar(e)$, we denote the set of variables that occur in e. Given a binary relation $\rightsquigarrow$, we define the usual *transitive* (resp., *transitive and reflexive*) closure of $\rightsquigarrow$ by $\rightsquigarrow^+$ (resp., $\rightsquigarrow^*$).

2.2 Rewrite Theories and Maude Modules

The static state structure as well as the dynamic behavior of a concurrent system can be formalized as a RWL specification that encodes a *conditional rewrite theory*. More specifically, a *conditional rewrite theory* (or simply *rewrite theory*) is a triple $\mathcal{R} = (\Sigma, E, R)$, where:

(i) (Σ, E) is a membership equational theory that allows us to define the system data types via equations, as well as algebraic and membership axioms. Σ is a signature that specifies the operators of $\mathcal{R}$, while $E = \Delta \cup B$ is the disjoint union of the set Δ, which contains conditional equations and conditional membership axioms, and the set B, which contains algebraic axioms associated with binary operators in Σ. The general Maude syntax of conditional equations and membership axioms is the following:

$$\texttt{ceq } [l] \; : \; \lambda = \rho \texttt{ if } C \; . \qquad \texttt{cmb } [l] \; : \; \lambda \; : \; s \texttt{ if } C \; .$$

where l is a label (i.e., a name that identifies the equation), $\lambda, \rho \in \tau(\Sigma, \mathcal{V})$, s is a sort and C is an *equational* condition, that is, a (possibly empty) conjunction of equations $\texttt{t} = \texttt{t}'$, matching equations $\texttt{p} := \texttt{t}$, and memberships $\texttt{t} : \texttt{s}'$ that is built using the binary conjunction connective $\bigwedge$, which is assumed to be associative. When C is empty, the syntax for equations and memberships is simplified as follows:

$$\texttt{eq } [l] \; : \; \lambda = \rho \; . \qquad \texttt{mb } [l] \; : \; \lambda \; : \; s \; .$$

A membership equational theory (Σ, E) is encoded in Maude through a *functional module* that is syntactically delimited by keywords $\texttt{fmod}$ and $\texttt{endfm}$. Functional modules provide executable models for the specified equational theories.

Example 1. The following Maude functional module[1] encodes an equational theory that defines the functional part of a simple, distributed banking system.

```
fmod BANK-EQ is inc BANK-INT+ID . pr SET{Id} .
  sorts Account PremiumAccount Status Msg State .
  subsort PremiumAccount < Account .
  subsorts  Account Msg < State .

  var ID : Id .            op <_|_|_> : Id Int Status -> Account [ctor] .
  var BAL : Int .          op active : -> Status [ctor] .
  var STS : Status .       op blocked : -> Status [ctor] .


  op Alice : -> Id [ctor] .        op Bob : -> Id [ctor] .
  op Charlie : -> Id [ctor] .      op Daisy : -> Id [ctor] .


  cmb < ID | BAL | STS > : PremiumAccount if ID in PreferredClients .


  op PreferredClients : -> Set{Id} .
  eq PreferredClients = Bob, Charlie .


  op updateStatus : Account -> Account .
  ceq updateStatus(< ID | BAL | active >) = < ID | BAL | blocked >
      if BAL < 0 .
  eq updateStatus(< ID | BAL | STS >) = < ID | BAL | STS > [owise] .
endfm
```

---

[1] BANK-EQ includes the functional module BANK-INT+ID, which (i) imports INT for integer manipulation and (ii) declares the sort Id that is used to parameterize SET{X :: TRIV}.

A bank account is represented as a term of the form < ID | BAL | STS > where ID is the owner of the account, BAL is the account balance, and STS is the account status, which can be blocked or active. The defined conditional membership axiom states that an account is a PremiumAccount if its owner is included in the PreferredClients set. Finally, the updateStatus operation updates the status account to blocked when the account balance is negative.

(ii) R is a set of conditional labeled rules whose Maude syntax is the following:

$$\texttt{crl } [l] \ : \ \lambda \ \texttt{=>} \ \rho \ \texttt{if } C \ .$$

where l is a label, $\lambda, \rho \in \tau(\Sigma, \mathcal{V})$, and C is a *rule* condition, i.e., an equational condition that may also contain rewrite expressions of the form $\mathbf{t} = \mathbf{t}'$. When a rule has no condition, we simply write $\texttt{rl } [l] : \lambda \ \texttt{=>} \ \rho$.

A rewrite theory $\mathcal{R} = (\Sigma, E, R)$ is specified in Maude by means of a *system module*, which is introduced by the syntax mod...endm. A system module may include both a functional representation of the equational theory (Σ, E) and the specification of the rewrite rules in R.

Example 2. The following Maude rewrite theory models the distributed behavior of the banking system of Example 1.

```
mod BANK is inc BANK-EQ .
  vars ID ID1 ID2 : Id .
  vars BAL BAL1 BAL2 M : Int .

  op empty-state : -> State [ctor] .
  op _;_ : State State -> State [ctor assoc comm id: empty-state] .
  ops credit debit : Id Int -> Msg [ctor] .
  op  transfer : Id Id Int -> Msg  [ctor] .

  rl [credit] : credit(ID,M) ; < ID | BAL | active > =>
                  updateStatus(< ID | BAL + M | active >) .
  rl [debit] : debit(ID,M) ; < ID | BAL | active > =>
                  updateStatus(< ID | BAL - M | active >) .
  rl [transfer] : transfer(ID1,ID2,M) ;
                    < ID1 | BAL1 | active > ; < ID2 | BAL2 | active >
                  => updateStatus(< ID1 | BAL1 - M | active >) ;
                     updateStatus(< ID2 | BAL2 + M | active >) .
endm
```

Each state of the system is modeled as a multiset (i.e., an associative and commutative list) of elements of the form $e_1; e_2; \ldots \ldots; e_n$. Each element e_i is either (i) a bank account; or (ii) a message modeling a debit, credit, or transfer operation. These account operations are implemented via three rewrite rules: namely, debit, credit, and transfer rules.

2.3   Rewriting and Generalization Modulo Equational Theories

Let us consider a conditional rewrite theory (Σ, E, R), with $E = \Delta \cup B$, where Δ is a set of conditional equations and membership axioms, and B is a set of equational axioms associated with some binary operators in Σ. The conditional rewriting modulo E relation (in symbols, $\to_{R/E}$) can be defined by lifting the usual conditional rewrite relation on terms [19] to the E-congruence classes $[t]_E$ on the term algebra $\tau(\Sigma, \mathcal{V})$ that are induced by $=_E$ [11]. In other words, $[t]_E$ is the class of all terms that are equal to t modulo E. Unfortunately, $\to_{R/E}$ is, in general, undecidable since a rewrite step $t \to_{R/E} t'$ involves searching through the possibly infinite equivalence classes $[t]_E$ and $[t']_E$.

The Maude interpreter implements conditional rewriting modulo E by means of two much simpler relations, namely $\to_{\Delta,B}$ and $\to_{R,B}$. These allow rules, equations and memberships to be intermixed in the rewriting process by simply using an algorithm of matching modulo B. We define $\to_{R\cup\Delta,B}$ as $\to_{R,B} \cup \to_{\Delta,B}$. Roughly speaking, the relation $\to_{\Delta,B}$ uses the equations of Δ (oriented from left to right) as simplification rules. Thus, by repeatedly applying the equations as simplification rules from a given term t, we eventually reach a term $t \downarrow_{\Delta,B}$ to which no further equations can be applied. The term $t \downarrow_{\Delta,B}$ is called a *canonical (or normal)* form of t w.r.t. Δ modulo B. An *equational simplification* of a term t in Δ modulo B is a rewrite sequence of the form $t \to^*_{\Delta,B} t \downarrow_{\Delta,B}$. Informally, the relation $\to_{R,B}$ implements rewriting with the rules of R, which might be non-terminating and non-confluent, whereas Δ is required to be terminating and Church-Rosser modulo B in order to guarantee the existence and unicity (modulo B) of a canonical form w.r.t. Δ for any term [14]. Terms are rewritten into canonical forms according to their sort structure, which is induced by the signature Σ and the membership axioms specified in Δ. In particular, through membership axioms of the form cmb [1] : λ : s if C, we can assert that any term B-matching λ has a specific sort s whenever a condition C holds. Equational simplification of terms is naturally lifted to substitutions as follows: given $\sigma = \{x_1/t_1, x_2/t_2, \ldots, x_n/t_n\}$, we define the *normalized* substitution $\sigma \downarrow_{\Delta,B} = \{x_i/(t_i \downarrow_{\Delta,B})\}_{i=1}^n$.

Formally, $\to_{R,B}$ and $\to_{\Delta,B}$ are defined as follows. Given a rewrite rule crl [r] : $\lambda \Rightarrow \rho$ if C $\in R$ (resp., an equation ceq [e] : $\lambda = \rho$ if C $\in \Delta$), a substitution σ, a term t, and a position w of t, $t \overset{r,\sigma,w}{\to}_{R,B} t'$ (resp., $t \overset{e,\sigma,w}{\to}_{\Delta,B} t'$) iff $\lambda\sigma =_B t_{|w}$, $t' = t[\rho\sigma]_w$, and C *evaluates to true* w.r.t. σ. When no confusion arises, we simply write $t \to_{R,B} t'$ (resp. $t \to_{\Delta,B} t'$) instead of $t \overset{r,\sigma,w}{\to}_{R,B} t'$ (resp. $t \overset{e,\sigma,w}{\to}_{\Delta,B} t'$).

Roughly speaking, a conditional rewrite step on the term t applies a rewrite rule/equation to t by replacing a *reducible* (sub-)*expression* of t (namely $t_{|w}$), called the *redex*, by its contracted version $\rho\sigma$, called the *contractum*, whenever the condition C is fulfilled. Note that the evaluation of a condition C is typically a recursive process since it may involve further (conditional) rewrites in order to normalize C to *true*. Specifically, an equation e *evaluates to true* w.r.t. σ if $e\sigma \downarrow_{\Delta,B} =_B true$; a matching equation p := t *evaluates to true* w.r.t. σ if $p\sigma =_B t\sigma \downarrow_{\Delta,B}$; a rewrite expression t $\Rightarrow$ p *evaluates to true* w.r.t. σ if there exists a rewrite sequence $t\sigma \rightarrow^*_{R\cup\Delta,B} u$, such that $u =_B p\sigma^2$; and, finally, a membership $t : s$ *evaluates to true* w.r.t. σ if $t\sigma$ has sort s.

Under appropriate conditions on the rewrite theory, a rewrite step $s \rightarrow_{R/E} t$ modulo E on a term s can be implemented without loss of completeness by applying a rewrite strategy that first simplifies the term s into its canonical form $s \downarrow_{\Delta,B}$, and then applies a rule $r \in R$ to $s \downarrow_{\Delta,B}$ [15].

A *computation* (trace) $\mathcal{C}$ for s_0 in the conditional rewrite theory $(\Sigma, \Delta \cup B, R)$ is then deployed as the (possibly infinite) rewrite sequence

$$s_0 \rightarrow^*_{\Delta,B} s_0 \downarrow_{\Delta,B} \rightarrow_{R,B} s_1 \rightarrow^*_{\Delta,B} s_1 \downarrow_{\Delta,B} \rightarrow_{R,B} \cdots$$

that interleaves $\rightarrow_{\Delta,B}$ rewrite steps and $\rightarrow_{R,B}$ rewrite steps following the strategy mentioned above. After each conditional rewriting step using $\rightarrow_{R,B}$, in general, the resulting term s_i, $i = 1,\ldots,n$, is not in canonical normal form. Therefore, it is normalized before the subsequent rewrite step with $\rightarrow_{R,B}$ is performed. Also, in the precise strategy adopted by Maude, the last term of a finite computation is finally normalized before the result is delivered. By ε, we denote the *empty* computation.

We define a *Maude step* from a given term s as any of the sequences $s \rightarrow^*_{\Delta,B} s \downarrow_{\Delta,B} \rightarrow_{R,B} t \rightarrow^*_{\Delta,B} t \downarrow_{\Delta,B}$ that head the non-deterministic Maude computations for s. Note that, for a canonical form s, a Maude step for s boils down to $s \rightarrow_{R,B} t \rightarrow^*_{\Delta,B} t \downarrow_{\Delta,B}$. We define $m\mathcal{S}(s)$ as the set of all the non-deterministic Maude steps from s.

A *generalization* of a pair of terms t_1, t_2 is a triple (g, θ_1, θ_2) such that $g\theta_1 = t_1$ and $g\theta_2 = t_2$. The triple (g, ϕ_1, ϕ_2) is the *least general generalization (lgg)* of the pair of terms t_1, t_2, written $lgg(t_1, t_2)$, if (1) (g, ϕ_1, ϕ_2) is a generalization of t_1, t_2 and (2) for every other generalization (g', ψ_1, ψ_2) of t_1, t_2, g' is more general than g. The *lgg* of a pair of terms is unique up to variable renaming [21].

In [9], the notion of least general generalization is extended to work modulo (order-sorted) equational theories, where function symbols can obey any combination of associativity, commutativity, and identity axioms (including the empty set of such axioms). Unlike the untyped case, for a pair of terms t_1, t_2 there is generally no single lgg, due to order-sortedness or to the equational axioms. Instead, there is a finite, minimal, and complete set of lggs (denoted by $lgg_E(t_1, t_2)$) so that any other equational generalizer has at least one of them as an instance. Given any element (g, ϕ_1, ϕ_2) of the set $lgg_E(t_1, t_2)$, we define the function π from $\mathcal{P}os(t)$ to $\mathcal{P}os(t_1)$ that provides an injective correspondence between (the position of) any variable in g and (the position of) the corresponding term in t_1; we need this because computing modulo algebraic axioms may cause the term structure of g to be different from both, t_1 and t_2. For instance, consider an associative and commutative symbol f and the terms $t_1 = f(b, c, a)$ and $t_2 = f(d, a, b)$. Then, a possible lgg modulo the associativity and commutativity

---

[2] Technically, to properly evaluate a rewrite expression $t \Rightarrow p$ or a matching condition $p := t$, the term p is required to be a Δ-pattern modulo B (i.e., a term p such that, for every substitution σ, if $x\sigma$ is a canonical form w.r.t. Δ modulo B for every $x \in Dom(\sigma)$, then $p\sigma$ is also a canonical form w.r.t. Δ modulo B).

of f is $(f(a, b, X), \{X/c\}, \{X/d\}) \in lgg_E(t_1, t_2)$, where X is a variable. Note that both t_1 and t_2 are syntactically different from $f(a, b, X)$, and the value $\pi(3) = 2$ indicates the subterm c of t_1 that is responsible for the mismatch with t_2. By $\widehat{lgg}_E(t_1, t_2)$ we denote the pair (G, π) where $G = (g, \phi_1, \phi_2)$ is arbitrarily chosen among those lggs in the set $lgg_E(t_1, t_2)$ that have fewer variables, and π is the corresponding position mapping from positions of g's variables to the relative subterms of t_1.

One of the main motivations of our work is to help automate as much as possible the validation and debugging of programs with respect to properties that are outside of Maude's typing system. Some of the properties we consider can arguably be expressed by means of sorts and memberships in Maude. Nevertheless, in the following section we deal with properties that these facilities cannot handle.

3   The Assertion Language

Assertions are linguistic constructions that formally express properties of a software system. Throughout this section, we consider a software system that is specified by a rewrite theory $\mathcal{R} = (\Sigma, \Delta \cup B, R)$. Without loss of generality, we assume that Σ includes at least the sort State. Terms of sort State are called *system states* (or simply *states*). State transitions are obtained by nondeterministically applying the rewrite rules in R to canonical forms of system states. A state s is simplified into its canonical form $s \downarrow_{\Delta,B}$ by using equations and algebraic/membership axioms in $\Delta \cup B$.

In our specification language, assertions are formulas built on user-defined functions. The meaning of such functions is specified by a user-defined program. Our framework supports two kinds of assertions: *functional* assertions and *system* assertions. Functional assertions allow properties to be logically defined on the equational component of the rewrite theory $\mathcal{R}$ while system assertions specify formal constraints on the possibly nondeterministic rule component of $\mathcal{R}$. The benefit of using a logic framework is that the definition and checking of all asserted properties can be performed in a uniform and familiar setting.

3.1   The Assertion Logic

The core of our assertion language is based on (order-sorted) predicate logic, where first order formulas are built over the signature Σ of the rewrite theory $\mathcal{R}$ enriched with a set of user-defined boolean function symbols (predicates). The truth values are given by the formulas true and false. The usual conjunction (and), disjunction (or), exclusive or (xor), negation (not), and implication (implies) logic operators are used to express composite properties. Variables in the formulas are not quantified.

Logic formulas can be defined in Maude by means of the predefined functional module BOOL [14], which specifies the built-in sort Bool, the truth values, the

```
mod BANK-PRED is inc BANK .
    var ACC : Account .              op isPremium : Account -> Bool .
    var ID : Id .                    op getBalance : Account -> Int .
    var BAL : Int .                  op getId : Account -> Id .
    var STS : Status .               op getStatus : Account -> Status .

    ceq isPremium(ACC) = true if ACC : PremiumAccount .
    eq isPremium(ACC) = false [owise] .

    eq getBalance(< ID | BAL | STS >) = BAL .
    eq getId(< ID | BAL | STS >) = ID .
    eq getStatus(< ID | BAL | STS >) = STS .
endm
```

Fig. 1. System properties specified by the BANK-PRED module.

logic operators, and the built-in operators for membership predicates _:: S for each sort S, and term equality _==_ and inequality _=/=_.

The built-in Boolean functions _==_ and _=/=_ have a straightforward operational meaning: given an expression u == v, then both u and v are simplified by the equations in the module (which are assumed to be Church-Rosser and terminating) to their canonical forms (perhaps modulo some axioms such as associativity) and these canonical forms are compared for equality. If they are equal, the value of u == v is true; if they are different, it is false. The predicate u =/= v is just the negation of u == v. In the module BOOL, valid formulas are reduced to the constant true, invalid formulas are reduced to the constant false, and all the others are reduced to a canonical form (modulo associativity and commutativity) consisting of an *exclusive or* of conjunctions.

Predicates that are not specified in BOOL are module-dependent and can be equationally defined as total Boolean functions over the domain formalized by $\mathcal{R}$. Therefore, we can define basic properties on a given rewrite theory $\mathcal{R}$ by means of a system module PRED($\mathcal{R}$) that

- imports the (Maude encoding of) the rewrite theory $\mathcal{R}$; and
- specifies predicates via user-defined operators that are associated with terminating and Church-Rosser equational definitions of some total Boolean function.

In this scenario, a well-formed formula is any term of sort Bool built using the operators and variables declared in the system module PRED($\mathcal{R}$).

We say that a formula φ *holds* in $\mathcal{R}$, iff φ can be reduced to true in PRED($\mathcal{R}$) (in symbols, $\mathcal{R} \models \varphi$).

Example 3. Consider the BANK system module of Example 2 and the new predicates given in the BANK-PRED module of Fig. 1. Then, within BANK-PRED we can specify the formula

```
not(isPremium(ACC:Account)) implies getBalance(ACC:Account) > 0
```

which is true for every nonpremium bank account ACC with a positive balance.

3.2   System and Functional Assertions

System assertions formalize properties over (portions of) system states. Formally, a *system assertion* (also called *constrained term* in [26]) is an expression of the form $S\{\varphi\}$ where S is a (possibly non ground) term in $\tau(\Sigma, \mathcal{V})$ of sort State, and φ is a well-formed formula such that $Var(\varphi) \subseteq Var(S)$.

System assertions are checked against states of the system specified by $\mathcal{R}$. Roughly speaking, a system assertion $S\{\varphi\}$ allows us to validate all system states s that match (modulo the equational theory E) the state "template" S w.r.t. the formula φ. More formally, we define the satisfaction of a system assertion in a system state as follows.

Definition 1 (System Assertion Satisfaction). *Let* $\mathcal{R} = (\Sigma, E, R)$ *be a rewrite theory. Let* $S\{\varphi\}$ *be a system assertion for* $\mathcal{R}$ *and* s *be a state in* $\mathcal{R}$. *Then,* $S\{\varphi\}$ *is* satisfied *in* s *(in symbols,* $s \models S\{\varphi\}$*) iff for each* $w \in \mathcal{P}os(s)$, *for each substitution* σ *if* $s_{|w} =_E S\sigma$ *then* $\varphi\sigma$ *holds in* $\mathcal{R}$.

Note that, if there is no subterm $s_{|w}$ of s that matches S (modulo E), we trivially have $s \models S\{\varphi\}$. This implies that $S\{\varphi\}$ is *not* satisfied in s (in symbols, $s \not\models S\{\varphi\}$) only in the case when there exist w and σ such that $s_{|w} =_E S\sigma$, and the formula $\varphi\sigma$ does not hold in $\mathcal{R}$. We call w a *system error symptom*. Roughly speaking, a system error symptom is the position of a subterm of the state s that is responsible for the violation of the considered assertion in s.

Definition 2 (System Error Symptoms). *The set of all system error symptoms for a state* s *and a system assertion* $S\{\varphi\}$ *is defined as follows:*

$$SysErr(s, S\{\varphi\}) = \{w \mid \exists\sigma.\ s_{|w} =_E S\sigma, w \in \mathcal{P}os(s),\ and\ \varphi\sigma \not\models \mathcal{R}\}.$$

Observe that $SysErr(s, S\{\varphi\}) = \emptyset$, *whenever* $s \models S\{\varphi\}$.

Example 4. Consider the extended rewrite theory of Example 3 together with the system assertion

$$\Theta = <\ \texttt{C:Id}\ |\ \texttt{B:Int}\ |\ \texttt{S:Status}\ >\ \{\ \texttt{not(isPremium(<\ C:Id\ |\ B:Int\ |}$$
$$\texttt{S:Status\ >))\ implies\ B:Int\ >\ 0\ }\}$$

Then, Θ is satisfied in the state

$$<\ \texttt{Alice}\ |\ \texttt{50}\ |\ \texttt{active}\ >\ ;\ <\ \texttt{Bob}\ |\ \texttt{40}\ |\texttt{active}\ >\ ;\ \texttt{debit(Alice,60)},$$

but it is not satisfied in $s_{err} = <$ Alice | -10 |blocked > ; < Bob | 40 | active >, since Alice's non-premium account has a negative balance. The detected error symptom for the considered state and assertion is the position 1 that refers to the subterm < Alice | -10 | blocked > of s_{err}.

The second type of assertions that we consider are *functional assertions*. Functional assertions allow one to specify the general pattern O of the canonical form for any input term t that matches a given template I, while allowing pre- and post-conditions $\varphi_{in}, \varphi_{out}$ over the equational simplification to also be declared. Their general form is $I\{\varphi_{in}\} \rightarrow O\{\varphi_{out}\}$ where $I, O \in \tau(\Sigma, V)$, $\varphi_{in}, \varphi_{out}$ well-formed formulas, $Var(\varphi_{in}) \subseteq Var(I)$ and $Var(\varphi_{out}) \subseteq Var(I) \cup Var(O)$. Intuitively, functional assertions allow us to specify the I/O behaviour of the equational simplification of a term t by providing

Input: an input template I that t can match and a pre-condition φ_{in} that t can meet;

Output: an output template O that the canonical form of t has to match and a post-condition φ_{out} that the computed canonical form of t has to meet (whenever the input term t matching I meets φ_{in}).

Note that, while system assertions $S\{\varphi\}$ resemble Matching Logic (ML) formulas $\pi \wedge \phi$ (called ML *patterns*), where π is a configuration term and ϕ is a first order logic formula, functional assertions $I\{\varphi_{in}\} \rightarrow O\{\varphi_{out}\}$ remind Reachability Logic (RL) formulas $\varphi \Rightarrow \varphi'$, where φ, φ' are ML patterns (for a survey on ML/RL, see [27]). Different from our functional assertions, which predicate on equational simplification sequences, RL formulas are evaluated on system computations. Namely, the semantics of a RL formula $\varphi \Rightarrow \varphi'$ is that any state satisfying φ transits (in zero or more steps) into a state satisfying φ', while ML formulas are used to express (and reason about) static state properties, similarly to our system assertions.

The notion of satisfaction for a functional assertion is given w.r.t. the equational simplification $\mu = t \rightarrow^*_{\Delta,B} t \downarrow_{\Delta,B}$ of term t into its canonical form $t \downarrow_{\Delta,B}$.

Definition 3 (Functional Assertion Satisfaction). *Let $\mathcal{R} = (\Sigma, E, R)$ be a rewrite theory, with $E = \Delta \cup B$. Let $I\{\varphi_{in}\} \rightarrow O\{\varphi_{out}\}$ be a functional assertion for $\mathcal{R}$, and μ be the equational simplification of the term t in $\tau(\Sigma, V)$ into its canonical form $t \downarrow_{\Delta,B}$ w.r.t. Δ modulo B. Then, $I\{\varphi_{in}\} \rightarrow O\{\varphi_{out}\}$ is satisfied in μ (in symbols, $\mu \models I\{\varphi_{in}\} \rightarrow O\{\varphi_{out}\}$) iff for each substitution σ_{in} s.t. $t =_B I\sigma_{in}$, if $\varphi_{in}\sigma_{in}$ holds in $\mathcal{R}$, then there exists σ_{out} such that $t \downarrow_{\Delta,B} =_B O(\sigma_{in} \downarrow_{\Delta,B})\sigma_{out}$ and $\varphi_{out}(\sigma_{in} \downarrow_{\Delta,B})\sigma_{out}$ holds in $\mathcal{R}$.*

Note that $I\{\varphi_{in}\} \rightarrow O\{\varphi_{out}\}$ is (trivially) satisfied in μ when either t does not match $I\sigma_{in}$ (modulo B) or $\varphi_{in}\sigma_{in}$ does not hold in $\mathcal{R}$. Intuitively, a functional error occurs in an equational simplification μ where the computed canonical form fails to match the structure or meet the properties of the output template O. In other words, $\Phi = I\{\varphi_{in}\} \rightarrow O\{\varphi_{out}\}$ is *not* satisfied in μ only in the case when there exists an *input* substitution σ_{in} s.t.

- $t =_B I\sigma_{in}$ and $\varphi_{in}\sigma_{in}$ holds in $\mathcal{R}$;
- $t \downarrow_{\Delta,B} \neq_B O(\sigma_{in} \downarrow_{\Delta,B})\sigma_{out}$ or $\varphi_{out}(\sigma_{in} \downarrow_{\Delta,B})\sigma_{out}$ does not hold in $\mathcal{R}$, for any substitution σ_{out}.

Definition 4 (Functional Error Symptoms). *Let $\mathcal{R} = (\Sigma, E, R)$ be a rewrite theory, with $E = \Delta \cup B$. Let $\Phi = I\{\varphi_{in}\} \rightarrow O\{\varphi_{out}\}$ be a functional assertion for $\mathcal{R}$. Let $\mu = t \rightarrow^*_{\Delta,B} t \downarrow_{\Delta,B}$ be an equational simplification such* *that* $\mu \not\models \Phi$ *with input substitution* σ_{in}. *Then, a* functional error symptom *for* μ *w.r.t.* Φ *is any position in* $\mathcal{P}os(t \downarrow_{\Delta,B})$ *that belongs to the following set:*

$$FunErr(\mu, \Phi) = \{\pi(w) \in \mathcal{P}os(t \downarrow_{\Delta,B}) \mid ((g, \sigma_1, \sigma_2), \pi)$$
$$= \widehat{lgg}_{\Delta \cup B}(t \downarrow_{\Delta,B}, O(\sigma_{in} \downarrow_{\Delta,B})) \ and \ g_{|w} \in \mathcal{V}ar(g), w \in \mathcal{P}os(g)\}$$

Roughly speaking, $FunErr(\mu, \Phi)$ is computed by "comparing" the canonical form $t \downarrow_{\Delta,B}$ with the instance $O(\sigma_{in} \downarrow_{\Delta,B})$ of the output template O by the (normalized) substitution $\sigma_{in} \downarrow_{\Delta,B}$ using a least general generalization algorithm modulo equational theories. More specifically, an arbitrarily-selected least general generalization (g, σ_1, σ_2) (modulo $\Delta \cup B$) between $t \downarrow_{\Delta,B}$ and $O(\sigma_{in} \downarrow_{\Delta,B})$ is chosen via $\widehat{lgg}_{\Delta \cup B}$, and potentially erroneous subterms of $t \downarrow_{\Delta,B}$ are detected by selecting every position $\pi(w) \in \mathcal{P}os(t \downarrow_{\Delta,B})$ in correspondence with a position $w \in \mathcal{P}os(g)$. The intuition behind this method is that variables in g reflect possible discrepancies between the canonical form and the instantiated output template, and, thus, subterms $(t \downarrow_{\Delta,B})_{|\pi(w)}$ represent, to some extent, a possible anomalous subterm of $t \downarrow_{\Delta,B}$.

It is worth noting that the use of $\widehat{lgg}_{\Delta \cup B}$ is generally preferable to the adoption of a pure syntactic lgg algorithm since it minimizes the number of variables in g (and, hence, the points of discrepancy between $t \downarrow_{\Delta,B}$ and $O(\sigma_{in} \downarrow_{\Delta,B})$, which facilitates isolating erroneous information. Let us see an example.

Example 5. Let us consider the equational simplification $f(0,0) \rightarrow^+_{\Delta,B} c(1,3)$ w.r.t. an equational theory $(\Sigma, \Delta \cup B)$ in which the operator c is declared commutative. Let $\Phi = f(X, Y) \{true\} \rightarrow c(Z, 1) \{even(Z)\}$ be a functional assertion, where predicate $even(Z)$ checks whether Z is an even number.

Then, $(f(0,0), c(1,3)) \not\models \Phi$ (with input substitution $\sigma_{in} = \{X/0, Y/0\}$), since variable Z in the output template $c(Z, 1)$ is bound to 3 and $even(3)$ is false. Then, $\widehat{lgg}_{\Delta \cup B}(c(1,3), c(Z,1))$ returns a pair $((g, \sigma_1, \sigma_2), \pi)$ such that g contains the minimum number of variables. For instance, $\widehat{lgg}_{\Delta \cup B}(c(1,3), c(Z,1)) = ((c(Z,1), \{Z/3\}, \{\}), \{1 \mapsto 2\})$ and $FunErr(\mu, \Phi) = \{2\}$, which precisely detects that the term $c(1,3)_{|2} = 3$ is what causes the violation of Φ.

By contrast, the computation of a purely syntactic least general generalization would have delivered the more general result $(c(W, Z), \{W/1, Z/3\}, \{\})$ and the larger functional error symptom set $\{1, 2\}$ (which represents the positions of both arguments of the canonical form $c(1,3)$), thereby hindering the isolation of the erroneous subterm of $c(1,3)$.

Example 6. Consider again the extended rewrite theory of Example 3. Then, the functional assertion

$$\Phi = \texttt{updateStatus(ACC:Account)}\{\texttt{isPremium(ACC:Account)}\} \rightarrow \texttt{ACC:Account}\{\texttt{true}\}$$

states that premium account statuses (as well as other information in the account) remain unchanged after updateStatus is invoked. Thus, Φ is not satisfied in the following equational simplification

$$\texttt{updateStatus(< Bob | 95-100 | active >)} \rightarrow^+_{\Delta,B} \texttt{< Bob | -5 | blocked >}$$

with input substitution $\sigma_{in} = \{$ACC/< Bob | 95-100 | active >$\}$ and $\sigma_{in} \downarrow_{\Delta,B} = \{$ACC/< Bob | -5 | active >$\}$. Hence, there is a single (syntatic) least general generalizer

$$\widehat{lgg}_{\Delta \cup B}(< \text{ Bob | -5 | blocked >},\text{ACC}(\sigma_{in} \downarrow_{\Delta,B})) =$$
$$= ((< \text{ Bob | -5 | X:Status >},\{\text{X/blocked}\},\{\text{X/active}\}),\{3 \mapsto 3\})$$

where $FunErr(\mu, \Phi) = \{3\}$ is the functional error symptom set that pinpoints the anomalous status on Bob's premium account < Bob | -5 | blocked >.

Finally, an *assertional specification* $\mathcal{A}$ for a rewrite theory $\mathcal{R} = (\Sigma, E, R)$ is a set of functional and system assertions for $\mathcal{R}$. By $\mathcal{F}(\mathcal{A})$, we denote the set of functional assertions in $\mathcal{A}$, while $\mathcal{S}(\mathcal{A})$ denotes the set of system assertions in $\mathcal{A}$. By $s \models \mathcal{S}(\mathcal{A})$ (resp. $\mu \models \mathcal{F}(\mathcal{A})$), we denote that s satisfies all assertions in $\mathcal{S}(\mathcal{A})$ (resp. μ satisfies all assertions in $\mathcal{F}(\mathcal{A})$).

In the following section, we outline our previous work on trace slicing for RWL theories.

4 Enhancing Trace Slicing

Trace slicing [1–5] is a transformation technique for RWL theories that can drastically reduce the size and complexity of entangled, textually-large execution traces by focusing on selected computation aspects. This is done by uncovering data dependences among related parts of the trace w.r.t. a user-defined slicing criterion (i.e., a set of symbols that the user wants to observe). This technique aims to improve the analysis, comprehension, and debugging of sophisticated rewrite theories by helping the user inspect involved traces in an easier way. By step-wisely reducing the amount of information in the simplified trace, it is easier for the user to locate program faults because pointless information or unwanted rewrite steps have been automatically removed. Roughly speaking, in our slices, the irrelevant subterms of a term are omitted, leaving "holes" that are denoted by special variable symbols $\bullet$.

A term *slice* of the term s is a term $s^{\bullet}$ that hides part of the information in s; that is, the irrelevant data in s that we are not interested in are simply replaced by (fresh) $\bullet$-variables of appropriate sort, denoted by $\bullet_i$, with $i = 0, 1, 2, \ldots$.

The next auxiliary definition formalizes the function $Tslice(t, P)$, which allows a term slice of t to be constructed w.r.t. a set of positions P of t. The function $Tslice$ relies on the function $fresh^{\bullet}$ whose invocation returns a (fresh) variable $\bullet_i$ of appropriate sort that is distinct from any previously generated variable $\bullet_j$.

Definition 5 (Term Slice). *Let $t \in \tau(\Sigma, \mathcal{V})$ be a term and let P be a set of positions s.t. $P \subseteq \mathcal{P}os(t)$. Then, the term slice $Tslice(t, P)$ of t w.r.t. P is computed as follows.*

$$Tslice(t, P) = recslice(t, P, \Lambda), \text{ where}$$

$$recslice(t, P, p) = \begin{cases} f(recslice(t_1, P, p.1), \ldots, & recslice(t_n, P, p.n)) \\ & \text{if } t = f(t_1, \ldots, t_n), n \geq 0, \text{ and } p \in \bar{P} \\ t & \text{if } t \in \mathcal{V} \text{ and } p \in \bar{P} \\ fresh^{\bullet} & \text{otherwise} \end{cases}$$

and $\bar{P} = \{u \mid u \leq p \wedge p \in P\}$ is the prefix closure of P.

Roughly speaking, the function $Tslice(t, P)$ yields a term slice of t w.r.t. a set of positions P that includes all symbols of t that occur within the paths from the root of t to any position in P, while the remaining information of t is abstracted by means of •-variables.

Example 7. Consider the specification of Example 1 and the state $t =$ < Alice | 50 | active > ; < Bob | 100 | active > ; debit(Alice,20). Consider the set $P = \{1.2, 3.1, 3.2\}$ of positions in t. Then,

$$Tslice(t, P) = < \; \bullet_1 \; | \; 50 \; | \; \bullet_2 \; > \; ; \; \bullet_3 \; ; \; \text{debit(Alice,20)}.$$

Trace slicing can be carried out forward or backward. While the forward trace slicing results in a form of impact analysis that identifies the scope and potential consequences of changing the program input, backward trace slicing allows provenance analysis to be performed; i.e., it shows how (parts of) a program output depend(s) on (parts of) its input and helps estimate which input data need to be modified to accomplish a change in the outcome. While dependency provenance provides information about the origins of (or influences upon) a given result, the notion of descendants is the key for impact evaluation. In the sequel, we focus on backward trace slicing.

Let us illustrate by means of an example how it can help the user *think backwards* (i.e., to deduce the conditions under which a program produces some observed data).

Example 8. Consider the BANK system module of Example 2 and the computation trace $\mathcal{C}_{bank}$ in program BANK that starts in the initial state

< Alice | 50 | active > ; < Bob | 20 | active > ; < Charlie | 20 |
active > ; < Daisy | 20 | active > ; debit(Alice,80) ; credit(Alice,20)

and ends in the final state

< Alice | - 10 | blocked > ; < Bob | 20 | active > ; < Charlie | 20 |
active > ; < Daisy | 20 | active >

Let us assume we manually define as the slicing criterion the negative balance −10 for client Alice, which is a possible malfunction of the BANK specification, since regular account balances must be non-negative numbers according to the semantics intended by the programmer. Therefore, we execute trace slicing on the trace $\mathcal{C}_{bank}$ w.r.t. the slicing criterion < $\bullet_1$ | −10 | $\bullet_2$ > ; $\bullet_3$; $\bullet_4$; $\bullet_5$

that observes the negative balance of Alice's account in order to determine the cause of such a disfunction. The output trace slice delivered by the trace slicing technique is as follows

$$< \bullet_1 \mid 50 \mid \bullet_2 > ; \ \bullet_3 ; \ \bullet_4 ; \ \bullet_5 ; \ \texttt{debit}(\bullet_1, 80) ; \ \texttt{credit}(\bullet_1, 20) \ \bullet \overset{\texttt{credit}}{\rightarrow}$$

$$< \bullet_1 \mid 70 \mid \bullet_2 > ; \ \bullet_3 ; \ \bullet_4 ; \ \bullet_5 ; \ \texttt{debit}(\bullet_1, 80) \ \bullet \overset{\texttt{debit}}{\rightarrow}$$

$$< \bullet_1 \mid \texttt{-10} \mid \bullet_2 > ; \ \bullet_3 ; \ \bullet_4 ; \ \bullet_5$$

which greatly simplifies the trace $\mathcal{C}_{bank}$ by only showing the origins of the observed negative balance while excluding all the bank accounts that are not related to Alice.

Throughout this paper, we assume the existence of a *backwardSlicing*$(s_0 \rightarrow^*_{\Delta \cup B} s_n, s_n^\bullet)$ function as defined in [8] that yields the backward trace slice $s_0^\bullet \rightarrow^*$ $s_n^\bullet$ of the computation trace $s_0 \rightarrow^*_{\Delta \cup B} s_n$ w.r.t. a term slice $s_n^\bullet$ of s_n. This function relies on an instrumentation technique for Maude steps that allows the relevant information of the step, such as the selected redex and the contractum produced by the step, to be traced explicitly despite the fact that terms are rewritten modulo a set B of equational axioms that may cause their components to be implicitly reordered in the original trace. Also, the dynamic dependencies exposed by backward trace slicing are exploited in [8] to provide a (preliminary) program slicing capability that can identify those parts of a Maude theory that can (potentially) affect the values computed at some point of interest.

The main idea of this work is to enhance backward trace slicing by using runtime assertion checking to automatically identify the relevant symbols to be traced back from the erroneous states of the trace, that is, those states where an assertion is falsified. In conventional program development environments, when a given assertion check fails, the programmer must thoughtfully identify which program statements impacted on the value(s) causing the assertion failure. An additional advantage of blending trace slicing and runtime checking together is that the runtime checking not only helps automate the trace slicing, but trace slicing also helps answer the question that immediately arises when an assertion is violated. This question is "What caused it?". By using our enhanced, backward trace slicing methodology, error diagnosis is greatly simplified because accurate criteria for slicing are automatically inferred from the computed error symptoms that immediately bootstrap the slicing process so that much of the irrelevant data that does not influence the falsified assertions is automatically cut off.

5   Integrating Assertion-Checking and Trace Slicing

Dynamic assertion-checking and trace slicing can be smoothly combined together to facilitate the debugging of ill-defined rewrite theories. In this section, we formulate an assertion-checking methodology to verify whether a given computation trace $\mathcal{C}$ meets the requirements formalized by an assertional specification $\mathcal{A}$. In the case when a functional or system assertion $A \in \mathcal{A}$ fails to be satisfied over $\mathcal{C}$, a fragment $\mathcal{P}$ of $\mathcal{C}$ (that exhibits the anomalous behaviour w.r.t. A) is returned together with the corresponding set of system/functional error symptoms. Then, we show how backward trace slicing can take advantage of the computed error symptoms to produce small, easy-to-inspect computation slices of all those fragments that have been proven to be erroneous by the assertion-checking methodology.

5.1 Dynamic Assertion-Checking

We first extend the notion of satisfaction of the functional assertions to state equational simplifications (i.e., equational simplifications that reduce a state into its canonical form), where the state may contain an arbitrary number of function calls that might eventually be simplified. For this purpose, we introduce the following auxiliary definitions. Given $\mathcal{R} = (\Sigma, E, R)$, with $E = \Delta \cup B$, the term t is an equational redex in $\mathcal{R}$ if there is $(\lambda = \rho \text{ if } C) \in \Delta$ and substitution σ such that $t =_B \lambda\sigma$. Given $\mathcal{R}$ and a system state s in $\mathcal{R}$, $Top(s)$ is the set of minimal positions $w \in \mathcal{P}os(s)$ such that $s_{|w}$ is an equational redex in $\mathcal{R}$. Formally,

$$Top(s) = \{w \in \mathcal{P}os(s) \,|\, s_{|w} \text{ is an equational redex and}$$
$$\not\exists w' \leq w \text{ s.t. } s_{|w'} \text{ is an equational redex}\}.$$

Roughly speaking, $Top(s)$ selects all the positions in $\mathcal{P}os(s)$ that identify those outermost subterms of s to be equationally simplified into their canonical form in order to compute $s\downarrow_{\Delta,B}$. In other words, given the equational simplification of the state s, $\mathcal{S} : s \to^+_{\Delta,B} s\downarrow_{\Delta,B}$, each subterm $s_{|w}$, with $w \in Top(s)$, is reduced to $(s\downarrow_{\Delta,B})_{|w}$ in $\mathcal{S}$. This allows functional assertions to be effectively checked over each equational simplification $s_{|w} \to^+_{\Delta,B} (s\downarrow_{\Delta,B})_{|w}$ such that $w \in Top(s)$.

Definition 6 (Extended Functional Assertion Satisfaction). *Let $\mathcal{R} = (\Sigma, E, R)$ be a rewrite theory, with $E = \Delta \cup B$, and let s be a system state in $\mathcal{R}$ such that $Top(s) \neq \{\Lambda\}$. Let $s \to^+_{\Delta,B} s\downarrow_{\Delta,B}$ be an equational simplification for the state s in $\mathcal{R}$. Let $\mathcal{A}$ be an assertional specification for $\mathcal{R}$. We say that $\mathcal{F}(\mathcal{A})$ is* satisfied *in $s \to^+_{\Delta,B} s\downarrow_{\Delta,B}$ (in symbols, $s \to^+_{\Delta,B} s\downarrow_{(\Delta,B)} \models \mathcal{F}(\mathcal{A})$), iff for each $w \in Top(s)$, $s_{|w} \to^+_{\Delta,B} (s\downarrow_{\Delta,B})_{|w} \models \mathcal{F}(\mathcal{A})$.*

System and functional error symptoms (whose definitions have been given in Sect. 3 for a single system/functional assertion) can be naturally extended to assertional specifications in the following way.

Definition 7 (State Error Symptoms). *Let $\mathcal{R} = (\Sigma, E, R)$, with $E = \Delta \cup B$, be a rewrite theory. Let $\mathcal{A}$ be an assertional specification for $\mathcal{R}$. Let s be a state of $\mathcal{R}$. Then,*

$$SysErr(s, \mathcal{A}) = \bigcup_{\Theta \in \mathcal{S}(\mathcal{A})} SysErr(s, \Theta)$$
$$FunErr(s \to^+_{\Delta,B} s\downarrow_{\Delta,B}, \mathcal{A}) = \bigcup_{\Phi \in \mathcal{F}(\mathcal{A}), w \in Top(s)} \{(s_{|w} \to^+_{\Delta,B} (s\downarrow_{\Delta,B})_{|w},$$
$$FunErr(s_{|w} \to (s\downarrow_{\Delta,B})_{|w}, \Phi))\}$$

The notion of satisfaction for an assertional specification in a given computation is then formalized as follows.

Definition 8 (Satisfaction of an Assertional Specification). *Let* $\mathcal{R} = (\Sigma, E, R)$, *with* $E = \Delta \cup B$, *be a rewrite theory and* $\mathcal{C}$ *be a computation in* $\mathcal{R}$. *Let* $\mathcal{A}$ *be an assertional specification for* $\mathcal{R}$. *Then the specification* $\mathcal{A}$ *is satisfied in* $\mathcal{C}$ *iff*

- *for each state* s *in* $\mathcal{C}$ *that is a canonical form w.r.t.* Δ *modulo* B, $s \models \mathcal{S}(\mathcal{A})$;
- *for each state* s *in* $\mathcal{C}$ *that is not a canonical form w.r.t.* Δ *modulo* B, $s \rightarrow^+_{\Delta,B}$ $s\!\downarrow_{(\Delta,B)} \models \mathcal{F}(\mathcal{A})$.

To check an assertional specification $\mathcal{A}$ in a given computation $\mathcal{C}$, we can simply traverse $\mathcal{C}$ and progressively evaluate system assertions over states and functional assertions over state equational simplifications, respectively. Definition 9 formalizes this methodology into the function $check(\mathcal{C}, \mathcal{A})$ that takes as input a computation $\mathcal{C}$ and an assertional specification $\mathcal{A}$ and delivers a triple $(\mathcal{P}, Err, flag)$ where $\mathcal{P}$ is a prefix of $\mathcal{C}$, Err is a set of functional or system error symptoms w.r.t. $\mathcal{A}$, and $flag \in \{none, sys, fun\}$.

Roughly speaking, function $check(\mathcal{C}, \mathcal{A})$ returns $(\mathcal{P}, Err, flag)$ as soon as it encounters either a state or a state equational simplification in which $\mathcal{A}$ is not satisfied: $\mathcal{P}$ represents a prefix of $\mathcal{C}$ that reaches a state in which a system/functional assertion is violated, Err specifies the associated error symptom set, and $flag$ declares the nature of the computed symptoms (fun stands for functional error symptoms, sys for system error symptoms, and the keyword $none$ indicates that no symptom has been identified).

Definition 9 (Assertion Checking). *Let* $\mathcal{R} = (\Sigma, E, R)$, *with* $E = \Delta \cup B$, *be a rewrite theory and* $\mathcal{C}$ *be a computation in* $\mathcal{R}$. *Let* $\mathcal{A}$ *be an assertional specification for* $\mathcal{R}$.

$$
check(\mathcal{C}, \mathcal{A}) = \begin{cases}
(\varepsilon, \emptyset, none) & \text{if } \mathcal{C} = \varepsilon \\
(\mu \rightarrow^*_{R,B} \mathcal{C}'', Err, flag) & \text{if } \mathcal{C} = \mu \rightarrow^*_{R,B} \mathcal{C}' \text{ and } \mu \models \mathcal{F}(\mathcal{A}) \\
 & \text{and } (\mathcal{C}'', Err, flag) = check(\mathcal{C}', \mathcal{A}) \\
(\mu, FunErr(\mu, \mathcal{F}(\mathcal{A})), fun) & \text{if } \mathcal{C} = \mu \rightarrow^*_{R,B} \mathcal{C}' \text{ and } \mu \not\models \mathcal{F}(\mathcal{A}) \\
(s \rightarrow_{R,B} \mathcal{C}'', Err, flag) & \text{if } \mathcal{C} = s \rightarrow^*_{R,B} \mathcal{C}', \ s = s \downarrow_{(\Delta,B)} \\
 & \text{and } s \models \mathcal{S}(\mathcal{A}) \\
 & \text{and } (\mathcal{C}'', Err, flag) = check(\mathcal{C}', \mathcal{A}) \\
(s, SysErr(s, \mathcal{S}(\mathcal{A})), sys) & \text{if } \mathcal{C} = s \rightarrow^*_{R,B} \mathcal{C}', \ s = s \downarrow_{(\Delta,B)} \\
 & \text{and } s \not\models \mathcal{S}(\mathcal{A})
\end{cases}
$$

where $\mu = s \rightarrow^+_{\Delta,B} s \downarrow_{\Delta,B}$ *is a non-empty equational simplification for* s.

The runtime checking methodology formalized in Definition 9 can be interpreted either as an asynchronous (and trace-storing) technique or as a synchronous one (by considering that the input trace $\mathcal{C}$ is lazily generated as successive Maude steps are incrementally consumed by the calculus). In the following section, we formalize a truly synchronous methodology where traces, or rather whole search trees, can be stepwisely examined in a forward direction, reporting a violation at the exact step where it occurs.

5.2 Runtime Assertion-Based Backward Trace Slicing

Given a conditional rewrite theory $\mathcal{R} = (\Sigma, E, R)$, with $E = \Delta \cup B$, the transition space of all computations in $\mathcal{R}$ from the initial state s_0 can be represented as a *computation tree*,[3] $\mathcal{T}_\mathcal{R}(s_0)$. RWL computation trees are typically large and complex objects that represent the highly-concurrent, nondeterministic nature of rewrite theories.

```
function analyze(s₀, (Σ, Δ ∪ B, R), A, depth)
1.   𝒯 = s₀
2.   d = 0
3.   while (d ≤ depth) do
4.     F = frontier(𝒯)
5.     for each s ∈ F
6.       for each M ∈ mS(s)
7.         (P, Err, flag) = check(M, A)
8.         case flag of
9.           none :
10.             𝒯 = expand(𝒯, s, M)
11.           sys :
12            w = selectSysSymptom(Err)
13            l• = TSlice(last(P), Posₒ(last(P)))
14.             return backwardSlice(s₀ →*ᴿ∪Δ,ᴮ P, l•)
15.           fun:
16.             (t →⁺Δ,ᴮ t↓Δ,ᴮ, L) = {selectFunSymptom(Err)})
17.             (t↓Δ,ᴮ)• = TSlice(t↓Δ,ᴮ, ⋃_{w∈L} Posₒ(t↓Δ,ᴮ))
18.             return backwardSlice(t →⁺Δ,ᴮ t↓Δ,ᴮ, (t↓Δ,ᴮ)•)
19.           end case
20.       end for
21.     end for
22.     d = d + 1
23. end while
24. return 𝒯
end
```

Fig. 2. The *analyze* function.

Our methodology checks rewrite theories w.r.t. an assertional specification $\mathcal{A}$ at runtime by incrementally generating and checking the computation tree

---

[3] In order to facilitate trace inspection, computations are visualized as trees, although they are internally represented by means of more efficient graph-like data structures that allow common subexpressions to be shared.

$\mathcal{T}_{\mathcal{R}}(s_0)$ until a fixed depth. In fact, the complete generation of $\mathcal{T}_{\mathcal{R}}(s_0)$ is generally not feasible since some of its branches may be infinite as they encode nonterminating computations. The general analysis algorithm, which is specified by the routine $analyze(s_0, \mathcal{R}, \mathcal{A}, depth)$, is given in Fig. 2. We use the following auxiliary notation: given a position w of a term t, $Pos_w(t) = \{w.w' \mid w.w' \in Pos(t)\}$. The computation tree is constructed breadth-first, starting from a tree $\mathcal{T}$ that consists of a single root node s_0. At each expansion stage, the leaf nodes of the current $\mathcal{T}$ are computed by the function $frontier(\mathcal{T})$. Expansion of an arbitrary node s is done by deploying all the possible Maude computation steps stemming from s that are given by $m\mathcal{S}(s)$. Whenever a Maude step $\mathcal{M}$ is produced, it is also checked w.r.t. the specification $\mathcal{A}$ by calling $check(\mathcal{M}, \mathcal{A})$ that computes the triple $(\mathcal{P}, Err, flag)$. According to the computed $flag$ value, the algorithm distinguishes the following cases:

$flag = none$. No error symptoms have been computed; hence, $\mathcal{A}$ is satisfied in the Maude step $\mathcal{M}$, and $\mathcal{M}$ can safely expand the node s by replacing s with the path represented by $\mathcal{M}$ (via the invocation of $expand(\mathcal{T}, s, \mathcal{M})$), thereby augmenting $\mathcal{T}$.

$flag = sys$. In this case, $check$ returns a set of system error symptoms Err together with a computation $\mathcal{P}$ (which is a prefix of the Maude step $\mathcal{M}$) that violates a system assertion of $\mathcal{A}$. The computation $s_0 \rightarrow^*_{R\cup\Delta,B} \mathcal{P}$ is then generated and backward sliced w.r.t. a term slice $l^\bullet$ of the last state of $\mathcal{P}$. This term slice conveys all the relevant information that we automatically retrieve by using Definition 5 from the (system) error symptom w selected by the function $selectSysSymptom(Err)$, while all other symbols in l are considered meaningless and simply pruned away. This way, the algorithm delivers a trace slice $s_0^\bullet \bullet\!\!\rightarrow \mathcal{P}^\bullet$ that removes from the computation all of the information that does not affect the production of the chosen error symptom.

$flag = fun$. Some functional assertions have been violated by the considered Maude step $\mathcal{M}$. Hence, the algorithm selects a functional error symptom $(t \rightarrow^+_{\Delta,B} t\!\downarrow_{\Delta,B}, L)$ and returns the backward trace slicing of $t \rightarrow^+_{\Delta,B} t\!\downarrow_{\Delta,B}$ w.r.t. a term slice of $t\!\downarrow_{\Delta,B}$ that includes all the subterms of $t\!\downarrow_{\Delta,B}$ that are rooted at positions in L. As explained in Sect. 3.2, these subterms indicate possible causes of the assertion violation.

It is worth noting that, in our framework, we do not attach any specific semantics to $selectSysSymptom$ and $selectFunSymptom$ functions since many selection strategies can be specified with different degrees of automation and associated tradeoffs. For instance, we can simply obtain a fully automatic selection strategy by selecting the first symptom in Err. On the other hand, a purely interactive strategy can be implemented by asking the user to choose a symptom at runtime.

Finally, if the $analyze$ function terminates without detecting any assertion violation, then a (verified) tree $\mathcal{T}$ is delivered that encodes the first $depth$ levels of the computation tree $\mathcal{T}_{\mathcal{R}}(s_0)$; otherwise, the trace slice of the first computation that is found to violate an assertion is delivered. When multiple assertions are violated, $analyze$ can be invoked iteratively.

5.3   The ABETS system

The assertion-based slicing methodology of Sect. 5.2 has been fully implemented in a prototype tool we call ABETS (*Assertion-BasEd Trace Slicer*), which is publicly available at http://safe-tools.dsic.upv.es/abets together with some documentation and examples. The implementation comprises: (i) a front-end consisting of a RESTful Web service written in Java, with an intuitive user interface based on AJAX technology written in HTML5, Canvas, and Javascript; and (ii) a back-end that implements the proposed trace analysis methodology in Maude. The implementation of the backend consists of about 350 Maude function definitions (approximately 2700 lines of source code) that partially reuse the slicing and exploration machinery developed in previous work [4,5,8], extending it with the constraint-checking capabilities described in this paper.

To perform dynamic analysis with ABETS, the user must provide (i) the Maude program to analyze together with an initial state, and (ii) the list of assertions to be checked together with the module that defines the extra predicates that are used in the assertions. In order to non-deterministically search for assertion violations, the tree expansion is carried out up to a given depth bound that is measured in Maude steps. Whenever an assertion fails to be satisfied in the computation tree, the user is given an automatically generated counterexample trace slice that he/she can fully inspect, query, and slice further.

In ABETS, the trace slices can be easily navigated and all of the relevant information of the rewrite steps involved (e.g., equation/rule applications, matching substitutions, redex/contractum positions) is recorded and made available to the user. Furthermore, by disregarding rules and equations that are not used in the computed trace slice, ABETS can also generate a dynamic program slice where only potentially faulty fragments of the code are kept. Our preliminary experience has shown that the synergistic capabilities of ABETS can provide a very powerful Swiss knife in error diagnosis and debugging. The system allows assertion checking to be disabled when the functions/states they refer to are no longer under consideration so that no overhead is incurred after program analysis.

For demonstration purposes, let us analyze the BANK system of Example 2 together with an assertional specification $\mathcal{A}_{BANK}$ that includes the system assertion Θ of Example 4 and the functional assertion Φ of Example 6. Let us consider the expansion of all Maude steps that originate from the initial state

```
s₀ = < Alice | 20 | active > ; < Bob | 50 | active > ; < Charlie | 10 |
     active > ; debit(Alice,30) ; transfer(Bob,Charlie,60)
```

This is achieved in ABETS by calling $analyze(\mathcal{R}_{BANK}, \mathcal{A}_{BANK}, 1)$, where $\mathcal{R}_{BANK}$ is the rewrite theory specified by the BANK system module.

The assertion checking algorithm immediately discovers that Θ is not satisfied in the following Maude step $\mathcal{M}$

$s_0 \overset{\text{debit}}{\longrightarrow}$ updateStatus(< Alice | 20 - 30 | active >) ; < Bob | 50 | active
   > ; < Charlie | 10 | active > ; transfer(Bob,Charlie,60)
 $\longrightarrow^+$ < Alice | -10 | blocked > ; < Bob | 50 | active > ; < Charlie |
   10 | active > ; transfer(Bob,Charlie,60)

since Alice's nonpremium account has a negative balance. Here, the transition $\longrightarrow^+$ represents the equational state simplification that follows the rewrite step from s_0 by using the debit rule.

Then, a system error symptom is automatically computed by the tool, which unambiguously signals the anomalous subterm < Alice | -10 | blocked > of the last state of $\mathcal{M}$, and produces the associated term slice

$$l^\bullet = \text{< Alice | -10 | blocked > ;} \ \bullet_1 \ ; \ \bullet_2 \ ; \ \bullet_3$$

Finally, the algorithm automatically generates the backward trace slice of $\mathcal{M}$ w.r.t. $l^\bullet$, that is,

$$\text{< Alice | 20 | active > ;} \ \bullet_1 \ ; \ \bullet_2 \ ; \ \text{debit(Alice,30) ;} \ \bullet_3 \longrightarrow^+$$
$$\text{< Alice | -10 | blocked > ;} \ \bullet_1 \ ; \ \bullet_2 \ ; \ \bullet_3$$

which suggests an erroneous implementation of the debit rule. Indeed, debit always authorizes withdrawals from a nonpremium account even when the intended payout exceeds the account balance, which is in contrast with the statement asserted by Θ.

Similarly, by re-executing the analysis algorithm on a mutation of the original BANK module that fixes the buggy debit rule, we can also discover a violation of the functional assertion Φ that detects an anomalous behaviour of function updateStatus: in fact, updateStatus blocks *every* bank account with a negative balance, while premium accounts should always be kept active. The delivered trace slice is shown in Fig. 3 toghether with the achieved reduction (87 %).

State	Label	Trace	Trace Slice
Trace information			✕
1	'Start	< Alice | 20 | active > ; < Bob | 50 | active > ; < Charlie | 10 | active > ; debit(Alice,30) ; transfer(Bob,Charlie,60)	
2	toBnf	debit(Alice,30) ; < Alice | 20 | active > ; < Bob | 50 | active > ; < Charlie | 10 | active > ; transfer(Bob,Charlie,60)	
3	fromBnf	< Bob | 50 | active > ; < Charlie | 10 | active > ; transfer(Bob, Charlie,60) ; debit(Alice,30) ; < Alice | 20 | active >	
4	debit	< Bob | 50 | active > ; < Charlie | 10 | active > ; transfer(Bob, Charlie,60) ; updateAccountStatus(< Alice | 20 - 30 | active >)	• ; • ; • ; **updateAccountStatus(< • | 50 - 60 | active >)**
5	builtIn	< Bob | 50 | active > ; < Charlie | 10 | active > ; transfer(Bob, Charlie,60) ; updateAccountStatus(< Alice | - 10 | active >)	• ; • ; • ; **updateAccountStatus(< • | - 10 | active >)**
6	Label-EQ43	< Bob | 50 | active > ; < Charlie | 10 | active > ; transfer(Bob, Charlie,60) ; < Alice | - 10 | blocked >	• ; • ; • ; < • | • | **blocked >**
Total size:		828	105
Reduction Rate: 87%			

Fig. 3. Trace Slice after refuting the functional assertion Φ of Example 6.

Finally, by running the *program slice* option of ABETS, all program statements related to the updateStatus function are automatically identified and isolated in a program slice, since they can (potentially) cause the erroneous program behavior.

6   Conclusions and Further Work

We have formalized a framework that integrates dynamic slicing and runtime assertion checking to help diagnose programming errors in rewriting logic theories. A key feature of our approach is that the assertions (or more precisely, their runtime checks) are used to automatically synthesize advisable slicing criteria from inferred error symptoms. Our methodology smoothly blends in with the general framework for the analysis and exploration of rewriting logic computations that we developed in previous research [8].

The techniques we have developed are adequately fast and usable when applied to programs of several hundred lines, yet there are certainly several ways that our prototype implementation can be improved. For instance, one issue of interest would be to properly extend the current linear representation of Maude steps in ABETS, which intentionally obviates recording the traces for the recursive evaluation of conditions for the sake of efficiency. Other planned improvements are to add more flexibility to the selection and processing of violated assertions and to refine the slicing criterion C that we infer from the falsified functional assertion $I \{\varphi_{in}\} \rightarrow O \{\varphi_{out}\}$, by further generalizing C using φ_{out} to reduce the number of variables of interest. Finally, we are also working on extending the system to deal with (object-oriented) Full Maude specifications.

References

1. Alpuente, M., Ballis, D., Espert, J., Romero, D.: Backward trace slicing for rewriting logic theories. In: Bjørner, N., Sofronie-Stokkermans, V. (eds.) CADE 2011. LNCS, vol. 6803, pp. 34–48. Springer, Heidelberg (2011)
2. Alpuente, M., Ballis, D., Frechina, F., Romero, D.: Backward trace slicing for conditional rewrite theories. In: Bjørner, N., Voronkov, A. (eds.) LPAR-18 2012. LNCS, vol. 7180, pp. 62–76. Springer, Heidelberg (2012)
3. Alpuente, M., Ballis, D., Frechina, F., Romero, D.: JULIENNE: a trace slicer for conditional rewrite theories. In: Giannakopoulou, D., Méry, D. (eds.) FM 2012. LNCS, vol. 7436, pp. 28–32. Springer, Heidelberg (2012)
4. Alpuente, M., Ballis, D., Frechina, F., Romero, D.: Using conditional trace slicing for improving Maude programs. Sci. Comput. Program. **80**, Part B:385–415 (2014)
5. Alpuente, M., Ballis, D., Frechina, F., Sapiña, J.: Slicing-based trace analysis of rewriting logic specifications with I JULIENNE. In: Felleisen, M., Gardner, P. (eds.) ESOP 2013. LNCS, vol. 7792, pp. 121–124. Springer, Heidelberg (2013)
6. Alpuente, M., Ballis, D., Frechina, F., Sapiña, J.: Inspecting rewriting logic computations (in a Parametric and Stepwise Way). In: Iida, S., Meseguer, J., Ogata, K. (eds.) Specification, Algebra, and Software. LNCS, vol. 8373, pp. 229–255. Springer, Heidelberg (2014)

7. Alpuente, M., Ballis, D., Frechina, F., Sapiña, J.: Debugging Maude programs via runtime assertion checking and trace slicing. Technical report, Department of Computer Systems and Computation, Universitat Politècnica de València (2015). http://safe-tools.dsic.upv.es/abets/abets-tr.pdf
8. Alpuente, M., Ballis, D., Frechina, F., Sapiña, J.: Exploring conditional rewriting logic computations. J. Symbolic Comput. **69**, 3–39 (2015)
9. Alpuente, M., Escobar, S., Espert, J., Meseguer, J.: A modular order-sorted equational generalization algorithm. Inf. Comput. **235**, 98–136 (2014)
10. Baader, F., Snyder, W.: Unification Theory. In: Robinson, J.A., Voronkov, A. (eds.) Handbook of Automated Reasoning, vol. I, pp. 447–533. Elsevier Science (2001)
11. Bruni, R., Meseguer, J.: Semantic foundations for generalized rewrite theories. Theor. Comput. Sci. **360**(1–3), 386–414 (2006)
12. Clarke, L.A., Rosenblum, D.S.: A historical perspective on runtime assertion checking in software development. ACM SIGSOFT Softw. Eng. Notes **31**(3), 25–37 (2006)
13. Clavel, M., Durán, F., Eker, S., Lincoln, P., Martí-Oliet, N., Meseguer, J., Talcott, C.: All About Maude - A High-Performance Logical Framework. LNCS. Springer, Heidelberg (2007)
14. Clavel, M., Durán, F., Eker, S., Lincoln, P., Martí-Oliet, N., Meseguer, J., Talcott, C.: Maude Manual (Version 2.6). Technical report, SRI International Computer Science Laboratory (2011). http://maude.cs.uiuc.edu/maude2-manual/
15. Durán, F., Meseguer, J.: A Maude coherence checker tool for conditional order-sorted rewrite theories. In: Ölveczky, P.C. (ed.) WRLA 2010. LNCS, vol. 6381, pp. 86–103. Springer, Heidelberg (2010)
16. Durán, F., Roldán, M., Vallecillo, A.: Invariant-driven strategies for Maude. Electron. Notes Theor. Comput. Sci. **124**(2), 17–28 (2005)
17. Goguen, J.A., Meseguer, J.: Equality, types, modules, and (why not?) generics for logic programming. J. Logic Program. **1**(2), 179–210 (1984)
18. Goguen, J.A., Meseguer, J.: Unifying functional, object-oriented and relational programming with logical semantics. In: Agha, G., Wegner, P., Yonezawa, A. (eds.), Research Directions in Object-Oriented Programming, pp. 417–478. The MIT Press (1987)
19. Klop, J.W.: Term rewriting systems. In: Abramsky, S., Gabbay, D., Maibaum, T. (eds.), Handbook of Logic in Computer Science, vol. I, pp. 1–112. Oxford University Press (1992)
20. Korel, B., Laski, J.: Dynamic program slicing. Inf. Process. Lett. **29**(3), 155–163 (1988)
21. Lassez, J.L., Maher, M.J., Marriott, K.: Unification Revisited. In: Minker, J. (ed.) Foundations of Deductive Databases and Logic Programming, pp. 587–625. Morgan Kaufmann, Los Altos, California (1988)
22. Leavens, G.T., Cheon, Y.: Design by Contract with JML (2005). http://www.eecs.ucf.edu/leavens/JML/jmldbc.pdf
23. Martí-Oliet, N., Palomino, M., Verdejo, A.: Rewriting logic bibliography by topic: 1990–2011. J. Logic Algebraic Program. **81**(7–8), 782–815 (2012)
24. Meseguer, J.: Conditional rewriting logic as a unified model of concurrency. Theoret. Comput. Sci. **96**(1), 73–155 (1992)
25. Meseguer, J.: Multiparadigm logic programming. In: Kirchner, H., Levi, G. (eds.) ALP 1992. LNCS, vol. 632, pp. 158–200. Springer, Heidelberg (1992)
26. Rocha, C., Meseguer, J., Muñoz, C.: Rewriting modulo SMT and open system analysis. In: Escobar, S. (ed.) WRLA 2014. LNCS, vol. 8663, pp. 247–262. Springer, Heidelberg (2014)

27. Roşu, G.: From Rewriting Logic, to Programming Language Semantics, to Program Verification. In: Martí-Oliet, N., Ölveczky, P.C., Talcott, C., (eds.) Logic, Rewriting, and Concurrency. LNCS, vol. 9200, pp. 598–616. Springer, Heidelberg (2015)
28. Roldán, M., Durán, F., Vallecillo, A.: Invariant-driven specifications in Maude. Sci. Comput. Program. **74**(10), 812–835 (2009)
29. TeReSe. Term Rewriting Systems. Cambridge University Press (2003)

Computer Modeling in Neuroscience:
From Imperative to Declarative Programming

Maude Modeling in Neuroscience

Thomas J. Anastasio$^{(\boxtimes)}$

Departments of Molecular and Integrative Physiology, and Computer Science,
Beckman Institute for Advanced Science and Technology,
University of Illinois at Urbana-Champaign, Urbana, IL 61801, USA
tja@illinois.edu
http://tja.beckman.illinois.edu/

Abstract. Theory and computational modeling have played important roles in neuroscience. Models of neural systems range from coolly abstract to scrupulously biologically detailed, but the overwhelming majority have been implemented using imperative programming languages. Very recently, declarative programming approaches have entered the realm of computational neuroscience, including models implemented in Maude. The declarative approach promises deeper insights into neurobiology, especially into the pathological processes that underlie neurological disorders. This chapter will provide a very short overview of imperative and declarative modeling in neuroscience, and will then describe a specific example of a model of a key neural process implemented in Maude. The Maude model provides potential new insights that would be difficult to obtain using an imperative approach.

Keywords: Neuroscience · Neurobiology · Emotion · Learning · Fear conditioning · Extinction · Marijuana · Endocannabinoids · Post-traumatic stress disorder

1 Introduction

Readers of this volume will be aware of the essential difference between imperative and declarative programming as tools for modeling systems, and for modeling the processes implemented by systems. They will also know that imperative tools are usually preferred for computationally intensive simulations of specific behaviors, while declarative tools are better for exploring and analyzing a system over a much fuller range of its possible behaviors. Declarative tools are particularly useful for the analysis of complicated systems and processes. To date, declarative tools have been applied mainly in the analysis of complicated engineered systems, but biological systems can be extraordinarily complicated and the analytical capabilities of declarative modeling tools can be brought profitably to bear on them as well.

The brain is arguably the most complicated biological system, and one to which declarative modeling tools should be applied. The main purpose of this chapter is to summarize a declarative model of an important neural process, in order to illustrate the

© Springer International Publishing Switzerland 2015
N. Martí-Oliet et al. (Eds.): Meseguer Festschrift, LNCS 9200, pp. 97–113, 2015.
DOI: 10.1007/978-3-319-23165-5_4

power of the declarative approach and to motivate its continued use in neuroscience. But computer modeling has been an important component of neuroscientific research for some time and the declarative approach is very new. Before describing the declarative modeling example that is the focus of this chapter, it is important briefly to outline previous, imperative computational modeling approaches in neuroscience so that the new declarative approach can be properly situated.

Whether imperative or declarative, computational models in neuroscience can be either abstract or biologically detailed, and they can be either concept-driven or data-driven. The declarative model to be described can be characterized as abstract and data-driven, but declarative models could take various forms just as the imperative models before them have done and continue to do.

It is difficult to draw a line separating abstract from biologically detailed models in computational neuroscience. Many of the most detailed models of single neurons have elegant theoretical underpinnings. Still, a distinction between them can be made. Some of the most influential abstract models are concept driven, in the sense that the impetus for model creation is an idea about the nature of the neural system being modeled, whether that system is a single neuron with many ion channels or an interconnected network of many neural units. Concept-driven models are usually abstract, but what they lack in terms of biological detail they make up in theoretical elegance. Despite that, concept-driven models have not contributed greatly to the kind of understanding that leads to treatments for neurological disorders.

At the other end of the spectrum, the impetus for the creation of biologically detailed models is a desire to see what happens when many isolated facts concerning a neural system are brought together within a consistent computational framework. Biologically detailed models are generally data-driven. They tend to focus more on single neurons or small groups of neurons rather than on neural networks, although hybrid models also exist in which detailed models of neurons are embedded within larger networks of abstract units. Data-driven models can provide insight into specific aspects of the behavior of a neural system and this insight often has relevance to possible treatments for neurological disorders. As mentioned, the great majority of models in computational neuroscience so far, whether concept- or data-driven, have been implemented using imperative programming languages. The next section provides a very brief overview of the types of models that have enlightened neuroscience to date.

2   Brief Survey of Imperative Models in Neuroscience

Biologically detailed modeling in computational neuroscience rests on the twin pillars of the Hodgkin-Huxley conductance-based model of neuronal action potentials [1], and the Rall cable-theory model of neuronal membranes [2]. Essentially, the Hodgkin-Huxley theory explains how specific ion channels (membrane-bound proteins) actively change their conductances as functions of membrane voltage and time and thereby bring about the action potentials that propagate down the axons of neurons (axons begin at neuronal somata and end in synaptic terminals; they provide the outputs from neurons). The Rall cable theory, based on the theory of transmission of electrical signals down long cables, explains how synaptic potentials propagate and integrate in the dendrites and somata of neurons as functions of the resistance and capacitance of neuronal membranes (dendrites convey synaptic potentials to the somata and provide the inputs to neurons). These seminal models have been greatly extended by adding new ion channels as they are discovered, increasing the realism and complexity of neuronal membrane geometries, exploring interactions between multiple neurons, and in many other ways [3].

Abstract modeling in computational neuroscience began with the work of Ratliff and Hartline [4] on the visual system of the *Limulus*, which resulted in the theory of lateral inhibition. This theory explains how a network of visual neurons that have inhibitory connections could compute a function approximating a second derivative in space that essentially separates the edges and contours from the uniform regions in an image [5]. Subsequently, similar mechanisms have been described in several other sensory systems including the auditory and somatosensory systems [6].

The inhibition in lateral inhibitory networks can be feed-forward or recurrent. In the feed-forward case an input that excites a given model neuron (unit) will inhibit the units to the side (lateral) of that unit, while in the recurrent case the units inhibit one another. In the recurrent case, closed loops and positive (excitatory) feedback can occur (because inhibition of inhibition is excitation), so recurrent lateral inhibition is generally used in conjunction with bounded nonlinear units. The resulting recurrent network can have multiple stable states as the feedback interactions push some units to the lower bound, others to the upper bound of the nonlinear unit activation function. Such networks can form circumscribed regions of active units and might underlie such functions in the brain as detection and selection [7].

Recurrent, nonlinear networks with excitatory or mixed excitatory/inhibitory connections can also have multiple stable states. The best known examples were described by Hopfield [8, 9]. Hopfield construed the stable states of his network as memories, and he showed that the stable states could correspond to a set of desired memory patterns when the strengths (or weights) of the connections between the units in the network equaled the covariations between the elements of the patterns. The Hopfield model has provided potential new insights into the operation of the hippocampus, a brain region closely associated with memory [10].

While the connection weights in Hopfield networks are set directly, many other types of neural networks require adaptive learning algorithms to set their weights. In so called unsupervised learning, a network learns to categorize its inputs on the basis of exposure to the input alone [11]. The learning is unsupervised in the sense that there are no constraints placed on the responses of the units, and the network is essentially free to choose its own categories. One very nice property of these networks is that they form a spatial structure in which units that respond to inputs of the same or similar category are near each other in the network, essentially forming a map over the network. Such synthetic maps have structures similar to the spatial structures found in many map-like brain regions [12].

In supervised learning, in contrast, each input pattern is matched with a required output pattern, and the network is trained on the basis of the error between the required output and its actual output over many learning trials [13]. Networks trained using supervised learning can develop nonuniform distributed representations in which the required input/output transformation is achieved, but the various components of the transformation are divided up and distributed in seemingly random ways over the units in the network. Such representations are counterintuitive but they resemble many of the non-map representations that are observed in the brain [14, 15].

It has been proven that units in networks trained using supervised learning algorithms, which may indeed perform a specific input/output transformation, actually also compute an estimate of the probability that a given input belongs to an output construed as a category [16]. Supervised training can produce units that estimate probabilities and in so doing resemble the actual behavior of neurons in certain brain regions [17]. The probabilistic view of neural computation has been extended to networks that update beliefs in terms of prior knowledge and current input [18]. This paradigm has been used to model the behavior of neurons in the visual system that recognize images [19] or predict the next location of a moving target [20].

All of the computational modeling very briefly outlined in this section was done using imperative programming. Declarative programming is beginning to catch on in biology [21], and some interesting models of biological systems have been implemented using Maude [22, 23]. (Maude, for those who may not know her, is the declarative programming language due to José Meseguer, who is the individual being honored at the Festschrift symposium from which this chapter proceeds.) Despite its growing popularity in biology more broadly, declarative programming has been, with a few exceptions [e.g. 24], nonexistent in neuroscience.

I myself (the author of this chapter) had been deeply involved in computational neuroscience of the more abstract variety for many years, and wrote an introductory textbook on it [25], but had also been searching for a computational modality better suited to analysis of the complicated details associated with real neurobiological and neuropathological processes. I then had the good fortune to audit a course taught by José Meseguer on Maude, during which he demonstrated to the class how Maude could marshal the forces of algebra to help us "reason about" complicated systems. José expressed great enthusiasm for my ideas on using Maude to study neurobiological processes, and he and his students were generous in providing guidance and help to me as I got started. Since then I have been active in applying Maude to various problems in neurobiology. Most of my work so far has been modeling Alzheimer Disease, which is by far the most lethal neurological killer [26–29]. My lab is currently modeling other neurobiological processes both normal and abnormal.

To provide an example of how we are applying declarative programming in computational neuroscience, the rest of this chapter provides a summary of a model of emotional learning implemented in Maude [30]. This specific example will show how Maude, and declarative programming in general, naturally provides analysis capabilities of great utility in the effort to understand complicated neurobiological processes. This example certainly will not exhaust the possibilities for declarative programming in the analysis of neurobiological processes, but it will illustrate its application to a class of problems that arises frequently in neuroscience.

Studies of a neurobiological process often uncover seemingly contradictory and inconsistent findings, where it appears that an essential component of the process does the opposite of what would be expected given the observed outcome of the process, or that a component that contributes to process outcome with one effect undoes its own contribution with another effect. Of course, findings on a neurobiological system can seem inconsistent because they are actually wrong, but another possibility is more interesting. It is that the findings are right but seem inconsistent because our view of the system is limited, whereas if we could expand our view we might see that the findings actually fit within a harmonious whole. Declarative tools including Maude are valuable in neurobiology because they facilitate an expanded view of a system by enumerating all possible system configurations. Through the use of such tools it is sometimes possible to see that component effects that seem antithetic given a limited view of a process are instead seen as consonant in many process configurations, and further, that analysis of those configurations can expand our view of the outcome of the neuro-biological process. The emotional learning example described throughout the rest of this chapter will demonstrate how Maude can be used to generate such an expanded view of a neurobiological process. In order to appreciate this example it will be nec-essary to describe the emotional learning context in some detail.

3   Declarative Model of Extinction of Fear Conditioning

It could be argued that fear is the mother of all emotions. It certainly confers survival advantage, by keeping animals away from threatening but avoidable situations. Our knowledge of the neurobiological basis of emotion, and particularly of fear, has been greatly increased by the neuroscientist Joseph LeDoux [31]. To study fear scientifically in lab animals, LeDoux and others pioneered the use of the fear conditioning paradigm [32–34]. The goal of psychological conditioning in general is to cause an animal to associate a normally neutral stimulus with a normally evocative stimulus so that, after conditioning, presentation of the previously neutral stimulus will evoke the same response as the normally evocative stimulus. In the parlance of psychological condi-tioning research, the normally neutral and evocative stimuli are known respectively as the conditioned stimulus (CS) and the unconditioned stimulus (US), while the responses to the CS and the US are known respectively as the conditioned response (CR) and the unconditioned response (UR). The fear conditioning paradigm is con-sistent with this general framework.

In fear conditioning, a laboratory animal (usually a rat) is placed in a cage, the floor of which can be electrified to deliver a mild shock. The natural response of a rat to a shock is to become immobile, or to freeze. To accomplish fear conditioning, a naïve rat (one without any prior experimental conditioning) is placed in a cage and a shock (US) and an audible tone (CS) are delivered together. The rat naturally freezes (UR). Thereafter the rat will freeze (CR) even if the tone is delivered without shock. Such successful fear conditioning requires only one trial in rats, but the conditioning is not necessarily permanent. Repeated presentation of the tone without the shock will cause response extinction, such that the CS by itself no longer evokes the CR. But extinction is not the same as forgetting. For example, if an extinguished rat is placed in a different cage it will again produce the CR in response to the CS. In extinction the rat has essentially learned that it need not respond to the CS in a specific context, but it has not forgotten the association of CS and US. All of this is of importance to the neuroscience of emotional learning and it also has practical implications.

Post-traumatic stress disorder (PTSD) has received renewed attention in this era of wars and atrocities, but a recent historical survey argues that it is not a new phenomenon (http://www.vva.org/archive/TheVeteran/2005_03/feature_HistoryPTSD. htm). PTSD is characterized as a stress response that persists long after the initial trauma that caused it has terminated (http://www.nimh.nih.gov/health/topics/post-traumatic-stress-disorder-ptsd/index.shtml). In important ways PTSD resembles a conditioned fear response that is resistant to extinction, and studies of fear conditioning and extinction are leading to potential new insights into this prevalent disorder [35]. One intriguing finding is that extinction of fear conditioning in mice is impaired by a genetic manipulation that blocks the normal functioning of the endogenous cannabinoid system [36]. Endogenous cannabinoids are the brain's own marijuana, or at least the main active constituent of marijuana, which is Δ^9-tetrahydrocannabinol [37]. Further work in rodents confirms that suppression of endocannabinoid functioning impairs extinction [38] while enhancement of endocannabinoid functioning facilitates extinction [39]. Clinical trails in humans indicate that oral Δ^9-tetrahydrocannabinol is effective in the treatment of PTSD [40].

On the neurobiological level the link between endocannabinoids and extinction presents a paradox, the nature of which cannot be appreciated without delving into the details. The brain region most closely associated with fear conditioning and extinction is the amygdala, an almond-shaped structure located in the temporal lobe of the cerebrum [41]. Seen from the outside, the amygdala appears to take input related to sensory signals including the US and CS of fear conditioning, and to produce output leading to fear responses such as the UR and CR of fear conditioning. Seen from the inside, the amygdala is a complicated structure composed of many sub-regions each containing many neural types. The lateral amygdala is one such sub-region, and among its neural types are excitatory projection neurons and inhibitory interneurons. The projection neurons, on the one hand, are responsible for transferring sensory signals from the input to the output stage of the amygdala. A key event in the acquisition of fear conditioning occurs when these projection neurons learn to respond to the CS as well as to the US, after which they can cause a CR to the CS alone. The interneurons, on the other hand, are responsible for inhibiting the projection neurons under certain circumstances, including those related to extinction of fear conditioning.

Here is the paradox: the function of endogenous cannabinoids is to *weaken* the inhibitory connection from the inhibitory to the projection neurons in the lateral amygdala. This would seem to oppose, rather than promote, extinction. The paradox is that endocannabinoids promote extinction on the behavioral level, but their neurobiological action is to suppress inhibition of (i.e. to excite) the very neurons that mediate fear conditioning. One way to solve this paradox is to take account of the many other connections in the amygdala, and to explore whether endocannabinoid suppression of inhibition of the projection neurons could still be compatible with extinction of fear conditioning. The purpose of the computational modeling study described in the rest of this section was to do just that. The approach was to write a specification characterizing the interconnections between the main neural types in the amygdala, and to search the space of interconnections for configurations that were compatible both with weakening of the inhibitory connections onto projection neurons and with extinction. As it turned out, there were many such configurations, and exploring them in the context of other facts concerning extinction neurobiology led to potential new insights.

3.1 Neurobiology of the Amygdala

There are several competing views of the neurobiology of fear conditioning, but they share a basic common structure. Probably with most well-established framework is the one by LeDoux and colleagues [42]. A diagram depicting this framework is shown in Fig. 1.

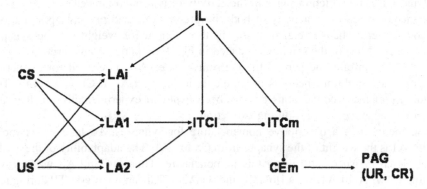

Fig. 1. Schematic of the neural pathway through the amygdala that mediates fear conditioning. CS, US, CR, and UR are conditioned or unconditioned stimulus or response. LA1 and LA2 are units representing each of two projection neurons, while LAi is a unit representing an inhibitory interneuron, in the lateral amygdala. Other abbreviations also are units that represent neurons in the corresponding brain region. IL is the infralimbic part of prefrontal cortex, ITCl and ITCm are the lateral and medial intercalated cell masses of the amygdala, CEm is the central nucleus of the amygdala, and PAG is the periaqueductal gray matter. Arrows and balls represent excitatory and inhibitory connections, respectively (this figure is reproduced from [30]).

The diagram shows the pathway from input to output in the amygdala that is represented computationally in the model. A great deal of data supports the structure, function, and adaptability, of the amygdala model. This data is summarized in review articles [32, 33, 42] and in the original modeling article [30]. Briefly, the input and output stages of the amygdala are, respectively, the lateral nucleus (LA) and the medial part of the central nucleus (CEm). The intercalated cell mass (ITC), which has lateral and medial parts (ITCl and ITCm), constitutes the bridge between LA and CEm. Neurons in ITC are inhibitory. Functionally, LA excites ITCl, ITCl inhibits ITCm, and ITCm inhibits CEm, so LA excites CEm over this pathway (since inhibition of inhibition is excitation).

The input can be either or both of CS and US, which in the case of fear conditioning are sensory signals related to sound and pain, respectively. Neural projections carrying these signals excite neurons in LA. The diagram depicts two LA projection neurons (LA1 and LA2) and one inhibitory interneuron (LAi); the reason for having two LA projection neurons is that some do, but others do not, receive input from LA inhibitory interneurons. Neurons in LA excite neurons in CEm over the ITC bridge described in the previous paragraph, and neurons in CEm command the fear response through activation of neurons in the periaqueductal gray matter (PAG). The fear response includes freezing but also includes increases in blood pressure, excretion of stress hormones, etc. This "fear pathway" is normally activated by pain but can be conditioned to stimuli, such as sound, by which it is not normally activated.

Fear conditioning, like other forms of conditioning, requires learning, and learning in the brain is mediated by changes in the strengths of the synapse between neurons. Such changes often take the form of long-term potentiation and long-term depression (LTP and LTD) by which changes in the activity of two interconnected neurons (e.g. simultaneous increases in activity of both the pre-synaptic and post-synaptic neuron) lead to long-term change (e.g. increase) in the strength (or weight) of the synapse between them. Most of the synapses depicted in Fig. 1 are known to undergo both LTP and LTD. The infralimbic part of the prefrontal cortex (IL) is critical for extinction [43]. It excites inhibitory neurons in amygdala including LAi, ITCl, and ITCm. The adaptability of those connections has not been explored experimentally but they are likely to be adaptive, at least during extinction.

The model uses a descriptive nomenclature for synapses. To take an example, wCStoLA1 is the weight of the synapse from CS to LA1. The adaptability of the model can be described succinctly using this nomenclature. Thus, wCStoLAi, wCStoLA1, wCStoLA2, wLAitoLA1, wLA1toITCl, and wLA2toITCl can undergo LTP during fear conditioning and LTD during extinction, while wILtoLAi, wILtoITCl, and wILtoITCm can undergo LTP during extinction with no weight modification allowed during fear conditioning. The weights wUStoLAi, wUStoLA1, wUStoLA2, wITCmtoCEm, wITCltoITCm, and wCEmtoPAG are not modifiable. That makes nine adaptable (i.e. modifiable) synapses in the model.

In the brain, fear conditioning is associated mainly with LTP of the synapses from CS to LA projection neurons. Since the pathway from LA to CEm (i.e. from input to output in the amygdala) is net excitatory, an increase in the connection strength from CS to LA will enable the CS to activate CEm leading to a fear response via PAG. Fear conditioning also could occur through increase in the connection strength from LA to ITCl. What about extinction? Extinction *could* occur through a *decrease* in the connection strength from CS to LA, essentially undoing any increase that brought fear conditioning about in the first place. However, crucial experiments show that CS continues to excite some neurons in LA (and inhibit none) even after extinction [34], [44, 45]. Then extinction *could* occur through *decrease* in the connection strength from LA to ITCl. The limitation here is that IL is known to be active during and after extinction and IL excites ITCl, which could override any decrease in LA to ITCl excitation. In fact, the contribution of IL poses a conundrum on top of the paradox posed by endocannabinoids.

IL is necessary for extinction, and during the process of extinction its activity increases. IL also contacts inhibitory neurons in amygdala and could produce extinction in at least two ways. By exciting ITCm, IL could produce net inhibition of CEm and so block the fear command. However, IL also excites ITCl, which inhibits ITCm, so by exciting ITCl it seems that IL would undo its own excitation of ITCm and fail to produce extinction. IL seems to pursue opposing goals. How does IL promote extinction if it can undo its own effects? This is the IL conundrum.

Also, by exciting LAi, IL could inhibit at least LA1 and reduce transmission of the fear signal, but this leads us back to the endocannabinoid paradox. In the context of the model, the effect of the endocannabinoid system, which is crucial to extinction, is to *decrease* the strength of the inhibitory connection from LAi to LA1. This would undermine the ability of IL, or of LAi itself, to suppress the response of LA1 and so promote extinction. This is again the endocannabinoid paradox.

The purpose of the modeling summarized in this section is to try to resolve both the IL conundrum and the endocannabinoid paradox. These are the findings that seem incongruous in the context of the extinction process. The computational approach to this problem begins with the realization that the connectivity of the amygdala is complex and there are many configurations of the weights (i.e. strengths) of its nine modifiable (i.e. adaptable) synapses. Thus, the expanded view of the process we wish to generate will consist of a complete enumeration of the synaptic-strength configurations of the amygdala fear-pathway system. The hope is that at least some of those configurations produce extinction, which is the process outcome, but also instantiate the seemingly incongruent combinations of connections that characterize the IL conundrum and/or the endocannabinoid paradox. This hope will be fulfilled, and it will lead to a new and fuller appreciation of what the actual outcome of extinction process could be. In order to appreciate this potential new view of extinction it will be necessary to describe the analysis in terms of specific model synapses.

For modeling purposes, synaptic weights are limited to only three discrete values (0, +1, +2 for excitatory synapses or 0, −1, −2 for inhibitory synapses). Even so there are 3^9 or 19,683 possible weight combinations. This number of configurations provides many possibilities for adaptation. The goal of the analysis is to search for weight combinations (i.e. configurations) that achieve extinction in conjunction with wLAitoLA1 LTD (i.e. decrease in the weight from LAi to LA1). By imposing the further constraint that CS must excite at least one of either LA1 or LA2 (and inhibit neither) after extinction (see above), potential insight into the role of the endocannabinoids can be obtained. Details of the computational model follow.

3.2 Computational Approach Specifics

Fear conditioning and extinction are learning processes that have LTP and LTD of specific synapses as neurobiological sub-processes. Together these processes can produce many different synaptic strength configurations, but only some of those will cause overall system behavior that is consistent with experimental observation. The approach taken here was first to specify the system of interactions that characterize the amygdala fear-conditioning pathway, and also to specify the allowed changes in synaptic connection weights, in Maude. Then the specification could be searched for specific synaptic weight configurations using the Maude search tool. The raw results of the analysis are numbers of synaptic weight configurations produced by the model that satisfy a series of increasingly restrictive, experimentally determined constraints. The results are summarized in Table 1. Interpretation of the raw results provides potential insights and delineates experimentally testable predictions from the model.

The fear pathway through the amygdala is essentially a neural network. Each model element (unit) represents a neuron from a specific brain region (e.g. unit ITCm represents a neuron in ITCm). The activity of each unit is meant to represent the activity of a neuron. Specifically, the activity level of a unit is proportional to the action potential firing rate of a neuron, but the unit activation functions are very simple, being simply the sum of the weighted inputs to a unit, bounded from below at 0 (because neurons cannot have negative firing rates). For example, the activity of element ITCm is computed as ITCm = bITCm + wITCltoITCm * ITCl + wILtoITCm * IL, where bITCm is a bias and wITCltoITCm is negative because the ITCl to ITCm connection is inhibitory. All units have initial activities of 1, except for US, CS, CEm, PAG, and IL, which are initially 0. Each unit has a bias, which ensures its proper initial activity given the initial connection weights, which are all of absolute value 1 except those from CS (wCStoLA1, wCStoLA2, and wCStoLAi) and IL (wILtoLAi, wILtoITCl, and wILtoITCm), which are initially 0. Switching US from 0 to 1 will activate the fear pathway, resulting in activation of PAG and production of UR. Initially, switching CS from 0 to 1 will have no effect, because the initial weights from CS to LA are all zero. Fear conditioning requires LTP of those connection weights.

In the model, fear conditioning is accomplished by setting CS to 1 and allowing weight modifications to occur until CS is able to activate PAG. When fear conditioning has ended, extinction is accomplished by keeping CS at 1 and again allowing weight modifications to occur until CS is no longer able to activate PAG. IL takes value 1 during extinction training but is otherwise 0. For both phases of learning, connection weights are modified one at a time, and every individual weight change is followed by evaluation of the effects of that weight change on the responses to CS (i.e. changes in activation) of all units in the model. Weight modification terminates if PAG goes to 1 during fear conditioning or to 0 during extinction.

In the Maude specification, rules produce allowed weight modifications while equations determine the effects of each weight change on unit activities (especially in responses to the CS). Thus rule executions cause the model system to transition from one connection weight configuration (i.e. model state) to another, while the equations elaborate the state by finding all the new unit activities. In Maude search, the number of states (i.e. connection weight configurations) that produce extinction following fear conditioning (see also next sub-section) is the same as the number of terminal states. Thus, the specification terminates for all possible paths and can achieve extinction following fear conditioning in all of them.

The model was also instantiated in MATLAB™, which is an imperative programming language widely used in neurobiology. The main reason for the MATLAB program was to serve as a crosscheck for the Maude specification, and herein lies an interesting difference between declarative modeling approaches in biology versus computer engineering. In computer engineering, the analyst typically specifies an engineered system (software, firmware, etc.), the details of which are all known, and the behavior of the specification can be thoroughly compared with that of the system to be analyzed to ensure the adequacy of the specification. In contrast in biology, the system to be analyzed can only be specified incompletely, and thorough comparison between the behavior of the specification and of the biological system is not practical.

It is therefore useful to have two versions of the same computational model, each written in a different programming language, to serve as crosschecks for each other.

For the model, the initial weight configuration and a battery of fear conditioned and extinguished weight configurations were checked for consistency between the programs written in the two different languages, Maude and MATLAB. The check was valid for the equational part of the Maude specification. In all cases the results agreed, providing assurance that the findings are not corrupted by programming error.

Another use for the MATLAB program was to use it for directed searches to find sets of weight changes that would produce extinction following fear conditioning. The directed search involved systematic exploration of the neighborhoods of randomly selecting start states for states of interest. Connection weight absolute values were restricted to the range between 0 and 2, inclusive, for both the directed searches in MATLAB and the state-space searches in Maude, but the latter allowed only integer-valued weight changes while the former allowed real-valued weight changes. In both cases, the number of connection weight configurations that produced extinction was compared with the number that did so in conjunction with LTD of wLAitoLA1, which represents endocannabinoid action in the model and is critical to this analysis. It was found that 56 % of all configurations found by the exhaustive Maude search met this criterion while only 18 % of all configurations found by the MATLAB directed search did so. Comparison of the results of exhaustive, integer-valued search versus random, real-valued search provided some assurance that the Maude search is not missing configurations that rely on non-integer weights.

3.3   Analysis of the Model

There are 3^9 or 19,683 possible weight combinations, as explained in Subsect. 3.1, given that each modifiable weight can take 1 of 3 integer levels and there are 9 modifiable weights. Of critical importance is to search for weight combinations (configurations) that achieve extinction in conjunction with LTD of the weight of the connection from LAi to LA1. To record whether wLAitoLA1 has undergone LTD during extinction, an additional weight-record parameter was added to record its value immediately following fear conditioning. Because wLAitoLA1 can either get more negative during fear conditioning (i.e. go to -2) or stay at its initial value (-1), the weight-record parameter has 2 possible values. This parameter is not a weight per se but it does increase the number of states of the system, so it increases the number of possible weight (and weight-record) configurations by a factor of 2 to 39,366.

Of the total of 39,366 weight (including weight-record) configurations, 8394 produce extinction following fear conditioning (Table 1 Row 1). Of those, 4659 do so in conjunction with LTD of wLAitoLA1 (Table 1 Row 2). Thus, roughly half of all extinguished weight configurations are consistent with the experimentally-documented decrease (i.e. LTD), due to endocannabinoids, of the inhibitory synapse onto LA1 (i.e. wLAitoLA1). This by itself is a useful finding. It indicates that, given what is known about the fear pathway through the amygdala, an endocannabinoid-induced decrease of inhibition onto some LA projection neurons, which should oppose extinction, at least does not preclude extinction in about half of the space of synaptic weight

Table 1. The results of searches of the model state space are tabulated according to conditions and numbers of compatible weight configurations.

Search conditions and numbers of compatible configurations

Conditions	Configurations
(1) No conditions other than that the 9 modifiable weights are each restricted to three absolute levels (0, +1, +2 or 0, −1, −2) and the 3 weights from IL are limited to LTP during extinction but the other 6 can undergo LTP during fear conditioning and LTD during extinction	8394
(2) 1 and LTD of wLAitoLA1	4659
(3) 2 but with LA1 or LA2 excited by CS	2335
(4) 3 but neither LA1 nor LA2 inhibited by CS	957
(5) 4 with weights constrained to be equal from CS (wCStoLAi = wCStoLA1 = wCStoLA2), LA (wLA1toITCl = wLA2toITCl), and IL (wILtoLAi = wILtoITCl = wILtoITCm)	0
(6) 4 with weights from CS (wCStoLAi, wCStoLA1, and wCStoLA2) free to be unequal but with these equality constraints: (wLA1toITCl = wLA2toITCl) and (wILtoLAi = wILtoITCl = wILtoITCm)	0
(7) 4 with weights from LA (wLA1toITCl and wLA2toITCl) free to be unequal but with these equality constraints: (wCStoLAi = wCStoLA1 = wCStoLA2) and (wILtoLAi = wILtoITCl = wILtoITCm)	21
(8) 4 with weights from IL (wILtoLAi, wILtoITCl, and wILtoITCm) free to be unequal but with these equality constraints: (wCStoLAi = wCStoLA1 wCStoLA2) and (wLA1toITCl = wLA2toITCl)	42
(9) 4 (no equality constraints) but no LTD of wLAitoLA1	303

configurations that produce extinction. The finding that there is a relatively large number of configurations that are consistent both with extinction and with wLAitoLA1 LTD does not exactly resolve the endocannabinoid paradox, but it does leave substantial room for further refinement.

Some but not all real LA neurons continue to respond to CS after actual extinction (see Subsect. 3.1), so at least one of LA1 and LA2 should have a CS response after simulated extinction. Of the 4659 configurations that achieve extinction in conjunction with LTD of wLAitoLA1, 2335 have LA1 or LA2 responding to CS (Table 1 Row 3). Although some real LA neurons do not respond to CS, none are actively inhibited by CS following extinction (see also Subsect. 3.1). The implication of this for the model is that neither LA1 nor LA2 should be inhibited by CS following extinction (note that CS can cause net inhibition of LA1 via LAi). The number of configurations that have this property and also achieve extinction in conjunction with wLAitoLA1 LTD is 957 (Table 1 Row 4). This is a rather special number. In these 957 synaptic weight configurations, extinction has occurred in conjunction with LTD of wLAitoLA1, which corresponds to the known contribution of endocannabinoids to extinction, and CS inhibits neither of LA1 nor LA2 but excites at least one of them, which corresponds to retention of the CS-US association despite extinction (remember that extinction is not forgetting). The full importance of this number of neurobiologically consistent configurations will be made apparent later in this sub-section.

With the endocannabinoid paradox on hold the IL conundrum can be addressed. The approach begins by asking whether the weights of all connections from the same brain area can all change by the same amount to accomplish extinction, or do they need the flexibility to be independently modifiable. The specific weights in question are those from CS (wCStoLAi, wCStoLA1, and wCStoLA2), from LA (wLA1toITCl and wLA2toITCl), and from IL (wILtoLAi, wILtoITCl, and wILtoITCm). Search results show that if the weights of the connections from CS are constrained to be equal (wCStoLAi = wCStoLA1 = wCStoLA2), and if the weights from LA projection neurons are constrained to be equal (wLA1toITCl = wLA2toITCl), and if the weights from IL are constrained to be equal (wILtoLAi = wILtoITCl = wILtoITCm), then the number of weight configurations that achieve extinction in conjunction with LTD of wLAitoLA1, and with LA1 or LA2 excited but neither inhibited by CS, is 0 (Table 1 Row 5). If the connection weights from CS can be differentially modified but not those from LA and IL, then the number of weight configurations that achieve extinction in conjunction with LTD of wLAitoLA1, and with LA1 or LA2 excited but neither inhibited by CS, is again 0 (Table 1 Row 6). These results show that differential modifiability of the weights of the connections from IL and/or LA will be necessary to accomplish extinction that is consistent with neurobiological constraints.

Further search results show that if the connection weights from LA alone, or if those from IL alone, can be differentially modified, with the others (from CS and IL, or from CS and LA, respectively) constrained to be equal, then there are 21 and 42 configurations, respectively, that achieve extinction in conjunction with LTD of wLAitoLA1 and LA1 or LA2 excited but neither inhibited by CS (Table 1 Rows 7 and 8). Thus, the model predicts that extinction in conjunction with LTD of the synapses of inhibitory interneurons onto LA projection neurons (endocannabinoid contribution), along with some retention of CS responding in LA (extinction is not forgetting), can occur but only if the synapses of LA projection neurons onto ITCl, or those of IL neurons onto LA inhibitory interneurons, ITCl, and ITCm, are differentially modifiable. This prediction should be testable experimentally.

This result also takes us some way toward solving the IL conundrum. Recall that in the model IL promotes extinction by activating ITCm but opposes extinction by activating ITCl. Of course, if IL can activate ITCm but not activate ITCl then it would not thwart its own contribution to extinction, but this requires differential modification of the weights of the connections from IL. What if IL cannot differentially modify its connection weights but they are all constrained to change together by the same amount? Then the fear pathway through the amygdala can *still* achieve extinction, *but* the connections from LA onto neurons in ITCl must be differentially modifiable! Thus, the analysis shows how the IL conundrum can be solved in principle. Whether it is solved that way in the actual amygdala awaits experimental verification of the predictions outlined in the previous paragraph.

This brings us back to the endocannabinoid paradox, or how can a *decrease* in inhibition onto an LA projection neuron, which should *oppose* extinction, actually contribute to the achievement of extinction along with its neurobiologically known concomitants. One more search was run in an effort to resolve this paradox. To make it robust, all equality constraints are dropped (so weights of connections from CS, LA, and IL can all be differentially modified again). Then LTD of wLAitoLA1 during simulated extinction is *disallowed* in the model. This would correspond to elimination of endocannabinoid-induced LTD in the amygdala.

With connection weights from CS, LA, and IL all differentially modifiable, and LA1 or LA2 excited but neither inhibited by CS after extinction, the number of configurations that achieve extinction *without* LTD of wLAitoLA1 is 303 (Table 1 Row 9). For direct comparison, a previous search revealed that, with connection weights from CS, LA, and IL all differentially modifiable, and LA1 or LA2 excited but neither inhibited by CS after extinction, the number of configurations that achieve extinction *with* LTD of wLAitoLA1 is 957 (Table 1 Row 4). These results show that LTD of inhibitory interneuron synapses is not only compatible with extinction, but it actually increases (in fact, triples) the number of weight configurations that achieve extinction while preserving some LA responses to CS in the model. The implications of this will be discussed more fully in the next sub-section.

3.4   Implications of Main Model Findings

The main finding of the analyses is that endocannabinoid-induced LTD of the synapses of LA inhibitory interneurons onto LA projection neurons is not only compatible with extinction, but it increases the number of connection weight configurations that achieve extinction while also preserving some fear memory in amygdala. By showing that the number of such configurations is increased threefold in the model, the analysis identifies preservation of fear memory during extinction as a highly likely outcome of endocannabinoid-induced LTD in the lateral amygdala.

The analysis was focused on numbers of model weight configurations, and these are relevant to potential explanations of extinction for two reasons. The first reason is that the actual mechanisms of synaptic weight (i.e. strength) modifications in the amygdala are still very incompletely understood. It is known that the amygdala brings about fear conditioning and extinction by making synaptic strength changes but, because it is not known exactly how those changes are made, it could justifiably be assumed that, given a well-established model based on known amygdala connectivity, the outcome associated with the most numerous synaptic strength configurations is the one most likely to occur in the real amygdala.

The second reason is that the actual mechanisms that produce synaptic strength modifications during fear conditioning and extinction may have substantial random components. A recent (imperative) model of fear conditioning and extinction suggests that possible adaptive mechanisms in the amygdala are stochastic [46]. In this view, the synapses in the fear pathway change weight at random as the whole neural system searches for a configuration that achieves fear conditioning and, subsequently, extinction. Given an essentially random search, the probability of a given outcome (e.g. extinction with some preservation of fear memory) should be proportional to the number of synaptic weight configurations that are compatible with that outcome. In the model, LTD of the synapses of inhibitory interneurons onto LA projection neurons increases the number of configurations that are compatible with extinction and some retention of CS memory. It is possible that the contribution of endocannabinoid-induced LTD in the real LA is to increases the amygdala's options for adaptation sufficiently for extinction with some retention of CS memory to occur.

4 New Approach in Computational Neuroscience

As noted at the beginning of this chapter, most computer models in neuroscience are implemented using imperative programming languages, and as such they are largely limited to simulation. Because models of neurobiological systems can have many elements and many potential states, the imperative explorations are usually limited to tiny fractions of their model state-spaces. The declarative approach brings powerful tools for the exploration of model state spaces and in that way it adds a critical new dimension to computational neuroscience. It is a highly data-driven approach that can be taken in order to synthesize a potentially very large quantity of facts and provide new insights into what those facts mean in the aggregate. Because neurobiological processes can seem prohibitively complicated, a declarative approach is essential in order to "reason about" them and ultimately characterize and understand them.

The model summarized in Sect. 3 took a lot of facts into account, and the analysis of its specification in Maude, involving searches for specific states that met an ever increasing set of exacting criteria, was admittedly intricate. To most neurobiologists, the Maude analysis of extinction along the amygdala fear pathway summarized in Sect. 3 would seem rather arcane, but it does lead to potential new insights into the seemingly paradoxical contribution of endocannabinoids to this process. Permitting a bit of speculation, the Maude analysis could suggest that marijuana and its derivatives are useful in the treatment of PTSD because they allow users to reduce their stressful reactions to stimuli that, through trauma, they associate with fear, but also allow them not to forget the original association. Permitting a bit more speculation, the analysis could even hint that many users find marijuana psychologically beneficial because it separates them from the stressful reactions they have to real or imagined threats but does not remove the associations, allowing users to view those associations from a new perspective. It is amusing to realize that something as technical and intricate as a Maude analysis of a neural pathway could lead to new ideas about why some people like to use marijuana. On a more serious level, it is exciting to imagine the new understandings of brain function that are made possible by the introduction of declarative programming techniques to computational neuroscience.

References

1. Hodgkin, A.L., Huxley, A.F.: A quantitative description of membrane current and its application to conduction and excitation in nerve. J. Physiol. **177**, 500–544 (1952)
2. Rall, W.: Theory of the physiological properties of dendrites. Ann. NY Acad. Sci. **2**, 1071–1092 (1962)
3. Koch, C.: Biophysics of Computation: Information Processing in Single Neurons. Oxford University Press, Oxford (2004)
4. Ratliff, F., Hartline, H.K.: The response of limulus optic nerve fibers to patterns of illumination on the receptor mosaic. J. Gen. Physiol. **42**, 1241–1255 (1959)
5. Ratliff, F.: Mach Bands: Quantitative Studies on Neural Networks in the Retina. Holden-Day, San Francisco (1965)

6. Shepherd, G.M.: Neurobiology, 3rd edn., pp. 239–242. Oxford University Press, New York (1994)
7. Haykin, S.: Neural Networks: A Comprehensive Foundation, pp. 680–709. Prentice Hall, Upper Saddle River (1999)
8. Hopfield, J.J.: Neural networks and physical systems with emergent collective computational abilities. PNAS **79**, 2554–2558 (1982)
9. Hopfield, J.J.: Neurons with graded response have collective computational properties like those of two-state neurons. PNAS **81**, 3088–3092 (1984)
10. Rolls, E.T.: An attractor network in the hippocampus: theory and neurophysiology. Learn. Mem. **14**, 714–731 (2007)
11. Kohonen, T.: Self-Organizing Maps, 2nd edn. Springer, Berlin (1997)
12. Knudsen, E.I., du Lac, S., Esterly, S.D.: Computational maps in the brain. Ann. Rev. Neurosci. **10**, 41–65 (1987)
13. Rumelhart, D.E., Hinton, G.E., Williams, R.J.: Learning internal representations by error propagation. In: Rumelhart, D.E., McClelland, J.L., PDP Research Group (eds.) Parallel Distributed Processing: Exploration in the Microstructure of Cognition, vol. 1: Foundations, pp. 318–362. MIT Press, Cambridge (1986)
14. Anastasio, T.J., Robinson, D.A.: The distributed representation of vestibulo-oculomotor signals by brainstem neurons. Biol. Cybern. **61**, 79–88 (1989)
15. Anastasio, T.J., Robinson, D.A.: Distributed parallel processing in the vertical vestibulo-ocular reflex: learning networks compared to tensor theory. Biol. Cybern. **63**, 161–167 (1990)
16. Richard, M.D., Lippmann, R.P.: Neural network classifiers estimate Bayesian a-posteriori probabilities. Neural Compt. **3**, 461–483 (1991)
17. Anastasio, T.J., Patton, P.E., Belkacem-Boussaid, K.: Using Bayes' rule to model multisensory enhancement in the superior colliculus. Neural Compt. **12**, 997–1019 (2000)
18. Lee, T.S., Mumford, D.: Hierarchical Bayesian inference in the visual cortex. J. Opt. Soc. Am. A: **20**, 1434–1448 (2003)
19. Rao, R.P.N.: Bayesian inference and attentional modulation in the visual cortex. NeuroReort **16**, 1843–1848 (2005)
20. Ma, R., Cui, H., Lee, S.-H., Anastasio, T.J., Malpeli, J.G.: Predictive encoding of moving target trajectory by neurons in the parabigeminal nucleus. J. Neurophysiol. **109**, 2029–2043 (2013)
21. Fisher, J., Henzinger, T.A.: Executable cell biology. Nat. Biotechnol. **25**, 1239–1249 (2007)
22. Eker, S., Knapp, M., Laderoute, K., Lincoln, P., Meseguer, J., Sonmez, K.: Pathway logic: symbolic analysis of biological signaling. Pac. Symp. Biocomput. **7**, 400–412 (2002)
23. Talcott, C.: Pathway logic. In: Bernardo, M., Degano, P., Zavattaro, G. (eds.) SFM 2008. LNCS, vol. 5016, pp. 21–53. Springer, Berlin (2008)
24. Tiwari, A., Talcott, C.: Analyzing a discrete model of *aplysia* central pattern generator. In: Heiner, M., Uhrmacher, A.M. (eds.) Computational Methods in Systems Biology, pp. 347–366. Springer, Berlin (2008)
25. Anastasio, T.J.: Tutorial on Neural Systems Modeling. Sinauer Associates, Sunderland (2010)
26. Anastasio, T.J.: Data-driven modeling of Alzheimer disease pathogenesis. J. Theor. Biol. **290**, 60–72 (2011)
27. Anastasio, T.J.: Exploring the contribution of estrogen to amyloid-beta regulation: a novel multifactorial computational modeling approach. Front. Pharmacol. **4**, 16 (2013)
28. Anastasio, T.J.: Computational identification of potential multitarget treatments for ameliorating the adverse effects of amyloid-β on synaptic plasticity. Front. Pharmacol. **5**, 1 (2014)

29. Anastasio, T.J.: Temporal-logic analysis of microglial phenotypic conversion with exposure to amyloid-β. Mol. BioSyst. **11**, 434–453 (2014)
30. Anastasio, T.J.: Computational search for hypotheses concerning the endocannabinoid contribution to the extinction of fear conditioning. Front. Comput. Neurosci. **7**, 74 (2013)
31. LeDoux, J.E.: The Emotional Brain: The Mysterious Underpinnings of Emotional Life. Touchstone, New York (1998)
32. Maren, S.: Long-term potentiation in the amygdala: a mechanism for emotional learning and memory. Trends Neurosci. **22**, 561–567 (1999)
33. LeDoux, J.E.: Emotion circuits in the brain. Annu. Rev. Neurosci. **23**, 155–184 (2000)
34. Herry, C., Ferraguti, F., Singewald, N., Letzkus, J.J., Ehrlich, I., Luthi, A.: Neuronal circuits of fear extinction. Eur. J. Neurosci. **31**, 599–612 (2010)
35. Mahan, A.L., Ressler, K.J.: Fear conditioning, synaptic plasticity and the amygdala: implications for posttraumatic stress disorder. Trends Neurosci. **35**, 24–35 (2012)
36. Marsicano, G., Wotjak, C.T., Azad, S.C., Bisogno, T., Rammes, G., Cascio, M.G., Hermann, H., Tang, J., Hofmann, C., Zieglgansberger, W., Di Marzo, V., Lutz, B.: The endogenous cannabinoid system controls extinction of aversive memories. Nature **418**, 530–534 (2002)
37. Martin, B.R., Mechoulan, R., Razdan, R.K.: Discovery and characterization of endogeneous cannabinoids. Life Sci. **65**, 573–595 (1999)
38. Ganon-Elazar, E., Akirav, I.: Cannabinoid receptor activation in the basolateral amygdala blocks the effects of stress on the conditioning and extinction of inhibitory avoidance. J. Neurosci. **29**, 11078–11088 (2009)
39. Chhatwal, J.P., Davis, M., Maguschak, K.A., Ressler, K.J.: Enhancing cannabinoid neurotransmission augments the extinction of conditioned fear. Neuropsychopharmacology **30**, 516–524 (2005)
40. Roitman, P., Mechoulan, R., Cooper-Kazaz, R., Shalev, A.: Preliminary, open-lable, pilot study of add-on oral Δ^9-tetrahydrocannabinol in chronic post-traumatic stress disorder. Clin. Drug Invest. **34**, 587–591 (2014)
41. Blair, H.T., Schafe, G.E., Bauer, E.P., Rodrigues, S.M., LeDoux, J.E.: Synaptic plasticity in the lateral amygdala: a cellular hypothesis of fear conditioning. Learn. Mem. **8**, 229–242 (2001)
42. Pare, D., Quirk, G.J., LeDoux, J.E.: New vistas on amygdala networks in conditioned fear. J. Neurophysiol. **92**, 1–9 (2004)
43. Quirk, G.J., Likhtik, E., Pelletier, J.G., Pare, D.: Stimulation of medial prefrontal cortex decreases the responsiveness of central amygdala output neurons. J. Neurosci. **23**, 8800–8807 (2003)
44. Repa, J.C., Muller, J., Apergis, J., Desrochers, T.M., Zhou, Y., LeDoux, J.E.: Two different lateral amygdala cell populations contribute to the initiation and storage of memory. Nat. Neurosci. **4**, 724–731 (2001)
45. Hobin, J.A., Goosens, K.A., Maren, S.: Context-dependent neuronal activity in the lateral amygdala represents fear memories after extinction. J. Neurosci. **23**, 8410–8416 (2003)
46. Vlachos, I., Herry, C., Luthi, A., Aertsen, A., Kumar, A.: Context-dependent encoding of fear and extinction memories in a large-scale network model of the basal amygdala. PLoS Comput. Biol. **7**, e1001104 (2011)

Hybrid Multirate PALS

Kyungmin Bae[1]([✉]) and Peter Csaba Ölveczky[2]

[1] Carnegie Mellon University, Pittsburgh, USA
kbae@andrew.cmu.edu
[2] University of Oslo, Oslo, Norway

Abstract. Multirate PALS reduces the design and verification of a virtually synchronous distributed real-time system to the design and verification of the underlying synchronous model. This paper introduces Hybrid Multirate PALS, which extends Multirate PALS to virtually synchronous distributed multirate hybrid systems, such as aircraft and power plant control systems. Such a system may have interrelated local physical environments, each of whose continuous behaviors may periodically change due to actuator commands. We define continuous interrelated local physical environments, and the synchronous and asynchronous Hybrid Multirate PALS models, and give a trace equivalence result relating a synchronous and an asynchronous model. Finally, we illustrate by an example how invariants can be verified using SMT solving.

1 Introduction

José Meseguer has recently proposed using *formal patterns* [13] to reduce the cost and difficulty of designing, implementing, and verifying state-of-the-art systems. Such a pattern formally specifies a generic solution to a frequently occurring distributed systems problem and comes with strong formal guarantees. The idea is that the significant effort invested in developing a formal pattern and proving its correctness can be amortized over the many instances of the pattern.

Inspired by concrete problems in avionics, José, in joint work with the second author and colleagues at Rockwell Collins and UIUC, developed the PALS ("physically asynchronous, logically synchronous") formal pattern for *virtually synchronous* distributed real-time systems [2,12]. Such a system consists of a number of distributed components with the same period that should logically proceed synchronously: at the beginning of each round, each component reads its inputs, performs a transition, and sends output. Virtually synchronous distributed systems are very hard to design correctly, due to concurrency, communication delays, execution times, skews of the local clocks, and so on. Furthermore, due to the state space explosion caused by the asynchrony of distributed real-time systems, their model checking verification quickly becomes unfeasible.

The point of PALS is that it is sufficient to design and model check the much simpler underlying idealized *synchronous design SD*, provided that the underlying infrastructure can guarantee bounds on the network delays, execution times, and imprecision of the local clocks. PALS then transforms a synchronous

© Springer International Publishing Switzerland 2015
N. Martí-Oliet et al. (Eds.): Meseguer Festschrift, LNCS 9200, pp. 114–134, 2015.
DOI: 10.1007/978-3-319-23165-5_5

design SD and performance parameters Γ into the distributed (and therefore asynchronous) real-time system $\mathcal{A}(SD, \Gamma)$. The formal guarantee of PALS is that the idealized design SD and its distributed real-time version $\mathcal{A}(SD, \Gamma)$ satisfy the same temporal logic properties [12].

The benefits of PALS were first demonstrated on an avionics system provided by Rockwell Collins. The synchronous design had 185 reachable states and could be model checked in a fraction of a second, whereas even an unrealistically simplified distributed version—with perfect local clocks and no network delays— had more than 3,000,000 reachable states; if the network delay could be 0 or 1, then no model checking was feasible.

PALS assumes that all components share the same period. However, different components often operate at different rates, yet need to synchronize. For example, the ailerons and the rudder of an airplane must synchronize to smoothly turn an airplane, even though the aileron controllers and the rudder controller typically operate with different frequencies. Together with José we therefore developed the *Multirate PALS* pattern for hierarchical multirate systems as an extension of PALS [4]. The effectiveness of the Multirate PALS pattern was illustrated with the modeling and analysis of an algorithm for turning an airplane [3]. To make the Multirate PALS methodology available also to domain-specific modelers, we have developed the *Multirate Synchronous AADL* language, which allows modelers to develop their synchronous designs using the industrial modeling standard AADL [7]. We have integrated the Real-Time Maude model checking of such designs inside the OSATE modeling environment for AADL [5,6].

On this festive occasion we extend the Multirate PALS methodology to the important class of virtually synchronous distributed *hybrid* systems. Many control systems can be seen as such systems. One example is the airplane turning control systems mentioned above; another is the well-known steam-boiler controller in a power plant: in each round, local sensors report their reading of continuous parameters to the main controller, which decides what action to take in order for the power plant to function safely and efficiently [1].

Extending the PALS framework to hybrid systems is challenging. PALS is based on abstracting from the time at which a message is sent (as long as it is sent within the appropriate time window); this is possible because the advance of time does not change the state of a system or its environment. However, if the environment is a physical component with a continuous dynamics, which, furthermore, may change because of actuator commands, then its state changes continuously with the elapse of time. Another significant challenge is that, because of the clock skews, different components will read their physical values at different times in the asynchronous system; this must be reflected in the synchronous model to have any hope of obtaining a PALS equivalence.

In this paper we introduce *Hybrid Multirate PALS*. The key idea is to start with a synchronous Multirate PALS model SD whose environment is fully nondeterministic. The different continuous trajectories of an environment, defined as functions of actuator commands, are specified as a *controlled physical environment E*. The synchronous model in Hybrid Multirate PALS is then the *environment restriction* $SD \restriction_\Pi E$, which considers only those behaviors of SD that are consistent with the continuous behaviors of the environment. To account for the timing imprecision of reading from, and writing to, the environment due to clock skews, the constraint E also takes clock skews into account. The corresponding asynchronous Hybrid Multirate PALS model is the environment restriction $\mathcal{MA}(SD, T, \Gamma) \restriction_\Pi E$, where $\mathcal{MA}(SD, T, \Gamma)$ is the multirate asynchronous model of SD with global period T and performance bounds Γ. This paper then presents a trace equivalence result relating $SD \restriction_\Pi E$ and $\mathcal{MA}(SD, T, \Gamma) \restriction_\Pi E$.

Different components may have different *local* environments, which, furthermore, can be *physically correlated* to each other. For example, if we consider two adjacent rooms, the temperature of one room will immediately affect the temperature of the other room. In the airplane turning control system, the controllers of the ailerons and rudder are physically correlated to its main controller, since any angular movement of the ailerons and rudder immediately changes the position of the airplane. Our environment constraints allow us to express such continuous interdependencies between multiple physical environments.

The first steps towards extending Multirate PALS with hybrid behaviors were taken in [3]. However, that work does not take into account the skews of the local clocks or the continuous interaction between different physical components. More significantly, [3] does not provide any relationship between the synchronous model and the asynchronous distributed models.

Adding hybrid behaviors and clock skews to the *synchronous* model raises the question of how to analyze it. Explicit-state model checking can no longer be used. Instead, we propose to symbolically verify *invariant properties* of the synchronous model using SMT solving, and illustrate our approach with an example. The verification of a synchronous model, with controlled physical environments and clock skews, is essentially reduced to the satisfaction of SMT formulas over the real numbers and ordinary differential equations. The satisfaction problem over such theories, which is undecidable in general, is decidable up to any given precision $\delta > 0$ [8]. We use the dReal SMT solver [9] to automatically verify invariants of the synchronous model up to some precision $\delta > 0$.

This paper is organized as follows. Section 2 gives some background on Multirate PALS. Section 3 explains how continuous environments can be specified. Section 4 presents Hybrid Multirate PALS, and gives a trace equivalence result relating the synchronous and the asynchronous Hybrid Multirate PALS models. Section 5 illustrates by an example how an invariant in the synchronous model can be verified by SMT solving, and Sect. 6 gives some concluding remarks.

2   Preliminaries on Multirate PALS

Multirate PALS transforms a *synchronous design* SD with global period T, together with bounds Γ on network delays, execution times, and local clock skews, into a distributed multirate real-time system $\mathcal{MA}(SD, T, \Gamma)$ that satisfies the same temporal logic properties. This section gives an overview of the synchronous models SD, the corresponding asynchronous models $\mathcal{MA}(SD, T, \Gamma)$, and the relationship between SD and $\mathcal{MA}(SD, T, \Gamma)$. We refer to [4,12] for details.

Discrete Synchronous Models. The synchronous model SD is formally specified as an *ensemble* $\mathcal{E}$ of nondeterministic *typed machines*. A typed machine is a tuple $M = (D_i, S, D_o, \delta_M)$, where (i) $D_i = D_{i_1} \times \cdots \times D_{i_n}$ is an input set (a value to the k-th *input port* is an element of D_{i_k}), (ii) S is a set of states, (iii) $D_o = D_{o_1} \times \cdots \times D_{o_m}$ is an output set (a value from the j-th *output port* is an element of D_{o_j}), and (iv) $\delta_M \subseteq (D_i \times S) \times (S \times D_o)$ is a total transition relation. This model has no explicit notion of time; the next state and the next outputs of a machine depend only on its current state and its inputs.

A collection of typed machines with different periods can be composed into a *multirate ensemble* as illustrated in Fig. 1, where the period of a slow machine is a multiple of the period of a fast machine. A multirate machine ensemble $\mathcal{E}$ is a tuple $\mathcal{E} = (J_S, J_F, e, \{M_j\}_{j \in J_S \cup J_F}, E, src, rate, adap)$, where: (i) J_S is a set of "slow" machine indices, and J_F is a set of "fast" machine indices ($J_S \cap J_F = \emptyset$); (ii) $e \notin J_S \cup J_F$ is the environment index; (iii) $\{M_j\}_{j \in J_S \cup J_F}$ is a family of typed machines; (iv) $E = (\mathcal{D}_i^e, \mathcal{D}_o^e)$ is the *environment* with $\mathcal{D}_i^e$ the *input set* and $\mathcal{D}_o^e$ the *output set*; (v) *src* is a *wiring diagram* assigning to each input port its "source" output port so that no connection exists between two "fast" machines; (vi) *rate* assigns to each fast machine its *rate*, denoting how many times faster than the slow machines it is; and (vii) *adap* assigns an *input adaptor* to each M_j for communication between components with different rates.

In each iteration of the system, all components in an ensemble $\mathcal{E}$ perform a transition in lockstep. A fast component $f \in J_F$ is *slowed down* and performs $k = rate(f)$ internal transitions in one such "global" synchronous step. Such a fast machine therefore produces k-tuples of outputs in one global synchronous step; however, since this output is read by a "slow" machine (which only executes *one* transition during a global synchronous step), the k-tuples must first be transformed to *single* values (e.g., the last value, or the average of the k values). Likewise, a single output from a slow machine must be transformed to a k-tuple of inputs (e.g., a k-tuple $(d, \perp, \ldots, \perp)$ for some "don't care" value $\perp$) for the fast machine M_f. These transformation are performed by *input adapters*.

The *synchronous composition* of a multirate ensemble $\mathcal{E}$ is equivalent to a single machine $M_{\mathcal{E}}$, whose state space is $S^{\mathcal{E}} = (\Pi_{j \in J} S_j) \times (\Pi_{j \in J} D_{OF}^j)$, consisting of the states S_j of its components, and the "feedback" outputs D_{OF}^j for $j \in J_S \cup J_F$ (that is, the outputs from machine j to some machine in $\mathcal{E}$). For example, the synchronous composition $M_{\mathcal{E}}$ of the ensemble $\mathcal{E}$ in Fig. 1 is the machine given by the outer box.

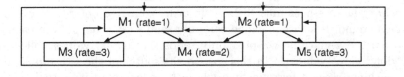

Fig. 1. A multirate ensemble $\mathcal{E}$, with M_1 and M_2 slow machines.

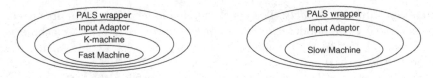

Fig. 2. The wrapper hierarchies for fast components (left) and slow components (right).

The behaviors of a multirate ensemble $\mathcal{E}$ can be specified as a transition system $ts(\mathcal{E}) = (S^{\mathcal{E}} \times D_i^{\mathcal{E}}, \longrightarrow_{\mathcal{E}})$, where $(\boldsymbol{s}, \boldsymbol{i}) \longrightarrow_{\mathcal{E}} (\boldsymbol{s}', \boldsymbol{i}')$ iff an ensemble in state $\boldsymbol{s}$ and with input $\boldsymbol{i}$ from the environment has a transition to state $\boldsymbol{s}'$ (i.e., $\exists \boldsymbol{o}.~((\boldsymbol{i}, \boldsymbol{s}), (\boldsymbol{s}', \boldsymbol{o})) \in \delta_{\mathcal{E}}$) and the environment can generate output $\boldsymbol{i}'$ in the next step. Although an environment could be modeled as a normal machine, the global ("slow") environment is made explicit for pragmatic reasons in Multirate PALS. In Multirate PALS, faster machines can have "local environments;" this is specified by composing the fast machine and its local environment and by modeling their composition as a single fast machine.

Asynchronous Models. In the distributed real-time model $\mathcal{MA}(\mathcal{E}, T, \Gamma)$, each distributed component consists of a machine M in $\mathcal{E}$ with some "wrappers" around it, as shown in Fig. 2. In this system, each machine M performs at its own rate. A wrapper has an input buffer, an output buffer, and access to the component's local clock, which deviates by less than ε from the perfect global clock. The outermost wrapper is the PALS wrapper, which encloses an input adaptor wrapper, which encloses either a (slow) machine or a k-machine wrapper, which encloses an ordinary (fast) machine.

The PALS wrappers share the same *slow period* T, and communicate with the other components by sending and receiving messages. An input adaptor wrapper reads the inputs from the PALS wrapper and applies an appropriate input adaptor function at the beginning of each *slow* period T according to its local clock. For a fast machine, its k-machine wrapper extracts each value from the k-tuple input and sends them to the enclosed typed machine at each fast period T/k. The k-machine wrapper also stores the outputs from the fast machine, and sends out the resulting k-*tuples* of outputs to its outer layer for each slow period T. The innermost typed machine runs at its given rate; at the beginning of its periods, it reads its input, performs a transition, and sends the outputs when the execution of the transition is finished.

Due to execution times and network delays, a fast machine in $\mathcal{MA}(\mathcal{E}, T, \Gamma)$ may *not* be able to finish all of its k internal transitions in a global ("slow") round *before* the messages must be sent in order to arrive before the beginning of the next round. If a fast machine can only send $k' < k$ outputs in each slow round, then the k-machine wrapper only sends the first k' values, followed by $k - k'$ "don't care" values $\bot$. In this case, if the source of the i-th input port of a slow machine M_j is a fast machine whose k' is less than its rate k, then its input adaptor function $adap(j)_i$ must be $(k' + 1)$-*oblivious*, that is, it "ignores" the values $v_{k'+1}, \ldots, v_k$ in a k-tuple $(v_1, \ldots, v_k)$.

Relating the Synchronous and Asynchronous Models. In the *stable states* of the asynchronous model $\mathcal{MA}(\mathcal{E}, T, \Gamma)$, all the input buffers of the PALS wrappers are full, and all other input and output buffers are empty. (A stable state is a snapshot of the system just before the components start performing local transitions.) The mapping $sync : Stable(\mathcal{MA}(\mathfrak{E}, T, \Gamma)) \rightarrow S^{\mathcal{E}} \times D_i^{\mathcal{E}}$ associates each stable state with the corresponding synchronous state of $M_{\mathcal{E}}$. We can relate two stable states by $t_1 \sim_{obi} t_2$, iff their corresponding input buffer contents *cannot* be distinguished by any input adaptors and their machine states are identical.

In the *big-step* transition system $(Stable(\mathcal{MA}(\mathcal{E}, T, \Gamma)), \longrightarrow_{st})$ between stable states, each "big-step transition" corresponds to a single step in the synchronous transition system $ts(\mathcal{E})$. The relation $(\sim_{obi} ; sync)$ is a bisimulation between the transition systems $(Stable(\mathcal{MA}(\mathcal{E}, T, \Gamma)), \longrightarrow_{st})$ and $ts(\mathcal{E})$. Furthermore, if a state labeling function $\mathcal{L} : S^{\mathcal{E}} \times \mathcal{D}_i^{\mathcal{E}} \rightarrow 2^{AP}$ for $\mathcal{E}$ cannot distinguish between $\sim_{obi}$-equivalent states, then the two Kripke structures for $Stable(\mathcal{MA}(\mathcal{E}, T, \Gamma))$ and $ts(\mathcal{E})$ are bisimilar and therefore satisfy the same CTL^* formulas.

3 Specifying Physical Environments

In Multirate PALS, a "local" environment of a component is assumed to be incorporated into the corresponding typed machine, and the "global" environment of the entire ensemble is given by another typed machine or by environment constraints [3,4]. Multirate PALS abstracts from the timing of events (as long they happen within certain time windows), and the environment constraints mean that the global environment has the same possible behaviors at any time. This model is therefore not suitable for distributed *hybrid* systems, where a (possibly "local") physical environment changes its state continuously with time elapse.

This section explains how physical environments with continuous behaviors can be specified using logical constraints over physical parameters. Such a specification must also take into account

- the imprecision of the local clocks, since environment "values" are read at times defined by the local clock;
- actuator/control commands that change the continuous behavior of the environment (e.g., "turn the heater off" or "turn the aileron θ degrees"); and
- the physical correlation between different physical environments.

A state of a (local) physical environment is given by a tuple $\boldsymbol{v} = (v_1, \ldots, v_l) \in \mathbb{R}^l$ of its physical parameters $\boldsymbol{x} = (x_1, \ldots, x_l)$. The behavior of the physical parameters $\boldsymbol{x}$ can be modeled using differential equations that specify *trajectories* $\tau_1, \ldots, \tau_l$ of the parameters $\boldsymbol{x}$ over time. A *trajectory* [10] of duration T is a function $\tau : [0, T] \rightarrow \mathbb{R}$ that defines the continuous behavior of a physical parameter. Let $\mathcal{T}$ be the set of all trajectories, and $\boldsymbol{\tau}(t) = (\tau_1(t), \ldots, \tau_l(t))$ for an l-tuple of trajectories $\boldsymbol{\tau} = (\tau_1, \ldots, \tau_l)$. The parameters $\boldsymbol{x} = (x_1, \ldots, x_l)$ can also be considered as trajectories in such a way that a state of the physical environment at time t is given by $\boldsymbol{x}(t)$.

3.1 Controlled Physical Environments

The local physical environment E_M of a typed machine M can be specified as a *controlled physical environment* that specifies every possible trajectory of its physical parameters, taking into account the control commands from M. We assume that the environment E_M and its local controller M are tightly integrated (no communication delay, etc.), and that the controller reads environment values of E_M and gives control commands to E_M at the beginning of each period. Since the local clock of M may differ from the perfect global clock by up to the maximal clock skew ϵ, we now take into account the local clock of M in our definition. Let $c_M : \mathbb{N} \to \mathbb{R}_{>0}$ denote a *periodic local clock* of M such that $c_M(0) = 0$ and $c_M(n) \in (nT - \epsilon, nT + \epsilon)$ for each $n > 0$; that is, $c_M(i)$ denotes the (global) time at the beginning of the $(i + 1)$-th period according to M's local clock.

Definition 1. *A controlled physical environment $E_M = (C, \boldsymbol{x}, \Lambda)$ consists of:*

- *C a set of control commands, representing "actuator outputs" from M;*
- *$\boldsymbol{x} = (x_1, \ldots, x_l)$ a vector of real-number-valued variables; and*
- *$\Lambda \subseteq (C \times \mathbb{R}_{\geq 0} \times \mathbb{R}^l) \times T^l$ a physical transition relation: $((a, t, \boldsymbol{v}), \boldsymbol{\tau}) \in \Lambda$ iff for a control command $a \in C$ from M that lasts for duration t, E_M's physical state follows the trajectory $\boldsymbol{\tau} \in T^l$ from a state $\boldsymbol{v} \in \mathbb{R}^l$ with $\boldsymbol{\tau}(0) = \boldsymbol{v}$.*

For a periodic local clock c_M, a physical transition of each i-th period defines a trajectory $\boldsymbol{\tau}$ of $\boldsymbol{x}$ during the time interval $[c_M(i) + \alpha_{M_{\max}}, c_M(i+1) + \alpha_{M_{\max}}]$, where the extra time value $\alpha_{M_{\max}}$ reflects the execution time of M (the actuator command will be "delayed" by $\alpha_{M_{\max}}$ as explained in Sect. 4). That is:

$$((a, c_M(i + 1) - c_M(i), \boldsymbol{v}), \boldsymbol{\tau}) \in \Lambda$$
$$\implies \forall t \in [c_M(i) + \alpha_{M_{\max}}, c_M(i+1) + \alpha_{M_{\max}}].\ \boldsymbol{x}(t) = \boldsymbol{\tau}(t - (c_M(i) + \alpha_{M_{\max}}))$$

For example, in the controlled physical environment E_M in Fig. 3, each physical transition defines a trajectory τ_i from the value v_i to v_{i+1} according to the control command a_i from M during the time interval $[c_M(i) + \alpha_{M_{\max}}, c_M(i+1) + \alpha_{M_{\max}}]$.

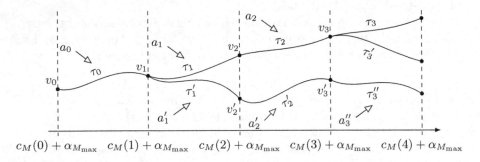

$$c_M(0) + \alpha_{M_{\max}} \quad c_M(1) + \alpha_{M_{\max}} \quad c_M(2) + \alpha_{M_{\max}} \quad c_M(3) + \alpha_{M_{\max}} \quad c_M(4) + \alpha_{M_{\max}}$$

Fig. 3. A controlled physical environment E_M of a controller M with a local clock c_M. E.g., $((a_0, c_M(1) - c_M(0), v_0), \tau_0) \in \Lambda$, $((a_1, c_M(2) - c_M(1), v_1), \tau_1) \in \Lambda$, and so on.

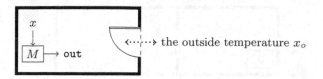

Fig. 4. Open room environment.

Example 1. Consider the thermostat controller M in Fig. 4 that controls the temperature of an open room. The temperature x of the room changes based on the mode out $\in \{on, off\}$ of M and the outside temperature x_o. This controlled physical environment can be specified by $E_M = (\{on, off\}, (x, x_o), \Lambda)$, where:

- $\{on, off\}$ is the set of control commands from the controller M;
- the physical parameters x and x_o denote, respectively, the room's temperature and the outside temperature; and
- $\Lambda \subseteq (\{on, off\} \times \mathbb{R}_{\geq 0} \times \mathbb{R}^2) \times \mathcal{T}^2$ is the physical transition relation such that $((\mathsf{out}, t, (v, v_o)), (\tau, \tau_o)) \in \Lambda$ iff there are two trajectories $\tau, \tau_o : [0, t] \to \mathbb{R}$, where $\tau(0) = v$, $\tau_o(0) = v_o$, and:

$$\frac{\mathrm{d}x}{\mathrm{d}t} = \begin{cases} K(h - ((1-k)x + kx_o)) & \text{if out} = on \\ -K((1-k)x + kx_o) & \text{if out} = off. \end{cases}$$

The values of the constants $K, h, k \in \mathbb{R}$ depend on the size of the room, the power of the heater, and the size of the open door, respectively. Given a trajectory τ_o of the outside temperature x_o, the temperature x rises according to the equation $\frac{\mathrm{d}x}{\mathrm{d}t} = K(h - ((1-k)x + kx_o))$ when the heater is turned on, and falls according to the equation $\frac{\mathrm{d}x}{\mathrm{d}t} = -K((1-k)x + kx_o)$ when the heater is turned off.   ♠

3.2 Correlating Physical Environments

Several physical environments can be physically correlated to each other, and one local environment may therefore immediately affect another environment. Such physical correlations can be naturally expressed as logical constraints over physical parameters (e.g., some parameter of one physical environment should always equal some other parameter of another physical environment).

Definition 2. *Consider n controlled physical environments $E_{M_i} = (C_i, \boldsymbol{x}_i, \Lambda_i)$ with $\boldsymbol{x}_i = (x_{i_1}, \ldots, x_{i_{l_i}})$ for $i = 1, \ldots, n$. Given a variable t for time and unary function symbols $x_{1_1}, \ldots, x_{1_{l_1}}, x_{2_1}, \ldots, x_{2_{l_2}}, \ldots, x_{n_1}, \ldots, x_{n_{l_n}}$ for trajectories, a time-invariant constraint is a first order logic formula of the form*

$$(\forall t)\ \psi(\boldsymbol{x}_1(t), \boldsymbol{x}_2(t), \ldots, \boldsymbol{x}_n(t)).$$

For instance, if a parameter x_1 of E_{M_1} must be equal to a parameter x_2 of E_{M_2}, then the time-invariant constraint is $(\forall t)\ x_1(t) = x_2(t)$. In practice, many physical correlations can be expressed as such *time-invariant equality constraints*.

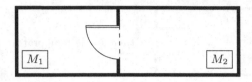

Fig. 5. Two rooms connected by an open door.

Example 2. Consider two *adjacent* rooms connected by an open door, illustrated in Fig. 5. The temperature of each room i is separately controlled by a normal thermostat controller M_i that turns the room's heater on or off.

The local physical environment of each controller M_i (for $i = 1, 2$) is just the open room environment $E_{M_i} = (\{on, off\}, (x_i, x_{o_i}), \Lambda_i)$ in Example 1. Each physical transition relation Λ_i states that the room's temperature x_i and the outside temperature x_{o_i} are governed by the differential equation:

$$\frac{\mathrm{d}x_i}{\mathrm{d}t} = \begin{cases} K_i(h_i - ((1-k)x_i + kx_{o_i})) & \text{if } \mathsf{out}_i = on \\ -K_i((1-k)x_i + kx_{o_i}) & \text{if } \mathsf{out}_i = off, \end{cases}$$

where $K_i, h_i \in \mathbb{R}$ are constants given by the size of each room and the heater's power, respectively, and $k \in \mathbb{R}$ depends on the size of the open door.

Because the rooms 1 and 2 are physically connected to each other by the open door, the outside temperature x_{o_1} of room 1 must be the same as the temperature x_2 of room 2, and the outside temperature x_{o_2} of room 2 must be the same as the temperature x_1 of room 1. This requirement can be specified by the time-invariant constraint $(\forall t)\ x_{o_1}(t) = x_2(t)\ \wedge\ x_{o_2}(t) = x_1(t)$. ♠

4   Hybrid Multirate PALS

This section introduces *Hybrid Multirate PALS*, which incorporates physical environment constraints, specified by controlled physical environments and time-invariant constraints, into the Multirate PALS framework. The idea is based on the methodology proposed in [3]: (i) the continuous parts of the system are specified by physical environments constraints, (ii) the "standard" Multirate PALS synchronous model is a *nondeterministic* model defined for *any* possible environment behavior; and (iii) the *environment restriction* defines the behavior of the system constrained by the behavior of a specific environment.

In contrast to [3], we now also accommodate continuous correlations between different physical environments as well as clock skews. The local clocks are now included also in the synchronous models, since the behaviors of a typed machine M are restricted by the controlled physical environment E_M, which gets commands and produces values at times given by the controller's local clock. To deal with the transition execution time of a controller M, the control output to E_M is "delayed" for the maximum execution time $\alpha_{\mathrm{max}M}$ of M.

Section 4.1 describes the Hybrid Multirate PALS synchronous model $\mathcal{E} \restriction_\Pi E_\mathcal{E}$, Sect. 4.2 defines the transition sequences possible in this environment-restricted synchronous model, Sect. 4.3 defines the Hybrid Multirate PALS asynchronous model, and Sect. 4.4 gives a trace equivalence result relating the synchronous and the asynchronous Hybrid Multirate PALS models.

4.1   Environment-Restricted Synchronous Ensembles

A controller machine $M = (D_i, S, D_o, \delta_M)$ is a *nondeterministic* typed machine parameterized by any possible observable behavior of its physical environment $E_M = (C, \boldsymbol{x}, \Lambda)$. A controller M is assumed to be tightly integrated with E_M: the controller can immediately observe and affect its environment.[1] That is, at the beginning of each period, a machine in the synchronous model (and therefore also in the asynchronous model) reads its inputs (which were the outputs in the previous iteration) and reads the *current* values of its local environment, performs a transition, sends output to other components (to be read in the next iteration), and gives a control command to its local environment, which takes effect after the maximum transition execution time $\alpha_{\max_M}$ of M has elapsed.

A machine M that integrates both a controller and its environment should have a state space of the form $S = S' \times \mathbb{R}^m$, where $\mathbb{R}^m$ is the state space of the m continuous environment parameters that the controller observes. The fact that the machine M is defined for *any* values of the environment follows directly from the definition of typed machines: their transition relations are total.

To compose a machine that already integrates a general *parameter environment* with its real environment, we must define the "interface" between the "controller part" and the "environment part" of a such an environment-parametric machine M. This interface is given by the following projection functions:

Definition 3. *For a state $s \in S$ of M and a physical state $\boldsymbol{v} \in \mathbb{R}^l$ of E_M, the relationship between the controller and its local environment is given by the following projection functions $\pi = (\pi_C, \pi_T, \pi_{\boldsymbol{x}}, \pi_S)$, where*

- $\pi_C(s) \in C$ *denotes the current control command of M to E_M;*
- $\pi_T(s) \in \mathbb{N}$ *denotes a "round number"; i.e., $\pi_T(s) = i$ means that the next iteration of the system is iteration i;*
- $\pi_{\boldsymbol{x}}(s) \in \mathbb{R}^m$, $m \leq l$, *denotes the "observed" parameters of E_M by M; and*
- $\pi_S(\boldsymbol{v}) \in \mathbb{R}^m$ *denotes the "observable" part of the physical state $\boldsymbol{v}$ by M.*

If a physical state $\boldsymbol{v}$ of E_M is observed by M in a certain state s, then the "observed" parameters $\pi_{\boldsymbol{x}}(s)$ by M must be identical to the "observable" parameters $\pi_S(\boldsymbol{v})$ of E_M; that is, $\pi_{\boldsymbol{x}}(s) = \pi_S(\boldsymbol{v})$.

The *environment restriction* of M by E_M with respect to the projection function π is defined as a typed machine $M \restriction_\pi E_M$ whose transition relation $\delta_{M \restriction_\pi E_M}$ is constrained by the observed behavior of E_M:

---

[1] "Remote" sensors and actuators that are not tightly integrated can be considered as parts of another controller that communicates with M through the network.

Definition 4. *The* environment restriction *of M by its physical environment E_M with respect to projection functions π and maximum transition time $\alpha_{\mathrm{max}M}$ is the typed machine $M \upharpoonright_\pi E_M = (D_i, S \times \mathbb{R}^l \times C, D_o, \delta_{M \upharpoonright_\pi E_M})$, where its transition $((\boldsymbol{i}, (s, \boldsymbol{v}, a)), ((s', \boldsymbol{v}', \pi_C(s)), \boldsymbol{o})) \in \delta_{M \upharpoonright_\pi E_M}$ holds iff:*

- *M has a transition $((\boldsymbol{i}, s), (s', \boldsymbol{o})) \in \delta_M$, where $\pi_T(s') = \pi_T(s) + 1$;*
- *E_M has two consecutive transitions with "middle" physical state $\boldsymbol{v}''$:*
  1. *using the previous actuator output $a \in C$, $((a, \alpha_{\mathrm{max}M}, \boldsymbol{v}), \boldsymbol{\tau}) \in \Lambda$, where $\boldsymbol{\tau}(0) = \boldsymbol{v}$ and $\boldsymbol{\tau}(\alpha_{\mathrm{max}M}) = \boldsymbol{v}''$,[2] and then*
  2. *using the current actuator output $\pi_C(s)$, $((\pi_C(s), T, \boldsymbol{v}''), \boldsymbol{\tau}') \in \Lambda$, where $T = c_M(\pi_T(s')) - (c_M(\pi_T(s)) + \alpha_{\mathrm{max}M})$, $\boldsymbol{\tau}(0) = \boldsymbol{v}''$ and $\boldsymbol{\tau}(T) = \boldsymbol{v}'$; and*
- *$\pi_{\boldsymbol{x}}(s) = \pi_S(\boldsymbol{v})$ and $\pi_{\boldsymbol{x}}(s') = \pi_S(\boldsymbol{v}')$.*

Since M needs some time $\alpha \le \alpha_{\mathrm{max}M}$ to compute the next actuator command, we assume for simplicity that the new actuator command $\pi_C(s)$ affects E_M after time $\alpha_{\mathrm{max}M}$ has elapsed in each round.

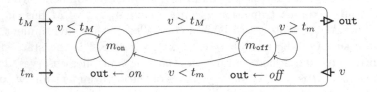

Fig. 6. A digital thermostat controller.

Example 3. Figure 6 illustrates a typed machine $M = (\mathbb{R}^2, S, \{*\}, \delta_M)$ for a digital thermostat controller that controls the temperature of a room, where:

- $\mathbb{R}^2$ is the input set, since M has two inputs $(t_M, t_m) \in \mathbb{R}^2$, with t_M a desired maximum temperature and t_m a desired minimum temperature;
- $S = \{m_{\mathrm{on}}, m_{\mathrm{off}}\} \times \mathbb{N} \times \mathbb{R}$, where each state (m, n, v) consists of the heater's mode m, a counter n denoting the current "round number," and the observed temperature v at time $c_M(n)$;
- $\{*\}$ is the singleton set, which indicates that M has no output port; and
- $\delta_M \subseteq (\mathbb{R}^2 \times S) \times (S \times \{*\})$ is M's transition relation defining next state $(m', n+1, v')$ from input (t_M, t_m) and a current state (m, n, v); that is, $(((t_M, t_m), (m, n, v)), ((m', n+1, v'), *)) \in \delta_M$ if

$$m' = \begin{cases} \text{if } v \le t_M \text{ then } m_{\mathrm{on}} \text{ else } m_{\mathrm{off}} \text{ fi} & \text{if } m = m_{\mathrm{on}} \\ \text{if } v \ge t_m \text{ then } m_{\mathrm{off}} \text{ else } m_{\mathrm{on}} \text{ fi} & \text{if } m = m_{\mathrm{off}}. \end{cases}$$

---

[2] From the physical state $\boldsymbol{v}$, the physical state $\boldsymbol{v}'$ is reachable through some trajectory $\boldsymbol{\tau}$ of duration $\alpha_{\mathrm{max}M}$ by the control command a.

In each step, M performs a transition based on the mode m, the inputs (t_M, t_m), and the observed temperature v at time $c_M(n)$. Notice that the *next* observed temperature of the nondeterministic controller M can be *any value v'*.

We now compose M with the environment E_M in Example 1 by defining the environment restriction $M \upharpoonright_\pi E_M = (\mathbb{R}^2, S \times \mathbb{R}^2 \times C, \{*\}, \delta_{M \upharpoonright_\pi E_M})$ with M's state space $S = \{m_{on}, m_{off}\} \times \mathbb{N} \times \mathbb{R}$, E_M's physical state space $\mathbb{R}^2$, and the delayed controller output $C = \{m_{on}, m_{off}\}$. The projection functions π are given by: $\pi_C(m_{on}, n, v) = on$ and $\pi_C(m_{off}, n, v) = off$; $\pi_T(m, n, v) = n$; $\pi_x(m, n, v) = v$, since the controller observes the temperature v; and $\pi_S(v_{room}, v_{outside}) = v_{room}$, since M observes the current room temperature.
$(((t_M, t_m), (m, n, v, v_e, v_o, a)), ((m', n', v', v'_e, v'_o, \pi_C(m, n, v)), *)) \in \delta_{M \upharpoonright_\pi E_M}$ iff:

- M has a transition $(((t_M, t_m), (m, n, v)), ((m', n+1, v'), *)) \in \delta_M$,
- E_M has two consecutive transitions with intermediate state (v''_e, v''_o)
  1. $((a, \alpha_{\max M}, (v_e, v_o)), (\tau, \tau_o)) \in \Lambda$ for some trajectories $(\tau, \tau_o) \in \mathcal{T}^2$ such that $(\tau, \tau_o)(0) = (v, v_o)$ and $(\tau, \tau_o)(\alpha_{\max M}) = (v''_e, v''_o)$, for a the control command given in the previous round;
  2. $((\pi_C(m, n, v), T, (v''_e, v''_o)), (\tau', \tau'_o)) \in \Lambda$ for some trajectories $(\tau', \tau'_o) \in \mathcal{T}^2$ of duration $T = c_M(n') - (c_M(n) + \alpha_{\max M})$ such that $(\tau, \tau_o)(0) = (v'', v''_o)$ and $(\tau', \tau'_o)(T) = (v', v'_o)$, for $\pi_C(m, n, v)$ the new control command; and
- M *observes* E_M: the observed temperature $\pi_x(m, n, v) = v$ by M is identical to the observable parameter $\pi_S(v_e, v_o) = v_e$ of E_M. ♠

The synchronous model in Hybrid Multirate PALS is specified as a multirate ensemble, where each component j has a local environment and projection functions π_j, and where physical correlations between their local physical environments are specified as time-invariant constraints:

Definition 5 (Hybrid Multirate Ensemble). *A hybrid multirate ensemble is a triple written $\mathcal{E} \upharpoonright_\Pi E_\mathcal{E}$, where*

- *$\mathcal{E}$ is a multirate ensemble $\mathcal{E} = (J_S, J_F, e, \{M_j\}_{j \in J_S \cup J_F}, E, src, rate, adap)$,*
- *$E_\mathcal{E} = \langle\{E_{M_j}\}_{j \in J_S \cup J_F}, (\forall t)\ \psi\rangle$ is a family of local environments, one for each machine, where $(\forall t)\ \psi$ specifies the time-invariant constraints over the physical parameters of the physical environments that define the immediate physical correlations between their local physical environments, and*
- *$\Pi = \{\pi_j\}_{j \in J_S \cup J_F}$ is a family of projection functions π_j.*

Example 4. Figure 7 shows a multirate ensemble $\mathcal{E}$ which controls the temperatures of two rooms. The ensemble $\mathcal{E}$ consists of three discrete components: the main controller Main sets a maximum temperature t_M and a minimum temperature t_m, and each thermostat controller M_i ($i = 1, 2$)—specified as a typed machine in Example 3—controls the room's heater. A controller M_i produces actuator output out_i, and its behavior depends on the temperature x_i of its room. The controllers M_1 and M_2 have different rates: $rate(1) = 1$ and $rate(2) = 2$.

The local environments of M_1 and M_2 are the two adjacent rooms connected by an open door in Example 2. The behavior of the environment restriction $M_i \restriction_{\pi_i} E_{M_i}$ also depends on the local clock c_{M_i} (M_i observes and affects the temperature based on c_{M_i}). Since the door between the rooms is open, the physical correlation between E_{M_1} and E_{M_2} is given by the time-invariant constraint $(\forall t)\ \psi \equiv (\forall t)\ x_{o_1}(t) = x_2(t) \ \wedge\ x_{o_2}(t) = x_1(t)$. The resulting hybrid multirate ensemble is $\mathcal{E} \restriction_{\Pi} \langle E_{M_1}, E_{M_2}, (\forall t)\ \psi \rangle$, for the obvious Π. ♠

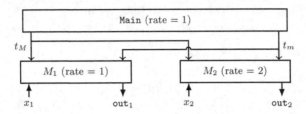

Fig. 7. A multirate ensemble $\mathcal{E}$.

4.2 Realizable Transition Sequences

Intuitively, the behaviors of a hybrid multirate ensemble $\mathcal{E} \restriction_{\Pi} E_{\mathcal{E}}$ are a *subset* of the behaviors of $\mathcal{E}$; namely, the behaviors restricted by the physical environment $E_{\mathcal{E}}$. We formally define the restricted behavior by means of *realizable transition sequences* of $\mathcal{E} \restriction_{\Pi} E_{\mathcal{E}}$. The idea is that a typed machine M corresponds to the transition system $ts(M)$, and a sequence of transitions in $ts(M)$ provides complete information about its local clocks, actuator control output, and observed physical environment states. A transition sequence is *realizable* iff its observed physical states are consistent with valid continuous environment behaviors with the given actuator control output and local clocks.

Formally, consider a sequence of transitions given by the transition system $ts(M)$: $\rho_M = (s_0, i_0) \rightarrow_M (s_1, i_1) \rightarrow_M (s_2, i_2) \rightarrow_M (s_3, i_3) \rightarrow_M \cdots$. The observed behavior of E_M is then provided by the projection functions: $\pi_{E_M}(\rho_M) = (\pi_C(s_0), \pi_T(s_0), \pi_{\boldsymbol{x}}(s_0)) \rightarrow_M (\pi_C(s_1), \pi_T(s_1), \pi_{\boldsymbol{x}}(s_1)) \rightarrow_M \cdots$. Each $\pi_T(s_i)$ records the beginning of the $(i+1)$-th period of M, and each $\pi_C(s_i)$ affects E_M in the $(i + 1)$-th period of M. If ρ is generated by M under the environment E_M, then $\pi_{E_M}(\rho)$ must follow E_M's physical transition relation Λ.

Definition 6. *A transition sequence $\rho_M = (s_0, i_0) \rightarrow_M (s_1, i_1) \rightarrow_M \cdots$ of M is realizable with respect to $E_M = (C, \boldsymbol{x}, \Lambda)$, a local clock c_M, maximum execution time $\alpha_{\max_M}$, initial control output a_{init}, and projection functions π, denoted by $E_M, \pi, c_M, \alpha_{\max_M}, a_{init} \models \rho_M$, iff for some trajectory of $\boldsymbol{x}$:*

– M correctly observes E_M: $\pi_{\boldsymbol{x}}(s_i) = \pi_S(\boldsymbol{x}(c_M(i)))$ for each $i \in \mathbb{N}$:, and
– E_M's two consecutive transition holds:

- $((a_i, \alpha_{\max_M}, v_i), \tau_i) \in \Lambda$ for each $i \in \mathbb{N}$, where if $i = 0$, then $a_i = a_{init}$, and if $i > 0$, then $a_i = \pi_C(s_{i-1})$, and $x(t) = \tau_i(t - c_M(\pi_T(s_i)))$ for $t \in [c_M(\pi_T(s_i)), c_M(\pi_T(s_i) + \alpha_{\max_M}]$.
- $((\pi_T(s_i), c_M(\pi_T(s_{i+1})) - t_i^0, \tau_i(\alpha_{\max_M}), \tau_i') \in \Lambda$ for each $i \in \mathbb{N}$, where $t_i^0 = c_M(\pi_T(s_i)) + \alpha_{\max_M}$ and $x(t) = \tau_i(t - t_i^0)$ for $t \in [t_i^0, c_M(\pi_T(s_{i+1}))]$.

It is easy to see that a realizable sequence of M by E_M and a sequence of the environment restriction $M \restriction_\pi E_M$ are in a one-to-one correspondence:

Lemma 1. *Given a machine M, a physical environment E_M, initial control output a_{init}, projection functions π, maximum execution time $\alpha_{\max_M}$, and a local clock c_M, for each sequence ρ_M such that $E_M, \pi, c_M, \alpha_{\max_M}, a_{init} \models \rho_M$, there exists a corresponding sequence $\rho_{M \restriction_\pi E_M}$ of $M \restriction_\pi E_M$, and vice versa.*

For a multirate ensemble $\mathcal{E} = (J_S, J_F, e, \{M_j\}_{j \in J_S \cup J_F}, E, src, rate, adap)$, its synchronous transition system $ts(\mathcal{E})$ defines a synchronous transition sequence $\rho_\mathcal{E} = (s_0, i_0) \to_\mathcal{E} (s_1, i_1) \to_\mathcal{E} (s_2, i_2) \to_\mathcal{E} (s_3, i_3) \to_\mathcal{E} \cdots$. Recall that each state s_i for $\mathcal{E}$ consists of the states $\{s_{j_i}\}_{j \in J_S \cup J_F}$ of its subcomponents and the feedback output. Therefore, there is a *local* transition sequence $s_{j_0} \to_j s_{j_1} \to_j s_{j_2} \to_j \cdots$ for each subcomponent M_j. Such local sequences are *slow-step* transitions that do not contain information about fast intermediate steps. But by construction of $\mathcal{E}$ with input adaptors, the feedback output in each synchronous state s_i contains all input values, including those for fast intermediate steps, used in the next step of the ensemble $\mathcal{E}$. Therefore, we can construct a complete local transition sequence $s_{j_0} = s_{j_{0,1}} \to s_{j_{0,2}} \cdots \to s_{j_{0,k}} \to_j s_{j_1} = s_{j_{1,1}} \to s_{j_{1,2}} \cdots \to s_{j_{1,k}} \to_j \cdots$. Hence, there exists a collection $\{\rho_{M_j}\}_{j \in J_S \cup J_F}$ of complete machine transition sequences, where each ρ_{M_j} is a transition sequence of M_j at *its own rate*, including all the intermediate fast steps.

Definition 7. *A sequence $\rho_\mathcal{E} = (s_0, i_0) \to_\mathcal{E} (s_1, i_1) \to_\mathcal{E} (s_2, i_2) \to_\mathcal{E} \cdots$ of $ts(\mathcal{E})$ is* realizable *w.r.t. $\{E_{M_j}, \pi_j, c_{M_j}, \alpha_{\max_{M_j}}, a_{init}^j\}_{j \in J_S \cup J_F}$ and a time-invariant constraint $(\forall t) \psi$, denoted by $\{E_{M_j}, \pi_j, c_M, \alpha_{\max_{M_j}}, a_{init}^j\}_{j \in J_S \cup J_F}, (\forall t) \psi \models \rho_\mathcal{E}$, iff for $j \in J_S \cup J_F$, for $\rho_\mathcal{E}$'s complete sequences $\{\rho_{M_j}\}_{j \in J_S \cup J_F}$, by some trajectory x_j, $E_{M_j}, \pi_j, c_{M_j}, \alpha_{\max_{M_j}}, a_{init}^j \models \rho_{M_j}$, and $\psi(\{x_j(t)\}_{j \in J_S \cup J_F})$ for any $t \in \mathbb{R}_{\geq 0}$.*

That is, the semantics of a hybrid multirate ensemble is given by the *realizable behaviors* subset of all transition sequences of the synchronous composition $ts(\mathcal{E})$. Each such realizable behavior consists of one realizable transition sequence for each machine such that these sequences together also satisfy the time-invariant constraints correlating the different local environments.

4.3   Hybrid Multirate PALS Distributed Models

Hybrid Multirate PALS maps a hybrid multirate ensemble $\mathcal{E} \restriction_\Pi E_\mathcal{E}$, together with performance bounds Γ, to the distributed real-time system

$$\mathcal{MA}(\mathcal{E}, T, \Gamma) \restriction_\Pi E_\mathcal{E}$$

whose behaviors are the subset of those of the Multirate PALS distributed system $\mathcal{MA}(\mathcal{E}, T, \Gamma)$ that can be realized by the physical environment(s) $E_{\mathcal{E}}$.

The realizable big-step transitions in $(Stable(\mathcal{MA}(\mathcal{E}, T, \Gamma)), \rightarrow_{st})$ can be defined as in the synchronous case. Consider a big-step transition sequence $\bar{\rho}_{\mathcal{E}} = (s_0, i_0) \rightarrow_{st} (s_1, i_1) \rightarrow_{st} (s_2, i_2) \rightarrow_{st} \cdots$ in $(Stable(\mathcal{MA}(\mathcal{E}, T, \Gamma)), \rightarrow_{st})$. Each stable state s_i in $\mathcal{MA}(\mathcal{E}, T, \Gamma)$ consists of the states of its components and the feedback output received in the input buffers. Similarly, there is a collection of $\{\bar{\rho}_{M_j}\}_{j \in J_S \cup J_F}$ of complete stable local transition sequences in $\mathcal{MA}(\mathcal{E}, T, \Gamma)$, where each sequence $\bar{\rho}_{M_j}$ is a stable transition sequence of M_j in $\mathcal{MA}(\mathcal{E}, T, \Gamma)$ at its own rate. Consider such a complete stable sequence $\bar{\rho}_{M_j} = (s_0^j, i_0^j) \rightarrow (s_1^j, i_1^j) \rightarrow (s_2^j, i_2^j) \rightarrow \cdots$ in $\mathcal{MA}(\mathcal{E}, T, \Gamma)$. Because M_j can sample E_{M_j}'s parameters at time $c_{M_j}(\pi_T(s_i^j))$ for $i \in \mathbb{N}$ and M_j's actuator outputs only depend on machine states, such sequences completely describe the interactions between M_j and E_{M_j} in $\mathcal{MA}(\mathcal{E}, T, \Gamma)$, given initial actuator outputs $\{a_{init}^j\}_{j \in J_F \cup J_S}$. The realizability of $\bar{\rho}_{M_j}$ in $\mathcal{MA}(\mathcal{E}, T, \Gamma)$ can be defined in the exactly same way, and so can the realizability of the big-step transition sequence $\bar{\rho}_{\mathcal{E}}$.

That is, the semantics of $\mathcal{MA}(\mathcal{E}, T, \Gamma) \upharpoonright_\Pi E_{\mathcal{E}}$ is a set of realizable behaviors in the big-step transition system $(Stable(\mathcal{MA}(\mathcal{E}, T, \Gamma)), \rightarrow_{st})$. Each such behavior consists of a realizable "locally-big" transition sequence for each machine, which together satisfy the time-invariant constraints correlating the physical behaviors.

4.4   Relating the Synchronous and Asynchronous Models

The correctness of Hybrid Multirate PALS immediately follows from the fact that a trace $\rho_{\mathcal{E}}$ for the synchronous model $\mathcal{E}$ and $\bar{\rho}_{\mathcal{E}}$ for the asynchronous model $\mathcal{MA}(\mathcal{E}, T, \Gamma)$ only depends on the discrete parts of the systems. Recall that there is a mapping $sync : Stable(\mathcal{MA}(\mathfrak{E}, T, \Gamma)) \rightarrow S^{\mathcal{E}} \times D_i^{\mathcal{E}}$ that associates each stable state with the corresponding synchronous state of $ts(\mathcal{E})$, so that $ts(\mathcal{E})$ and $Stable(\mathcal{MA}(\mathfrak{E}, T, \Gamma))$ are bisimilar to each other by $\sim_{obi}; sync$. Therefore, a big-step transition $\rho_{\mathcal{E}}$ for $\mathcal{E}$ and a big-step stable transition $\bar{\rho}_{\mathcal{E}}$ for $\mathcal{MA}(\mathcal{E}, T, \Gamma)$ are in a one-to-one correspondence, Furthermore, if a machine state s in $\rho_{\mathcal{E}}$ corresponds to a machine state s' in $\bar{\rho}_{\mathcal{E}}$, they must be identical, since $sync$ maps the same machine states. By construction, local time $\pi_T(s)$ at the beginning of its period, actuator output $\pi_C(s)$, and observed environment parameter $\pi_x(s)$ only depends on machine state s. This means that all physical measurements and physical activation happen at the same time in both synchronous and asynchronous models, and thus their continuous behaviors must be identical. Consequently:

Theorem 1. $\mathcal{E} \upharpoonright_\Pi E_{\mathcal{E}}$ and $\mathcal{MA}(\mathcal{E}, T, \Gamma) \upharpoonright_\Pi E_{\mathcal{E}}$ have the same set of realizable transition sequences (for $\sim_{obi}; sync(s)$-equivalent initial states and the same initial actuator output).

Let $\mathcal{L} : S^{\mathcal{E}} \times D_i^{\mathcal{E}} \rightarrow 2^{AP}$ be a state labeling function on $\mathcal{E}$. By the correctness of Multirate PALS, if $\mathcal{L}$ cannot distinguish $\sim_{obi}$-equivalent states, two Kripke structures for $Stable(\mathcal{MA}(\mathcal{E}, T, \Gamma))$ and $ts(\mathcal{E})$ are bisimilar and satisfy the same

CTL^* formulas. By Theorem 1, $\mathcal{E} \upharpoonright_\Pi E_\mathcal{E}$ and $\mathcal{MA}(\mathcal{E}, T, \Gamma) \upharpoonright_\Pi E_\mathcal{E}$ have the same set of realizable transition sequences. Consider an $ACTL^*$ formula φ (recall that $ACTL^*$ is an universal fragment of CTL^* whose counterexample is given by a sequence). If there exists a *realizable* counterexample φ in $\mathcal{E} \upharpoonright_\Pi E_\mathcal{E}$, then there also exists a realizable counterexample in $\mathcal{MA}(\mathcal{E}, T, \Gamma) \upharpoonright_\Pi E_\mathcal{E}$:

Theorem 2. *Given a multirate ensemble $\mathcal{E}$, its physical environment $E_\mathcal{E} = \langle\{E_j\}_{j \in J_S \cup J_F}, (\forall t)\ \psi\rangle$, and a labeling function $\mathcal{L} : S^\mathcal{E} \times \mathcal{D}_i^\mathcal{E} \to 2^{AP}$ that cannot distinguish $\sim_{obi}$-equivalent states. Then the Kripke structures for $\mathcal{E} \upharpoonright_\Pi E_\mathcal{E}$ and $\mathcal{MA}(\mathcal{E}, T, \Gamma) \upharpoonright_\Pi E_\mathcal{E}$ satisfy the same $ACTL^*$ formulas (for $(\sim_{obi}; sync(s))$-related initial states and the same initial actuator output).*

5   Verifying Invariants Using SMT Solving

This section shows how an invariant of a hybrid multirate ensemble can be verified, for *all* possible local clocks, using SMT solving. The idea is to express a synchronous transition in $\mathcal{E} \upharpoonright_\Pi E_\mathcal{E}$ as a logical formula $\Psi(\boldsymbol{x}, \boldsymbol{x}')$, where $\boldsymbol{x}$ denotes the states of its physical environments at the beginning of the round and $\boldsymbol{x}'$ denotes those at the end of the round. Suppose that a safety property of the system is expressed as a formula $\Phi(\boldsymbol{x})$. If we can prove

$$\Phi(\boldsymbol{x}) \wedge \Psi(\boldsymbol{x}, \boldsymbol{x}') \implies \Phi(\boldsymbol{x}')$$

then the safety property $\Phi(\boldsymbol{x})$ holds in each synchronous state if it holds in the initial state. This implication can often be checked automatically using SMT solving by checking the unsatisfiability of the negation $\Phi(\boldsymbol{x}) \wedge \Psi(\boldsymbol{x}, \boldsymbol{x}') \wedge \neg\Phi(\boldsymbol{x}')$.

5.1   Logical Representations

A controlled physical environment E_M can also be expressed as logical constraints in first-order logic formulas over reals that involve differential equations. This representation allows us to analyze the system using SMT solvers.

Definition 8. *If $E_M = (C, \boldsymbol{x}, \Lambda)$ is a controlled physical environment with physical parameters $\boldsymbol{x} = (x_1, \ldots, x_l)$, for unary function symbols $x_1, \ldots, x_l$, then the physical transition relation Λ can be expressed as a first order logic formula of the form $\varphi_{E_M}(y_C; y_t, y_t'; y_{v_1}, \ldots, y_{v_l})$, where:*

- *the variable y_C denotes control commands from M;*
- *the variables y_t and y_t' denote times at the beginning and the end of the trajectory duration, respectively; and*
- *the variables $y_{v_1}, \ldots, y_{v_l}$ denote the initial values of $\boldsymbol{x}$ at time y_t.*

The formula $\varphi_{E_M}(y_{C_1}, \ldots, y_{C_j}; y_t, y_t'; y_{v_1}, \ldots, y_{v_l})$ specifies the trajectories of the physical parameters $\boldsymbol{x} = (x_1, \ldots, x_l)$ from time y_t to time y_t'. That is:

$$\varphi_{E_M}(a; t_0, t_1; \boldsymbol{v}) \iff ((a, t_1 - t_0, \boldsymbol{v}), \boldsymbol{\tau}) \in \Lambda \wedge \forall t \in [t_0, t_1].\ \boldsymbol{x}(t) = \boldsymbol{\tau}(t - t_0).$$

Example 5. Consider the open room environment $E_M = (\{on, off\}, (x, x_o), \Lambda)$ in Example 1. The formula $\varphi_{E_M}(y_{\text{out}}; y_t, y_t'; y_v, y_{v_o})$ is given by:

$$x_o(y_t) = y_{v_o} \;\wedge$$

$$\left[\begin{array}{l} (y_{\text{out}} = on \wedge \forall t \in [y_t, y_t'].\; x(t) = y_v + \displaystyle\int_0^{t-y_t} K(h - ((1-k)x + kx_o))\,\mathrm{d}t) \\[2mm] \vee\; (y_{\text{out}} = off \wedge \forall t \in [y_t, y_t'].\; x(t) = y_v + \displaystyle\int_0^{t-y_t} (-K((1-k)x + kx_o))\,\mathrm{d}t) \end{array}\right]$$

♠

The *behavior* of a number of controlled physical environments can be expressed as logical formulas. For E_M, the behavior up to its N-th period can be expressed by the formula: $\bigwedge_{0 \le i < N} \varphi_{E_M}(a_i; c_M(i), c_M(i+1); \boldsymbol{x}(c_M(i)))$, given controller inputs $a_0, \ldots, a_{k-1}$. Consider two environments E_{M_i}, for $i = 1, 2$, with time-invariant constraint $(\forall t)\, \psi(\boldsymbol{x}_1(t), \boldsymbol{x}_2(t))$. Suppose that M_2 is k times faster than M_1. The combined behavior of E_{M_1} and E_{M_2} is then given by the formula

$$\varphi_{E_{M_1}}(a_1; c_{M_1}(0), c_{M_1}(1); \boldsymbol{x}(c_{M_1}(0)))$$
$$\wedge\; \bigwedge_{0 \le j < k} \varphi_{E_{M_2}}(a_{2_j}; c_{M_2}(j), c_{M_2}(j+1); \boldsymbol{x}(c_{M_2}(j))) \;\wedge\; (\forall t)\, \psi(\boldsymbol{x}_1(t), \boldsymbol{x}_2(t)),$$

given controller inputs a_1 from M_1 and $a_{2_0}, \ldots, a_{2_{k-1}}$ from M_2. Equality constraints, such as $x_1(t) = x_2(t)$, can be removed from the formula by replacing equal with equal (for example, by replacing each function symbol x_1 with x_2).

For a typed machine M, its transition relation δ_M can be expressed as a logic formula of the form $\varphi_M(\boldsymbol{y}_i; \boldsymbol{y}_s; \boldsymbol{y}_s'; \boldsymbol{y}_o)$, with the variables $\boldsymbol{y}_i$ for the inputs, $\boldsymbol{y}_s$ for the current state, $\boldsymbol{y}_s'$ for the next state, and $\boldsymbol{y}_o$ for the outputs, so that $\varphi_M(\boldsymbol{d}_i; s; s'; \boldsymbol{d}_o) \iff ((\boldsymbol{d}_i, s), (s', \boldsymbol{d}_o)) \in \delta_M$. A formula for an environment restriction or a connection between input and output ports can be expressed by adding appropriate equality conditions between the corresponding variables.

Example 6. For the digital thermostat controller in Example 3, the logic formula $\varphi_M(y_{t_M}, y_{t_m}; y_m, y_n, y_v; y_m', y_n', y_v'; *)$ can be defined by:

$$y_n' = y_n + 1 \;\wedge\; \left(y_m' = \begin{cases} \textbf{if } y_v \le t_M \textbf{ then } m_{\text{on}} \textbf{ else } m_{\text{off}} \textbf{ fi} & \textbf{if } y_m = m_{\text{on}} \\ \textbf{if } y_v \ge t_m \textbf{ then } m_{\text{off}} \textbf{ else } m_{\text{on}} \textbf{ fi} & \textbf{if } y_m = m_{\text{off}} \end{cases}\right).$$

The formula $\varphi_{M \restriction_\pi E_M}(y_{t_M}, y_{t_m}; y_m, y_n, y_v, y_{v_o}, y_a; y_t, y_t'; y_m', y_n', y_v', y_{v_o}', y_a')$ for the environment restriction $M \restriction_\pi E_M$ is then given by:

$$\exists y_v'', y_{v_o}''.\; \varphi_M(y_{t_M}, y_{t_m}; y_m, y_n, y_v; y_m', y_n', y_v'; *) \;\wedge$$
$$y_t = c_M(\pi_T(y_m, y_n, y_v)) \;\wedge\; y_t' = c_M(\pi_T(y_m', y_n', y_v')) \;\wedge$$
$$\varphi_{E_M}(y_a; y_t, y_t + \alpha_{\text{max}M}; y_v, y_{v_o}) \;\wedge$$
$$y_v = x(y_t) \;\wedge\; y_{v_o} = x_o(y_t) \;\wedge$$
$$\varphi_{E_M}(\pi_C(y_m, y_n, y_v); y_t + \alpha_{\text{max}M}, y_t'; y_v'', y_{v_o}'') \;\wedge$$
$$y_v'' = x(y_t + \alpha_{\text{max}M}) \;\wedge\; y_{v_o}'' = x_o(y_t + \alpha_{\text{max}M}) \;\wedge$$
$$y_v' = x(y_t') \;\wedge\; y_{v_o}' = x_o(y_t'),$$

with the interface projection functions $\pi_C(y_m, y_n, y_v) = y_m$, $\pi_T(y_m, y_n, y_v) = n$, $\pi_x(y_m, y_n, y_v) = y_v$, and $\pi_S(y_v, v_o) = y_v$. ♠

5.2   Local Clocks and SMT Solving

It is hard or impossible to predict the concrete values of a local clock c_M, since the values of local clocks are determined on-the-fly by clock synchronization mechanisms [2,11]. We therefore represent c_M by logical formulas. Assuming PALS bounds Γ, a periodic local clock c_M satisfies $\exists t \in (\tau - \epsilon, \tau + \epsilon).\ c_M(n) = t$ for any time $\tau > \epsilon$ when a global round begins. We only need the values of c_M up to its rate k for one synchronous step. Hence, we add the formula $\exists t_M^n \in (\tau + n(T/k) - \epsilon, \tau + n(T/k) + \epsilon).\ c_M(n) = t_M^n$ with a fresh variable t_M^n for $n = 1, \dots, k$. We can define a logical formula $\Psi(x, x')$ that expresses a synchronous transition of $\mathcal{E}$ during time $[\tau - \epsilon, \tau + T - \epsilon)$ using the above constructs (e.g., φ_M and φ_{E_M}), where t_M^n is written in $\Psi(x, x')$ instead of c_M.

Example 7. For the multirate ensemble $\mathcal{E}$ for two adjacent rooms in Example 2, the formula $\Psi(y_{v_1}, y_{v_2}; y'_{v_1}, y'_{v_2})$ for duration $[\tau - \epsilon, \tau + T - \epsilon)$ is the conjunction of the following formulas:

1. symbolic local clocks of M_1 and M_2 for one synchronous step:

$$\tau - \epsilon < t_{M_1} < \tau + \epsilon \ \wedge$$
$$\tau - \epsilon < t_{M_2}^1 < \tau + \epsilon \ \wedge \qquad \tau + T/2 - \epsilon < t_{M_2}^2 < \tau + T/2 + \epsilon$$

2. the behavior of the physical environments E_{M_1} and E_{M_2} before their rounds begin (at t_{M_1} and t_{M_2}, respectively):

$$\varphi_{E_{M_1}}(y_{m_1}; \tau - \epsilon, t_{M_1}; y_{v_1}, y_{v_{o_1}}) \ \wedge \ \varphi_{E_{M_2}}(y_{m_2}; \tau - \epsilon, t_{M_2}^1; y_{v_2}, y_{v_{o_2}})$$

3. the behavior of the environment restrictions (see Example 6), where the ensemble connections are reflected in the input variable names:

$$\varphi_{M_1 \upharpoonright_{\pi_1} E_{M_1}}(y_{t_M}, y_{t_m}; \ y_{m_1}, y_{n_1}, y_{v_1}, y_{v_{o_1}}; \ t_{M_1}, \tau + T - \epsilon; y'_{m_1}, y'_{n_1}, y'_{v_1}, y'_{v_{o_1}}) \wedge$$
$$\varphi_{M_2 \upharpoonright_{\pi_2} E_{M_2}}(y_{t_M}, y_{t_m}; \ y_{m_2}, y_{n_2}, y_{v_2}, y_{v_{o_2}}; \ t_{M_2}^1, t_{M_2}^2; \qquad y''_{m_2}, y''_{n_2}, y''_{v_2}, y''_{v_{o_2}}) \wedge$$
$$\varphi_{M_2 \upharpoonright_{\pi_2} E_{M_2}}(y_{t_M}, y_{t_m}; \ y''_{m_2}, y''_{n_2}, y''_{v_2}, y''_{v_{o_2}}; \ t_{M_2}^2, \tau + T - \epsilon; \ y'_{m_2}, y'_{n_2}, y'_{v_2}, y'_{v_{o_2}})$$

4. the time-invariant constraint $(\forall t)\ x_{o_1}(t) = x_2(t) \ \wedge \ x_{o_2}(t) = x_1(t)$, which can be removed from Ψ by replacing each x_{o_1} by x_2 and each x_{o_2} by x_1.

That is, the whole formula $\Psi(y_{v_1}, y_{v_2}; y'_{v_1}, y'_{v_2})$ has the form $1 \wedge 2 \wedge 3 \wedge 4$.

We have verified the two-room thermostat system in Example 4 using the dReal SMT solver [9]. The dReal solver can check the satisfiability of a formula Ψ over the real numbers—involving non-linear real functions, such as polynomials, trigonometric functions, and solutions of Lipschitz-continuous ordinary differential equations (ODEs)—up to a given precision $\delta > 0$ If dReal returns

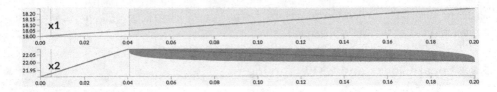

Fig. 8. The counterexample trajectories of x_1 and x_2, where $\tau = 0.02$.

false, then Ψ is unsatisfiable. If dReal returns true, then Ψ's *syntactic numerical perturbation*[3] by δ is satisfiable, although Ψ *could* still be unsatisfiable.

In the analysis, we let $\alpha_{\max M_i} = 0$ and use the values $h_1 = 100$, $h_2 = 200$, $K_1 = 0.015$, $K_2 = 0.025$, $k = 0.01$, and $T = 0.1$ for the constants in the differential equations. The two outputs from Main are always $t_M = 21$ and $t_m = 19$; that is, the desired temperature is between 19 and 21° in both rooms. This is obviously not achievable, so we analyze the invariant

$$\Phi(x_1, x_2) \equiv (18 < x_1 < 22) \wedge (18 < x_2 < 22).$$

If the maximal local clock skew $\epsilon = 0.005$, then $\Phi(y_{v_1}, y_{v_2}) \wedge \Psi(y_{v_1}, y_{v_2}; y'_{v_1}, y'_{v_2})$ $\wedge \neg \Phi(y'_{v_1}, y'_{v_2})$ is unsatisfiable,[4] which means that $\Phi(y_{v_1}, y_{v_2}) \wedge \Psi(y_{v_1}, y_{v_2}; y'_{v_1}, y'_{v_2})$ $\implies \Phi(y'_{v_1}, y'_{v_2})$ holds, and that Φ therefore is an invariant from any initial state satisfying Φ. However, if $\epsilon = 0.02$, then the inductive condition is violated by the trajectories in Fig. 8, which were generated by dReal, using precision $\delta = 0.001$.[5] Figure 8 shows that the temperature of x_2 rises for extra $0.02 + \delta$ time units (where the round begins at time $\tau = 0.02$), so that the value of x_2 at time 0.2 (i.e., $\tau + 0.2 - \epsilon$) can be greater than 22.   ♠

6   Concluding Remarks

We have extended the complexity-reducing Multirate PALS design and verification methodology to the class of virtually synchronous distributed multirate *hybrid* systems, which includes avionics, automotive, and power plant control systems. In such systems, the local continuous environments may be interrelated and may change trajectories due to actuator commands. We have formalized Hybrid Multirate PALS by defining its synchronous and asynchronous models, and have given a trace equivalence result relating a synchronous model and its asynchronous counterpart; these therefore satisfy the same ACTL* formulas.

A main difference between Multirate PALS and Hybrid Multirate PALS is that time elapse does *not* affect the behavior of the system in Multirate PALS, as long as events take place inside certain time intervals. This is no longer the

---

[3] E.g., if $\psi \equiv x > 3 \wedge y = z$, then for $\delta = 0.1$, its syntactic numerical perturbation by δ is $x - 3 > -0.1 \wedge y - z \geq -0.1 \wedge z - y \geq -0.1$.

[4] The analysis took 28 min on Intel Xeon 2.0 GHz with 64 GB memory.

[5] The analysis took 10 h 27 min on the same machine.

case when the environment has continuous behaviors. To have an equivalence between a synchronous and an asynchronous model, both must read environment values, and give actuator commands, at the same time. This means that also the *synchronous* model must take local clocks and execution times into account. Hybrid Multirate PALS should nevertheless significantly reduce the complexity of designing and verifying a virtually synchronous CPS: it allows us to abstract from asynchronous communication (and the resulting interleavings), message buffering, network delays, backoff timers, and so on.

Since the synchronous models now include both clock skews and differential equations, explicit-state model checking is no longer practical. We therefore show how system invariants can be symbolically verified using the dReal SMT solver.

Much work remains, including: investigating the effectiveness of Hybrid Multirate PALS on larger case studies, including the airplane turning system that was treated in an *ad hoc* way in [3]; developing verification techniques for properties beyond invariants; investigating whether a synchronous model and its asynchronous counterpart are bisimilar and hence satisfy the same CTL* properties; and making the Hybrid Multirate PALS methodology available to domain-specific modeling by extending industry-standard modeling languages, such as AADL [7], to specify Hybrid Multirate PALS (synchronous) models, and by integrating automatic verification into their modeling environments.

Acknowledgments. José Meseguer has been a mentor and role model for both of us. We would like to thank you, José, for showing us how beautiful, and at the same time practical, theoretical computer science can be. We thank you for your exemplary guidance and your support throughout our research careers. But first and foremost, we thank you for many years of true friendship.

References

1. Abrial, J.R., Börger, E., Langmaack, H. (eds.): Formal Methods for Industrial Applications: Specifying and Programming the Steam Boiler Control. LNCS, vol. 1165. Springer, Heidelberg (1996)
2. Al-Nayeem, A., Sun, M., Qiu, X., Sha, L., Miller, S.P., Cofer, D.D.: A formal architecture pattern for real-time distributed systems. In: RTSS. IEEE (2009)
3. Bae, K., Krisiloff, J., Meseguer, J., Ölveczky, P.C.: Designing and verifying distributed cyber-physical systems using Multirate PALS: an airplane turning control system case study. Sci. Comput. Program. **103**, 13–50 (2015)
4. Bae, K., Meseguer, J., Ölveczky, P.C.: Formal patterns for multirate distributed real-time systems. Sci. Comput. Program. **91**, 3–44 (2014)
5. Bae, K., Ölveczky, P.C., Meseguer, J., Al-Nayeem, A.: The SynchAADL2Maude tool. In: de Lara, J., Zisman, A. (eds.) FASE 2012. LNCS, vol. 7212, pp. 59–62. Springer, Heidelberg (2012)
6. Bae, K., Ölveczky, P.C., Meseguer, J.: Definition, semantics, and analysis of Multirate Synchronous AADL. In: Jones, C., Pihlajasaari, P., Sun, J. (eds.) FM 2014. LNCS, vol. 8442, pp. 94–109. Springer, Heidelberg (2014)
7. Feiler, P.H., Gluch, D.P.: Model-Based Engineering with AADL. Addison-Wesley, Boston (2012)

8. Gao, S., Avigad, J., Clarke, E.M.: δ-complete decision procedures for satisfiability over the reals. In: Gramlich, B., Miller, D., Sattler, U. (eds.) IJCAR 2012. LNCS, vol. 7364, pp. 286–300. Springer, Heidelberg (2012)
9. Gao, S., Kong, S., Clarke, E.M.: dReal: an SMT solver for nonlinear theories over the reals. In: Bonacina, M.P. (ed.) CADE 2013. LNCS, vol. 7898, pp. 208–214. Springer, Heidelberg (2013)
10. Lynch, N., Segala, R., Vaandrager, F.: Hybrid I/O automata. Inf. Comput. **185**(1), 105–157 (2003)
11. Lynch, N.A.: Distributed Algorithms. Morgan Kaufmann, San Francisco (1996)
12. Meseguer, J., Ölveczky, P.C.: Formalization and correctness of the PALS architectural pattern for distributed real-time systems. Theoret. Comput. Sci. **451**, 1–37 (2012)
13. Meseguer, J.: Taming distributed system complexity through formal patterns. Sci. Comput. Program. **83**, 3–34 (2014)

Debits and Credits in Petri Nets and Linear Logic

Massimo Bartoletti[1]([✉]), Pierpaolo Degano[2], Paolo Di Giamberardino[1], and Roberto Zunino[3]

[1] Dipartimento di Matematica e Informatica,
Università degli Studi di Cagliari, Cagliari, Italy
bart@unica.it
[2] Dipartimento di Informatica, Università di Pisa, Pisa, Italy
[3] Dipartimento di Matematica, Università degli Studi di Trento,
Trento, Italy

Abstract. Exchanging resources often involves situations where a participant gives a resource without obtaining immediately the expected reward. For instance, one can buy an item without paying it in advance, but contracting a debt which must be eventually honoured. Resources, credits and debits can be represented, either implicitly or explicitly, in several formal models, among which Petri nets and linear logic. In this paper we study the relations between two of these models, namely intuitionistic linear logic with mix and Debit Petri nets. In particular, we establish a natural correspondence between provability in the logic, and marking reachability in nets.

Gracias, Pepe. Ha sido un privilegio para mi poder observar cómo tu agudo razonamiento consiga caracterizar los aspectos relevantes de un problema con precisión y exactitud, desvelando su funcionalidad y rol — Pierpaolo

1 Introduction

The exchange of resources (both physical and virtual) is a natural aspect of many kinds of interactions, e.g., those involving participants distributed over an open network. To reason about these interactions, it is beneficial to model them using a formal system, e.g., a logic, a process calculus, or a Petri net. Typically, parties exchange resources in a circular way: one provides the other with a resource, and waits for something in return. If not dealt with properly, circularity can lead to a deadlock: this is especially the case when parties are mutually distrusting, and so no one is willing to do the first move. This is a classical issue, discussed by philosophers at least since Hobbes' Leviathan [1], and more recently dealt with by several works in the area of concurrency theory, e.g. [2–7].

As an example, consider the following scenario: Alice wants a birthday cake (*cb*), but she only has the ingredients to make an apple cake (*ia*); Bob wants an apple cake (*ca*), but he only has the ingredients to make a birthday cake (*ib*). They

© Springer International Publishing Switzerland 2015
N. Martí-Oliet et al. (Eds.): Meseguer Festschrift, LNCS 9200, pp. 135–159, 2015.
DOI: 10.1007/978-3-319-23165-5_6

make a deal: each one will cook for the other, and then they will exchange cakes (and eat them). We want to model this situation using logic: in particular, since we deal with resources to be consumed and produced, we will make use of linear logic (LL) [8]. Each resource will correspond to an atomic formula; a linear implication $A \multimap B$ represents a process which consumes the resource A to produce the resource B, while a tensor product $A \otimes B$ stands for the conjunction of the resources A, B. The neutral element 1 of $\otimes$ denotes the absence of resources.

A first attempt to model the deal between Alice and Bob could be the following, where we represent Alice's and Bob's proposals as multisets of linear logic formulas:

- Alice's proposal: $\Gamma_{Alice} = ia, (ia \otimes cb) \multimap ca$
- Bob's proposal: $\Gamma_{Bob} = ib, (ib \otimes ca) \multimap cb$

After a correct interaction between Alice and Bob, we expect that all resources have been consumed: in the logical model, this corresponds to deducing $\Gamma_{Alice}, \Gamma_{Bob} \vdash 1$ (where $\vdash$ is the entailment relation of the logic). However, this sequent is not provable in linear logic. The reason why the entailment fails is that both Alice and Bob wait for the other to deliver something before starting to cook, but since no one starts, no cake can be made.

Cancellative linear logic [9] is a logic where Alice and Bob reach an agreement on their respective proposals, as the sequent $\Gamma_{Alice}, \Gamma_{Bob} \vdash 1$ is provable. An explanation of this fact is that the previous proposals are equivalent, in cancellative linear logic, to those below:

- Alice's proposal: $\Gamma'_{Alice} = ia, ia \multimap (ca \otimes cb^{\perp})$
- Bob's proposal: $\Gamma'_{Bob} = ib, ib \multimap (cb \otimes ca^{\perp})$

From the second proposals, it is evident that $\Gamma'_{Alice}, \Gamma'_{Bob} \vdash ca \otimes cb \otimes ca^{\perp} \otimes cb^{\perp}$ (this latter fact also holds in LL). Differently from LL, cancellative linear logic proves the *annihilation* principle $A \otimes A^{\perp} \vdash 1$, allowing dual formulas to cancel out. A consequence of this fact is that $\Gamma'_{Alice}, \Gamma'_{Bob} \vdash 1$, as wanted. Intuitively, negative atoms, like $ca^{\perp}$, act as *debits*, while positive ones, like ca, act as *credits*.

However, the first and the second proposals have a slightly different flavour. In the first, Alice is using cb in the left-hand side of the implication, and this could be interpreted as requiring Bob's cake *in advance*, in order to produce her apple cake. Instead, in the second proposal Alice can start producing her cake *without* Bob's one, but she records the debit $cb^{\perp}$ in the right-hand side of the implication.

The reason why cancellative linear logic does not capture the difference between the two proposals is that it proves the *inverse annihilation* principle $1 \vdash A \otimes A^{\perp}$, making the two proposals equivalent. This principle models the fact that it is always possible to generate *from scratch* a pair debit, credit. However, we find this principle not always realistic: in our scenario, it would allow Alice to prepare her cake even in the absence of the needed ingredients (which would be recorded as debits, though).

We therefore look for a refinement of the logic, wherein debits can be generated in a controlled manner: formally, we want annihilation, but not its inverse. To this purpose, we consider a linear logic which comprises a rule, usually called [MIX], which makes it possible to prove annihilation [10]. In particular, we focus on intuitionistic linear logic with [MIX] (ILL^{mix}) [11].

In this logic, the second proposal of Alice and Bob leads to an agreement, while the first one does not: formally, $\Gamma'_{Alice}, \Gamma'_{Bob} \vdash 1$, but $\Gamma_{Alice}, \Gamma_{Bob} \nvdash 1$. This is because, in the second proposal, the deadlock situation is avoided by allowing Alice to give an apple cake to Bob, provided that contextually Bob is charged with a *debit* to give her a birthday cake.

Contribution. In this paper we focus on the Horn fragment of ILL^{mix}, which only admits negation, tensor products, linear implications, and exponentials. We provide this fragment with a big-step operational semantics, as well as a small-step one, and we prove that Horn ILL^{mix} is sound and complete with respect to these semantics (Theorems 2 and 3).

Leveraging on these semantics, we proceed to prove our main result, i.e., that provability in Horn ILL^{mix} is equivalent to the reachability problem for Debit nets [12] with delayed annihilation (Theorem 4), which is known to be decidable.

2   Intuitionistic Linear Logic with Mix

In this section we recall from [11] the syntax, the sequent calculus, and some facts about ILL^{mix}.

Definition 1 (Syntax of ILL^{mix}). *Assume a denumerable set* $\mathbf{A}$ *of atoms, ranged over by* $a, b, \ldots$. *The formulas* $A, B, \ldots$ *of ILL^{mix} are defined as follows:*

$$A ::= a \mid A^{\perp} \mid A \otimes A \mid A \multimap A \mid A \& A \mid A \oplus A \mid {!}A \mid 1 \mid 0 \mid \top \mid \perp$$

The sequent calculus of ILL^{mix} is depicted in Fig. 1; the symbol γ in the expression $\Gamma \vdash \gamma$ may stand either for empty or for a formula A. We say that A is provable whenever $\vdash A$ can be deduced with the rules in Fig. 1.

We observe that $A^{\perp}$ can be defined as $A \multimap \perp$; in this case the rules [NEGR] and [NEGL] become derivable from the other rules of ILL^{mix}. However, we prefer to regard $A^{\perp}$ as a primitive, rather than syntactic sugar.

Theorem 1 (Cut elimination [11]). *Every provable sequent of ILL^{mix} admits a proof without the* [CUT] *rule.*

Example 1. As an example of an ILL^{mix} proof, we provide the proof of the annihilation principle $A \otimes A^{\perp} \vdash 1$:

$$
\cfrac{
  \cfrac{
    \cfrac{
      \cfrac{}{A \vdash A}\ {\scriptstyle[\text{AX}]}
    }{A, A^{\perp} \vdash}\ {\scriptstyle[\text{NEGL}]}
  }{A \otimes A^{\perp} \vdash}\ {\scriptstyle[\otimes\,\text{L}]}
  \qquad
  \cfrac{}{\vdash 1}\ {\scriptstyle[\text{1R}]}
}{A \otimes A^{\perp} \vdash 1}\ {\scriptstyle[\text{MIX}]}
$$

$$\frac{}{A \vdash A}\ \text{[\textsc{ax}]} \qquad \frac{\Gamma \vdash A \qquad \Gamma', A \vdash \gamma}{\Gamma, \Gamma' \vdash \gamma}\ \text{[\textsc{cut}]} \qquad \frac{\Gamma \vdash \qquad \Gamma' \vdash \gamma}{\Gamma, \Gamma' \vdash \gamma}\ \text{[\textsc{mix}]}$$

$$\frac{\Gamma, A \vdash}{\Gamma \vdash A^{\perp}}\ \text{[\textsc{negr}]} \qquad \frac{\Gamma \vdash A}{\Gamma, A^{\perp} \vdash}\ \text{[\textsc{negl}]}$$

$$\frac{\Gamma \vdash}{\Gamma \vdash \perp}\ \text{[$\perp$R]} \qquad \frac{}{\perp \vdash}\ \text{[$\perp$L]} \qquad \frac{}{\vdash 1}\ \text{[1R]} \qquad \frac{\Gamma \vdash \gamma}{\Gamma, 1 \vdash \gamma}\ \text{[1L]} \qquad \frac{}{\Gamma \vdash \top}\ \text{[$\top$]} \qquad \frac{}{\Gamma, 0 \vdash A}\ \text{[0L]}$$

$$\frac{\Gamma, A \vdash \gamma \qquad \Gamma, B \vdash \gamma}{\Gamma, A \oplus B \vdash \gamma}\ \text{[$\oplus$L]} \qquad \frac{\Gamma \vdash A}{\Gamma \vdash A \oplus B}\ \text{[$\oplus$R1]} \qquad \frac{\Gamma \vdash B}{\Gamma \vdash A \oplus B}\ \text{[$\oplus$R2]}$$

$$\frac{\Gamma \vdash A \qquad \Gamma \vdash B}{\Gamma \vdash A \& B}\ \text{[\&R]} \qquad \frac{\Gamma, A \vdash \gamma}{\Gamma, A \& B \vdash \gamma}\ \text{[\&L1]} \qquad \frac{\Gamma, B \vdash \gamma}{\Gamma, A \& B \vdash \gamma}\ \text{[\&L2]}$$

$$\frac{\Gamma, A, B \vdash \gamma}{\Gamma, A \otimes B \vdash \gamma}\ \text{[$\otimes$L]} \qquad \frac{\Gamma \vdash A \qquad \Gamma' \vdash B}{\Gamma, \Gamma' \vdash A \otimes B}\ \text{[$\otimes$R]}$$

$$\frac{\Gamma \vdash A \qquad \Gamma', B \vdash \gamma}{\Gamma, \Gamma', A \multimap B \vdash \gamma}\ \text{[$\multimap$L]} \qquad \frac{\Gamma, A \vdash B}{\Gamma \vdash A \multimap B}\ \text{[$\multimap$R]}$$

$$\frac{\Gamma, A \vdash \gamma}{\Gamma, !A \vdash \gamma}\ \text{[!L]} \qquad \frac{!\Gamma \vdash A}{!\Gamma \vdash !A}\ \text{[!R]} \qquad \frac{\Gamma \vdash \gamma}{\Gamma, !A \vdash \gamma}\ \text{[\textsc{weakl}]} \qquad \frac{\Gamma, !A, !A \vdash \gamma}{\Gamma, !A \vdash \gamma}\ \text{[\textsc{col}]}$$

Fig. 1. Sequent calculus of ILL^{mix}.

We remark that introducing the rule [\textsc{mix}] is equivalent to adding the principle $\perp \vdash 1$. In fact, if $\perp \vdash 1$ is assumed, the rule [\textsc{mix}] becomes derivable:

$$\frac{\dfrac{\dfrac{\Gamma \vdash \gamma}{\Gamma, 1 \vdash \gamma}\ \text{[1L]} \qquad \dfrac{\Delta \vdash}{\Delta \vdash \perp}\ \text{[$\perp$ R]}}{\Gamma, \Delta, \perp \multimap 1 \vdash \gamma}\ \text{[$\multimap$ L]} \qquad \dfrac{\perp \vdash 1}{\vdash \perp \multimap 1}\ \text{[$\multimap$ R]}}{\Gamma, \Delta \vdash \gamma}\ \text{[\textsc{cut}]}$$

Inversely, this principle is derivable using the rule [\textsc{mix}], as follows:

$$\frac{\dfrac{}{\perp \vdash}\ \text{[$\perp$ L]} \qquad \dfrac{}{\vdash 1}\ \text{[1R]}}{\perp \vdash 1}\ \text{[\textsc{mix}]}$$

In *cancellative linear logic* [9], instead, $1 \dashv\vdash \perp$ and $A \multimap B \dashv\vdash A^{\perp} \otimes B$. As a consequence both the principles $1 \vdash A \otimes A^{\perp}$ and $A \otimes A^{\perp} \vdash 1$ are valid.

2.1 Simple Products and Multisets

We denote by $\mathbf{A}^{\perp}$ the set $\{a^{\perp} \mid a \in \mathbf{A}\}$. We call *literals* the elements of $\mathbf{L} = \mathbf{A} \cup \mathbf{A}^{\perp}$. A *simple product* (resp. *positive simple product*) is a tensor product

$a_1 \otimes \ldots \otimes a_n$ where $a_i \in \mathbf{L} \cup \{1\}$ (resp. $a_i \in \mathbf{A} \cup \{1\}$) for all i. We will use X, Y, W, Z as metavariables for simple products.

A *multiset* over $\mathbf{L}$ is a function from $\mathbf{L}$ to the set $\mathbb{N}$ of natural numbers. The union $\uplus$ of multisets is defined as expected; by $\varnothing$ we denote the empty multiset (that is, the constant function 0). The *support* $set(\Sigma)$ of a multiset Σ is the set $\{a \mid \Sigma(a) > 0\}$. For each simple product X, we define the multiset $mset_X$ as:

– if $X = 1$ then $mset_1$ is the constant function 0;
– if $X = x \in \mathbf{L}$, then $mset_x$ is the function: $mset_x(y) = \begin{cases} 1 & \text{if } y = x \\ 0 & \text{otherwise} \end{cases}$
– if $X = X_1 \otimes X_2$, then $mset_{X_1 \otimes X_2} = mset_{X_1} \uplus mset_{X_2}$.

Given a multiset of simple products $\Omega = \{W_1, \ldots, W_n\}$, we denote by $\bigotimes \Omega$ the simple product $W_1 \otimes \ldots \otimes W_n$. Hereafter we will often exploit implicitly this correspondence, and we will use the same metavariables to denote both multisets of literals and simple products: the difference will be clear from the context.

2.2 Horn ILL^{mix} Sequents

Horn implications, ranged over by H, H', are formulas of the form $X \multimap Y$, where X is positive. *Horn sequents* are sequents of the form:

$$\Omega, \Gamma, !\Delta \vdash Z$$

where Ω is a multiset of simple products, and Γ and Δ are multisets of Horn implications; when Z is positive, we say that the Horn sequent is *honoured*. A Horn theory is a pair $(\Gamma, !\Delta)$.

Example 2. Figure 2 shows a proof in ILL^{mix} of the Horn sequent corresponding to the second proposal of Sect. 1:

$$ia, \; ib, \; ia \multimap (ca \otimes cb^{\perp}), \; ib \multimap (cb \otimes ca^{\perp}) \; \vdash \; 1$$

The Horn sequent corresponding to the first proposal, instead, is not provable.

3 Debit Nets

In this section we present a minor variant of the debit nets in [12], in order to obtain a correspondence with Horn ILL^{mix}.

We assume the reader familiar with Petri nets, and only recall here some basic notions [13]. A Petri net is a tuple (S, T, F), where S is a set of *places*, T is a set of *transitions* (with the constraint that $S \cap T = \emptyset$), and $F \colon (S \times T) \cup (T \times S) \to \mathbb{N}$ is a *weight function*. The state of a net is given by a *marking*, that is a function $m : S \to \mathbb{N}$ assigning to each place a certain number of tokens. The behaviour of a Petri net is described by a transition relation between *markings*: if $m(s)$ contains at least $F(s, t)$ tokens for all s, then the transition t can fire, decreasing $m(s)$ by $F(s, t)$ tokens and increasing it by $F(t, s)$.

$$
\cfrac{
  \cfrac{
    \cfrac{
      \cfrac{
        \cfrac{
          \cfrac{
            \cfrac{\dfrac{\overline{ca \vdash ca}\ [\text{AX}]}{ca, ca^{\perp} \vdash}\ [\text{NEGL}] \qquad \dfrac{\overline{cb \vdash cb}\ [\text{AX}]}{cb, cb^{\perp} \vdash}\ [\text{NEGL}]}{ca, cb^{\perp}, cb, ca^{\perp} \vdash}\ [\text{MIX}]
          }{ca, cb^{\perp}, cb \otimes ca^{\perp} \vdash}\ [\otimes\text{L}]
        }{ca \otimes cb^{\perp}, cb \otimes ca^{\perp} \vdash}\ [\otimes\text{L}] \qquad \overline{\vdash 1}\ [\text{1R}]
      }{ca \otimes cb^{\perp}, cb \otimes ca^{\perp} \vdash 1}\ [\text{MIX}] \qquad \overline{ib \vdash ib}\ [\text{AX}]
    }{ib, ca \otimes cb^{\perp}, ib \multimap (cb \otimes ca^{\perp}) \vdash 1}\ [\multimap\text{L}] \qquad \overline{ia \vdash ia}\ [\text{AX}]
  }{ia, ib, ia \multimap (ca \otimes cb^{\perp}), ib \multimap (cb \otimes ca^{\perp}) \vdash 1}\ [\multimap\text{L}]
}{}
$$

Fig. 2. Proof of the sequent corresponding to the second proposal of Sect. 1.

Debit nets (DPN) [12] extend Petri nets by allowing places to give tokens "on credit", so that transitions can fire even in the absence of the required number of tokens. Technically, each place s contains a number of tokens $m(s)$ (modelling credits) and of *antitokens* $d(s)$ (modelling debits). In general, token and antitokens can co-exist in a place. Transitions affect m in the standard way: for a transition t to be fired the marking $m(s)$ must contain at least $F(s,t)$ tokens for all s, and after the firing $m(s)$ will be decreased by $F(s,t)$ tokens and increased by $F(t,s)$. Instead, upon the firing of t, the number of antitokens $d(s)$ is increased by $L(s,t)$, where the *lending function* $L : S \times T \to \mathbb{N}$ specifies how many tokens are borrowed at each time. Note the differences between m, F and d, L. First, F and m are used to check whether a transition can be fired, while L and d are not. This renders the fact that a debit can neither prevent nor cause a transition to fire. Second, $F(t,s)$ is defined while $L(t,s)$ is not: this is because the generation of antitokens is already obtained through $L(s,t)$. Hence, debits can only be increased by firing transitions.

At any time tokens and antitokens can cancel out through a special *annihilation* step. More precisely, both $m(s)$ and $d(s)$ can be simultaneously decremented when non-zero (this is the *delayed annihilation* policy of [12]).

Definition 2 (Debit net). *A debit net is a tuple $N = (S, T, F, L)$ where:*

- *(S, T, F) is a Petri net,*
- *$L \colon S \times T \to \mathbb{N}$ is the lending function.*

We now formalise the notion of marking and of *honoured markings*, i.e. those where all debits have been honoured.

Definition 3 (Marking). *A marking of a debit net $N = (S, T, F, L)$ is a pair (m, d) of functions such that*

- *$m \colon S \to \mathbb{N}$ is the token function*
- *$d \colon S \to \mathbb{N}$ is the antitoken function*

A marking (m, d) of N is honoured iff $d(s) = 0$ for all places s of N.

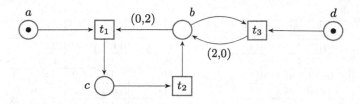

Fig. 3. A Debit Petri net.

Example 3. In Fig. 3 we represent a DPN with places $\{a, b, c, d\}$, having initial marking $\{a \mapsto (1,0), b \mapsto (0,0), c \mapsto (0,0), d \mapsto (1,0)\}$. When representing DPNs, we adopt the following drawing conventions: places are depicted as circles, transitions as squares, and arcs connecting transition to places are decorated with their weights. In case of arcs connecting places to transitions we have a pair of natural numbers, the first representing the weight of the *standard* arcs (possibly 0) and the second the weight of the lending ones. We do not write the weight $(1,0)$ (from places to transitions) or 1 (from transitions to places), and do not draw any arc between a place and a transition if both standard and lending arcs have null weights. Tokens are depicted as filled bullets, while antitokens as empty bullets.

Definition 4 (Step and computation). *Let $N = (S, T, F, L)$ be a DPN, and let (m, d) be a marking of N. We say that:*

- *$t \in T$ is enabled at (m, d) iff $m(s) \geq F(s, t)$ for all s;*
- *annihilation is enabled at (m, d) for s iff $m(s) > 0$ and $d(s) > 0$.*

A step from (m, d) to (m', d') — in symbols $(m, d) \rightarrow_N (m', d')$ — can occur whenever one of the following holds:

- *t is enabled at (m, d), and, for all $s \in S$:*

$$m'(s) = m(s) - F(s, t) + F(t, s) \qquad d'(s) = d(s) + L(s, t)$$

- *annihilation is enabled at (m, d) for $\tilde{s}$, and, for all $s \in S$:*

$$m'(s) = \begin{cases} m(s) - 1 & \text{if } s = \tilde{s} \\ m(s) & \text{otherwise} \end{cases} \qquad d'(s) = \begin{cases} d(s) - 1 & \text{if } s = \tilde{s} \\ d(s) & \text{otherwise} \end{cases}$$

A computation is a finite sequence of steps. As usual, we denote with $\rightarrow_N^*$ the reflexive and transitive closure of $\rightarrow_N$.

Example 4. In the DPN in Fig. 3 there are two possible computations, depending on when annihilation occurs. A computation is represented as a sequence of vectors, the elements of which are pairs $(m(s), d(s))$.

	a	b	c	d
	(1,0)	(0,0)	(0,0)	(1,0)
t_1	(0,0)	(0,2)	(1,0)	(1,0)
t_2	(0,0)	(1,2)	(0,0)	(1,0)
annihil	(0,0)	(0,1)	(0,0)	(1,0)

	a	b	c	d
	(1,0)	(0,0)	(0,0)	(1,0)
t_1	(0,0)	(0,2)	(1,0)	(1,0)
t_2	(0,0)	(1,2)	(0,0)	(1,0)
t_3	(0,0)	(2,2)	(0,0)	(0,0)
annihil	(0,0)	(1,1)	(0,0)	(0,0)
annihil	(0,0)	(0,0)	(0,0)	(0,0)

The computation depicted in the left part of the previous table follows an *instantaneous* annihilation policy: annihilation occurs as soon as possible (the leftmost column records the transition taken to obtain the marking in the row). Indeed, as soon as we find $(1, 2)$ in place b, we annihilate it to $(0, 1)$. Note that this computation leaves a debit in place b, which can not be honoured by any further transitions, since the net is stuck. Instead, under the *delayed annihilation* policy, settling debits is not prioritized. Hence, when b reaches $(2, 1)$ we can either annihilate, obtaining the previous computation, or instead perform transition t_3, obtaning the computation in the right part of the previous table, where every debit is eventually honoured.

4   Debit Nets as a Model of Horn ILL^{mix}

In this section we reduce provability of honoured Horn sequents in ILL^{mix} to reachability in DPNs. As an intermediate step, we will endow Horn ILL^{mix} with two operational semantics: a big-step and a small-step one. The proof proceeds as follows: in Theorem 2 we show that the big-step semantics coincides with Horn ILL^{mix} provability when applied to honoured sequents; in Proposition 3 we show that small-step semantics simulate faithfully computations in DPNs. The equivalence between the two semantics (Proposition 1) then allows to derive our principal result (Theorem 4) as a corollary. A similar proof technique is used by Kanovich in [14], to prove the equivalence between reachability in Petri nets and provability in Horn LL (without mix and negation).

4.1   Big-Step Semantics

The big-step semantics of Horn ILL^{mix} is formalised as a relation $\Downarrow$ between triples $(W, \Gamma, !\Delta)$ and simple products Z. The intuition is that $(W, \Gamma, !\Delta) \Downarrow Z$ holds in the big-step semantics if and only if the sequent $(W, \Gamma, !\Delta) \vdash Z$ is provable in ILL^{mix}, whenever Z is positive. Note that here we are interpreting W as a multiset of literals.

Definition 5 (Big-step semantics of Horn ILL^{mix}). *We inductively define the relation* $(W, \Gamma, !\Delta) \Downarrow Z$ *by the rules in Fig. 4.*

Intuitively, the axiom [$\Downarrow$ **H**] models the consumption of the resource X to produce Y, by using an implication $X \multimap Y$. Rule [$\Downarrow$ **S**] models the settlement

$$\frac{}{(X, \varnothing, \varnothing) \Downarrow X} \; [\Downarrow \mathbf{I}] \qquad \frac{(W, \Gamma_1, !\Delta_1) \Downarrow U \qquad (U, \Gamma_2, !\Delta_2) \Downarrow Z}{(W, \Gamma_1 \uplus \Gamma_2, !(\Delta_1 \uplus \Delta_2)) \Downarrow Z} \; [\Downarrow \text{CUT}]$$

$$\frac{}{(X, X \multimap Y, \varnothing) \Downarrow Y} \; [\Downarrow \mathbf{H}] \qquad \frac{}{(a \otimes a^{\perp}, \varnothing, \varnothing) \Downarrow 1} \; [\Downarrow \mathbf{S}]$$

$$\frac{(X, \Gamma, !\Delta) \Downarrow Z}{(X, \Gamma, !(\Delta \uplus \{H\})) \Downarrow Z} \; [\Downarrow \mathbf{W_!}] \qquad \frac{(X, \Gamma, !\Delta) \Downarrow Y}{(X \otimes V, \Gamma, !\Delta) \Downarrow Y \otimes V} \; [\Downarrow \mathbf{M}]$$

$$\frac{(X, (\Gamma \uplus \{H\}), !\Delta) \Downarrow Z}{(X, \Gamma, !(\Delta \uplus \{H\})) \Downarrow Z} \; [\Downarrow \mathbf{L_!}] \qquad \frac{(X, \Gamma, !(\Delta \uplus \{H, H\})) \Downarrow Z}{(X, \Gamma, !(\Delta \uplus \{H\})) \Downarrow Z} \; [\Downarrow \mathbf{C_!}]$$

Fig. 4. Big-step semantics of Horn ILL^{mix}.

of a debit $a^{\perp}$ with the corresponding credit a. Rules $[\Downarrow \text{CUT}]$ and $[\Downarrow \mathbf{M}]$ deal with composition of computations; rules $[\Downarrow \mathbf{C_!}]$, $[\Downarrow \mathbf{W_!}]$, and $[\Downarrow \mathbf{L_!}]$ are the counterpart of the structural rules of LL, stating that !-ed implications may be re-used at will.

Theorem 2. *Let* $\Omega, \Gamma, !\Delta \vdash Z$ *be an honoured Horn sequent. Then:*

$$\left(\bigotimes \Omega, \Gamma, !\Delta \right) \Downarrow Z \iff \Omega, \Gamma, !\Delta \vdash Z$$

4.2 Small-Step Semantics

We now introduce a small-step semantics of Horn ILL^{mix}, which we will show equivalent to the big-step one in Proposition 1. Together with Theorem 2, we will obtain a correspondence between the small-step semantics and provability in Horn ILL^{mix}.

$$(X \otimes W, \Gamma \uplus \{X \multimap Y\}, !\Delta) \rightsquigarrow (Y \otimes W, \Gamma, !\Delta) \qquad [\rightsquigarrow_H]$$

$$(X \otimes W, \Gamma, !\Delta \uplus \{!(X \multimap Y)\}) \rightsquigarrow (Y \otimes W, \Gamma, !\Delta \uplus \{!(X \multimap Y)\}) \qquad [\rightsquigarrow_{!H}]$$

$$(a \otimes a^{\perp} \otimes W, \Gamma, !\Delta) \rightsquigarrow (1 \otimes W, \Gamma, !\Delta) \qquad [\rightsquigarrow_S]$$

Fig. 5. Small step semantics of Horn ILL^{mix}.

Definition 6 (Small-step semantics of Horn ILL^{mix}). *We define the transition system* $(\Omega, \rightsquigarrow)$ *as follows:*

- Ω *is the set of all triples of the form* $(W, \Gamma, !\Delta)$*, where* W *is a multiset of literals, and* Γ, Δ *are multisets of Horn implications.*
- $\rightsquigarrow$ *is defined by the rules in Fig. 5. As usual, we denote with* $\rightsquigarrow^*$ *the transitive and reflexive closure of the relation* $\rightsquigarrow$*.*

$$T \quad = set(\Gamma) \uplus set(\Delta)$$

$$S \quad = S^{atm} \cup S^{ctrl} \qquad \text{where } S^{atm} = \langle set(\Gamma \uplus \Delta) \rangle \text{ and } S^{ctrl} = set(\Gamma)$$

$$F(s,t) = \begin{cases} mset_X(s) & \text{if } s \in S^{atm} \text{ and } t = X \multimap Y \\ 1 & \text{if } s \in S^{ctrl} \text{ and } t = in_0(s) \\ 0 & \text{otherwise} \end{cases}$$

$$F(t,s) = \begin{cases} mset_Y(s) & \text{if } s \in S^{atm} \text{ and } t = X \multimap Y \\ 0 & \text{otherwise} \end{cases}$$

$$L(s,t) = \begin{cases} mset_Y(s^{\perp}) & \text{if } s \in S^{atm} \text{ and } t = X \multimap Y \\ 0 & \text{otherwise} \end{cases}$$

Fig. 6. Encoding of Horn ILL^{mix} theories into Debit nets.

The multisets W and Γ in the small-step semantics play a role similar to that of markings in DPNs. In particular, W takes into account the tokens and antitokens in places, while Γ is used to bound how many times a transition can be fired. This intuition will be exploited in the following section, to establish a correspondence between the small-step semantics and computations in DPNs.

We now briefly comment on the rules in Fig. 5. Rule $[\leadsto_H]$ applies an implication in Γ, and then discharges one of its occurrences; rule $[\leadsto_{!H}]$ is similar, except that it does not discharge any occurrence; finally, rule $[\leadsto_S]$ annihilates a token with an antitoken.

Proposition 1. $(W, \Gamma, !\Delta) \Downarrow Z \iff (W, \Gamma, !\Delta) \leadsto^* (Z, \varnothing, !\Delta)$

The following theorem directly follows from Theorem 2 and Proposition 1.

Theorem 3. *Let $\Omega, \Gamma, !\Delta \vdash Z$ be an honoured Horn sequent. Then*

$$\left(\bigotimes \Omega, \Gamma, !\Delta \right) \leadsto^* (Z, \varnothing, !\Delta) \iff \Omega, \Gamma, !\Delta \vdash Z$$

4.3   Encoding Horn ILL^{mix} into Debit Nets

We now provide an encoding of Horn theories into Debit nets. We start by defining a function $\langle A \rangle$ associating a Horn formula A with the set of atoms occurring in it (this function is extended to sets of formulas as usual):

$$\langle 1 \rangle = \emptyset \qquad \langle a^{\perp} \rangle = \langle a \rangle = \{a\} \qquad \langle A \otimes B \rangle = \langle A \multimap B \rangle = \langle A \rangle \cup \langle B \rangle$$

Given two multisets of Horn implications Γ, Δ, the Horn theory $(\Gamma, !\Delta)$ can be encoded as a Debit net as follows.

Definition 7. *For a pair $(\Gamma, !\Delta)$, we define the DPN $\mathcal{N}(\Gamma, !\Delta)$ in Fig. 6.*

We now comment on Definition 7. For every Horn implication in Γ and Δ we generate a transition in T. We keep the transitions coming from Γ and Δ separate, using disjoint union $\uplus$ (with left and right injections denoted by in_0 and in_1, respectively). We use the *set* function to ignore multiplicity: e.g., if the multiset Γ contains two equal implications, only a single transition is generated. Then, for any atom occurring in Δ or Γ (disregarding multiplicity), we generate an *atom place* in the net (in S^{atm}). Tokens and antitokens in this place represent the credits and debits for that atom, respectively. Further, we generate a *control place* (in S^{ctrl}) for each implication in Γ. During a computation, the number of tokens in this place corresponds to the multiplicity of an implication in Γ. Since implications in Δ are under a !, their multiplicity is immaterial, hence we do not generate control places for Δ.

The function $F(s,t)$ specifies how many tokens from s are consumed by firing a transition t. Assume t corresponds to an implication $X \multimap Y$ (either in Γ or in Δ). Then, firing t consumes tokens in two different ways. First, for each occurrence of a literal a in X, it consumes a token ($mset_X(a)$ tokens removed from place a). Second, it consumes a single token from the control place associated to the implication $X \multimap Y$ in Γ. Technically, t is the left injection in_0 of the control place $s = X \multimap Y$; with a little abuse of notation, when writing $t = X \multimap Y$ we mean that $t = in_i(X \multimap Y)$ for $i = 0$ or $i = 1$. The function $F(t,s)$ specifies how many tokens are produced in place s by firing t. When firing a transition for $X \multimap Y$ (either in Γ or in Δ), we generate a token for each occurrence of a positive literal a of Y ($mset_Y(a)$ tokens added to place a). Finally, the function $L(s,t)$ specifies how many *antitokens* are produced: the transition for $X \multimap Y$ generates an antitoken for each occurrence of a *negative* literal $a^\perp$ in Y ($mset_Y(a^\perp)$ antitokens added to place a).

Given a DPN $N = \mathcal{N}(\Gamma_0, !\Delta_0)$, we say that a pair (W, Γ) of a simple product W and a multiset of Horn implications Γ is *compatible* with N iff $\langle W \rangle \subseteq S^{atm}$ (the set of atom places of N), and $set(\Gamma) \subseteq set(\Gamma_0)$. Every pair (W, Γ) compatible with N can be represented with the marking $[W, \Gamma]$ of N, defined below. Roughly, the marking counts the multiplicity of each positive and negative literal in W, as well as the multiplicity of the implications in Γ.

Definition 8. *Let* $N = \mathcal{N}(\Gamma_0, !\Delta_0)$ *for some Horn theory* $(\Gamma_0, !\Delta_0)$, *and let* (W, Γ) *be compatible with* N. *We define the marking* $[W, \Gamma] = (m, d)$ *of* N *as:*

$$m(s) = \begin{cases} mset_W(s) & if\, s \in S^{atm} \\ \Gamma(in_0(s)) & if\, s \in S^{ctrl} \end{cases} \qquad d(s) = \begin{cases} mset_W(s^\perp) & if\, s \in S^{atm} \\ 0 & otherwise \end{cases}$$

Note that the above operator $[W, \Gamma]$ is not injective, since e.g. $W' = a \otimes b$ and $W'' = b \otimes a$ will lead to the same marking. However, injectivity can be recovered considering simple products up to commutativity, associativity, and 1 identities. From now on, we will consider simple products up to this equivalence. The following proposition ensures that the operator is also surjective.

Proposition 2. *For all markings* (m, d) *of* $N = \mathcal{N}(\Gamma_0, !\Delta_0)$ *there exists a unique* (W, Γ) *compatible with* N *such that* $(m, d) = [W, \Gamma]$.

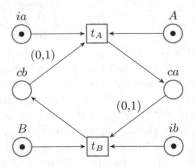

Fig. 7. The second proposal of Sect. 1 as a DPN ($t_A = in_0(A), t_B = in_0(B)$).

Example 5. Consider the second proposal of Alice and Bob's in Sect. 1, modelled as the Horn theory $(\Gamma, !\varnothing)$, where:

$$\Gamma = \{A, B\} \qquad \text{with } A = ia \multimap ca \otimes cb^{\perp} \text{ and } B = ib \multimap cb \otimes ca^{\perp}$$

In Fig. 7 we show the DPN $\mathcal{N}(\Gamma, \varnothing)$, with initial marking $[\{ia, ib\}, \Gamma]$. Note that after firing t_A and t_B (in any order), followed by two annihilation steps, the DPN reaches the empty marking.

We now establish a strict correspondence between the small-step semantics of ILL^{mix} and DPN computations. First, we relate the states in the semantics to markings in the DPN, through the $[\cdot]$ operator. Then, we show that each step in the semantics corresponds to a step in the DPN, and *vice versa*.

Proposition 3. *Let $N = \mathcal{N}(\Gamma_0, !\Delta)$, and let (W, Γ) be compatible with N. Then:*

$$(W, \Gamma, !\Delta) \rightsquigarrow (W', \Gamma', !\Delta) \qquad \Longleftrightarrow \qquad [W, \Gamma] \rightarrow_N [W', \Gamma']$$

Note that, when taking the $\Leftarrow$ direction in the above statement, assuming markings of the form $[W, \Gamma]$ is not a restriction, because Proposition 2 guarantees surjectivity.

Combining the above correspondence with the one in Theorem 3, we obtain our main result: the provability of an honoured Horn sequent in ILL^{mix} is equivalent to reachability of certain honoured markings in DPNs.

Theorem 4. *Let $N = \mathcal{N}(\Gamma, !\Delta)$. An honoured Horn sequent $\Omega, \Gamma, !\Delta \vdash Z$ of ILL^{mix} is provable iff $[\bigotimes \Omega, \Gamma] \rightarrow_N^* [Z, \varnothing]$.*

Example 6. Consider the marked DPN N in Fig. 3. A Horn theory $(\Gamma, !\Delta)$ such that $\mathcal{N}(\Gamma, !\Delta) = N$ is the following:

$$\Gamma = \varnothing \qquad \Delta = \{a \multimap b^{\perp} \otimes b^{\perp} \otimes c, \; c \multimap b, \; b \otimes d \multimap b \otimes b\}$$

The unique pair (W_0, Γ_0) corresponding to the marking in Fig. 3 is $W_0 = a \otimes d$, $\Gamma_0 = \varnothing$. Consider again the computation in the rightmost table of Example 4. According to Theorem 4, the following Horn ILL^{mix} sequent is provable:

$$a \otimes d, \; !(a \multimap b^{\perp} \otimes b^{\perp} \otimes c), \; !(c \multimap b), !(b \otimes d \multimap b \otimes b) \vdash 1$$

Remark 1. Note that the $\Rightarrow$ direction of Theorem 4 does not hold, in general, when the sequent is not assumed to be honoured. For instance, the Horn sequent $b, a \multimap b^{\perp} \vdash a^{\perp}$ is provable in ILL^{mix}, but the corresponding DPN has no computations leading to a marking with an antitoken in place a. In a certain sense, in ILL^{mix} we can *reverse* an implication $a \multimap b^{\perp}$, by using it as $b \multimap a^{\perp}$, whereas transitions in DPNs can be taken in only one direction. In honoured Horn sequents, we can no more reverse transitions, since the right-hand side of a sequent can only contain *positive* atoms.

5 Related Work and Conclusions

The starting point of this paper has been cancellative linear logic [9], an extension of linear logic where $\perp$ is identified with 1, and $\otimes$ with $\otimes$. As a consequence, both the annihilation principle $a \otimes a^{\perp} \vdash 1$ and its inverse $1 \vdash a \otimes a^{\perp}$ are valid: interpreting a as a credit and $a^{\perp}$ as a debit, the annihilation principle states that debits and credits cancel out. The inverse annihilation principle, instead, allows for generating a resource along with its corresponding debit. In [9] an extension of the token game of Petri nets is introduced, called *financial game*, where a pair token-antitoken can be either generated or annihilated. Building on this intuition, in this paper we have shown that adding [Mix] to the Horn fragment of LL is enough to permit cancelling debits, without identifying $\otimes$ with

$$\frac{}{(X, X \multimap Y, \varnothing) \Downarrow Y} \; [\Downarrow \mathbf{H}]$$

(nor allowing to freely generate credits and debits). Our main result is that provability in Horn ILL^{mix} corresponds to reachability in Stott and Godfrey's Debit nets [12].

Relations between linear logic and Petri nets have been studied by several authors, using both syntactical [14,15] and semantical methods [16–20]. Most of the papers in the semantical side connect Petri nets and LL within suitable algebraic frameworks. In particular, Meseguer and Martí-Oliet [18] compare Petri nets with multiplicative-additive linear logic, using a common categorical model, based on symmetric monoidal categories. Using semantics as a bridge, they show how linear logic can be used as a "specification language" for Petri nets. To do that, they define a satisfiability relation between Petri nets and linear logic sequents. The fragment of linear logic we have considered does not include some linear operators, e.g. internal and external choice, which are instead dealt with by [18]. To extend our correspondence to $\&$, it seems enough to share control places in DPNs between $\&$-ed transitions. The operator $\oplus$ could be dealt with by considering *non-deterministic* DPNs, in the same way as [15] relates Horn LL theories (extended with $\oplus$) with non-deterministic Petri nets.

On a more syntactical level, Kanovich [14,15] studies the computational power of the Horn fragment of LL, comparing it with Petri nets and Minsky machines. In particular, reachability in Petri nets and provability in Horn LL are shown equivalent. The strategy used to prove our main results is similar to Kanovich's; nevertheless, some differences are worth noticing. First, in this paper we have considered a different fragment of LL, featuring linear implication, tensor product, negation, 1 and [Mɪx], as well as DPNs instead of Petri nets (we have sketched above how to extend our correspondence to & and ⊕). Second, the intermediate objects used by Kanovich to connect the logic with nets are a sort of dags (called *Horn programs*), while we have used two operational semantics. Third, in Kanovich's encoding of Horn LL in Petri nets, all implications are under a !, while we have also allowed transitions to be consumed.

Another variant of Petri nets where tokens can be taken "on credit" has been presented in [21]. This model, called Lending Petri nets (LPNs), is similar to our version of DPNs: a main difference is that we have adopted a *delayed* annihilation policy, while that of LPNs is *instantaneous*, i.e. tokens and anti-tokens cannot coexist in the same place. While the instantaneous policy makes Debit nets Turing powerful [12], the delayed annihilation policy makes DPNs equi-expressive to Petri nets.

Reasoning about mutual commitments in a non-linear logic has been addressed in [4], by extending intuitionistic propositional logic with a *contractual implication* connective ⇸. Roughly, a contractual implication $a \to (b \to c)$ can be interpreted as a non-linear variant of $a \multimap (b^\perp \otimes c)$. This logic is related to Lending Petri nets: indeed, Lending Petri nets form a sound and complete model of the Horn fragment of the logic [21], analogously to the relation between Horn ILL^{mix} and DPNs studied in this paper. In [22] the correspondence between PCL and LPNs is pushed further, by showing that proof traces [6] of a Horn PCL theory Δ are exactly the honoured firing sequences in $\mathcal{N}(\Delta)$.

Acknowledgments. This work has been partially supported by Aut. Reg. of Sardinia grants L.R.7/2007 CRP-17285 (TRICS) and P.I.A. 2010 ("Social Glue"), by MIUR PRIN 2010-11 project "Security Horizons", and by EU COST Action IC1201 "Behavioural Types for Reliable Large-Scale Software Systems" (BETTY).

A   Proofs

A.1   Proofs for Sect. 4.1

Lemma 1. *If* $(W, \Gamma, !\Delta) \Downarrow Z$ *then* $W, \Gamma, !\Delta \vdash Z$.

Proof. We proceed by induction on the height of the derivation of $(W, \Gamma, !\Delta) \Downarrow Z$. We have the following cases, according on the last rule used:

- [⇓ **I**], [⇓ CUT], [⇓ **W**!], [⇓ **C**!], or [⇓ **L**!]. Straightfoward.
- [⇓ **H**]. We have:

$$\cfrac{\cfrac{}{X \vdash X}{}^{[\text{AX}]} \quad \cfrac{}{Y \vdash Y}{}^{[\text{AX}]}}{X, X \multimap Y \vdash Y}{}^{[\multimap\text{L}]}$$

and we obtain the thesis as follows:

$$\cfrac{}{(a \otimes a^\perp, \varnothing, \varnothing) \Downarrow 1}{}^{[\Downarrow \text{ S}]}$$

− [⇓ S]. We have:

$$\cfrac{\cfrac{\cfrac{\cfrac{}{a \vdash a}\ {}_{[\text{AX}]}}{a, a^{\perp} \vdash}\ {}_{[\text{NEGL}]}}{a \otimes a^{\perp} \vdash}\ {}_{[\otimes\,\text{L}]} \qquad \cfrac{}{\vdash 1}\ {}_{[1\text{R}]}}{a \otimes a^{\perp} \vdash 1}\ {}_{[\text{MIX}]}$$

and we obtain the thesis as follows:

$$\cfrac{(X, \Gamma, !\Delta) \Downarrow Y}{(X \otimes V, \Gamma, !\Delta) \Downarrow Y \otimes V}\ {}_{[\Downarrow\,\text{M}]}$$

− [⇓ M]. We have:

$$\cfrac{\cfrac{X, \Gamma, !\Delta \vdash Y \qquad \cfrac{}{V \vdash V}\ {}_{[\text{AX}]}}{X, V, \Gamma, !\Delta \vdash Y \otimes V}\ {}_{[\otimes\text{R}]}}{X \otimes V, \Gamma, !\Delta \vdash Y \otimes V}\ {}_{[\otimes\text{L}]}$$

and by using the induction hypothesis, we obtain the thesis as follows:

$$\cfrac{\cfrac{}{a \vdash a}\ {}_{[\text{AX}]}}{a, a^{\perp} \vdash}\ {}_{[\text{NEGL}]}$$

□

Definition 9 (Almost-Horn honoured sequent). *We say that a sequent $\Omega, \Gamma, !\Delta \vdash \gamma$ is almost-Horn honoured if Ω is a multiset of simple products, Γ, Δ are multisets of Horn implications, and γ is a positive simple product or empty.*

Definition 10. (Clean proof). *We say that proof π of an almost-Horn honoured sequent $\Omega, \Gamma, \Delta \vdash \gamma$ is clean when all the applications of a rule* [NEGL] *in π are placed just below an* [AX] *rule, as follows:*

$$\cfrac{\cfrac{\Omega_1, \Gamma_1, !\Delta_1 \vdash a \qquad \Omega_2, \Gamma_2, !\Delta_2 \vdash}{\Omega_1, \Omega_2, \Gamma_1, \Gamma_2, !\Delta_1, !\Delta_2 \vdash a}\ {}_{[\text{MIX}]}}{\Omega_1, \Omega_2, a^{\perp}, \Gamma_1, \Gamma_2, !\Delta_1, !\Delta_2 \vdash}\ {}_{[\text{NEGL}]}$$

Lemma 2. *Any provable almost-Horn honoured sequent admits a clean cut-free proof.*

Proof. Let $\Omega, \Gamma, \Delta \vdash \gamma$ be a provable almost-Horn honoured sequent. By Theorem 1 it admits a cut-free proof π. We show that every occurrence of the rule [NEGL] which does not respect the pattern of Definition 10, can be moved upwards in the proof. We reason by cases, depending on the rule r just above [NEGL] in π. Since π is cut-free and all its sequents are almost-Horn honoured, we can restrict to the following cases:

- [Ax], [1L], [⊗L], [!L], [weakL], [coL]. Straightforward.
- [Mix]. We have:

$$\cfrac{\cfrac{\Omega_1, \Gamma_1, !\Delta_1 \vdash a}{\Omega_1, a^\perp, \Gamma_1, !\Delta_1, \vdash} \text{[negL]} \qquad \Omega_2, \Gamma_2, !\Delta_2 \vdash}{\Omega_1, \Omega_2, a^\perp, \Gamma_1, \Gamma_2, !\Delta_1, !\Delta_2 \vdash} \text{[Mix]}$$

and we obtain the thesis as follows:

$$\cfrac{\cfrac{\Omega_1, Y, \Gamma_1, !\Delta_1 \vdash a \qquad \Omega_2, \Gamma_2, !\Delta_2 \vdash X}{\Omega_1, \Omega_2, \Gamma_1, \Gamma_2, X \multimap Y, !\Delta_1, !\Delta_2 \vdash a} \text{[⊸L]}}{\Omega_1, \Omega_2, a^\perp, \Gamma_1, \Gamma_2, X \multimap Y, !\Delta_1, !\Delta_2 \vdash} \text{[negL]}$$

- [⊸ L]. We have:

$$\cfrac{\cfrac{\Omega_1, Y, \Gamma_1, !\Delta_1 \vdash a}{\Omega_1, Y, a^\perp, \Gamma_1, !\Delta_1, \vdash} \text{[negL]} \qquad \Omega_2, \Gamma_2, !\Delta_2 \vdash X}{\Omega_1, \Omega_2, a^\perp, \Gamma_1, \Gamma_2, X \multimap Y, !\Delta_1, !\Delta_2 \vdash} \text{[⊸L]}$$

and we obtain the thesis as follows:

$$\cfrac{\cfrac{\vdots}{a, a^\perp \vdash} \qquad \cfrac{}{\vdash 1} \text{[1R]}}{a, a^\perp \vdash 1} \text{[Mix]}$$

$\square$

Definition 11 (Proper proof). *For a proof π of ILL^{mix}, we say that an application of* [Mix] *rule is* proper, *whenever it has the following form:*

$$\cfrac{\Gamma \vdash W \qquad \Gamma', W \vdash Z}{\Gamma, \Gamma' \vdash Z} \text{[Cut]}$$

We say that a proof π of an honoured almost-Horn sequent is proper *if it is clean, and every occurrence of the* [Mix] *rule in π is proper.*

Definition 12 (Harmless cut). *Given a proof π of ILL^{mix}, we say that the application of a* [Cut] *rule is* harmless *whenever it has the following form:*

$$\cfrac{\cfrac{}{a \vdash a} \text{[Ax]}}{a, a^\perp \vdash} \text{[negL]}$$

where W is a positive simple product.

Lemma 3. *A provable Horn honoured sequent admits a proper proof where all the applications of the* [CUT] *rule are harmless.*

Proof. We prove the following stronger statement. Assume that an almost-Horn honoured sequent $\Omega, \Gamma, !\Delta \vdash \gamma$ is provable. Then:

(a) if $\gamma = Z$, then there exists a proper proof of $\Omega, \Gamma, !\Delta \vdash Z$ where all [CUT] rules are harmless.

(b) if γ is empty, then there exists a proper proof of $\Omega, \Gamma, !\Delta \vdash 1$ where all [CUT] rules are harmless.

By Lemma 2, consider a clean cut-free proof π of the sequent. We proceed by induction on the height of π. Since π is cut-free, we can restrict to the following cases, according to the last rule used in π:

- [Ax], [1R]. In these cases there is nothing to prove.
- [1L], [⊗L], [⊗R], [⊸L], [!L], [WEAKL], [COL]. Straightforward by the induction hypothesis.
- [NEGL]. Since π is clean, is must have the following form:

$$\cfrac{\cfrac{\cfrac{}{a \vdash a}\text{[Ax]}}{a, a^{\perp} \vdash}\text{[NEGL]} \qquad \cfrac{}{\vdash 1}\text{[1R]}}{a, a^{\perp} \vdash 1}\text{[MIX]}$$

which we replace with the following proof:

$$\cfrac{\Omega_2, \Gamma_2, !\Delta_1 \vdash \qquad \Omega_1, \Gamma_1, !\Delta_1 \vdash Z}{\Omega_1, \Omega_2, \Gamma_1, \Gamma_2, !\Delta_1, !\Delta_2 \vdash Z}\text{[MIX]}$$

- [NEGR]. This case is not possible, since π is cut-free and γ is either a positive simple product or empty.
- [MIX]. We have the following two subcases:
  1. π has the form:

$$\cfrac{\Omega_1, \Gamma_1, !\Delta_1 \vdash Z \qquad \cfrac{\Omega_2, \Gamma_2, !\Delta_1 \vdash 1}{\Omega_1, \Omega_2, \Gamma_1, \Gamma_2, !\Delta_1, !\Delta_2 \vdash Z \otimes 1}\text{[⊗R]} \qquad \cfrac{\cfrac{\cfrac{}{Z \vdash Z}\text{[Ax]}}{Z, 1 \vdash Z}\text{[1L]}}{Z \otimes 1 \vdash Z}\text{[⊗L]}}{\Omega_1, \Omega_2, \Gamma_1, \Gamma_2, !\Delta_1, !\Delta_2 \vdash Z}\text{[CUT]}$$

By the induction hypothesis (applied on both premises), we obtain:

$$\cfrac{\Omega_1, \Gamma_1, !\Delta_1 \vdash \qquad \Omega_2, \Gamma_2, !\Delta_1 \vdash}{\Omega_1, \Omega_2, \Gamma_1, \Gamma_2, !\Delta_1, !\Delta_2 \vdash}\text{[MIX]}$$

where the application of the [CUT] rule is harmless.

2. π has the form:

$$
\cfrac{
\cfrac{\Omega_1, \Gamma_1, !\Delta_1 \vdash 1 \qquad \Omega_2, \Gamma_2, !\Delta_1 \vdash 1}{\Omega_1, \Omega_2, \Gamma_1, \Gamma_2, !\Delta_1, !\Delta_2 \vdash 1 \otimes 1} \; [\otimes R]
\qquad
\cfrac{\cfrac{\cfrac{}{1 \vdash 1} \; [Ax]}{1, 1 \vdash 1} \; [1L]}{1 \otimes 1 \vdash 1} \; [\otimes L]
}{\Omega_1, \Omega_2, \Gamma_1, \Gamma_2, !\Delta_1, !\Delta_2 \vdash 1} \; [Cut]
$$

By the induction hypothesis and harmless application of [Cut], we obtain:

$$
\cfrac{\Omega, \Gamma, H, \Delta \vdash Z}{\Omega, \Gamma, !H, !\Delta \vdash Z} \; [!L]
$$

where the application of the [Cut] rule is harmless.   □

Lemma 4. *Let $\Omega, \Gamma, !\Delta \vdash Z$ be a provable honoured Horn sequent. Then:*

$$
\left(\bigotimes \Omega, \Gamma, !\Delta \right) \Downarrow Z
$$

Proof. By Lemma 3, there exists a proper proof π of $\Omega, \Gamma, !\Delta \vdash Z$ containing only harmless applications of the [Cut] rule. We proceed by induction on the height of π, and then by cases on the last rule used.

- [Ax]. Trivial by rule [$\Downarrow$ I].
- [!L]. We have:

$$
\cfrac{\left(\bigotimes \Omega, (\Gamma \uplus \{H\}), !\Delta \right) \Downarrow Z}{\left(\bigotimes \Omega, \Gamma, !(\Delta \uplus \{H\}) \right) \Downarrow Z} \; [\Downarrow \mathbf{L}_!]
$$

By the induction hypothesis, we obtain:

$$
\cfrac{\Omega, X, Y, \Gamma, !\Delta \vdash Z}{\Omega, X \otimes Y, \Gamma, !\Delta \vdash Z} \; [\otimes L]
$$

- [negL]. This case is not possible, since the righthand side of the final sequent of π cannot be empty by hypothesis.
- [negR]. This case is not possible, since the righthand side of the final sequent of π must be a positive simple product.
- [⊗L]. We have:

$$
\cfrac{\Omega_1, \Gamma_1, !\Delta_1 \vdash Z_1 \qquad \Omega_2, \Gamma_2, !\Delta_2 \vdash Z_2}{\Omega_1, \Omega_2, \Gamma_1, \Gamma_2, !\Delta_1, !\Delta_2 \vdash Z_1 \otimes Z_2} \; [\otimes R]
$$

By the induction hypothesis, $\left(\bigotimes (\Omega \uplus \{X\} \uplus \{Y\}), \Gamma, !\Delta \right) \Downarrow Z$. The thesis follows because $\bigotimes (\Omega \uplus \{X\} \uplus \{Y\}) = \bigotimes (\Omega \uplus \{X \otimes Y\})$.

- [⊗R]. We have:

$$\dfrac{\dfrac{(\bigotimes \Omega_1, \Gamma_1, !\Delta_1) \Downarrow Z_1}{(\bigotimes \Omega_1 \otimes \bigotimes \Omega_2, \Gamma_1, !\Delta_1) \Downarrow Z_1 \otimes \bigotimes \Omega_2} \ [\Downarrow \mathbf{M}] \quad \dfrac{(\Omega_2, \Gamma_2, !\Delta_2) \Downarrow Z_2}{(Z_1 \otimes \bigotimes \Omega_2, \Gamma_2, !\Delta_2) \Downarrow Z_1 \otimes Z_2} \ [\Downarrow \mathbf{M}]}{(\bigotimes \Omega_1 \otimes \bigotimes \Omega_2, (\Gamma_1 \uplus \Gamma_2), !(\Delta_1 \uplus \Delta_2)) \Downarrow Z_1 \otimes Z_2} \ [\Downarrow \text{Cut}]$$

By applying the induction hypothesis on both premises, we obtain:

$$\dfrac{\Omega_1, \Gamma_1, !\Delta_1 \vdash X \quad \Omega_2, Y, \Gamma_2, !\Delta_2 \vdash Z}{\Omega_1, \Omega_2, \Gamma_1, X \multimap Y, \Gamma_2, !\Delta_1, !\Delta_2 \vdash Z} \ [\multimap\text{L}]$$

- [⊸L]. We have:

$$\dfrac{\dfrac{\dfrac{(\bigotimes \Omega_1, \Gamma_1, !\Delta_1) \Downarrow X \quad \dfrac{}{(X, X \multimap Y, \varnothing) \Downarrow Y} \ [\Downarrow \mathbf{H}]}{(\bigotimes \Omega_1, (\Gamma_1 \uplus \{X \multimap Y\}), !\Delta_1) \Downarrow Y} \ [\Downarrow \text{Cut}]}{(\bigotimes \Omega_1 \otimes \bigotimes \Omega_2, (\Gamma_1 \uplus \{X \multimap Y\}), !\Delta_1) \Downarrow Y \otimes \bigotimes \Omega_2} \ [\Downarrow \mathbf{M}] \quad (Y \otimes \bigotimes \Omega_2, \Gamma_2, !\Delta_2) \Downarrow Z}{(\bigotimes \Omega_1 \otimes \bigotimes \Omega_2, (\Gamma_1 \uplus \{X \multimap Y\} \uplus \Gamma_2), !(\Delta_1 \uplus \Delta_2)) \Downarrow Z} \ [\Downarrow \text{Cut}]$$

Since $X \multimap Y$ is a Horn implication, then X must be a positive simple product, and so $\Omega_1, \Gamma_1, !\Delta_1 \vdash X$ is an honoured Horn sequent. Therefore we can apply the induction hypothesis, from which we obtain:

$$\dfrac{\dfrac{\vdots}{a, a^{\perp} \vdash} \quad \dfrac{}{\vdash 1} \ [\text{1R}]}{a, a^{\perp} \vdash 1} \ [\text{Mix}]$$

- [Mix]. Since π is proper, it must be:

$$\dfrac{\Omega_1, \Gamma_1, !\Delta_1 \vdash Z_1 \quad \Omega_2, Z_1, \Gamma_2, !\Delta_2 \vdash Z}{\Omega_1, \Omega_2, \Gamma_1, \Gamma_2, !\Delta_1, !\Delta_2 \vdash Z} \ [\text{Cut}]$$

and the thesis follows from rule [⇓ **s**].

- [Cut]. Since π contains only harmless cuts, by Definition 12 there exist $\Omega_1, \Omega_2, Z_1, \Gamma_1, \Gamma_2$, and Δ_1, Δ_2 such that Z_1 is honoured, and the last rule in π is:

$$\dfrac{\dfrac{\bigotimes \Omega_1, \Gamma_1, !\Delta_1 \Downarrow Z_1}{\bigotimes \Omega_1 \otimes \bigotimes \Omega_2, \Gamma_1, !\Delta_1 \Downarrow Z_1 \otimes \bigotimes \Omega_2} \ [\Downarrow \mathbf{M}] \quad \bigotimes \Omega_2 \otimes Z_1, \Gamma_2, !\Delta_2 \Downarrow Z}{(\bigotimes \Omega_1 \otimes \bigotimes \Omega_2, \Gamma_1 \uplus \Gamma_2, !(\Delta_1 \uplus \Delta_2)) \Downarrow Z} \ [\Downarrow \text{Cut}]$$

By applying the induction hypothesis on both premises, we obtain:

$$\bigotimes \Omega_1, \Gamma_1, !\Delta_1 \Downarrow Z_1 \qquad \bigotimes \Omega_2 \otimes Z_1, \Gamma_2, !\Delta_2 \Downarrow Z$$

Therefore, we obtain:

$$\frac{}{\vdash 1} \; [1\text{R}]$$

– [1R]. We have:

$$\frac{}{(1, \varnothing, \varnothing) \Downarrow 1} \; [\Downarrow \mathbf{I}]$$

Since the empty multiset is associated with the simple product 1, we obtain:

$$\frac{\Omega, \Gamma, !\Delta \vdash Z}{1, \Omega, \Gamma, !\Delta \vdash Z} \; [1\text{L}]$$

– [1L]. We have:

$$\frac{\dfrac{\dfrac{}{(X, X \multimap Y, \varnothing) \Downarrow Y} \; [\Downarrow \mathbf{H}]}{(X \otimes V, X \multimap Y, \varnothing) \Downarrow Y \otimes V} \; [\Downarrow \mathbf{M}] \qquad (Y \otimes V, \Gamma', !\Delta) \Downarrow Z}{(X \otimes V, (\Gamma' \uplus \{X \multimap Y\}), !\Delta) \Downarrow Z} \; [\Downarrow \text{CUT}]$$

By the induction hypothesis, we know that:

$$\left(\bigotimes \Omega, \Gamma, !\Delta\right) \Downarrow Z$$

Since $mset_{\otimes} \Omega = mset_{1 \otimes \bigotimes \Omega}$, we conclude that:

$$\left(1 \otimes \bigotimes \Omega, \Gamma, !\Delta\right) \Downarrow Z \qquad\qquad \square$$

Theorem 2. Let $\Omega, \Gamma, !\Delta \vdash Z$ be an honoured Horn sequent. Then:

$$\left(\bigotimes \Omega, \Gamma, !\Delta\right) \Downarrow Z \iff \Omega, \Gamma, !\Delta \vdash Z$$

Proof. The ($\Longrightarrow$) direction follows from Lemma 1; the ($\Longleftarrow$) direction follows from Lemma 4. $\qquad\qquad \square$

A.2   Proofs for Sect. 4.2

Lemma 5. *The following facts hold:*

1. *If $(W_1, \Gamma_1, !\Delta_1) \rightsquigarrow^* (W_2, \Gamma_1', !\Delta_1)$ and $(W_2, \Gamma_2, !\Delta_2) \rightsquigarrow^* (W_3, \Gamma_2', !\Delta_2)$, then $(W_1, \Gamma_1 \uplus \Gamma_2, !(\Delta_1 \uplus \Delta_2)) \rightsquigarrow^* (W_3, \Gamma_1' \uplus \Gamma_2', !(\Delta_1 \uplus \Delta_2))$.*
2. *If $(W, \Gamma, !\Delta) \rightsquigarrow^* (W', \Gamma', !\Delta)$ and V is a simple product, then $(W \otimes V, \Gamma, !\Delta) \rightsquigarrow^* (W' \otimes V, \Gamma', !\Delta)$.*
3. *If $(W, \Gamma \uplus \{H\}, !\Delta) \rightsquigarrow^* (W', \Gamma', !\Delta)$ where H is a Horn implication, then $(W, \Gamma, !(\Delta \uplus \{H\})) \rightsquigarrow^* (W', \Gamma', !(\Delta \uplus \{H\}))$.*

4. *If* $(W, \Gamma, !\Delta) \rightsquigarrow^* (W', \Gamma', !\Delta)$ *where* H *is a Horn implication, then* $(W, \Gamma, !(\Delta \uplus \{H\})) \rightsquigarrow^* (W', \Gamma', !(\Delta \uplus \{H\}))$.
5. *If* $(W, \Gamma, !(\Delta \uplus \{H, H\})) \rightsquigarrow^* (W', \Gamma', !(\Delta \uplus \{H, H\}))$ *where* H *is a Horn implication, then* $(W, \Gamma, !(\Delta \uplus \{H\})) \rightsquigarrow^* (W', \Gamma', !(\Delta \uplus \{H\}))$.

Proof. Straightforward.   □

Proposition 1. $(W, \Gamma, !\Delta) \Downarrow Z \iff (W, \Gamma, !\Delta) \rightsquigarrow^* (Z, \varnothing, !\Delta)$

Proof. For the ($\implies$) direction, we proceed by induction on the height of the derivation of $(W, \Gamma, !\Delta) \Downarrow Z$, and then by cases on the last rule applied.

For the base case, we have three possible subcases:

- [$\Downarrow$ I]. By reflexivity of $\rightsquigarrow^*$ $(X, \varnothing, \varnothing) \rightsquigarrow^* (X, \varnothing, \varnothing)$.
- [$\Downarrow$ H]. By rule [$\rightsquigarrow_H$], $(X, X \multimap Y, \varnothing) \rightsquigarrow (Y, \varnothing, \varnothing)$.
- [$\Downarrow$ S]. By rule [$\rightsquigarrow_S$], $(a \otimes a^\perp, \varnothing, \varnothing) \rightsquigarrow (1, \varnothing, \varnothing)$.

For the inductive case, we have the following subcases:

- [$\Downarrow$CUT]. By applying the induction hypothesis on both premises, we obtain:

$$(W, \Gamma_1, !\Delta_1) \rightsquigarrow^* (U, \varnothing, !\Delta_1) \qquad\qquad (U, \Gamma_2, !\Delta_2) \rightsquigarrow^* (Z, \varnothing, !\Delta_2)$$

By item (1) of Lemma 5 we obtain the thesis:

$$(W, \Gamma_1 \uplus \Gamma_2, !\Delta_1 \uplus !\Delta_2) \rightsquigarrow^* (Z, \varnothing, !\Delta_1 \uplus !\Delta_2)$$

- [$\Downarrow$ M]. By the induction hypothesis and item (2) of Lemma 5.
- [$\Downarrow$ L$_!$]. By the induction hypothesis and item (3) of Lemma 5.
- [$\Downarrow$ W$_!$]. By the induction hypothesis and item (4) of Lemma 5.
- [$\Downarrow$ C$_!$]. By the induction hypothesis and item (5) of Lemma 5.

For the ($\impliedby$) direction, we proceed by induction on the length n of the computation $(W, \Gamma, !\Delta) \rightsquigarrow^n (Z, \varnothing, !\Delta)$.

- $n = 0$: then the computation consists of the single state $(Z, \varnothing, !\Delta)$, and $(Z, \varnothing, !\Delta) \Downarrow Z$ is derivable by rule [$\Downarrow$ I] followed by as many applications of [$\Downarrow$ W$_!$] as the cardinality of Δ.
- $n > 0$: Let $s \rightsquigarrow s'$ be the first transition of the computation. Let us call t_0 the sub-computation of lenght $n - 1$ starting from s'. We have the following three subcases, depending on the rule used to deduce $s \rightsquigarrow s'$:
  - [$\rightsquigarrow_H$]. By definition, there exist X, Y, V such that $s' = (Y \otimes V, \Gamma', !\Delta)$, $W = X \otimes V$, $\Gamma = \Gamma' \uplus \{X \multimap Y\}$ and
    $(X \otimes V, \Gamma' \uplus \{X \multimap Y\}, !\Delta) \rightsquigarrow (Y \otimes V, \Gamma', !\Delta) \rightsquigarrow^* (Z, \varnothing, !\Delta)$
    The induction hypothesis gives us $(Y \otimes V, \Gamma', !\Delta) \Downarrow Z$, so we can build the following:

$$((X \otimes V, \Gamma, !\Delta) \rightsquigarrow (Y \otimes V, \Gamma, !\Delta) \rightsquigarrow^* (Z, \varnothing, !\Delta)$$

- $[\leadsto_{!H}]$. By definition, there exist X, Y, V such that $s' = (Y \otimes V, \Gamma, !\Delta)$, $W = X \otimes V$, $X \multimap Y \in \Delta$ and

$$((X \otimes V, \Gamma, !\Delta) \leadsto (Y \otimes V, \Gamma, !\Delta) \leadsto^* (Z, \varnothing, !\Delta)$$

The induction hypothesis gives us $(Y \otimes V, \Gamma, !\Delta) \Downarrow Z$, so we can build the following:

$$\cfrac{\cfrac{\cfrac{\overline{(X, X \multimap Y, \varnothing) \Downarrow Y}^{[\Downarrow \mathbf{H}]}}{(X \otimes V, X \multimap Y, \varnothing) \Downarrow Y \otimes V}^{[\Downarrow \mathbf{M}]} \quad (Y \otimes V, \Gamma, !\Delta) \Downarrow Z}{(X \otimes V, (\Gamma \uplus \{X \multimap Y\}), !\Delta) \Downarrow Z}^{[\Downarrow \mathrm{CUT}]}}{\cfrac{(X \otimes V, \Gamma, !(\Delta \uplus \{X \multimap Y\})) \Downarrow Z}{(X \otimes V, \Gamma, !\Delta) \Downarrow Z}^{[\Downarrow \mathbf{C}_!]}}^{[\Downarrow \mathbf{L}_!]}$$

- $[\leadsto_S]$. By definition, there exist V and an atom a such that $s' = (1 \otimes V, \Gamma, !\Delta)$, $W = a \otimes a^\perp \otimes V$, and:

$$(a \otimes a^\perp \otimes V, \Gamma, !\Delta) \leadsto (1 \otimes V, \Gamma, !\Delta) \leadsto^* (Z, \varnothing, !\Delta)$$

The induction hypothesis gives us $(1 \otimes V, \Gamma', !\Delta) \Downarrow Z$, so we can build the following:

$$\cfrac{\cfrac{\overline{(a \otimes a^\perp, \varnothing, \varnothing) \Downarrow 1}^{[\Downarrow \mathbf{S}]}}{(a \otimes a^\perp \otimes V, \varnothing, \varnothing) \Downarrow 1 \otimes V}^{[\Downarrow \mathbf{M}]} \quad (1 \otimes V, \Gamma, !\Delta) \Downarrow Z}{(a \otimes a^\perp \otimes V, \Gamma, !\Delta) \Downarrow Z}^{[\Downarrow \mathrm{CUT}]}$$

Theorem 3. *Let* $\Omega, \Gamma, !\Delta \vdash Z$ *be an honoured Horn sequent. Then:*

$$(\bigotimes \Omega, \Gamma, !\Delta) \leadsto^* (Z, \varnothing, !\Delta) \quad \Longleftrightarrow \quad \Omega, \Gamma, !\Delta \vdash Z$$

Proof. Straightforward by Proposition 1 and Theorem 2. □

A.3   Proofs for Sect. 4.3

Proposition 2. *For all markings (m, d) of $N = \mathcal{N}(\Gamma_0, !\Delta_0)$ there exists a unique (W, Γ) compatible with N such that $(m, d) = [W, \Gamma]$.*

Proof. We prove that $[\cdot]$ is injective and surjective over (W, Γ) compatible with $\mathcal{N}(\Gamma_0, \Delta_0)$; the result then follows straightforwardly. Let us assume $(W_1, \Gamma_1) \neq (W_2, \Gamma_2)$; then either $mset_{W_1}(s) \neq mset_{W_2}(s)$ for some $s \in S^{atm}$, or $\Gamma_1(in_0(s)) \neq \Gamma_2(in_0(s))$ for some $s \in S^{ctrl}$, or $mset_{W_1}(s^\perp) \neq mset_{W_2}(s^\perp)$ for some $s \in S^{atm}$; but then by compatibility and by definition of $[\cdot]$, $[W_1, \Gamma_1] \neq [W_2, \Gamma_2]$. This proves injectivity.

For surjectivity, if (m, d) is a marking of $\mathcal{N}(\Gamma_0, \Delta_0)$, we can build (W, Γ) compatible with $\mathcal{N}(\Gamma_0, \Delta_0)$ s.t. $(m, d) = [W, \Gamma]$ as follows: to retrieve W we observe that m, d define a multiset M of occurences of literals as observed in Sect. 3; we take W to be the tensor product of all the elements of M. Let Γ comprise every implication $s \in S^{ctrl}$ with multiplicity $m(s)$. By construction, $\langle W \rangle \subseteq S^{atm}$ and $\Gamma \subseteq \Gamma_0$, so they are compatible with $\mathcal{N}(\Gamma_0, \Delta_0)$ and always by construction $(m, d) = [W, \Gamma]$. $\qquad\qquad\qquad\qquad\qquad\qquad\qquad\qquad\qquad\qquad\qquad\qquad$ □

Proposition 3. *Let $N = \mathcal{N}(\Gamma_0, !\Delta)$, and let (W, Γ) be compatible with N. Then:*

$$(W, \Gamma, !\Delta) \rightsquigarrow (W', \Gamma', !\Delta) \quad\iff\quad [W, \Gamma] \to_N [W', \Gamma']$$

Proof. From left to right we reason by cases, depending on the small-step rule we are using:

- $[\rightsquigarrow_{!H}]$: then $W = X \otimes V$, $W' = Y \otimes V$, $\Gamma = \Gamma'$ and $((X \otimes V, \Gamma, !\bar{\Delta} \cup \{!(X \multimap Y)\}) \rightsquigarrow (Y \otimes V, \Gamma, !\bar{\Delta} \cup \{!(X \multimap Y)\}))$ where $!\Delta = !\bar{\Delta} \cup \{!(X \multimap Y)\}$
  By Definition 8 we know that $[W, \Gamma] = (m, d)$ for some marking (m, d) such that for all $s \in S^{atm}$ we have $m(s) = mset_W(s) = mset_X(s) + mset_V(s)$ and $r = in_1(X \multimap Y)$ for some transition r of N; then for all s, $mset_X(s) = F(s, r)$ by Definition 7; since $mset_X(s) + mset_V(s) = mset_W(s) = m(s)$, r is enabled in (m, d); moreover, for all s, we know by Definition 7 that $mset_Y(s) = F(r, s)$ and $mset_Y(s^\perp) = L(s, r)$ so after firing r, $mset_{W'}(s) = mset_Y(s) + mset_V(s) = m'(s)$ and $mset_Y(s^\perp) + mset_V(s^\perp) = d'(s)$ by Definition 7. Further $m(s) = m'(s)$ when $s \in S^{ctrl}$. Therefore, by Definition 8, then $[W', \Gamma'] = (m', d')$.

- $[\rightsquigarrow_H]$: then $W = X \otimes V$, $W' = Y \otimes V$, $\Gamma = \Gamma' \uplus \{X \multimap Y\}$ and $(X \otimes V, \Gamma' \uplus \{X \multimap Y\}, !\Delta) \rightsquigarrow (Y \otimes V, \Gamma', !\Delta)$
  By Definition 8 we know that $[W, \Gamma] = (m, d)$ for some marking (m, d) such that for all $s \in S^{atm}$ we have $m(s) = mset_W(s) = mset_X(s) + mset_V(s)$ and $r = in_0(X \multimap Y)$ for some transition r of N; then for all s, $mset_X(s) = F(s, r)$ by Definition 7; since $mset_X(s) + mset_V(s) = mset_W(s) = m(s)$, r is enabled in (m, d); moreover, for all s, we know by Definition 7 that $mset_Y(s) = F(r, s)$ and $mset_Y(s^\perp) = L(s, r)$ so after firing r, $mset_{W'}(s) = mset_Y(s) + mset_V(s) = m'(s)$ and $mset_Y(s^\perp) + mset_V(s^\perp) = d'(s)$ by Definition 7; moreover r has been fired in (m', d') (so its control place has one fewer token). By Definition 8, then $[W', \Gamma'] = (m', d')$.

- $[\rightsquigarrow_S]$: then $W = (a \otimes a^\perp) \otimes V$, $W' = 1 \otimes V$, $\Gamma = \Gamma'$, and $((a \otimes a^\perp) \otimes V, \Gamma, !\Delta) \rightsquigarrow (1 \otimes V, \Gamma, !\Delta)$
  By Definition 8 we know that $[W, \Gamma] = (m, d)$ for some marking (m, d) such that for all $s \in S^{atm}$ we have $m(s) = mset_W(s)$; now since $mset_W(a)$ and $mset_W(a^\perp) > 0$, $m(a) > 0$ and $d(a) > 0$, so annihilation is enabled at (m, d). After firing annihilation, we know that $m'(a) = m(a) - 1$ and $d'(a) = d(a) - 1$, while for all other $s \neq a$, we have $m'(s) = m(s)$ and $d'(s) = d(s)$. Now it is easy to verify that $mset_{W'}(a) = mset_W(a) - 1$ and $mset_{W'}(a^\perp) = mset_W(a^\perp) - 1$ and for all $s \neq a$ $mset_{W'}(s) = mset_W(s)$ (resp. $mset_{W'}(s^\perp) = mset_W(s^\perp)$). Control places are unaffected, so $m(s) = m'(s)$ when $s \in S^{ctrl}$. We conclude that $(m', d') = [W', \Gamma']$.

From right to left we reason by cases, depending on the type of step:

- Suppose we fire $r = in_i(X \multimap Y)$ at $(m, d) = [W, \Gamma]$. We know that $(m, d) \to (m', d')$ by firing r for some m', d' and $(m', d') = [W', \Gamma']$ for some W', Γ' by Proposition 2; since r is enabled in (m, d), for all $s \in S^{atm}$ such that $mset_X(s) \geq 0$, we have that $m(s) \geq mset_X(s)$ and since $(m, d) = [W, \Gamma]$, we have that $mset_W(s) \geq mset_X(s)$. This means that, $W = X \otimes V$ for some V (so when $s \in S^{atm}$ we have $m(s) = mset_X(s) + mset_V(s)$ and $d(s) = mset_X(s^\perp) + mset_V(s^\perp)$). Now we have two subcases:

  - if $i = 1$, then for all $s \in S^{ctrl}$, $m(s) = m'(s)$ so, since $(m, d) = [W, \Gamma]$ and $(m', d') = [W', \Gamma']$, $\Gamma = \Gamma'$; moreover for all $s \in S^{atm}$, $m'(s) = mset_Y(s) + mset_V(s)$ and $d'(s) = mset_Y(s^\perp) + mset_V(s^\perp)$ (by definition of $\to$, and Definition 8). But then, $W' = Y \otimes V$ and $((X \otimes V, \Gamma, !\Delta \cup \{!(X \multimap Y)\}) \rightsquigarrow (Y \otimes V, \Gamma, !\Delta \cup \{!(X \multimap Y)\}))$

  - otherwise, $i = 0$ which implies $(m, d) = [W, \Gamma]$ and $(m', d') = [W', \Gamma']$, with $\Gamma' = \Gamma \setminus \{X \multimap Y\}$. Moreover $m'(s) = mset_Y(s) + mset_V(s)$ and $d'(s) = mset_Y(s^\perp) + mset_V(s^\perp)$ (by definition of $\to$, and Definition 8). But then $W' = Y \otimes V$ and $(X \otimes V, \Gamma, !\Delta) \rightsquigarrow (Y \otimes V, \Gamma', !\Delta)$

- annihilation is enabled in $(m, d) = [W, \Gamma]$. We know that $(m, d) \to (m', d')$ through an annihilation step and $(m', d') = [W', \Gamma']$ for some W', Γ' by Proposition 2. Since an annihilation is enabled, for some atom a, $m(a) \geq 1, d(a) \geq 1$; now for all s, $m(s) = mset_W(s)$, and $d(s) = mset_W(s^\perp)$, so for some a, V, $W = (a \otimes a^\perp) \otimes V$. Then $(W, \Gamma, !\Delta) \rightsquigarrow (1 \otimes V, \Gamma', !\Delta)$ (where $\Gamma = \Gamma'$). It remains to prove that $W' = 1 \otimes V$; this follows from the fact that for $s = a$, $m'(s) = mset_W(s) - 1$ and $d'(s) = mset_W(s^\perp) - 1$ (by definition of $\to$, and Definition 8).   □

Theorem 4. *Let $N = \mathcal{N}(\Gamma, !\Delta)$. An honoured Horn sequent $\Omega, \Gamma, !\Delta \vdash Z$ of ILL^{mix} is provable iff $[\bigotimes \Omega, \Gamma] \to_N^* [Z, \varnothing]$.*

Proof. By Theorem 3, we know that $\Omega, \Gamma, !\Delta \vdash Z$ is provable if and only if $(\bigotimes \Omega, \Gamma, !\Delta) \rightsquigarrow^* (Z, \emptyset, !\Delta)$. By Proposition 3, $(\bigotimes \Omega, \Gamma, !\Delta) \rightsquigarrow^* (Z, \emptyset, !\Delta)$ iff $[\bigotimes \Omega, \Gamma] \to_N^* [Z, \emptyset]$.   □

References

1. Hobbes, T.: The Leviathan, chap. XIV (1651)
2. Viswanathan, M., Viswanathan, R.: Foundations for circular compositional reasoning. In: Orejas, F., Spirakis, P.G., van Leeuwen, J. (eds.) ICALP 2001. LNCS, vol. 2076, pp. 835–847. Springer, Heidelberg (2001). doi:10.1007/3-540-48224-5_68
3. Maier, P.: Compositional circular assume-guarantee rules cannot be sound and complete. In: Gordon, A.D. (ed.) FOSSACS 2003. LNCS, vol. 2620, pp. 343–357. Springer, Heidelberg (2003). doi:10.1007/3-540-36576-1_22
4. Bartoletti, M., Zunino, R.: A calculus of contracting processes. In: Proceedings of the LICS, pp. 332–341 (2010). doi:10.1109/LICS.2010.25
5. Bartoletti, M., Cimoli, T., Pinna, G.M., Zunino, R.: Circular causality in event structures. Fundamenta Informaticae **134**(3–4), 219–259 (2014). doi:10.3233/FI-2014-1101
6. Bartoletti, M., Cimoli, T., Giamberardino, P.D., Zunino, R.: Vicious circles in contracts and in logic. Sci. Comput. Program. (to appear). doi:10.1016/j.scico.2015.01.005

7. Bartoletti, M., Cimoli, T., Pinna, G.M., Zunino, R.: Contracts as games on event structures, JLAMP (to appear). doi:10.1016/j.jlamp.2015.05.001

8. Girard, J.-Y.: Linear logic. Theoret. Comput. Sci. **50**, 1–102 (1987). doi:10.1016/0304-3975(87)90045-4

9. Martí-Oliet, N., Meseguer, J.: An algebraic axiomatization of linear logic models. In: Topology and Category Theory in Computer Science, pp. 335–355. Oxford University Press Inc. (1991)

10. Fleury, A., Retoré, C.: The Mix rule. Mathematical Structures in Computer Science **4**(2), 273–285 (1994). doi:10.1017/S0960129500000451

11. Kamide, N.: Linear logics with communication-merge. J. Logic Comput. **15**(1), 3–20 (2005). doi:10.1093/logcom/exh029

12. Stotts, P.D., Godfrey, P.: Place/transition nets with debit arcs. Inf. Process. Lett. **41**(1), 25–33 (1992). doi:10.1016/0020-0190(92)90076-8

13. Reisig, W.: Petri Nets: An Introduction. Monographs in Theoretical Computer Science, vol. 4. Springer, Heidelberg (1985). doi:10.1007/978-3-642-69968-9

14. Kanovich, M.I.: Linear logic as a logic of computations. Ann. Pure Appl. Logic **67**, 183–212 (1994). doi:10.1016/0168-0072(94)90011-6

15. Kanovich, M.I.: Petri nets, Horn programs, linear logic and vector games. Ann. Pure Appl. Logic **75**(1–2), 107–135 (1995). doi:10.1016/0168-0072(94)00060-G

16. Asperti, A., Ferrari, G.L., Gorrieri, R.: Implicative formulae in the "proofs as computations" analogy. In: Proceedings of the POPL, pp. 59–71 (1990). doi:10.1145/96709.96715

17. Engberg, U., Winskel, G.: Completeness results for linear logic on Petri nets. Ann. Pure Appl. Logic **86**(2), 101–135 (1997). doi:10.1016/S0168-0072(96)00024-3

18. Martí-Oliet, N., Meseguer, J.: From Petri nets to linear logic. Math. Struct. Comput. Sci. **1**(1), 69–101 (1991). doi:10.1017/S0960129500000062

19. Ishihara, K., Hiraishi, K.: The completeness of linear logic for Petri net models. Logic J. IGPL **9**(4), 549–567 (2001)

20. Kanovich, M.I., Okada, M., Scedrov, A.: Phase semantics for light linear logic. ENTCS **6**, 221–234 (1997). doi:10.1016/S1571-0661(05)80159-8

21. Bartoletti, M., Cimoli, T., Pinna, G.M.: Lending Petri nets and contracts. In: Arbab, F., Sirjani, M. (eds.) FSEN 2013. LNCS, vol. 8161, pp. 66–82. Springer, Heidelberg (2013). doi:10.1007/978-3-642-40213-5_5

22. Bartoletti, M., Cimoli, T., Pinna, G.M.: Lending Petri nets. Sci. Comput. Program. (2015). doi:10.1016/j.scico.2015.05.006

Alice and Bob Meet Equational Theories

David Basin, Michel Keller, Saša Radomirović, and Ralf Sasse$^{(\boxtimes)}$

Department of Computer Science, Institute of Information Security,
ETH Zurich, Zurich, Switzerland
`ralf.sasse@inf.ethz.ch`

Dedicated to José Meseguer on his 65th Birthday.

Abstract. Cryptographic protocols are the backbone of secure communication over open networks and their correctness is therefore crucial. Tool-supported formal analysis of cryptographic protocol designs increases our confidence that these protocols achieve their intended security guarantees. We propose a method to automatically translate textbook style Alice&Bob protocol specifications into a format amenable to formal verification using existing tools. Our translation supports specification modulo equational theories, which enables the faithful representation of algebraic properties of a large class of cryptographic operators.

1 Introduction

Internet security builds on cryptographic protocols that achieve properties such as secrecy, entity authentication, and privacy. The correct operation of these protocols is critical and manual analysis is not up to the task. Indeed, some protocols were used for years before flaws were detected using symbolic analysis tools [4,19]. Today, there are many such tools available based on different formalisms and specification languages: ProVerif [6] uses the applied pi calculus, Scyther [13] uses role scripts, Maude-NPA [15] uses strands [17], and TAMARIN [29] specifies protocols using multiset rewriting. Unfortunately the input languages of all of these tools are difficult for non-expert users to master, which hinders the widespread acceptance and use of these tools.

The starting point for this paper is our work on the TAMARIN tool, which has been used successfully to analyze many cryptographic protocols [5,25,30]. TAMARIN uses multiset rewriting as its input language, which is very general; but this generality makes it difficult to use, especially for non-experts. Hence an attractive proposition is to support, additionally, a simpler and more intuitive language, closer to the text-book notation that many users know from their studies. The most popular language of this kind is generally known as Alice&Bob protocol specifications. Due to its popularity, versions of it have been considered for other analysis tools [1]. It is indeed simple, but it suffers from ambiguities and imprecision.

We propose a new way to specify protocols in an Alice&Bob style that supports specification *modulo user-specifiable equational theories*. This enables

© Springer International Publishing Switzerland 2015
N. Martí-Oliet et al. (Eds.): Meseguer Festschrift, LNCS 9200, pp. 160–180, 2015.
DOI: 10.1007/978-3-319-23165-5_7

one to specify, along with a protocol, the algebraic properties of the cryptographic operators used. To support this, we analyze the protocol specification's executability and translate executable specifications into an intermediate language based on role scripts. We then determine which checks should be made on the messages by the participants, based on their current knowledge and the equational theory, and insert those checks into the compiled role scripts. Nonexecutable specifications are rejected, and warnings to the user are displayed when checks cannot be inserted as expected, for example, when the same name is used repeatedly, but due to encryption the principal cannot verify if all occurrences are instantiated in the same way. The role scripts can then be further translated to any protocol analysis tool's input language, as our theoretical results are general. We have implemented this first general-purpose translation step and, for TAMARIN's input language, we have also implemented the second step. Taken together, this provides an automatic translation from Alice&Bob specifications to the TAMARIN tool's input language [18].

As mentioned, equational theories are used to model the algebraic properties of cryptographic operators. Support for different equational theories therefore allows a more precise analysis of protocols. Specifying suitable equational theories is fundamental for the symbolic analysis of cryptographic protocols as otherwise attacks may be missed. Moreover, for some protocols, their execution would even be impossible without algebraic properties. For example, for Diffie-Hellman key exchange, the partners cannot establish the same key without exponentiation being commutative. We focus on subterm-convergent theories, which we require for the results presented in this paper. Note that these theories have the finite variant property [12,16], which is a prerequisite for many automated protocol analysis tools as well.

To handle the imprecision inherent in basic Alice&Bob notation, we build upon previous work that makes explicit many of this notation's assumptions. In particular, Caleiro, Viganò, and Basin [9] provide an operational semantics based on the spi calculus that formalizes how principals construct and parse messages and makes explicit what checks should be made by honest principals. Mödersheim [26] studies a similar problem in the context of equational theories. See Sect. 5 for a more detailed comparison.

We also build upon and take inspiration in our work from the research and tools of José Meseguer. TAMARIN uses Maude [11] as a back-end for unification modulo equational theories. Moreover, inspired by Maude-NPA [15], TAMARIN computes variants following [16]. This motivated the design decision of supporting user-specified subterm-convergent equational theories in addition to built-in theories with the finite variant property that is used both in TAMARIN and in the translation we present in this paper.

In Sect. 2 we describe Alice&Bob notation. In Sect. 3 we present the role-script notation for protocols, explain how to decide the executability of given Alice&Bob protocols and how to add appropriate checks, and provide an example. We describe our automated translation to TAMARIN in Sect. 4 and compare to related work in Sect. 5, before we draw conclusions in Sect. 6.

2 Alice&Bob Protocol Notation

In this section we formalize Alice&Bob notation. First, we highlight ambiguities with the text-book version of this notation due to its inherent imprecision, and we explain the general idea of our formalization. Afterwards we specify this notation in more detail.

2.1 Overview

In Alice&Bob notation, a protocol is specified as a list of *message exchange steps* of the form

$$A \rightarrow B: \quad msg.$$

These steps describe the actions that are performed by honest principals in a protocol run. The semantics of an Alice&Bob specification defines the behavior of the principals running the protocol. Our semantics is based on the work of Caleiro, Viganò, and Basin [9].

To illustrate the need for a formal semantics, consider the simplified basic Kerberos authentication protocol [27] shown in Fig. 1.

At first glance, this protocol's meaning seems clear. The principal in role C sends his identity, the name of a resource V, and the nonce n to the authentication server S. A *nonce* is an arbitrary number to be used only once in a security protocol. The principal in role S then responds by returning two ciphertexts, the first one generated with a shared key k_{CS} between C and S and the second generated with a key k_{VS}, shared between V and S. The principal C then decrypts the first ciphertext to obtain the key k and verifies the nonce n and the intended communication partner V. If these checks succeed, then C encrypts fresh nonces t and t' under k and sends this ciphertext along with the second ciphertext to V. The principal V then responds with the encryption of t under the key k, which is obtained by decrypting the second ciphertext $\{k, C\}_{k_{VS}}$.

While this account provides a high-level explanation of the protocol's workings, the precise actions the principals must take are not fully spelled out. In order to send the message $\{k, V, n\}_{k_{CS}}, \{k, C\}_{k_{VS}}$ in Step 2, S must first construct it. Intuitively, one would assume that S knows k_{CS} and k_{VS} and generates a fresh key k and can therefore construct the message. But this is not stated explicitly. It is possible that the protocol's designer had in mind that k is known only to C and V, while S only knows the two ciphertexts in Step 2. The specification as given does not resolve this ambiguity. It is also unclear whether n is

1. $C \rightarrow S$: C, V, n
2. $S \rightarrow C$: $\{k, V, n\}_{k_{CS}}, \{k, C\}_{k_{VS}}$
3. $C \rightarrow V$: $\{t, t'\}_k, \{k, C\}_{k_{VS}}$
4. $V \rightarrow C$: $\{t\}_k$

Fig. 1. An intuitive but ambiguous description of an authentication protocol.

actually a nonce, even though the choice of name suggests this. It could just as well be a constant or a publicly known value, neither of which we would assume of a nonce.

Another aspect left implicit in Fig. 1 is what the principals do with the messages they receive. If we assume that k_{CS} actually denotes the shared secret key between C and S and that both know the key and C has no other prior knowledge, then C should extract the key k by decrypting the first ciphertext of Step 2 with the secret key k_{CS}. This is the only way C can construct the message $\{t, t'\}_k$ in Step 3. For this reason, we must formalize the new information gained by analyzing incoming messages based on the knowledge that a principal has.

We formalize Alice&Bob notation based on the notion of knowledge, given by a set of messages, which grows during protocol execution when new messages are received or fresh nonces are generated. We define what information is stored, how incoming messages are parsed and compared to existing knowledge, and how messages are composed for sending.

In the rest of this section, we provide a complete formalization of Alice&Bob notation, which is the basis of our Alice&Bob input language and our translation to the intermediate representation format that follows. Note that Alice&Bob notation is independent of the adversary model. We do not define the capabilities of the adversary as they can be independently specified.

2.2 Messages and Message Model

We use a signature Σ containing three sorts, *Fresh*, *Public*, and *Msg*, where both *Fresh* and *Public* are subsorts of *Msg*. Each operator $f \colon s_1 \ldots s_n \to s$ defined on any of the sorts has a top sort overloading, i.e., $f \colon Msg \ldots Msg \to Msg$. We assume disjoint sets of countably infinite variables X_s for each sort, with $X = \bigcup_s X_s$. $\mathcal{T}_{\Sigma,s}$ is the set of ground terms of sort s and $\mathcal{T}_{\Sigma,s}(X)$ is the set of terms of sort s. We use $\mathcal{T}_\Sigma$ and $\mathcal{T}_\Sigma(X)$ for the corresponding term algebras.

By default, Σ includes the following operators: pairing, projections (first or second element of a pair), symmetric and asymmetric encryption and decryption, digital signing, and hashing. The set of equations $\mathcal{E}$ defining these operators gives rise to an equational theory $(\Sigma, \mathcal{E})$.

In addition to the default operators, users can specify further operators together with the equations defining them. The user-defined equations must be subterm-convergent: directed from left to right, the resulting rewrite system must be convergent and the right-hand side of any equation is either (1) a constant that is in normal form with respect to the rewrite system or (2) a strict subterm of the left-hand side. Combining the default theory with such additional user-supplied operators and their equational specification yields a theory that is also subterm-convergent. We say that a term t is *derivable* from a set of terms M if it can be constructed by repeated application of operators in Σ to the terms in M under the equational theory $\mathcal{E}$.

We define positions in terms as sequences $p = [i_1, \ldots, i_n]$ of positive natural numbers. We use $t|_p$ to denote the subterm of t at position p, and for the empty sequence $[\,]$ we define $t = t|_{[\,]}$. We use the operator $\cdot$ to concatenate sequences.

Two positions p, p' are siblings if $|p| = |p'|$ and there is an immediate parent position p'' such that $p = p'' \cdot [i]$ and $p' = p'' \cdot [i']$ with $i \neq i'$ for two natural numbers i and i'.

All other standard notation follows the account of Baader and Nipkow [2].

2.3 Alice&Bob in Detail

Prior to defining Alice&Bob notation, we first describe the principal's initial knowledge and the knowledge that a principal has during a protocol's execution.

Knowledge and Basic Sets. Principals remember messages they acquire during protocol execution. We call the set of messages that are derivable from acquired messages the principal's *knowledge*. Knowledge is essential for constructing messages to be sent, analyzing received messages, and comparing the messages received in a protocol step to messages received in an earlier step.

A principal's knowledge is in general infinite; if Alice knows a message m, she can immediately derive infinitely many messages, for example by concatenating arbitrarily many copies of m. To finitely represent the knowledge of the principals participating in a protocol, *basic sets* proposed by Caleiro et al. [9] can be used. A basic set for M is a minimal set of terms from which all terms in M can be derived.

Initial Knowledge. A principal's initial knowledge is the basic set of messages that the principal knows prior to performing any protocol actions, that is, before the first sending or receiving action. A default initial knowledge can be generated from the protocol's context, i.e., the protocol and all the roles appearing in it, or the initial knowledge can be explicitly given.

Alice&Bob Protocol Specifications. We now define Alice&Bob protocol specifications.

Definition 1. *An* Alice&Bob *protocol specification is a quadruple* $(Spec, \rho, \Sigma, \mathcal{E})$, *where:*

- *Spec is a finite sequence* $step_1, \ldots, step_n$ *of message exchange steps where, for* $t \in \{1, \ldots, n\}$, $step_t$, *has the form*

$$\mathbf{label_t}. \quad S \to R\colon (n_1, \ldots, n_v).m.$$

  Here R and S are distinct role names (terms of sort Public), $n_1, \ldots, n_v$ are distinct variable names of sort Fresh, $\mathbf{label_t}$ is a unique name given to this message exchange step, and $m \in \mathcal{T}_\Sigma(X)$ is a message.
- *ρ is a partial map $\mathcal{T}_{\Sigma, Public}(X) \to \mathcal{P}(\mathcal{T}_\Sigma(X))$ from role names to sets of messages representing that role's explicit initial knowledge in the protocol. Note that this explicit initial knowledge may very well be empty and can then be omitted. The notational conventions below explain the standard initial knowledge that is always assumed to be available to each role.*

– $(\Sigma, \mathcal{E})$ *is a subterm-convergent equational theory specifying operators and their defining equations.*

Note that we require fresh nonces $(n_1, \ldots, n_v)$ to be stated explicitly. They are assumed to be generated randomly by the sender at the beginning of the step, before the message is constructed and sent. This information is actually redundant: since we know the initial knowledge, we can find out if a nonce is fresh. Nevertheless, we explicitly declare fresh nonces to improve readability and to help catch specification errors.

Our definition says that the fresh nonces must have *distinct* variable names. This not only includes the current message exchange step, but also the complete protocol. Two fresh variables with the same name may not appear in a protocol and a variable that appears in the initial knowledge of a principal may not be redefined as a fresh variable. Moreover, a fresh variable name must not coincide with any role name; this is ensured by having different sorts for role names and fresh variables.

The labels are given just for reference and we omit them in many cases. Also, we drop the parentheses enclosing the fresh variable names when there are none.

Notational Conventions. Alice&Bob protocol specifications rely heavily on implicit notational conventions. For instance, the notation k_{CS} suggests that this term is a shared key between C and S. There is also no need to mention that C is the client and S is the server. We need to be a bit more formal in a computer-interpretable input language, but we also use the following notational conventions and some short-hands to keep our Alice&Bob notation compact yet still precise:

– Variables representing fresh terms (of sort $Fresh$) and general message terms (of sort Msg) are denoted by lower case letters, possibly with subscripts.
– Variables representing public terms (of sort $Public$), including role names, are denoted by capital letters. In the following example, the principal in role A sends her own name to the principal in role B,

$$A \rightarrow B: \ A.$$

Note that the 'A' before the colon denotes the name of a role while the 'A' after the colon denotes the name of the principal that executes role A during an execution of the protocol.
– Constants are public terms that are denoted as strings in single quotes. Below, the principal in role A sends the constant 'Hello!' to the principal in role B:

$$A \rightarrow B: \ \text{'Hello!'}.$$

– The asymmetric encryption of a message m with the public key $pk(A)$ of principal A is denoted by $\{m\}_{pk(A)}$. This is syntactic sugar for $\text{enc}(m, \text{pk}(k))$, where $\text{enc}(\_, \_)$, $\text{dec}(\_, \_)$, and $\text{pk}(\_)$ are operators defined by the equation

$$\text{dec}(\text{enc}(m, \text{pk}(k)), k) = m. \tag{1}$$

Thus $pk(A)$ is syntactic sugar for $\texttt{pk}(k)$, where k is a private key (a fresh term) that A knows.

- Digital signatures are a special case of asymmetric encryption, with $sk(A)$ denoting the secret key k of the principal in role A with the associated public key denoted as $pk(A)$. Signature verification is defined by the equation

$$\texttt{sigverify}(\texttt{sign}(m, sk(A)), m, pk(A)) = \texttt{True}. \tag{2}$$

This equation is treated as a predicate by principals. In protocol analysis tools such as TAMARIN, it is possible to restrict the protocol executions analyzed to those that fulfill the predicate.

- The symmetric encryption of a message m with the secret key $k(A, B)$ shared by the principals A and B is denoted by $\{m\}_{k(A,B)}$. This is syntactic sugar for $\texttt{senc}(m, k)$, where $\texttt{senc}(\_, \_)$ and $\texttt{sdec}(\_, \_)$ are operators defined by the equation

$$\texttt{sdec}(\texttt{senc}(m, k), k) = m \tag{3}$$

and $k(A, B) = k$ for a shared secret key k (a fresh term) that both A and B know.

- For each principal A, we assume that A's initial knowledge contains its own private key $sk(A)$. Moreover, A's initial knowledge includes for each principal B that principal's public key $pk(B)$, and the shared secret key $k(A, B)$. These keys need not be explicitly specified as a principal's initial knowledge. The corresponding encryption functions and their Definitions (1) and (3) are included in the equational theory by default. Similarly, pairing and projection of terms are also included and they satisfy the following two equations.

$$\texttt{fst}(\texttt{pair}(a, b)) = a$$
$$\texttt{snd}(\texttt{pair}(a, b)) = b$$

For $n \geq 2$, we write $(a_1, \ldots, a_n)$ for the repeated, left-associative application of the pairing operator to the terms $a_1, \ldots, a_n$. We drop the parentheses whenever the resulting expression is unambiguous.

With these conventions, Fig. 2 shows a specification of the simplified Kerberos authentication protocol in our Alice&Bob syntax.

1.	$C \rightarrow S:$	$(n). \, C, V, n$
2.	$S \rightarrow C:$	$(k). \, \{k, V, n\}_{k(C,S)}, \{k, C\}_{k(V,S)}$
3.	$C \rightarrow V:$	$(t, t'). \, \{t, t'\}_k, \{k, C\}_{k(V,S)}$
4.	$V \rightarrow C:$	$\{t\}_k$

Fig. 2. The simplified Kerberos authentication protocol in our Alice&Bob specification.

3   From Equations to Rewriting Rules

We translate Alice&Bob protocol scripts to a tool's input language via an intermediate representation called *role scripts*. A role script represents a principal's view of the protocol specification. It consists of the principal's send and receive actions, which each take a message as an argument. The principal sends and receives messages to and from a channel, without information on who the actual communication partner is.

3.1   Role Scripts for Protocols

Given input in Alice&Bob notation, we first check its executability under the fine interpretation of Caleiro et al. [9]. We explain this in Sect. 3.2. Then we translate the protocol to role scripts, one role script for each protocol role. We add appropriate checks to be taken by the principals receiving messages, in Sect. 3.3. In Sect. 3.4 we provide an example that illustrates our algorithms. Afterwards we can generate output in the input language of any suitable protocol verification tool, and we have implemented this for TAMARIN. Output for other tools could be generated from the intermediate role scripts in a similar fashion.

Role scripts have the same syntax as Alice&Bob messages described in Sect. 2.2. They also include the knowledge principals acquire from the messages they receive; in this way, the role scripts make all information explicit.

We represent a protocol by the equational theory its operators support, the security goals of interest, and a collection of role scripts that specify the message exchanges. Each role script consists of a name, an initial knowledge, and an (ordered) list of actions to be performed. The actions are sending or receiving a message, creating nonces, and updating the principal's knowledge afterwards. Both incoming and outgoing messages contain the name of the designated partner role. In the derived role script specification, our algorithm explicitly states which generated names are fresh, i.e., nonces, and the checks that need to be performed by roles on received messages. Moreover, the specification of the secrecy of a given term, as well as non-injective and injective agreement [20] between roles is supported.

3.2   Deciding Executability

An Alice&Bob specification is a list of message exchange steps. A step is *executable* if the sender's knowledge (at that point in the protocol's execution) is sufficient to create the message specified to be sent. Message creation here refers to the capability of the principal to (i) generate new nonces and add them to his knowledge, and (ii) apply operators to messages in his knowledge. Note that to derive a principal's knowledge based on the messages he previously received, it is necessary to apply operators (like decryption) and use their algebraic properties. An Alice&Bob specification is executable if all steps of all roles are executable.

Thus to decide whether an Alice&Bob specification is executable, we must determine for each step of every role whether the term specified to be sent in the step can be derived from the principal's knowledge at that point in the protocol.

Without equational theories, we have a simple separation of knowledge derivation rules into construction and deconstruction rules, reminiscent of Paulson's inductive approach, where they are called synthesis and analysis rules. A similar procedure can be used with equational theories that are subterm-convergent and we now give the high-level description of this. All operators can be applied as *construction* rules that produce, bottom-up, constructed terms. From each equation of the theory, we extract several *deconstruction* rules that produce, top-down, (de-)constructed terms.

To prevent endless loops, we split an agent's knowledge into two sets, the set of constructed terms $\mathcal{C}$ and the set of terms to be deconstructed $\mathcal{D}$. Construction rules may only be applied to terms in the set $\mathcal{C}$ and produce terms that we add to $\mathcal{C}$. Deconstruction rules take a term from the set $\mathcal{D}$ and zero or more terms from $\mathcal{C}$ and produce terms that we add to $\mathcal{D}$. Moreover, we have one rule that may add any term from $\mathcal{D}$ to $\mathcal{C}$. As a result, constructed terms never need be deconstructed later because deconstruction yields a subterm (due to the subterm-convergence property) that must have been used in the construction of the term, and thus is known already and this shorter derivation can be used.

More precisely, construction rules are created as follows: Let $\mathcal{C}^n$ denote the n-fold Cartesian product of $\mathcal{C}$. Let $f \in \Sigma$ be an n-ary function symbol. Then we add the rule $(t_1, t_2, \ldots, t_n) \in \mathcal{C}^n \vdash f(t_1, \ldots, t_n) \in \mathcal{C}$ to the set of construction rules. In particular, there is exactly one construction rule for each function symbol.

For deconstruction rules, let $l = r$ be an equation oriented such that r is a constant or a strict subterm of l. If r is ground, no deconstruction rule is necessary, as the specifications of ground terms are public knowledge. Otherwise, we obtain a set of deconstruction rules for every occurrence of r in l as follows. (See *drules* and *cprems* functions in [28, Section 3.2.3].)

1. Consider the set *positions* of all positions p in l that mark a subterm equal to r, i.e., all p such that $l|_p = r$.
2. For each $p \in positions$, consider all positions D_p that are strictly above p and not equal to $[\,]$, i.e., subterms $l|_{p'}$ of l that contain the term r at position p'' in $l|_{p'}$, i.e., $l|_{p'.p''} = r$ for some $p'' \neq [\,]$, except for $l|_p$ and l.
3. For each $p' \in D_p$, consider the set $C_{p'}$ of positions that have a sibling above or equal to p'.
4. The deconstruction rules are

$$l|_{p'} \in \mathcal{D} \wedge \left(\bigwedge_{q \in C_{p'}} l|_q \in \mathcal{C} \right) \vdash l|_p \in \mathcal{D}, \tag{4}$$

for each $p' \in D_p$, where $l|_{p'}$ must be a term in $\mathcal{D}$ and the terms corresponding to positions in $C_{p'}$ are in $\mathcal{C}$. The deconstructed term $l|_p$ is added to $\mathcal{D}$.

In the following, we call these rules *drules*, and we denote such a rule as $drule(l|_{p'}, p'')$, where p'' is as in Step 2 above and $l|_{p'}$ is the left-most term in

Rule (4). By applying a drule $drule(x, q)$ to a term t we obtain $t|_q$ if x matches t and there exist terms in $\mathcal{C}$ fulfilling the remaining requirements of the left-hand side of the rule, and $\bot$ otherwise. We can now use these rules with the following effective, but computationally expensive, procedure to compute the closure of $\mathcal{D}$ under application of drules to terms in $\mathcal{D}$. The closure algorithm is as follows.

Closure Algorithm. Figure 3 (left) provides pseudo-code for the algorithm `close`(M). This algorithm repeatedly applies all drules to all terms in M, adding the resulting terms to M, until a fixed point is reached.

This algorithm always terminates on finite input sets $\mathcal{D}$. Let $\overline{\mathcal{D}}$ be the set of all (sub-)terms of elements in $\mathcal{D}$. Let $\widehat{\mathcal{D}}$ be the result of applying the closure algorithm to $\mathcal{D}$. All terms derivable from $\mathcal{D}$ are in $\overline{\mathcal{D}}$ due to subterm-convergence, so $\mathcal{D} \subseteq \widehat{\mathcal{D}} \subseteq \overline{\mathcal{D}}$. The closure algorithm monotonically increases its set of derived terms M. It terminates when no element is added to M in an iteration. As the resulting set $\widehat{\mathcal{D}}$ is bounded above by the finite set of (sub-)terms, i.e., $\overline{\mathcal{D}}$, termination is ensured.

Derivability. To decide whether a term t is derivable from a given knowledge set M, we use the algorithm `derivable`(M, t), given in Fig. 3 (right). The algorithm first computes the closure of M and the size of t. It then uses the procedure `constructall`, with the closure and t's size as input, to generate all terms, up to the given size, which are built from elements of the closure with any operator application. The term t is then derivable if and only if it is in this set of constructed terms. Since t is finite, the derivability check terminates.

Our derivability algorithm is sound: the above derivability procedure only returns `true` for terms that are actually derivable because it only uses drules (in the closure algorithm) and construction rules on the initial set of terms. Our algorithm is also complete as it returns the correct answer `true` for all derivable terms. We sketch a proof by contradiction of this: Assume there is a term t that is derivable from M for which our algorithm returns `false`. That term is derived using operator application and simplification with the equational theory on the terms in M. If t is built using only operator application, then it would trivially be derivable according to our algorithm, which is a contradiction. Otherwise an equation in the equational theory must be used in the derivation. Pick the smallest subterm s of t on which some equation is applicable, i.e., there is no proper subterm of s for which any equation can be applied. As the equations

```
        Input: M.
        M' := M
  (*)   ∀m ∈ M, ∀drule(l, q) ∈ drules:
            t := apply drule(l, q) to m
                if t ≠ ⊥ ∧ t ∉ M
                    then M' := M' ∪ {t}
            if M' ≠ M then M := M' goto (*)
        return M
```

```
        Input: M, t.
        D := close(M)
        C := M ∪ D
        if t ∈ constructall(C, size(t))
            then return true
            else return false
```

Fig. 3. Algorithm `close`(M) (left) and algorithm `derivable`(M, t) (right).

are subterm-convergent, applying an equation results in a subterm $s|_p$ at some position p (or a constant, which is known anyway) which must be constructed using only operator applications. Then we can replace s by $s|_p$ in the derived term t to create a term t'. Now there is one less equation application possible on t'. We can repeat this until no more equation can be applied as there is a finite number of equation applications to start with. Thus our algorithm returns true for this term t as well, which is a contradiction.

3.3   Checking Received Messages

For cryptographic protocols, it is not only important that all participants can generate all the outgoing messages, but also that each participant checks each incoming message as thoroughly as possible, to make sure it is as expected and that the other principals have not deviated from the protocol.

An obvious example of what to check is that a message authentication code received matches the intended message. Similarly, if a principal generates a nonce, sends it, and expects to receive it in a subsequent message, he should check that the nonce he receives actually matches the nonce he generated. These are sensible checks that should be included in principals' role scripts; but one must take care to avoid unrealistic or even infeasible checks.

Example 1. Consider again the Kerberos authentication protocol shown in Fig. 2. In this protocol, it is not possible for the agent C to verify in Step 2 that the key k appearing in the first ciphertext is equal to the key k in the second ciphertext. We therefore do not add such a check to C. However, C can verify that the nonce n generated and sent to S as well as the identity of the intended communication partner V in Step 1 are equal to the nonce n and the identity V appearing in the first ciphertext in Step 2. Hence we add these two checks to C.

In general, whenever the same name appears multiple times in a role script, all its instances must be identical, except for those that theprincipal in question cannot actually analyze and check. An example of this is a key inside a ticket created for a different principal, such as is the case for the principal C in the second ciphertext in Step 2 of Kerberos. The principal cannot see the content of that ticket and thus one must not add checks to this principal on this opaque data.

The key ideas of the algorithms, which we describe in detail in the rest of this section, are as follows. We compute the closure of the agent's knowledge with the above closure algorithm, but track additional information on how terms are derived. Then, for each received message, we check if any of the *accessible* parts of the received message were previously known to the agent. If so, we generate a check that compares the received message part in question to the previously known terms. To determine which parts are accessible, we use the closure algorithm as well. Essentially, we store all the received terms and the terms derived from them in a marked set. We test if any term pattern in the marked set can be generated in different ways from the current knowledge (which includes other received messages), and if so we generate a check that the resulting role script should make. Note that this includes the possibility of comparing (parts of) two received messages with each other if they are supposed to be the same. We now delve into the formal details.

Let M be the agent's knowledge. We annotate every term in M with its provenance and we distinguish between $\mathcal{D}$-terms and $\mathcal{C}$-terms with different annotations. We can thus partition M into the two subsets $\mathcal{D}$ and $\mathcal{C}$. The annotation is constructed as follows. If $m \in M$ was received in the i-th protocol step, it is in $\mathcal{D}$ and annotated with $[i]$ and denoted $m^{[i]}$. The subterms of a term $m^{[i]}$ are additionally annotated with their position in $m^{[i]}$, while for the term itself we use $m^{[i]}$ as shorthand for $m^{[i:[\ ]]}$. We call these annotations *locators*. These terms are obtained by applying a drule. For example, suppose that $m^{[3]} = \langle t, t \rangle$, then the two subterms of $m^{[3]}$ are $m^{[3:[1]]} = t$, $m^{[3:[2]]} = t$. Terms in $\mathcal{C}$ are annotated by sets of locators rather than a single locator. When a term $m^{[i:p]} \in \mathcal{D}$ is added to the set $\mathcal{C}$, its annotation is changed from $[i : p]$ to the one-element set $\{[i : p]\}$ and the term is written as $m^{\{[i:p]\}}$. Terms that are constructed from terms with locators are annotated with the set of locators consisting of the union of the locators of all terms used in the construction. I.e., the term $f(t_1^{j_1}, \ldots, t_k^{j_k})$ is annotated with the set $j_1 \cup \ldots \cup j_k$. The need for locators will become clear in the algorithm that computes the checks on the received messages.

Definition 2. *We say that a term $m^{[i:p]}$ can be* verified *with knowledge set M, if it can be constructed from M without using its subterms. Formally, $m^{[i:p]} \in \mathcal{D}$ is equal to some $m^L \in \mathcal{C}$, where $[i : p]$ is not a prefix of any element in L (and not equal to one). We define the function* $\mathtt{verifies}(M, m^{[i:p]})$ *to return the witness m^L if it exists and $\perp$ otherwise.*

Note that the only terms that must be verified by an agent in a protocol execution are $\mathcal{D}$-terms.

While generating an agent's role script from an Alice&Bob specification we insert all possible checks that can be performed on received messages. In some cases, a received message can only be checked after further messages have been received, for instance in commitment schemes. In such a scheme, a principal receives an encrypted message first and the decryption key afterwards. The principal cannot check the first message until he receives the key. We use the set Γ to store messages for which checks may still be needed.

We start with the set $\Gamma = \emptyset$ and the set M equal to the agent's initial knowledge. All terms in the initial knowledge are annotated with $[0]$. Fresh nonces generated by the agent are also annotated with $[0]$ and added to M, but not to Γ. We incrementally build the checks for messages in the order they are received. Let m be a message received in the i-th protocol step. We generate the possible checks by applying the algorithm $\mathtt{agent\text{-}checks}(m^{[i]}, M, \Gamma)$ described below on the message $m^{[i]}$, M, and Γ and afterwards update M and Γ with the algorithm's output. We define $\mathtt{agent\text{-}checks}$ in Fig. 4. It uses the algorithm $\mathtt{check}$, defined in Fig. 5, that checks individual messages.

The algorithm $\mathtt{agent\text{-}checks}$ takes as input a message $m^{[i]}$, a set of unchecked terms Γ, and a knowledge set M. It first adds $m^{[i]}$ to both Γ and M. It then closes

Input: $m^{[i]}$, M, Γ.
$\Gamma := \Gamma \cup m^{[i]}$
$M := M \cup m^{[i]}$
$M' := M$
(*) $\forall m^{[j:p]} \in M$, $\forall drule(l,q) \in drules$:
$\qquad t^{[j:p \cdot q]} := $ apply $drule(l,q)$ to $m^{[j:p]}$
$\qquad\qquad$ if $t^{[j:p \cdot q]} \neq \bot \wedge t^{[j:p \cdot q]} \notin M$
$\qquad\qquad\qquad$ then $\Gamma := \Gamma \cup \{t^{[j:p \cdot q]}\}$, $M' := M' \cup \{t^{[j:p \cdot q]}\}$
$\quad$ if $M' \neq M$ then $M := M'$ goto (*)

$\Gamma' := \emptyset$
$\forall \gamma \in \Gamma$,
$\qquad$ if check$(\gamma, M) = \bot$
$\qquad\qquad$ then $\Gamma' := \Gamma' \cup \gamma$
$\qquad\qquad$ else print check(γ, M)
$\Gamma := \Gamma'$
return M, Γ

Fig. 4. Algorithm agent-checks$(m^{[i]}, M, \Gamma)$

the knowledge set M with our knowledge closure algorithm modified to respect locators. Afterwards, it calls the check algorithm for all terms in the resulting set Γ. The result from check is used to add checks to the generated role script. Terms for which checks were added are removed from Γ.

The check algorithm creates checks for individual messages as follows. Its input is an annotated message $m^{[i:p]}$ (i.e., i-th protocol step, position p), and knowledge set M. If $m^{[i:p]}$ can be verified with M (see Definition 2) then check returns the corresponding check $m^{[i:p]} =^? m^L$, otherwise it returns $\bot$ to signify that no check is possible for this term with the given knowledge set.

Note that for the term $m^{[i:p]}$, the locator implicitly keeps track of how the term was derived from $m^{[i]}$. That way the check created in check compares the term derived from the received message $m^{[i]}$ with the term m^L constructed by the agent.

We briefly sketch why the checking algorithms are sound and complete. By construction, all checks $m^{[i:p]} =^? m^L$ generated by the check algorithm are computable. Both the left-hand term and the right-hand term are in the agent's knowledge. The checks are correct in the sense that $m^{[i:p]}$ is a subterm of a

Input: $m^{[i:p]}$, M.
$m^L := $ verifies$(M, m^{[i:p]})$
if $m^L \neq \bot$
$\qquad$ then return $m^{[i:p]} =^? m^L$
$\qquad$ else return $\bot$

Fig. 5. Algorithm check$(m^{[i:p]}, M)$

received message term and none of the subterms of m^L are derived from $m^{[i:p]}$. That is, the terms used to compute the left-hand and right-hand side have different origins. Finally, we generate all possible checks in the `agent-checks` algorithm since every subterm that can be derived from a received message is added to the set Γ. Each term in Γ that can be constructed from terms that have a different origin is checked against such a construction.

3.4 Putting It All Together

For a given protocol, we execute the above algorithms for each role. If the algorithms decide that the role is executable, they create a role script for the role's actions in the protocol. This includes the checks the role needs to make and warnings that are returned to the user. Essentially, everything left in Γ at the end of the protocol gives rise to a warning. Of course, if some term t cannot be checked and a warning is issued, then all terms that use t as a proper subterm cannot be checked either, and we therefore do not issue warnings for these terms.

The warnings are intended to inform the protocol designer about which message (sub-)terms the specified roles must accept as valid without having a way to check them. The protocol designer should then ensure that these unchecked parts are intentionally given as part of the specification. In particular, (i) they could be irrelevant (and should be dropped), or (ii) they are tickets that are just forwarded and checked by another role (and hence are there for a good reason), or (iii) they are important and protected under encryption with appropriate authentication. In this last case, the unchecked message part contains *new terms for this role* that the role must trust, and have confidence in, due to the overall protocol run. This could be, for example, a fresh key from a key distribution server, and generally speaking a term that is only received and not confirmed in any manner. Hence our approach not only creates explicit checks, but it determines which terms are not actually checked, and it calls the protocol designer's attention to this by issuing warnings.

Consider again the Kerberos authentication protocol previously given in Fig. 2. By the conventions stated in Sect. 2.3, the protocol implicitly uses Eq. (3) for the symmetric encryption and decryption of terms. Thus Σ contains the symbols `senc` and `sdec`. By Sect. 3.2 we first obtain the following two construction rules:

$$(x, y) \in \mathcal{C}^2 \vdash \mathsf{senc}(x, y) \in \mathcal{C} \tag{5}$$
$$(x, y) \in \mathcal{C}^2 \vdash \mathsf{sdec}(x, y) \in \mathcal{C}. \tag{6}$$

Next we apply the four step procedure in Sect. 3.2 to obtain a set of deconstruction rules for Eq. (3).

1. We have $l = \mathsf{sdec}(\mathsf{senc}(m, k), k)$ and $r = m$. As illustrated in Fig. 6, there is one position p in l such that $l|_p = r$, namely $p = [1, 1]$.
2. By Fig. 6, there is exactly one position that is above $p = [1, 1]$ and not equal to $[\ ]$, namely $[1]$, so $D_{[1,1]} = \{[1]\}$.

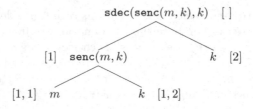

Fig. 6. Tree of subterms of $\mathtt{sdec}(\mathtt{senc}(m,k),k)$ and their positions.

3. Figure 6 shows that $C_{[1]}$, the set of positions that have a sibling above or equal to $[1]$, is $\{[2]\}$.
4. Our deconstruction rule is therefore:

$$l|_{[1]} \in \mathcal{D} \wedge l|_{[2]} \in \mathcal{C} \vdash l|_{[1,1]} \in \mathcal{D},$$

that is

$$\mathtt{senc}(m,k) \in \mathcal{D} \wedge k \in \mathcal{C} \vdash m \in \mathcal{D}.$$

Similarly, we obtain three construction rules for the pairing operator **pair** and the two projections **fst** and **snd**, and two deconstruction rules for the two projections. We omit the details.

We now turn to the role scripts. For readability, we omit irrelevant terms from a role's knowledge. The three role scripts in the Kerberos protocol are then given as follows:

– Role script for role S

$$
\begin{array}{ll}
 & S \text{ knows: } \quad S^{[0]}, C^{[0]}, V^{[0]}, k(C,S)^{[0]}, k(V,S)^{[0]} \\
\mathbf{1.} & \quad C \to S: \ (n).((C,V),n) \\
 & S \text{ checks: } \quad (C,V)^{[1:[1]]} = (C^{[0]}, V^{[0]}) \\
 & S \text{ checks: } \quad C^{[1:[1,1]]} = C^{[0]} \\
 & S \text{ checks: } \quad V^{[1:[1,2]]} = V^{[0]} \\
 & S \text{ knows: } \quad S^{[0]}, C^{[0]}, V^{[0]}, k(C,S)^{[0]}, k(V,S)^{[0]}, n^{[1:[2]]} \\
\mathbf{2.} & \quad S \to C: \ (k).\{k,V,n\}_{k(C,S)}, \{k,C\}_{k(V,S)}, \\
 & S \text{ warns: } \quad n^{[1:[2]]}
\end{array}
$$

This role script is executable because S can generate the message in Step 2 from its knowledge shown above Step 2 by applying pairing and encryption. Our algorithm obtains the check after Step 1 as the first subterm (C,V) of the received pair $((C,V),n)$ is constructable from the initial knowledge. Also, both subterms of this pair, namely C and V, are constructable from the initial knowledge as well, so our algorithm adds an additional check each. Note that our algorithm produces redundant checks here (those for $C^{[1:[1,1]]}$ and $V^{[1:[1,2]]}$), but these can be filtered out afterwards as they are subterms of the term $(C,V)^{[1:[1]]}$ checked against the same right-hand side. The only warning produced is for $n^{[1:[2]]}$, which is indeed a term that this principal cannot check anything about.

- Role script for role C

$$C \text{ knows: } S^{[0]}, C^{[0]}, V^{[0]}, k(C,S)^{[0]}$$

    1. $C \to S:\ (n).C, V, n$

    2. $S \to C:\ (k).\{k, (V,n)\}_{k(C,S)}, \{k, C\}_{k(V,S)}$

        $C \text{ checks}:\ (V,n)^{[2:[1,1,2]]} = (V^{[0]}, n^{[0]})$

        $C \text{ checks}:\ V^{[2:[1,1,2,1]]} = V^{[0]}$

        $C \text{ checks}:\ n^{[2:[1,1,2,2]]} = n^{[0]}$

        $C \text{ knows}:\ S^{[0]}, C^{[0]}, V^{[0]}, k(C,S)^{[0]}, n^{[0]}, k^{[2:[1,1,1]]}$

    3. $C \to V:\ (t,t').\{t,t'\}_k, \{k, C\}_{k(V,S)}$

    4. $V \to C:\ \{t\}_k$

        $C \text{ checks}:\ (\{t\}_k)^{[4]} = \{t^{[0]}\}_{k^{[2:[1,1,1]]}}$

        $C \text{ checks}:\ t^{[4:[1]]} = t^{[0]}$

        $C \text{ warns}:\ k^{[2:[1,1,1]]}$

This role script is executable because C can generate the first message from its initial knowledge and the message in Step 3 from its knowledge shown above Step 3. The term k is obtained by applying Eq. (3) to the first component of the pair received in Step 2 and picking the first element of the resulting pair. The only resulting warning is for the received key, which cannot be checked.

- Role script for role V

$$V \text{ knows: } S^{[0]}, C^{[0]}, V^{[0]}, k(V,S)^{[0]}$$

    3. $C \to V:\ (t,t').\{t,t'\}_k, \{k, C\}_{k(V,S)}$

        $V \text{ checks}:\ C^{[3:[2,1,2]]} = C^{[0]}$

        $V \text{ knows}:\ S^{[0]}, C^{[0]}, V^{[0]}, k(V,S)^{[0]}, k^{[3:[2,1,1]]}, t^{[3:[1,1,1]]}, t'^{[3:[1,1,2]]}$

    4. $V \to C:\ \{t\}_k$

        $V \text{ warns}:\ k^{[3:[2,1,1]]}, t^{[3:[1,1,1]]}, t'^{[3:[1,1,2]]}$

This role script is executable because V can generate the message $\{t\}_k$ from its knowledge. The resulting warning is about the three terms for which no check can be included: the key k and the two nonces t and t'.

4   Automated Translation to Tamarin

An instance of our Alice&Bob translation to a tool-supported protocol specification language is described in more detail in [18]. The translation goes from Alice&Bob via an *intermediate representation* format to TAMARIN's input language. The intermediate representation is functionally the same as the role scripts described in this paper, with minor syntactic differences. TAMARIN uses the tool-generated input to analyze the given protocol.

On loading a protocol theory, TAMARIN detects when a protocol rule is not executable; however our automatic translation only produces theories that meet the executability requirement. TAMARIN supports user-specified subterm-convergent equational theories. Our translation therefore simply copies the equational theory with some minor syntactic changes. Checks on received messages are implemented by pattern matching on the premises of rules. The security goals of secrecy, non-injective agreement, and injective agreement are translated into their canonical definition used when specifying protocols for TAMARIN.

More detail and further examples are available in [18] and the tool is available at the webpage [3]. Due to space constraints, we do not list the TAMARIN specification for our running example produced by the translation.

There is also a prototype implementation of an explicit check generator, following our algorithms described above. It is available at [3].

5   Related Work

We will first discuss other research related to Alice&Bob notation. Afterwards we consider different tools' input languages, which are the target languages for our translation effort. Finally, we discuss other proposed translation mechanisms.

Formalization of Alice&Bob Notation. Alice&Bob notation, while intuitive, suffers from ambiguities and imprecision as shown in Sect. 2.1. To clarify what protocol specification notation actually means, Caleiro et al. [8,9] and Mödersheim [26] have investigated the semantics of Alice&Bob notation.

Caleiro et al. work with a fixed message model and consider how principals' knowledge increases during a protocol run as principals receive messages. Moreover, they provide an operational semantics, based on the spi calculus, that makes explicit the actions that a principal must execute. The key aspect of the operational semantics is that it provides detailed checks to be performed by principals to ensure there was no adversary involvement. Our semantics of Alice&Bob protocol specifications are based on this work.

In contrast to Caleiro et al.'s semantics, which is based on a fixed message model, Mödersheim gives a formalization of Alice&Bob notation that is defined over an arbitrary algebraic theory. However, his method does not directly yield the actions that must be taken by honest principals.

Input Languages. Each automated security protocol verification tool defines its own input language. Most of these languages look rather different from Alice&Bob notation. However, their underlying concepts are often similar to the core idea of Alice&Bob notation: communication is modeled by specifying the messages that are sent and received by principals participating in a protocol run. Most input languages however do not explicitly pair the sender and the receiver as is done in Alice&Bob notation.

In Maude-NPA [15] protocols are specified by defining *strands*, which are similar to the *roles* used in our work. A strand specifies a sequence of sending and receiving messages, from one participant's point of view. Using Maude's [11] unification capabilities, Maude-NPA then reasons modulo equational theories.

Similarly, protocols in Scyther [13] are specified by explicitly stating which actions (sending, receiving, generation of fresh numbers, and claims of security properties) must be taken by principals. Scyther-proof [23] is a tool based on a proof-generating variant [24] of the verification theory underlying Scyther. Its input language uses proper Alice&Bob-style notation for specifying protocols.

In ProVerif [7] protocols are specified in the applied pi calculus as the parallel compositions of processes that correspond to roles in Alice&Bob notation. Checks on received messages must be explicitly stated.

TAMARIN's [29] input language is based on specifying rewriting rules for multisets of so-called facts. State is usually expressed with the help of user-defined facts and communication by the predefined In and Out facts, which represent sending and receiving actions. Hence, TAMARIN also works by stating the send and receive actions of principals.

Even though all of the tools' input languages have aspects in common with Alice&Bob notation, they all heavily rely on the specification of additional information, such as algebraic properties and typing rules, which must be stated explicitly. Mödersheim uses an elegant Alice&Bob-style language called AnB [26] where the algebraic properties of the messagemodel are assumed to be fixed, and consequently need not be included in the protocol specification itself. In our work, we fuse the different approaches into an input language that has a pre-defined message model, similar to [26], but which is extensible with user-specifiable subterm-convergent equations while leaving the adversary model unspecified.

Existing Translations. There are two steps that a translation from an Alice&Bob input language into a tool-specific language must perform. The first is to verify that a given Alice&Bob specification is executable and the second is to extract the security checks that a role must perform on received messages. It is in these two steps that existing translations differ. Executability is important, as otherwise protocol participants cannot carry out their steps and run the protocol. Non-executability indicates either a mistake in the protocol or its formalization, for example, a missing setup assumption. Checks on incoming messages are also important, for example, to ensure that the message authentication code a participant received refers to the message actually received.

Chevalier et al. [10] translate protocol narrations to strand-like role scripts that are annotated with explicit unifiability conditions. Their protocol narrations are similar to our Alice&Bob input, i.e., they specify the messages exchanged as well as the sending and receiving agent, and the initial knowledge of each agent. They verify executability during the translation to role scripts. Their security checks are such that each participant verifies received messages as far as possible. Namely, a message item that is reused in the description must be the same for the receiver in all incoming messages. This imposes checks that are practically infeasible for agents to perform and we give an example in Sect. 3.3, Example 1. For these cases, we do not add such infeasible checks; instead we warn the protocol designer that something might not be working as intended due to the inherent imprecision of such protocol narrations or Alice&Bob specifications. The output of their translation is not directly analyzable by any existing tool, unlike ours, which can directly output descriptions that can be analyzed by TAMARIN, in addition to producing role scripts.

Another alternative translation is based on endpoint projections [22]. The protocol, given in a language close to Alice&Bob notation, is transformed in multiple steps to a role script-like language. Along the way, the implicit assumptions in the input are made explicit by making choices. The endpoint projection must, in particular, deal with the asymmetries introduced by security protocols, for example, the receiver might not (yet) know the key being used in an encryption. Making these choices to get to an explicit presentation is similar to what we do. The main restrictions in their work, compared to ours, is that their translation cannot handle equational theories beyond equations formalizing symmetric and asymmetric encryption and it does not allow principals to transmit two or more messages in a row without receiving a message in between.

The *Common Authentication Protocol Specification Language* CAPSL [14] uses message list input similar to our Alice&Bob specifications, extended with Casper's % operator [21] to indicate receiver patterns. The use of these extended patterns makes protocol specification more cumbersome, but it substantially simplifies the problem of determining which checks should be made when receiving messages, which we do in our work without resorting to such patterns. CAPSL translates the input message lists into an intermediate language CIL, given in multiset rewriting. Subsequent translations from CIL to different analysis tools, such as Maude, have also been carried out [14].

6   Conclusions

We have presented an analysis of Alice&Bob style protocol specifications, yielding a translation into an intermediate role script-like format that handles executability concerns and generates appropriate checks for correct message reception. We have also implemented a further translation of this intermediate format to the input of TAMARIN, a cryptographic protocol verification tool that uses multiset rewriting for protocol specifications. This translation is automated and allows Alice&Bob protocol specifications to be analyzed and verified with TAMARIN.

Alice&Bob notation is simple and can be used by novices. Indeed we believe that our work could help to teach undergraduate students about protocol specification and analysis in a formal methods course, with some caveats. The students can specify protocols nicely using Alice&Bob notation, and, when the back-end verification succeeds, they have proven the security property. However, when verification fails, the tool's counter-example is still in a format that is not very meaningful for students. To make this viable for teaching, one would need to define and implement a back-translation from the tool's output representation to Alice&Bob like syntax that is easier to understand. This back-translation should benefit from knowledge gained in the initial translation from Alice&Bob notation to the tool's input. In particular, for TAMARIN, this counter-example output is in the form of a constraint system of dependency graphs. We are investigating the conversion of these back to Alice&Bob notation as future work.

References

1. Armando, A., Basin, D., Boichut, Y., Chevalier, Y., Compagna, L., Cuellar, J., Drielsma, P.H., Heám, P.C., Kouchnarenko, O., Mantovani, J., Mödersheim, S., von Oheimb, D., Rusinowitch, M., Santiago, J., Turuani, M., Viganò, L., Vigneron, L.: The AVISPA tool for the automated validation of internet security protocols and applications. In: Etessami, K., Rajamani, S.K. (eds.) CAV 2005. LNCS, vol. 3576, pp. 281–285. Springer, Heidelberg (2005)
2. Baader, F., Nipkow, T.: Term Rewriting and All That. Cambridge University Press, Cambridge (1998)
3. Basin, D., Keller, M., Radomirović, S., Sasse, R.: Alice&Bob protocols. http://www.infsec.ethz.ch/research/software/anb.html
4. Basin, D., Cremers, C., Meier, S.: Provably repairing the ISO/IEC 9798 standard for entity authentication. J. Comput. Secur. **21**(6), 817–846 (2013)
5. Basin, D., Cremers, C., Kim, T.H.-J., Perrig, A., Sasse, R., Szalachowski, P.: ARPKI: attack resilient public-key infrastructure. In: Ahn, G.-J., Yung, M., Li, N. (eds.) Proceedings of the 2014 ACM SIGSAC Conference on Computer and Communications Security, 3–7 November 2014, Scottsdale, AZ, USA, pp. 382–393. ACM (2014)
6. Blanchet, B.: Proverif automatic cryptographic protocol verifier user manual. CNRS, Departement d'Informatique, Ecole Normale Superieure, Paris (2005)
7. Blanchet, B.: An efficient cryptographic protocol verifier based on prolog rules. In: Computer Security Foundations Workshop (CSFW), pp. 82–96. IEEE (2001)
8. Caleiro, C., Viganò, L., Basin, D.: Deconstructing Alice and Bob. Electron. Notes Theoret. Comput. Sci. **135**(1), 3–22 (2005). Proceedings of the Workshop on Automated Reasoning for Security Protocol Analysis, ARSPA 2005
9. Caleiro, C., Viganò, L., Basin, D.: On the semantics of Alice&Bob specifications of security protocols. Theor. Comput. Sci. **367**(1–2), 88–122 (2006)
10. Chevalier, Y., Rusinowitch, M.: Compiling and securing cryptographic protocols. Inf. Process. Lett. **110**(3), 116–122 (2010)
11. Clavel, M., Durán, F., Eker, S., Lincoln, P., Martí-Oliet, N., Meseguer, J., Talcott, C. (eds.): All About Maude - A High-Performance Logical Framework. LNCS, vol. 4350. Springer, Heidelberg (2007)
12. Comon-Lundh, H., Delaune, S.: The finite variant property: how to get rid of some algebraic properties. In: Giesl, J. (ed.) RTA 2005. LNCS, vol. 3467, pp. 294–307. Springer, Heidelberg (2005)
13. Cremers, C.: Unbounded verification, falsification, and characterization of security protocols by pattern refinement. In: ACM Conference on Computer and Communications Security (CCS), pp. 119–128. ACM (2008)
14. Denker, G., Millen, J.K.: CAPSL intermediate language. In: Proceedings of FMSP 1999 (1999). http://www.csl.sri.com/users/millen/capsl/
15. Escobar, S., Meadows, C., Meseguer, J.: Maude-NPA: cyptographic protocol analysis modulo equational properties. In: Aldini, A., Barthe, G., Gorrieri, R. (eds.) FOSAD. LNCS, vol. 5705, pp. 1–50. Springer, Heidelberg (2007)
16. Escobar, S., Sasse, R., Meseguer, J.: Folding variant narrowing and optimal variant termination. J. Logic Algebraic Program. **81**(7–8), 898–928 (2012)
17. Fabrega, F.J.T., Herzog, J., Guttman, J.: Strand spaces: what makes a security protocol correct? J. Comput. Secur. **7**, 191–230 (1999)
18. Keller, M.: Converting Alice and Bob protocol specifications to Tamarin. Bachelor's thesis, ETH Zurich (2014). http://www.infsec.ethz.ch/research/software/anb.html

19. Lowe, G.: Breaking and fixing the Needham-Schroeder public-key protocol using FDR. In: TACAS, pp. 147–166 (1996)
20. Lowe, G.: A hierarchy of authentication specifications. In: Proceedings of the 10th IEEE Workshop on Computer Security Foundations, CSFW 1997, pp. 31–43, Washington, DC, USA. IEEE Computer Society (1997)
21. Lowe, G.: Casper: a compiler for the analysis of security protocols. J. Comput. Secur. 6(1), 53–84 (1998)
22. McCarthy, J., Krishnamurthi, S.: Cryptographic protocol explication and end-point projection. In: Jajodia, S., Lopez, J. (eds.) ESORICS 2008. LNCS, vol. 5283, pp. 533–547. Springer, Heidelberg (2008)
23. Meier, S.: GitHub repository of scyther-proof Project. https://github.com/meiersi/scyther-proof
24. Meier, S., Cremers, C., Basin, D.: Strong invariants for the efficient construction of machine-checked protocol security proofs. In: CSF, pp. 231–245. IEEE Computer Society (2010)
25. Meier, S., Schmidt, B., Cremers, C., Basin, D.: The TAMARIN prover for the symbolic analysis of security protocols. In: Sharygina, N., Veith, H. (eds.) CAV 2013. LNCS, vol. 8044, pp. 696–701. Springer, Heidelberg (2013)
26. Mödersheim, S.: Algebraic properties in Alice and Bob notation. In: ARES, pp. 433–440. IEEE Computer Society (2009)
27. Neuman, B.C., Ts'o, T.: Kerberos: an authentication service for computer networks. Communications 32(9), 33–38 (1994)
28. Schmidt, B.: Formal analysis of key exchange protocols and physical protocols. Ph.D. dissertation, ETH Zurich (2012)
29. Schmidt, B., Meier, S., Cremers, C., Basin, D.: Automated analysis of Diffie-Hellman protocols and advanced security properties. In: Computer Security Foundations Symposium (CSF), pp. 78–94. IEEE (2012)
30. Schmidt, B., Sasse, R., Cremers, C., Basin, D.: Automated verification of group key agreement protocols. In: 2014 IEEE Symposium on Security and Privacy, SP 2014, 18–21 May 2014, Berkeley, CA, USA, pp. 179–194. IEEE Computer Society (2014)

On First-Order Model-Based Reasoning

Maria Paola Bonacina[1]([⊠]), Ulrich Furbach[2],
and Viorica Sofronie-Stokkermans[2]

[1] Dipartimento di Informatica, Università Degli Studi di Verona, Verona, Italy
mariapaola.bonacina@univr.it
[2] Fachbereich Informatik, Universität Koblenz-Landau, Koblenz, Germany
{furbach,sofronie}@uni-koblenz.de

Dedicated to
José Meseguer,
friend and colleague.

Abstract. Reasoning semantically in first-order logic is notoriously a challenge. This paper surveys a selection of *semantically-guided* or *model-based* methods that aim at meeting aspects of this challenge. For first-order logic we touch upon *resolution-based* methods, *tableaux-based* methods, *DPLL-inspired* methods, and we give a preview of a new method called SGGS, for *Semantically-Guided Goal-Sensitive* reasoning. For first-order theories we highlight *hierarchical* and *locality-based* methods, concluding with the recent *Model-Constructing satisfiability calculus*.

1 Introduction

Traditionally, automated reasoning has centered on *proofs* rather than *models*. However, models are useful for applications, intuitive for users, and the notion that *semantic guidance* would help proof search is almost as old as theorem proving itself. In recent years there has been a surge of *model-based* first-order reasoning methods, inspired in part by the success of model-based solvers for propositional satisfiability (SAT) and satisfiability modulo theories (SMT).

The core procedure of these solvers is the *conflict-driven clause learning* (CDCL) version [52,60,62,88] of the Davis-Putnam-Logemann-Loveland (DPLL) procedure for propositional logic [32]. The original Davis-Putnam (DP) procedure [33] was proposed for first-order logic, and featured propositional, or ground, resolution. The DPLL procedure replaced propositional resolution with *splitting*, initially viewed as breaking disjunctions apart by *case analysis*, to avoid the growth of clauses and the non-determinism of resolution. Later, splitting was understood as *guessing*, or *deciding*, the truth value of a propositional variable, in order *to search for a model* of the given set of clauses. This led to read DPLL as a *model-based* procedure, where all operations are centered around a candidate partial model, called *context*, represented by a sequence, or *trail*, of literals.

DPLL-CDCL brought back propositional resolution as a mechanism to generate *lemmas*, and achieve a better balance between *guessing* and *reasoning*. The

N. Martí-Oliet et al. (Eds.): Meseguer Festschrift, LNCS 9200, pp. 181–204, 2015.
DOI: 10.1007/978-3-319-23165-5_8

model-based character of the procedure became even more pronounced: when the current candidate model falsifies a clause, this *conflict* is *explained* by a heuristically controlled series of resolution steps, a resolvent is added as lemma, and the candidate partial model is repaired in such a way to remove the conflict, satisfy the lemma, and backjump as far away as possible from the conflict. SMT-solvers integrate in DPLL-CDCL a decision procedure for satisfiability in a theory or combination of theories $\mathcal{T}$: the $\mathcal{T}$-satisfiability procedure raises a $\mathcal{T}$-*conflict* when the candidate partial model is not consistent with $\mathcal{T}$, and generates $\mathcal{T}$-*lemmas* to add theory reasoning to the inference component [7, 34].

While SAT and SMT-solvers offer fast decision procedures, they typically apply to sets of propositional or ground clauses, without quantifiers. Indeed, decidability of the problem and termination of the procedure descend from the fact that the underlying language is the *finite* set of the input atoms.

ATP (Automated Theorem Proving) systems offer theorem-proving strategies that are designed for the far more expressive language of first-order logic, but are only *semi-decision procedures* for validity, as the underlying language, and search space, are *infinite*. This trade-off between *expressivity* and *decidability* is ubiquitous in logic and artificial intelligence. First-order satisfiability is not even semi-decidable, which means that first-order model-building cannot be mechanized in general. Nevertheless, there exist first-order reasoning methods that are *semantically-guided* by a *fixed* interpretation, and even *model-based*, in the sense that the state of a derivation contains a representation of a candidate partial model that evolves with the derivation.

In this survey, we illustrate a necessarily incomplete selection of such methods for first-order logic (Sect. 2) or first-order theories (Sect. 3). In each section the treatment approximately goes from syntactic or axiomatic approaches towards more semantic ones, also showing connections with José Meseguer's work. All methods are described in expository style, and the interested reader may find the technical details in the references. Background material is available in previous surveys, such as [18, 59, 67–69] for theorem-proving strategies, [19] for decision procedures based on theorem-proving strategies or their integration with SMT-solvers, and books such as [17, 70, 76].

2   Model-Based Reasoning in First-Order Logic

In this section we cover *semantic resolution*, which represents the early attempts at injecting semantics in resolution; *hypertableaux*, which illustrates model-based reasoning in tableaux, with applications to fault diagnosis and description logics; the *model-evolution calculus*, which lifts DPLL to first-order logic, and a new method called *SGGS*, for *Semantically-Guided Goal-Sensitive* reasoning, which realizes a first-order CDCL mechanism.

2.1   Semantic Resolution

Soon after the seminal article by Alan Robinson introducing the resolution principle [75], James R. Slagle presented *semantic resolution* in [79]. Let S be the finite set of first-order clauses to be refuted. Slagle's core idea was to use a given Herbrand interpretation I to avoid generating resolvents that are true in I, since expanding a consistent set should not lead to a refutation. The following example from [31] illustrates the concept in propositional logic:

Example 1. Given $S = \{\neg A_1 \vee \neg A_2 \vee A_3,\ A_1 \vee A_3,\ A_2 \vee A_3\}$, let I be all-negative, that is, $I = \{\neg A_1, \neg A_2, \neg A_3\}$. Resolution between $\neg A_1 \vee \neg A_2 \vee A_3$ and $A_1 \vee A_3$ generates $\neg A_2 \vee A_3$, after merging identical literals. Similarly, resolution between $\neg A_1 \vee \neg A_2 \vee A_3$ and $A_2 \vee A_3$ generates $\neg A_1 \vee A_3$. However, these two resolvents are true in I. Semantic resolution prevents generating such resolvents, and uses all three clauses to generate only A_3, which is false in I.

Formally, say that we have a clause N, called *nucleus*, and clauses $E_1, \ldots, E_q$, with $q \geq 1$, called *electrons*, such that the electrons are *false* in I. Then, if there is a series of clauses $R_1, R_2, \ldots, R_q, R_{q+1}$, where R_1 is N, R_{i+1} is a resolvent of R_i and E_i, for $i = 1, \ldots, q$, and R_{q+1} is *false* in I, semantic resolution generates only R_{q+1}. The intuition is that electrons are used to resolve away literals in the nucleus until a clause false in I is generated.

Example 2. In the above example, $\neg A_1 \vee \neg A_2 \vee A_3$ is the nucleus N, and $A_1 \vee A_3$ and $A_2 \vee A_3$ are the electrons E_1 and E_2, respectively. Resolving N and E_1 gives $\neg A_2 \vee A_3$, and resolving the latter with E_2 yields A_3: only A_3 is retained, while the intermediate resolvent $\neg A_2 \vee A_3$ is not.

Semantic resolution can be further restricted by assuming a precedence $>$ on predicate symbols, and stipulating that in each electron the predicate symbol of the literal resolved upon must be maximal in the precedence. The following example also from [31] is in first-order logic:

Example 3. For $S = \{Q(x) \vee Q(a) \vee \neg R(y) \vee \neg R(b) \vee S(c),\ \neg Q(z) \vee \neg Q(a),\ R(b) \vee S(c)\}$, let I be $\{Q(a), Q(b), Q(c), \neg R(a), \neg R(b), \neg R(c), \neg S(a), \neg S(b), \neg S(c)\}$, so that $I \not\models \neg Q(z) \vee \neg Q(a)$ and $I \not\models R(b) \vee S(c)$. Assume the precedence $Q > R > S$. Thus, $Q(x) \vee Q(a) \vee \neg R(y) \vee \neg R(b) \vee S(c)$ is the nucleus N, and $\neg Q(z) \vee \neg Q(a)$ and $R(b) \vee S(c)$ are the electrons E_1 and E_2, respectively. Resolution between N and E_1 on the Q-literals produces $\neg R(y) \vee \neg R(b) \vee S(c)$, which is not false in I, and therefore it is not kept. Note that this resolution step is a binary resolution step between a factor of N and a factor of E_1. Resolution between $\neg R(y) \vee \neg R(b) \vee S(c)$ and E_2 on the R-literals yields $S(c)$. This second resolution step is a binary resolution between a factor of $\neg R(y) \vee \neg R(b) \vee S(c)$ and E_2. Resolvent $S(c)$ is false in I and it is kept.

In these examples I is given by a finite set of literals: Example 1 is propositional, and in Example 3 the Herbrand base is finite, because there are no function symbols. The examples in [79] include a theorem from algebra, where the interpretation is given by a multiplication table and hence is really of semantic nature. The crux of semantic resolution is the *representation* of I. In theory, a Herbrand interpretation is given by a subset of the Herbrand base of S. In practice, one needs a *finite* representation of I, which is a non-trivial issue, whenever the Herbrand base is not finite, or a mechanism to test the truth of a literal in I. Two instances of semantic resolution that aimed at addressing this issue are *hyperresolution* [74] and the *set-of-support strategy* [86].

Hyperresolution assumes that I contains either all negative literals or all positive literals. In the first case, it is called *positive* hyperresolution, because electrons and all resolvents are positive clauses: positive electrons are used to resolve away all negative literals in the nucleus to get a positive hyperresolvent. In the second case, it is called *negative* hyperresolution, because electrons and all resolvents are negative clauses: negative electrons are used to resolve away all positive literals in the nucleus to get a negative hyperresolvent. Example 1 is an instance of positive hyperresolution.

The set-of-support strategy assumes that $S = T \uplus SOS$, where SOS (for Set of Support) contains initially the clauses coming from the negation of the conjecture, and $T = S \setminus SOS$ is consistent, for some I such that $I \models T$ and $I \not\models SOS$. A resolution of two clauses is permitted, if at least one is from SOS, in order to avoid expanding the consistent set T. All resolvents are added to SOS. Thus, all inferences involve clauses descending from the negation of the conjecture: a method with this property is deemed *goal-sensitive*.

In terms of implementation, positive hyperresolution is often implemented in contemporary theorem provers by resolution with *selection of negative literals*. Indeed, resolution can be restricted by a *selection function* that selects negative literals [4]. A clause can have all, some, or none of its negative literals selected, depending on the selection function. In *resolution with negative selection*, the negative literal resolved upon must be selected, and the other parent must not contain selected literals. If some negative literal is selected for each clause containing one, one parent in each resolution inference will be a positive clause, that is, an electron for positive hyperresolution. Thus, a selection function that selects some negative literal in each clause containing one induces resolution to simulate hyperresolution as a macro inference involving several steps of resolution.

The set-of-support strategy is available in all theorem provers that feature the *given-clause loop* [61], which is a de facto standard for resolution-based provers. This algorithm maintains two lists of clauses, named *to-be-selected* and *already-selected*, and at each iteration it extracts a *given clause* from *to-be-selected*. In its simplest version, with only resolution as inference rule, it performs all resolutions between the given clause and the clauses in *already-selected*; adds all resolvents to *to-be-selected*; and adds the given clause to *already-selected*. If one initializes these lists by putting the clauses in T in *already-selected*, and the clauses in SOS in *to-be-selected*, this algorithm implements the set-of-support strategy. Indeed, in the original version of the given-clause algorithm, *to-be-selected* was called SOS, and *already-selected* was called *Usable*.

State-of-the-art resolution-based theorem provers implement more sophisticated versions of the given clause algorithm, which also accomodate *contraction rules*, that delete (e.g., *subsumption, tautology deletion*) or simplify clauses (e.g., *clausal simplification, equational simplification*). The compatibility of contraction rules with semantic strategies is not obvious, as shown by the following:

Example 4. Let $T = \{\neg P,\ P \vee Q\}$ and $SOS = \{\neg Q\}$. Clausal simplification, which is a combination of resolution and subsumption, applies $\neg Q$ to simplify $P \vee Q$ to P. If the result is $T = \{\neg P,\ P\}$ and $SOS = \{\neg Q\}$, the consistent set T becomes inconsistent, and the refutational completeness of resolution with set-of-support collapses, since the set-of-support strategy does not allow us to resolve P and $\neg P$, being both in T. The correct application of clausal simplification yields $T = \{\neg P\}$ and $SOS = \{\neg Q,\ P\}$, so that the refutation can be found.

In other words, if a clause in SOS simplifies a clause, whether in T or in SOS, the resulting clause must be added to SOS. The integration of contraction rules and other enhancements, such as *lemmaizing*, in semantic strategies was investigated in general in [22].

Semantic resolution, hyperresolution, and the set-of-support strategy exhibit *semantic guidance*. We deem a method *semantically guided*, if it employs a *fixed* interpretation to drive the inferences. We deem a method *model-based*, if it builds and transforms a *candidate model*, and uses it to drive the inferences.

A beginning of the evolution from being semantically guided to being model-based can be traced back to the SCOTT system [80], which combined the finite model finder FINDER, that searches for small models, and the resolution-based theorem prover OTTER [61]. As the authors write "SCOTT brings semantic information gleaned from the proof attempt into the service of the syntax-based theorem prover." In SCOTT, FINDER provides OTTER with a *guide model*, which is used for an extended set-of-support strategy: in each resolution step at least one of the parent clauses must be false in the guide model. During the proof search FINDER updates periodically its model to make more clauses true. Thus, inferences are controlled as in the set-of-support strategy, but the guide model is *not fixed*, which is why SCOTT can be seen as a forerunner of model-based methods. Research on the cooperation between theorem prover and finite model finder continued with successors of OTTER, such as Prover9, and successors of FINDER, such as MACE4 [87]. This line of research has been especially fruitful in applications to mathematics (e.g., [3,38]).

2.2   Hypertableaux

Tableau calculi offer an alternative to resolution and they have been discussed abundantly in the literature (e.g., Chap. 3 in [76]). Their advantages include no need for a clause normal form, a single proof object, and an easy extendability to other logics. The disadvantage, even in the case of clause normal form tableaux, is that variables are *rigid*, which means that substitutions have to be applied to all occurrences of a variable within the entire tableau. The *hypertableau calculus* [10] offers a more liberal treatment of variables, and borrows the concept of hyperinference from positive hyperresolution.

In this section, we adopt a Prolog-like notation for clauses: $A_1 \vee \ldots \vee A_m \vee \neg B_1 \vee \ldots \vee \neg B_n$ is written $A_1, \ldots, A_m \Leftarrow B_1, \ldots, B_n$, where $A_1, \ldots, A_m$ form the *head* of the clause and are called *head literals*, and $B_1, \ldots, B_n$ form the *body*. There are two rules for constructing a hypertableau (cf. [10]): the *initialization*

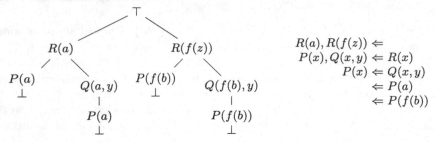

Fig. 1. A sample hypertableaux refutation with the clause set on the right.

rule gives a tableau consisting of a single node labeled with $\top$; this one-element branch is *open*. The *hyperextension* rule selects an open branch and a clause $A_1, \ldots, A_m \Leftarrow B_1, \ldots, B_n$, where $m, n \geq 0$, from the given set S, such that there exists a most general unifier σ which makes all the $B_i \sigma$'s *follow logically* from the model given by the branch. If there is a variable in the clause that has an occurrence in more than one head literal A_i, a *purifying substitution* π is used to ground this variable. Then the branch is extended by new nodes labeled with $A_i \sigma \pi, \ldots, A_m \sigma \pi$. A branch is *closed* if it can be extended by a clause without head literals. S is unsatisfiable if and only if there is a hypertableau for S whose branches are all closed.

Two major advantages of hyperextension are that it avoids unnecessary branching, and only variables in the clauses are universally quantified and get instantiated, while variables in the branches are treated as *free* variables (except those occurring in different head literals). The latter feature allows a superposition-like handling of equality [11], while the former is relevant for hypertableaux for description logic [78], which we shall return to in the next section. Hypertableaux were implemented in the *Hyper* theorem prover for first-order logic, followed by *E-Hyper* implementing also the handling of equality.

Example 5. An example refutation is given in Fig. 1. The initial tableau is set up with the only positive clause. Extension at $R(a)$ with the second clause uses $\sigma = \{x \leftarrow a\}$: since y appears only once in the resulting head, $\pi = \varepsilon$ and y remains as a free variable. In the right subtree $R(f(z))$ is extended with the second clause and $\sigma = \{x \leftarrow f(z)\}$. In the head $P(f(z)), Q(f(z), y)$ of the resulting clause z is repeated: an *instance generation* mechanism produces $\pi = \{z \leftarrow b\}$, or the instance $P(f(b)), Q(f(b), y) \Leftarrow R(f(b))$, to find a refutation. Note how the tableau contains by construction only positive literals, and the interpretation given by a branch is used to control the extension steps very much like in hyperresolution.

2.3 Model-Based Transformation of Clause Sets

Hypertableaux use partial models, that is, models for parts of a clause set, control the search space. An open branch that cannot be expanded further represents a model for the entire clause set. In this section we present a transformation method, borrowed from model-based diagnosis and presented in [8], which is based on a given model and therefore can be installed on top of hypertableaux. In applications to diagnosis, one has a set of clauses S which corresponds to a description of a system, such as an electrical circuit. Very often there is a model I of a correctly functioning system available; in case of an electrical circuit it may be provided by the design tool itself. If the actual circuit is fed with an input and does not show the expected output, the task is to find a diagnosis, or those parts of the circuit which may be broken. Instead of doing reasoning with the system description S and its input and output in order to find the erroneous parts, the idea is to compute only *deviations* from the initially given model I.

Assume that S is a set of propositional clauses and I a set of propositional atoms; as a very simple example take

$$S = \{B \Leftarrow, \; C \Leftarrow A, B\} \text{ and } I = \{A\}.$$

Each clause in S is transformed by replacing a positive literal L by $\neg neg\_L$ and a negative literal $\neg L$ by neg_L, if L is contained in I. In other words, a literal which is contained in the initial model moves to the other side of the arrow and is renamed with the prefix $neg\_$ as in

$$S' = \{B \Leftarrow, \; C, neg\_A \Leftarrow B\}.$$

This transformation is model-preserving, as every model of S is a model of S'. For this it suffices to assign true to neg_L if and only if L is false, for every $L \in I$, and keep truth values unchanged for atoms outside of I. This property is independent of I, and it holds even if I is not a model of S. In our example, after initialization, first hyperextension with $B \Leftarrow$, and then hyperextension with $C, neg\_A \Leftarrow B$, yield the open branches $\{B, C\}$ and $\{B, neg\_A\}$. Hyperextension with $C, neg\_A \Leftarrow B$ can be applied because only B occurs in the body. Since A is assumed to be true in I, it can be added: adding A to $\{B, C\}$ yields model $\{A, B, C\}$; adding A to $\{B, neg\_A\}$ yields model $\{B\}$. If deriving A in S is very expensive, it pays off to save this derivation by moving A as neg_A to the body of the clause. In this example a Horn clause becomes non-Horn, introducing the case where A is false, and neg_A holds, although A is in I. Symmetrically, a non-Horn clause may become Horn. This transformation technique enabled a hypertableau prover to compute benchmarks from electrical engineering [8], and was also applied to the view update problem in databases [2].

Although this transformation mechanism only works in the propositional case, it can be extended to description logic [39]. Indeed, most description logic reasoners are based on tableau calculi, and a hypertableau calculus was used in [78] as a basis for an efficient reasoner for the description logic $\mathcal{SHIQ}$. For this purpose, the authors define *DL-clauses* as clauses without occurrences of function symbols, and such that the head is allowed to include disjunctions of atoms, which may contain existential role restrictions as in

$$\exists repairs.Car(x) \Leftarrow Mechanic(x).$$

In other words, a given $\mathcal{SHIQ}$-Tbox is translated to a large extent into first-order logic; only existential role restrictions are kept as positive "literals." Given a Tbox in the form of a set of DL-clauses, if we have in addition an Abox, or a set of ground assertions, we can use the interpretation given by the ABox as initial model for the model-based transformation [39]. On this basis, the already mentioned *E-Hyper* reasoner was modified to become *E-KRHyper*, which was shown to be a decision procedure for $\mathcal{SHIQ}$ in [16].

2.4   The Model Evolution Calculus

The practical success of DPLL-based SAT solvers suggested the goal of lifting features of DPLL to the first-order level. Research focused on *splitting first-order clauses*, seen as a way to improve the capability to handle non-Horn clauses. Breaking first-order clauses apart is not as simple as in propositional logic, because a clause stands for all its ground instances, and literals share variables that are implicitly universally quantified. Decomposing disjunction is a native feature in tableaux, whose downside is represented by rigid variables, as already discussed in Sect. 2.2, where we saw how hypertableaux offer a possible answer.

The quest for ways to split efficiently clauses such as $A(x) \lor B(x)$ led to the *model evolution calculus* [13]. In this method splitting $A(x) \lor B(x)$ yields a branch with $A(x)$, meaning $\forall x A(x)$, and one with $\neg A(c)$, the Skolemized form of $\neg \forall x A(x) \equiv \exists x \neg A(x)$. Splitting in this way has the disadvantage that the signature changes, and Skolem constants, being new, do not unify with other non-variable terms. Thus, the model evolution calculus employs *parameters*, in place of Skolem constants, to replace existentially quantified variables. These parameters are similar to the free variables of hypertableaux.

The similarity between the model evolution calculus and DPLL goes beyond splitting, as the model evolution calculus aims at being a faithful lifting of DPLL to first-order logic. Indeed, a central feature of the model evolution calculus is that it maintains a *context* Λ, which is a finite set of literals, representing a Herbrand interpretation I_Λ, seen as a candidate partial model of the input set of clauses S. Thus, the model evolution calculus is a *model-based* first-order method. Literals in Λ may contain variables, implicitly universally quantified as in clauses, and parameters. Clauses are written in the form $\Lambda \vdash C$, so that each clause carries the context with itself.

In order to determine whether $I_\Lambda \models L$, for L an atom in the Herbrand base of S, one looks at the most specific literal in Λ that subsumes L; in case of a tie, L is picked with positive sign. If I_Λ is not a model of S, the inference system unifies input clauses against Λ to find instances that are not true in I_Λ: these instances are subject to splitting, to modify Λ and repair I_Λ. Otherwise, the system recognizes that Λ cannot be fixed and declares S unsatisfiable. As DPLL uses *depth-first search with backtracking*, the model evolution calculus uses depth-first search with backtracking and *iterative deepening* on term depth, which however may skew the search towards big proofs with small term depth.

The model evolution calculus was implemented in the *Darwin* prover [9], and extended to handle equality on its own [12] and with superposition [14].

2.5 SGGS: Semantically-Guided Goal-Sensitive Reasoning

SGGS, for *Semantically-Guided Goal-Sensitive* reasoning, is a new theorem-proving method for first-order logic [26–29], which inherits features from several of the strategies that we surveyed in the previous sections. SGGS is *semantically guided* by a fixed initial interpretation I like semantic resolution; and it is *goal-sensitive* like the set-of-support strategy. With hyperresolution and hyper-tableaux, it shares the concept of hyperinference, although the hyperinference in SGGS, as we shall see, is an instance generation inference, and therefore its closest ancestor is *hyperlinking* [58,71], an inference rule that uses the most general unifier of a hyperresolution step to generate instances of the parents, rather than a hyperresolvent.

Most importantly, SGGS is *model-based* at the first-order level, in the sense of working by representing and transforming a candidate partial model of the given set S of first-order clauses. This fundamental characteristic is in common with the model evolution calculus, but while the latter lifts DPLL, SGGS lifts DPLL-CDCL to first-order logic, and it combines the model-based character with the semantic guidance and the goal sensitivity. Indeed, SGGS was motivated by the quest for a method that is simultaneously first-order, model-based, semantically-guided, and goal-sensitive. Furthermore, SGGS is *proof confluent*, which means it does not need backtracking, and it does not necessarily reduce to either DPLL or DPLL-CDCL, if given a propositional problem.

In DPLL-CDCL, if a literal L appears in the trail that represents the candidate partial model, all occurrences of $\neg L$ in the set of clauses are false. If all literals of a clause C are false, C is in *conflict*; if all literals of C except one, say Q, are false, Q is an *implied literal* with C as *justification*. The status of C *depends* on the *decision levels* where the complements of its literals were either guessed (decision) or implied (Boolean propagation). SGGS generalizes these concepts to first-order logic. Since variables in first-order literals are implicitly universally quantified, if L is true, $\neg L$ is false, but if L is false, we only know that a ground instance of $\neg L$ is true. SGGS restores the symmetry by introducing the notion of *uniform falsity*: L is uniformly false, if all its ground instances are false, or, equivalently, if $\neg L$ is true. A first rôle of the given interpretation I is to provide a *reference model* where to evaluate the truth value of literals: a literal is I-*true*, if it is true in I, and I-*false*, if it is uniformly false in I.

An *SGGS clause sequence* Γ is a sequence of clauses, where every literal is either I-true or I-false, so that it tells the truth value in I of all its ground instances. In every clause C in Γ a literal is *selected*: if $C = L_1 \vee \ldots \vee L_n$ and L_n is selected, we write the clause as $L_1 \vee \ldots \vee [L_n]$, or, more compactly, $C[L_n]$, with a slight abuse of the notation. SGGS tries to modify I into a model of S (if I is a model of S the problem is solved). Thus, I-false literals are preferred for selection, and an I-true literal is selected only in a clause whose literals are all I-true, called I-*all-true* clause. A second rôle of the given interpretation I is to provide a *starting point* for the search of a model for S.

An SGGS clause sequence Γ represents a *partial interpretation* $I^p(\Gamma)$: if Γ is the empty sequence, denoted by ε, $I^p(\Gamma)$ is empty; if Γ is $C_1[L_1], \dots, C_i[L_i]$, and $I^p(\Gamma|_{i-1})$ is the partial interpretation represented by $C_1[L_1], \dots, C_{i-1}[L_{i-1}]$, then $I^p(\Gamma)$ is $I^p(\Gamma|_{i-1})$ plus the ground instances $L_i\sigma$ of L_i, such that $C_i\sigma$ is ground, $C_i\sigma$ is not satisfied by $I^p(\Gamma|_{i-1})$, and $\neg L_i\sigma$ is not in $I^p(\Gamma|_{i-1})$, so that $L_i\sigma$ can be added to satisfy $C_i\sigma$. In other words, each clause adds the ground instances of its selected literal that satisfy ground instances of the clause not satisfied thus far.

An *interpretation* $I[\Gamma]$ is obtained by consulting first $I^p(\Gamma)$, and then I: for a ground literal L, if its atom appears in $I^p(\Gamma)$, its truth value in $I[\Gamma]$ is that in $I^p(\Gamma)$; otherwise, it is that in I. Thus, $I[\Gamma]$ is I modified to satisfy the clauses in Γ by satisfying the selected literals, and since I-true selected literals are already true in I, the I-false selected literals are those that matter. For example, if Γ is $[P(x)]$, $\neg P(f(y)) \vee [Q(y)]$, $\neg P(f(z)) \vee \neg Q(g(z)) \vee [R(f(z), g(z))]$, and I is all negative like in positive hyperresolution, $I[\Gamma]$ satisfies all ground instances of $P(x)$, $Q(y)$, and $R(f(z), g(z))$, and no other positive literal.

SGGS generalizes Boolean, or clausal, propagation to first-order logic. Consider an I-false (I-true) literal M selected in clause C_j in Γ, and an I-true (I-false) literal L in C_i, $i > j$: if all ground instances of L appear negated among the ground instances of M added to $I^p(\Gamma)$, L is uniformly false in $I[\Gamma]$ because of M, and *depends* on M, like $\neg L$ *depends* on L in propositional Boolean propagation, when L is in the trail. If this happens for *all* its literals, clause $C[L]$ is in *conflict* with $I[\Gamma]$; if this happens for all its literals except L, L is an *implied literal* with $C[L]$ as *justification*. SGGS employs *assignment functions* to keep track of the *dependencies* of I-true literals on selected I-false literals, realizing a sort of *first-order propagation modulo semantic guidance* by I. SGGS ensures that I-all-true clauses in Γ are either conflict clauses or justifications.

The main inference rule of SGGS, called *SGGS-extension*, uses the current clause sequence Γ and a clause C in S to generate an instance E of C and add it to Γ to obtain the next clause sequence Γ'. SGGS-extension is a hyperinference, because it unifies literals $L_1, \dots, L_n$ of C with I-false selected literals $M_1, \dots, M_n$ of opposite sign in Γ. The hyperinference is *guided* by $I[\Gamma]$, because I-false selected literals contribute to $I[\Gamma]$ as explained above. Another ingredient of the instance generation mechanism ensures that every literal in E is either I-true or I-false. SGGS-extension is also responsible for selecting a literal in E.

The *lifting theorem* for SGGS-extension shows that if $I[\Gamma] \not\models C'$ for some ground instance C' of a clause $C \in S$, SGGS-extension builds an instance E of C such that C' is an instance of E. There are three kinds of SGGS-extension: (1) add a clause E which is in conflict with $I[\Gamma]$ and is I-all-true; (2) add a clause E which is in conflict with $I[\Gamma]$ but is not I-all-true; and (3) add a clause E which is not in conflict with $I[\Gamma]$. In cases (1) and (2), it is necessary to *solve the conflict*: it is here that SGGS lifts the conflict-driven clause learning (CDCL) mechanism of DPLL-CDCL to the first-order level.

In DPLL-CDCL a conflict is *explained* by resolving a conflict clause C with the justification D of a literal whose complement is in C, generating a new conflict clause. Typically resolution continues until we get either the empty clause $\bot$ or an *asserting clause*, namely a clause where only one literal Q is falsified in the current decision level. DPLL-CDCL learns the asserting clause and backjumps to the shallowest level where Q is undefined and all other literals in the asserting clause are false, so that Q enters the trail with the asserting clause as justification. SGGS *explains* a conflict by resolving the conflict clause E with an I-all-true clause $D[M]$ in Γ which is the justification of the literal M that makes an I-false literal L in E uniformly false in $I[\Gamma]$. Resolution continues until we get either $\bot$ or a conflict clause $E[L]$ which is I-all-true. If $\bot$ arises, S is unsatisfiable. Otherwise, SGGS *moves* the I-all-true clause $E[L]$ to the left of the clause $B[M]$, whose I-false selected literal M makes L uniformly false in $I[\Gamma]$. The effect is to *flip* at once the truth value of *all* ground instances of L in $I[\Gamma]$, so that the conflict is solved, L is implied, and $E[L]$ satisfied.

In order to simplify the presentation, up to here we omitted that clauses in SGGS may have *constraints*. For example, $x \not\equiv y \triangleright P(x,y) \vee Q(y,x)$ is a *constrained clause*, which represents its ground instances that satisfy the constraints: $P(a,b) \vee Q(b,a)$ is an instance, while $P(a,a) \vee Q(a,a)$ is not. The reason for constraints is that selected literals of clauses in Γ may *intersect*, in the sense of having ground instances with the same atoms. Since selected literals determine $I^p(\Gamma)$, whence $I[\Gamma]$, non-empty intersections represent *duplications*, if the literals have the same sign, and *contradictions*, otherwise. SGGS removes duplications by deletion of clauses, and contradictions by resolution. However, before doing either, it needs to *isolate* the shared ground instances in the selected literal of *one* clause. For this purpose, SGGS features inference rules that replace a clause by a *partition*, that is, a set of clauses that represent the same ground instances and have *disjoint* selected literals. This requires constraints. For example, a partition of $[P(x,y)] \vee Q(x,y)$ is $\{true \triangleright [P(f(z),y)] \vee Q(f(z),y),\ top(x) \neq f \triangleright [P(x,y)] \vee Q(x,y)\}$, where the constraint $top(x) \neq f$ means that variable x cannot be instantiated with a term whose topmost symbol is f. If L and M in $C[L]$ and $D[M]$ of Γ intersect, SGGS partitions $C[L]$ by $D[M]$: it partitions $C[L]$ into $A_1 \triangleright C_1[L_1], \ldots, A_n \triangleright C_n[L_n]$ so that only L_j, for some j, $1 \leq j \leq n$, intersects with M, and $A_j \triangleright C_j[L_j]$ is either deleted or resolved with $D[M]$.

The following example shows an SGGS-refutation:

Example 6. Given $S = \{\neg P(f(x)) \vee \neg Q(g(x)) \vee R(x),\ P(x),\ Q(y),\ \neg R(c)\}$, let I be all negative. An SGGS-derivation starts with the empty sequence. Then, four SGGS-extension steps apply:

$\Gamma_0 : \varepsilon$
$\Gamma_1 : [P(x)]$
$\Gamma_2 : [P(x)],\ [Q(y)]$
$\Gamma_3 : [P(x)],\ [Q(y)],\ \neg P(f(x)) \vee \neg Q(g(x)) \vee [R(x)]$
$\Gamma_4 : [P(x)],\ [Q(y)],\ \neg P(f(x)) \vee \neg Q(g(x)) \vee [R(x)],\ [\neg R(c)]$

At this stage, the selected literals $R(x)$ and $\neg R(c)$ intersect, and therefore SGGS partitions $\neg P(f(x)) \vee \neg Q(g(x)) \vee [R(x)]$ by $[\neg R(c)]$:

$$\Gamma_5\colon [P(x)], \ [Q(y)], \ x \not\equiv c \triangleright \neg P(f(x)) \vee \neg Q(g(x)) \vee [R(x)],$$
$$\neg P(f(c)) \vee \neg Q(g(c)) \vee [R(c)], \ [\neg R(c)]$$

Now the I-all-true clause $\neg R(c)$ is in conflict with $I[\Gamma_5]$. Thus, SGGS moves it left of the clause $\neg P(f(c)) \vee \neg Q(g(c)) \vee [R(c)]$ that makes $\neg R(c)$ false in $I[\Gamma_5]$, in order to amend the induced interpretation. Then, it resolves these two clauses, and replaces the parent that is not I-all-true, namely $\neg P(f(c)) \vee \neg Q(g(c)) \vee [R(c)]$, by the resolvent $\neg P(f(c)) \vee \neg Q(g(c))$:

$$\Gamma_6\colon [P(x)], \ [Q(y)], \ x \not\equiv c \triangleright \neg P(f(x)) \vee \neg Q(g(x)) \vee [R(x)], \ [\neg R(c)],$$
$$\neg P(f(c)) \vee \neg Q(g(c)) \vee [R(c)]$$
$$\Gamma_7\colon [P(x)], \ [Q(y)], \ x \not\equiv c \triangleright \neg P(f(x)) \vee \neg Q(g(x)) \vee [R(x)], \ [\neg R(c)],$$
$$\neg P(f(c)) \vee [\neg Q(g(c))]$$

Assuming that in the resolvent the literal $\neg Q(g(c))$ gets selected, there is now an intersection between selected literals $\neg Q(g(c))$ and $Q(y)$, so that SGGS partitions $Q(y)$ by $\neg P(f(c)) \vee \neg Q(g(c))$:

$$\Gamma_8\colon [P(x)], \ top(y) \neq g \triangleright [Q(y)], \ z \not\equiv c \triangleright [Q(g(z))], \ [Q(g(c))],$$
$$x \not\equiv c \triangleright \neg P(f(x)) \vee \neg Q(g(x)) \vee [R(x)], \ [\neg R(c)], \ \neg P(f(c)) \vee [\neg Q(g(c))]$$

At this point, the I-all-true clause $\neg P(f(c)) \vee [\neg Q(g(c))]$ is in conflict with $I[\Gamma_8]$. As before, SGGS moves it left of the clause that makes its selected literal $\neg Q(g(c))$ false, namely $[Q(g(c))]$, in order to fix the candidate model, and then resolves $\neg P(f(c)) \vee [\neg Q(g(c))]$ and $[Q(g(c))]$, replacing the latter by the resolvent $\neg P(f(c))$:

$$\Gamma_9\colon \ [P(x)], \ top(y) \neq g \triangleright [Q(y)], \ z \not\equiv c \triangleright [Q(g(z))], \ \neg P(f(c)) \vee [\neg Q(g(c))],$$
$$[Q(g(c))], \ x \not\equiv c \triangleright \neg P(f(x)) \vee \neg Q(g(x)) \vee [R(x)], \ [\neg R(c)]$$
$$\Gamma_{10}\colon [P(x)], \ top(y) \neq g \triangleright [Q(y)], \ z \not\equiv c \triangleright [Q(g(z))], \ \neg P(f(c)) \vee [\neg Q(g(c))],$$
$$[\neg P(f(c))], \ x \not\equiv c \triangleright \neg P(f(x)) \vee \neg Q(g(x)) \vee [R(x)], \ [\neg R(c)]$$

The resolvent has only one literal which gets selected; since $[\neg P(f(c))]$ intersects with $[P(x)]$, the next inference partitions $[P(x)]$ by $[\neg P(f(c))]$:

$$\Gamma_{11}\colon top(x) \neq f \triangleright [P(x)], \ y \not\equiv c \triangleright [P(f(y))], \ [P(f(c))], \ top(y) \neq g \triangleright [Q(y)],$$
$$z \not\equiv c \triangleright [Q(g(z))], \ \neg P(f(c)) \vee [\neg Q(g(c))], \ [\neg P(f(c))],$$
$$x \not\equiv c \triangleright \neg P(f(x)) \vee \neg Q(g(x)) \vee [R(x)], \ [\neg R(c)]$$

The next step moves the I-all-true clause $[\neg P(f(c))]$, which is in conflict with $I[\Gamma_{11}]$, to the left of the clause $[P(f(c))]$ that makes $[\neg P(f(c))]$ false in $I[\Gamma_{11}]$, and then resolves these two clauses to generate the empty clause:

$$\Gamma_{12}\colon top(x) \neq f \triangleright [P(x)], \ y \not\equiv c \triangleright [P(f(y))], \ [\neg P(f(c))], \ [P(f(c))],$$
$$top(y) \neq g \triangleright [Q(y)], \ z \not\equiv c \triangleright [Q(g(z))], \ \neg P(f(c)) \vee [\neg Q(g(c))],$$
$$x \not\equiv c \triangleright \neg P(f(x)) \vee \neg Q(g(x)) \vee [R(x)], \ [\neg R(c)]$$
$$\Gamma_{13}\colon top(x) \neq f \triangleright [P(x)], \ y \not\equiv c \triangleright [P(f(y))], \ [\neg P(f(c))], \ \bot,$$
$$top(y) \neq g \triangleright [Q(y)], \ z \not\equiv c \triangleright [Q(g(z))], \ \neg P(f(c)) \vee [\neg Q(g(c))],$$
$$x \not\equiv c \triangleright \neg P(f(x)) \vee \neg Q(g(x)) \vee [R(x)], \ [\neg R(c)]$$

This example only illustrates the basic mechanisms of SGGS. This method is so new that it has not yet been implemented: the hope is that its conflict-driven model-repair mechanism will have on first-order theorem proving an effect similar to that of the transition from DPLL to DPLL-CDCL for SAT-solvers. If this were true, even in part, the benefit could be momentous, considering that CDCL played a key rôle in the success of SAT technology. Another expectation is that non-trivial semantic guidance (i.e., not based on sign like in hyperresolution) pays off in case of many axioms or large knowledge bases.

3   Model-Based Reasoning in First-Order Theories

There are basically two ways one can think about a theory presented by a set of axioms: as the set of all theorems that are logical consequences of the axioms, or as the set of all interpretations that are models of the axioms. The two are obviously connected, but may lead to different styles of reasoning, that we portray by the selection of methods in this section. We cover approaches that *build axioms* into resolution, *hierarchical* and *locality-based* theory reasoning, and a recent method called *Model-Constructing satisfiability calculus* or *MCsat*.

3.1   Building Theory Axioms into Resolution and Superposition

The early approaches to theory reasoning emphasized the axioms, by building them into the inference systems. The first analyzed theory was *equality*: since submitting the equality axioms to resolution, or other inference systems for first-order logic, leads to an explosion of the search space, *paramodulation*, *superposition*, and *rewriting* were developed to build equality into resolution (e.g., [4, 21, 48, 73, 77] and Chaps. 7 and 9 in [76]).

Once equality was conquered, research flourished on building-in theories (e.g., [30, 36, 40, 49, 53, 54, 66, 72]). *Equational theories*, that are axiomatized by sets of equalities, and among them *permutative theories*, where the two sides of each axiom are permutations of the same symbols, as in *associativity* and *commutativity*, received the most attention. A main ingredient is to replace syntactic unification by unification *modulo* a set E of equational axioms, a concept generalized by José Meseguer to *order-sorted E-unification* (e.g., [37, 43, 46]). This kind of approach was pursued further, by building into superposition axioms for *monoids* [42], *groups* [85], *rings* and *modules* [84], or by generalizing superposition to embed *transitive relations* other than equality [5]. The complexities and limitations of these techniques led to investigate the methods for *hierarchical* theory reasoning that follow.

3.2   Hierarchical Reasoning by Superposition

Since José Meseguer's work with Joe Goguen (e.g., [44]), it became clear that a major issue at the cross-roads of reasoning, specifying, and programming, is that theories, or specifications, are built by *extension* to form *hierarchies*.

A *base theory* T_0 is defined by a set of *sorts* S_0, a *signature* Σ_0, possibly a set of axioms N_0, and the class C_0 of its *models* (e.g., term-generated Σ_0-algebras). An *extended* or *enriched* theory T adds new sorts ($S_0 \subseteq S$), new function symbols ($\Sigma_0 \subseteq \Sigma$), called *extension functions*, and new axioms ($N_0 \subseteq N$), specifying properties of the new symbols. For the base theory the class of models is given, while the extension is defined axiomatically. A pair (T_0, T) as above forms a *hierarchy* with *enrichment axioms* N.

The crux of extending specifications was popularized by Joe Goguen and José Meseguer as *no junk and no confusion*: an interpretation of S and Σ, which is a model of N, is a model of T only if it extends a model in C_0, without collapsing its sorts, or making distinct elements equal (*no confusion*), or introducing new elements of base sort (*no junk*). A sufficient condition for the latter is *sufficient completeness*, a property studied also in inductive theorem proving, which basically says that every ground non-base term t' of base sort is equal to a ground base term t. Sufficient completeness is a strong restriction, violated by merely adding a constant symbol: if $\Sigma_0 = \{a, b\}$, $N = N_0 = \{a \not\simeq b\}$, and $\Sigma = \{a, b, c\}$, where a, b, and c are constants of the same sort, the extension is not sufficiently complete, because c is junk, or a model with three distinct elements is not isomorphic to one with two. Although sufficient completeness is undecidable in general (e.g., [57]), sufficient completeness analyzers exist (e.g., [45, 47, 56]), with key contributions by José Meseguer.

Hierarchic superposition was introduced in [6] and developed in [41] to reason about a hierarchy (T_0, T) with enrichment axioms N, where N is a set of clauses. We assume to have a decision procedure to detect that a finite set of Σ_0-clauses is T_0-unsatisfiable. Given a set S of Σ-clauses, the problem is to determine whether S is false in all models of the hierarchic specification, or, equivalently, whether $N \cup S$ has no model whose reduct to Σ_0 is a model of T_0. The problem is solved by using the T_0-reasoner as a black-box to take care of the base part, while superposition-based inferences apply only to non-base literals.[1] First, for every clause C, whenever a subterm t whose top symbol is a base operator occurs immediately below a non-base operator symbol (or vice versa), t is replaced by a new variable x and the equation $x \simeq t$ is added to the antecedent of C. This transformation is called *abstraction*. Then, the inference rules are modified to require that all substitutions are *simple*, meaning that they map variables of base sort to base terms. A meta-rule named *constraint refutation* detects that a finite set of Σ_0-clauses is inconsistent in T_0 by invoking the T_0-reasoner. Hierarchic superposition was proved refutationally complete in [6], provided T_0 is compact, which is a basic preliminary to make constraint refutation mechanizable, and $N \cup S$ is *sufficiently complete with respect to simple instances*, which means that for every model I of all simple ground instances of the clauses in $N \cup S$, and every ground non-base term t', there exists a ground base term t (which may depend on I) such that $I \models t' \simeq t$.

---

[1] Other approaches to subdivide work between superposition and an SMT-solver appeared in [20, 25].

There are situations where the enrichment adds *partial* functions: Σ_0 contains only total function symbols, while $\Sigma \setminus \Sigma_0$ may contain partial functions and total functions having as codomain a new sort. Hierarchic superposition was generalized to handle both total and partial function symbols, yielding a *partial hierarchic superposition calculus* [41]. To have an idea of the difficulties posed by partial functions, consider that replacement of equals by equals may be unsound in their presence. For example, $s \not\simeq s$ may hold in a partial algebra (i.e., a structure where some function symbols are interpreted as partial), if s is undefined. Thus, the equality resolution rule (e.g., resolution between $C \vee s \not\simeq s$ and $x \simeq x$) is restricted to apply only if s is guaranteed to be defined. Other restrictions impose that terms replaced by inferences may contain a partial function symbol only at the top; substitutions cannot introduce partial function symbols; and every ground term made only of total symbols is smaller than any ground term containing a partial function symbol in the ordering used by the inference system. The following example portrays the partial function case:

Example 7. Let $\mathcal{T}_0$ be the base theory defined by $\mathcal{S}_0 = \{\mathsf{data}\}$, $\Sigma_0 = \{b: \rightarrow \mathsf{data}, f: \mathsf{data} \rightarrow \mathsf{data}\}$, and $N_0 = \{\forall x\, f(f(x)) \simeq f(x)\}$. We consider the extension with a new sort list, total functions $\{\mathsf{cons}: \mathsf{data}, \mathsf{list} \rightarrow \mathsf{list}, \mathsf{nil}: \rightarrow \mathsf{list}, d :\rightarrow \mathsf{list}\}$, partial functions $\{\mathsf{car} : \mathsf{list} \rightarrow \mathsf{data}, \mathsf{cdr} : \mathsf{list} \rightarrow \mathsf{list}\}$, and the following clauses, where $N = \{(1), (2), (3)\}$ and $S = \{(4), (5)\}$:

$$
\begin{array}{ll}
(1) & \mathsf{car}(\mathsf{cons}(x, l)) \simeq x \\
(2) & \mathsf{cdr}(\mathsf{cons}(x, l)) \simeq l \\
(3) & \mathsf{cons}(\mathsf{car}(l), \mathsf{cdr}(l)) \simeq l \\
(4) & f(b) \simeq b \\
(5) & f(f(b)) \simeq \mathsf{car}(\mathsf{cdr}(\mathsf{cons}(f(b), \mathsf{cons}(b, d))))
\end{array}
$$

The partial hierarchic superposition calculus deduces:

$$
\begin{array}{lll}
(6) & x \not\simeq f(f(b)) \vee y \not\simeq f(b) \vee z \not\simeq b \vee x \not\simeq \mathsf{car}(\mathsf{cdr}(\mathsf{cons}(y, \mathsf{cons}(z, d)))) & \text{Abstr. (5)} \\
(7) & x \not\simeq f(f(b)) \vee y \not\simeq f(b) \vee z \not\simeq b \vee x \not\simeq \mathsf{car}(\mathsf{cons}(z, d)) & \text{Superp. (2),(6)} \\
(8) & x \not\simeq f(f(b)) \vee y \not\simeq f(b) \vee z \not\simeq b \vee x \not\simeq z & \text{Superp. (1),(7)} \\
(9) & \bot & \text{Constraint refutation (4),(8)}
\end{array}
$$

Under the assumption that $\mathcal{T}_0$ is a universal first-order theory, which ensures compactness, the partial hierarchic superposition calculus was proved sound and complete in [41]: if a contradiction cannot be derived from $N \cup S$ using this calculus, then $N \cup S$ has a model which is a partial algebra. Thus, if the unsatisfiability of $N \cup S$ does not depend on the totality of the extension functions, the partial hierarchic superposition calculus can detect its inconsistency. In certain problem classes where partial algebras can always be made total, the calculus is complete also for total functions. Research on hierarchic superposition continued in [1], where an implementation for extensions of linear arithmetic was presented, and in [15], where the calculus was made "more complete" in practice.

3.3   Hierarchical Reasoning in Local Theory Extensions

A series of papers starting with [81] identified a class of theory extensions $(\mathcal{T}_0, \mathcal{T})$, called *local*, which admit a complete hierarchical method for checking satisfiability of *ground* clauses, without requiring either sufficient completeness or that T_0 is a universal first-order theory. The enrichment axioms in N do not have to be clauses: if they are, we have an extension *with clauses*; if N consists of formulæ of the form $\forall \bar{x}\,(\Phi(\bar{x}) \lor D(\bar{x}))$, where $\Phi(\bar{x})$ is an *arbitrary* Σ_0-formula and $D(\bar{x})$ is a Σ-clause, with at least one occurrence of an extension function, we have an extension *with augmented clauses*. The basic assumption that T_0, or a fragment thereof, admits a decision procedure for satisfiability clearly remains.

As we saw throughout this survey, instantiating universally quantified variables is crucial in first-order reasoning. Informally, a theory extension is *local*, if it is sufficient to consider only a *finite* set of instances. Let G be a set of ground clauses to be refuted in T, and let $N[G]$ denote the set of instances of the clauses in N where every term whose top symbol is an extension function is a ground term occurring in N or G. Theory T is a *local extension* of T_0, if $N[G]$ suffices to prove the T-unsatisfiability of G [81]. Subsequent papers studied variants of locality, including those for extensions with augmented clauses, and for combinations of local theories, and proved that locality can be recognized by showing that certain partial algebras embed into total ones [50,51,81,82].

If T is a local extension, it is possible to check the T-satisfiability of G by hierarchical reasoning [50,51,81,82], allowing the introduction of new constants by *abstraction* as in [64]. By locality, G is T-unsatisfiable if and only if there is no model of $N[G] \cup G$ whose restriction to Σ_0 is a model of T_0. By abstracting away non-base terms, $N[G] \cup G$ is transformed into an equisatisfiable set $N_0 \cup G_0 \cup D$, where N_0 and G_0 are sets of Σ_0-clauses, and D contains the definitions introduced by abstraction, namely equalities of the form $f(g_1, \ldots, g_n) \simeq c$, where f is an extension function, $g_1, \ldots, g_n$ are ground terms, and c is a new constant. The problem is reduced to that of testing the T_0-satisfiability of $N_0 \cup G_0 \cup \mathsf{Con}_0$, where Con_0 contains the instances of the congruence axioms for the terms in D:

$$\mathsf{Con}_0 = \{\bigwedge_{i=1}^{n} c_i \simeq d_i \Rightarrow c \simeq d \mid f(c_1, \ldots, c_n) \simeq c, f(d_1, \ldots, d_n) \simeq d \in D\},$$

which can be solved by a decision procedure for T_0 or a fragment thereof.

In the following example T_0 is the theory of *linear arithmetic* over the real numbers, and T is its extension with a monotone unary function f, which is known to be a local extension [81]:

Example 8. Let G be $(a \leq b \land f(a) = f(b) + 1)$. The enrichment $N = \{x \leq y \Rightarrow f(x) \leq f(y)\}$ consists of the monotonicity axiom. In order to check whether G is T-satisfiable, we compute $N[G]$, omitting the redundant clauses $c \leq c \Rightarrow f(c) \leq f(c)$ for $c \in \{a, b\}$:

$$N[G] = \{a \leq b \Rightarrow f(a) \leq f(b), \ b \leq a \Rightarrow f(b) \leq f(a)\}.$$

The application of abstraction to $N[G] \cup G$ yields $N_0 \cup G_0 \cup D$, where:

$$N_0 = \{a \leq b \Rightarrow a_1 \leq b_1, \ b \leq a \Rightarrow b_1 \leq a_1\}, \qquad G_0 = \{a \leq b, \ a_1 \simeq b_1 + 1\},$$

$D = \{a_1 \simeq f(a),\ b_1 \simeq f(b)\}$, and a_1 and b_1 are new constants. Thus, Con_0 is $\{a \simeq b \Rightarrow a_1 \simeq b_1\}$. A decision procedure for linear arithmetic applied to $N_0 \cup G_0 \cup \mathsf{Con}_0$ detects unsatisfiability.

3.4   Beyond SMT: Satisfiability Modulo Assignment and MCsat

Like SGGS generalizes conflict-driven clause learning (CDCL) to first-order logic and Herbrand interpretations, the *Model-Constructing satisfiability calculus*, or *MCsat* for short, generalizes CDCL to decidable fragments of first-order theories and their models [35,55].

Recall that in DPLL-CDCL the trail that represents the candidate partial model contains only propositional literals; the inference mechanism that explains conflicts is propositional resolution; and learnt clauses are made of input atoms. These three characteristics are true also of the DPLL($\mathcal{T}$) paradigm for SMT-solvers [7], where an *abstraction function* maps finitely many input first-order ground atoms to finitely many propositional atoms. In this way, the method bridges the gap between the first-order language of the theory $\mathcal{T}$ and the propositional language of the DPLL-CDCL core solver. In DPLL($\mathcal{T}$), also $\mathcal{T}$-lemmas are made of input atoms, and the guarantee that no new atoms are generated is a key ingredient of the proof of termination of the method in [65].

Also when $\mathcal{T}$ is a union of theories $\mathcal{T} = \bigcup_{i=1}^{n} \mathcal{T}_i$, the language of atoms remains finite. The standard method to combine satisfiability procedures for theories $\mathcal{T}_1, \ldots, \mathcal{T}_n$ to get a satisfiability procedure for their union is *equality sharing* [64], better known as *Nelson-Oppen scheme*, even if equality sharing was the original name given by Greg Nelson, as reconstructed in [63]. Indeed, a key feature of equality sharing is that the combined procedures only need to share equalities between constant symbols. These equalities are mapped by the abstraction function to *proxy variables*, that is, propositional variables that stand for the equalities. As there are finitely many constant symbols, there are also finitely many proxy variables.

MCsat generalizes both model representation and inference mechanism beyond satisfiability modulo theories (SMT), because it is designed to decide a more general problem called *satisfiability modulo assignment* (SMA). An SMA problem consists of determining the satisfiability of a formula S in a theory $\mathcal{T}$, given an initial assignment I to some of the variables occuring in S, including *both* propositional variables and *free first-order variables*. SMT can be seen as a special case of SMA where I is empty. Also, since an SMT-solver builds partial assignments during the search for a satisfying one, an intermediate state of an SMT search can be viewed as an instance of SMA. A first major generalization of MCsat with respect to DPLL-CDCL and DPLL($\mathcal{T}$) is to allow the trail to contain also *assignments to free first-order variables* (e.g., $x \leftarrow 3$). Such assignments can be *semantic decisions* or *semantic propagations*, thus called to distinguish them from the Boolean decisions and Boolean propagations that yield the standard Boolean assignments (e.g., $L \leftarrow true$).

The answer to an SMA problem is either a model of S including the initial assignment I, or "unsatisfiable" with an *explanation*, that is, a formula S' that follows from S and is inconsistent with I. This notion of explanation is a generalization of the explanation of conflicts by propositional resolution in DPLL-CDCL. Indeed, a second major generalization of MCsat with respect to DPLL-CDCL and DPLL($\mathcal{T}$) is to allow the inference mechanism that explains conflicts to generate *new atoms*, as shown in the following example in the quantifier-free fragment of the theory of equality:

Example 9. Assume that S is a conjunction of literals including $\{v \simeq f(a),\ w \simeq f(b)\}$, where a and b are constant symbols, f is a function symbol, and v and w are free variables. If the trail contains the assignments $a \leftarrow \alpha,\ b \leftarrow \alpha,\ w \leftarrow \beta_1,\ v \leftarrow \beta_2$, where α, β_1, and β_2 denote distinct values of the appropriate sorts, there is a conflict. The explanation is the formula $a \simeq b \Rightarrow f(a) \simeq f(b)$, which is an instance of the substitutivity axiom, or congruence axiom, for function f. Note how the atoms $a \simeq b$ and $f(a) \simeq f(b)$ need not appear in S, and therefore such a lemma could not be generated in DPLL($\mathcal{T}$).

In order to apply MCsat to a theory $\mathcal{T}$, one needs to give clausal inference rules to *explain* conflicts in $\mathcal{T}$. These inference rules generate clauses that may contain *new* (i.e., non-input) ground atoms in the signature of the theory. New atoms come from a *basis*, defined as the closure of the set of input atoms with respect to the inference rules. The proof of termination of the MCsat transition rules in [35] requires that the basis be *finite*. The following example illustrates the importance of this finiteness requirement:

Example 10. Given $S = \{x \geq 2,\ \neg(x \geq 1) \vee y \geq 1,\ x^2 + y^2 \leq 1 \vee xy > 1\}$, and starting with an empty trail $M = \emptyset$, a Boolean propagation puts $x \geq 2$ in the trail. Theory propagation adds $x \geq 1$, because $x \geq 2$ implies $x \geq 1$ in the theory, and $x \geq 1$ appears in S. A Boolean propagation over clause $\neg(x \geq 1) \vee y \geq 1$ adds $y \geq 1$, so that we have $M = x \geq 2,\ x \geq 1,\ y \geq 1$. If a Boolean decision guesses next $x^2 + y^2 \leq 1$ and then a semantic decision adds $x \leftarrow 2$, we have $M = x \geq 2,\ x \geq 1,\ y \geq 1,\ x^2 + y^2 \leq 1,\ x \leftarrow 2$ and a conflict, as there is no value for y such that $4 + y^2 \leq 1$. Learning $\neg(x = 2)$ as an explanation of the conflict does not work, because the procedure can then try $x \leftarrow 3$, and hit another conflict. Clearly, we do not want to learn the infinite sequence $\neg(x = 2),\ \neg(x = 3),\ \neg(x = 4) \ldots$.

Similarly, also a systematic application of the inference rules to enumerate all atoms in a finite basis would be too inefficient. The key point is that the inference rules are applied only to explain conflicts and amend the current partial model, so that the generation of new atoms is *conflict-driven*. This concept is connected with that of *interpolation* (e.g., [83] for interpolation and locality, [23] for a survey on interpolation of ground proofs, and [24] for an approach to interpolation of non-ground proofs): given two inconsistent formulæ A and B, a formula that follows from A and is inconsistent with B is an interpolant of A and B, if it is made only of symbols that appear in both A and B. In a theory $\mathcal{T}$, the notions of being inconsistent and being logical consequence are relative to $\mathcal{T}$, and the interpolant is allowed to contain theory symbols even if they are not common to A and B. Since an explanation is a formula S' that follows from S and is inconsistent with I, an interpolant of S and I (written as a formula) is an explanation. We illustrate these ideas continuing Example 10:

Example 11. The solution is to observe that $x^2 + y^2 \le 1$ implies $-1 \le x \wedge x \le 1$, which is inconsistent with $x = 2$. Note that $-1 \le x \wedge x \le 1$ is an interpolant of $x^2 + y^2 \le 1$ and $x = 2$, as x appears in both. Thus, a desirable explanation is $(x^2 + y^2 \le 1) \Rightarrow x \le 1$, or $\neg(x^2 + y^2 \le 1) \vee x \le 1$ in clausal form, which brings the procedure to update the trail to $M = x \ge 2$, $x \ge 1$, $y \ge 1$, $x^2 + y^2 \le 1$, $x \le 1$. At this point, $x \ge 2$ and $x \le 1$ cause another theory conflict, which leads the procedure to learn the lemma $\neg(x \ge 2) \vee \neg(x \le 1)$. A first step of explanation by resolution between $\neg(x^2 + y^2 \le 1) \vee x \le 1$ and $\neg(x \ge 2) \vee \neg(x \le 1)$ yields $\neg(x^2 + y^2 \le 1) \vee \neg(x \ge 2)$. A second step of explanation by resolution between $\neg(x^2 + y^2 \le 1) \vee \neg(x \ge 2)$ and $x \ge 2$ yields $\neg(x^2 + y^2 \le 1)$, so that the trail is amended to $M = x \ge 2$, $x \ge 1$, $y \ge 1$, $\neg(x^2 + y^2 \le 1)$, finally repairing the decision (asserting $x^2 + y^2 \le 1$) that caused the conflict.[2]

In summary, MCsat is a fully model-based procedure, which lifts CDCL to SMT and SMA. Assignments to first-order variables and new literals are involved in decisions, propagations, conflict detections, and explanations, on a par with Boolean assignments and input literals. The theories covered in [35,55] are the quantifier-free fragments of the theories of equality, linear arithmetic, and boolean values, and their combinations. MCsat is also the name of the implementation of the method as described in [55].

4   Discussion

We surveyed model-based reasoning methods, where inferences build or amend partial models, which guide in turn further inferences, balancing search with inference, and search for a model with search for a proof. We exemplified these concepts for first-order clausal reasoning, and then we lifted them, sort of speak, to theory reasoning. Automated reasoning has made giant strides, and state of the art systems are very sophisticated in working with mostly syntactic information. The challenge of model-based methods is to go towards a semantically-oriented style of reasoning, that may pay off for hard problems or new domains.

Acknowledgments. The first author thanks David Plaisted, for starting the research on SGGS and inviting her to join in August 2008; and Leonardo de Moura, for the discussions on MCsat at Microsoft Research in Redmond in June 2013. The third author's work was partially supported by the German Research Council (DFG) as part of the Transregional Collaborative Research Center "Automatic Verification and Analysis of Complex Systems" (SFB/TR 14 AVACS, www.avacs.org).

---

[2] The problem in Examples 10 and 11 appeared in the slides of a talk entitled "Arithmetic and Optimization @ MCsat" presenting joint work by Leonardo de Moura, Dejan Jovanović, and Grant Olney Passmore, and given by Leonardo de Moura at a Schloss Dagsthul Seminar on "Deduction and Arithmetic" in October 2013.

References

1. Althaus, E., Kruglov, E., Weidenbach, C.: Superposition modulo linear arithmetic SUP(LA). In: Ghilardi, S., Sebastiani, R. (eds.) FroCoS 2009. LNCS (LNAI), vol. 5749, pp. 84–99. Springer, Heidelberg (2009)
2. Aravindan, C., Baumgartner, P.: Theorem proving techniques for view deletion in databases. J. Symbolic Comput. **29**(2), 119–148 (2000)
3. Arthan, R., Oliva, P.: (Dual) Hoops have unique halving. In: Bonacina, M.P., Stickel, M.E. (eds.) Automated Reasoning and Mathematics: Essays in Memory of William W. McCune. LNCS (LNAI), vol. 7788, pp. 165–180. Springer, Heidelberg (2013)
4. Bachmair, L., Ganzinger, H.: Rewrite-based equational theorem proving with selection and simplification. J. Log. Comput. **4**(3), 217–247 (1994)
5. Bachmair, L., Ganzinger, H.: Ordered chaining calculi for first-order theories of transitive relations. J. ACM **45**(6), 1007–1049 (1998)
6. Bachmair, L., Ganzinger, H., Waldmann, U.: Refutational theorem proving for hierarchic first-order theories. Appl. Algebra Eng. Commun. Comput. **5**, 193–212 (1994)
7. Barrett, C., Sebastiani, R., Seshia, S.A., Tinelli, C.: Satisfiability modulo theories. In: Biere, A., Heule, M., Van Maaren, H., Walsh, T. (eds.) Handbook of Satisfiability, chap. 26, pp. 825–886. IOS Press, Amsterdam (2009)
8. Baumgartner, P., Fröhlich, P., Furbach, U., Nejdl, W.: Semantically guided theorem proving for diagnosis applications. In: Proceedings of IJCAI-16, vol. 1, pp. 460–465 (1997)
9. Baumgartner, P., Fuchs, A., Tinelli, C.: Implementing the model evolution calculus. Int. J. Artif. Intell. Tools **15**(1), 21–52 (2006)
10. Baumgartner, P., Furbach, U., Niemelä, I.: Hyper tableaux. In: Alferes, J.J., Moniz Pereira, L., Orłowska, E. (eds.) JELIA 1996. LNCS (LNAI), vol. 1126, pp. 1–17. Springer, Heidelberg (1996)
11. Baumgartner, P., Furbach, U., Pelzer, B.: The hyper tableaux calculus with equality and an application to finite model computation. J. Logic Comput. **20**(1), 77–109 (2008)
12. Baumgartner, P., Pelzer, B., Tinelli, C.: Model evolution calculus with equality - revised and implemented. J. Symbolic Comput. **47**(9), 1011–1045 (2012)
13. Baumgartner, P., Tinelli, C.: The model evolution calculus as a first-order DPLL method. Artif. Intell. **172**(4–5), 591–632 (2008)
14. Baumgartner, P., Waldmann, U.: Superposition and model evolution combined. In: Schmidt, R.A. (ed.) CADE-22. LNCS (LNAI), vol. 5663, pp. 17–34. Springer, Heidelberg (2009)
15. Baumgartner, P., Waldmann, U.: Hierarchic superposition with weak abstraction. In: Bonacina, M.P. (ed.) CADE-24. LNCS (LNAI), vol. 7898, pp. 39–57. Springer, Heidelberg (2013)
16. Bender, M., Pelzer, B., Schon, C.: System description: E-KRHyper 1.4 – Extensions for unique names and description logic. In: Bonacina, M.P. (ed.) CADE-24. LNCS (LNAI), vol. 7898, pp. 126–134. Springer, Heidelberg (2013)
17. Bibel, W., Schmitt, P.H. (eds.): Automated Deduction - A Basis for Applications (in 2 volumes). Kluwer Academic Publishers, Dordrecht (1998)
18. Bonacina, M.P.: A taxonomy of theorem-proving strategies. In: Veloso, M.M., Wooldridge, M.J. (eds.) Artificial Intelligence Today - Recent Trends and Developments. LNCS (LNAI), vol. 1600, pp. 43–84. Springer, Heidelberg (1999)

19. Bonacina, M.P.: On theorem proving for program checking - Historical perspective and recent developments. In: Fernández, M. (eds.): Proceedings of PPDP-12, pp. 1–11. ACM Press (2010)
20. Bonacina, M.P., Echenim, M.: Theory decision by decomposition. J. Symbolic Comput. **45**(2), 229–260 (2010)
21. Bonacina, M.P., Hsiang, J.: Towards a foundation of completion procedures as semidecision procedures. Theoret. Comput. Sci. **146**, 199–242 (1995)
22. Bonacina, M.P., Hsiang, J.: On semantic resolution with lemmaizing and contraction and a formal treatment of caching. New Gener. Comput. **16**(2), 163–200 (1998)
23. Bonacina, M.P., Johansson, M.: Interpolation of ground proofs in automated deduction: a survey. J. Autom. Reasoning **54**(4), 353–390 (2015)
24. Bonacina, M.P., Johansson, M.: On interpolation in automated theorem proving. J. Autom. Reasoning **54**(1), 69–97 (2015)
25. Bonacina, M.P., Lynch, C.A., de Moura, L.: On deciding satisfiability by theorem proving with speculative inferences. J. Autom. Reasoning **47**(2), 161–189 (2011)
26. Bonacina, M.P., Plaisted, D.A.: Constraint manipulation in SGGS. In: Kutsia, T., Ringeissen, C. (eds.) Proceedings of UNIF-28, RISC Technical Reports, pp. 47–54. Johannes Kepler Universität, Linz (2014)
27. Bonacina, M.P., Plaisted, D.A.: SGGS theorem proving: an exposition. In: Schulz, S., de Moura, L., Konev, B. (eds.) Proceedings of PAAR-4, EasyChair Proceedings in Computing (EPiC), pp. 25–38 (2015)
28. Bonacina, M.P., Plaisted, D.A.: Semantically guided goal-sensitive reasoning: inference system and completeness (2015, in preparation)
29. Bonacina, M.P., Plaisted, D.A.: Semantically-guided goal-sensitive reasoning: model representation. J. Autom. Reasoning **55**(1), 1–29 (2015). doi:10.1007/s10817-015-9334-4
30. Boy de la Tour, T., Echenim, M.: Permutative rewriting and unification. Inf. Comput. **205**(4), 624–650 (2007)
31. Chang, C.-L., Lee, R.C.-T.: Symbolic logic and mechanical theorem proving. Academic Press, New York (1973)
32. Davis, M., Logemann, G., Loveland, D.: A machine program for theorem-proving. Commun. ACM **5**(7), 394–397 (1962)
33. Davis, M., Putnam, H.: A computing procedure for quantification theory. J. ACM **7**, 201–215 (1960)
34. de Moura, L., Bjørner, N.: Satisfiability modulo theories: introduction and applications. Commun. ACM **54**(9), 69–77 (2011)
35. de Moura, L., Jovanović, D.: A model-constructing satisfiability calculus. In: Giacobazzi, R., Berdine, J., Mastroeni, I. (eds.) VMCAI 2013. LNCS, vol. 7737, pp. 1–12. Springer, Heidelberg (2013)
36. Dershowitz, N., Jouannaud, J.-P.: Rewrite systems. In: van Leeuwen, J. (ed.) Handbook of Theoretical Computer Science, vol. B, pp. 243–320. Elsevier, Amsterdam (1990)
37. Escobar, S., Meseguer, J., Sasse, R.: Variant narrowing and equational unification. Electron. Notes Theoret. Comput. Sci. **238**, 103–119 (2009)
38. Fitelson, B.: Gibbard's collapse theorem for the indicative conditional: an axiomatic approach. In: Bonacina, M.P., Stickel, M.E. (eds.) Automated Reasoning and Mathematics: Essays in Memory of William W. McCune. LNCS (LNAI), vol. 7788, pp. 181–188. Springer, Heidelberg (2013)

39. Furbach, U., Schon, C.: Semantically guided evolution of $\mathcal{SHI}$ ABoxes. In: Galmiche, D., Larchey-Wendling, D. (eds.) TABLEAUX 2013. LNCS (LNAI), vol. 8123, pp. 17–34. Springer, Heidelbeg (2013)

40. Gallier, J.H., Snyder, W.: Designing unification procedures using transformations: a survey. EATCS Bull. **40**, 273–326 (1990)

41. Ganzinger, H., Sofronie-Stokkermans, V., Waldmann, U.: Modular proof systems for partial functions with Evans equality. Inf. Comput. **240**(10), 1453–1492 (2006)

42. Ganzinger, H., Waldmann, U.: Theorem proving in cancellative abelian monoids. In: McRobbie, M.A., Slaney, J.K. (eds.) CADE-13. LNCS, vol. 1104, pp. 388–402. Springer, Heidelberg (1996)

43. Goguen, J., Meseguer, J.: Order-sorted unification. J. Symbolic Comput. **8**(4), 383–413 (1989)

44. Goguen, J., Meseguer, J.: Order-sorted algebra I: equational deduction for multiple inheritance, overloading, exceptions and partial operations. Theoret. Comput. Sci. **105**(2), 217–273 (1992)

45. Hendrix, J., Clavel, M., Meseguer, J.: A sufficient completeness reasoning tool for partial specifications. In: Giesl, J. (ed.) RTA 2005. LNCS, vol. 3467, pp. 165–174. Springer, Heidelberg (2005)

46. Hendrix, J., Meseguer, J.: Order-sorted equational unification revisited. Electron. Notes Theoret. Comput. Sci. **290**(3), 37–50 (2012)

47. Hendrix, J., Meseguer, J., Ohsaki, H.: A sufficient completeness checker for linear order-sorted specifications modulo axioms. In: Furbach, U., Shankar, N. (eds.) IJCAR 2006. LNCS (LNAI), vol. 4130, pp. 151–155. Springer, Heidelberg (2006)

48. Hsiang, J., Rusinowitch, M.: Proving refutational completeness of theorem proving strategies: the transfinite semantic tree method. J. ACM **38**(3), 559–587 (1991)

49. Hsiang, J., Rusinowitch, M., Sakai, K.: Complete inference rules for the cancellation laws. In: Proceedings of IJCAI-10, pp. 990–992 (1987)

50. Ihlemann, C., Jacobs, S., Sofronie-Stokkermans, V.: On local reasoning in verification. In: Ramakrishnan, C.R., Rehof, J. (eds.) TACAS 2008. LNCS, vol. 4963, pp. 265–281. Springer, Heidelberg (2008)

51. Ihlemann, C., Sofronie-Stokkermans, V.: On hierarchical reasoning in combinations of theories. In: Giesl, J., Hähnle, R. (eds.) IJCAR 2010. LNCS (LNAI), vol. 6173, pp. 30–45. Springer, Heidelberg (2010)

52. Marques-Silva, J.P., Sakallah, K.A.: GRASP: a new search algorithm for satisfiability. In: Proceedings of ICCAD 1996, pp. 220–227 (1997)

53. Jouannaud, J.-P., Kirchner, C.: Solving equations in abstract algebras: a rule-based survey of unification. In: Lassez, J.-L., Plotkin, G. (eds.) Computational Logic - Essays in Honor of Alan Robinson, pp. 257–321. MIT Press, Cambridge (1991)

54. Jouannaud, J.-P., Kirchner, H.: Completion of a set of rules modulo a set of equations. SIAM J. Comput. **15**(4), 1155–1194 (1986)

55. Jovanović, D., Barrett, C., de Moura, L.: The design and implementation of the model-constructing satisfiability calculus. In: Jobstman, B., Ray, S. (eds.) Proceedings of FMCAD-13. ACM and IEEE (2013)

56. Kapur, D.: An automated tool for analyzing completeness of equational specifications. In: Proceedings of ISSTA-94, pp. 28–43. ACM (1994)

57. Kapur, D., Narendran, P., Rosenkrantz, D.J., Zhang, H.: Sufficient-completeness, ground-reducibility and their complexity. Acta Informatica **28**(4), 311–350 (1991)

58. Lee, S.-J., Plaisted, D.A.: Eliminating duplication with the hyperlinking strategy. J. Autom. Reasoning **9**, 25–42 (1992)

59. Lifschitz, V., Morgenstern, L., Plaisted, D.A.: Knowledge representation and classical logic. In: van Harmelen, F., Lifschitz, V., Porter, B. (eds.) Handbook of Knowledge Representation, vol. 1, pp. 3–88. Elsevier, Amsterdam (2008)
60. Malik, S., Zhang, L.: Boolean satisfiability: from theoretical hardness to practical success. Commun. ACM **52**(8), 76–82 (2009)
61. McCune, W.W.: OTTER 3.3 reference manual. Technical Report ANL/MCS-TM-263, MCS Division, Argonne National Laboratory, Argonne (2003)
62. Moskewicz, M.W., Madigan, C.F., Zhao, Y., Zhang, L., Malik, S.: Chaff: Engineering an efficient SAT solver. In: Blaauw, D., Lavagno, L. (eds.) Proceedings of DAC-39, pp. 530–535 (2001)
63. Nelson, G.: Combining satisfiability procedures by equality sharing. In: Bledsoe, W.W., Loveland, D.W. (eds.) Automatic Theorem Proving: After 25 Years, pp. 201–211. American Mathematical Society, Providence (1983)
64. Nelson, G., Oppen, D.C.: Simplification by cooperating decision procedures. ACM Trans. Program. Lang. **1**(2), 245–257 (1979)
65. Nieuwenhuis, R., Oliveras, A., Tinelli, C.: Solving SAT and SAT modulo theories: from an abstract Davis-Putnam-Logemann-Loveland procedure to DPLL(T). J. ACM **53**(6), 937–977 (2006)
66. Peterson, G.E., Stickel, M.E.: Complete sets of reductions for some equational theories. J. ACM **28**(2), 233–264 (1981)
67. Plaisted, D.A.: Mechanical theorem proving. In: Banerji, R.B. (ed.) Formal Techniques in Artificial Intelligence, pp. 269–320. Elsevier, Amsterdam (1990)
68. Plaisted, D.A.: Equational reasoning and term rewriting systems. In: Gabbay, D.M., Hogger, C.J., Robinson, J.A. (eds.) Handbook of Logic in Artificial Intelligence and Logic Programming. Vol. I: Logical Foundations, pp. 273–364. Oxford University Press, Oxford (1993)
69. Plaisted, D.A.: Automated theorem proving. Wiley Interdisciplinary Reviews: Cognitive Science **5**(2), 115–128 (2014)
70. Plaisted, D.A., Zhu, Y.: The Efficiency of Theorem Proving Strategies. Friedrich Vieweg & Sohns, Braunschweig (1997)
71. Plaisted, D.A., Zhu, Y.: Ordered semantic hyper linking. J. Autom. Reasoning **25**, 167–217 (2000)
72. Plotkin, G.: Building in equational theories. Mach. Intell. **7**, 73–90 (1972)
73. Robinson, G.A., Wos, L.: Paramodulation and theorem proving in first order theories with equality. Mach. Intell. **4**, 135–150 (1969)
74. Robinson, J.A.: Automatic deduction with hyper-resolution. Int. J. Comput. Math. **1**, 227–234 (1965)
75. Robinson, J.A.: A machine oriented logic based on the resolution principle. J. ACM **12**(1), 23–41 (1965)
76. Robinson, J.A., Voronkov, A. (eds.): Handbook of Automated Reasoning. Elsevier and MIT Press, Amsterdam (2001)
77. Rusinowitch, M.: Theorem-proving with resolution and superposition. J. Symbolic Comput. **11**(1 & 2), 21–50 (1991)
78. Shearer, R., Motik, B., Horrocks, I.: HermiT: a highly efficient OWL reasoner. In: Ruttenberg, A., Sattler, U., Dolbear, C. (eds.) Proceedings of OWLED-5, vol. 432 of CEUR (2008)
79. Slagle, J.R.: Automatic theorem proving with renamable and semantic resolution. J. ACM **14**(4), 687–697 (1967)
80. Slaney, J., Lusk, E., McCune, W.: SCOTT: semantically constrained Otter system description. In: Bundy, A. (ed.) CADE-12. LNCS (LNAI), vol. 814, pp. 764–768. Springer, Heidelberg (1994)

81. Sofronie-Stokkermans, V.: Hierarchic reasoning in local theory extensions. In: Nieuwenhuis, R. (ed.) CADE-20. LNCS (LNAI), vol. 3632, pp. 219–234. Springer, Heidelberg (2005)
82. Sofronie-Stokkermans, V.: Hierarchical and modular reasoning in complex theories: the case of local theory extensions. In: Konev, B., Wolter, F. (eds.) FroCos 2007. LNCS (LNAI), vol. 4720, pp. 47–71. Springer, Heidelberg (2007)
83. Sofronie-Stokkermans, V.: Interpolation in local theory extensions. Logical Methods in Computer Science, **4**(4), 1–31, Paper 1 (2008)
84. Stuber, J.: Superposition theorem proving for abelian groups represented as integer modules. Theoret. Comput. Sci. **208**(1–2), 149–177 (1998)
85. Waldmann, U.: Superposition for divisible torsion-free abelian groups. In: Kirchner, C., Kirchner, H. (eds.) CADE-15. LNCS (LNAI), vol. 1421, pp. 144–159. Springer, Heidelberg (1998)
86. Wos, L., Carson, D., Robinson, G.: Efficiency and completeness of the set of support strategy in theorem proving. J. ACM **12**, 536–541 (1965)
87. Zhang, H., Zhang, J.: MACE4 and SEM: a comparison of finite model generators. In: Bonacina, M.P., Stickel, M.E. (eds.) Automated Reasoning and Mathematics: Essays in Memory of William W. McCune. LNCS (LNAI), vol. 7788, pp. 101–130. Springer, Heidelberg (2013)
88. Zhang, L., Malik, S.: The quest for efficient boolean satisfiability solvers. In: Voronkov, A. (ed.) CADE-18. LNCS (LNAI), vol. 2392, pp. 295–313. Springer, Heidelberg (2002)

A Normal Form for Stateful Connectors

Roberto Bruni[1]([✉]), Hernán Melgratti[2], and Ugo Montanari[1]

[1] Department of Computer Science, University of Pisa, Pisa, Italy
bruni@di.unipi.it
[2] Departamento de Computación, FCEyN, Universidad de Buenos Aires - Conicet,
Buenos Aires, Argentina

Abstract. In this paper we consider a calculus of connectors that allows for the most general combination of synchronisation, non-determinism and buffering. According to previous results, this calculus is tightly related to a flavour of Petri nets with interfaces for composition, called Petri nets with boundaries. The calculus and the net version are equipped with equivalent bisimilarity semantics. Also the buffers (the net places) can be one-place (C/E nets) or with unlimited capacity (P/T nets). In the paper we investigate the idea of finding normal form representations for terms of this calculus, in the sense that equivalent (bisimilar) terms should have the same (isomorphic) normal form. We show that this is possible for finite state terms. The result is obtained by computing the minimal marking graph (when finite) for the net with boundaries corresponding to the given term, and reconstructing from it a canonical net and a canonical term.

Keywords: Algebras of connectors · Petri nets with boundaries

1 Introduction

One of the foci of our long-standing collaboration with José Meseguer has been concerned with the algebraic properties of Petri nets and their computations, exploiting suitable symmetric (strict) monoidal categories [13,14,22,23]. In the context of the ASCENS project[1], we have recently investigated a flavour of composable Petri nets, called *Petri nets with boundaries*, originally proposed by Pawel Sobocinski in [28]. Petri nets with boundaries should not be confused with bounded nets: the former come equipped with left/right interfaces for composition, the latter require the existence of a bound on the number of tokens that can be present in the same place during the computation. Petri nets with boundaries allow to conveniently model stateful connectors in component-based systems and have been related to other widely adopted component-based frameworks, like BIP [4], in [10]. In particular we have shown in [12] that they are

---

This research was supported by the EU project IP 257414 (ASCENS), EU 7th FP under grant agreement no. 295261 (MEALS), by the Italian MIUR Project CINA (PRIN 2010/11), and by the UBACyT Project 20020130200092BA.
[1] http://www.ascens-ist.eu/.

© Springer International Publishing Switzerland 2015
N. Martí-Oliet et al. (Eds.): Meseguer Festschrift, LNCS 9200, pp. 205–227, 2015.
DOI: 10.1007/978-3-319-23165-5_9

equivalent to the algebra of stateless connectors from [8] extended with one-place buffers. In this paper we consider an algebra of connectors that allow for the most general combination of synchronisation, non-determinism and buffering and investigate the idea of finding a normal form representation for terms of this algebra, under some finiteness hypotheses.

Component-based design is a modular engineering practice that relies on the separation of concerns between coordination and computation. Component-based systems are built from loosely coupled computational entities, the *components*, whose interfaces comprise the number, kind and peculiarities of communication ports. The term *connector* denotes entities that glue the interaction of components [25], by imposing suitable constraints on the allowed communications. The evolution of a network of components and connectors is as if played in rounds: At each round, the components try to interact through their ports and the connectors allow/disallow some of the interactions selectively. A connector is called *stateless* when the interaction constraints it imposes are the same at each round; *stateful* otherwise.

In the case of the algebra of stateless connectors [8], terms are assigned input-output sorts, written $P : (n, m)$ or $P : n \to m$, where n is the arity (i.e., the number of ports) of the left-interface and m of the right-interface. Terms are constructed by composing in series and in parallel five kinds of basic connectors (and their duals, together with identities $I : (1,1)$) that express basic forms of (co) monoidal synchronisation and non-determinism: (self-dual) symmetry $X : (2,2)$, synchronisation $\nabla : (1,2)$ and $\Delta : (2,1)$, mutual exclusion $\wedge : (1,2)$ and $\vee : (2,1)$, hiding $\perp : (1,0)$ and $\top : (0,1)$, and inaction $\perp : (1,0)$ and $\top : (0,1)$. The parallel composition $P_1 \otimes P_2$ of two terms $P_1 : (n_1, m_1)$ and $P_2 : (n_2, m_2)$ has sort $(n_1 + n_2, m_1 + m_2)$ and corresponds to put the two connectors side by side, without interaction constraints between them. The sequential composition $P_1; P_2 : (n, m)$ is defined only if the right-interface k of $P_1 : (n, k)$ matches with the left-interface of $P_2 : (k, m)$ and corresponds to plug together such interfaces, enforcing port-wise synchronisation. It is immediate to see that each term $P : (n, m)$ has a corresponding dual $P^c : (m, n)$ (defined recursively by letting $(P_1 \otimes P_2)^c = P_1^c \otimes P_2^c$ and $(P_1; P_2)^c = P_2^c; P_1^c$) and a normal form axiomatisation is provided in [8] whose equivalence classes form a symmetric strict monoidal category (PROduct and Permutation category, PROP [16,21]) of so-called tick-tables. All such connectors are stateless.

The simplest extension to stateful connectors consists of adding one-place buffers as basic terms: $\bigcirc : (1, 1)$ denotes the empty buffer, willing to receive a "token" when an action is executed on its left port; and $\odot : (1, 1)$ denotes the full buffer, willing to give the "token" away when an action is executed on its right port. This way, certain interactions can be dynamically enabled or disabled depending on the presence or absence of "tokens" in the buffers. Such stateful connectors can be put in correspondence with Petri nets with boundaries up to bisimilarity [9,12,28]. In fact, the operational semantics of connectors and Petri nets with boundaries can be expressed in terms of labelled transition systems (LTS) whose labels are pairs (a, b) with a being a string that describes the actions observed on the ports of the left-interface and b those on the right-interface. In our case a basic action observed on a single port is a natural number, describing the number of firings on which that port is involved, or equivalently, the number of "tokens" travelling on that port; therefore a and b are strings of natural numbers. A transition with such an observation is written $P \xrightarrow{a}{b} P'$. In the case of connectors, states are terms of the algebra, while in the case of nets states are markings. In both cases the "sizes" of the interfaces are preserved by transitions, e.g., if $P \xrightarrow{a}{b} P'$ and $P : (n, m)$, then $|a| = n$, $|b| = m$ and $P' : (n, m)$. Interestingly, the abstract semantics induced by ordinary bisimilarity over such LTS is a congruence w.r.t. sequential and parallel composition. Regarding the correspondence, first, it is shown that any net $N : m \to n$ with initial marking X can be associated with a connector $\boldsymbol{T}_{N_X} : (m, n)$ that preserves and reflects the semantics of N. Conversely, for any connector $P : (m, n)$ there exists a bisimilar net $\{\!|P|\!\} : m \to n$ defined by structural recursion on P. Roughly, in both cases, the one-place buffers of the connector correspond to the places of the Petri net.

The problem of finding an axiomatisation for stateful connectors such that normal forms can be found for bisimilarity classes is complicated by the fact that the number of buffers is not preserved by bisimilarity: the same "abstract state" can be described by a different combination of places. As a simple example, take a net with two transitions α and β and a place p whose pre-set is $\{\alpha\}$ and whose post-set is $\{\beta\}$. Clearly if p is substituted by any number of places connected in the same way to α and β, then the overall behaviour is not changed.

The solution provided here is to translate a term P to the corresponding net $\{\!|P|\!\}$. Then we build the marking graph of $\{\!|P|\!\}$. It must be finite because only a finite number of markings exist. Moreover we observe that marking graphs can be represented up to bisimilarity by a Petri net with boundaries that has one place for each reachable marking of $\{\!|P|\!\}$ (i.e., one place for each state of the marking graph). Finally, the translation of such net to the corresponding connector gives a canonical representation of P, in the sense that any other term P' bisimilar to P will yield the same term (up to suitable permutations).

The same procedure can be followed when Place/Transition (P/T) Petri nets with boundaries are considered. In this case, places capacity is unconstrained, i.e., a place can contain any number of tokens. Correspondingly, we start from terms of the P/T Petri calculus, where the basic constructors ◯ and ⊙ are replaced by a denumerable set of constructors $(\!|n|\!)$ for any natural number n, each representing a buffer with n tokens. Given the correspondence in [12], between P/T Petri calculus terms and P/T nets with boundaries, we can again translate a term P to the corresponding net $\{\!|P|\!\}$, but building a finite marking graph of $\{\!|P|\!\}$ requires the net to be bounded.[2] This is equivalent to require that only a finite set of terms is reachable from the term P via transitions. The marking graph can then be minimised (w.r.t. the number of states, up to bisimilarity) and translated to an equivalent P/T Petri calculus term.

---

[2] Formally, a net is *bounded* if $\exists k \in \mathbb{N}$ such that in any reachable marking the number of tokens in any place is less than or equal to k, i.e., k is a bound for the capacity of places. Note that the marking graph of a net is finite iff the net is bounded.

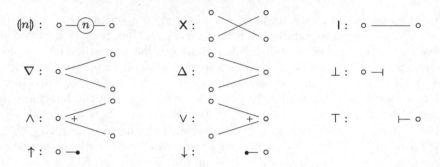

Fig. 1. Graphical representation of terms

Structure of the Paper. Section 2 introduces the P/T and the C/E Petri calculi, together with their bisimilarity semantics. Section 3 recalls Petri nets with
boundaries and their tight correspondence with Petri calculi. Section 4 shows
how to obtain a normal form for a P/T Petri calculus term P by computing
the minimal marking graph for $\{[P]\}$ and from it a canonical P/T Petri net N.
Finally, the canonical form of P is obtained by mapping N back into a term of
the P/T calculus. A similar process is outlined in Sect. 5 for terms of the C/E
Petri calculus. Section 6 concludes the paper.

2   Petri Calculi

As a matter of presentation, along the paper we find it convenient to present
first the more general version (P/T case) of the definition and constructions,
because it can be largely reused in the simpler case (C/E).

2.1   The P/T Petri Calculus

The P/T calculus is an algebra of connectors that mixes freely elementary synchronization constraints with mutual exclusion and (unbounded) memory. It is
obtained by extending the algebra of stateless connectors with a denumerable
set of constants $(\!|n|\!)$ (one for any $n \in \mathbb{N}$), each of them representing a buffer that
currently contains n data items, aka tokens.

The syntax of terms of the P/T Calculus is below, where $n \in \mathbb{N}$.

$$
\begin{aligned}
P ::= & \ (\!|n|\!) & & \text{buffer with } n \text{ data items} \\
& | \ \mathsf{I} & & \text{identity wire} & & | \ \mathsf{X} & & \text{twist} \\
& | \ \nabla \ | \ \Delta & & \text{duplicator and its dual} & & | \ \bot \ | \ \top & & \text{hiding and its dual} \\
& | \ \wedge \ | \ \vee & & \text{mutex and its dual} & & | \ \downarrow \ | \ \uparrow & & \text{inaction and its dual} \\
& | \ P \otimes P & & \text{parallel composition} & & | \ P ; P & & \text{sequential composition}
\end{aligned}
$$

The diagrammatical representation of terms is shown in Fig. 1. Any term P
has a unique associated *sort* (k, l) with $k, l \in \mathbb{N}$, that fixes the size k of the left
interface and the size l of the right interface of P (see Fig. 2).

$$(\![n]\!) : (1,1) \qquad \mathsf{I} : (1,1) \qquad \mathsf{X} : (2,2) \qquad \nabla : (1,2) \qquad \Delta : (2,1)$$

$$\bot : (1,0) \qquad \top : (0,1) \qquad \wedge : (1,2) \qquad \vee : (2,1) \qquad \downarrow : (1,0) \qquad \uparrow : (0,1)$$

$$\frac{P_1 : (k,l) \quad P_2 : (m,n)}{P_1 \otimes P_2 : (k+m,l+n)} \qquad\qquad \frac{P_1 : (k,n) \quad P_2 : (n,l)}{P_1 \,;\, P_2 : (k,l)}$$

Fig. 2. Sort inference rules

$$\frac{n,h,k \in \mathbb{N} \quad k \le n}{(\![n]\!) \xrightarrow[k]{h} (\![n+h-k]\!)} \;(\textsc{Tkio}_{n,h,k}) \qquad \frac{k \in \mathbb{N}}{\mathsf{I} \xrightarrow[k]{k} \mathsf{I}} \;(\textsc{Id}_k) \qquad \frac{h,k \in \mathbb{N}}{\mathsf{X} \xrightarrow[kh]{hk} \mathsf{X}} \;(\textsc{Tw}_{h,k})$$

$$\frac{k \in \mathbb{N}}{\nabla \xrightarrow[kk]{k} \nabla} \;(\nabla_k) \qquad \frac{k \in \mathbb{N}}{\Delta \xrightarrow[k]{kk} \Delta} \;(\Delta_k) \qquad \frac{k \in \mathbb{N}}{\bot \xrightarrow{k} \bot} \;(\bot_k) \qquad \frac{k \in \mathbb{N}}{\top \xrightarrow[k]{} \top} \;(\top_k)$$

$$\frac{h,k \in \mathbb{N}}{\wedge \xrightarrow[hk]{h+k} \wedge} \;(\wedge_{h,k}) \qquad \frac{h,k \in \mathbb{N}}{\vee \xrightarrow[h+k]{hk} \vee} \;(\vee_{h,k}) \qquad \frac{}{\downarrow \xrightarrow{0} \downarrow} \;(\downarrow) \qquad \frac{}{\uparrow \xrightarrow[0]{} \uparrow} \;(\uparrow)$$

$$\frac{P \xrightarrow[b]{a} Q \quad R \xrightarrow[d]{c} S}{P \otimes R \xrightarrow[bd]{ac} Q \otimes S} \;(\textsc{Ten}) \qquad\qquad \frac{P \xrightarrow[c]{a} Q \quad R \xrightarrow[b]{c} S}{P \,;\, R \xrightarrow[b]{a} Q \,;\, S} \;(\textsc{Cut})$$

Fig. 3. Operational semantics of P/T calculus

The operational semantics is defined by means of the LTS in Fig. 3 whose states are terms P of the algebra and whose transitions are labelled by pairs $(a,b) \in \mathbb{N}^* \times \mathbb{N}^*$, written $P \xrightarrow[b]{a} P'$, where if $P : (k,l)$ then $|a| = k$, $|b| = l$ and $P' : (k,l)$. For each $i \in \{1 \dots k\}$, a_i is the number of actions executed on the i-th port of the left interface. Analogously, for each $j \in \{1 \dots l\}$, b_j is the number of actions executed on the j-th port of the right interface. Since data items can be created and deleted, but all connectors are maintained by the rules, the target P' preserves the overall structure of P (i.e., P and P' can differ only for sub-terms of the form $(\![n]\!)$).

We remark that some of the rules are more precisely schemes. For instance, there is one particular instance of rule $(\textsc{Tkio}_{n,h,k})$ for any possible choice of n, h and k. We think the rules are self-explanatory: Rule $(\textsc{Tkio}_{n,h,k})$ models the case where a buffer with n tokens releases $k \le n$ tokens and receives h new tokens in the same step; at the end $n + h - k$ tokens are left in the buffer. Rule $(\textsc{Id}_k)$ and $(\textsc{Tw}_{h,k})$ just (re)wire the observation on the left interface to the one on the right. Rules (∇_k) and (Δ_k) enforce action synchronization on all ports. Rules $(\bot_k)$ and $(\top_k)$ hide any action on its interface. Rules $(\wedge_{h,k})$ and $(\vee_{h,k})$ mix the actions observed on the interface with two ports. Rules $(\downarrow)$ and $(\uparrow)$ enforce inaction on their (single) ports. Finally, rules $(\textsc{Ten})$ and $(\textsc{Cut})$ deal with parallel and sequential composition.

Notably, the induced bisimilarity is a congruence w.r.t. $\otimes$ and $;$ [12].

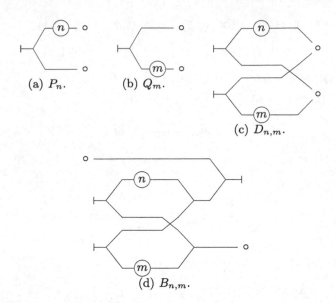

(a) P_n.   (b) Q_m.

(c) $D_{n,m}$.

(d) $B_{n,m}$.

Fig. 4. Petri calculus term for a buffer of capacity n

Example 1. As an example, we show one possible way to represent a buffer with capacity n. First, let $P_n = \top \; ; \nabla \; ; ((\!|n|\!) \otimes \mathsf{I}) : (0,2)$ and $Q_m = \top \; ; \nabla \; ; (\mathsf{I} \otimes (\!|m|\!)) : (0,2)$ shown in Fig. 4(a) and (b). It is immediate to check that, for any $h \leq n$ the only transitions for P_n are of the form $P_n \xrightarrow[hk]{} P_{n+k-h}$ and symmetrically, for Q_m and $k \leq m$, are of the form $Q_m \xrightarrow[hk]{} Q_{m+h-k}$. Let $C = (\mathsf{I} \otimes \mathsf{X} \otimes \mathsf{I}) \; ; (\Delta \otimes \Delta) : (4,2)$. Again, it is immediate to check that the only transitions for C are of the form $C \xrightarrow[hk]{hkhk} C$. Then, let $D_{n,m} = (P_n \otimes Q_m) \; ; C : (0,2)$ shown in Fig. 4(c). We have that $D_{n,m} \xrightarrow[hk]{} D_{n+k-h,m+h-k}$ with $h \leq n$ and $k \leq m$. Note that $(n-h+k) + (m-k+h) = n-m$, i.e., the numbers of tokens in the connector is invariant under transitions. Thus, the term $B_{n,m} = (\mathsf{I} \otimes D_{n,m}) \; ; ((\Delta \; ; \bot) \otimes \mathsf{I}) : (1,1)$ shown in Fig. 4(d) has transitions $B_{n,m} \xrightarrow[k]{h} B_{n+k-h,m+h-k}$ with $h \leq n$ and $k \leq m$ and $B_{n,0}$ is a buffer of capacity n (the sub-term P_n counts the free positions of the buffer, while Q_0 the busy ones).

2.2   The C/E Petri Calculus

It is quite common to impose some capacity over buffers. For example, we could think to consider only buffers of the form $(\!|c,n|\!)$ with $n \leq c$, where n is the number of tokens in the buffer and c is its maximal capacity. In this case, the transition $(\!|c,n|\!) \xrightarrow[k]{h} (\!|c,m|\!)$ would be possible only if $k \leq n$ and $h \leq c-n$ with $m = n+h-k$. ($(\!|c,n|\!)$ roughly corresponds to the process $B_{c-n,n}$ from Example 1).

In this section we focus on the simplest such case, where buffers have capacity one, also called one-place buffers. The corresponding calculus, originally introduced in [28], can be seen as the consequent restriction of the P/T Petri calculus

$$\overline{}\,(\text{TкI}) \qquad \overline{}\,(\text{TкO}) \qquad \overline{}\,(\text{TкE}) \qquad \overline{}\,(\text{TкF})$$
$$\bigcirc \xrightarrow[0]{1} \odot \qquad\quad \odot \xrightarrow[1]{0} \bigcirc \qquad\quad \bigcirc \xrightarrow[0]{0} \bigcirc \qquad\quad \odot \xrightarrow[0]{0} \odot$$

Fig. 5. Operational semantics for the one-place buffer (of the C/E Petri Calculus)

to operate over one-place buffers; in Petri net terminology, this restriction is called Condition/Event (C/E). Terms of the C/E Petri Calculus are defined by the grammar:

$$P ::= \bigcirc \mid \odot \mid \mathsf{I} \mid \mathsf{X} \mid \nabla \mid \Delta \mid \bot \mid \top \mid \wedge \mid \vee \mid \downarrow \mid \uparrow \mid P \otimes P \mid P\,;P$$

The constructors are the same as the ones of P/T calculus except for $\bigcirc$ and $\odot$ that respectively mimic the behaviour of $(\!|0,1|\!)$ and $(\!|1,1|\!)$. As before, any term P has a unique associated *sort*, with $\bigcirc : (1,1)$ and $\odot : (1,1)$ (remaining cases are defined as in Fig. 2).

The operational semantics is then defined by replacing Rule $(\text{TкIO}_{n,h,k})$ in Fig. 3 with the four rules in Fig. 5, representing respectively: the arrival of a token in the empty buffer (rule (TкI)); the release of a token from the full buffer (rule (TкO)); the inactivity of the empty/full buffer (rules (TкE), (TкF)).

Remark 1. The semantics of the C/E Petri calculus presented here slightly differs from the original one in [28] and all its variants considered in [12]. If we restrict to consider stateless connectors, i.e., terms not involving $\bigcirc$ and $\odot$, then their semantics is the one called 'weak' in [12], whereas the 'strong' semantics would allow only one action at a time to take place in a port, e.g., only transitions $\wedge \xrightarrow[00]{0} \wedge$, $\wedge \xrightarrow[10]{1} \wedge$ and $\wedge \xrightarrow[01]{1} \wedge$ would be considered for the connector $\wedge$. Differently from the weak case, here we forbid tokens to traverse buffers during a step, in agreement with the classical C/E semantics where a loop cannot fire. However, other variants can be nicely accounted for by changing the rules for $\bigcirc$ and $\odot$. For example, consume/produce loops can be dealt with by adding the transition $\odot \xrightarrow{1} \odot$. On the one hand, we think the semantics proposed here improves the correspondence between C/E Petri calculus and C/E Petri nets with boundaries (avoiding the use of the 'contention' relation from [12]) and, on the other hand, it yields a more uniform definition with the P/T case, preserving all good properties, like bisimilarity being a congruence w.r.t. $\otimes$ and ;.

Example 2. A buffer with capacity n can be represented by combining n buffers of capacity 1: we just let $B_1 = \bigcirc : (1,1)$ and $B_{n+1} = \wedge\,;\,(B_n \otimes \bigcirc)\,;\,\vee : (1,1)$.

3 Nets with Boundaries

Nets with boundaries extends ordinary Petri nets by equipping them with left and right interfaces made of ports. Ports are different from places in that places in the pre-set of a transition α impose a bound on the number of instances of α that can be fired concurrently, while ports do not. In fact ports can account for an unbounded number of instances of transitions attached to them to fire concurrently. This is desirable, not an anomaly, because we can account for any execution context in which the nets with boundaries are plugged in.

3.1   P/T Petri Nets with Boundaries

Petri nets [26] consist of *places*, which are repositories of *tokens*, and *transitions* that remove and produce tokens. Places of a *Place/Transition net* (P/T net) can hold zero, one or more tokens and arcs are weighted. The state of a P/T net is described in terms of *(P/T) markings*, i.e., (finite) multisets of tokens.

A multiset on a set X is a function $X \to \mathbb{N}$. The set of multisets on X is denoted $\mathcal{M}_X$. We let $\mathcal{U}, \mathcal{V}$ range over $\mathcal{M}_X$. For $\mathcal{U}, \mathcal{V} \in \mathcal{M}_X$, we write $\mathcal{U} \subseteq \mathcal{V}$ iff $\forall x \in X : \mathcal{U}(x) \leq \mathcal{V}(x)$ and we use the usual multiset operations for union $(\cup)$, difference $(-)$ and scalar multiplication $(\cdot)$. We use $\varnothing \in \mathcal{M}_X$ for the empty multiset s.t. $\varnothing(x) = 0$ for all $x \in X$ and we write x for the singleton multiset $\mathcal{U}$ such that $\mathcal{U}(x) = 1$ and $\mathcal{U}(y) = 0$ for all $y \neq x$. Given a finite X, if $f : X \to \mathcal{M}_Y$ and $\mathcal{U} \in \mathcal{M}_X$ then we shall abuse notation and write $f(\mathcal{U}) = \bigcup_{x \in X} \mathcal{U}(x) \cdot f(x)$.

Definition 1 (P/T net). *A P/T net is a 4-tuple* $(P, T, ^\circ-, -^\circ)$ *where:* P *is a set of* places; T *is a set of* transitions; *and* $^\circ-, -^\circ : T \to \mathcal{M}_P$ *are functions assigning* pre- *and* post-sets *to transitions.*

Let $\mathcal{X} \in \mathcal{M}_P$, we write $N_\mathcal{X}$ for the marked P/T net N with marking $\mathcal{X}$.

Definition 2 (P/T step semantics). *Let* $N = (P, T, ^\circ-, -^\circ)$ *be a P/T net,* $\mathcal{X}, \mathcal{Y} \in \mathcal{M}_P$. *For* $\mathcal{U} \in \mathcal{M}_T$ *a multiset of transitions, we write:*

$$N_\mathcal{X} \to_\mathcal{U} N_\mathcal{Y} \quad \overset{\text{def}}{=} \quad {}^\circ\mathcal{U} \subseteq \mathcal{X},\ \mathcal{U}^\circ \subseteq \mathcal{Y}\ \&\ \mathcal{X} - {}^\circ\mathcal{U} = \mathcal{Y} - \mathcal{U}^\circ.$$

The remaining of this section recalls the composable nets proposed in [28]. Due to space limitation, we refer to [12] for a detailed presentation. In the following we let $\underline{n}$ range over finite ordinals, i.e., $\underline{n} \overset{\text{def}}{=} \{0, 1, \ldots, n - 1\}$.

Definition 3 (P/T net with boundaries). *Let* $m, n \in \mathbb{N}$. *A (finite) P/T net with boundaries* $N : m \to n$ *is a tuple* $N = (P, T, ^\circ-, -^\circ, {}^\bullet-, -^\bullet)$, *where:*

- $(P, T, ^\circ-, -^\circ)$ *is a finite P/T net;*
- ${}^\bullet- : T \to \mathcal{M}_{\underline{m}}$ *and* $-^\bullet : T \to \mathcal{M}_{\underline{n}}$ *are functions that bind transitions to the left and right boundaries of* N;

Let $\mathcal{X} \in \mathcal{M}_P$, we write $N_\mathcal{X}$ for the P/T net N with boundaries whose current marking is $\mathcal{X}$. Note that, for any $k \in \mathbb{N}$, there is a bijection $\ulcorner - \urcorner : \mathcal{M}_{\underline{k}} \to \mathbb{N}^k$ between multisets on $\underline{k}$ and strings of natural numbers of length k, defined by $\ulcorner \mathcal{U} \urcorner_i \overset{\text{def}}{=} \mathcal{U}(i)$, namely, the i-th natural number in the string $\ulcorner \mathcal{U} \urcorner$ assigned to the multiset $\mathcal{U}$ is the multiplicity of the i-th port in $\mathcal{U}$. For example, given the multiset $\mathcal{U} = \{0, 0, 2\} \in \mathcal{M}_{\underline{4}}$ we have $\ulcorner \mathcal{U} \urcorner = 2\ 0\ 1\ 0$.

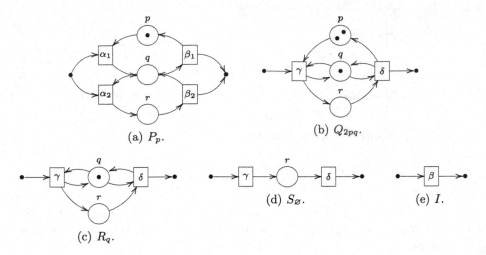

(a) P_p.

(b) Q_{2pq}.

(c) R_q.

(d) $S_\varnothing$.

(e) I.

Fig. 6. Five marked P/T nets with boundaries

Definition 4 (P/T Labelled Semantics). *Let $N = (P, T, {}^\circ-, -^\circ, {}^\bullet-, -^\bullet)$ be a P/T net with boundaries and $\mathcal{X}, \mathcal{Y} \in \mathcal{M}_P$. We write*

$$N_\mathcal{X} \xrightarrow{a}{b} N_\mathcal{Y} \quad \overset{\text{def}}{=} \quad \exists \mathcal{U} \in \mathcal{M}_T \text{ s.t. } N_\mathcal{X} \to_\mathcal{U} N_\mathcal{Y}, \ a = \ulcorner{}^\bullet\mathcal{U}\urcorner \ \& \ b = \ulcorner\mathcal{U}^\bullet\urcorner. \quad (1)$$

Example 3. Figure 6 shows five different marked P/T nets with boundaries. Places are circles and a marking is represented by the presence or absence of tokens; rectangles are transitions and arcs stand for pre- and post-set relations. The left (respectively, right) interface is depicted by points situated on the left (respectively, on the right). Figure 6(a) shows the marked net $P_p : 1 \to 1$ containing three places, four transitions and initially marked with one token in place p. Figure 6(b) shows the marked net $Q_{2pq} : 1 \to 1$ containing three places, two transitions and initially marked with two tokens in p and one in q. These two nets are bisimilar: they both model a buffer with capacity two, in which messages are produced over the left interface and consumed over the right interface. Figure 6(c) and (d) show two different models for unbounded buffers. They are not bisimilar: while R_q serialises all operations on the buffer, $S_\varnothing$ allows for the concurrent production/consumption of messages. Note that transition γ in Fig. 6(d) has an empty pre-set and δ has an empty post-set. Figure 6(e) shows the net $I : 1 \to 1$ that contains no places. The sole transition β has empty pre and post-sets. This net can forward any quantity of tokens received on its left port to the right port and, hence, it is neither bisimilar to R_q nor to $S_\varnothing$.

While from the point of view of ordinary Petri nets having empty pre-/post-sets is quite a peculiar feature, which makes life harder when defining the operational semantics, we emphasize that in our context of decomposing nets into their minimal components this is a highly valuable property. In fact, the interfaces of

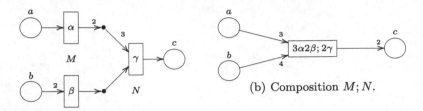

(a) Two P/T with boundaries M and N.

Fig. 7. Composition of P/T with boundaries

nets with boundaries have the role of synchronizing the transitions of different components. In this perspective, it is natural to have nets without places as basic components.

Nets with boundaries can be composed in parallel and in series.

Given $N_\mathcal{X} : m \to n$ and $M_\mathcal{Y} : k \to l$, their tensor product is the net $N_\mathcal{X} \otimes M_\mathcal{Y} :$ $m + k \to n + l$ whose sets of places and transitions are the disjoint union of the corresponding sets in N and M, whose maps $^\circ{-}, {-}^\circ, {}^\bullet{-}, {-}^\bullet$ are defined according to the maps in N and M and whose initial marking is $\mathcal{X} \cup \mathcal{Y}$. Intuitively, the tensor product corresponds to put the nets N and M side-by-side.

The sequential composition $N_\mathcal{X}; M_\mathcal{Y} : m \to n$ of $N_\mathcal{X} : m \to k$ and $M_\mathcal{Y} : k \to n$ is slightly more involved. Intuitively, transitions attached to the left or right boundaries can be seen as transition fragments, that can be completed by attaching other complementary fragments to that boundary. When two transition fragments in N share a boundary node, then they are two mutually exclusive options for completing a fragment of M attached to the same boundary node. Thus, the idea is to combine the transitions of N with those of M when they share a common boundary, as if their firings were synchronised. As in general (infinitely) many combinations are possible, the composed nets is defined by selecting a minimal (multi-)set of synchronisations that suffices to represent any other possible synchronisation as a linear combinations of the chosen ones (i.e., as the concurrent firing of several transitions). The initial marking is $\mathcal{X} \cup \mathcal{Y}$ (formal definition can be found at [12]). As an example, Fig. 7(b) shows the sequential composition of the nets $M : 0 \to 2$ and $N : 2 \to 0$ from Fig. 7(a). A firing of α produces two tokens on the port to which γ is also attached, while a firing of γ requires three tokens from the same port and one from the other port, to which β is attached to. Therefore the minimal multi-set of transitions that allows the synchronization between α, β and γ contains three instances of α and two instances of β and γ.

3.2   From P/T Nets with Boundaries to P/T Calculus and Back

The contribution in [12] enlightens a tight semantics correspondence between P/T calculus and P/T nets with boundaries. Concretely, two translations are defined. The first encoding $\boldsymbol{T}_{-}$ shows that each net $N_\mathcal{X}$ can be mapped into a P/T calculus process $T_{N_{\mathcal{X}}}$ that preserves and reflects operational semantics (and thus also bisimilarity). The second encoding $\{\!\!\{-\}\!\!\}$ provides the converse translation, from a P/T Petri calculus process P to a P/T net with boundaries $\{\!\!\{P\}\!\!\}$, defined by structural induction. We recall here the two main correspondence results and omit the details due to space constraints.

Theorem 1. *Let P be a term of P/T calculus.*

(i) if $P \xrightarrow[b]{a} P'$ then $\{\!\!\{P\}\!\!\} \xrightarrow[b]{a} \{\!\!\{P'\}\!\!\}$.
(ii) if $\{\!\!\{P\}\!\!\} \xrightarrow[b]{a} N_{\mathcal{X}}$ then $\exists P'$ such that $P \xrightarrow[b]{a} P'$ and $\{\!\!\{P'\}\!\!\} = N_{\mathcal{X}}$.

Theorem 2. *Let N be a finite P/T net with boundaries, then*

(i) if $N_{\mathcal{X}} \xrightarrow[b]{a} N_{\mathcal{Y}}$ then $T_{N_{\mathcal{X}}} \xrightarrow[b]{a} T_{N_{\mathcal{Y}}}$.
(ii) if $T_{N_{\mathcal{X}}} \xrightarrow[b]{a} Q$ then $\exists \mathcal{Y}$ such that $N_{\mathcal{X}} \xrightarrow[b]{a} N_{\mathcal{Y}}$ and $Q = T_{N_{\mathcal{Y}}}$.

3.3 C/E Nets with Boundaries

A well-known subclass of bounded P/T nets are C/E nets. In C/E nets, places have maximum capacity 1 and pre- and post-set of transitions are restricted to sets (instead of multisets). Formally,

Definition 5 (C/E net). *A C/E net is a P/T net $N = (P, T, {}^{\circ}-, -^{\circ})$ where:[3] P is a set of* places; *T is a set of* transitions; *and ${}^{\circ}-, -^{\circ} : T \to 2^P$ are functions.*

In addition, a *C/E marking* is just a subset of places $X \subseteq P$ (not a multiset). We let N_X denote the net N with marking X.

Definition 6 (C/E step semantics). *Let $N = (P, T, {}^{\circ}-, -^{\circ})$ be a C/E net, $X, Y \subseteq P$ and $\mathcal{U} \subseteq \mathcal{M}_T$ a multiset of transitions s.t. ${}^{\circ}\mathcal{U}$ and $\mathcal{U}^{\circ}$ are sets, write:*

$$N_X \to_{\mathcal{U}} N_Y \quad \overset{\text{def}}{=} \quad {}^{\circ}\mathcal{U} \subseteq X, \ \mathcal{U}^{\circ} \cap X = \varnothing \ \& \ Y = (X \backslash {}^{\circ}\mathcal{U}) \cup \mathcal{U}^{\circ}.$$

We remark that the constraint on ${}^{\circ}\mathcal{U}$ and $\mathcal{U}^{\circ}$ to be sets ensures that every pair of transitions in $\mathcal{U}$ has disjoint pre- and post-sets. This definition allows the concurrent firing of several instances of the same transition when its pre- and post-sets are both empty: As explained before, even if places are bounded this will allow for ports of unbounded capacity (w.r.t. the number of actions that can take place concurrently) in C/E nets with boundaries.

Definition 7 (C/E nets with boundaries). *A P/T net with boundaries $N = (P, T, {}^{\circ}-, -^{\circ}, {}^{\bullet}-, -^{\bullet})$ is a C/E net with boundaries if $(P, T, {}^{\circ}-, -^{\circ})$ is a C/E net.*

A marking of a C/E net with boundaries is just a set of places of the net, i.e., $X \subseteq P$. Note that while pre- and post-set of transitions are sets and not multisets, multiplicity are maintained by ${}^{\bullet}-$ and $-^{\bullet}$ w.r.t. left and right ports: many tokens can be exchanged concurrently over a single port in one step.

---

[3] In the context of C/E nets some authors call places *conditions* and transitions *events*.

Definition 8 (C/E Labelled Semantics). *Let* $N = (P, T, {}^\circ-, -{}^\circ, {}^\bullet-, -{}^\bullet)$ *be a C/E net with boundaries and* $X, Y \subseteq P$. *Write:*

$$N_X \xrightarrow[b]{a} N_Y \stackrel{\text{def}}{=} \exists \mathcal{U} \subseteq \mathcal{M}_T \text{ s.t. } N_X \rightarrow_{\mathcal{U}} N_Y, \ a = \ulcorner {}^\bullet\mathcal{U}\urcorner \ \& \ b = \ulcorner \mathcal{U}^\bullet\urcorner \quad (2)$$

Remark 2. Following the presentation of the C/E Petri Calculus in Sect. 2.2 (see Remark 1), we have presented here a slightly different definition for C/E Petri nets with boundaries w.r.t. [12] by allowing richer observations over interfaces (strings of natural numbers instead of just 0/1).

Example 4. All nets in Fig. 6 except from Q_{2pq} (Fig. 6(b)) can be interpreted as C/E nets with boundaries. We remark that P_p has the same behaviour when considering both the P/T net and the C/E labelled semantics (because P_p is a 1-bounded P/T net). Similarly, the semantics of $I_\varnothing$ is also invariant under both views. Differently, the behaviour of R_q and $S_\varnothing$ changes when considering the C/E semantics. The former is deadlocked, because of the self-looping transitions, while the latter models a buffers of capacity one that alternates the production and consumption of tokens.

The correspondence results in Sect. 3.2 can be restated also for the case of the Petri Calculus and C/E nets with boundaries along the lines shown in [12].

4   Normal Forms for Finite State P/T terms

This section shows how to obtain normal forms for finite state connectors. We will take advantage of the mutual encodings between P/T calculus terms and P/T nets with boundaries summarised in Sect. 3.2. In order to obtain the normal for a connector, we will proceed as follows: (i) we translate a P/T calculus term into an equivalent P/T net with boundaries by using the encoding $\{\!|\_|\!\}$, (ii) we compute a canonical representation (up to isomorphism) for the corresponding net with boundaries, (iii) we map back the canonical representation of the net into a term of the P/T calculus by using the encoding $T_{\_}$. The canonical representation of the net is obtained by analysing its associated marking graph.

Definition 9 (Reachable marking). *Let* $N_{\mathcal{X}} : n \rightarrow m$ *be a P/T net with boundaries. Then,* $\mathcal{Y}$ *is a reachable marking of* $N_{\mathcal{X}}$ *if there exists a (possibly empty) finite sequence of transitions* $N_{\mathcal{X}} \xrightarrow[b_1]{a_1} N_{\mathcal{X}_1} \xrightarrow[b_2]{a_2} \cdots \xrightarrow[b_k]{a_k} N_{\mathcal{Y}}$ *with* $a_i \in \mathbb{N}^n$ *and* $b_i \in \mathbb{N}^m$. *We write* $\mathcal{RM}(N_{\mathcal{X}})$ *for the set of all reachable markings of* $N_{\mathcal{X}}$.

Definition 10 (Marking graph of a net with boundaries). *Let* $N : n \rightarrow m$ *be a P/T net with boundaries with initial marking* $\mathcal{X}$. *The marking graph of* $N_{\mathcal{X}}$ *is the state transition graph* $\mathcal{MG}(N_{\mathcal{X}}) = (\mathcal{RM}(N_{\mathcal{X}}), \mathbb{T})$ *where* $\mathbb{T} \subseteq \mathcal{M}_P \times \mathbb{N}^n \times \mathbb{N}^m \times \mathcal{M}_P$ *is as follows:* $\mathbb{T} = \{(\mathcal{Y}, a, b, \mathcal{Z}) \mid \mathcal{Y}, \mathcal{Z} \in \mathcal{RM}(N_{\mathcal{X}}) \wedge N_{\mathcal{Y}} \xrightarrow[b]{a} N_{\mathcal{Z}}\}$.

We say $\mathcal{MG}(N_{\mathcal{X}})$ is *(in)finite state* when $\mathcal{RM}(N_{\mathcal{X}})$ is (in)finite. We say $\mathcal{MG}(N_{\mathcal{X}})$ is *finite* when it is finite state and $\mathbb{T}$ is also finite, we say it is *infinite* otherwise.

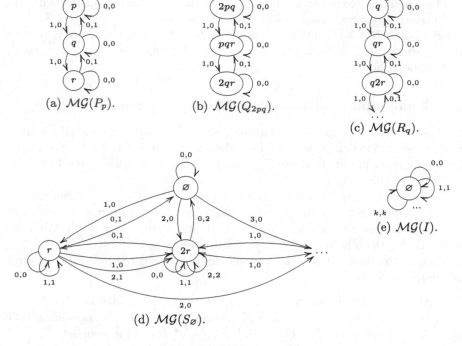

(a) $\mathcal{MG}(P_p)$.

(b) $\mathcal{MG}(Q_{2pq})$.

(c) $\mathcal{MG}(R_q)$.

(d) $\mathcal{MG}(S_\varnothing)$.

(e) $\mathcal{MG}(I)$.

Fig. 8. Marking graphs

$\mathcal{MG}(N_\chi)$ is finitely branching if for any $\mathcal{Y} \in \mathcal{RM}(N_\chi)$ it holds that $\mathbb{T}_\mathcal{Y} = \{(\mathcal{V}, a, b, \mathcal{Z}) \mid (\mathcal{V}, a, b, \mathcal{Z}) \in \mathbb{T} \wedge \mathcal{V} = \mathcal{Y}\}$ is finite.

Note that for any $N_\chi : n \to m$, it holds that $N_\chi \xrightarrow[0^m]{0^n} N_\chi$. Therefore, every node in a marking graph of the net has a self-loop with label $(0^n, 0^m)$.

Example 5. Figure 8 shows the marking graphs for the nets in Fig. 6. We remark that the marking graphs for P_p and Q_{2pq} are finite and isomorphic. On the contrary, the remaining three are infinite. The marking graph for R_q and $S_\varnothing$ are infinite state (because the corresponding nets are unbounded). Nevertheless, while $\mathcal{MG}(R_q)$ is finitely branching, $\mathcal{MG}(S_\varnothing)$ is not (e.g., any state in $\mathcal{MG}(S_\varnothing)$ has a transition labelled $(k, 0)$ for any $k \in \mathbb{N}$). Although $\mathcal{MG}(I)$ is finite state, it is infinitely branching.

Remark 3. The marking graph of a net with boundaries is finite state only if the underlying net is bounded. Note that the marking graph of a net containing a transition with empty pre-set and non-empty post-set is unbounded (for instance, the net $S_\varnothing$ in Fig. 8(d)).

Remark 4. The marking graph of a P/T net with boundaries containing a transition with an empty preset is infinitely branching (e.g., the nets $I_\varnothing$ in Fig. 6(e) and $S_\varnothing$ in Fig. 8(d)). On the contrary, when every transitions in the net has a non-empty preset, the marking graph is finitely branching because each marking constraints the number of concurrently fireable instances of each transition.

The remaining of the section is devoted to the definition of the normal form of (finite state) connectors. We deal with the general case by using a *divide et impera* approach. We solve two sub-problems first: (i) the encoding of nets with finite marking graphs (Sect. 4.1) and (ii) the encoding of infinitely branching stateless nets (Sect. 4.2).

4.1   Finite (State and Transition) Marking Graphs

In this section we show how to obtain the normal form for P/T nets with boundaries whose marking graph is finite, i.e., when it is bounded and every transition has a non-empty preset. We leave this as an implicit assumption for all the nets considered in this section.

We first note that for a finite graph we can apply, e.g., a partition refinement algorithm [18,24] to obtain the smallest (up-to iso) (in terms of states and transitions) automaton amongst all those bisimilar to the given graph. We write $min(MG)$ for the minimal graph in the equivalence class of MG.

We note that any finite marking graph can be represented by a P/T net with boundaries as follows:

Definition 11 (Marking graph as a net with boundaries). *Let $MG = (S,T)$ with $T \in S \times \mathbb{N}^n \times \mathbb{N}^m \times S$ be a marking graph. The corresponding P/T net with boundaries is $\mathcal{NB}(MG) = (S,T',{}^\circ-,-^\circ,{}^\bullet-,-^\bullet) : n \to m$ s.t.*

- $T' = T \setminus \{(s,0^n,0^m,s) \mid s \in S\}$ *(we can safely omit self-looping transitions that are not attached to ports);*
- ${}^\circ(s,a,b,t) = s$ *and* $(s,a,b,t)^\circ = t$*;*
- ${}^\bullet(s,a,b,t) = \mathcal{U}$ *where* $\mathcal{U} \in \mathcal{M}_{\underline{n}}$ *and* $\ulcorner\mathcal{U}\urcorner = a$*;*
- $(s,a,b,t)^\bullet = \mathcal{V}$ *where* $\mathcal{V} \in \mathcal{M}_{\underline{m}}$ *and* $\ulcorner\mathcal{V}\urcorner = b$*.*

We let $can(N_\mathcal{X}) \overset{\text{def}}{=} \mathcal{NB}(min(\mathcal{MG}(N_\mathcal{X})))_{\{\mathcal{X}\}}$.

Lemma 1 (Minimal net with boundaries). *Let $N : n \to m$ be a net with boundaries, then we have that $N_\mathcal{X}$ and $can(N_\mathcal{X})$ are bisimilar.*

Proof. It follows by noting that $N_\mathcal{X}$ and $\mathcal{MG}(N_\mathcal{X})$ are bisimilar by construction; $\mathcal{MG}(N_\mathcal{X})$ and $min(\mathcal{MG}(N_\mathcal{X}))$ are bisimilar by definition; and $min(\mathcal{MG}(N_\mathcal{X}))$ and $\mathcal{NB}(min(\mathcal{MG}(N_\mathcal{X})))_{\{\mathcal{X}\}}$ are bisimilar by construction.

Corollary 1. *$can(N)$ is unique (up-to iso) because $\mathcal{NB}(-)$ and $\mathcal{MG}(-)$ are functions and the minimal automaton is also unique (up-to iso).*

Corollary 2. *Given two bisimilar nets with boundaries $N_\mathcal{X}$ and $M_\mathcal{X}$, the nets $can(N_\mathcal{X})$ and $can(M_\mathcal{X})$ are isomorphic.*

Example 6. Consider the P/T term $Q = (\nabla\otimes(\top;\nabla));(\mathsf{I}\otimes T\otimes\mathsf{I});((\Delta;\bot)\otimes\Delta)$ with $T = \mathsf{X};(\nabla\otimes\nabla);(\langle\!\langle2\rangle\!\rangle\otimes(\vee;\langle\!\langle1\rangle\!\rangle);\wedge)\otimes\langle\!\langle0\rangle\!\rangle);(\Delta\otimes\Delta)$ depicted in Fig. 9(a). The equivalent P/T net with boundaries $\{\![Q]\!\}$ is the net Q_{2pq} shown in Fig. 6(b). The corresponding marking graphs is in Fig. 8(b). This graph is minimal, i.e., there does not exist a bisimilar graph with a smaller number of states and/or transitions. Therefore, $min(\mathcal{MG}(Q_{2pq})) = \mathcal{MG}(Q_{2pq})$ and $can(Q_{2pr}) = \mathcal{NB}(\mathcal{MG}(Q_{2pr}))_{2pr}$, which is shown in Fig. 9(b). Then, the normal form $nf(Q)$ is obtained by encoding back $can(Q_{2pr})$ as a P/T term (shown in Fig. 9(c)).

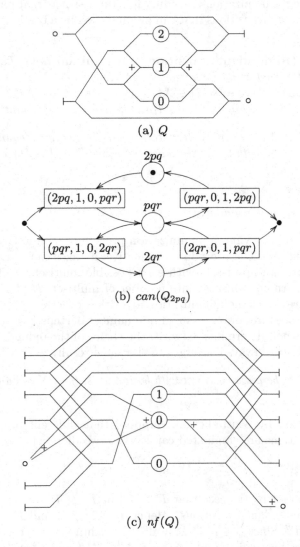

(a) Q

(b) $can(Q_{2pq})$

(c) $nf(Q)$

Fig. 9. Normal form of a term with finite marking graph

We remark that the marking graph $\mathcal{MG}(P_p))$ (Fig. 8(a)), corresponding to P_p in Fig. 6(a), is isomorphic to the marking graph of Q_{2pq}. This implies that both P_p and Q_{2pq} have the same normal form.

4.2 Stateless, Infinitely Branching Marking Graphs

The simplest case of finite state, but infinite branching marking graph, is a net without places, like the net I in Fig. 6(e), whose marking graph is (partially) depicted in Fig. 8(e).

We introduce a minimization procedure for stateless nets that removes redundant transitions, i.e., transitions that can be mimicked by a combination of other transitions in the net.

Definition 12 (Redundant transition and minimal net). *Let $N : m \to n$ be the stateless P/T net with boundaries $N = (\varnothing, T, {}^{\circ}-, -^{\circ}, {}^{\bullet}-, -^{\bullet})$. A transition $t \in T$ is redundant if there exists $\mathcal{U} \in \mathcal{M}_{T-\{t\}}$ s.t. ${}^{\bullet}t = {}^{\bullet}\mathcal{U}$ and $t^{\bullet} = \mathcal{U}^{\bullet}$. We say that a stateless net is minimal if every transition is not redundant.*

Lemma 2. *Let $N : m \to n$ be the stateless P/T net with boundaries $N = (\varnothing, T, {}^{\circ}-, -^{\circ}, {}^{\bullet}-, -^{\bullet})$ with $t \in T$ redundant. Define $T' = T - \{t\}$ and*

$$N' = (\varnothing, T', {}^{\circ}-|_{T'}, -^{\circ}|_{T'}, {}^{\bullet}- |_{T'}, -^{\bullet}|_{T'}).$$

Then, $N_{\varnothing}$ and $N'_{\varnothing}$ are bisimilar.

The above result provides a minimization procedure by iteratively removing redundant transitions. The procedure is effective: it takes each transition $t \in T$ and compares pre- and post-sets with each possible multisets $\mathcal{U}$ of T. Since ${}^{\bullet}t$ and $t^{\bullet}$ are finite, there is just a finite number of multisets $\mathcal{U}$ of T to consider. We note $\tilde{N}$ the result of the minimization algorithm over N.

The above procedure converges in a finite number of steps, because T is finite. The procedure is non-deterministic (w.r.t. the choice of the redundant transition t to eliminate) but it always converges to the same result.

Lemma 3. *Let N be a stateless net with boundaries, then $\tilde{N}$ is uniquely defined (up-to iso).*

Proof. We proceed by contradiction. Suppose that different orders in which redundant transitions are eliminated can lead to two different outcomes

$$N' = (\varnothing, T', {}^{\circ}-, -^{\circ}, {}^{\bullet}-, -^{\bullet}) \text{ and } N'' = (\varnothing, T'', {}^{\circ}-, -^{\circ}, {}^{\bullet}-, -^{\bullet}).$$

Clearly it cannot be the case that $T' \subset T''$ or $T'' \subset T'$ (otherwise T' or T'' would contain redundant transitions). Hence $T'' \setminus T' \neq \varnothing$ and $T' \setminus T'' \neq \varnothing$.

Let $t' \in T' \setminus T''$. Since $t' \in T' \subseteq T$, it must be redundant w.r.t. the transitions in T'', i.e., there must exist $\mathcal{U}' \in \mathcal{M}_{T''}$ s.t. ${}^{\bullet}t' = {}^{\bullet}\mathcal{U}'$ and $t'^{\bullet} = \mathcal{U}'^{\bullet}$. Following a similar reasoning, any transition t'' in $T'' \setminus T'$ must be redundant w.r.t. the transitions in T' and expressible as a suitable $\mathcal{U}'' \in \mathcal{M}_{T'}$.

Moreover, there must be at least one transition $t'' \in \mathcal{U}'$, non isomorphic to t', such that $t'' \in T'' \setminus T'$ (otherwise t' would be redundant w.r.t. transitions in T'). Then, since any such t'' can be expressed in terms of $\mathcal{U}'' \in \mathcal{M}_{T'}$, it follows that t' can be expressed as a multiset $\mathcal{U} \in \mathcal{M}_{T'}$. Now there are two cases:

 – $t' \notin \mathcal{U}$, but this is absurd, because t' would be redundant;
 – $t' \in \mathcal{U}$, but this is absurd, because we would have $\mathcal{U} = t'$ isomorphic to t''.

Lemma 4. *Let N be a stateless net, then $\tilde{N}_{\varnothing}$ and $N_{\varnothing}$ are bisimilar.*

Lemma 5. *Let N and M be two stateless bisimilar nets, then $\tilde{N}_{\varnothing} = \tilde{M}_{\varnothing}$ (up-to iso).*

Proof. The proof follows by contradiction. Assume that there is a transition t in $\tilde{N}$ that is not matched by a transition in $\tilde{M}$. Let $\tilde{N}_{\varnothing} \to_t \tilde{N}_{\varnothing}$. Then, $N_{\varnothing} \to_t N_{\varnothing}$. Since $N_{\varnothing}$ and $M_{\varnothing}$ are bisimilar, $M_{\varnothing} \to_{\mathcal{U}} M_{\varnothing}$ with $^{\bullet}t = {}^{\bullet}\mathcal{U}$ and $t^{\bullet} = \mathcal{U}^{\bullet}$. Consequently, $M_{\varnothing} \to_{\mathcal{U}} M_{\varnothing}$. If $|\mathcal{U}| = 1$, we are done. Otherwise, $\mathcal{U} = k_1 \cdot t_1 \cup \ldots \cup k_n \cdot t_n$ with $n > 1$. Then, for any transition t_i we conclude that $\tilde{N}_{\varnothing} \to_{t_i} \tilde{N}_{\varnothing}$. Hence, t is redundant in $\tilde{N}$, which contradicts the assumption that $\tilde{N}$ is minimal.

Example 7. Consider the stateless term $Sl = (\wedge \otimes \wedge); (\mathsf{I} \otimes (\Delta; \nabla) \otimes \mathsf{I}); (\vee \otimes \vee)$ depicted in Fig. 10(a). The corresponding net with boundaries $\{[Sl]\}$ is shown in Fig. 10(b). Note that the transition β is redundant because it can be expressed as the concurrent firing of α and β; consequently, it is removed by the minimization algorithm, which produces the minimal net $\widetilde{\{[Sl]\}}$ shown in Fig. 10(c). Finally, the normal form $nf(Sl)$ for the net Sl is obtained by encoding back $\widetilde{\{[Sl]\}}$ as the Petri calculus term shown in Fig. 10(d).

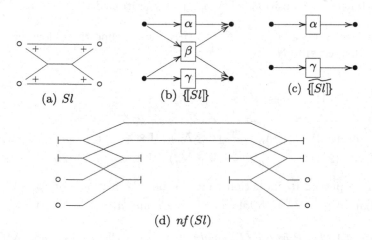

(a) Sl (b) $\{[Sl]\}$ (c) $\widetilde{\{[Sl]\}}$

(d) $nf(Sl)$

Fig. 10. A stateless term of the P/T calculus Sl

4.3   Finite State and Infinitely Branching Marking Graph

When the marking graph is finite state but infinitely branching, the associated net has both transitions with empty pre- and post-set and transitions with non-empty post-set (by Remark 3, the net cannot contain transitions with empty pre-set and non-empty post-set). We show first that the behaviour of a net can be described by combining the behaviour of two subnets containing respectively the stateless and stateful behaviours.

Definition 13 (Stateless and Stateful subnets). *Let $N : m \to n$ be the P/T net with boundaries $N = (P, T, {}^\circ-, -{}^\circ, {}^\bullet-, -{}^\bullet)$. A transition $t \in T$ is stateless if ${}^\circ t = t^\circ = \varnothing$. We write T^{sl} for the set of all stateless transitions and $T^{\mathsf{sf}} = T \backslash T^{\mathsf{sl}}$ denote the set of stateful transitions. The stateless subnet of N is*

$$N^{\mathsf{sl}} = (\varnothing, T^{\mathsf{sl}}, {}^\circ - |_{T^{\mathsf{sl}}}, -{}^\circ|_{T^{\mathsf{sl}}}, {}^\bullet - |_{T^{\mathsf{sl}}}, -{}^\bullet|_{T^{\mathsf{sl}}})$$

Similarly, the stateful subnet is

$$N^{\mathsf{sf}} = (P, T^{\mathsf{sf}}, {}^\circ - |_{T^{\mathsf{sf}}}, -{}^\circ|_{T^{\mathsf{sf}}}, {}^\bullet - |_{T^{\mathsf{sf}}}, -{}^\bullet|_{T^{\mathsf{sf}}})$$

We can now tightly relate the behaviour of N with those of N^{sl} and N^{sf}.

Lemma 6. *Let N_X be a marked P/T net with boundaries. Then,*

- *If $N_X \xrightarrow[b]{a} N_Y$, then there exist a_1, a_2, b_1 and b_2 such that $a = a_1 + a_2$, $b = b_1 + b_2$, $N^{\mathsf{sf}}_X \xrightarrow[b_1]{a_1} N^{\mathsf{sf}}_Y$ and $N^{\mathsf{sl}}_\varnothing \xrightarrow[b_2]{a_2} N^{\mathsf{sl}}_\varnothing$.*
- *If $N^{\mathsf{sf}}_X \xrightarrow[b_1]{a_1} N^{\mathsf{sf}}_Y$ and $N^{\mathsf{sl}}_\varnothing \xrightarrow[b_2]{a_2} N^{\mathsf{sl}}_\varnothing$, then $N_X \xrightarrow[b_1+b_2]{a_1+a_2} N_Y$.*

Proof. The proof follows by definition of the subnets and the operational semantics of P/T nets, as transitions of N are just partitioned into N^{sl} and N^{sf}.

In the following we let $\mathsf{I}_n \overset{\text{def}}{=} \bigotimes_n \mathsf{I} : (n, n)$ and define the following terms of the P/T calculus, $\forall n \in \mathbb{N}$:

$$
\begin{aligned}
&\mathsf{X}_0 \overset{\text{def}}{=} \mathsf{I} : (1,1) \qquad && \Lambda_0 = \mathsf{V}_0 \overset{\text{def}}{=} \uparrow; \downarrow : (0,0) \\
&\mathsf{X}_1 \overset{\text{def}}{=} \mathsf{X} : (2,2) \qquad && \mathsf{X}_{n+1} \overset{\text{def}}{=} (\mathsf{X}_n \otimes \mathsf{I}) ; (\mathsf{I}_n \otimes \mathsf{X}) : (n+2, n+2) \\
&\Lambda_1 \overset{\text{def}}{=} \Lambda : (1,2) \qquad && \Lambda_{n+1} \overset{\text{def}}{=} (\Lambda \otimes \Lambda_n) ; (\mathsf{I} \otimes \mathsf{X}_n \otimes \mathsf{I}_n) : (n+1, 2n+2) \\
&\mathsf{V}_1 \overset{\text{def}}{=} \mathsf{V} : (2,1) \qquad && \mathsf{V}_{n+1} \overset{\text{def}}{=} (\mathsf{V} \otimes \mathsf{V}_n) ; (\mathsf{I} \otimes \mathsf{X}_n \otimes \mathsf{I}_n) : (2n+2, n+1)
\end{aligned}
$$

It can be proved by induction that the only transitions for Λ_n and V_n are $\Lambda_n \xrightarrow[b]{a} \Lambda_n$ and $\mathsf{V}_n \xrightarrow[a]{b} \mathsf{V}_n$ with $|a| = n$, $|b| = 2n$, and $a_i = b_i + b_{n+i}$ for all $i < n$.

Definition 14. *Let P be a P/T calculus term s.t. $\{\!|P|\!\} : m \to n$ and $\mathcal{MG}(\{\!|P|\!\})$ is finite state. The normal form of P, written $nf(P)$, is as follows*

$$nf(P) = \Lambda_m; (\boldsymbol{T}_{can(\{\!|P|\!\}^{\mathsf{sf}})} \otimes \boldsymbol{T}_{\widetilde{\{\!|P|\!\}^{\mathsf{sl}}}}); \mathsf{V}_n$$

Lemma 7. *Let P be a P/T calculus term s.t. $\mathcal{MG}(\{\!|P|\!\})$ is finite state. Then, P and $nf(P)$ are bisimilar.*

Proof. It follows from the behaviour of Λ_n and V_n, Lemmata 1, 4 and 6 and the correspondence Theorems 1 and 2.

Lemma 8. *Let P and Q be two bisimilar P/T calculus terms s.t. $\mathcal{MG}(\{[P]\})$ and $\mathcal{MG}(\{[Q]\})$ are finite state. Then, $nf(P) = nf(Q)$ (up-to iso).*

Proof. It follows by contradiction. Assume that $min(\{[P]\}^{\mathsf{sf}}) = min(\{[Q]\}^{\mathsf{sf}})$ and $\widetilde{\{[P]\}}^{\mathsf{sl}} = \widetilde{\{[Q]\}}^{\mathsf{sl}}$ does not hold. Therefore, it should be the case that either i) $min(\{[P]\}^{\mathsf{sf}})$ and $min(\{[Q]\}^{\mathsf{sf}})$ are not bisimilar; or ii) $\widetilde{\{[P]\}}^{\mathsf{sl}} = \widetilde{\{[Q]\}}^{\mathsf{sl}}$ are not bisimilar. In both cases we conclude that $min(\{[P]\})$ and $min(\{[Q]\})$ (and therefore P and Q) are not bisimilar. For (i), we note that the marking graphs differ in a transition connecting two different states (and this cannot be mimicked by stateless transitions); for (ii) every state will miss at least a self-loop transition (since $\mathcal{MG}(\{[P]\})$ and $\mathcal{MG}(\{[Q]\})$ are finite state, all infinite self-loops in the marking graphs are originated by stateless transitions).

Corollary 3 (Idempotency). $nf(P) = nf(nf(P))$ *(up-to iso).*

5   Normal Forms for the C/E Petri Calculus

The case of C/E Petri calculus is quite interesting, because now any term P models a finite state connector, so that we can reduce to normal form any term.

Lemma 9. *Let P be a Petri calculus term. Then $\mathcal{MG}(\{[P]\})$ is finite state.*

Proof. The C/E net with boundaries $\mathcal{MG}(P)$ has as many places as the number of subterms $\bigcirc$ and $\odot$ in P and the reachable states of $\mathcal{MG}(P)$ are just subsets of the places in $\mathcal{MG}(P)$, thus they are finitely many.

Now by using the approach for P/T nets we can obtain the normal form for every Petri calculus term. The only subtlety to deal with is when mapping a marking graph into a C/E net, because marking graphs can contain self-loops, as illustrated by the following example.

Example 8. Consider the C/E net C_p in Fig. 11(a). The corresponding marking graph is in Fig. 11(b) and the corresponding minimal automaton is in Fig. 11(c). If we apply $\mathcal{NB}$ we obtain the net in Fig. 11(d). Note that transition α cannot be fired under the C/E semantics because it inhibits consume/produce loops. Hence, the obtained net is not bisimilar to C_p. In order to translate back the minimal marking graph to a C/E net, we need to handle self-loops differently. While $\mathcal{NB}$ already removes any trivial self-loop (i.e., with empty observation) from the minimal marking graph, non-trivial self-loops are handled by duplicating states, as illustrated in Fig. 11(e). Then, the normal form is obtained by using the C/E corresponding to the minimal marking graph without non-trivial self-loops.

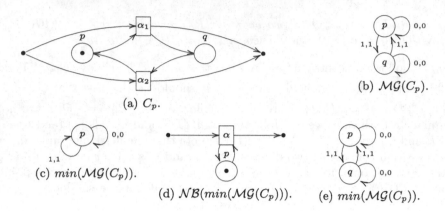

Fig. 11. Minimisation of C/E nets

6   Concluding Remarks

In this paper we have considered a calculus of connectors that allows for the most general combination of synchronisation, non-determinism and buffering. The touchstone of its generality is its ability of modeling a variety of Petri nets compositionally, up to bisimilarity. Often bisimilarity implies the existence of a minimal representative, but such a construction has not been exhibited yet for Petri nets, at least directly. Thus in the paper we interpret the case graph of a net as a transition system labelled with the synchronizations observable on its boundaries. Then we can minimize such a LTS and reinterpret it univocally as a net and as a term of the calculus. Thus minimality is restricted to a case graph (step) semantics, which we might say observes parallelism but not concurrency.

Related Work. An algebra consisting of five kinds of basic stateless connectors (plus their duals) is presented in [8], together with the operational, observational and denotational semantics and a complete normal-form axiomatisation. The behaviour of connectors ∧ and ∨ is slightly different from the one considered here, because in [8] only one action can take place at the time, e.g., only transitions $\wedge \xrightarrow{\frac{1}{10}} \wedge$ and $\wedge \xrightarrow{\frac{1}{01}} \wedge$ are considered instead of $\wedge \xrightarrow{\frac{n+m}{n\,m}} \wedge$.

The Tile Model [15] offers a semantic framework for concurrent systems, of which the algebra of stateless connectors is just a particular instance. Roughly, the semantics of component-based systems can be expressed via tiles when configurations and observations form two monoidal categories with the same objects. Tiles define LTSs whose labels are pairs ⟨trigger, effect⟩. In this context, the usual notion of equivalence is called *tile bisimilarity*, which is a congruence (w.r.t. sequential and parallel composition) when a suitable rule format is met [15].

Reo [1] is an exogenous coordination model based on channel-like connectors that mediate the flow of data among components. Notably, a small set of point-to-point primitive connectors is sufficient to express a large variety of interesting interaction patterns, including several forms of mutual exclusion, synchronisation, alternation, and context-dependency. Components and primitive connectors can be composed into larger Reo circuits by disjoint union up-to the merging of shared nodes. The semantics of Reo has been formalised in many ways, tile model included [2]. See [17] for a recent survey.

BIP [4] is a component framework for constructing systems by superposing three layers of modelling: (1) Behaviour, representing the sequential computation of individual components; (2) Interaction, defining the handshaking mechanisms between these components; and (3) Priority, assigning a partial order of privileges to interactions. In absence of priorities, the interaction layer admits the algebraic presentation given in [5] and has been related to connectors in [10].

The wire calculus [27] takes inspiration from [19,20] but shares similarities with the tile model. It is presented as a process algebra where each process comes with a sort, written $P : (n, m)$ for a process P with n ports on the left and m on the right. The usual action prefixes $a.P$ of process algebras are extended by allowing the simultaneous input of a trigger a and output of an effect b, written $\frac{a}{b}.P$, where a (resp. b) is a string of actions, one for each port of the process. The Petri calculus [9,28] can be regarded as a dialect of the wire calculus.

Nets with boundaries [28] take inspiration from the open nets of [3], whose interfaces consist of places instead of ports.

Future Work. Some recent work [6,7] exploits an algebra of connectors similar to ours to define a relational denotational semantics and a structural operational semantics for *signal flow graphs*, a classical structure in control theory and signal processing. We plan to investigate connections between Petri nets with boundaries and signal flow graphs. We might also consider extending the results of this paper to other more expressive semantics, observing e.g. causality. Another direction in which our results could be extended is dealing with systems with a higher degree of dynamism, that adapt their behavior to evolving environments: e.g., systems whose structure and interaction capabilities can change at runtime. Some recent progresses in this direction are discussed in [11].

Acknowledgements. We thank the anonymous reviewers for their careful reading and helpful suggestions for improving the presentation. We would like to express infinite gratitude to José, for his guidance, support and friendship during our long-standing collaboration.

References

1. Arbab, F.: Reo: a channel-based coordination model for component composition. Math. Struct. Comp. Sci. **14**(3), 329–366 (2004)
2. Arbab, F., Bruni, R., Clarke, D., Lanese, I., Montanari, U.: Tiles for Reo. In: Corradini, A., Montanari, U. (eds.) WADT 2008. LNCS, vol. 5486, pp. 37–55. Springer, Heidelberg (2009)
3. Baldan, P., Corradini, A., Ehrig, H., Heckel, R.: Compositional semantics for open Petri nets based on deterministic processes. Math. Struct. Comp. Sci. **15**(1), 1–35 (2005)

4. Basu, A., Bozga, M., Sifakis, J.: Modeling heterogeneous real-time components in BIP. In: SEFM 2006, pp. 3–12. IEEE Computer Society (2006)
5. Bliudze, S., Sifakis, J.: The algebra of connectors - structuring interaction in BIP. IEEE Trans. Comput. **57**(10), 1315–1330 (2008)
6. Bonchi, F., Sobociński, P., Zanasi, F.: A categorical semantics of signal flow graphs. In: Baldan, P., Gorla, D. (eds.) CONCUR 2014. LNCS, vol. 8704, pp. 435–450. Springer, Heidelberg (2014)
7. Bonchi, F., Sobocinski, P., Zanasi, F.: Full abstraction for signal flow graphs. In: POPL 2015, pp. 515–526. ACM (2015)
8. Bruni, R., Lanese, I., Montanari, U.: A basic algebra of stateless connectors. Theor. Comput. Sci. **366**(1–2), 98–120 (2006)
9. Bruni, R., Melgratti, H., Montanari, U.: A connector algebra for P/T nets inter-actions. In: Katoen, J.-P., König, B. (eds.) CONCUR 2011. LNCS, vol. 6901, pp. 312–326. Springer, Heidelberg (2011)
10. Bruni, R., Melgratti, H., Montanari, U.: Connector algebras, Petri nets, and BIP. In: Clarke, E., Virbitskaite, I., Voronkov, A. (eds.) PSI 2011. LNCS, vol. 7162, pp. 19–38. Springer, Heidelberg (2012)
11. Bruni, R., Melgratti, H., Montanari, U.: Behaviour, interaction and dynamics. In: Iida, S., Meseguer, J., Ogata, K. (eds.) Specification, Algebra, and Software. LNCS, vol. 8373, pp. 382–401. Springer, Heidelberg (2014)
12. Bruni, R., Melgratti, H.C., Montanari, U., Sobocinski, P.: Connector algebras for C/E and P/T nets' interactions. Log. Methods Comput. Sci. **9**(3), 1–65 (2013)
13. Bruni, R., Meseguer, J., Montanari, U., Sassone, V.: Functorial models for Petri nets. Inf. Comput. **170**(2), 207–236 (2001)
14. Degano, P., Meseguer, J., Montanari, U.: Axiomatizing the algebra of net compu-tations and processes. Acta Inf. **33**(7), 641–667 (1996)
15. Gadducci, F., Montanari, U.: The tile model. In: Proof, Language, and Interaction, pp. 133–166. The MIT Press (2000)
16. Hackney, P., Robertson, M.: On the category of props (2012). arXiv:1207.2773
17. Jongmans, S.S.T., Arbab, F.: Overview of thirty semantic formalisms for Reo. Sci. Ann. Comput. Sci. **22**(1), 201–251 (2012)
18. Kanellakis, P.C., Smolka, S.A.: CCS expressions, finite state processes, and three problems of equivalence. In: PODC 1983, pp. 228–240. ACM (1983)
19. Katis, P., Sabadini, N., Walters, R.F.C.: Representing place/transition nets in Span(Graph). In: Johnson, M. (ed.) AMAST 1997. LNCS, vol. 1349. Springer, Heidelberg (1997)
20. Katis, P., Sabadini, N., Walters, R.F.C.: Representing place/transition nets in Span(Graph). In: Johnson, M. (ed.) AMAST 1997. LNCS, vol. 1349, pp. 322–336. Springer, Heidelberg (1997)
21. MacLane, S.: Categorical algebra. Bull. AMS **71**(1), 40–106 (1965)
22. Meseguer, J., Montanari, U.: Petri nets are monoids. Inf. Comp. **88**(2), 105–155 (1990)
23. Meseguer, J., Montanari, U., Sassone, V.: On the semantics of place/transition Petri nets. Math. Struct. Comp. Sci. **7**(4), 359–397 (1997)
24. Paige, R., Tarjan, R.E.: Three partition refinement algorithms. SIAM J. Comput. **16**(6), 973–989 (1987)
25. Perry, D.E., Wolf, E.L.: Foundations for the study of software architecture. ACM SIGSOFT Soft. Eng. Notes **17**, 40–52 (1992)
26. Petri, C.: Kommunikation mit Automaten. Ph.D. thesis, Institut für Instrumentelle Mathematik, Bonn (1962)

27. Sobocinski, P.: A non-interleaving process calculus for multi-party synchronisation. In: ICE 2009, EPTCS, vol. 12, pp. 87–98 (2009)
28. Sobociński, P.: Representations of Petri net interactions. In: Gastin, P., Laroussinie, F. (eds.) CONCUR 2010. LNCS, vol. 6269, pp. 554–568. Springer, Heidelberg (2010)

Enlightening Ph.D. Students
with the Elegance of Logic

My Personal Memory About Prof. José Meseguer

Shuo Chen[✉]

One Microsoft Way, Redmond, WA 98052, USA
shuochen@microsoft.com

Abstract. This article provides my recollection about how Prof. José Meseguer enlightened me to study security problems from the logic perspective. His lectures and advices are having a long term influence on my research career.

1 Preface

For many of us in the academia, the years in the Ph.D. program are the period when our minds started to open up to the wonder of science and waited to be inspired by some great pioneers in the field. It is one of the most memorable periods in one's life.

The Ph.D. program of computer science in the University of Illinois at Urbana-Champaign (UIUC) consists of three milestones – the qualifying exam, the thesis proposal and the final defense. The first stage, the qualifying exam, has a nice property of "fail-fast". Students are required to pass the exam in the second or the third semester, or they will be asked to quit. The final stage is also short (typically about one year) and often predictable (since the advisor understands the time frame of the student's new job). What I would describe as a long and dark tunnel is the second stage – the baffling period before the thesis proposal. Unlike the first stage, which is mainly to know the field, the second stage is all about knowing yourself – finding out something deep about your true passion. Very fortunately, I got to know Prof. José Meseguer, an enlightening professor who brought the elegance of logic into my mind.

2 My Research Direction Before Knowing José

My advisor is Prof. Ravi Iyer. He is a renowned expert in the field of fault tolerance and dependable computing. Ravi always encouraged students to study real-world systems to get first-hand experience. He also put great emphasis on empirical measurements of operational data on these systems. A really unique advantage of our group is that Ravi has a strong background in statistics and probability, because his own Ph.D. degree was in statistics. Our group didn't have much difficulty publishing papers in the most prestigious conference in the field, IEEE DSN.

Ravi envisioned that the fault tolerance expertise in our group could be extended to cover security topics. Under his guidance, a more senior Ph.D. student and I decided to

© Springer International Publishing Switzerland 2015
N. Martı́-Oliet et al. (Eds.): Meseguer Festschrift, LNCS 9200, pp. 228–231, 2015.
DOI: 10.1007/978-3-319-23165-5_10

devote our effort to security. Looking back, I see this as a very important growth period in my career. It helped me step into the security area and gave me the confidence to work on real-world systems. I enjoyed so much the real-world systems work, such as modifying network server programs, or even hacking into the OS kernel, to change program behaviors and observe many interesting security consequences caused by these changes. We built automation technologies to conduct large-scale experiments, and systematized the findings by classification and statistics. These studies continued to be published in the fault tolerance community.

Despite the progress, it was difficult to go deeper in my research because there is a fundamental difference between fault tolerance and security: in fault tolerance research, it is often valid to consider faults as a natural phenomenon, and model their arrivals as a stochastic process. Theories of probability and statistics can be nicely applied under this problem setting. Security research, however, usually needs to consider the threat as the action of a deliberate and malicious adversary. It is hard to apply probabilistic approaches to measure the "likelihood of security". In fact, it is even questionable whether such a likelihood can be objectively defined while still being useful. How to formulate a scientific problem then?

This situation baffled me for a few years. On one hand, I was passionate about studying real-world systems to build up my knowledge and collect many empirical insights. On the other hand, these didn't turn into deeper scientific research ideas.

3   José's Course on Formal Methods

Although I had fulfilled all the course requirements of the Ph.D. program, Ravi strongly suggested that I should take a few more courses to broaden my scope. He believed that this might help me walk out of the fog. I don't remember why I decided to take José's course on formal methods. It was surely a wise decision, because it opened up a whole new horizon for me.

The lectures in the first few weeks were about natural numbers, arithmetic operations and basic algebras. It amazed me that these elementary-school concepts are so interesting when they are viewed from the perspective of logic. Even more impressively, they were all concretely expressed in rewriting logic, of which the primitive is nothing but matching-and-replacing substrings. The elegance of logic resonated with something hidden in my heart: I like validating claims that can be resolved in a binary manner, rather than discussing the less exact arguments that are common in other computer science courses.

The course went on. José started to teach how to model an algorithm using rewriting logic and prove its correctness using the Maude theorem prover. At this point, I realized that formal methods are directly relevant to security research. Many important security problems can be defined as program correctness problems, as long as researchers concretely understand the program semantics and the security goals to achieve. In other words, if I was able to define semantics for some insights that I obtained over the years, I would able to bring scientific rigor to my research.

4   Face-to-Face Discussions with José

The most beneficial part of taking José's course was not the lectures, but the invaluable opportunity to discuss ideas with him face-to-face.

It wasn't easy for me to set up the first meeting with José. I was concerned that my ideas were too rudimentary in terms of formal reasoning, and my logic background was quit lacking, so how could José be interested? Eventually, I settled down to one idea and wrote a few pages about it. One day, outside the lecture room, I tried to briefly explain the idea to him. To my surprise, it ended up being a long discussion! José easily understood what I was trying to do, and offered many valuable comments. More importantly, he encouraged me to move forward. This discussion meant a lot for me, because I really needed the encouragement from an expert in the field.

José and I continued the discussion throughout the rest of the semester. We not only discussed specific ideas, but also our thoughts about research in general, such as how theoretical research and empirical research are related, how to show success of a research idea, etc. Through these discussions, it became clear that formal analysis should be a component in my dissertation.

Under the guidance of Ravi and José, I spent one more semester successfully developing the framework for my thesis proposal. I am very grateful for José's enlightenment and encouragement that helped me find my passion and go through the baffling period.

5   Our Collaborations After My Graduation

I joined Microsoft Research Redmond as a security researcher after graduation, and continued to discuss extensively with José. Our project was to use rewriting logic to model the logic of Internet Explorer's graphic user interface, in order to find logic flaws that allow a malicious webpage to spoof the contents in the address bar and the status bar. José brought one of his best students, Ralf Sasse, into this project. The three of us collaborated tightly, and spent a tremendous amount of effort understanding and modeling the source code of Internet Explorer. We flew between Urbana and Redmond several times to expedite the progress. I still remember the night when José waited in the Urbana-Champaign Willard Airport for my flight, which was delayed due to a snowstorm. José brought me to his home and prepared a meal for me. I viewed this as a very special honor that only a close collaborator could have. Yes, José Meseguer, the inventor of rewriting logic, cooked for me!

Our effort was paid off, as we discovered 13 security bugs before Internet Explorer 7 was shipped. Because of the severity, Microsoft asked us to withhold the paper submission. Withholding good new results from publication is a difficult situation for scientists, but José was very supportive, because he understood the societal aspect of security research. Eventually, the project had a happy ending: 11 out of the 13 vulnerabilities were fixed when Internet Explorer 7 was shipped, and our paper was accepted to the top security conference, IEEE Symposium on Security and Privacy.

José and I continued to discuss research ideas because of our shared interest. He invited me to serve on the thesis committee of Ralf Sasse. I know that Ralf and José developed a number of innovative technologies to prove security for a real browser and many cryptographic protocols.

6   Long-Term Influence

I have been working on many security research projects across different areas, including memory safety, browser security, web/mobile application security, and security protocols. Interestingly, program semantics always come into the picture from one angle or another. I continue to be curious about what a program tries to do and what it actually does. Many of my papers demonstrate that logic flaws are causing realistic security and privacy breaches in today's cloud and mobile systems, so formal methods are a valuable solution.

It is clear that the inspiration I got from José many years ago is having a long term influence on my research focus and methodology. I feel very honored to know José as a teacher, a collaborator and a friend. Of course, in his career, José must have inspired many other scholars in different ways. That's why he receives so much respect from the research community. I believe that the respect comes not only from his intellectual contributions, but are a result of his nature of being open, caring, friendly and enlightening. It is really joyful that we are celebrating his achievements. Happy 65th Birthday, José!

Acknowledgement.   The author thanks Cormac Herley for proofreading this article.

Two Decades of Maude

Manuel Clavel[1], Francisco Durán[2](✉), Steven Eker[3], Santiago Escobar[4],
Patrick Lincoln[3], Narciso Martí-Oliet[5], and Carolyn Talcott[3]

[1] IMDEA Software, Madrid, Spain
[2] Universidad de Málaga, Málaga, Spain
duran@lcc.uma.es
[3] CSL, SRI International, Menlo Park, CA, USA
[4] DSIC-ELP, Universitat Politècnica de València, Valencia, Spain
[5] Facultad de Informática, Universidad Complutense de Madrid, Madrid, Spain

Dedicado a José Meseguer, con ocasión de su 65 cumpleaños, con cariño, amistad y agradecimiento por todo el trabajo realizado conjuntamente en estas dos décadas.

Abstract. This paper is a tribute to José Meseguer, from the rest of us in the Maude team, reviewing the past, the present, and the future of the language and system with which we have been working for around two decades under his leadership. After reviewing the origins and the language's main features, we present the latest additions to the language and some features currently under development. This paper is not an introduction to Maude, and some familiarity with it and with rewriting logic are indeed assumed.

1   The Origins

The story of Maude does not begin on a dark and stormy night as many stories do, but on a sunny Californian day. Maude was conceived at the Logic and Specification Group, part of the Computer Science Laboratory at SRI International, in Menlo Park, California, in the Spring of 1990. José Meseguer was leading that group, after working for several years with J. A. Goguen and other colleagues on order-sorted equational logic [47] and its implementation in the OBJ3 language [48], among many other topics. At that time he was proposing a new computational logic which could provide on the one hand a unified model of concurrency [63,64], and on the other hand declarative support for (concurrent) object-oriented programming [62,66]. This new logic was thought of as an extension of (order-sorted) equational logic with rules (understood either as logical inference rules or as transitions in a concurrent system) which, as the equations, would also be executed by rewriting, and for this reason was called *rewriting logic*. The good properties of the logic for unifying several models of computation, including concurrent ones, were soon generalized to representing other models of computation and also other logics, so that rewriting logic was proposed as a logical and semantic framework [58,59].

In the same way that order-sorted equational logic was implemented as a specification and programming language in OBJ3, behind rewriting logic there

© Springer International Publishing Switzerland 2015
N. Martí-Oliet et al. (Eds.): Meseguer Festschrift, LNCS 9200, pp. 232–254, 2015.
DOI: 10.1007/978-3-319-23165-5_11

was a language waiting to be implemented. But the language, soon to be called *Maude*, was there from the beginning. In addition to the papers [62,66] devoted to the rewriting logic support of object-oriented programming, in those years several other papers were produced from the programming point of view [65,75, 83], emphasizing the multiparadigm and parallel programming characteristics of the proposal. Moreover, even the more theoretical papers, such as the ones about the logical framework [58,59], used the Maude notation for presenting examples in such a way that, with slight changes, most of them could be executed in the current implementation of the language. Those examples already used primitive versions of constructs for parameterized specification and some of them were associated with a name, *Maudelog*, for a version of Maude with logical variables and unification, a feature which has taken much longer to be fully implemented (see Sect. 3). The way "from OBJ to Maude and beyond" has been well explained by José Meseguer himself in [70].

Although the implementation of the Maude language and system were not yet underway, techniques for compilation of rewriting onto parallel architectures were studied as part of the Rewrite Rule Machine (RRM) project [54,55] in the early nineties. A sublanguage of Maude, called Simple Maude, which included term rewriting, graph rewriting, and object-oriented rewriting, was proposed as part of this project.

The prospects for an implementation of Maude greatly improved in the mid nineties, with the arrival at SRI International of Steven Eker, as a postdoc expert on term rewriting implementation, and Manuel Clavel, as a PhD student to work on rewriting logic reflection [12]. This is indeed the reason for the title of this paper: it is around this date when the Maude team was born. The work developed in that period was shown in the first public presentation of the Maude system [25], which took place at the first Workshop on Rewriting Logic and its Applications (WRLA) in Asilomar, California, in 1996 [67]. Another presentation at the same event showed the first realization of the reflection ideas in rewriting logic and Maude [26].

Coincidentally, Francisco Durán also joined the group during that event, as a PhD student, to work on the Maude module algebra [29], which led to the development of *Full Maude*. Although the advances of all the work being done by the Maude team in all these areas were shown at the second WRLA in Pont-à-Mousson, France, in 1998 [53] (implementation [17], reflection [13,15], module algebra [34]), the first public release of Maude had to wait yet another year until the end of 1999 [16]. Maude 1 was presented in RTA 1999 [18], in FASE 2000 [21], in an ETAPS 2000 tutorial [19], and in a journal paper published in 2002 [22].

However, that first public release of Maude was a proof-of-concept. Although it already had many interesting features, there were so many other missing features that it was not the end, but the beginning of much more work, as discussed by the "Towards Maude 2.0" paper [20] presented at the third WRLA in Kanazawa, Japan, in 2000 [45]. It required a lot of effort to complete the implementation and also to write a good manual for Maude 2.0, publicly released in the Summer of 2003, with a presentation in RTA 2003 [23]. Among other new features, this new version provided support for membership equational logic, support for rewrite expressions in rule conditions, new predefined modules, a new version of its metalevel, and an LTL model checker.

The Maude 2 features kept increasing and improving along the following years, when we managed to have yearly releases, helped by intense meetings of the Maude team, at that time distributed in different locations both in the US and in Europe, after each edition of the WRLA, in Pisa (2002), Barcelona (2004), and Vienna (2006). We reached an important milestone with the publication in 2007 of the book "All About Maude" [24]. The book coincided with the release of Maude 2.3, where the main features of the language and its implementation stabilized, including parameterized modules, interaction with external objects, and a greater catalogue of predefined—some of them parameterized—modules and views, among others.

Since then, the Maude team has produced several additional releases until the recent Maude 2.7. Most of the new features in Maude 2.4, presented at RTA 2009 [14], and subsequent versions after it, have been related to order-sorted unification and narrowing, which is the subject of Sect. 3. In this brief summary of the work related to the origins of Maude and rewriting logic we cannot do justice to all the work done by many people around the world in this area; instead, we direct the reader to the survey written by José Meseguer himself on twenty years of rewriting logic, published in 2012 [72], together with an annotated bibliography [60] compiling all the papers on rewriting logic and its applications written in the period 1990–2012.

2 The Language

The close contact with many specification and programming applications has served as a good stimulus for a substantial increase in expressive power of the rewriting logic formalism in general, and of its Maude realization in particular. Maude is a high-performance language and system supporting both equational and rewriting logic computation for a wide range of applications, including development of theorem-proving tools, language prototyping, executable specification and analysis of concurrent and distributed systems, and logical framework applications in which other logics are represented, translated, and executed.

2.1 Generalized Rewrite Theories in Maude

Maude's functional modules are theories in *membership equational logic* [9,69], a Horn logic whose atomic sentences are either equalities $t = t'$ or membership assertions of the form $t : s$, stating that a term t has a certain sort s. Such a logic extends OBJ3's [48] order-sorted equational logic and supports sorts, subsorts, subsort polymorphic overloading of operators, and definition of partial functions with equationally defined domains.

A Maude (system) module is a generalized rewrite theory, defined as a 4-tuple $\mathcal{R} = (\Sigma, E \cup Ax, \phi, R)$, where $(\Sigma, E \cup Ax)$ is a membership equational theory, Ax is a set of equational axioms for which rewriting modulo is available,

R is a set of labeled conditional rewrite rules, and ϕ is a function assigning to each operator $f : k_1 \ldots k_n \to k$ in Σ the subset $\phi(f) \subseteq \{1, \ldots, n\}$ of its frozen arguments. Rewriting in $(\Sigma, E \cup Ax, \phi, R)$ happens modulo the equational axioms Ax. Maude supports rewriting modulo different combinations of associativity (A), commutativity (C), identity (U), left identity (Ul), right identity (Ur), and idempotence axioms. Computationally, rules are interpreted as local transition rules in a possibly concurrent system. Logically, they are interpreted as inference rules in a logical system. This makes rewriting logic both a general semantic framework to specify concurrent systems and languages [68], and a general logical framework to represent and execute different logics [59]. The combination of evaluation strategies and frozen arguments allows Maude to perform context-sensitive rewriting [57] with both equations E and rules R modulo Ax.

Maude accepts module hierarchies of functional and system modules with user-definable mixfix syntax. The Maude system is implemented in C++ and is highly modular. Maude's core is its rewrite engine, which is extensible, and indeed has been extended, in many different ways since its inception. For instance, new equational theories can be "plugged in" and new built-in symbols with special rewriting (equation or rule) semantics may be easily added. To date, rewriting modulo all combinations of associativity, commutativity, left and right identity, and idempotence have been implemented apart from those that contain both associativity and idempotence.

Over the years, the development of Maude has been guided by the goal of providing a better support for both rewriting logic and its underlying membership equational logic. For instance, the duality between its logical and operational views was completed with the addition of the **nonexec** attribute in Maude 2.0. The point is that efficient and complete computation by rewriting is only possible for equational theories that satisfy properties such as confluence, sort-decreasingness, and termination. Similarly, to be efficiently executable, a generalized rewrite theory $\mathcal{R} = (\Sigma, E \cup Ax, \phi, R)$ should first of all have $(\Sigma, E \cup Ax)$ satisfying the above executability requirements, and should furthermore be coherent [36].

Executability is of course what we want for programming; but it is too restrictive for specification, transformation, and reasoning purposes. For this reason, there is a linguistic distinction between modules, that are typically used for programming as executable theories, and theories, which need not be executable and are used for specification purposes, for example, to specify the semantic requirements of actual parameters of parameterized modules, or for theorem-proving purposes. Maude supports specification of arbitrary membership equational logic theories and of arbitrary rewrite theories, while at the same time keeping a sharp distinction between executable and non-executable statements (i.e., equations, memberships, or rules) by means of the **nonexec** attribute. Fully executable equational and rewrite theories are called admissible, and satisfy the above-mentioned executability requirements. This support for a disciplined coexistence of executable and non-executable statements allows not only a seamless

integration of specification and code, but also a seamless integration of Maude
with its formal tools.

Maude includes some built-in functional modules providing convenient high-
performance functionality within the Maude system. In particular, the built-
in modules of integers, natural, rational, and floating-point numbers, quoted
identifiers, and strings provide a minimal set of efficient operations for Maude
programmers.

2.2 Reflection in Maude

Informally, a reflective logic is a logic in which important aspects of its metatheory can be represented at the object level in a consistent way, so that the object-
level representation correctly simulates the relevant meta-theoretic aspects.

Rewriting logic is reflective [12] in the precise sense of having a universal
theory U that can represent any finitely presented rewrite theory T (including
U itself) and any terms t, t' in T as terms $\overline{T}$ and $\overline{t}, \overline{t'}$ in U, so that we have the
following equivalence

$$T \vdash t \rightarrow t' \iff U \vdash \langle \overline{T}, \overline{t} \rangle \rightarrow \langle \overline{T}, \overline{t'} \rangle.$$

Since U is representable in itself, we can then achieve a "reflective tower" with
an arbitrary number of levels of reflection.

Maude efficiently supports this reflective tower through its META-LEVEL module, where Maude terms and modules are reified as elements of a data types Term
and Module, respectively. The processes of reducing a term to normal form in
a functional module and of rewriting a term in a system module using Maude's
default interpreter are respectively reified by *descent* functions metaReduce and
metaRewrite. Similarly, the process of applying a rule of a system module to a
subject term is reified by a function metaApply. Furthermore, parsing and pretty
printing of a term in a signature, as well as key sort operations are also reified by
corresponding metalevel functions, and up and down functions to move terms,
modules, and views between levels.

The reflective capabilities of Maude provide a great range of possibilities,
many of which have been exploited with different purposes. It has been used, for
example, to define alternative rewriting strategies, to define strategy languages,
to define module operations, and in general to extend Maude in different ways.
This extensibility by reflection is exploited in Maude's design and implementation. Full Maude is an extension of Maude written in Maude itself which has
been used since the beginnings of Maude as a place in which to design and
experiment with new features. For example, a module algebra of parameterized
modules, views, and module expressions in the OBJ style was available in Maude
through Full Maude [30,34,35] long before it was implemented in C++ for (Core)
Maude 2.4. Object-oriented modules, with convenient syntax for object-oriented
applications, or parameterized views are currently available in Full Maude but
not yet in Core Maude. In summary, we have been 'using our own medicine',
using Maude to specify our system before facing the effort of implementing it.

Indeed, Full Maude has been more than a place in which to experiment with new features, it has provided basic infrastructure on which to define extensions of Maude such as Real-Time Maude [76].

The reflective capabilities of Maude have also been key for the development of executable implementations of very different formal models and programming languages. See, e.g., the coordination models for distributed objects in [82] and [73], the definition of Mobile Maude [32] and its socket-based distributed implementation [37], or the Maude Action Tool [28], which provides an executable environment for action semantics.

2.3 Maude's Formal Tools

In addition to its core functionality for rewriting, Maude comes with a number of tools. Some of these tools are directly integrated in the core system, which provides specific commands for them, as for the `search` command, for searching for terms satisfying given pattern and condition reachable from a given initial term, or through operators in predefined modules, as in the case of its LTL model checker. Other tools are provided as extensions by different authors, some using the infrastructure provided by Full Maude, such as the Church-Rosser checker, the coherence checker [36], the termination tool [33], the explicit-state model checker for linear temporal logic of rewriting (LTLR) [4,71], or the LTL logical model checker [2]; and some independently, such as the Maude inductive theorem prover [12], or the sufficient completeness checker [50]. An attempt to bring these tools under a common environment so that they can be used together to keep track of pending proof obligations and help in their interaction to discharge these proof obligations is currently under development in what is being called the Maude Formal Environment [38].

3   The Present: Unification and Narrowing

As mentioned before, Maude inherited many features from its predecessors, such as order-sorted equational logic and the use of commonly occurring attributes like associativity and commutativity, but other features of its predecessors were left behind, e.g., Eqlog [46] envisioned an integration of order-sorted equational logic with Horn logic, providing logical variables, constraint solving, and automated reasoning capabilities on top of order-sorted equational logic; and MaudeLog [65] envisioned an integration of order-sorted rewriting logic with queries including logical variables. The paper [40] revisited this topic and showed how many modern programming features can be implemented using Maude.

Unification is a fundamental deductive mechanism used in many automated deduction tasks and it is essential for programming languages with logical variables. Many functional and logic programming languages use an evaluation mechanism called *narrowing* [1], which is a generalization of term rewriting allowing free variables in terms (as in logic programming) and replacing pattern matching by unification in order to (non-deterministically) reduce these terms.

Unification and narrowing were introduced in Maude in 2009 as part of the Maude 2.4 release [14]. In that version of Maude, unification worked for any combination of symbols being either free or associative-commutative (AC), and it was developed by Eker as a built-in feature in Core Maude. Narrowing worked for system modules without equations and relied on the built-in unification algorithm. It supported the concept of symbolic reachability analysis of terms with logical variables, computing suitable substitutions for the variables in both the origin and the destination terms. Narrowing was first implemented in Full Maude, which allowed us to carry on research on its reasoning capabilities. The latest developments in Maude 2.6 were presented at RTA 2011 [31]. First, Eker improved the built-in unification algorithms to allow any combination of symbols being either free, commutative (C), associative-commutative (AC), or associative-commutative with an identity symbol (ACU). The performance was dramatically improved, allowing further development of other techniques in Maude. Second, the concepts of *variant* [27] and *variant-based unification* [43] led to a significant improvement in the reasoning capabilities. Given an equational theory $(\Sigma, E \cup Ax)$, the E, Ax-variants of a term t are the set of all pairs consisting of a substitution σ and the E, Ax-canonical form of $t\sigma$. Variant generation, variant-based unification, and symbolic reachability based on variant-based unification were all implemented in Full Maude.

In the most recent Maude 2.7 version, Eker has extended the available capabilities. First, the built-in unification algorithm allows any combination of symbols being free, C, AC, ACU, CU, U, Ul, Ur. Second, variant generation and variant-based unification are implemented in C++ at the Core Maude level with excellent performance. Note that the former version of variant generation and variant-based unification in Maude 2.6 was implemented for very simple equational theories called *strongly right irreducible*, but the new implementation in Maude 2.7 got rid of this restriction, allowing really complex equational theories and their combinations.

The classical application of narrowing modulo an equational theory is to perform $E \cup Ax\text{-}unification$ by narrowing with oriented equations E modulo axioms Ax. Indeed, the variant-based equational order-sorted unification algorithm implemented in Maude 2.7 is based on a narrowing strategy, called *folding variant narrowing* [43], that terminates when $E \cup Ax$ has the finite variant property [27], even though unrestricted narrowing typically does not terminate when Ax contains AC axioms [27,43].

An interesting example of the flexibility of folding variant narrowing, even beyond equational unification, is given for the classic *missionaries and cannibals* problem. In this problem, three missionaries and three cannibals must cross a river using a boat. The boat cannot cross the river with no people on board, and cannot carry more than two people. In any of the banks, the missionaries cannot be outnumbered by cannibals, otherwise the cannibals would eat the missionaries.

A solution for this problem was presented by Goguen and Meseguer in [49] as an equational logic program, requiring constraint-solving features, logical variables, order-sorted types, and axioms. Features of the original solution have been adapted to the current equational variant-based programming features available in Maude using the ideas of [40], and the resulting module is shown in Fig. 1.

The imported module TRIPLIST defines trip lists (TripList), with concatenation operator _*_ as constructor[1] and a length operation #_. Module PSET defines sorts Elem of people and PSet of multisets of people. Multisets are constructed with union operator _+_, which is declared as associative, commutative and with an identity symbol, and come with operators _-_ for removal and _/_ for intersection.

Our aim is to find a list of trips, where each trip is a term rooted by a predicate boat with a set of missionaries and cannibals. Odd positions in the list represent trips from the left bank to the right bank, and even positions trips from right to left. The MAC module defines constants for the missionaries (taylor, helen, and william) and the cannibals (umugu, nzwave, and amoc). lb(L) (resp., rb(L)) represents the people set in the left (resp., right) bank after the sequence of trips L. mset(PS) (resp., cset(PS)) gives the subset of missionaries (resp., cannibals) in PS. The function boatok checks whether a trip is ok — one or two people in the boat, where these are from the set of defined cannibals and missionaries — and solve is the general predicate for checking/generating the trip list solution — a trip list is a solution if it is a 'good' list, and the sequence of trips leaves the left bank empty. A trip list L * T is good if T is a valid trip (boatok), the sublist L is good, and the number of cannibals in each bank is smaller than the number of missionaries in the same bank for each trip in the sequence.

The key is therefore in the definition of the Success sort in the SUCCESS module. The success sort has a constant success, an operator _>>_ that defines the conditional evaluation of constraints, such that the left side is evaluated before the right side, and _=:=_, which represents unification between two terms. For each sort, _=:=_ is defined only in the positive cases, returning success; see below for its definition for sort Bool.

```
op _>>_ : [Success] [Success] -> [Success] [frozen(2)] .
rl success >> X:[Success] => X:[Success] .

op _=:=_ : Bool Bool -> [Success] [comm] .
rl X:Bool =:= X:Bool => success .
```

Folding variant narrowing is performed using those equations labeled with the variant flag, while the remaining equations are used as usual in Maude. Here we can ask for solutions to the very general problem of the names of missionaries and cannibals carried from one side to the other of the river.

---

[1] The original solution assumes that lists are created using an associative symbol, but unification modulo associativity is infinitary and it is not available in Maude. The _*_ operator is therefore not declared associative.

```
fmod MAC is
  pr SUCCESS + TRIPLIST + PSET .
  ops taylor helen william : -> Elem [ctor] .
  ops umugu nzwawe amoc : -> Elem [ctor] .
  var L : TripList .    var T : Trip .    var PS : PSet .

  op gen : Elem -> [Success] .
  eq gen(taylor) = success [variant] .    eq gen(taylor) = success .
  eq gen(helen) = success [variant] .     eq gen(helen) = success .
  eq gen(william) = success [variant] .   eq gen(william) = success .
  eq gen(umugu) = success [variant] .     eq gen(umugu) = success .
  eq gen(nzwawe) = success [variant] .    eq gen(nzwawe) = success .
  eq gen(amoc) = success [variant] .      eq gen(amoc) = success .

  op m0 : -> [PSet] .                     op c0 : -> [PSet] .
  eq m0 = taylor helen william .          eq c0 = umugu nzwawe amoc .
  op mset : PSet -> [PSet] .              op cset : PSet -> [PSet] .
  eq mset(PS) = PS /\ m0 .                eq cset(PS) = PS /\ c0 .

  op boatok : Trip -> [Success] .         op boat : PSet -> Trip [ctor] .
  eq boatok(boat(X:Elem)) = gen(X:Elem) .
  eq boatok(boat(X1:Elem X2:Elem))
   = gen(X1:Elem) >> gen(X2:Elem) >> ((X1:Elem =/= X2:Elem) =:= true) .

  ops lb rb : TripList -> [PSet] .
  eq lb(nil) = m0 c0 .
  eq lb(L * boat(PS))
    = if (even # L) then (lb(L) - PS) else (lb(L) /\PS) fi .
  eq rb(nil) = empty .
  eq rb(L * boat(PS))
    = if (even # L) then (rb(L) /\ PS) else (rb(L) - PS) fi .

  op good : TripList -> [Success] .
  eq good(nil) = success .
  eq good(L * T)
   = boatok(T)
     >> good(L)
     >> ( (# cset(lb(L * T)) =< # mset(lb(L * T))
            or (# mset(lb(L * T)) == 0))
          and
          (# cset(rb(L * T)) =< # mset(rb(L * T))
            or (# mset(rb(L * T)) == 0)) ) =:= true .

  op solve : TripList -> [Success] .
  eq solve(L) = good(L) >> (lb(L) == empty) =:= true .
endfm
```

Fig. 1. Missionaries and cannibals example

```
Maude> variant unify
        solve(nil * boat(E1:Elem E2:Elem) *
            boat(E1:Elem) * boat(E3:Elem E4:Elem) *
            boat(E2:Elem) * boat(E5:Elem E6:Elem) *
            boat(E6:Elem E3:Elem) * boat(E1:Elem E6:Elem) *
            boat(E4:Elem) * boat(E2:Elem E4:Elem) *
            boat(E6:Elem) * boat(E6:Elem E3:Elem))   =?    success   .

  Unifier #1
  E1:Elem --> helen     E2:Elem --> amoc      E3:Elem --> umugu
  E4:Elem --> nzwawe    E5:Elem --> william   E6:Elem --> taylor
  Unifier #2
  E1:Elem --> william   E2:Elem --> amoc      E3:Elem --> umugu
  E4:Elem --> nzwawe    E5:Elem --> helen     E6:Elem --> taylor
  ...
  Unifier #36
  E1:Elem --> helen     E2:Elem --> umugu     E3:Elem --> amoc
  E4:Elem --> nzwawe    E5:Elem --> taylor    E6:Elem --> william
```

Enumerating all thirty-six solutions takes only a few minutes thanks to the efficient implementation of folding variant narrowing in Core Maude. The most general question is `variant unify solve(L) =? success`, which enumerates all the necessary boat movements from one bank to the other and the missionaries and cannibals moved each time. However, we would have to add more variant equations, for recursive instantiation of a variable of sort `Triplist`, and for recursive instantiation of a variable of sort `PSet`, apart of the current variant equations for instantiation of variables of sort `Elem`. The current implementation in Maude is not able to handle the folding variant narrowing search space associated to a unification problem like that, though it will enumerate the solutions given enough resources.

The modern application of narrowing with rules R modulo $E \cup Ax$ is that of *symbolic reachability analysis* [74]. In this case, the rules R are understood as transition rules instead of equations. Narrowing is a complete deductive method [74] for symbolic reachability analysis, that is, for solving existential queries of the form $\exists \overline{x} \; t(\overline{x}) \rightarrow^* t'(\overline{x})$ in the sense that the formula holds for R iff there is a narrowing sequence $t \leadsto^*_{R, E \cup Ax} u$ such that u and t' have an $E \cup Ax$-unifier. Furthermore, in symbolic reachability analysis, we may be interested in verifying properties more general than existential properties of the form $\exists X \; t \rightarrow^* t'$, since one can generalize the above reachability property to properties of the form $\mathcal{R}, t \models \varphi$, for φ a temporal logic formula. The papers [2, 42] show how narrowing can be used (again, both at the level of transitions with rules R and at the level of equations E) to perform *logical model checking*. Two distinctive features are: (i) the term t does not describe a single initial state, but a possibly infinite set of instances of t (i.e., a possibly infinite set of initial states); and (ii) the set of reachable states does not have to be finite. Therefore, standard model-checking techniques may not be usable, because of a possible double infinity: in the number of initial states, and in the number of states reachable for each of those initial states.

So far, the most successful story about rewriting logic with narrowing is the Maude-NPA protocol analyzer [41], where cryptographic protocols are formally specified as order-sorted rewrite theories and the security analysis is performed in a backwards way, from an attack state to an initial state.

4   The Near Future: Rewriting Modulo SMT

The rapid progress of *satisfiability modulo theories* (SMT) solvers [7] has been one of the most important developments in automated verification and reasoning. A feature recently added to Maude (in an internal version, not publicly released at the time of the publication of this paper) is support for *rewriting modulo SMT* [78], so that functional and system modules can have conditions dealing with SMT data types, which are then solved by the usually more effective SMT solvers.

SMT solvers are decision procedures for an existential fragment of first-order logic with equality, where variables range over SMT data types, such as Booleans, integers, and reals. After presenting the way in which rewriting modulo SMT is being implemented in Maude in Sect. 4.1, we describe a sample application in Sect. 4.2.

4.1   Maude SMT

When performing rewriting modulo SMT, the object being rewritten is a symbolic representation of a (possibly infinite) family of terms. In its current Maude implementation, the representation of such family of terms is an ordered pair, where the first component is a term which may include variables ranging over data types supported by an SMT solver and SMT operators on those variables, and the second component is a constraint on those SMT variables. Rewriting proceeds as a search where each rewrite rule may have a condition, interpreted as an SMT constraint. In order to make a rewrite step, the accumulated constraints must be satisfiable, as checked by an SMT solver; when a conditional rule succeeds, the constraint it enforced on the SMT variables in the new term is 'conjuncted' with the existing constraint. Since Maude has no built-in knowledge of the SMT theories, no simplification of the accumulated constraint is performed.

Maude's interface to SMT data types closely follows the SMT-LIB standard [6]. In particular there are functional modules BOOLEAN, INTEGER, REAL, and REAL-INTEGER which provide signatures for the SMT-LIB theories of Booleans, integers, reals, and reals combined with integers, respectively. Although the current implementation has some restrictions, we expect to have a full implementation of rewriting modulo SMT in a near future release of Maude.

An SMT rewriting search is initiated with the smt-search command, which has a syntax similar to the syntax of the search command. The start term may only include SMT variables, which may also appear in the pattern term, and condition.

To give a flavor of how rewriting modulo SMT works, let us consider a small example. We define the *gcd* function using state transitions on a pair of SMT integers that encode Euclid's algorithm.

```
mod EUCLID is protecting INTEGER .
  sort State .
  op gcd : Integer Integer -> State .
  op return : Integer -> State .

  vars I J K X Y Z : Integer .

  crl gcd(X, Y) => gcd(X - Y, Y) if X > Y = true .
  crl gcd(X, Y) => gcd(X, Y - X) if X < Y = true .
  crl gcd(X, Y) => return(X) if X = Y .
endm
```

We then ask about the existence of a pair of integers X, Y such that $gcd(X, Y) = 3$ and $X + Y = 27$.

```
Maude> smt-search [1] gcd(X, Y) =>* return(Z)
            such that Z = 3 /\ X + Y = 27 .

Solution 1
empty substitution
where Z === 3 and X + Y === 27 and X > Y and X - Y > Y and
  X - Y - Y > Y and X - Y - Y - Y < Y and X - Y - Y - Y ===
  Y - (X - Y - Y - Y) and Z === X - Y - Y - Y
```

Maude searches the (typically infinite) tree of SMT rewrites on (term, constraint) pairs for a state that matches the pattern and satisfies the constraint on the variables given in the command. In the success case, it returns a substitution for non-SMT variables in the pattern and a satisfiable constraint on SMT variables. If the search graph is infinite as in this case, the command will not terminate in the failure case unless a depth bound is given.

Like most other Maude commands, smt-search is reflected at the metalevel by a corresponding descent function.

Currently, Maude uses CVC4 [5] as its backend SMT solver, however calls to the SMT solver are implemented via an abstract interface and wrappers for other SMT libraries could easily be added in the future.

4.2 Symbolic Analysis of Distance-Bounding Protocols

Having the possibility of using constrained variables gives us the opportunity of making finite potentially infinite search spaces. As for a more interesting application of rewriting modulo SMT, let us consider the case of *distance-bounding protocols* [10], a class of security protocols that infer an upper bound on the distance between two agents from the round trip time of messages. This is used, for example, for controlling some kinds of access and for clock synchronization. In a distance-bounding protocol session, the verifier (V) and the prover (P) exchange messages:

$$V \to P : m$$
$$P \to V : m'$$

where m is a challenge and m' is a response message (constructed using the components of m such as nonces in m). In order to infer the distance to the prover, the verifier remembers the time, t_0, when the message m was sent, and the time, t_1, when the message m' returns. From the difference $t_1 - t_0$ and the assumptions on the speed of the transmission medium, v, the verifier can compute an upper bound on the distance to the prover, namely $(t_1 - t_0) \times v$.

In [52], a novel attack on distance bounding called *attack in-between-ticks* is presented. The attack is formalized using a model in which provers, verifiers, and attackers may have different clock rates, processing speeds, or observation granularity. The model is based on a multiset-rewriting formalism called *timed local state transition systems* [51] which supports both discrete and dense time. The key insight for the attack is that an attacker can mask his location by exploiting the fact that a message may be sent at any point between two clock ticks of the verifier's clock, while the verifier measures the time at discrete clock ticks. For example, if the time bound is 3, a message could start at time 1.7, which is 2 on the verifier's clock, and the reply received at 4.9, which is 5 on the verifier's clock. From the verifier's perspective the attacker is within range, since $5 - 2 = 3$, but in fact the round trip time was $4.9 - 1.7 = 3.2$.

The model was formalized in Maude SMT and smt-search was used to find a symbolic representation of a family of attacks. Using Maude SMT the potentially infinite search space becomes finite, by treating the distance between the verifier and the prover as a constrained variable.

To illustrate the use of SMT, we show the Tick rule which advances system time following the approach of Real-Time Maude [76].

```
var  S : Soup .
vars T T1 T2 : Real .

crl [Tick] : { S (Time @ T) (vTime @ T1) }
   => { S (Time @ T2) (vTime @ T1) }
   if (T2 > T and (T2 < T1 + 1/1)) = true
   [nonexec] .
```

Here, a system state is a soup of timed facts (F @ T) enclosed in brackets. There is a unique fact, Time @ T, representing the physical time. The fact vTime @ T1 represents time as perceived by the verifier. In Real-Time Maude execution and search use a time sampling strategy. In contrast, using Maude SMT the new time is left symbolic, with constraints on its range. Here the constraint says that the next time should be greater than the current time, but should not advance beyond the next verifier time $(T1 + 1)$.

The following command searches for a state in which the verifier, v, accepts a reply from the prover, p, (Ok(< p, M:Msg >) @ T:Real), where the distance bound is 3 (the allowed round trip time is 2 * 3) and the distance from the verifier to the prover, dvp, (or to an attacker, dva) is greater than 3.

```
Maude> smt-search [1] { gensym(0)
                        dist(dva, dvp)
                        (Time @ 0/1)
                        (vTime @ 0/1)
                        (VO(p) @ 0/1)
                        (PO(v) @ 0/1) }
              =>+ { (Ok(< p, M:Msg >) @ T:Real)
                    S:Soup }
              such that (dva > 3/1 and dvp > 3/1 ) = true .
```

One (simplified) solution is:

```
S --> ...
   (Time @ toInteger(dvp + dvp) + 1/1)
   (RStart(< p, n(0) >) @ 0/1)                *** the real start
   (RStop(< p, n(0) >) @ dvp + dvp)           *** the real stop
   (V1(< p, n(0) >) @ toInteger(dvp + dvp) + 1/1) *** the verifiers view
  where dva > 3/1 and dvp > 3/1 and
  toInteger(dvp + dvp) <= 2/1 * 3/1    *** the discrete time looses here
  and T === toInteger(dvp + dvp) + 1/1
  ...
M --> n(0)
```

The verifier's start time is $1/1$ thus the elapsed time is `toInteger(dvp + dvp)` $<=$ $2/1 * 3/1$.

5   Pathway Logic

There have been many applications of Maude in many different areas, including some as diverse as security [41], cyber-physical systems [3,44], and model-driven engineering [8,77] (see, for example, the survey [72]). We very briefly discuss in this section one of them, Pathway Logic, which innovates by modeling nature rather that the usual digital artifacts, very nicely illustrating the modeling capabilities of Maude.

Pathway Logic (PL) is a system for modeling and reasoning about cellular processes such as signal transduction, metabolism, and cell-cell communication in the immune system. The semantic underpinnings of PL is Maude, and José Meseguer was part of the original PL team that developed the key ideas [39]. The first instance of a PL model was a model of a cancer-related signaling pathway crafted in Maude in 2000. In order to facilitate scaling up and interacting with the PL models, the executable Maude model has been augmented with the Pathway Logic Assistant (PLA), an interactive graphical interface that allows a user to easily create specific models, and browse and query them [81]. In addition, a mechanism for semantically grounding the language with links to standard databases has been put in place, and a substantial collection of formal models has been developed [56,80].

5.1   About Pathway Logic

Signal transduction is the mechanism by which cells sense their environment, process this information, and make decisions: what proteins to produce, what metabolic pathways to activate, whether to replicate, move, or possibly die. Typically, the signal is a chemical or protein in the cells environment that binds to a receptor protein (in the cell membrane). The receptor becomes *active*, initiating the signaling process. The signal is transmitted by change in state and location of proteins involved.

In PL a cell state is represented as a soup of occurrences, where each occurrence has three components: a protein or other biomolecule (gene, metabolite, etc.), a modifier, and a location. The modifier indicates the state of the protein, including binding of small molecules or phosphates, or ability to act on other proteins (enzyme activity). For example, the term < [Hras - GTP], CLi > is the occurrence of the protein Hras modified by binding to the small molecule GTP (Guanosine TriPhosphate), attached to the inside of the cell membrane (CLi). [2] Signal transduction steps are formalized as local rewrite rules operating on the relevant part of the cell state.

As an example, rule 1.EgfR.act formalizes the initiation of signaling in response to the presence of Egf (Epidermal growth factor) in a cell's exterior.

```
rl [1.EgfR.act]
  < ?ErbB1L:ErbB1L, XOut > < EgfR, EgfRC >
  =>
  < ?ErbB1L:ErbB1L :[ EgfR - Yphos], EgfRC >
```

Here, ?ErbB1L:ErbB1L is a variable of sort ErbB1L, XOut is the cells external environment, and the infix operator _:_ represents complex formation. In the model there are two proteins of sort ErbB1L: Egf and Tgfa (Transforming growth factor alpha).

How does the Egf signal propagate? To answer the question, we use the PLA. Figure 2 shows the subnet containing all rules relevant to activation (binding to GTP) of Hras in response to Egf. The subnet is generated by backwards collection from the goal H = < [Hras - GTP], CLi > in the Egf response network.

A specific execution path can be found by mapping the subnet and goal to the language of the LoLA model checker [79] asserting that the goal cannot be reached. If there is a counter example, LoLA returns a list of transitions that can be fired to reach the goal. Maude then converts this to a network and generates the expression to display the interactive graph.

Figure 2 also shows the result of comparing two different execution paths, the second is the result of removing one of the occurrences from the subnet (simulating a knockout) before asking LoLA.

---

[2] There are in fact two internal syntactic forms for representing cell state: a soup of locations; and a soup of occurrences. We restrict attention to the latter as occurrences correspond to places in a Petri net.

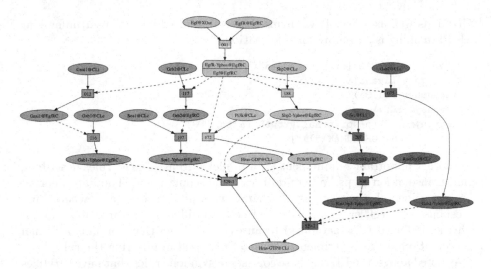

Fig. 2. Activation of Hras in response to Egf. The full subnet with two execution paths compared. Pink belongs to both paths, blue and cyan to different paths. Ovals represent occurrences, rectangles represent rules, with input ovals connected by incoming arrows and output connected by outgoing arrows. Dashed arrows connect occurrences that both input and output (enzymes) (Color figure online).

5.2   Maude's Role in Pathway Logic

The integration of the Maude executable model and PLA is achieved using the IOP platform for communication amongst a group of *actors*. Maude's *loop mode and reflection* are key to turning Maude into an actor [61]. The PL Maude actor extends the IMaude actor [61] with rules for handling PL specific requests. IMaude provides data structures for representing state, managing asynchronous interaction, and saving and restoring state. The latter makes essential use of Maude's capabilities for pretty-printing and parsing. PLA is an actor built using an interpreter of a Scheme-like language, JLambda, layered on top of Java. The Maude PL actor listens for requests from PLA and generates expressions in the JLambda language to instruct PLA to construct and render interactive network graphs.

Proteins, modifications, and locations are given different names by different biologists. Thus, to understand what a biological model is talking about it is important to link the names (constants) used to reference databases that provide canonical names and additional information. This is accomplished using the metadata attribute of Maude operator declarations. PL operator metadata is a string encoding an S-expression that maps key words to values such as database access identifiers, synonym lists, and biological classifiers. With the aid of a meta-model (componentInfoSpec) the metadata is rendered as a menu of information and active links that are presented when the user clicks on a graph element. The following is the operator declaration for Egf. The metadata includes identifiers used by two reference databases (spnumber for UniProt and hugosym for

HGNC), as well as a list of synonyms that can be used when a user unfamiliar with PL naming is searching for information.

```
op Egf : -> ErbB1L [ctor metadata"(\
  (category Ligand)\
  (spnumber P01133)\
  (hugosym EGF)\
  (synonyms \"Pro-epidermal growth factor\"\
            \"EGF_HUMAN\"))"] .
```

Another important requirement for a model is to justify the rules describing signal transduction steps. Where do these rules come from? They are inferred from experimental observations of what is present in a cell (and where) and of response of a cell or a population of cells to different perturbations. In PL each rule is linked (via metadata) to an evidence page that contains a formal representation of the experimental observations used in inferring the rule. [3]

As noted above, the PL rule base consists of symbolic rules that have variables whose sort consists of a finite set of proteins or modifications or locations. Petri net tools (and graphical representations) need concrete instances. One possibility is simply to generate all possible instances. This generates many useless rules, because proteins of a given sort behave similarly in some cases, and differently in other cases. Thus we take advantage of the Maude function to generate all matches of a rule to a given state to generate only concrete rules that are possibly reachable from initial states of interest.

In the spirit of *May I borrow your logic* [11], PL supports multiple representations of a PL knowledge base: Maude signature and rules, Petri nets, JSON, and SBML (Systems Biology Markup Language). The Petri net representation is used for efficient analysis of large networks that takes advantage of the restricted nature of PL rules. SBML is an exchange format used to share models between different systems biology tools including simulators and visualizers. The JSON representation is used to treat a PL knowledge base as a database with efficient query of static relations; for example, finding all rules that involve the protein whose UniProt identifier is P01112 (Hras). Maude could be programmed to answer such queries, but putting the information in a database makes it more widely accessible. In addition, the JSON representation is an easily parsable exchange format that is being used by researchers developing modeling and analysis tools.

Transformations to different representation systems is done by reflecting the PL model to the metalevel and transforming it to a representation in the target system, which can then be written to a file with the help of the PLA actor.

6   Further Ahead

We have presented our view of the two decades we have been involved in the development of Maude. We have gone from the first years of Maude to the

---

[3] Rules are currently inferred by a human curator.

current state of the system and the new features currently under development. What features will Maude users be expecting in the future? Or what features will have a bigger impact for Maude in the future? As we explained before, a current trend in programming languages is to become multi-paradigm, offering flexibility and simplicity for problem specification and solving, and we believe our efforts will go into that direction.

On the one hand, all the logical and symbolic features will be boosted. For instance, other SMT libraries such as veriT could be added to Maude. Also, variant generation and variant-based unification, as well as narrowing in system theories, consider only unconditional rules and equations: conditional narrowing, both at the level of equational logic and rewriting logic, should be added. Moreover, more built-in unification algorithms will be included in Core Maude: we have explored unification algorithms for associativity, for homomorphic encryption, for exclusive-or, etc. Furthermore, we envision conditional narrowing combined with SMT solvers, so that many different reasoning facilities are seamlessly combined.

On the other hand, tool support will be incremented. Maude would not be such a good logical framework without its metalevel capabilities. The Maude Formal Environment will be improved with better tool integration. And we should not forget about tools built on top of Maude, such as the Pathway Logic Assistant, Real-Time Maude, or the Maude-NPA protocol analyzer.

Acknowledgements. Francisco Durán was partially supported by Universidad de Málaga, Campus de Excelencia Internacional Andalucía Tech and Spanish Ministry for Economy and Competitiveness (MINECO) and the European Union (FEDER) under grant POLYCIMS (ref. TIN2014-52034-R). Santiago Escobar was partially supported by the EU (FEDER) and the Spanish MINECO under grant TIN2013-45732-C4-1-P. Narciso Martí-Oliet was partially supported by Spanish MINECO under grant StrongSoft (TIN2012-39391-C04-04) and Comunidad de Madrid program N-GREENS Software (S2013/ICE-2731).

References

1. Antoy, S., Echahed, R., Hanus, M.: A needed narrowing strategy. J. ACM **47**(4), 776–822 (2000)
2. Bae, K., Escobar, S., Meseguer, J.: Abstract logical model checking of infinite-state systems using narrowing. In: van Raamsdonk, F. (ed.) 24th International Conference on Rewriting Techniques and Applications, RTA 2013, Eindhoven, The Netherlands, 24–26 June 2013. LIPIcs, vol. 21, pp. 81–96. Schloss Dagstuhl - Leibniz-Zentrum fuer Informatik (2013)
3. Bae, K., Krisiloff, J., Meseguer, J., Ölveczky, P.C.: Designing and verifying distributed cyber-physical systems using multirate PALS: an airplane turning control system case study. Sci. Comput. Program. **103**, 13–50 (2015)
4. Bae, K., Meseguer, J.: Model checking linear temporal logic of rewriting formulas under localized fairness. Sci. Comput. Program. **99**, 193–234 (2015)

5. Barrett, C., Conway, C.L., Deters, M., Hadarean, L., Jovanović, D., King, T., Reynolds, A., Tinelli, C.: CVC4. In: Gopalakrishnan, G., Qadeer, S. (eds.) CAV 2011. LNCS, vol. 6806, pp. 171–177. Springer, Heidelberg (2011)
6. Barrett, C., Stump, A., Tinelli, C.: The SMT-LIB standard: Version 2.0. Technical report, Department of Computer Science, The University of Iowa (2010). http:// smt-lib.org
7. Barrett, C.W., Sebastiani, R., Seshia, S.A., Tinelli, C.: Satisfiability modulo theories. In: Biere, A., Heule, M., van Maaren, H., Walsh, T. (eds.) Handbook of Satisfiability. Frontiers in Artificial Intelligence and Applications, vol. 185, pp. 825–885. IOS Press (2009)
8. Boronat, A., Meseguer, J.: An algebraic semantics for MOF. Formal Asp. Comput. **22**(3–4), 269–296 (2010)
9. Bouhoula, A., Jouannaud, J.-P., Meseguer, J.: Specification and proof in membership equational logic. Theor. Comput. Sci. **236**(1), 35–132 (2000)
10. Brands, S., Chaum, D.: Distance bounding protocols. In: Helleseth, T. (ed.) EUROCRYPT 1993. LNCS, vol. 765, pp. 344–359. Springer, Heidelberg (1994)
11. Cerioli, M., Meseguer, J.: May i borrow your logic? (Transporting logical structure along maps). Theor. Comput. Sci. **173**, 311–347 (1997)
12. Clavel, M.: Reflection in General Logics and in Rewriting Logic, with Applications to the Maude Language. Ph.D. thesis, Universidad de Navarra, Spain, February 1998
13. Clavel, M.: Reflection in general logics, rewriting logic, and Maude. In: Kirchner, Kirchner (eds.) [53], pp. 71–82
14. Clavel, M., Durán, F., Eker, S., Escobar, S., Lincoln, P., Martí-Oliet, N., Meseguer, J., Talcott, C.: Unification and narrowing in Maude 2.4. In: Treinen, R. (ed.) RTA 2009. LNCS, vol. 5595, pp. 380–390. Springer, Heidelberg (2009)
15. Clavel, M., Durán, F., Eker, S., Lincoln, P., Martí-Oliet, N., Meseguer, J.: Metalevel computation in Maude. In: Kirchner and Kirchner [53], pp. 331–352
16. Clavel, M., Durán, F., Eker, S., Lincoln, P., Martí-Oliet, N., Meseguer, J., Quesada, J.F.: Maude: specification and programming in rewriting logic. SRI International, January 1999. http://maude.cs.uiuc.edu/maude1/manual/
17. Clavel, M., Durán, F., Eker, S., Lincoln, P., Martí-Oliet, N., Meseguer, J., Quesada, J.F.: Maude as a metalanguage. In: Kirchner and Kirchner [53], pp. 147–160
18. Clavel, M., Durán, F., Eker, S., Lincoln, P., Martí-Oliet, N., Meseguer, J., Quesada, J.F.: The Maude system. In: Narendran, P., Rusinowitch, M. (eds.) RTA 1999. LNCS, vol. 1631, pp. 240–243. Springer, Heidelberg (1999)
19. Clavel, M., Durán, F., Eker, S., Lincoln, P., Martí-Oliet, N., Meseguer, J., Quesada, J.F.: A Maude tutorial. Tutorial distributed as documentation of the Maude system, Computer Science Laboratory, SRI International. Presented at the European Joint Conference on Theory and Practice of Software, ETAPS 2000, Berlin, Germany, 25 March 2000
20. Clavel, M., Durán, F., Eker, S., Lincoln, P., Martí-Oliet, N., Meseguer, J., Quesada, J.F.: Towards Maude 2.0. In: Futatsugi [45], pp. 294–315
21. Clavel, M., Durán, F., Eker, S., Lincoln, P., Martí-Oliet, N., Meseguer, J., Quesada, J.F.: Using Maude. In: Maibaum, T. (ed.) FASE 2000. LNCS, vol. 1783, pp. 371–374. Springer, Heidelberg (2000)
22. Clavel, M., Durán, F., Eker, S., Lincoln, P., Martí-Oliet, N., Meseguer, J., Quesada, J.F.: Maude: specification and programming in rewriting logic. Theor. Comput. Sci. **285**(2), 187–243 (2002)

23. Clavel, M., Durán, F., Eker, S., Lincoln, P., Martí-Oliet, N., Meseguer, J., Talcott, C.L.: The Maude 2.0 system. In: Nieuwenhuis, R. (ed.) RTA 2003. LNCS, vol. 2706, pp. 76–87. Springer, Heidelberg (2003)

24. Clavel, M., Durán, F., Eker, S., Lincoln, P., Martí-Oliet, N., Meseguer, J., Talcott, C. (eds.): All About Maude - A High-Performance Logical Framework. LNCS, vol. 4350. Springer, Heidelberg (2007)

25. Clavel, M., Eker, S., Lincoln, P., Meseguer, J.: Principles of Maude. In: Meseguer [67], pp. 65–89

26. Clavel, M., Meseguer, J.: Reflection and strategies in rewriting logic. In: Meseguer [67], pp. 126–148

27. Comon-Lundh, H., Delaune, S.: The finite variant property: how to get rid of some algebraic properties. In: Giesl, J. (ed.) RTA 2005. LNCS, vol. 3467, pp. 294–307. Springer, Heidelberg (2005)

28. Braga, C.O., Haeusler, E.H., Meseguer, J., Mosses, P.D.: Maude action tool: using reflection to map action semantics to rewriting logic. In: Rus, T. (ed.) AMAST 2000. LNCS, vol. 1816, pp. 407–421. Springer, Heidelberg (2000)

29. Durán, F.: A Reflective Module Algebra with Applications to the Maude Language. Ph.D. thesis, Universidad de Málaga, Spain, June 1999

30. Durán, F.: The extensibility of Maude's module algebra. In: Rus, T. (ed.) AMAST 2000. LNCS, vol. 1816, pp. 422–437. Springer, Heidelberg (2000)

31. Durán, F., Eker, S., Escobar, S., Meseguer, J., Talcott, C.L.: Variants, unification, narrowing, and symbolic reachability in Maude 2.6. In: Schmidt-Schauß, M. (ed.) Proceedings of the 22nd International Conference on Rewriting Techniques and Applications, RTA 2011, Novi Sad, Serbia, 30 May - 1 June, 2011. LIPIcs, vol. 10, pp. 31–40. Schloss Dagstuhl - Leibniz-Zentrum fuer Informatik (2011)

32. Durán, F., Eker, S., Lincoln, P., Meseguer, J.: Principles of mobile Maude. In: Kotz, D., Mattern, F. (eds.) MA 2000, ASA/MA 2000, and ASA 2000. LNCS, vol. 1882, pp. 73–85. Springer, Heidelberg (2000)

33. Durán, F., Lucas, S., Meseguer, J.: MTT: the Maude termination tool (system description). In: Armando, A., Baumgartner, P., Dowek, G. (eds.) IJCAR 2008. LNCS (LNAI), vol. 5195, pp. 313–319. Springer, Heidelberg (2008)

34. Durán, F., Meseguer, J.: An extensible module algebra for Maude. In: Kirchner and Kirchner [53], pp. 174–195

35. Durán, F., Meseguer, J.: Maude's module algebra. Sci. Comput. Program. $66(2)$, 125–153 (2007)

36. Durán, F., Meseguer, J.: On the Church-Rosser and coherence properties of conditional order-sorted rewrite theories. J. Logic Algebraic Program. $81(7\text{--}8)$, 816–850 (2012)

37. Durán, F., Riesco, A., Verdejo, A.: A distributed implementation of mobile Maude. Electr. Notes Theor. Comput. Sci. $176(4)$, 113–131 (2007)

38. Durán, F., Rocha, C., Álvarez, J.M.: Towards a Maude formal environment. In: Agha, G., Danvy, O., Meseguer, J. (eds.) Formal Modeling: Actors, Open Systems, Biological Systems. LNCS, vol. 7000, pp. 329–351. Springer, Heidelberg (2011)

39. Eker, S., Knapp, M., Laderoute, K., Lincoln, P., Meseguer, J., Sonmez, K.: Pathway logic: symbolic analysis of biological signaling. In: Proceedings of the Pacific Symposium on Biocomputing, pp. 400–412, January 2002

40. Escobar, S.: Functional logic programming in Maude. In: Iida, S., Meseguer, J., Ogata, K. (eds.) Specification, Algebra, and Software. LNCS, vol. 8373, pp. 315–336. Springer, Heidelberg (2014)

41. Escobar, S., Meadows, C., Meseguer, J.: Maude-NPA: cryptographic protocol analysis modulo equational properties. In: Aldini, A., Barthe, G., Gorrieri, R. (eds.) FOSAD 2007/2008/2009. LNCS, vol. 5705, pp. 1–50. Springer, Heidelberg (2007)
42. Escobar, S., Meseguer, J.: Symbolic model checking of infinite-state systems using narrowing. In: Baader, F. (ed.) RTA 2007. LNCS, vol. 4533, pp. 153–168. Springer, Heidelberg (2007)
43. Escobar, S., Sasse, R., Meseguer, J.: Folding variant narrowing and optimal variant termination. J. Logic Algebraic Program. **81**(7–8), 898–928 (2012)
44. Fadlisyah, M., Ölveczky, P.C., Ábrahám, E.: Formal modeling and analysis of interacting hybrid systems in HI-Maude: what happened at the 2010 sauna world championships? Sci. Comput. Program. **99**, 95–127 (2015)
45. Futatsugi, K. (ed.): Proceedings of the Third International Workshop on Rewriting Logic and its Applications, WRLA 2000, Kanazawa, Japan, 18–20 September 2000. Electronic Notes in Theoretical Computer Science. Elsevier, Amsterdam (2000)
46. Goguen, J., Meseguer, J.: Eqlog: equality, types and generic modules for logic programming. In: DeGroot, D., Lindstrom, G. (eds.) Logic Programming, Functions, Relations and Equations, pp. 295–363. Prentice-Hall (1986)
47. Goguen, J., Meseguer, J.: Order-sorted algebra I: equational deduction for multiple inheritance, overloading, exceptions and partial operations. Theor. Comput. Sci. **105**, 217–273 (1992)
48. Goguen, J., Winkler, T., Meseguer, J., Futatsugi, K., Jouannaud, J.-P.: Introducing OBJ. In: Goguen, J.A., Malcolm, G. (eds.) Software Engineering with OBJ: Algebraic Specification in Action, pp. 3–167. Kluwer Academic Publishers (2000)
49. Goguen, J.A., Meseguer, J.: Equality, types, modules, and (why not?) generics for logic programming. J. Logic Program. **1**(2), 179–210 (1984)
50. Hendrix, J., Meseguer, J., Ohsaki, H.: A sufficient completeness checker for linear order-sorted specifications modulo axioms. In: Furbach, U., Shankar, N. (eds.) IJCAR 2006. LNCS (LNAI), vol. 4130, pp. 151–155. Springer, Heidelberg (2006)
51. Kanovich, M., Kirigin, T.B., Nigam, V., Scedrov, A., Talcott, C., Perovic, R.: A rewriting framework for activities subject to regulations. In: Tiwari, A. (ed.) 23rd International Conference on RewritingTechniques and Applications (RTA). LIPIcs, vol. 15, pp. 305–322. Schloss Dagstuhl - Leibniz-Zentrum fuer Informatik (2012)
52. Kanovich, M., Kirigin, T.B., Nigam, V., Scedrov, A., Talcott, C.: Discrete vs. dense times in the analysis of cyber-physical security protocols. In: Focardi, R., Myers, A. (eds.) POST 2015. LNCS, vol. 9036, pp. 259–279. Springer, Heidelberg (2015)
53. Kirchner, C, Kirchner, H. (eds.) Proceedings of the Second International Workshop on Rewriting Logic and its Applications, WRLA 1998, Pont-à-Mousson, France, 1–4 September 1998. Electronic Notes in Theoretical Computer Science, vol. 15. Elsevier, Amsterdam (1998)
54. Lincoln, P., Martí-Oliet, N., Meseguer, J.: Specification, transformation, and programming of concurrent systems in rewriting logic. In: Blelloch, G.E., Chandy, K.M., Jagannathan, S. (eds.) Specification of Parallel Algorithms, DIMACS Workshop, 9–11 May 1994. DIMACS Series in Discrete Mathematics and Theoretical Computer Science, vol. 18, pp. 309–339. American Mathematical Society (1994)
55. Lincoln, P., Martí-Oliet, N., Meseguer, J., Ricciulli, L.: Compiling rewriting onto SIMD and MIMD/SIMD machines. In: Halatsis, C., Philokyprou, G., Maritsas, D., Theodoridis, S. (eds.) PARLE 1994. LNCS, vol. 817, pp. 37–48. Springer, Heidelberg (1994)
56. Lincoln, P.D., Talcott, C.: Symbolic systems biology and pathway logic. In: Iyengar, S. (eds.) Symbolic Systems Biology, pp. 1–29. Jones and Bartlett (2010)

57. Lucas, S.: Context-sensitive rewriting strategies. Inf. Comput. **178**(1), 294–343 (2002)
58. Martí-Oliet, N., Meseguer, J.: Action and change in rewriting logic. In: Pareschi, R., Fronhöfer, B. (eds.) Dynamic Worlds: From the Frame Problem to Knowledge Management. Applied Logic Series, vol. 12, pp. 1–53. Kluwer Academic Publishers (1999)
59. Martí-Oliet, N., Meseguer, J.: Rewriting logic as a logical and semantic framework. In: Gabbay, D.M., Guenthner, F. (eds.) Handbook of Philosophical Logic, 2nd edn., vol. 9, pp. 1–87. Kluwer Academic Publishers (2002)
60. Martí-Oliet, N., Palomino, M., Verdejo, A.: Rewriting logic bibliography by topic: 1990–2011. J. Logic Algebraic Program. **81**(7–8), 782–815 (2012)
61. Mason, I.A., Talcott, C.L.: IOP: the interoperability platform & IMaude: an interactive extension of Maude. In: Martí-Oliet, N. (ed.) Proceedings Fifth International Workshop on Rewriting Logic and its Applications, WRLA 2004, Barcelona, Spain, 27–28 March 2004. Electronic Notes in Theoretical Computer Science, vol. 117, pp. 315–333. Elsevier (2005). http://www.sciencedirect.com/science/journal/15710661
62. Meseguer, J.: A logical theory of concurrent objects. In: Meyrowitz, N. (ed.) Proceedings of the ECOOP-OOPSLA 1990 Conference on Object-Oriented Programming, Ottawa, Canada, 21–25 October 1990, pp. 101–115. ACM Press (1990)
63. Meseguer, J.: Rewriting as a unified model of concurrency. In: Baeten, J.C.M., Klop, J.W. (eds.) CONCUR 1990. LNCS, vol. 458, pp. 384–400. Springer, Heidelberg (1990)
64. Meseguer, J.: Conditional rewriting logic as a unified model of concurrency. Theor. Comput. Sci. **96**(1), 73–155 (1992)
65. Meseguer, J.: Multiparadigm logic programming. In: Kirchner, H., Levi, G. (eds.) ALP 1992. LNCS, vol. 632, pp. 158–200. Springer, Heidelberg (1992)
66. Meseguer, J.: A logical theory of concurrent objects and its realization in the Maude language. In: Agha, G., Wegner, P., Yonezawa, A. (eds.) Research Directions in Concurrent Object-Oriented Programming, pp. 314–390. The MIT Press (1993)
67. Meseguer, J. (ed.) Proceedings of the First International Workshop on Rewriting Logic and its Applications, WRLA 1996, Asilomar, California, 3–6 September 1996. Electronic Notes in Theoretical Computer Science, vol. 4. Elsevier, Amsterdam (1996)
68. Meseguer, J.: Rewriting logic as a semantic framework for concurrency: a progress-report. In: Sassone, V., Montanari, U. (eds.) CONCUR 1996. LNCS, vol. 1119, pp. 331–372. Springer, Heidelberg (1996)
69. Meseguer, J.: Membership algebra as a logical framework for equationalspecification. In: Parisi-Presicce, F. (ed.) WADT 1997. LNCS, vol. 1376, pp. 18–61. Springer, Heidelberg (1998)
70. Meseguer, J.: From OBJ to Maude and beyond. In: Futatsugi, K., Jouannaud, J.-P., Meseguer, J. (eds.) Algebra, Meaning, and Computation. LNCS, vol. 4060, pp. 252–280. Springer, Heidelberg (2006)
71. Meseguer, J.: The temporal logic of rewriting: a gentle introduction. In: Degano, P., De Nicola, R., Meseguer, J. (eds.) Concurrency, Graphs and Models. LNCS, vol. 5065, pp. 354–382. Springer, Heidelberg (2008)
72. Meseguer, J.: Twenty years of rewriting logic. J. Logic Algebraic Program. **81**(7–8), 721–781 (2012)
73. Meseguer, J., Talcott, C.: Semantic models for distributed object reflection. In: Magnusson, B. (ed.) ECOOP 2002. LNCS, vol. 2374, pp. 1–36. Springer, Heidelberg (2002)

74. Meseguer, J., Thati, P.: Symbolic reachability analysis using narrowing and its application to verification of cryptographic protocols. High.-Order Symbolic Comput. **20**(1–2), 123–160 (2007)
75. Meseguer, J., Winkler, T.C.: Parallel programmming in Maude. In: Banâtre, J.-P., Le Métayer, D. (eds.) Research Directions in High-Level Parallel Programming Languages 1991. LNCS, vol. 574. Springer, Heidelberg (1992)
76. Ölveczky, P.C., Meseguer, J.: Semantics and pragmatics of real-time Maude. High.-Order Symbolic Comput. **20**(1–2), 161–196 (2007)
77. Rivera, J.E., Durán, F., Vallecillo, A.: Formal specification and analysis of domain specific models using Maude. Simulation **85**(11–12), 778–792 (2009)
78. Rocha, C., Meseguer, J., Muñoz, C.: Rewriting modulo SMT and open system analysis. In: Escobar, S. (ed.) WRLA 2014. LNCS, vol. 8663, pp. 247–262. Springer, Heidelberg (2014)
79. Schmidt, K.: LoLA: a low level analyser. In: Nielsen, M., Simpson, D. (eds.) ICATPN 2000. LNCS, vol. 1825, pp. 465–474. Springer, Heidelberg (2000)
80. Talcott, C.: Symbolic modeling of signal transduction in pathway logic. In: Perrone, L.F., Wieland, F.P., Liu, J., Lawson, B.G., Nicol, D.M., Fujimoto, R.M. (eds.) 2006 Winter Simulation Conference, pp. 1656–1665 (2006)
81. Talcott, C., Dill, D.L.: Multiple representations of biological processes. In: Priami, C., Plotkin, G. (eds.) Transactions on Computational Systems Biology VI. LNCS (LNBI), vol. 4220, pp. 221–245. Springer, Heidelberg (2006)
82. Talcott, C.L.: Coordination models based on a formal model of distributed object reflection. Electr. Notes Theor. Comput. Sci. **150**(1), 143–157 (2006)
83. Winkler, T.C.: Programming in OBJ and Maude. In: Lauer, P.E. (ed.) Functional Programming, Concurrency, Simulation and Automated Reasoning. LNCS, vol. 693, pp. 229–277. Springer, Heidelberg (1993)

Formal Universes

Erwin Engeler[✉]

ETH Zurich, Zürich, Switzerland
`engeler@math.ethz.ch`

Abstract. This essay addresses the concerns of the foundations of mathematics of the early 20th century which led to the creation of formally axiomatized universes. These are confronted with contemporary developments, particularly in computational logic and neuroscience. Our approach uses computational models of mental experiments with the infinite in set-theory and symbol-manipulation systems, in particular models of combinatory logic.

1 Introduction

To guide the perplexed, philosophers used to propose their systems, physicists their Grand Unified Theories, and theologians their faith and scriptures. What about mathematicians? Formal universes.

In this essay[1] I try to tell this story as I see it, distinguishing two kinds of universes: One is centered in set theory, the other in symbol-manipulating systems. Both claim universality for mathematics. For evidence both adduce developments of vast stretches of mathematics, reducing it to the respective "foundations". I permit myself some skeptical comments on the whole enterprise.

2 Universalism I : Paradise Lost?

Mathematics creates its own universe, not out of chaos or *tohu-wabohu* as in Genesis, but out of nothingness, the empty set. Start with the empty set $V_0 = \emptyset$ and repeat taking the set of all subsets: $V_1 = P(\emptyset), V_2 = P(P(\emptyset)), \ldots$, all the while collecting what you have obtained so far. At the beginning all is finite; the first time an infinite set obtains is when collecting up after infinitely many steps. This is the set V_ω of hereditarily finite sets, a small universe of mathematical objects. In it we may recognize (or fashion) the natural numbers, the integers, the rationals (individually, not as totalities). According to the famous 19th century algebraist Kronecker this is all that God created, the rest is "Menschenwerk". Works of man are for example the real numbers which arise if we take the power-set of the hereditarily finite sets and, continuing the process, functions, function spaces, ... the whole menagerie, the universe of mathematics, called the

---

[1] In memory and appreciation of a quarter century of discussion on the foundation and philosophy of mathematics and computer science with José Meseguer.

© Springer International Publishing Switzerland 2015
N. Martí-Oliet et al. (Eds.): Meseguer Festschrift, LNCS 9200, pp. 255–263, 2015.
DOI: 10.1007/978-3-319-23165-5_12

cumulative hierarchy, due to von Neumann and Zermelo [18]. All of this out of nothingness, insubstantial.

Indeed? What is the substance of mathematics; is there one? Substance is something that you have to reckon with, can reckon with. Mathematicians can manipulate, reckon with the objects of their domain, the universe of sets, named "Cantor's Paradise" by Hilbert [12].

In this view, the problem of foundations is to capture the kind of reckoning that goes on in the construction of the universe we just described. Around 1900 this meant to axiomatize: Postulate the empty set, pairing, power set, union, etc. in fact the axioms of Zermelo. These axioms arguably constitute the basis of mathematics today, they are still taken as "axiomatic" by most. Zermelo also produced an axiomatization of thermodynamics, long since forgotten. Axiomatization at that time meant mainly to describe compactly and plausibly the knowledge of a subject matter—with the intention to support logical conclusions, theorems, covering the field—*more geometrico* (Spinoza.)

Hilbert was one to recognize that axiomatization did not by itself create the universe of discourse, nor does it automatically fix it. His axiomatization of real numbers arguably fixes them,[2] but their "reality" had to be seen in the system of logical deductions from these axioms. Consequently, foundations as an axiomatic discipline became based on formal logic. The problem of foundation was reduced to proving deductive consistency by finite universally accepted means, and existence became identified with being free of formal logical contradictions. In the words of the main adversary, the founding intuitionist Brouwer, this meant that mathematics consisted only of "marks on paper".—And, as is well known, Hilbert's program was shown to fail by Gödel.—Banned from the Garden of Eden we now follow Voltaire's advice and cultivate our own gardens, outside.

This, of course, is not the whole story. In fact, the challenge was taken up by an impressive and ongoing development of proof theory (e.g. locating incompleteness of Peano arithmetic in the formal limitations of induction), and of axiomatic set theory (showing up the limitations of "platonic insight" into the transfinite by set-theoretic independence results), and much more. In the following we describe a small section of these developments from a somewhat personal perspective.

3   Universalism II : Paradise Regained?

Of course, nowadays numbers, reals, sets, functions do not only live on paper, they also live in our computers: as the data types of programming languages, supported by interpreters and compilers on sophisticated hardware. This universe of objects "to be reckoned with",[3] the types of objects available for computing is ever expanding. The definition of data types in most programing languages proceeds in a more or less pragmatic manner, guided by emerging applications, including the "internet of things". This is a valid approach to universality, and

---

[2] With respect to the "naive" set theory in which its structure is discussed.
[3] Cf. German "rechnen" = to compute.

important efforts are being made to provide it with systematic foundations as we shall see.

To illustrate this second kind of approach to universalism in mathematics, I shall use its historically first worked-out, and technically simplest, example. It is based on a development that originated in the 1930s in answer to the same "crisis of foundations" sketched above: Combinatory logic, lambda calculus, and type theories. Today these theories can also be seen as a foundational approach to computer science. Since Turing, in particular since universal Turing Machines, and also since von Neumann-Machines, inputs and programs are of the same nature and which we may simply call "data". They are perhaps marks on the Turing tape or bits in the computer memory. The basic operation is application: Programs may be applied to input data and of course result in data, which may again be programs. Programs may also be applied to programs, again resulting in data, etc. Indeed, we may admit that all combinations of applications on data result again in data, (e.g. in the model of Sect. 4 below).

Formally we have a the universe D of data which admits a binary operation on all its elements: if a and b are in D, then so is a applied to b, written $a \cdot b$. Any combination of data is results in data. For example $(x \cdot z) \cdot (y \cdot z)$ is a kind of data: If the program y is used to modify input z to a new input $y \cdot z$, and x uses the input z to obtain a new program $x \cdot z$, then $(x \cdot z) \cdot (y \cdot z)$ is the data resulting from applying the new program to the new input. Obtaining this data is in fact a program $\mathbf{S}$ which is applied to data x, y and z in sequence: $((\mathbf{S} \cdot x) \cdot y) \cdot z)$ equals $(x \cdot z) \cdot (y \cdot z)$.

Universality is expressed by the following axiom scheme: For every combination $\phi(x_1, \dots x_n)$ of data there is a data t_ϕ such that $t_\phi x_1 \dots x_n = \phi(x_1, \dots x_n)$, (association to the left is understood tacitly.) Such t_ϕ are called "combinators". A set D with an application operation satisfying the axiom scheme is called a combinatory algebra. Combinatory Logic, was invented by H.B. Curry in his 1929 Göttingen thesis [3] directed by Paul Bernays, also Doktorvater of the present author as well as of Saunders MacLane and Gerhard Gentzen. It is the formal theory of equations between combinators as its objects. As Schönfinkel had already shown, two combinators suffice for expressing all combinators, namely the above $\mathbf{S}$, characterized by its equation, together with $\mathbf{K}$, characterized by $\mathbf{K}xy = x$. These equations may be understood as rewrite rules of the logic. The two combinators come out in the proof of the axiom scheme.

For Combinatory Logic the question of consistency arises again. But while Hilbert's program failed for the first formal universe (of set theory and Peano arithmetic), Curry's formal system is consistent, as proved by Church and Rosser using the same finitist proof-theoretic tools that failed in the first case. This proof considers combinatory logic and algebra as a rewrite system. As a rewrite system, with rules such as $\mathbf{S}xyz \to (xz)(yz)$, it is surprisingly rich: it admits natural numbers (as combinations of $\mathbf{S}$ and $\mathbf{K}$), partial recursive functions (and thereby a theory of computability equivalent to Turing Machines), exhibits the phenomenon of undecidability (of termination of rewriting), etc. It is for this reason that we may consider combinatory logic as a foundation for computer science. It is open to formal extensions. One important such extension results from the introduction of types (e.g. to Lambda Calculus).

As a symbol manipulation system, Combinatory Logic is the prototype of a development that has been mastery reviewed by Meseguer [14]. The extensive development of rewrite logic supports the claim that the concept of mathematics as a symbol-manipulating enterprise is sustainable.

Rewrite systems prefer to live on computers. The currently successful computational system COQ offers such an environment. We remarked above that types are introduced in the classical programming languages in a restricted manner. To be part of a mathematical, or computational universe, containing the structures on which mathematics is "really" done, it would need types. COQ introduces basic types such as natural numbers and allows definition of new types by recursion, products (of dependent types) and continues through a hierarchy of classical mathematical structures, groups, fields, etc. to contemporary structures such as homotopy groups; all of this by uniform definitional templates. Moreover, it embodies procedures which support construction, and checking, of mathematical proofs—after a process of formalization that is empowered by its structural richness. The mathematics on which this rests is Homotopy Type Theory.[4]

A quite different example of universalism akin to rewrite systems is Wolfram's proposal to understand the mathematical sciences by rewrite rules in the shape of rules for cellular automata [17]. As a remarkable single-handed effort it challenges by its suggestiveness and scope. It should not be blamed for not completely and convincingly reaching its goal.[5]

4   Reductionism

Reducing one corpus of knowledge (e.g. chemistry) to another (e.g. quantum theory) is more a kind of (successful) mind-set than a true methodology. Called "reductionism", it is a well-recognized topic in the philosophy of science and has been the subject of an extensive literature. Whether it has ever been fully successful in the natural sciences is open to doubt.[6] How about mathematics?

Universalism and reductionism go hand in hand. This is the case of the set-theoretic universe. Indeed it is the tacit understanding of most mathematicians that all mathematics can "in principle" be reduced to set theory and logic. But this comes with steep costs: Mathematics, when fully reduced to its set-theoretic basic components may become quite opaque to mere humans. But not always.

Let us return to the second kind of universe, combinatory logic, which we introduced as a specimen of a symbol-manipulating systems and whose "existence" relies on a formal consistency proof. If the first universe, of set theory, is "good enough", then it should be possible to reduce the second universe to it, that is: to create a pocket universe for combinatory logic in set theory.

---

[4] An ongoing project accessible online at HomotopyTypeTheory.org.

[5] For a critique cf. [11].

[6] Among others by my late colleague, the quantum chemist H. Primas, in [1].

This would be an algebraic structure with elements corresponding to the combinators and their laws. My favorite construction[7] starts with an arbitrary non-empty set A, the basic data (e.g. marks on the Turing tape or bits in computer memory), and builds upon it a set $G(A)$ of formal expressions. It is recursively defined by $G(A) = A \cup \{\alpha \longrightarrow x : \alpha \subseteq G(A), x \in G(A)\}$, α finite. It thus consist of expressions $\alpha \longrightarrow x$ obtained by starting with basic data by iterating replacement of components x and $y \in \alpha$.—From this we define the combinatory algebra $\mathcal{D}_A$: Its set D of elements is the set of all subsets of $G(A)$, which stands as the set of data, $D = P(G(A))$. The application operation is defined by $X \cdot Y = \{x : \exists \alpha \subseteq Y, \alpha \longrightarrow x \in X\}$ for arbitrary data X and Y in D.[8]

This author's Combinatory Program [8] shows in detail that much of mathematics can be developed by suitably enriching $\mathcal{D}_A$ with additional operations and admitting specific basic sets A.

The extension to the theory of dependent types as in Homotopy Type Theory, mentioned above, also calls for a reduction to the set-theoretic universe. This is accomplished by Voevodsky in the current univalent foundation project,[9] which we cannot discuss here.

5   Criticism

Universalist tendencies are present in many fields and for many reasons; they are obviously attractive intellectually. But universality claims can also be the source of fundamentalism. In religion and politics unspeakable extremes have resulted throughout history. If such fundamentalism is applied to the natural sciences it may be strange (geology: disputing the age of the world; biology: disputing evolution). But occasionally, and frighteningly, there are even cases concerning mathematics (symbolic logic, banned as "bourgeois idealism" under Stalin; some "decadent" mathematical developments to be prosecuted, banned and replaced by "Deutsche Mathematik" by the Nazi).

So there are reasons to be skeptical towards universalism, even if it is esthetically attractive such as in the search for a Grand Unified Theory (for physics) and in the formal universes for mathematics and computer science of which we talked above. There are many others, equally attractive such as Feferman and Jäger's explicit mathematics and their universes [13].

How could one be skeptical about mathematics? Using the Zermelo-Fraenkel axioms of set theory ZFC, are we really sure that we fully understand infinite sets and their properties? Isn't set theory a kind of story that mathematicians tell among themselves, somewhat akin to fairy tales: Infinite sets are like the Easter Bunny which we have never actually seen but about some of whose properties

---

[7] The Plotkin-Scott-Engeler model.

[8] Understanding the model may be helped by considering sets of expressions $\alpha \longrightarrow x$ as partial and many-valued function from $G(A)$ to $G(A)$, namely as sets of pairs of arguments and values in $G(A)$. (If we so wish, we may also see these expressions as lists with head x and tail α).

[9] Reviewed in a recent survey [15].

we all agree; set theory also hides its precious gifts for us to search for them. Now, of course this is a doubtful kind of doubting. But still, on what basis rests the undeniable consent of mathematicians on these axioms? Some even doubt the power-set axiom for infinite sets, for example Gödel at one time, in his 1933 talk "The present situation in the foundations of mathematics" [9].

Mathematics is what all sufficiently patient learners and inventive practitioners of mathematics agree on. Let's fix on that and make it into a concrete model. Imagine mathematicians making thought-experiments on that part of the universe of sets that is immediately available to finite minds: the set of hereditarily finite sets (the first limit in the cumulative hierarchy, cf. Sect. 2). The goal of experiments is to verify the axioms of ZFC as expressed in first-order logic. Experiments are performed on the hereditarily finite sets by mentally executing programs based on the basic operations and basic predicates used in the axioms. The patience of such a mathematician may be measured by her or his willingness to test a universal quantifier by running through all sets up to rank n in the cumulative hierarchy. The inventiveness may be measured by the complexity n of programs that this mathematician creates to test an existential quantifier (by some measurement of the complexity of programs, only finitely many programs having smaller complexity than n). A mathematician of strength n in patience and inventiveness admits his individual theory ZFC_n of sets. Now consider ZFC_∞, the statements accepted by all sufficiently strong mathematicians. It turns out to contain ZFC.[10] Unfortunately (for logic?—for the model?), the set ZFC_∞, while it does not contain any theorem together with its negation, is not closed under classical logical deduction [16]. As were the intuitionists, we may be lead to doubt the reliability of classical logic when we approach infinite objects. How confident can we be?

Here is an extreme viewpoint: Some years ago van Dantzig [4] asked the question whether $10^{10^{10}}$ is a finite number, doubting that our mind is able to conceive this number as being built up by continued attaching individual marks to a given object (the intuitive basis of the number concept as proposed by Peano, Whitehead and Russell, Hilbert), or by a series of mental acts. Still, we seem to be content that $10^{10^{10}} + 10^{10^{20}} = 10^{10^{20}} + 10^{10^{10}}$, trusting a proof by induction that $a + b = b + a$ for all natural numbers. But does not such a proof beg the question in that it presumes in the induction step that local laws, verified in the small, persist in the large?

Again, this doubt is rather doubtful. But it points in another direction, that of the limits of the mind, of mental imagination.

In view of the exciting advanced of neurology, e.g. in explaining mechanisms of numerical abilities[11] it may be of interest to use mathematical models to approach the question on how and why the human brain can deal with natural numbers and conceive of them as a totality.

Let me present this in a somewhat idiosyncratic fashion; let us talk about thinking. Thinking means to apply thoughts to thoughts, thoughts being things

---

[10] Worked out in a 1971/78 paper by the author, reprinted in his collection Algorithmic Properties of Structures, World Publ.Co., 1990, pp. 87–95.

[11] Cf. the review by an originator of the idea of a number sense [7].

like concepts, impressions, memories, activities, projects—anything that you can think about, including mathematics. And, of course, the results of applying thoughts to thoughts. For example, applying the thought ⌜this object is even⌝ to the thought ⌜the number 14⌝ results in the thought ⌜14 is an even number⌝, a thought that happens to be true, (but this is of no concern.) We can also apply the thought ⌜this object is green⌝ to ⌜the number 14⌝, resulting in the possible, but rather unusual synesthetic thought ⌜the number 14 is green⌝. Thinking is free, all combinations of thoughts are admitted into the universe of thoughts. We arrive again at combinators, now talking about "thoughts" instead of "data".

The totality of thoughts may thus be understood as a combinatory algebra. In the following we let neurology suggest a model for the algebra of thoughts. Consider the brain as a set of connected neurons, firing at discrete times. The Neural Algebra sketched below describes the total activity of this neural net. Recall that for the graph model $\mathcal{D}_A$ (Sect. 4 above), we were free to choose the base set A. Let A therefore consist here of statements $a(t)$ expressing the fact that neuron a fires at time t. Assume we have knowledge of the firing history of some brain and of the underlying firing laws. These laws fix the causality of the activation of, say, neuron a_n at time t_n by neurons $a_0, \ldots, a_{n-1}$ connected to it and firing at (earlier) times $t_0, \ldots, t_{n-1}$ respectively. Let the expression $\{a_0(t_0), \ldots, a_{n-1}(t_{n-1})\} \longrightarrow a_n(t_n)$ denote this fact.

The construction analog to that of $\mathcal{D}_A$ above, now performed on these givens, produces a combinatory algebra $\mathcal{N}_A$, our Neural Algebra. Its elements are sets X each of which describes a time segment of the distributed activation history in respective parts of the brain. The sets X by construction consist of expressions denoting causal cascades of firings.[12]

By Crick's neurological hypothesis [2] all mental activities, perceptions, concepts, etc., "thoughts" for short, correspond to such activation histories and are therefore modeled here by elements in the algebra $\mathcal{N}_A$. Moreover, the operation of application $X \cdot Y$ describes the causal functioning of the brain: the processing X of an input Y, e.g. a visual input, produces $X \cdot Y$, the perception of an observed object. Remark: This is a much simplified version of the model used in the author's ongoing project on Neural Algebra. The full model ties more closely to the basic neural net, and allows to pass from objects in the algebra to underlying structures of the net.[13]

The brain model $\mathcal{N}_A$ allows us to speculate about the presence of the natural numbers in the human brain. First proposed by Dedekind [6], the infinity of natural numbers may be constructed "psychologically" as the following thought: Taking any thought object, e.g. the thought ⌜I am thinking of a number⌝. Reflecting on this thought is again a thought; and so on, yielding an infinity of thoughts.[14] In the Neural Algebra context there is no problem: We may assume that the element N_0

---

[12] To interpret expression in X as cascades, transform subexpressions such as $\{a\} \longrightarrow (\{b, c\} \longrightarrow d)$ successively by absorption into $\{a, b, c\} \longrightarrow d$.

[13] Diverse papers available online from the author's website, cf. "Neural Algebra".

[14] Proof of theorem 66; dismissed by the critical comment of Emmy Noether, one of the editors.

of $\mathcal{N}_\mathcal{A}$ represents thinking of the number zero, and the element R represents the mental operation of ⌜reflect on a thought⌝. Then Dedekind's construction goes like this: $N_{i+1} = N_i \cup R \cdot N_i, i = 0, 1, \ldots$. Collecting up, this gives the thought N of the totality, namely $N = \bigcup_i N_i$ which solves the recursion equation $N = R \cdot N$ with initial condition N_0. This N is a legitimate object of $\mathcal{N}_\mathcal{A}$ and stands for ⌜I am thinking of the totality of numbers⌝. It is a recursively defined object; not an easy subject, for many people it is too close to a vicious circle. The difficulty is that N is an infinite set and can therefore not be totally present in a finite brain restricted to finite time; the brain would be "lost in thought". At best N is present as an approximate thought, an illusion. Jumping a few steps to include other concepts and operations of arithmetic as objects in $\mathcal{N}_\mathcal{A}$, what we then are "really" able to conceive about arithmetic remains a sort of illusion, strangely convincing.

Shall we be content with that? Will we have to accept the judgement of the Erdgeist in Goethe's Faust?[15]

The Spirit of Nature, called up by the polymath Doctor Faustus by powerful formulas, spoke thus:

> "You grasp the mind
> you understand,
> not mine."[16]

With all respect for (Goethe's) Nature: certainly not!

6   Rejoinder

It is a good policy, unfortunately rarely followed, to be skeptical of one's own models.

In view of the missing consistency proof, is set theory and indeed most of the mathematical enterprise, just dogmata implanted in the brains of those students that we did not fail in our courses? Are we truly only guided by experiences in the strictly finite and reachable? This goes counter to all experience. The thriving industry of mathematics, pure, applied and expanding into all sciences is evidence enough. What seems to be at work here is what Martin Davis calls "Pragmatic Platonism" [5], which guides the mathematician's "Anschauung" into uncharted realms.

Mathematicians, like all scientists, are legitimized to use whatever technical means are available to aid the naked eye and brain. Otherwise they would be like savages, frightened by the telescope. Of course there are limits. Proof-finding algorithms (first satirized by Swift in Gulliver's Travels) are largely chimeric, and checking out otherwise inaccessible structures and highly complex and long proofs on their computer may leave some colleagues wondering about the reliability of coding and supporting software. But it pays to push these limits.

I refuse to believe that numbers are some sort of evanescent ghosts in our biological brains. Such beliefs were called "the most ridiculous" by Gödel.[17] The

---

[15] J.W.v.Goethe, Faust, Der Tragödie erster Teil, Nacht.

[16] Free translation by this author. Original: "Du gleichst dem Geist den Du begreifst, nicht mir".

[17] In a letter to A. Robinson, [10].

interpretation of N proposed above as the view we have of the natural numbers (as an object in $\mathcal{N}_A$) is simplistic: By combining thought objects we may form the thought object of composing this N. This is an example of generating an abstract entity as a manipulable object in $\mathcal{N}_A$. Such objects are present in our minds in a vein similar to the thoughts about the thoughts of another person: objects that are clearly present in our mind but not completed there. This is what I think the Pragmatic Platonist experiences as the "strangely convincing" presence of N mentioned above.

To conclude: Mathematics is a wonderful cultural treasure, shared and expanding by many and pursued pragmatically, without prejudice and mindfully. Universalist tendencies, with moderation and caution, help us to appreciate its coherence and beauty.

References

1. Agazzi, E. (ed.): The Problem of Reductionism in Science. Kluwer Academic Publishers, Dordrecht (1991)
2. Crick, F.: The Astonishing Hypothesis. Simon and Schuster Ltd., London (1994)
3. Curry, H.B.: Grundlagen der kombinatorischen Logik, Am. J. Math. 51, 509–536, 789–834 (1930)
4. van Dantzig, D.: Is $10^{10^{10}}$ a finite number? Dialectica 9, 273–277 (1955)
5. Davis, M.: Pragmatic Platonism. In: Friedman, M., Tennant, N. (eds.) Foundational Adventures Essays in Honor of Harvey. College Publications, London (2014)
6. Dedekind, R.: Was sind und was sollen die Zahlen? reprinted in Dedekind's collected works, Braunschweig (1932)
7. Dehaene, S., et al.: Arithmetic and the brain. Curr. Opin. Neurobiol. 14, 218–224 (2004)
8. Engeler, E., et al.: The Combinatory Programme. Birkhäuser Basel, Boston (1995)
9. Feferman, S., Gödel, K. (eds.): Collected Works, vol. III, p. 50. Oxford University Press, Oxford (2003). comments by S.F. p. 39
10. Feferman, S., Gödel, K. (eds.): Collected Works, vol. V, p. 204. Oxford University Press, Oxford (2003)
11. Gray, L.: A mathematician looks at Wolfram's new kind of science. Notices AMS 50, 200–211 (2003)
12. Hilbert, D.: Ueber das unendliche. Math. Ann. 95, 161–190 (1926)
13. Jäger, G., et al.: Universes in explicit mathematics. Ann. Pure Appl. Logic 109, 141–162 (2001)
14. Meseguer, J.: Twenty years of rewriting logic. J. Logic Algebraic Program. 81, 721–778 (2012)
15. Pelayo, A., Warren, M.A.: Homotopy type theory and Voevodsky's univalent foundations. Bulletin A.M.S. 51, 597–648 (2014)
16. Welti, E.: Die Philosophie des Strikten Finitismus. Peter Lang Verlag, Bern (1981)
17. Wolfram, S.: A New Kind of Science. Wolfram Media, Champaign (2002)
18. Zermelo, E.: Ueber Grenzzahlen und Mengenbereiche. Fund. Math. 16, 29–47 (1930)

When Is a Formula a Loop Invariant?

Stephan Falke and Deepak Kapur$^{2(\boxtimes)}$

^{1} Institute for Theoretical Computer Science, KIT, Karlsruhe, Germany
stephan.falke@gmail.com
^{2} Department of Computer Science, University of New Mexico, Albuquerque, USA
kapur@cs.unm.edu
^{3} Aicas GmbH, Karlsruhe, Germany

Dedicated to José Meseguer on his 65th birthday.

Abstract. Invariant properties at various program locations play a critical role in enhancing confidence in the reliability of software. A procedure for checking whether a given set of formulas associated with various program locations is an invariant or not is proposed. The procedure attempts to check whether the formulas are preserved by various program paths, in which case it declares the formulas to be invariant; otherwise, it attempts to strengthen them based on verification conditions generated from the program paths that did not preserve the formulas. This iterative process is continued until a verification condition along some path cannot be satisfied or an initial state of the program violates the formulas strengthened thus far. It is shown that under certain conditions, for certain theories including conjunctions of polynomial equalities, the procedure terminates, either declaring the input set of formulas to be not invariant, or generating a strengthened set of inductive invariant formulas. For other theories including Presburger arithmetic, the procedure may not terminate in general; heuristics are proposed for such cases for approximating strengthening of formulas to ensure termination. There is a direct relationship between this approach for checking formulas to be invariant and the k-induction approach for verifying properties to be k-inductive. This relationship is explored in depth.

1 Introduction

Specifying and ensuring program invariants are an excellent way to enhance confidence in the reliability of software. Static type checking is one way to check simple but useful invariants of expressions in a program for many programming languages. Static program analysis, which attempts to detect errors such as divide by zero, array index out of bounds, null pointer dereference, buffer

Preliminary results reported in this paper were first presented at *International Conference on Mathematics Mechanization–in Honor of Prof. Wen-Tsun Wu's 90th birthday (ICMM), 2009, Beijing, China,* and also at the Dagstuhl Seminar *Deduction and Arithmetic,* October 2013. This research has been partially supported by the NSF awards CCFf-1248069 and DMS-1217054 as well as a grant for Visiting Professorship for Senior International Scientists, the Chinese Academy of Sciences.

N. Martí-Oliet et al. (Eds.): Meseguer Festschrift, LNCS 9200, pp. 264–286, 2015.
DOI: 10.1007/978-3-319-23165-5_13

overflow, etc., is considered particularly useful since many system crashes are caused by such bugs in software. Loop invariants are of particular interest as they help in understanding the program behavior. Even though automatically generating invariants of programs is considered a very difficult problem, considerable progress has been made in developing heuristics for automatically generating loop invariants both by static and dynamic analyses of software, when invariants are formulas from certain logical theories.

There are at least two kinds of program invariants associated with specific program locations: (i) simple invariant properties which hold whenever control reaches such a program location, and (ii) inductive invariants, which are properties that are true the first time the location is reached and are preserved by every program path through that location. While inductive invariants are easier to check, simple invariants are often easier to state by a programmer and include in a specification.

Given a program written in a programming language for which an axiomatic semantics in the form of Floyd-Hoare style proof rules can be given, checking whether a formula is an inductive invariant is decidable under the assumption that the theory in which verification conditions get expressed is decidable. In general, this is however a semidecidable problem since the underlying logical theory may only be semi-decidable. Much to one's surprise, checking whether a formula is an invariant seems to be inherently even more difficult.

In this paper, we propose an approach for checking whether a given formula is an invariant at a program location by attempting to strengthen it so that the strengthened formula is an inductive invariant. This also enables verifying whether a programmer indeed annotated a program correctly, or whether the program behavior correctly captures the programmer's intent. We focus on invariants expressed as conjunctions of atomic formulas from a logical theory. The proposed procedure is iterative: it attempts to check whether a given formula associated with a program location is an inductive invariant or not by examining whether every basic cycle through that location preserves the formula. If the procedure succeeds, it declares the formula to be invariant; otherwise, for every cycle that does not preserve the formula, it attempts to strengthen the formula by adding additional atomic formulas from the logical theory in an attempt to fix lack of preservation of the formula through the cycle(s) violating it. Depending upon how this strengthening is attempted, the procedure can also sometimes determine whether the original formula is not an invariant. In case the procedure declares an input formula to be an invariant for a given program location, it also outputs a possibly strengthened formula which is an inductive invariant for that program location and implies the original input formula.

For logical theories including propositional calculus, quantifier-free theories of polynomial equalities and UTVPI [11] (also called octagonal constraints), it is proved that the procedure is a decision procedure under certain conditions, i.e., it declares whether an input formula is an invariant or not. When an invariant is a conjunction of linear inequalities (over rationals, integers or reals), the proposed procedure may not terminate. Heuristics are proposed to ensure termination of the procedure by computing weak strengthenings of hypothesized invariants. The second half of the paper is focused on relating the proposed procedure for strengthening a formula in an attempt to make it to be inductive to the k-induction method for verifying invariants. It is shown that under certain conditions, if the procedure for strengthening a formula does not use approximations, then the proposed procedure terminates if and only if the k-induction method for checking whether the formula is an invariant succeeds.

2   Examples

We illustrate the problem formulation and the proposed approach using simple illustrative examples informally presented at the source code level.

Example 1. Consider the following program fragment:

$(x, y, z) \leftarrow (1, 1, 0)$
while $x \leq n$
   do $(x, y, z) \leftarrow (x + y + 2, y + 2, z + 1)$

The formula $x = (z+1)^2$ can be shown to be a loop invariant. It is, however, not an inductive loop invariant since the verification condition $(x = (z+1)^2 \wedge x \leq n) \implies x + y + 2 = (z+2)^2$ does not hold. The left-hand side of this implication can be used to simplify its right-hand side by replacing z^2 by $x - 2*z - 1$ and subsequent algebraic simplifications. The right-hand side then becomes $y = 2*z + 1$. Adding this equality as a conjunct to the loop invariant produces $x = (z+1)^2 \wedge y = 2*z + 1$. Since

$$(x = 1 \wedge y = 1 \wedge z = 0) \implies (x = (z+1)^2 \wedge y = 2*z + 1),$$
$$(x = (z+1)^2 \wedge y = 2*z + 1) \implies (x + y + 2 = (z+2)^2 \wedge y + 2 = 2*(z+1) + 1),$$

this conjunction, which is a strengthening of the original formula $x = (z+1)^2$, is an inductive loop invariant implying that $x = (z+1)^2$ as well as $y = 2*z + 1$ are indeed invariants.   ◇

As the reader would have noticed, it was possible in this case to strengthen the original formula such that the strengthened formula is inductive. The strengthening was computed from the simplification of the verification condition arising from the loop body with the expectation that this strengthening would establish the verification condition. This process ended in 2 iterations but that need not be the case as illustrated by the example below.

Example 2. Consider the following program fragment:

$(x, y, z) \leftarrow (0, 0, 0)$
while TRUE
   do $(x, y, z) \leftarrow (x + 1, y + 1, z + x - y)$

Then $z \leq 0$ is a loop invariant. Since $z \leq 0 \not\Rightarrow z+x-y \leq 0$, it is not inductive. It can, however, be strengthened to $z \leq 0 \wedge x - y \leq 0$ which is true initially as well as preserved by the body of the loop. ◇

The reader would notice that in the first example, under the assumption $(x = (z + 1)^2 \wedge x \leq n)$, $x + y + 2 = (z + 2)^2$ is equivalent to $y = 2 * z + 1$. However, in the second example, under the assumption $z \leq 0$, $z + x - y \leq 0$ is not equivalent to $x - y \leq 0$ but rather $x - y \leq 0 \implies z + x - y \leq 0$. The use of $x - y \leq 0$ is an example of weak (or approximate) strengthening of the original formula $z \leq 0$. How such an approximate strengthening can be obtained automatically is discussed in Sect. 4. If we had just strengthened the original formula with the conclusion in the verification condition: $z + x - y \leq 0$ without approximating it by $x - y \leq 0$, adding such strengthenings $z + k(x - y) \leq 0, k \geq 1$ would not terminate.

Another observation to make from the above example is that $z = 0$, which is stronger than $z \leq 0$ is also an invariant; this can be checked as in the first example. The proposed procedure would compute a strengthening $z = 0 \wedge x - y = 0$, which is stronger than $x - y \leq 0$. But it is unable to first generate the strengthening $z = 0$ of the original input formula $z \leq 0$. As discussed later, the proposed approach attempts to generate the weakest strengthening of the input formula that is inductive.

Consider slight variations of the above examples which illustrate the cases when the input formulas are not invariants.

Example 3. Consider a slight variation of the first example in which the assignment to y is changed to $y + 1$.

$$(x, y, z) \leftarrow (1, 1, 0)$$
while $x \leq n$
$\quad$ **do** $(x, y, z) \leftarrow (x + y + 2, y + 1, z + 1)$

The formula $x = (z + 1)^2$ is not invariant any more. The verification condition of the body of the loop generated using $x = (z + 1)^2$ remains the same; thus $x = (z+1)^2$ is strengthened as in the first example to $x = (z+1)^2 \wedge y = 2 * z + 1$. However, the verification condition using this formula now is:

$$(x = (z+1)^2 \wedge y = 2*z+1 \wedge x \leq n) \implies (x+y+2 = (z+2)^2 \wedge y+1 = 2*(z+1)+1),$$

which is not valid. It is then possible to claim that the input formula is not an invariant of the above program. In fact, a test case can be generated from a satisfying assignment for the negated verification condition which invalidates the alleged invariant. ◇

Similarly, consider the following variation of the second example.

Example 4. Is $x + y + z = 0$ an invariant of the loop from Example 2? The verification condition generated from the body of the loop is

$$x + y + z = 0 \implies x + 1 + y + 1 + z + x - y = 0$$

which is not valid. Simplification leads to a strengthening of the input formula by $x - y + 2 = 0$; however, the initial state $x = 0, y = 0, z = 0$ does not satisfy the strengthened formula any longer, implying that it is not an invariant and hence the input formula $x + y + z = 0$ is also not an invariant.   ◇

3   Preliminaries

Instead of using a specific programming language, we use a more abstract framework of a transition system to which many programs can be translated easily.[1]

We use a signature Σ consisting of function symbols Σ^F and predicate symbols Σ^P. A Σ-model $\mathcal{A}$ consists of a non-empty set A (the *universe*) and a mapping $(\_)^{\mathcal{A}}$ assigning to each constant symbol $a \in \Sigma^F$ an element $a^{\mathcal{A}} \in A$, to each function symbol $f \in \Sigma^F$ of arity $n > 0$ a total function $f^{\mathcal{A}} : A^n \to A$, to each propositional symbol $B \in \Sigma^P$ and element $b^{\mathcal{A}} \in \{\mathbf{true}, \mathbf{false}\}$, and to each $p \in \Sigma^P$ of arity $n > 0$ a total function $p^{\mathcal{A}} : A^n \to \{\mathbf{true}, \mathbf{false}\}$. The mapping $(\_)^{\mathcal{A}}$ extends to an interpretation of terms and formulas in the usual way.

Given a signature Σ, we fix a Σ-theory Th, which is just a Σ-model $\mathcal{A}$. We denote by Φ a (yet undetermined) quantifier-free logic (i.e., subset of quantifier-free formulas over Σ) for this theory.

Definition 5 (Transition Systems). *A transition system (over the logic Φ) $\mathcal{S} = \langle X, \mathcal{L}, l^0, \tau^0, \mathcal{T} \rangle$ consists of:*

- *a set $X = \{x_1, \ldots, x_n\}$ of variables,*
- *a set $\mathcal{L}$ of locations,*
- *a start location $l^0 \in \mathcal{L}$,*
- *a formula $\tau^0 \in \Phi$ with $\mathcal{V}(\tau^0) \subseteq X$ that restricts the initial values of the variables, and*
- *a set $\mathcal{T}$ of transitions of the form $\langle l_1, \tau, \sigma, l_2 \rangle$, where $l_1, l_2 \in \mathcal{L}$ are locations, $\tau \in \Phi$ is a formula with $\mathcal{V}(\tau) \subseteq X$, and $\sigma : X \to \mathit{Terms}(\Sigma^F, X)$ is a variable update.*

The transition system $\mathcal{S}$ is conjunctive if τ^0 and τ (for all $\langle l_1, \tau, \sigma, l_2 \rangle \in \mathcal{T}$) are conjunctions of atomic formulas in Φ.

A configuration of $\mathcal{S}$ has the form $\langle l, \boldsymbol{v} \rangle$ for a location $l \in \mathcal{L}$ and a valuation $\boldsymbol{v} : X \to A$.[2] We write $\langle l_1, \boldsymbol{v_1} \rangle \to_t \langle l_2, \boldsymbol{v_2} \rangle$ for an evaluation step with a transition $t = \langle l_1, \tau, \sigma, l_2 \rangle$ if $\boldsymbol{v_1}$ satisfies τ and $\boldsymbol{v_2} = \sigma \boldsymbol{v_1}$ (i.e., $\boldsymbol{v_2}(x) = \boldsymbol{v_1}(\sigma(x))$ for all $x \in X$. We drop the subscript if we do not care about the used transition. A run of $\mathcal{S}$ is a (finite or infinite) sequence $\langle l_0, \boldsymbol{v_0} \rangle \to \langle l_1, \boldsymbol{v_1} \rangle \to \langle l_2, \boldsymbol{v_2} \rangle \ldots$ of evaluation steps such that $l_0 = l^0$ is the start location and the valuation $\boldsymbol{v_0}$ satisfies the formula τ^0.

The transition system $\mathcal{S}$ is unconditional if $\tau = \top$ for all $\langle l_1, \tau, \sigma, l_2 \rangle \in \mathcal{T}$. In this paper, $\top$ denotes true and $\bot$ denotes false.

---

[1] Provided a sufficiently expressive theory is used.
[2] Here, $\boldsymbol{v}$ extends to terms in the usual way.

A transition system $\mathcal{S}' = \langle X, \mathcal{L}, l^0, \tau'^0, \mathcal{T}' \rangle$ is an *abstraction* of another transition system $\mathcal{S} = \langle X, \mathcal{L}, l^0, \tau^0, \mathcal{T} \rangle$ iff (i) $\tau^0 \implies \tau'^0$ and (ii) for every $\langle l_1, \tau, \sigma, l_2 \rangle \in \mathcal{T}$, there is a corresponding $\langle l_1, \tau', \sigma, l_2 \rangle \in \mathcal{T}$ such that $\tau \implies \tau'$. In particular, it is possible to define an unconditional abstraction of any transition system by just dropping all the guards τ.

Definition 6 (Invariant Map). *For a given transition system, a formula map is a function $\mathcal{IM} : \mathcal{L} \to \Phi$ such that $\mathcal{V}(\mathcal{IM}(l)) \subseteq X$ for each location l.*

A formula map $\mathcal{IM}_2$ strengthens a formula map $\mathcal{IM}_1$ if $\mathcal{IM}_1(l) \implies \mathcal{IM}_2(l)$ for all locations l.

A formula map $\mathcal{IM}$ is an invariant map *if for each run $\langle l_0, \boldsymbol{v_0} \rangle \to \langle l_1, \boldsymbol{v_1} \rangle \to \ldots$ and each configuration $\langle l_i, \boldsymbol{v_i} \rangle$ occurring in this run, the valuation $\boldsymbol{v_i}$ satisfies the formula $\mathcal{IM}(l_i)$.*

An invariant map $\mathcal{IM}$ is inductive *if*

- $\tau^0 \implies \mathcal{IM}(l^0)$, *implying that initial states satisfy the invariant map and*
- $\mathcal{IM}(l_1) \wedge \tau \implies \mathcal{IM}(l_2)\,\sigma$, *the verification condition, for each transition $\langle l_1, \tau, \sigma, l_2 \rangle$, where $\mathcal{IM}(l_2)\,\sigma$ stands for updating variables in $\mathcal{IM}(l_2)$ by σ.*

It is easy to see that if $\mathcal{IM}$ is an inductive invariant map of an unconditional $\mathcal{S}$, then it is also an inductive invariant map of any $\mathcal{S}'$ of which $\mathcal{S}$ is an unconditional abstraction.

4   Inductive Strengthening Procedure

We present below a procedure for strengthening a given input formula map $\mathcal{IM}$ with the objective of computing an inductive invariant map that strengthens $\mathcal{IM}$ or declaring that $\mathcal{IM}$ is not an invariant map. The procedure is recursively invoked on a strengthened formula map until

- executions of all transitions preserve it, thus declaring the strengthened formula map to be an inductive invariant map and hence the original input formula map to be an invariant map, or either
- the strengthened formula map does not hold at the initial location (i.e., for the initial state(s)) or
- a verification condition for some transition is invalid.

In the last two cases, the strengthened formula map is not an invariant map, thus suggesting that the original input formula map may also not be an invariant map in case approximations are made in strengthening the formula map generated during the procedure. The strengthening procedure is not guaranteed to terminate either.

STRENGTHEN($\mathcal{IM}$)

1   **if** $\tau^0 \not\implies \mathcal{IM}(l^0)$
2   **then return** $\bot$ ▷ cannot compute inductive strengthening
3   **elseif** $\mathcal{IM}(l_1) \wedge \tau \not\implies \mathcal{IM}(l_2)\,\sigma$ for some transition $\langle l_1, \tau, \sigma, l_2 \rangle$
4   **then** construct ψ such that $\mathcal{IM}(l_1) \wedge \tau \wedge \psi \implies \mathcal{IM}(l_2)\,\sigma$
5   **return** STRENGTHEN($\mathcal{IM}[l_1 \mapsto \mathcal{IM}(l_1) \wedge \psi]$)   ▷ recursive call
6   **else return** $\mathcal{IM}$ ▷ $\mathcal{IM}$ is inductive

In the above procedure, line 1 tests whether the input formula map $\mathcal{IM}$ holds in the initial state(s); line 3 tests the verification condition corresponding to a transition from a location l_1 to a location l_2 (where l_2 could be l_1). The reader should notice that the procedure is always attempting to strengthen its input by conjoining an additional formula since the objective is to show that the original formula map is invariant. We thus do not attempt to try formulas obtained by weakening the input by disjuncting with formulas. If the procedure terminates resulting in a formula map as its output, the formula map is then an inductive invariant of the transition system; further, it implies all formula maps, including the original input, in the recursive calls and thus they are also invariants.

The main challenge in the strengthening procedure is to compute appropriate formulas ψ in line 4 with the following objectives:

1. The procedure should terminate on a large class of input formula maps, i.e., for every location l, the sequence of formulas $\mathcal{IM}(l), (\mathcal{IM}(l) \wedge \psi_1), (\mathcal{IM}(l) \wedge \psi_1) \wedge \psi_2, \ldots$ terminates, and it is desired as well that it takes only a few steps to converge.
2. **Soundness:** If the input to the procedure is not an invariant, it should terminate declaring so.
3. **Completeness:** If the input is an invariant, then it should terminate declaring so, as well as produce an inductive invariant map.

Below we discuss some possible ways to compute ψ from $\mathcal{IM}(l_1)$, τ, and $\mathcal{IM}(l_2)\,\sigma$. Leaving aside the case of making ψ false, which does not give any information, the following possibilities are related to each other.

I. *Trivial (I)*: Choose $\psi = \mathcal{IM}(l_2)\,\sigma$. While this strategy is attractive given that ψ is likely to have the same form as $\mathcal{IM}$, for most logical theories in which the formula map is specified, this strategy is not likely to terminate. However, as discussed below, many interesting properties about STRENGTHEN($\mathcal{IM}$) can be proved using this strategy.

II. *Conditional (II)*: $\psi = (\tau \implies \mathcal{IM}(l_2)\,\sigma)$. This strategy, while quite powerful, is not attractive since it requires strengthening using Horn clauses instead of atomic formulas. If invariants are specified as Horn clauses, then this strategy can be attractive as well. However, this strategy also suffers like the previous strategy that it is not likely to terminate.

III. *Equivalent (III)*: ψ is an equivalent simplified form of $\mathcal{IM}(l_2)\,\sigma$ in the concrete logic Φ under the context of $\mathcal{IM}(l_1) \wedge \tau$, i.e., $\mathcal{IM}(l_1) \wedge \tau \implies (\mathcal{IM}(l_2)\,\sigma \iff \psi)$. Making $\psi = \mathcal{IM}(l_2)\,\sigma$ as in Strategy I or $(\tau \implies \mathcal{IM}(l_2)\,\sigma$ as in Strategy II in the absence of any simplification is always possible, implying that this strategy subsumes Strategies I and II.

In a concrete logic Φ, $\mathcal{IM}(l_2)\,\sigma$ can typically be simplified using the context, but the kind of equivalent simplified forms as well as the simplification process are dependent upon Φ. It is hoped that simplification would be helpful especially in establishing the termination of STRENGTHEN($\mathcal{IM}$).

For these first three strategies, we will show later that under certain conditions, if the execution of STRENGTHEN($\mathcal{IM}$) leads to a strengthening of the input formula map that does not hold for the initial state, then the strengthened formula map is not an invariant; furthermore, it implies that the original formula map given as the input as well as subsequent inputs to the recursive calls STRENGTHEN($\mathcal{IM}$) are not invariants.

The fourth strategy has a different goal:

IV. *Approximating* $\mathcal{IM}(I_2)$ *w.r.t.* $\mathcal{IM}(l_1) \wedge \tau$ *(IV)*: This strategy is conservative in contrast to the first three strategies with the main focus on ensuring termination of STRENGTHEN($\mathcal{IM}$) by sacrificing completeness. Instead of computing a formula equivalent to $\mathcal{IM}(l_2)$, the context $\mathcal{IM}(l_1) \wedge \tau$ is used in the concrete logic Φ to derive a nontrivial formula ψ (other than false) such that $\mathcal{IM}(l_1) \wedge \tau \implies (\psi \implies \mathcal{IM}(l_2)\sigma)$. It is also desired that computations of approximate ψ's take only a few steps to converge.
It is especially useful when Φ is a quantifier-free theory of conjunctive linear inequalities. This is discussed later in Sects. 5.2 and 6.1.
Because of approximations made in this strategy, it may not be possible to declare from the approximated formula not being true in the initial state that the original formula map is not an invariant.

The procedure STRENGTHEN($\mathcal{IM}$) is said to succeed on $\mathcal{IM}$ iff (i) it terminates and (ii) either (a) declares $\mathcal{IM}$ to not be an invariant, or (b) produces a strengthened formula map $\mathcal{IM}'$ that implies $\mathcal{IM}$ and is inductive; the outcome (ii (b) implies that $\mathcal{IM}$ is an invariant. STRENGTHEN($\mathcal{IM}$) is said to fail on $\mathcal{IM}$ if (i) it either does not terminate or (ii) terminates at Line 2 because a formula map does not hold in the initial state but it cannot be asserted that the formula map is not an invariant due to approximations made.

For a given concrete logic Φ, it is assumed below that for each transition, the hypotheses and the conclusion in a verification condition generated due to any transition in a transition system is also in Φ; in particular, we will assume that each τ_i as well as each $\mathcal{IM}(l_i)$ associated with any location l_i is a conjunction of atomic formulas in Φ; furthermore, $\mathcal{IM}(l_i)\sigma$ is also a conjunction of atomic formulas. Otherwise, it would become necessary to approximate (abstract) guards as well as verification conditions.[3]

5   Application to Specific Logical Theories

In this section, we show how STRENGTHEN($\mathcal{IM}$) can be implemented for some logics. It is easy to see that in a logic in which there are only finitely many nonequivalent formulas constructed from a fixed vocabulary, e.g., propositional logic, the strengthening procedure can be made to terminate using any of the strategies. For example, for a boolean program for a hardware circuit description [5], even when $\mathcal{IM}(l_2)\sigma$ itself is used to strengthen a formula map (i.e., ψ in the strengthening procedure is $\mathcal{IM}(l_2)\sigma$), the procedure terminates given that

---

[3] What approximations are necessary and in what cases they must be performed depends on the manipulation needed to compute ψ in the strengthening procedure.

there are only finitely many states. It is further possible to simplify $\mathcal{IM}(l_2)\,\sigma$ using Strategy III [15].

Below, we discuss two logics which have infinitely many nonequivalent formulas: (i) the quantifier-free logic of conjunctive polynomial equalities over an algebraically closed field, and (ii) the quantifier-free logic of linear arithmetic. It can be proved that for the quantifier-free logic of UTVPI constraints (also known as octagonal domains [18]), the STRENGTHEN($\mathcal{IM}$) procedure terminates; we do not provide details because of space limitations.

5.1 Polynomial Equalities

Consider the theory Φ of the quantifier-free conjunctive logic of polynomial equalities over $\mathbb{Q}$. Thus, atoms have a form $p = 0$ for a polynomial p over program variables. Invariants under consideration are conjunctions of polynomial equalities. For this theory, algorithms from computer algebra can be effectively used; satisfiability check as well as simplification needed for various strengthening strategies above can be performed using Buchberger's Gröbner Basis algorithm [6]. We briefly review basic concepts related to Gröbner basis computation; more details can be found in [8]. For the application of Gröbner basis algorithm to automated reasoning, the reader may consult [12,15]; the use of Gröbner basis algorithm for generating loop invariants is discussed in [14,21,22].

A Gröbner basis G of a finite set F of multivariate polynomials over the rationals is a special finite basis of the ideal generated by F with the following properties: (i) every polynomial simplifies to a unique normal form when the polynomials in G are viewed as rewrite rules with their left hand sides being the largest monomial with respect to an admissible term ordering; the normal form of every polynomial in the ideal generated by F (and hence G) is 0, (ii) the polynomials in F do not have a common solution (i.e., the formula ($f_1 = 0 \wedge \cdots \wedge f_k = 0$), where $F = \{f_1, \cdots, f_k\}$, is unsatisfiable) over complex numbers if and only if G includes 1. Most importantly, G can be computed from F using Buchberger's completion algorithm. G can also be used to compute the number of common solutions of F.

A Gröbner basis algorithm decides satisfiability of a finite conjunction of polynomial equalities over the complex numbers. If a formula is shown to be unsatisfiable by a Gröbner basis algorithm, it is unsatisfiable over integers, reals or rationals as well but the converse is not true. Further, ($f_1 = 0 \wedge \cdots \wedge f_k = 0$) $\implies$ ($g = 0$) can be simplified to and is equivalent to ($g_1 = 0 \wedge \cdots \wedge g_l = 0$) $\implies$ ($\bar{g} = 0$), where $G = \{g_1, \cdots, g_l\}$ is a Gröbner basis of $\{f_1, \cdots, f_k\}$ and $\bar{g}$ is the normal form of g using G. Strategies requiring simplification of $\mathcal{IM}(l_2)\,\sigma$ (w.r.t. $\mathcal{IM}(l_1) \wedge \tau$ or just τ) can be effectively implemented using a Gröbner basis algorithm.

We assume below that in a transition system, all formulas – the initial state, guards, formula maps are conjunctions of polynomial equalities, and assignments are assumed to be polynomials expressions.

A formula map holds in an initial state iff the polynomial in every polynomial equality at l_0 reduces to 0 using a Gröbner basis of the polynomials in the polynomial equalities specifying the initial state τ^0. Similarly, a verification condition $\mathcal{IM}(l_1) \wedge \tau \implies \mathcal{IM}(l_2)\,\sigma$ is valid iff the polynomials in polynomial equalities in $\mathcal{IM}(l_2)\,\sigma$ reduce to 0 using a Gröbner basis of polynomials appearing in $\mathcal{IM}(l_1) \wedge \tau$.[4]

To illustrate, let us revisit the first example discussed in the introduction. Since the loop condition is not a polynomial equality, it is abstracted to be true. The formula map $x = (z+1)^2$ gives the verification condition:

$$x = (z+1)^2 \implies (x+y+2) = ((z+1)+1)^2.$$

The Gröbner basis of $\{x = (z+1)^2\}$ is itself using the lexicographic ordering in which $x > y > z$; it can be used to simplify the conclusion giving $y - 2*z - 1 = 0$ which cannot be simplified any further. The second call to STRENGTHEN takes the input $x = (z+1)^2 \wedge y = 2*z + 1$, producing the verification condition:

$$(x = (z+1)^2 \wedge y = 2*z+1) \implies (x+y+2 = ((z+1)+1)^2) \wedge (y+2 = 2*(z+1)+1).$$

The validity of this verification condition follows from the Gröbner basis of $\{x = (z+1)^2, y = 2*z+1\}$ which is again itself; simplifying both polynomial equations in the conclusion gives true.

Theorem 7. *If Strategy III is implemented using Gröbner Bases as described above, then* STRENGTHEN$(\mathcal{IM})$ *terminates.*

Proof (sketch). In order to simplify presentation, we assume for now that $|\mathcal{L}| = |\mathcal{T}| = 1$. The proof extends to the general case without technical complications.

Assume that STRENGTHEN$(\mathcal{IM})$ does not terminate. Then, we obtain an infinite sequence

$$\mathcal{IM}(l)$$
$$\mathcal{IM}(l) \wedge \psi_1$$
$$\mathcal{IM}(l) \wedge \psi_1 \wedge \psi_2$$
$$\vdots$$

of conjunctions of polynomial equalities. Now consider the ideals generated by these polynomial equalities (by abuse of notation, we write $\langle \varphi \rangle$ for the ideal generated by the polynomials p occurring in an atom $p = 0$ in the conjunctive formula φ):

$$\langle \mathcal{IM}(l) \rangle$$
$$\langle \mathcal{IM}(l) \wedge \psi_1 \rangle$$
$$\langle \mathcal{IM}(l) \wedge \psi_1 \wedge \psi_2 \rangle$$
$$\vdots$$

---

[4] Strictly speaking, Gröbner basis of the radical ideal of the polynomials is computed for satisfiability check, since $x^2 = 0 \implies x = 0$ is valid.

Clearly, $\langle \mathcal{IM}(l) \rangle \subseteq \langle \mathcal{IM}(l) \wedge \psi_1 \rangle \subseteq \langle \mathcal{IM}(l) \wedge \psi_1 \wedge \psi_2 \rangle \subseteq \dots$. By construction, for each polynomial p occurring in an atom $p = 0$ in the conjunction ψ_i, we have $p \notin \langle \mathcal{IM}(l) \wedge \psi_1 \wedge \dots \wedge \psi_{i-1} \rangle$. Therefore, $\langle \mathcal{IM}(l) \rangle \subset \langle \mathcal{IM}(l) \wedge \psi_1 \rangle \subset \langle \mathcal{IM}(l) \wedge \psi_1 \wedge \psi_2 \rangle \subset \dots$ and we obtain an infinite ascending chain of ideals. This is impossible by Hilbert's basis theorem.   □

Corollary 8. *Any strengthening strategy that adds a finite set of polynomial equalities to any of the formulas in $\mathcal{IM}$ terminates.*

Proof. The proof follows from the same argument as used in the Theorem above using Hilbert's basis theorem.   □

From the above corollary, it follows that all strategies terminate for conjunctive theory of polynomials equalities.

Corollary 9. *If S is unconditional, then Strategy III using Gröbner Bases as described above decides whether $\mathcal{IM}$ is an invariant map.*

Proof. Easy consequence of Theorems 15 and 7. A formula map $\mathcal{IM}$ is not an invariant map iff during the execution of STRENGTHEN($\mathcal{IM}$), a formula map is generated which the initial state does not satisfy.   □

As shown later during discussing the relationship between STRENGTHEN and k-induction, if S is not unconditional, then if STRENGTHEN fails at Line 2, it cannot be asserted that the input formula is not an invariant.

5.2   Linear Arithmetic

Consider Φ to be the quantifier-free logic of linear arithmetic over $\mathbb{Z}$ (or $\mathbb{Q}$ or $\mathbb{R}$) with atomic formulas of the form $\left(\sum_{x \in X} a_x * x \right) + c \geq 0$, where $a_x, c \in \mathbb{Z}$. Furthermore, each x also ranges over $\mathbb{Z}$. It is unclear how simplification can be performed in this logic to implement Strategy III; augmenting $\mathcal{IM}(l_2)\,\sigma$ to the formula map as in Strategy I often leads to STRENGTHEN not terminating as is evident from the second example in the introduction: strengthening after k iterations generates a formula map $\bigwedge_{k=0}^{n} z + k(x - y) \leq 0$.

There can be, however, many different heuristics to perform approximation of $\mathcal{IM}(l_2)\,\sigma$ w.r.t. $\mathcal{IM}(l_1) \wedge \tau$, i.e., Strategy IV.

Assume two conjunctive formulas ξ and χ (w.l.o.g we can assume that $\xi = (p_1 \geq 0 \wedge \dots \wedge p_n \geq 0)$ and $\chi = (q_1 \geq 0 \wedge \dots \wedge q_m \geq 0)$ for linear polynomials p_i and q_j). To find ψ such that $\xi \wedge \psi \implies \chi$, consider a linear template for $\psi = \left(\left(\sum_{x \in X} a_x * x \right) + c \geq 0 \right)$ where the a_x and c are parameters whose value needs to be determined.[5] The objective is to find values for the a_x and c such that

$$p_1 \geq 0 \wedge \dots \wedge p_n \geq 0 \wedge \left(\sum_{x \in X} a_x * x \right) + c \geq 0 \implies q_1 \geq 0 \wedge \dots \wedge q_m \geq 0,$$

---

[5] It would also be possible to let ψ be a conjunction of such atoms.

This implication is equivalent to:

$$p_1 \geq 0 \wedge \ldots \wedge p_n \geq 0 \implies (q_1 \geq 0 \wedge \ldots \wedge q_m \geq 0) \vee \neg \left(\left(\sum_{x \in X} a_x * x \right) + c \geq 0 \right) \quad \text{and}$$

$$p_1 \geq 0 \wedge \ldots \wedge p_n \geq 0 \implies (q_1 \geq 0 \wedge \ldots \wedge q_m \geq 0) \vee \left(\left(\sum_{x \in X} -a_x * x \right) - c > 0 \right).$$

As discussed in [13,14], quantifier elimination in which program variables $x \in X$ are eliminated can be performed, resulting in constraints on parameters c, α_x such that every value of these parameters satisfying these constraints give a candidate ψ. Often, it is possible to find the strongest possible ψ if all solutions to the constraints on parameters can be finitely described.

Below we discuss a simple heuristic based on Farkas' lemma. Since there are versions of Farkas' lemma dealing with strict linear inequalities, this heuristic works on linear inequalities over rationals ($\mathbb{Q}$) and reals ($\mathbb{R}$) as well.

Computing Approximation Using Farkas' Lemma. The above implication follows from a conjunction of the following implications for $1 \leq i \leq m$.

$$p_1 \geq 0 \wedge \ldots \wedge p_n \geq 0 \implies q_i \geq 0 \vee \left(\sum_{x \in X} -a_x * x \right) - c > 0$$

The above problem can be further simplified using another heuristic by replacing a disjunction $a \geq 0 \vee b \geq 0$ by $a + b \geq 0$ since $a + b \geq 0 \implies (a \geq 0 \vee b \geq 0)$ (as done in [17]). Applying this heuristic on the above formulas results in stronger requirements (for each $1 \leq i \leq m$):

$$p_1 \geq 0 \wedge \ldots \wedge p_n \geq 0 \implies q_1 + \left(\sum_{x \in X} -a_x * x \right) - c > 0$$

$$\vdots$$

$$p_1 \geq 0 \wedge \ldots \wedge p_n \geq 0 \implies q_m + \left(\sum_{x \in X} -a_x * x \right) - c > 0$$

For illustration, consider the second example in the introduction. Starting with the formula $z \leq 0$, the verification condition corresponding to the body of the loop is $z \leq 0 \implies (z + x - y \leq 0)$ which is not valid. We wish to compute as general a ψ as possible such that $(z \leq 0 \wedge \psi) \implies (z + x - y \leq 0)$. Applying the above heuristic, we have $z \leq 0 \implies ((z + x - y \leq 0) \vee (-az - bx - cy - d < 0))$, where $\psi : (az + bx + cy + d) \leq 0$. Eliminating $\vee$ using the heuristic that $(m + n \leq 0) \implies (m \leq 0 \vee n \leq 0)$, we have $z \leq 0 \implies (((-a+1)z + (-b+1)x + (-c-1)y) - d < 0$, from which $a = 0, b = 1, c = -1, d > 0$ can be derived as a possible solutions, giving $\psi : x - y \leq 0$. Examples in [3] which cannot be solved using the IC3 approach [2], can be done using the proposed approach and the above heuristic for approximating strengthenings.

We are currently investigating various heuristics for computing ψ's so that STRENGTHEN converges and works for most cases, as well as comparing their performance vis a vis invariant checking.

6 k-Induction and Strengthening

The above definition of an invariant map being inductive is based on the principle of mathematical induction on the length of a sequence of transitions in computations. Further, a single hypothesis is used in a proof of an induction step: namely, if an invariant map holds at a particular location l_1, and there is a transition at that location to another location l_2, then the invariant map holds in the newly updated state at l_2. Assuming that an invariant map holds in the initial state l_0 and using such a single step transition, it follows that the invariant map holds at all locations reachable from l_0 in finitely many transitions.

There are, however, more general versions of an induction proof rule on natural numbers in which a proof of an induction step is attempted using many induction hypotheses: the principle of complete induction (also called strong induction or total induction) uses a property for all numbers $< m+1$ as induction hypotheses (or just k numbers immediately preceding $m + 1$) while attempting a proof of the property for $m + 1$. For loop programs, this amounts to proving a property holds for first k iterations (including no execution) of the body of the loop), and then to prove that the property holds for the $(m + 1)^{\text{th}}$ iteration of the body of the loop assuming that the property held for the previous k iterations. In the context of a transition system, this amounts to first showing that a formula map is preserved for all transition sequences of length $\leq k$, and then proving that it holds for an arbitrary transition in a transition sequence of length $> k$, assuming it held for the previous k transitions.

Definition 10 (k-Inductive Invariant Map). *Let $k \geq 1$. Then an invariant map $\mathcal{IM}$ is k-inductive if*

- *(base cases):*
  for each sequence $\langle l_1, \tau_1, \sigma_1, l_2 \rangle, \langle l_2, \tau_2, \sigma_2, l_3 \rangle, \ldots, \langle l_k, \tau_k, \sigma_k, l_{k+1} \rangle$ of transitions with $l_1 = l^0$,

$$\tau^0 \implies \mathcal{IM}(l_1)$$
$$\tau^0 \wedge \tau_1 \implies \mathcal{IM}(l_2)\vartheta_1$$
$$\tau^0 \wedge \tau_1 \wedge \tau_2\vartheta_1 \implies \mathcal{IM}(l_3)\vartheta_2$$
$$\vdots$$
$$\tau^0 \wedge \tau_1 \wedge \tau_2\vartheta_1 \wedge \ldots \wedge \tau_{k-1}\vartheta_{k-2} \implies \mathcal{IM}(l_k)\vartheta_{k-1}$$

- *(step case):*
  for each sequence $\langle l_1, \tau_1, \sigma_1, l_2 \rangle, \langle l_2, \tau_2, \sigma_2, l_3 \rangle, \ldots, \langle l_k, \tau_k, \sigma_k, l_{k+1} \rangle$ *of transitions,*

$$\mathcal{IM}(l_1) \wedge \tau_1 \wedge \mathcal{IM}(l_2)\vartheta_1 \wedge \tau_2\vartheta_1 \wedge \mathcal{IM}(l_3)\vartheta_2 \wedge \ldots \wedge \mathcal{IM}(l_k)\vartheta_{k-1} \wedge \tau_k\vartheta_{k-1}$$
$$\implies \mathcal{IM}(l_{k+1})\underset{k}{\vartheta}$$

Here, ϑ_i for $1 \leq i \leq k$ denotes the variable update with $\vartheta_i(x) = x\,\sigma_1 \cdots \sigma_i$. Note that the notions of "inductive" and "1-inductive" coincide.

The following procedure can determine if a given formula map $\mathcal{IM}$ is k-inductive for some $k \geq 0$ for a given transition system; if such a k is found, then $\mathcal{IM}$ is indeed an invariant map. It is proved below that a formula map $\mathcal{IM}$ could be an invariant but it is not k-inductive for any $k \geq 0$.

K-INDUCTION($\mathcal{IM}$)

```
1   k ← 1
2   while TRUE
3       do if IM is k-inductive
4           then return k
5       elseif a base case fails
6           then return ⊥              ▷ IM is not an invariant map
7       else k ← k + 1
```

The procedure K-INDUCTION is said to succeed if it either declares that $\mathcal{IM}$ is not an invariant map or it returns a positive number k declaring that $\mathcal{IM}$ is k-inductive. The procedure may not terminate however. Unlike STRENGTHEN which can fail also because of approximations, the only way the above procedure fails is if it does not terminate.

If a transition system allows only transition sequences up to some predetermined fixed length n, then the above procedure is a decision procedure since it always terminates in $\leq n$ steps as it would have exhausted all possible transition sequences of the transition system. In contrast, STRENGTHEN attempts to compute the weakest inductive invariant map of a given transition system that implies the input formula map under the assumption the input is indeed an invariant map of the transition system. If the relation computed by a transition system is a relatively small finite table, then K-INDUCTION is likely to perform better than STRENGTHEN because of the exhaustive search being performed by K-INDUCTION.

The following theorem shows that there is a direct relationship between checking whether a formula map $\mathcal{IM}$ being k-inductive for some k and strengthening of $\mathcal{IM}$ using STRENGTHEN with respect to various strategies.

Theorem 11. *Using Strategy I in* STRENGHTEN,

1. K-INDUCTION($\mathcal{IM}$) *succeeds if* STRENGTHEN($\mathcal{IM}$) *succeeds.*
2. *If S is unconditional,* K-INDUCTION($\mathcal{IM}$) *succeeds iff* STRENGTHEN($\mathcal{IM}$) *succeeds.*

3. *If S is unconditional and* STRENGTHEN($\mathcal{IM}$) = ⊥, *then $\mathcal{IM}$ is* not *an invariant map.*

The above theorem is generalized further to Theorem 15 using Strategy III by instantiating ψ to be $\mathcal{IM}(l_2)\,\sigma$. Its proof is thus omitted.

The following example shows that the converse of 1 in the above theorem does not hold, i.e., STRENGTHEN may fail in cases where K-INDUCTION succeeds if $\psi = \mathcal{IM}(l_2)\,\sigma$ is taken. The reason is that a loop body is never executed for a nonempty subset of initial states for which the guard associated with the loop does not hold.

Example 12. Consider the transition system with $X = \{x, y\}$, $\mathcal{L} = \{l\}$, $l^0 = l$, $\tau^0 = (x = y \lor x = y + 2)$, and $\mathcal{T} = \{\langle l, x \leq y + 1, \sigma, l\rangle\}$ with $\sigma = \{x \mapsto x + 2\}$. Let $\mathcal{IM}$ be the invariant map with $\mathcal{IM}(l) = (y \leq x \land x \leq y + 2)$. Then $\mathcal{IM}$ is 2-inductive since

$$(x = y \lor x = y + 2) \implies \mathcal{IM}(l)$$
$$(x = y \lor x = y + 2) \land x \leq y + 1 \implies \mathcal{IM}(l)\,\sigma$$
$$\mathcal{IM}(l) \land x \leq y + 1 \land \mathcal{IM}(l)\,\sigma \land x + 2 \leq y + 1 \implies \mathcal{IM}(l)\sigma^2$$

are valid (where the last implication is valid since its left-hand side is unsatisfiable). On the other hand STRENGTHEN($\mathcal{IM}$) fails. Since

$$\mathcal{IM}(l) \land x \leq y + 1 \not\Longrightarrow \mathcal{IM}(l)\,\sigma$$

STRENGTHEN($\mathcal{IM}$) causes a recursive call to STRENGTHEN($\mathcal{IM}[l \mapsto \mathcal{IM}(l) \land y \leq x + 2 \land x + 2 \leq y + 2]$). This recursive call returns ⊥ since

$$(x = y \lor x = y + 2) \not\Longrightarrow \mathcal{IM}(l) \land y \leq x + 2 \land x + 2 \leq y + 2$$

is not valid (take, e.g., $x = 2$ and $y = 0$). The reader would notice that the guard $x \leq y + 1$ is true only for a subset of initial states satisfying $x = y \lor x = y + 2$. If those states are pruned, the subset of initial states is $x = y$ for which STRENGTHEN does terminate giving an inductive invariant $x = y$ which implies $y \leq x \land x \leq y + 2$.   ◇

Example 12 can also be slightly simplified in order to use only polynomial equalities for similar reasons as in the previous example.

Example 13. Consider the transition system with $X = \{x, y\}$, $\mathcal{L} = \{l\}$, $l^0 = l$, $\tau^0 = ((x - y) * (x - y - 2) = 0)$, and $\mathcal{T} = \{\langle l, (x - y) * (x - y - 1) = 0, \sigma, l\rangle\}$ with $\sigma = \{x \mapsto x + 2\}$. Let $\mathcal{IM}$ be the invariant map with $\mathcal{IM}(l) = ((x - y) * (x - y - 1) * (x - y - 2) = 0)$. Then $\mathcal{IM}$ is 2-inductive since

$$(x - y) * (x - y - 2) = 0 \implies \mathcal{IM}(l)$$
$$(x - y) * (x - y - 2) = 0 \land (x - y) * (x - y - 1) = 0 \implies \mathcal{IM}(l)\,\sigma$$
$$\mathcal{IM}(l) \land (x - y) * (x - y - 1) = 0 \land \mathcal{IM}(l)\,\sigma \land (x - y + 2) * (x - y + 1) = 0 \implies \mathcal{IM}(l)\overset{2}{\sigma}$$

are valid (where the last implication is valid since its left-hand side is unsatisfiable). On the other hand STRENGTHEN($\mathcal{IM}$) fails. Since

$$\mathcal{IM}(l) \wedge (x - y) * (x - y - 1) = 0 \not\Rightarrow \mathcal{IM}(l)\sigma$$

STRENGTHEN($\mathcal{IM}$) causes a recursive call to STRENGTHEN($\mathcal{IM}[l \mapsto \mathcal{IM}(l) \wedge (x - y + 2) * (x - y + 1) * (x - y) = 0]$). This recursive call returns $\bot$ since

$$(x - y) * (x - y - 2) = 0 \not\Rightarrow \mathcal{IM}(l) \wedge (x - y + 2) * (x - y + 1) * (x - y) = 0$$

is not valid (take, e.g., $x = 2$ and $y = 0$). As in Example 12, in this case, if the initial state(s) are pruned by the guard, then STRENGTHEN succeeds: $(x - y) * (x - y - 2) = 0 \wedge (x - y) * (x - y - 1) = 0$ simplifies to $x - y = 0$. The invariant map generated then is $x - y = 0$ implying $((x - y) * (x - y - 1) * (x - y - 2) = 0)$.

$\diamond$

The same example also shows that STRENGTHEN may fail in cases where K-INDUCTION succeeds if ψ is obtained from $\mathcal{IM}(l_2)\,\sigma$ by nontrivial equivalent simplification w.r.t $\mathcal{IM}(l_1) \wedge \tau$, i.e., if Strategy III is followed. However, once again, if initial states are pruned using the guards, then the following example works if Strategy III is followed. As shown in a later theorem, if the guard is also incorporated in strengthening as is the case in conditional Strategy II, then the two approaches are equivalent.

Example 14. Consider the transition system from Example 12 again. Simplifying the formula $\mathcal{IM}(l)\sigma = (y \leq x + 2 \wedge x + 2 \leq y + 2)$ w.r.t. $\mathcal{IM}(l) \wedge \tau = (y \leq x \wedge x \leq y + 2 \wedge x \leq y + 1)$ gives, e.g., $x = y$. But $(x = y \vee x = y + 2) \not\Rightarrow x = y$.

$\diamond$

Theorem 11 is generalized below where ψ is computed using Strategy III since it can be obtained as a corollary if ψ below is instantiated to be $\mathcal{IM}(l_2)\,\sigma$.

Theorem 15. *Assume that ψ is chosen in line 4 of* STRENGHTEN *such that $\mathcal{IM}(l_1) \wedge \tau \implies (\mathcal{IM}(l_2)\sigma \iff \psi)$, i.e., Strategy III is followed. Then*

1. K-INDUCTION($\mathcal{IM}$) *succeeds if* STRENGTHEN($\mathcal{IM}$) *succeeds.*
2. *If S is unconditional,* K-INDUCTION($\mathcal{IM}$) *succeeds iff* STRENGTHEN($\mathcal{IM}$) *succeeds.*
3. *If S is unconditional and* STRENGTHEN($\mathcal{IM}$) $= \bot$, *then $\mathcal{IM}$ is not an invariant map.*

Proof (sketch). In order to simplify presentation, we again assume that $|\mathcal{L}| = |\mathcal{T}| = 1$. The proof extends to the general case without technical complications.

Thus, let $\mathcal{T} = \langle l, \tau, \sigma, l \rangle$. For the first statement, we show that the obligations generated at recursion depth k in STRENGTHEN($\mathcal{IM}$) are stronger than the obligations generated in iteration k of K-INDUCTION($\mathcal{IM}$).

For iteration k of K-INDUCTION($\mathcal{IM}$) to succeed,

$$\tau^0 \implies \mathcal{IM}(l)$$
$$\tau^0 \wedge \tau \implies \mathcal{IM}(l)\sigma$$
$$\tau^0 \wedge \tau \wedge \tau\sigma \implies \mathcal{IM}(l)\sigma^2$$

$$\vdots$$

$$\tau^0 \wedge \tau \wedge \tau\sigma \wedge \ldots \wedge \tau\sigma^{k-1} \implies \mathcal{IM}(l)\sigma^{k-1}$$
$$\mathcal{IM}(l) \wedge \tau \wedge \mathcal{IM}(l)\sigma \wedge \tau\sigma \wedge \mathcal{IM}(l)\sigma^2 \wedge \ldots \wedge \mathcal{IM}(l)\sigma^{k-1} \wedge \tau\sigma^{k-1}$$
$$\implies \mathcal{IM}(l)\sigma^k$$

need to be valid. For STRENGTHEN($\mathcal{IM}$) to succeed at recursion depth k,

$$\tau^0 \implies \mathcal{IM}(l) \wedge \psi_1 \wedge \ldots \wedge \psi_{k-1}$$
$$\mathcal{IM}(l) \wedge \psi_1 \wedge \ldots \wedge \psi_{k-1} \wedge \tau \implies \mathcal{IM}(l)\sigma \wedge \psi_1\sigma \wedge \ldots \wedge \psi_{k-1}\sigma$$

need to be valid, where $\mathcal{IM}(l) \wedge \psi_1 \wedge \ldots \wedge \psi_{i-1} \wedge \tau \implies (\mathcal{IM}(l)\sigma \wedge \psi_1\sigma \wedge \ldots \wedge \psi_{i-1}\sigma \iff \psi_i)$.

Using the conditions on the ψ_i imposed during their construction, it is now relatively easy to see that the first k implications for K-INDUCTION($\mathcal{IM}$) are valid if the first implication for STRENGTHEN($\mathcal{IM}$) is valid. Similarly, validity of the last implication for K-INDUCTION($\mathcal{IM}$) follows from validity of the second implication for STRENGTHEN($\mathcal{IM}$). For unconditional systems, $\tau = \top$ and validity of the implications for K-INDUCTION($\mathcal{IM}$) implies validity of the implications for STRENGTHEN($\mathcal{IM}$) and the second statement follows.

For the third statement, assuming that $\tau = \top$, we show below that if ψ is $\mathcal{IM}(l_2)\sigma$ in every step, the statement holds; that would imply that the statement also holds if ψ is a simplification of $\mathcal{IM}(l_2)\sigma$ using $\mathcal{IM}(l_1)$.

Let STRENGTHEN($\mathcal{IM}$) $= \bot$. Thus, $\tau^0 \not\implies \mathcal{IM}(l) \wedge \mathcal{IM}(l)\sigma \wedge \mathcal{IM}(l)\sigma^2 \wedge \ldots \wedge \mathcal{IM}(l)\sigma^{k-1}$ for some $k \geq 1$ but $\tau^0 \implies \mathcal{IM}(l) \wedge \mathcal{IM}(l)\sigma \wedge \mathcal{IM}(l)\sigma^2 \wedge \ldots \wedge \mathcal{IM}(l)\sigma^{k-i}$ for all $i \geq 2$. Thus, there is a run of $\mathcal{S}$ (of length $k-1$) leading to a state not satisfying $\mathcal{IM}(l)$, i.e., $\mathcal{IM}(l)$ is not an invariant of $\mathcal{S}$. $\qquad\square$

Theorem 16. *Assume that $\psi = (\tau \implies \mathcal{IM}(l_2)\sigma)$ is chosen in line 4 of* STRENGTHEN, *i.e., Strategy II is followed. Then*

1. K-INDUCTION($\mathcal{IM}$) *succeeds iff* STRENGTHEN($\mathcal{IM}$) *succeeds.*
2. *If* STRENGTHEN($\mathcal{IM}$) $= \bot$, *then* $\mathcal{IM}$ *is not an invariant map.*

Proof (sketch). As before, in order to simplify presentation, we assume for now that $|\mathcal{L}| = |\mathcal{T}| = 1$. The proof extends to the general case without technical complications.

Thus, let $\mathcal{T} = \langle l, \tau, \sigma, l \rangle$. For the first statement, we show that the obligations generated at recursion depth k in STRENGTHEN($\mathcal{IM}$) following Strategy II are equivalent to the obligations generated in iteration k of K-INDUCTION($\mathcal{IM}$).

For iteration k of K-INDUCTION($\mathcal{IM}$) to succeed,

$$\tau^0 \implies \mathcal{IM}(l)$$
$$\tau^0 \wedge \tau \implies \mathcal{IM}(l)\,\sigma$$
$$\tau^0 \wedge \tau \wedge \tau\sigma \implies \mathcal{IM}(l)\sigma^2$$

$$\vdots$$

$$\tau^0 \wedge \tau \wedge \tau\sigma \wedge \ldots \wedge \tau\sigma^{k-2} \implies \mathcal{IM}(l)\sigma^{k-1}$$
$$\mathcal{IM}(l) \wedge \tau \wedge \mathcal{IM}(l)\,\sigma \wedge \tau\sigma \wedge \mathcal{IM}(l)\sigma^2 \wedge \ldots \wedge \mathcal{IM}(l)\sigma^{k-1} \wedge \tau\sigma^{k-1}$$
$$\implies \mathcal{IM}(l)\sigma^k$$

need to be valid. For STRENGTHEN($\mathcal{IM}$) to succeed at recursion depth k,

$$\tau^0 \implies \xi_0 \wedge \xi_1 \wedge \xi_2 \wedge \ldots \wedge \xi_{k-1}$$
$$\xi_0 \wedge \xi_1 \wedge \xi_2 \wedge \ldots \wedge \xi_{k-1} \wedge \tau \implies \xi_0\sigma \wedge \xi_1\sigma \wedge \xi_2\sigma \wedge \ldots \wedge \xi_{k-1}\sigma$$

need to be valid where $\xi_0 = \mathcal{IM}(l)$ and $\xi_{i+1} = (\tau \implies \xi_i\sigma)$ for all $i \geq 0$. Here, the first implication can easily be seen to be equivalent to the conjunction of the first k implications for K-INDUCTION($\mathcal{IM}$). Since $\xi_{i+1} \wedge \tau \implies \xi_i\,\sigma$, the second implication is equivalent to

$$\xi_1 \wedge \xi_2 \wedge \ldots \wedge \xi_{k-1} \wedge \tau \implies \xi_{k-1}\sigma$$

It is now easy to see that this implication is equivalent to the final implication for K-INDUCTION($\mathcal{IM}$).

For the second statement, the proof is similar to the proof of the third statement in Theorem 15 (with ξ_i instead of $\mathcal{IM}(l)\,\sigma^i$).   □

6.1   Strategy Approximation (IV) Can Often Be Effective

As shown above, STRENGTHEN (as well as K-INDUCTION) need not terminate using the first three strategies if invariants are specified in richer logics including Presburger arithmetic. Interestingly, as the following example shows, STRENGTHEN may succeed in cases where K-INDUCTION fails if ψ is approximated instead of using $\mathcal{IM}(l_2)\,\sigma$ when simplified using the context $\mathcal{IM}(l_1) \wedge \tau$, i.e., if Strategy IV is followed. We consider this as one of the major advantages of using STRENGTHEN instead of K-INDUCTION even for program analysis.

Example 17. Consider the transition system with $X = \{x, y\}$, $\mathcal{L} = \{l\}$, $l^0 = l$, $\tau^0 = (x = 0 \wedge y = 0)$, and $\mathcal{T} = \{\langle l, \top, \{x \mapsto x + y, y \mapsto 2 * y\}, l \rangle\}$. Let $\mathcal{IM}$ be the invariant map with $\mathcal{IM}(l) = (x \geq 0)$. Then $\mathcal{IM}$ is not k-inductive for any $k \geq 1$ since

$$x \geq 0 \wedge x + y \geq 0 \wedge x + 3 * y \geq 0 \wedge \ldots \wedge x + (2^{k-1} - 1) * y \geq 0$$
$$\implies x + (2^k - 1) * y \geq 0$$

is not valid for any $k \geq 1$ (e.g., if $x = 2^{k-1} - 1$ and $y = -1$).

Using approximation (Strategy IV), STRENGTHEN($\mathcal{IM}$) succeeds after one recursive call if ψ is obtained by approximating $\mathcal{IM}(l)\,\sigma$ by w.r.t. $\mathcal{IM}(l) \wedge \tau$: $\mathcal{IM}(l)\,\sigma = (x+y \geq 0)$ and $\mathcal{IM}(l) \wedge \tau = (x \geq 0)$. Using the approach described in Sect. 5.2, $\psi = (y \geq 0)$. The strengthened invariant map $\mathcal{IM}(l) = (x \geq 0 \wedge y \geq 0)$ is inductive.

It is obvious that the transition system is not doing anything interesting since in every configuration (state), $x = 0 \wedge y = 0$. From $x = 0$, the inductive invariant $x = 0 \wedge y = 0$ can be easily generated by STRENGTHEN using Strategy I or II as well K-INDUCTIVE. Starting with $x \geq 0$ causes nontermination of both STRENGTHEN and K-INDUCTIVE however because neither procedure is able to strengthen $x \geq 0$ to $x \geq 0 \wedge -x \geq 0$ first. $\diamond$

All examples in [3] in which invariants are expressed in linear arithmetic can be easily done using Strategy IV; this is in contrast to the IC3 approach which cannot handle such examples.

This observation raises many interesting issues for further research. Particularly, how can approximations be developed for various logical theories to generate formulas which are not necessarily equivalent to a given formula subject to a context, such that STRENGTHEN terminates, i.e., successive approximations converge?

7   Related Work

We were unable to locate any work in literature that focussed on the issue of deciding whether a given formula is a program invariant at one of its locations even though there appears to be hints of posing this problem in [5]. There is, however, a strong relationship between this problem and the automatic invariant generation problem, especially if the strongest possible invariants of certain shape (equivalently from a particular logical theory) can be generated by an automatic invariant generation procedure; in that case, for a formula sharing the shape can be an invariant only if it is implied by the strongest invariant. Instead of reviewing the literature on automatic invariant generation for programs, we focus below on approaches that start with a nontrivial candidate invariant in an attempt to generate an inductive invariant from it as that is a useful heuristic for showing whether the candidate invariant is an invariant.

In [20], Pasareanu and Visser presented an incremental strengthening method for generating inductive invariants from post conditions. They combined predicate abstraction and symbolic execution for strengthening, which employs generalization by considering only those predicates which are common across loop iterations. They seemed to be the first one to have suggested iterative strengthening for inductive invariant generation even though there are similar hints for a strategy based on strengthening in [16].

The research reported in this paper was triggered after reading [5].[6] We however addressed a different problem, namely, how it can be determined that an alleged property (or annotation) about a program is indeed an invariant at a given location and whether we could develop a decision procedure for this problem for a subclass of logical theories. Apparently unaware of [20], Bradley proposed a strengthening approach for generating inductive invariants in [4] in which inductive strengthenings are computed relative to a property (from the previous loop iteration); as in [20], generation of strengthenings could also take into account a post condition. Bradley subsequently proposed a related approach for hardware circuit analysis and implemented it in IC3 [2]. The new idea requires generating a counter-example as the negation of a boolean clause, during an attempt to show the validity of a verification condition, generalize that counter-example by successively dropping literals from the clause by checking whether all such counter-examples can be eliminated, thus computing a strengthening which is propagated forward (since counter-examples may have to be traced back for elimination). This approach is claimed by Bradley[7] to be an incremental invariant generation method using a mixture of bottom-up and top-down approaches for generating strengthening using both forward and backward propagation. Bradley's method was improved upon in [19] and called property directed reachability framework.

The proposed approach in contrast is different; it is based on forward reasoning much like abstract interpretation. In contrast to IC3, it does not use any counter-examples from verification conditions that cannot be established to rule out states one by one; rather it repeatedly tries to strengthen candidate invariants so that the verification conditions from the previous round can be established, to work toward the strengthened candidate becoming inductive. The following example from Bradley's survey paper [3] illustrates the difference between the power of the proposed approach in contrast to the IC3 approach.[8]

Example 18. The objective is to decide whether $P : y \geq 1$ is an invariant for the following loop:

$$(x, y) \leftarrow (1, 1)$$
while $*$
$\quad$ **do** $(x, y) \leftarrow (x + y, y + x)$

P holds in the initial state. The only verification condition is: $y \geq 1 \implies y + x \geq 1$; the approximation method discussed in Sect. 4 will generate $x \geq 0$ as the strengthening;[9] however, we consider other strategies, particularly Strategy I.

---

[6] We thank John Cochran for bringing that paper to our attention and collaborating with us during the initial stages of this research activity.

[7] See http://theory.stanford.edu/~arbrad/ic3.txt.

[8] It is shown in [3] how a template based approach for generating linear inequalities can be augmented to generate strengthings from counter-examples. Using this extension, it is possible to Example 18.

[9] All other examples in [3] can also be successfully done using Strategy IV.

The candidate invariant is strengthened to $(y \geq 1 \wedge y + x \geq 1)$; it holds in the initial state; the verification condition using this strengthened candidate is: $(y \geq 1 \wedge y + x \geq 1) \implies (y + x \geq 1 \wedge 2(x + y) \geq 1)$ which is valid, which establishes that $(y \geq 1 \wedge y + x \geq 1)$ and hence $y \geq 1$ are invariants. $\diamond$

There have recently been proposals to extend the IC3 framework to deal with QF_{LA} [1,7,10] and apply it for software analysis. As commented in [7], extending the IC3 framework to infinite state systems is nontrivial unless quantifier elimination to remove the primed variables (variables in the next transition) is used. Most extensions thus use interpolants and (approximate) quantifier elimination to compute preimages and counter-examples. Termination of the extended procedures is also an issue. Comparison of these extensions with the proposed approach needs to be investigated.

Dillig et al. [9] proposed an abduction based lazy guess-and-check approach and used quantifier-elimination over Presburger arithmetic to generate strengthenings. Their approach is also goal-directed since they start backwards from a given post condition of a program and generate candidate invariants using post condition and the loop test. Their approach is also iterative. However, instead of computing the weakest precondition of the loop body, they compute strengthenings by performing quantifier-elimination of as large a subset of variables as possible from a verification condition that does not hold along some path. Since there is no guarantee that the result can lead to an invariant, their procedure must backtrack and try different subsets of variables for quantifier elimination. Further, their procedure need not terminate; particularly there is no termination proof for any logical theory in [9]. It is also unclear whether their procedure can detect whether a post condition could never be met by a program. Since they perform quantifier elimination, their procedure can generate invariants of arbitrary boolean structure if it succeeds. As shown in [14], quantifier elimination can be used to generate linear invariants in which program variables are eliminated to identify constraints on parameters used in a template to specify candidate invariants; so it is no surprise that quantifier elimination can be used for strengthening as well. In contrast, the proposed approach does not need any backtracking. Most importantly, it gives a decision procedure for theories which do not admit infinite chains of stronger formulas (ordered by implication ordering). In certain cases (since polynomial equalities can simulate disjunctions (see Example 13)), the proposed approach can also generate disjunctive invariants.

Most related works also have heuristics for strengthening which can be used in the proposed STRENGTHEN procedure to approximate ψ in Strategy IV.

Except for Zhihai and Kapur [23] which proposed the notion of inner-invariant, which is the formula after the loop test in a loop that is invariant, we were unable to find any literature on generation of invariants that are not inductive. An inner-invariant formula associated with a loop, if strong enough, is typically not an inductive invariant since it is not preserved during the last iteration of a terminating loop. The paper discussed how such noninductive invariants can be generated using template based approach first introduced in [13] (see also [14]).

8 Conclusions

This paper addresses a fundamental question in program analysis: whether a formula associated with a program location in a program is invariant or not. It is obvious that this check cannot be performed by testing; a test can only tell if a formula is not an invariant whereas it cannot tell whether it is an invariant.

It is proved that for the quantifier-free theory of conjunctive polynomial equalities, there is a decision procedure under certain cases. For other theories, we are unable to get a decision procedure; however, we propose heuristics using which it can often be determined whether a given formula is an invariant or not. While performing this check, the proposed procedure also generates an inductive invariant which implies the formula in case of it being an invariant. The procedure is iterative: it attempts to check whether a given formula is inductive invariant; if it succeeds, then it terminates declaring the formula to be an invariant; otherwise, it attempts to strengthen the given formula so that a verification condition along a program path involving the location of the invariant can be established. Different strengthening strategies are discussed and their relationship to k-inductive method for checking formulas to be invariant is explored.

The focus in the paper is on formulas which are conjunctions of atomic formulas from a given logical theory. The approach is general and can, in principle, handle formulas with any boolean structure insofar as it is possible to generate suitable strengthenings in STRENGTHEN.

The proposed approach is discussed starting with a formula alleged to be an invariant by forward reasoning. It should however be possible to extend the approach so that it can be directed to achieve a post condition associated with a program by incorporating backward reasoning.

References

1. Bjørner, N., Gurfinkel, A.: Property directed polyhedral abstraction. In: D'Souza, D., Lal, A., Larsen, K.G. (eds.) VMCAI 2015. LNCS, vol. 8931, pp. 263–281. Springer, Heidelberg (2015)
2. Bradley, A.R.: SAT-based model checking without unrolling. In: Jhala, R., Schmidt, D. (eds.) VMCAI 2011. LNCS, vol. 6538, pp. 70–87. Springer, Heidelberg (2011)
3. Bradley, A.R.: Understanding IC3. In: Cimatti, A., Sebastiani, R. (eds.) SAT 2012. LNCS, vol. 7317, pp. 1–14. Springer, Heidelberg (2012)
4. Bradley, A.R., Manna, Z.: Verification constraint problems with strengthening. In: Barkaoui, K., Cavalcanti, A., Cerone, A. (eds.) ICTAC 2006. LNCS, vol. 4281, pp. 35–49. Springer, Heidelberg (2006)
5. Bradley, A.R., Manna, Z.: Property-directed incremental invariant generation. Formal Aspects Comput. **20**(4–5), 379–405 (2008)
6. Buchberger, B.: Gröbner bases: an algorithmic method in polynomial ideal theory. In: Bose, N.K. (ed.) Multidimensional Systems Theory-progress, Directions And Open Problems In Multidimensional Systems, pp. 184–232. Reidel Publishing Co, Holland (1985)

7. Cimatti, A., Griggio, A.: Software model checking via IC3. In: Madhusudan, P., Seshia, S.A. (eds.) CAV 2012. LNCS, vol. 7358, pp. 277–293. Springer, Heidelberg (2012)

8. Little, J., Cox, D., O'Shea, D.: Ideals, Varieties, and Algorithms: an Introduction to Computational Algebraic Geometry and Commutative algebra, 2nd edn. Springer, New York (1997)

9. Dillig, I., Dillig, T., Li, B., McMillan, K.: Inductive invariant generation via abductive inference. In: Lopes, C.V. (ed.) Proceedings of the 2013 ACM SIGPLAN International Conference on Object Oriented Programming Systems Languages and Applications. ACM (2013)

10. Hoder, K., Bjørner, N.: Generalized property directed reachability. In: Cimatti, A., Sebastiani, R. (eds.) SAT 2012. LNCS, vol. 7317, pp. 157–171. Springer, Heidelberg (2012)

11. Stuckey, P.J., Jaffar, J., Maher, M.J., Yap, R.H.C.: Beyond finite domains. In: Borning, Alan (ed.) PPCP 1994. LNCS, vol. 874. Springer, Heidelberg (1994)

12. Kapur, D.: A refutational approach to geometry theorem proving. Artif. Intell. J. **37**(1), 61–93 (1988)

13. Kapur, D.: Automatically generating loop invariants using quantifier elimination. Technical report TR-CS-2003-58, Department of Computer Science, University of New Mexico, Albuquerque, NM, USA, (2003)

14. Kapur, D.: A quantifier-elimination based heuristic for automatically generating inductive assertions for programs. J. Syst. Sci. Complexity **19**(3), 307–330 (2006)

15. Kapur, D., Narendran, P.: An Equational Approach to Theorem Proving in First-Order Predicate Calculus. General Electric Corporate Research and Development, Los Angeles (1985)

16. Manna, Z., Pnueli, A.: Temporal verification of reactive systems:safety. Technical report, ISBN 0-387-94459-1, Springer-Verlag, New York (1995)

17. Heizmann, M., Hoenicke, J., Leike, J., Podelski, A.: Linear ranking for linear lasso programs. In: Van Hung, D., Ogawa, M. (eds.) ATVA 2013. LNCS, vol. 8172, pp. 365–380. Springer, Heidelberg (2013)

18. Miné, A.: Weakly relational numerical abstract domains. Ph.D. thesis, Ecole Polytechnique X, France (2004)

19. Mishchenko, A., Eén, N., Brayton, R.K.: Efficient implementation of property directed reachability. In: Bjesse, P., Slobodová, A. (ed.) Proceedings of the 11th International Conference on Formal Methods in Computer-Aided (FMCAD 2011), pp. 125–134. IEEE (2011)

20. Păsăreanu, C.S., Visser, W.: Verification of Java programs using symbolic execution and invariant generation. In: Graf, S., Mounier, L. (eds.) SPIN 2004. LNCS, vol. 2989, pp. 164–181. Springer, Heidelberg (2004)

21. Rodríguez-Carbonell, E., Kapur, D.: Automatic generation of polynomial invariants of bounded degree using abstract interpretation. Sci. Comput. Program. **64**(1), 54–75 (2007)

22. Rodríguez-Carbonell, E., Kapur, D.: Generating all polynomial invariants in simple loops. J. Symbolic Comput. **42**(4), 443–476 (2007)

23. Zhang, Z., Kapur, D.: On invariant checking. J. Syst. Sci. Complexity **26**(3), 470–482 (2013)

Generic Proof Scores for Generate & Check Method in CafeOBJ

Kokichi Futatsugi[✉]

Research Center for Software Verification (RCSV), Japan Advanced Institute of
Science and Technology (JAIST), 1-1 Asahidai, Nomi, Ishikawa 923-1292, Japan
futatsugi@jaist.ac.jp

Abstract. Generic proof scores for the generate & check method in
CafeOBJ are described. The generic proof scores codify the generate &
check method as parameterized modules in the CafeOBJ language independently of specific systems to which the method applies. Basic proof
scores for a specific system can be obtained by instantiating the formal
parameter modules of the parameterized modules with the actual specification modules of the specific system. The effectiveness of the generic
proof scores is demonstrated by applying them to a couple of non-trivial
examples.

1 Introduction

Constructing specifications and verifying them in the upstream of system development are still most important challenges in system development and engineering. It is because many critical defects are caused at the phases of domain,
requirement, and design specifications. Proof scores are intended to meet these
challenges [8,9].

A system and the system's properties are specified in an executable algebraic
specification language (CafeOBJ [3] in our case). Proof scores are described also in
the same specification language for checking whether the specifications imply the
supposed properties. Specifications and proof scores are expressed in equations,
and the checks are done by reduction with the equations. The logical soundness
of the checks is guaranteed by the fact that the reductions are consistent with
the equational reasoning with the equations [11].

The concept of proof supported by proof scores is similar to that of LP [14].
Proof scripts written in tactic languages provided by theorem provers such as
Coq [6] and Isabelle/HOL [18] have similar nature as proof scores. However,
proof scores are written uniformly with specifications in an executable algebraic
specification language and can enjoy a transparent, simple, executable, and efficient logical foundation based on the equational and rewriting logics [11,17].

The generate & check method is a theorem proving method for transition
systems based on (1) generation of finite state patterns that cover all possible
infinite states, and (2) checking the validity of verification conditions for each
element of the finite state patterns [10]. The state space of a transition system is

© Springer International Publishing Switzerland 2015
N. Martí-Oliet et al. (Eds.): Meseguer Festschrift, LNCS 9200, pp. 287–310, 2015.
DOI: 10.1007/978-3-319-23165-5_14

defined as a quotient set (i.e. a set of equivalence classes) of terms of a topmost sort *State*, and the transitions are defined with conditional rewrite rules over the quotient set. A property to be verified is either (1) an invariant (i.e. a state predicate that is valid for all reachable states) or (2) a (p leads-to q) property for two state predicates p and q, where (p leads-to q) means that from any reachable state s with (p(s) = true) the system will get to a state t with (q(t) = true) no matter what transition sequence is taken.

The generate & check method is already described in [10], and this paper's contribution is the development of the generic proof scores as parameterized modules that codify the method. Modularization via parameterization of proof scores is crucial because (a) it helps to identify reusable proof scores, (b) it helps to give good structures to proof scores, and (c) (a) & (b) make proof scores easy to understand and flexible enough for transparent interactive deduction via reductions and modifications (i.e. interactive verification).

The rest of the paper is organized as follows. Section 2 digests preliminary materials from [10]. Section 3 presents the generic proof scores for the generate & check method. Section 4 presents the system and property specifications of QLOCK (a mutual exclusion protocol). Section 5 describes the development of the QLOCK proof scores using the generic proof scores presented in Sect. 3. Section 6 presents related works, some features of the method and the generic proof scores including a sketch of the ABP case, and future issues.

2   Preliminaries

2.1   Transition Systems

A **transition system** is defined as a three tuple (St, Tr, In). St is a set of states, $Tr \subseteq St \times St$ is a set of transitions on the states, and $In \subseteq St$ is a set of initial states. $(s, s') \in Tr$ denotes a transition from state s to state s'. A sequence of states $s_1 s_2 \cdots s_n$ with $(s_i, s_{i+1}) \in Tr$ for each $i \in \{1, \cdots, n-1\}$ is defined to be a **transition sequence**. Note that any $s \in St$ is defined to be a transition sequence of length 1. A state $s^r \in St$ is defined to be **reachable** iff (if and only if) there exists a transition sequence $s_1 s_2 \cdots s_n$ with $s_n = s^r$ for $n \in \{1, 2, \cdots\}$ such that $s_1 \in In$. A state predicate p (i.e. a function from St to Bool) is defined to be an **invariant** (or an invariant property) iff $(p(s^r) = \text{true})$ for any reachable state s^r. Let $\Sigma = (S, \leq, F)$ be a regular order-sorted signature [13] with a set of sorts S, and let $X = \{X_s\}_{s \in S}$ be an S-sorted set of variables. Let $T_\Sigma(X)$ be an S-sorted set of $\Sigma(X)$-terms, let $T_\Sigma(X)_s$ be a set of $\Sigma(X)$-terms of sort s, let E be a set of $\Sigma(X)$-equations, and let (Σ, E) be an equational specification with a unique topmost sort (i.e. a sort without subsorts) State. Let $\theta \in T_\Sigma(Y)^X$ be a substitution (i.e. a map) from X to $T_\Sigma(Y)$ for disjoint X and Y then θ extends to a morphism from $T_\Sigma(X)$ to $T_\Sigma(Y)$, and $t\theta$ is the term obtained by substituting $x \in X$ in t with $x\theta$. Let $tr = (\forall X)(l \rightarrow r \text{ if } c)$ be a rewrite rule with $l, r \in T_\Sigma(X)_{\text{State}}$ and $c \in T_\Sigma(X)_{\text{Bool}}$, then tr is called a transition rule and defines the one step transition relation $\rightarrow_{tr} \in T_\Sigma(Y)_{\texttt{State}} \times T_\Sigma(Y)_{\texttt{State}}$ for Y being disjoint from X as follows.

$$(s \rightarrow_{tr} s') \stackrel{\text{def}}{=} (\exists \theta \in T_\Sigma(Y)^X)((s =_E l\,\theta) \text{ and } (s' =_E r\,\theta) \text{ and } (c\,\theta =_E \texttt{true}))$$

Note that $=_E$ is understood to be defined with $((\Sigma \cup Y), E)$ by considering $y \in Y$ as a fresh constant if Y is not empty. Let $TR = \{tr_1, \cdots, tr_m\}$ be a set of transition rules, let $\rightarrow_{TR} \stackrel{\text{def}}{=} \bigcup_{i=1}^m \rightarrow_{tr_i}$, and let $In \subseteq (T_\Sigma)_{\texttt{State}}/(=_E)_{\texttt{State}}$. In is assumed to be defined via a state predicate $init$ that is defined with E, i.e. $(s \in In)$ iff $(init(s) =_E \texttt{true})$. Then (Σ, E, TR) defines a transition system $((T_\Sigma)_{\texttt{State}}/(=_E)_{\texttt{State}}, \rightarrow_{TR}, In).$[1] A specification (Σ, E, TR) is called a transition specification.

2.2 Verification of Invariant Properties

Let $TS = (St, Tr, In)$ be a transition system, let $p_1, p_2, \cdots, p_n$ $(n \in \{1, 2, \cdots\})$ be state predicates of TS, and let $inv(s) \stackrel{\text{def}}{=} (p_1(s) \text{ and } p_2(s) \text{ and } \cdots \text{ and } p_n(s))$ for $s \in St$.

Lemma 1 (Invariant Lemma). The following three conditions are sufficient for a state predicate p^t to be an invariant.

 (1) $(\forall s \in St)(inv(s) \text{ implies } p^t(s))$
 (2) $(\forall s \in St)(init(s) \text{ implies } inv(s))$
 (3) $(\forall (s, s') \in Tr)(inv(s) \text{ implies } inv(s'))$ □

A predicate that satisfies the conditions (2) and (3) like inv is called an **inductive invariant**. If p^t itself is an inductive invariant then taking $p_1 = p^t$ and $n = 1$ is enough. However, $p_1, p_2, \cdots, p_n$ $(n > 1)$ are almost always needed to be found for getting an inductive invariant, and to find them is the most difficult part of the invariant verification.

2.3 Verification of (p Leads-to q) Properties

Invariants are fundamentally important properties of transition systems. They are asserting that something bad will not happen (i.e. safety property). However, it is sometimes also important to assert that something good will surely happen (i.e. liveness property).

Let $TS = (St, Tr, In)$ be a transition system, and let p, q be predicates with arity $(St, Data)$ of TS, where $Data$ is a data sort needed to specify p, q.[2] A transition system is defined to have the $(p$ **leads-to** $q)$ **property** iff the system will get to a state t with $q(t, d)$ from a state s with $p(s, d)$ no matter what

---

[1] $(T_\Sigma)_{\texttt{State}}/(=_E)_{\texttt{state}}$ is better to be understood as $T_\Sigma/=_E$, for usually the sort $\texttt{State}$ can only be understood together with other related sorts like $\texttt{Bool}$, $\texttt{Nat}$, $\texttt{Queue}$, etc.

[2] We may need some $Data$ for specifying a predicate on a transition system like "the agent with the name N is working" where N is $Data$.

transition sequence is taken.[3] The (p leads-to q) property is a liveness property, and is adopted from the UNITY logic [4].

Lemma 2 (p leads-to q). Based on the original transition system $TS = (St, Tr, In)$, let $\widehat{St} \overset{\text{def}}{=} St \times Data$, let $(((s,d),(s',d)) \in \widehat{Tr}) \overset{\text{def}}{=} ((s,s') \in Tr)$, let $\widehat{In} \overset{\text{def}}{=} In \times Data$, and let $\widehat{TS} \overset{\text{def}}{=} (\widehat{St}, \widehat{Tr}, \widehat{In})$. Let $inv1, inv2, inv3, inv4$ be invariants of $\widehat{TS}$ and let m be a function from $\widehat{St}$ to $\texttt{Nat}$ (the set of natural numbers), then the following 4 conditions are sufficient for the (p leads-to q) property to be valid for $\widehat{TS}$. Here $\widehat{s} \overset{\text{def}}{=} (s,d)$ for any $d \in Data$, $p(\widehat{s}) \overset{\text{def}}{=} p(s,d)$ and $q(\widehat{s}) \overset{\text{def}}{=} q(s,d)$.

(1) $(\forall(\widehat{s},\widehat{s'}) \in \widehat{Tr})$
      $((inv1(\widehat{s})$ and $p(\widehat{s})$ and $(\texttt{not } q(\widehat{s})))$ implies $(p(\widehat{s'})$ or $q(\widehat{s'})))$
(2) $(\forall(\widehat{s},\widehat{s'}) \in \widehat{Tr})$
      $((inv2(\widehat{s})$ and $p(\widehat{s})$ and $(\texttt{not } q(\widehat{s})))$ implies $(m(\widehat{s}) > m(\widehat{s'})))$
(3) $(\forall\widehat{s} \in \widehat{St})$
      $((inv3(\widehat{s})$ and $p(\widehat{s})$ and $(\texttt{not } q(\widehat{s})))$ implies $(\exists\widehat{s'} \in \widehat{St})((\widehat{s},\widehat{s'}) \in \widehat{Tr}))$
(4) $(\forall\widehat{s} \in \widehat{St})$
      $((inv4(\widehat{s})$ and $(p(\widehat{s})$ or $q(\widehat{s}))$ and $(m(\widehat{s}) = 0))$ implies $q(\widehat{s}))$ $\square$

2.4 Generate and Check for $\forall st \in St$

A term $t' \in T_\Sigma(Y)$ is defined to be an **instance** of a term $t \in T_\Sigma(X)$ iff there exits a substitution $\theta \in T_\Sigma(Y)^X$ such that $t' = t\,\theta$.

A finite set of terms $C \subseteq T_\Sigma(X)$ is defined to **subsume** a (possibly infinite) set of ground terms (i.e. terms without variables) $G \subseteq T_\Sigma$ iff for any $t' \in G$ there exits $t \in C$ such that t' is an instance of t. Note that $T_\Sigma \overset{\text{def}}{=} T_\Sigma(\phi)$.

Lemma 3 (Subsume Lemma). Let a finite set of state terms $C \subseteq T_\Sigma(X)_\texttt{State}$ subsume the set of all ground state terms $(T_\Sigma)_\texttt{State}$, and let p be a state predicate, then the following holds.

$$((\forall s \in C)(p(s) \twoheadrightarrow^*_E \texttt{true})) \texttt{ implies } ((\forall t \in (T_\Sigma)_\texttt{State})(p(t) \twoheadrightarrow^*_E \texttt{true})) \quad \square$$

Lemma 3 implies the validity of following **Generate&Check-S**.

[Generate&Check-S] Let $((T_\Sigma)_\texttt{State}/(=_E)_\texttt{State}, \rightarrow_{TR}, In)$ be a transition system defined by a transition specification (Σ, E, TR) (see Sect. 2.1). Then, for a state predicate p_{st}, doing the following **Generate** and **Check** are sufficient for verifying $(\forall t \in (T_\Sigma)_\texttt{State})(p_{st}(t) =_E \texttt{true})$.

  Generate a finite set of state terms $C \subseteq T_\Sigma(X)_\texttt{State}$ that subsumes $(T_\Sigma)_\texttt{State}$.
  Check $(p_{st}(s) \twoheadrightarrow^*_E \texttt{true})$ for each $s \in C$. $\square$

Note that $(t_1 \twoheadrightarrow^*_E t_2)$ means that the term t_1 is reduced to the term t_2 by the CafeOBJ's reduction engine, and $(t_1 \twoheadrightarrow^*_E t_2)$ implies $(t_1 \rightarrow^*_E t_2)$ but not necessarily $(t_1 \rightarrow^*_E t_2)$ implies $(t_1 \twoheadrightarrow^*_E t_2)$.

---
[3] See [10] for a more precise definition.

2.5 Built-in Search Predicate

The verification conditions (3) of Lemma 1 and (1), (2) of Lemma 2 contain universal quantifications over the set of transitions Tr. CafeOBJ's built-in search predicate makes it possible to translate a universal quantification over Tr into a universal quantification over St.

The built-in search predicate is a generic operator like the `if_then_else_fi` operator and is declared as follows.

```
op _=(_,_)=>+_if_suchThat_{_} :
    *Cosmos* NzNat* NzNat* *Cosmos* Bool Bool *Cosmos* -> Bool .
```

`*Cosmos*` can be any sort, provided that the 1st and the 2nd `*Cosmos*`s belong to the same connected component of ordered sorts. `NzNat*` is a sort consists of non-zero natural numbers plus `*` (infinity).

By partially instantiating this built-in search predicate, the predicate for one step search over sort `State` with the following rank can be assumed to exit.

```
op _=(*,1)=>+_if_suchThat_{_} : State State Bool Bool Info -> Bool.
```

`(*,1)`, i.e. instantiated 2nd and 3rd arguments, represent one step searches with no limit on the number of objects to be searched. The 1st `State` is assumed to be given as the input. The 2nd `State` and the 1st `Bool` are bound by CafeOBJ system to the searched next (i.e. one step later) state and the condition of the transition used (a condition of an unconditional transition is `true`), respectively. The 2nd `Bool` is declare by a user to be a Boolean expression that usually includes the 1st `State` (current state), the 2nd `State` (next state), and the 1st `Bool` (condition). `Info` is also declare by a user to be any expression. If the reduced value of the Boolean expression bound to the 2nd `Bool` is `true`, the reduced value of the expression bound to `Info` is printed out for inspection. The built-in search predicate returns `true` if printout exits for some searched next state, and returns `false` otehwise. The precise definition of the behavior of the built-in search predicate for one step search over sort `State` can be found in Sect. 4.2 of [10].

Let q be a predicate "pred q : State State" for stating some relation of the current state and the next state, like $(inv(s)$ implies $inv(s'))$ in the condition (3) of Lemma 1. Note that this q has nothing to do with q of (p leads-to q). Let the predicates `_then_` and `valid-q` be defined as follows in CafeOBJ using the built-in search predicate. Note that `_then_` is different from `_implies_` because (`B:Bool implies true = true`) for `_implies_` but only (`true then true = true`) for `_then_`.

```
-- information constructor
[Infom] op (ifm _ _ _ _) : State State Bool Bool -> Infom {constr}
-- for checking conditions of ctrans rules
pred _then_ : Bool Bool .
eq (true then B:Bool) = B . eq (false then B:Bool) = true .
-- predicate to be checked for a State
pred valid-q : State State Bool .
eq valid-q(S:State,SS:State,CC:Bool) =
    not(S =(*,1)=>+ SS if CC suchThat
        not((CC then q(S, SS)) == true) {(ifm S SS CC q(S,SS))}) .
```

Note that defining `valid-q(s,SS:State,CC:Bool)`using the built-in search predicate is just possible because it is a built-in predicate. That is, the above definition with an equation in the object level is impossible with the search command in Maude [16].

For a state term $s \in T_\Sigma(Y)_{\text{State}}$, the reduction of the Boolean term:

`valid-q(s,SS:State,CC:Bool)`

with $\twoheadrightarrow_E^* \cup \rightarrow_{TR}$ behaves as follows based on the definition of the behavior of the built-in search predicate (Sect. 4.2 of [10]). Note that the $\rightarrow_{TR}$ part is effective only for determining the behavior of the built-in search predicate.

1. Search for evey pair (tr_j, θ) of a transition rule $tr_j = (\forall X)(l_j \rightarrow r_j \text{ if } c_j)$ in Tr and a substitution $\theta \in T_\Sigma(Y)^X$ such that $\mathtt{s} = l_j\,\theta$.
2. For each found (tr_j, θ), let ($\mathtt{SS} = r_j\,\theta$) and ($\mathtt{CC} = c_j\,\theta$) and print out (ifm s SS CC q(s,SS)) and tr_j if (not((CC then q(s,SS)) == true) $\twoheadrightarrow_E^*$ true).
3. Returns `false` if any printout exits, and returns `true` otherwise.

Note that for each found pair (tr_j, θ) with $tr_j = (\forall X)(l_j \rightarrow r_j \text{ if } c_j)$ either (1) no printout if $((c_j\,\theta \twoheadrightarrow_E^* \text{ false})$ or $((c_j\,\theta \twoheadrightarrow_E^* \text{ true})$ and $(\mathtt{q}(l_j\,\theta, r_j\,\theta) \twoheadrightarrow_E^* \text{ ture})))$ or (2) print out if $(\text{not}(c_j\,\theta \twoheadrightarrow_E^* \text{ false})$ and $(\text{not}(c_j\,\theta \twoheadrightarrow_E^* \text{ true})$ or $\text{not}(\mathtt{q}(l_j\,\theta, r_j\,\theta) \twoheadrightarrow_E^* \text{ ture})))$.

2.6   Generate & Check for $\forall tr \in Tr$

Definition 4 (Cover). Let $C \subseteq T_\Sigma(Y)$ and $C' \subseteq T_\Sigma(X)$ be finite sets. C is defined to **cover** C' iff for any ground instance $t'_g \in T_\Sigma$ of any $t' \in C'$, there exits $t \in C$ such that t'_g is an instance of t and t is an instance of t'.   □

Note that C subsumes $(T_\Sigma)_{\text{State}}$ if C coveres $\{S : \texttt{State}\}$ (a singleton of a variable of sort `State`).

Lemma 5 (Cover Lemma). Let $C' \subseteq T_\Sigma(X)_{\text{State}}$ be the set of all the left-hand sides of the transition rules in TR, and let $C \subseteq T_\Sigma(Y)$ cover C', then the following holds.

$$(\forall t \in C)(\texttt{valid} - \texttt{q}(\texttt{t}, \texttt{SS} : \texttt{State}, \texttt{CC} : \texttt{Bool}) \twoheadrightarrow_E^* \cup \rightarrow_{TR} \textbf{true})$$
$$\texttt{implies}$$
$$(\forall(s, s') \in ((T_\Sigma \times T_\Sigma) \cap \rightarrow_{TR}))(\mathtt{q}(s, s') \rightarrow_E^* \textbf{true})) \quad □$$

Lemma 5 implies the validity of following Generate&Check-T1/T2.

[Generate&Check-T1] Let $((T_\Sigma)_{\text{State}}/(=_E)_{\text{State}}, \rightarrow_{TR}, In)$ be a transition system defined by a transition specification (Σ, E, TR) (see Sect. 2.1), and let $C' \subseteq T_\Sigma(X)$ be the set of all the left-hand sides of the transition rules in TR. Then doing the following **Generate** and **Check** is sufficient for verifying

$$(\forall(s, s') \in ((T_\Sigma \times T_\Sigma) \cap \rightarrow_{TR}))(\mathtt{q_{tr}}(s, s') =_E \textbf{true})$$

for a predicate "`pred q`$_{\mathtt{tr}}$ `: State State`".

Generate a finite set of state terms $C \subseteq T_\Sigma(Y)_{\texttt{State}}$ that covers C'.
Check (`valid-q`$_{\texttt{tr}}$`(t,SS:State,CC:Bool)` $\twoheadrightarrow^*_E \cup \to_{TR}$ **true**) for each $\texttt{t} \in C$. $\square$

Generate&Check-T1 can be modified into Generate&Check-T2.

[Generate&Check-T2] Let $TR = \{tr_1, \cdots, tr_m\}$ be a set of transition rules, and let $tr_i = (\forall X)(l_i \to r_i \text{ if } c_i)$ for $i \in \{1, \cdots, m\}$. Then doing the following **Generate** and **Check** for all of $i \in \{1, \cdots, m\}$ is sufficient for verifying

$$(\forall(s, s') \in ((T_\Sigma \times T_\Sigma) \cap \to_{TR}))(\mathsf{q}_{\texttt{tr}}(s, s') =_E \textbf{true})$$

for a predicate "**pred q**$_{\texttt{tr}}$ **: State State**".

Generate a finite set of state terms $C_i \subseteq T_\Sigma(Y)_{\texttt{State}}$ that covers $\{l_i\}$.
Check (`valid-q`$_{\texttt{tr}}$`(t,SS:State,CC:Bool)` $\twoheadrightarrow^*_E \cup \to_{tr_i}$ **true**) for each $\texttt{t} \in C_i$. $\square$

2.7 Generate&Check for Verification of Invariant Properties

The conditions (1) and (2) of Lemma 1 can be verified by using Generate&Check-S with $\mathsf{p}_{\texttt{st}-1}(s)$ and $\mathsf{p}_{\texttt{st}-2}(s)$ defined as follows respectively.

(1) $\mathsf{p}_{\texttt{st}-1}(s) = (inv(s) \text{ implies } p^t(s))$
(2) $\mathsf{p}_{\texttt{st}-2}(s) = (init(s) \text{ implies } inv(s))$

Note that if $inv \overset{\text{def}}{=} (p_1 \text{ and} \cdots \text{and } p_n)$ and $p^t = (p_{i_1} \text{ and} \cdots \text{and } p_{i_m})$ for $\{i_1, \cdots, i_m\} \subseteq \{1, \cdots, n\}$ then condition (1) is directly obtained.

The condition (3) of Lemma 1 can be verified by using Generate & Check-T1 or T2 with $\mathsf{q}_{\texttt{tr}-3}(s, s')$ defined as follows.

(3) $\mathsf{q}_{\texttt{tr}-3}(s, s') = (inv(s) \text{ implies } inv(s'))$

2.8 Generate&Check for Verification of (p Leads-to q) Properties

The conditions (1) and (2) of Lemma 2 can be verified by using Generate&Check-T1 or T2 in Sect. 2.6 with $\mathsf{q}_{\texttt{tr}-1}(\widehat{s}, \widehat{s'})$ and $\mathsf{q}_{\texttt{tr}-2}(\widehat{s}, \widehat{s'})$ defined as follows respectively.

(1) $\mathsf{q}_{\texttt{tr}-1}(\widehat{s}, \widehat{s'}) = ((inv1(\widehat{s}) \text{ and } p(\widehat{s}) \text{ and } (\text{not } q(\widehat{s}))) \text{ implies } (p(\widehat{s'}) \text{ or } q(\widehat{s'})))$
(2) $\mathsf{q}_{\texttt{tr}-2}(\widehat{s}, \widehat{s'}) = ((inv2(\widehat{s}) \text{ and } p(\widehat{s}) \text{ and } (\text{not } q(\widehat{s}))) \text{ implies } (m(\widehat{s}) > m(\widehat{s'})))$

The conditions (3) and (4) of Lemma 2 can be verified by using Generate&Check-S in Sect. 2.4 with $\mathsf{p}_{\texttt{st}-3}(\widehat{s})$ and $\mathsf{p}_{\texttt{st}-4}(\widehat{s})$ defined as follows respectively.

(3) $\mathsf{p}_{\texttt{st}-3}(\widehat{s}) = ((inv3(\widehat{s}) \text{ and } p(\widehat{s}) \text{ and } (\text{not } q(\widehat{s}))) \text{ implies } (\widehat{s} = (*, 1) \Rightarrow + \texttt{ SS : State}))$
(4) $\mathsf{p}_{\texttt{st}-4}(\widehat{s}) = ((inv4(\widehat{s}) \text{ and } (p(\widehat{s}) \text{ or } q(\widehat{s})) \text{ and } (m(\widehat{s}) = 0)) \text{ implies } q(\widehat{s}))$

Note that $(s \texttt{ =(*,1)=>+ SS:State})$ is a simplified built-in search predicate that returns **true** if there exits $s' \in St$ such that $(s, s') \in Tr$.

2.9   System and Property Specifications, and Proof Scores

For verifying a system, a model of the system should be formalized and described as system specifications that are formal specifications of the behavior of the system. In conjunction with the system specifications, functions and predicates that are necessary for expressing the system's supposed properties are formalized and described as property specifications. Note that the supposed property we are considering is either (1) invariant property or (2) (p leads-to q) property. Proof scores are developed to verify that the system's supposed properties are deduced from the system and property specifications.

The effectiveness of the generic proof scores presented in this paper (Sect. 3) is demonstrated by applying them to two non-trivial examples ABP (Alternating Bit Protocol) and QLOCK (Mutual Exclusion Protocol by Locking with Queue). Because of the limit of space, this paper concentrates on explaining QLOCK specifications and how QLOCK proof scores can be obtained by applying the generic proof scores to the QLOCK specifications. The ABP case is sketched in Sect. 6.

All of the generic proof scores, the ABP specifications and proof scores, and the QLOCK specifications and proof scores are organized as three directories of files in the CafeOBJ language and posted at the following web page.

http://www.jaist.ac.jp/~kokichi/misc/1505gpsgcmco/

Interested readers are encouraged to look into the web page, for the posted Cafe-OBJ codes contain quite a few comments including the comments on the CafeOBJ language itself that are not included in this paper.

3   Generic Proof Scores for Generate & Check Method

This section presents the seven parameterized CafeOBJ modules that codify the three verification conditions of Sect. 2.7 for invariant properties and the four verification conditions of Sect. 2.8 for (p leads-to q) properties.[4]

The seven parameterized modules specify the seven verification conditions in an executable way, and only by instantiating the formal parameters of the parameterized modules with an actual specification modules of a specific system, the basic proof scores for the specific system are completed.

3.1   GENcases: Generating Patterns and Checking on Them

The module GENcases[5] specifies the pattern generation and the validity checking of predicates on the generated patterns.

[check_, [_], (_||_), (_,_), (_;_)]: Function check_ is specified as follows and performs the validity checks on all the patterns defined by SST. If all the validity checks are

---

[4] The file genCheck.cafe on the web page contains the seven parameterized modules. The files exState.cafe, genCases.cafe, pnat.cafe and predCj.cafe are used in genCheck.cafe. Note that each file without suffix "abp-" or "qlock-" in its name is not depend on QLOCK or ABP and generic for the generate & check method.

[5] The module GENcases is in the file genCases.cafe on the web page.

successful, mmi(SST) disappears and check(SST) returns ($):Ind. Otherwise, each case where checked condition does not reduce to true is included in the reduced value of mmi(SST). Function mmi_ is explained later in this section (Sect. 3.1).

```
op check_ : SqSqTr -> IndTr .   eq check(SST:SqSqTr) = ($ | mmi(SST)) .
```

Sort SqSqTr is specified as follows, and an SqSqTr (i.e. an element of sort SqSqTr) is (1) an SqSqEn or (2) a tree (or a sequence) of SqSqEns (i.e. elements of sort SqSqEn) composed with the associative binary operator _||_. An SqSqEn is an SqSq enclosed with [and].

```
[SqSqEn < SqSqTr]
op [_] : SqSq -> SqSqEn .   op _||_ : SqSqTr SqSqTr -> SqSqTr {assoc}
```

Sort SqSq is specified as follows, and an SqSq is (1) a ValSq, (2) a VlSq, or (3) a sequence of ValSqs or VlSqs composed with the associative binary operator _,_that has empSS as an identity (id: empSS). A ValSq is (1) a Val or (2) a sequence of Vals composed with the associative binary operator _,_. A VlSq is (1) a Val or (2) a sequence of VlSqs composed with the associative binary operator _;_. Note that the operator _,_ is overloaded (i.e. denotes two different operations), and a term composed with the same type of associative binary operators inductively is called a sequence for SqSq, ValSq, and VlSq while a SqSqTr is called a tree.

```
[Val < ValSq] op _,_ : ValSq ValSq -> ValSq {assoc}
[Val < VlSq] op _;_ : VlSq VlSq -> VlSq {assoc}
[ValSq VlSq < SqSq]
op empSS : -> SqSq .  op _,_ : SqSq SqSq -> SqSq {assoc id: empSS}
```

The operator _;_ specifies possible alternatives and the following equation reduces the alternatives into a term composed with the operator _||_.

```
eq [(SS1:SqSq,(V:Val;VS:VlSq),SS2:SqSq)]
   = [(SS1,V,SS2)] || [(SS1,VS,SS2)] .
```

The equation applies recursively and any subterm with the alternative operator _;_ is reduced into a term with _||_. That is, for any term $sqSq$ of sort SqSq the term [$sqSq$] is reduced to the term composed with operator _||_ of terms [$valSq_i$] ($i = 1,2,\cdots$) for $valSq_i$ of sort ValSq. This kind of reductions are called *alternative expansions*.

For example, if terms v1, v2, v3 are of sort Val, the following reduction happens. Note that, because empSS is declared to be an identity for operator "_,_ : SqSq SqSq -> SqSq", the equation covers the cases in which SS1 and/or SS2 in the left-hand side of the equation are/is empSS.

```
[(v1;v2;v3),(v1;v2)]
=red=>
[ (v1 , v1) ] || [ (v2 , v1) ] || [ (v3 , v1) ] ||
[ (v1 , v2) ] || [ (v2 , v2) ] || [ (v3 , v2) ]
```

[t__, g__]: To make the alternative expansions with _;_ more versatile, functions t__ and g__ are introduced as follows. String is a sort from CafeOBJ built-in module STRING and denotes the set of character strings like "abc", "v1", "_%_". By using t__, a user is assumed to specify term constructors with appropriate identifiers in the first argument, and the constructors determined by accompanying g__ can be used to specify the alternative expansions with _;_. The following two equations for g__ make the expansion of a nested expression with [_]s and _;_s possible, and reduce "g *strg sqSqTr*" to "t *strg sqSqTr*" if *sqSqTr* is of sort ValSq.

```
op t__ : String ValSq -> Val .
op g__ : String SqSqTr -> VlSq .
eq g(S:String)(SST1:SqSqTr || SST2:SqSqTr) = (g(S) SST1);(g(S) SST2) .
eq g(S:String)[VSQ:ValSq] = t(S)(VSQ) .
```

For example, let the following equations for t__ be given.[6]

```
[Qu Aid Label Aobs State < Val]
eq t("lb[_]:__")(A:Aid,L:Label,AS:Aobs) = ((lb[A]: L) AS) .
eq t("_$_")(Q:Qu,AS:Aobs) = (Q $ AS) .
```

Then the following expansion by reduction of alternatives is possible for QLOCK state terms if we assume q is of sort Qu, a1 and a2 are of sort Aid, and as is of sort Abos.

```
[(g("_$_")[(empQ;(a1 & q)),(g("lb[_]:__")[a2,(rs;ws;cs),as])])]
=red=>
[(empQ $ ((lb[a2]: rs) as))] || [((a1 & q) $ ((lb[a2]: rs) as))] ||
[(empQ $ ((lb[a2]: ws) as))] || [((a1 & q) $ ((lb[a2]: ws) as))] ||
[(empQ $ ((lb[a2]: cs) as))] || [(((a1 & q) $ ((lb[a2]: cs) as))]
```

The specifications of alternative expansions with _;_, [_], g__ are called **alternative scripts** or **alternative expansion scripts**. Alternative scripts are simple but powerful enough to specify a fairly large number of necessary patterns. Note that an alternative script is a term of sort SqSqTr.

[IndTr, mmi_, mi_, v_]: Sort IndTr and function mmi_ are specified as follows, and mmi_ translates a SqSqTr to a IndTr and mmi[*sqSq*] reduces to mi(*sqSq*) if *sqSq* is of sort ValSq.

```
-- indicator and indicator tree
[Ind < IndTr]
op $ : -> Ind .
op _|_ : IndTr IndTr -> IndTr {assoc}
-- make make indicator
op mmi_ : SqSqTr -> IndTr .
eq mmi(SST1:SqSqTr || SST2:SqSqTr) = (mmi SST1) | (mmi SST2) .
eq mmi[VSQ:ValSq] = mi(VSQ) .
```

---

[6] These equations are in the file qlock-genStTerm.cafe on the web page.

Indicator i__ and making indicator function mi_ are specified as follows. Function ii_ (information indicator) and predicate v_ (value) to be checked on ValSq are assumed to be defined by a user. mi(*valSq*) reduces to "(i v(*valSq*) ii(*valSq*))", and disappears if the first argument v(*valSq*) reduces to true. This implies that predicate v_ is valid for all the ValSqs specified by SST if check(SST) returns ($):Ind.

```
-- indicator
[Info] op i__ : Bool Info -> Ind .
-- making any indicator with'true' disappear
eq (i true II:Info) | IT:IndTr = IT .
eq IT:IndTr | (i true II:Info) = IT .
-- information constructor
op ii_ : ValSq -> Info .
-- the predicate to be checked
pred v_ : ValSq .
-- make indicator for v_
op mi_ : ValSq -> Ind .
eq mi(VSQ:ValSq) = (i v(VSQ) ii(VSQ)) .
```

3.2   Three Parameterized Modules for Invariant Properties

[PREDcj]: For defining conjunctions of predicates flexibly, the following parameterized module PREDcj[7] is prepared.

```
-- defining the conjunction of predicates
-- via the sequence of the names of the predicates
mod! PREDcj (X :: TRIV) {
-- names of predicates on Elt.X and the sequences of the names
[Pname < PnameSeq]
-- associative binary operator for constructing non nil sequences
op _ _ : PnameSeq PnameSeq -> PnameSeq {constr assoc}
-- cj(pns,e) defines the conjunction of predicates
-- whose names constitute the sequence pns
op cj : PnameSeq Elt.X -> Bool .
eq cj((PN:Pname PNS:PnameSeq),E:Elt) = cj(PN,E) and cj(PNS,E) . }
```

By using the cj (conjunction) operator of PREDcj, a conjunction of predicates can be expressed just as a sequence of the names of the predicates. This helps prompt modifications of component predicates of *inv* in the checks of the conditions (1),(2),(3) of Sect. 2.7 and of *inv1*, *inv2*, *inv3*, *inv4* in the checks of the conditions (1),(2),(3),(4) of Sect. 2.8.

[INV-1v, INV-2v][8]: The following two parameterized modules INV-1v and INV-2v codify the verification conditions (1) and (2) of Sect. 2.7 directly. The theory module STEpcj specifies the modules with sorts corresponding to Ste, Pname,

---

[7] The module PREDcj is in the file predCj.cafe on the web page.

[8] The modules INV-1v, INV-2v are in the file genCheck.cafe on the web page.

PnameSeq and a function corresponding to cj that make a predicate be presented as cj(*pNameSeq*,*ste*).

By defining the predicate v_ from the module GENcases as the predicate p_{st-1} of the condition (1) or the predicate p_{st-2} of the condition (2), necessary checks are done on all state patterns. The PnameSeqs p-iinv (twice), p^t, and p-init are assumed to be reified after the parameter modules are substituted with actual specification modules (i.e. after the instantiation of parameter modules).

```
mod* STEpcj {[Ste] [Pname < PnameSeq] pred cj : PnameSeq Ste .}
mod! INV-1v (ST :: STEpcj) {ex(GENcases)
  -- possible inductive invariant and target predicate
  ops p-iinv p^t : -> PnmSeq .
  [Ste < Val] eq v(S:Ste) = cj(p-iinv,S:Ste) implies cj(p^t,S) . }
mod! INV-2v (ST :: STEpcj) {ex(GENcases)
  ops p-init p-iinv : -> PnmSeq .
  [Ste < Val] eq v(S:Ste) = cj(p-init,S) implies cj(p-iinv,S) . }
```

[VALIDq, G&C-Tv, INV-3q][9]: The following parameterized module VALIDq directly specifies valid-q of Sect. 2.5. inc(RWL) declares the importation of the built-in module RWL that is necessary for using the built-in search predicate.

```
mod* STE {[Ste]}
mod! VALIDq (X :: STE) {inc(RWL)
  -- predicate to be checked for all the transitions
  pred q : Ste Ste .
  -- information constructor
  [Infom] op (ifm _ _ _ _) : Ste Ste Bool Bool -> Infom {constr}
  pred _then _ : Bool Bool . pred valid-q : Ste Ste Bool .
  eq (true then B:Bool) = B . eq (false then B:Bool) = true .
  eq valid-q(S:Ste,SS:Ste,CC:Bool) =
    not(S =(*,1)=>+ SS if CC suchThat
        not((CC then q(S, SS)) == true) {(ifm S SS CC q(S,SS))}) . }
```

The following module G&C-Tv defines v(S:Ste,SS:Ste,CC:Bool) as valid-q(S, SS,CC). Note that S:Ste,SS:Ste,CC:Bool in the left-hand side is of sort ValSq but S,SS,CC in the right-hand side is of sort Ste,Ste,Bool that is the sort list (or arity) of the standard form (i.e. without _) operator valid-q. In the module INV-3q, by defining q of the module VALIDq as q_{tr-3} of the condition (3) of Sect. 2.7, necessary checks are done on all state patterns. The PnameSeq p-iinv is assumed to be reified after the instantiation of the parameter module "ST :: STEpcj".

```
mod! G&C-Tv (S :: STE) {ex(VALIDq(S) + GENcases)
  [Ste Bool < Val] eq v(S:Ste,SS:Ste,CC:Bool) = valid-q(S,SS,CC) . }
mod! INV-3q (ST :: STEpcj) {ex(G&C-Tv(ST))
  op p-iinv : -> PnmSeq .
  eq q(S:Ste,SS:Ste) = (cj(p-iinv,S) implies cj(p-iinv,SS)) . }
```

Note that the three parameterized modules INV-1v, INV-2v, INV-3q have the same parameter declaration "ST :: STEpcj". It indicates that the modules

---

[9] The modules VALIDq, G&C-Tv, INV-3q are in the file genCheck.cafe on the web page.

obtained by applying the parameterized module PREDcj to appropriate modules can be substituted for the parameter modules of these three parameterized modules.

3.3 Four Parameterized Modules for (p Leads-to q) Properties

[EX-STATE, PCJ-EX-STATE][10] : For specifying the four verification conditions for (p leads-to q) properties, the states are needed to be extend with data. The following parameterized module EX-STATE specifies the state extension following Lemma 2 directly. The theory module ST-DT requires functions p, q, m for (p leads-to q) properties, and cj for defining predicates via sequences of their names. The functions p, q, m on ExState are specified based on the functions p, q, m on State and Data. The transitions over ExState are specified based on the transitions over State by declaring two equations with the built-in search predicates _=(*,1)=>+_if_suchThat_{_} and _=(*,1)=>+_. This succinct and powerful way to define transitions of extended system based on transitions of base system is possible because the built-in search predicates are generic and can present transitions over any sort (i.e. over *Cosmos*, see Sect. 2.5). The equation for t__ is for composing a term of sort ExState with the constructor _%_ in the alternative expansion script.

```
-- theory module with state and data
mod* ST-DT {ex(PNAT)
  [Ste Data] ops p q : Ste Data -> Bool . op m : Ste Data -> Nat.PNAT .
  [Pnm < PnmSeq] op cj : PnmSeq Ste -> Bool . }
mod! EX-STATE (SD :: ST-DT) {inc(RWL) ex(GENcases)
  [ExState Infom]
  -- state constructor for extended states
  op _%_ : Ste Data -> ExState {constr}
  -- the transitions on ExState is the same as the transitons on Ste
  eq ((S:Ste % D:Data) =(*,1)=>+ (SS:Ste % D)
      if CC:Bool suchThat B:Bool {I:Infom})
    = (S =(*,1)=>+ SS if CC suchThat B {I}) .
  eq ((S:Ste % D:Data) =(*,1)=>+ (SS:Ste % D)) = (S =(*,1)=>+ SS) .
  -- predicates p and q on ExState
  ops p q : ExState -> Bool .
  eq p(S:Ste % D:Data) = p(S,D) . eq q(S:Ste % D:Data) = q(S,D) .
  -- measure function on ExState
  op m : ExState -> Nat.PNAT . eq m(S:Ste % D:Data) = m(S,D) .
  -- t__ is introduced in the module GENcases
  [Ste Data ExState < Val] eq t("_%_")(S:Ste,D:Data) = (S % D) . }
```

The following parameterized module PCJ-EX-STATE makes the cj available on ExState and relate that to the cj on Ste.

```
mod! PCJ-EX-STATE (SD :: ST-DT) {
  ex((PREDcj((EX-STATE(SD)){sort Elt -> ExState}))
```

---

[10] The modules EX-STATE, PCJ-EX-STATE are in the file exState.cafe on the web page.

```
*{sort Pname -> ExPname, sort PnameSeq -> ExPnameSeq})
[Pnm < ExPname] [PnmSeq < ExPnameSeq]
eq cj(PN:Pnm,(S:Ste % D:Data)) = cj(PN,S) . }
```

[PQ-1q, PQ-2q, PQ-3v, PQ-4v][11] : The four parameterized modules for the four
verification conditions for (p leads-to q) properties are specified as follows. These
are direct translation from the four conditions of Sect. 2.8. The parameterized
modules PQ-1q and PQ-2q are using Generate&Check-T1 or Generate & Check-
T2, and the parameterized module G&C-Tv is necessary for reifying the predicate
q. The parameterized modules PQ-3v, PQ-4v are using Generate&Check-S, and
only the module GENcases is necessary for reifying the predicate v_.

```
-- theory module with p,q,m,cj on states
mod* STPQpcj {ex(PNAT)
  [Ste] ops p q : Ste -> Bool . op m : Ste -> Nat.PNAT .
  [Pnm < PnmSeq] op cj : PnmSeq Ste -> Bool . }
mod! PQ-1q (SQ :: STPQpcj) {ex(G&C-Tv(SQ))
  op pq-1-inv : -> PnmSeq .
  eq q(S:Ste,SS:Ste) =
    (cj(pq-1-inv,S) and p(S) and not(q(S))) implies (p(SS) or q(SS)) . }
mod! PQ-2q (SQ :: STPQpcj) {ex(G&C-Tv(SQ))
  op pq-2-inv : -> PnmSeq .
  eq q(S:Ste,SS:Ste) =
    (cj(pq-2-inv,S) and p(S) and not(q(S))) implies (m(S) > m(SS)) . }
mod! PQ-3v (SQ :: STPQpcj) {inc(RWL) ex(GENcases)
  op pq-3-inv : -> PnmSeq . [Ste < Val]
  eq v(S:Ste,SS:Ste) =
    (cj(pq-3-inv,S) and p(S) and not(q(S))) implies (S =(*,1)=>+ SS) . }
mod! PQ-4v (SQ :: STPQpcj) {pr(GENcases)
  op pq-4-inv : -> PnmSeq . [Ste < Val]
  eq v(S:Ste) =
    (cj(pq-4-inv,S) and (p(S) or q(S)) and (m(S) = 0)) implies q(S) . }
```

Note that the four parameterized modules PQ-1q, PQ-2q, PQ-3v, PQ-4v have
the same parameter declaration "(SQ :: STPQpcj)". It indicates that the modules obtained by applying the parameterized module PCJ-EX-STATE to appropriate
modules can be substituted for the parameter modules of these four parameterized modules.

4   QLOCK Specifications

This section gives formal specifications in CafeOBJ of a simple but non-trivial
example. The specifications are going to be used in Sect. 5 for showing the effectiveness of the generic proof scores presented in Sect. 3.

The example is a mutual exclusion protocol QLOCK. A mutual exclusion
protocol can be described as follows:

---

[11] The modules PQ-1q, PQ-2q, PQ-3v, PQ-4v are in the file genCheck.cafe on the web
  page.

Assume that many agents (or processes) are competing for a common equipment (e.g. a printer or a file system), but at any moment of time only one agent can use the equipment. That is, the agents are mutually excluded in using the equipment. A protocol (concurrent mechanism or algorithm) which can achieve the mutual exclusion is called "mutual exclusion protocol".

QLOCK is realized by using a unique global queue (first-in first-out storage) of agent names (or identifiers) as follows.

- Each of an unbounded number of agents who participates in the protocol behaves as follows:
  - If an agent wants to use the common equipment and its name is not in the queue yet, put its name at the bottom of the queue.
  - If an agent wants to use the common equipment and its name is already in the queue, check if its name is on the top of the queue. If its name is on the top of the queue, start to use the common equipment. If its name is not on the top of the queue, wait until its name is on the top of the queue.
  - If the agent finishes to use the common equipment, remove its name from the top of the queue.
- The protocol starts from the state with the empty queue.

4.1 QLOCK System Specifications[12]

The QLOCK protocol is specified by the following three modules WT, TY, EXc.

A state configuration (term of sort State) of QLOCK is modeled as a pair (term composed with a operator _$_) of the global queue (term of sort Qu) and the set of terms (lb[A:Aid]: L:Label) for all the agents. Term (lb[A:Aid]: L:Label) of sort Aob (agent observation) represents that an agent A is at a section denoted by a label (lb) L. A label is either rs (remainder section), ws (waiting section), or cs (critical section). A set is composed with associative, commutative, and idempotent binary operators "_ _". A term representing a set of terms of sort Aob is of sort Aobs (agent observation set).

The transition rule of the module TY indicates that if the top element of the queue is A:Aid (i.e. Qu is (A:Aid & Q:Qu)) and agent A is at waiting section ws (i.e. the label of A is ws; (lb[A:Aid]: ws)) then A gets into cs (i.e. (lb[A]: cs)) without changing contents of the queue (i.e. Qu is (A & Q)). The other two transition rules can be read similarly. Note that the modules WT, TY, EXc formulate the three actions explained above precisely and succinctly. QLOCKsys1 is just combining the three modules.

```
-- wt: want transition
mod! WT {pr(STATE)
trans[wt]:    (Q:Qu    $ ((lb[A:Aid]: rs) AS:Aobs))
```

---

[12] The specifications explained in this section are in the file qlock-sys.cafe on the web page.

```
         => ((Q & A) $ ((lb[A    ]: ws) AS)) . }
-- ty: try transition
mod! TY {pr(STATE)
trans[ty]:   ((A:Aid & Q:Qu) $ ((lb[A]: ws) AS:Aobs))
          => ((A     & Q)    $ ((lb[A]: cs) AS)) . }
-- exc: exit transition with a condition
mod! EXc {pr(STATE)
ctrans[exc]:   ((A1:Aid & Q:Qu) $ ((lb[A2:Aid]: cs) AS:Aobs))
            => (          Q     $ ((lb[A2    ]: rs) AS)) if (A1 = A2) . }
-- system specification of QLOCK
mod! QLOCKsys1{pr(WT + TY + EXc)}
```

An unconditional transition rule starts with `trans`, contains the rule's name `[_]:`, a current state term, `=>`, a next state term, and should end with "$_\sqcup$.". A conditional transition rule starts with `ctrans`, contains same components as `trans`, and `if` followed by a condition (a predicate) before "$_\sqcup$.".

Note that the term `Q:Qu` matches any term of sort `Qu` and the term `(lb[A:Aid]: rs)` matches any term `(lb[aid]: rs)` with `aid` of sort `Aid`. Note also that the second component of a state configuration is a set (i.e. a term composed with associative, commutative, and idempotent binary constructors "`_ _`") These implies that the left-hand side of the transition rule `wt` matches to a state in multiple ways depending on how many agents with `rs` are in the state, and an unbounded number of transitions may be defined by the rule `wt`. The rules `ty` and `exc` have a similar nature.

For STATEn with `Aid` $= \{a_1, \cdots, a_n\}$, the transition rules `wt,ty,exc` define the one step transition relations $\rightarrow_{\text{wt}}, \rightarrow_{\text{ty}}, \rightarrow_{\text{exc}}$ respectively on the state space `State` $= T_{\Sigma_{\text{STATE}n}}$. QLOCKsys1$n$ with STATEn defines a set of transitions $Tr_{\text{QLOCKsys1}n} \overset{\text{def}}{=} (\rightarrow_{\text{wt}} \cup \rightarrow_{\text{ty}} \cup \rightarrow_{\text{exc}}) \subseteq (T_{\Sigma_{\text{STATE}n}} \times T_{\Sigma_{\text{STATE}n}})$ (see Sect. 2.1).

It is easily seen that the rule `ty` can be translated to a conditional rule, and the rule `exc` can be translated to an unconditional rule.[13]

4.2 QLOCK Property Specifications[14]

The property specifications for the generate & check method are assumed to specify the following predicates.

1. The possible inductive invariant predicate inv, the target state predicate p^t, and the initial state predicate $init$ for verifying invariant properties (Sect. 2.7).
2. The invariant predicates $inv1$, $inv2$, $inv3$, $inv4$, the predicates p, q and the measure function m for verifying (p leads-to q) properties (Sect. 2.8).

Usually, the predicates are specified as conjunctions of elemental predicates.

---

[13] The file `qlock-sys-ex.cafe` on the web page contains the translated `tyc` and `ex` rules.

[14] The modules in this section is in the file `qlock-prop.cafe` unless otherwise stated.

For QLOCK, we adopt a strategy of formalizing necessary functions and elemental predicates based on the Peano style natural numbers. The strategy works well especially for specifying the measure function m for a (p leads-to q) property as demonstrated in Sect. 5.2.

[Predicates and Measure Function for (p Leads-to q) Property]: Following module PQMonState specifies the p and q predicates and a measure function m for defining a (p leads-to q) property of QLOCK. The function lags denotes the label of an agent in a state (see the module STATEfuns). Function #dms is specified in the following module STATEfuns-pq and intended to denote the properly decreasing natural numbers according to the state transitions of QLOCK. Note that function #ls (the number of a label in a state) is specified in module STATEfuns. PNAT*[15] is PNAT with the _*_ (times) operation.

```
mod! STATEfuns-pq {ex(STATEfuns + PNAT*)
  op #daq : Qu Aid -> Nat .  op #ccs : State -> Nat .
  ...
  -- decreasing Nat measure for the (p leads-to q) property
  op #dms : State Aid -> Nat .
  eq #dms(S:State,A:Aid)
    = ((s s s 0) * #daq(qu(S),A)) + #ls(S,rs) + #ccs(S) . }
mod! PQMonState {ex(STATEpcj + STATEfuns-pq)
  ops p q : State Aid -> Bool . eq p(S:State,A:Aid) = (lags(S,A) = ws) .
                                eq q(S:State,A:Aid) = (lags(S,A) = cs) .
  op m : State Aid -> Nat.PNAT . eq m(S:State,A:Aid) = #dms(S,A) . }
```

Based on this specification, the (q leads-to q) property of QLOCK is verified in Sect. 5.2.

4.3 Extended State (State % Aid) and Possible Inductive Invariants

For using Generate&Check-S or Generate&Check-T1/T2 with p(S:State,A:Aid), q(S:State,A:Aid), and m(S:State,A:Aid) of the module PQMonState, State needs to be extended with Aid. The following module EX-PQMonST extends State of PQMonState with Aid for constructing ExState (= (State % Aid)) by applying the parameterized module PCJ-EX-STATE to the module PQMonState.

```
-- ExState with PQMonST/State and Aid/Date
-- and a new invariant qas on ExState = (State % Date)
mod! EX-PQMonST {
  pr((PCJ-EX-STATE(PQMonST{sort Ste -> State, sort Data -> Aid,
                          sort Pnm -> Pname, sort PnmSeq -> PnameSeq})))
  -- if an agent is in the Qu then the agent is in Aobs
  op qas : -> ExPname .
  eq cj(qas,((Q:Qu $ AS:Aobs) % A:Aid))
    = ((#aq(Q, A) = s 0) implies not(#ass(AS, A) = 0)) . }
```

---

[15] The module PNAT* is in the file qlock-natQuSet.cafe on the web page.

A new elemental predicate is defined with the name qas of sort ExPname in EX-PQMonST. By using the predicate names defined in STATEpred-inv and qas, possible inductive invariants are specified in the following module INV.

```
-- possible inductive invariant predicates
mod! INV {ex(EX-PQMonST + STATEpred-inv)
  ops inv1 inv2 : -> PnameSeq .  ops inv3 : -> ExPnameSeq .
  --'wfs mx qep rs ws cs' are defined in STATEpred-inv
  eq inv1 = wfs .  eq inv2 = mx qep rs ws cs . eq inv3 = qas . }
```

In Sect. 5.1, the conjunction "inv1 inv2 inv3" is proved to be an inductive invariant. As a matter of fact, any of inv1, inv2, or inv3 can be proved to be an inductive invariant.

5   QLOCK Proof Scores

The three base modules Q-INV-1v, Q-INV-2v, Q-INV-3q for the three invariant verification conditions and the four base modules Q-PQ-1q, Q-PQ-2q, Q-PQ-3v, Q-PQ-4v for the four (p leads-to q) property verification conditions of QLOCK are obtained by instantiating the formal parameter modules of the parameterized modules INV-1v, INV-2v, INV-3q, and PQ-1q, PQ-2q, PQ-3v, PQ-4v with the module EX-PQMonST.[16]

5.1   Proof Scores for Invariant Properties

The following proof scores prove the three verification conditions for invariant properties under the condition in which each predicate inv, p^t, or $init$ is defined with the predicate name sequence "inv1 inv2 inv3", mx, or init respectively. [Q-INV-1-genCheck[17]]: The reduction "red ck ." in the following proof score proves the first verification condition "(cj(inv1 inv2 inv3,S) implies cj(mx, S)) for any state S of sort ExState" if it returns ($):Ind.

```
mod! Q-INV-1-genCheck {ex(Q-INV-1)
  op ck : -> IndTr . eq ck = check(['s:ExState]) . }
open Q-INV-1-genCheck . pr(INV)
  eq p-iinv = inv1 inv2 inv3 . eq p^t = mx . red ck . close
```

open opens the module Q-INV-1-genCheck; pr(INV) imports a necessary module INV; two equations reify predicate names p-iinv and p^t as "inv1 inv2 inv3" and mx respectively; "red ck ." reduces check(['s:ExState]) and returns the result, and close closes the opened tentative module.

cj(inv1 inv2 inv3,S) trivially implies cj(mx,S) for any S because inv2 includes mx, hence the most general trivial state pattern 's:ExState is enough.

---

[16] The base modules Q-INV-1v, Q-INV-2v, Q-INV-3q, and Q-PQ-1q, Q-PQ-2q, Q-PQ-3v, Q-PQ-4v are in the file qlock-genCheck.cafe on the web page.

[17] The module Q-INV-1-genCheck is in the file qlock-inv-1-ps.cafe on the web.

[`Q-INV-2-genCheck`[18], `Q-INV-3-genCheck`[19]]: The proof score for the second
verification condition using the module `Q-INV-2-genCheck` is constructed similarly
to the proof score for the third verification condition using `Q-INV-3-genCheck`, and
is omitted.

The reduction "`red ck .`" in the following proof score proves the third verification condition "(`cj(inv1 inv2 inv3,S) implies cj(inv1 inv2 inv3,SS)`) for
any transitions (`S,SS`) of $QLOCK$ with `ExState` (i.e. `S` and `SS` are of sort `ExState`)"
if it returns (`$`):`Ind`.

```
mod! Q-INV-3-genCheck {ex(Q-INV-3 + GENstTerm + CONSTandLITL)
  ops sst1 sst2 sst3 : -> SqSqTr .
  eq sst1 = [(g("_%_")[(g("_$_")[empQ,(g("lb[_]:__")[b1,rs,as])]),
                      (b1;b2)]),SS:ExState,CC:Bool]  ||
            [(g("_%_")[(g("_$_")[(b1 & q),(g("lb[_]:__")[(b1;b2),rs,as])]),
                      (b1;b2;b3)]),SS,CC] .
  eq sst2 = [(g("_%_")[(g("_$_")[(b1 & q),(g("lb[_]:__")[b1,ws,as])]),
                      (b1;b2)]),SS:ExState,CC:Bool] .
  eq sst3 = [(g("_%_")[(g("_$_")[(b1 & q),(g("lb[_]:__")[(b1;b2),cs,as])]),
                      (b1;b2;b3)]),SS:ExState,CC:Bool] .
  op ck : -> IndTr .  eq ck = check(sst1 || sst2 || sst3) . }
-- Generate&Check-T1
open Q-INV-3-genCheck . pr(QLOCKsys1 + INV + FACTtbu)
  eq p-iinv = inv1 inv2 inv3 . red ck . close
```

The module `GENstTerm`[20] is necessary to use `g__` in the alternative script (the
argument of `check_`). The alternative script `sst1` specifies three state patterns
that cover the left-hand side (`Q:Qu $ ((lb[A:Aid]: rs) as:Aobs)`) of the transition rule `wt`. `b1`, `b2`, `b3` are fresh constant literals declared in the module
`CONSTandLITL`[21]. Note that all possibilities of tree occurrences of `Aid` are covered by (`...[(...[(b1 ...),(...[(b1;b2)...])]),(b1;b2;b3)]`). Note also that
`SS:State,CC:Bool` in `sst1` are necessary for using the built-in search predicate "`_=(*,1)=>+_if_suchThat_{_}`". The script `sst2` specifies the state pattern
(`(b1 & q) $ ((lb[b1]: ws) as)`) that directly cover the left-hand side (`(A:Aid
& Q:Qu) $ ((lb[A]: ws) AS:Aobs)`) of the transition rule `ty`. The alternative
script `sst3` specifies state patterns that cover the left-hand side (`(A1:Aid &
Q:Qu) $ ((lb[A2: Aid]: cs) as:Aobs)`) of the transition rule `exc`. Hence, by
Generate&Check-T1 in Sect. 2.6, the correctness of the above proof score is
implied.

Note that the module `FACTtbu`[22] that declare the fundamental facts like "`eq
[NatGt1]: ((N:Nat.PNAT = 0) and (N > 0)) = false .`" is necessary here. Facts
declared in `FACTtbu` can be proved with other proof scores usually by using induction on term structures.

---

[18] The module `Q-INV-2-genCheck` is in the file `qlock-inv-2-ps.cafe` on the web page.
[19] The module `Q-INV-3-genCheck` is in the file `qlock-inv-3-ps.cafe` on the web page.
[20] The module `GENstTerm` is in the file `qlock-genStTerm.cafe` on the web page.
[21] The module `CONSTandLITL` is in the file `qlock-constAndLitl.cafe` on the web page.
[22] The module `FACTtbu` is in the file `qlock-factTbu.cafe` on the web page.

By using Generate&Check-T2 instead of Generate&Check-T1, checking the transition rules one by one is possible.[23] Generate&Check-T2 is an effective way for detecting errors in specifications and for finding necessary lemmas during proof score developments.

5.2   Proof Scores for (p Leads-to q) Property

This section presents proof scores for the four verification conditions of the (p(_,A:Aid) leads-to q(_,A:Aid)) property, where the predicates p and q are defined in the module PQMonState. The property says "if an agent gets the ws label then it will surely get the cs label".
[Q-PQ-1-genCheck[24], Q-PQ-2-genCheck[25]]: The reduction "red ck ." in the following proof score proves the first verification condition "(cj(pq-1-inv,S) and p(S) and not(q(S))) implies (p(SS) or q(SS)) for any transition (S,SS) of QLOCK with ExState" if it returns ($):Ind.

```
mod! Q-PQ-1-genCheck {ex(Q-PQ-1 + GENstTerm + CONSTandLITL)
  op ck : -> IndTr .
  eq ck = check(
  [(g("_%_")[(g("_$_")[empQ,(g("lb[_]:__")[b1,rs,as])]),
           (b1;b2)]),SS:ExState,CC:Bool] ||
  [(g("_%_")[(g("_$_")[(b1 & q),(g("lb[_]:__")[(b1;b2),rs,as])]),
           (b1;b2;b3)]),SS,CC]           ||
   [(g("_%_")[(g("_$_")[(b1 & q),(g("lb[_]:__")[b1,ws,as])]),
           (b1;b2)]),SS,CC]           ||
   [(g("_%_")[(g("_$_")[(b1 & q),(g("lb[_]:__")[(b1;b2),cs,as])]),
           (b1;b2;b3)]),SS,CC] ) . }
open Q-PQ-1-genCheck . pr(QLOCKsys1) red ck . close
```

pq-1-inv is not reified in the open...close clause, and it means cj(pq-1-inv,S) in the premises of the first verification condition is not necessary.
The alternative script at the argument position of check_ specifies state patterns of sort ExState. The same discussion about the module Q-INV-3-genCheck applies and the alternative script of module Q-PQ-1-genCheck covers the left-hand sides of the three transition rules on ExState. Note that the three transition rules on ExState are not defined directly, but induced from the three transition rules on State with the following two equations defined in the module EX-STATE.

```
eq ((S:State % D:Data) =(*,1)=>+ (SS:State % D)
      if CC:Bool suchThat B:Bool {I:Infom})
    = (S =(*,1)=>+ SS if CC suchThat B {I}) .
eq ((S:State % D:Data) =(*,1)=>+ (SS:State % D)) = (S =(*,1)=>+ SS) .
```

---

[23] You can see the proof score using Generate&Check-T2 in the file qlock-inv-3-ps.cafe on the web page.

[24] The module Q-PQ-1-genCheck is in the file qlock-pq-1-ps.cafe on the web page.

[25] The module Q-PQ-2-genCheck is in the file qlock-pq-2-ps.cafe on the web page.

Hence, by Generate&Check-T1 in Sect. 2.6, the correctness of the above proof score is implied.

The proof score for the second verification condition using the module Q-PQ-2-genCheck can be constructed in a similar way and is omitted.

[Q-PQ-3-genCheck[26], Q-PQ-4-genCheck[27]]: The reduction "red ck ." in the following proof score proves the third verification condition "((cj(pq-3-inv,S) and p(S) and not(q (S))) implies (S =(*,1)=>+ SS))" for any state S of QLOCK with ExState" if it returns ($):Ind. Note that the verification condition is defined on the transitions (S,SS) but the second argument of the built-in search predicate _=(*,1)=>_ is searched automatically by the CafeOBJ system when the first argument is fixed.

```
mod! Q-PQ-3-genCheck {
ex(Q-PQ-3 + CONSTandLITL + GENstTerm)
  op ck : -> IndTr .
  eq ck = check (
    [(g("_%_")[(g("_$_")[q,empty]),b1]),SS:ExState] ||
    [(g("_%_")[(g("_$_")[empQ,(g("lb[_]:__")[b1,(rs;ws;cs),as])]),
              (b1;b2)]),SS] ||
    [(g("_%_")[(g("_$_")[(b1 & q),(g("lb[_]:__")[b1,(rs;ws;cs),as])]),
              (b1;b2)]),SS] ) . }
open Q-PQ-3-genCheck . pr(QLOCKsys1 + INV)
  eq pq-3-inv = inv1 inv2 . red ck . close
```

Because the predicate defined via "inv3 = qas" is proved to be an invariant by the proof score including the module Q-INV-3-genCheck, any Aid in the queue is in Aobs for any reachable state. Therefore, any reachable state is an instance of the state pattern (((A1:Aid & Q:Qu) $ ((lb[A1]: L:Label) AS:Aobs)) % A2:Aid) of sort ExState. This fact implies that the alternative script at the argument position of check_ expands to the term of sort SqSqTr that includes the state patterns subsuming all the reachable states.[28] This, in turn, implies the correctness of the above proof score. The proof score for the fourth verification condition using Q-PQ-4-genCheck is constructed similarly and omitted.

6   Conclusion

6.1   Related Works

There are many researches on verifications of transition systems, and we only give a brief general view and point out most related works based on Maude [16].

---

[26] The module Q-PQ-3-genCheck is in the file qlock-pq-3-ps.cafe on the web page.

[27] The module Q-PQ-4-genCheck is in the file qlock-pq-4-ps.cafe on the web page.

[28] You can see the expanded term after the eof in the file qlocik-pq-3-ps.cafe on the web page.

Verification methods for transition systems are largely classified into deductive and algorithmic ones. The majority of the deductive methods are applications of theorem proving methods/systems [6,15,18,20] to verifications of concurrent systems or distributed protocols with infinite states. Most dominant algorithmic methods are based on model checking methods/systems [2,5] and are targeted on automatic verifications of temporal properties of finite state transition systems. The generate & check method is a deductive method with algorithmic combinatorial generations of cover sets. Moreover reduction with equations is only one deduction mechanism.

Maude [16] is a sister language of CafeOBJ and both languages share many important features. The idea that underlies the transition specification (Σ, E, TR) and the transition system $((T_\Sigma)_{\texttt{State}}/(=_E)_{\texttt{State}}, \rightarrow_{TR}, In)$ in Sect. 2.1 is the same as the one for the topmost rewrite theory [17,21,22]. Maude's basic logic is rewriting logic [17] and verification of transition systems with Maude focuses on sophisticated model checking with a powerful associative and/or commutative rewriting engine. There are recent attempts to extend the model checking with Maude for verifying infinite state transition systems [1,7]. They are based on narrowing with unification, whereas the generate & check method is based on cover sets with ordinary matching and reduction.

6.2   Some Features of Generic Proof Scores, Generate & Check Method, and CafeOBJ

Besides the QLOCK example, we have already developed ABP proof scores by applying the generic proof scores. Three base modules ABP-INV-1v, ABP-INV-2v, ABP-INV-3q for three ABP invariant verification conditions are obtained by instantiating the formal parameter modules of the parameterized modules INV-1v, INV-2v, INV-3q with module STATE.[29] The remarkable structural similarity between the three base modules ABP-INV-1v, ABP-INV-2v, ABP-INV-3q and Q-INV-1v, Q-INV-2v, Q-INV-3q for the three invariant verification conditions of ABP and QLOCK shows that the generic proof scores work quite nicely.[30]

Most difficult parts in constructing proof scores using the generate & check method include (1) finding a cover set in an appropriate abstraction level and (2) finding necessary lemmas. These inherently require human insight and it is hard to expect easy solutions. The generation of elements of a cover set, however, can be guided by inspecting the lefthand sides of transition rules. It also helps to find lemmas to identify an element of the cover set for which a verification condition can not be reduced to true.

Several theorem proving methods have been developed in CafeOBJ including (a) OTS (Observational Transition System) method [19] (b) the generate & check method [10] and (c) CITP (Constructor based Inductive Theorem Proving) method [12]. The generate & check method makes significant use of state

---

[29] The base modules ABP-INV-1v, ABP-INV-2v, ABP-INV-3q are in the file abp-genCheck.cafe on the web page.

[30] You can see the similarity by looking into the files abp-genCheck.cafe and qlock-genCheck.cafe on the web page.

configurations and transition (rewriting) rules over them, and is a complement to the OTS method. The CITP method is another recent development and is now implemented in CafeOBJ system.[31] Current proof scores in CafeOBJ have a potential to combine these three methods in an appropriate way.

6.3   Future Issues

The generate & check method is more important for large and/or complex systems, for it is difficult to do case analyses manually for them. Once a state configuration is properly designed, a large number of patterns (i.e. elements of a cover set) that cover all possible cases can be generated and checked. The generic proof scores (i.e. the seven parameterized modules) in Sect. 3 has the potential to make the applications of the generate & check method easy and transparent. Although, we are still in an early stage of applying the generic proof scores, the following can be expected.

- The verifications of larger and/or more complex systems become possible.
- The achieved proof scores are well structured and transparent, and have a high potential to be good quality verification documents.

Investigating to what extent the above expectations will be realized is interesting and important future issue.

Acknowledgments. It is a great pleasure for the author (KF) to have the chance to prepare this paper for the Festschrift in honor of Professor José Meseguer who has originated the rewriting logic and been leading the development of the area. The work reported in this paper is based on that development.

Comments from anonymous reviewers help to improve the quality of the paper and are appreciated.

This work was supported in part by Grant-in-Aid for Scientific Research (S) 23220002 from Japan Society for the Promotion of Science (JSPS).

References

1. Bae, K., Escobar, S., Meseguer, J.: Abstract logical model checking of infinite-state systems using narrowing. In: van Raamsdonk, F. (ed.) RTA. LIPIcs, vol. 21, pp. 81–96. Schloss Dagstuhl - Leibniz-Zentrum fuer Informatik (2013)
2. Baier, C., Katoen, J.P.: Principles of Model Checking. MIT Press, Cambridge (2008)
3. CafeOBJ (2015). http://cafeobj.org/
4. Chandy, K.M., Misra, J.: Parallel Program Design - a Foundation. Addison-Wesley, Boston (1989)
5. Clarke, E.M., Grumberg, O., Peled, D.: Model Checking. MIT Press, Cambridge (2001)

---

[31] Examples of CITP usage (i.e. `:goal` and `:ctf`) can be found in file `abp-factTbu-ps.cafe` on the web page.

6. Coq (2015). http://coq.inria.fr
7. Escobar, S., Meseguer, J.: Symbolic model checking of infinite-state systems using narrowing. In: Baader, F. (ed.) RTA 2007. LNCS, vol. 4533, pp. 153–168. Springer, Heidelberg (2007)
8. Futatsugi, K.: Verifying specifications with proof scores in CafeOBJ. In: Proceedings of 21st IEEE/ACM International Conference on Automated Software Engineering (ASE 2006), pp. 3–10. IEEE Computer Society (2006)
9. Futatsugi, K.: Fostering proof scores in CafeOBJ. In: Dong, J.S., Zhu, H. (eds.) ICFEM 2010. LNCS, vol. 6447, pp. 1–20. Springer, Heidelberg (2010)
10. Futatsugi, K.: Generate & check method for verifying transition systems in CafeOBJ. In: Nicola, R.D., Hennicker, R. (eds.) Software, Services, and Systems. LNCS, vol. 8950, pp. 171–192. Springer, Switzerland (2015)
11. Futatsugi, K., Găină, D., Ogata, K.: Principles of proof scores in CafeOBJ. Theor. Comput. Sci. **464**, 90–112 (2012)
12. Găină, D., Lucanu, D., Ogata, K., Futatsugi, K.: On automation of OTS/CafeOBJ method. In: Iida, S., Meseguer, J., Ogata, K. (eds.) Specification, Algebra, and Software. LNCS, vol. 8373, pp. 578–602. Springer, Heidelberg (2014)
13. Goguen, J.A., Meseguer, J.: Order-sorted algebra I: equational deduction for multiple inheritance, overloading, exceptions and partial operations. Theor. Comput. Sci. **105**(2), 217–273 (1992)
14. Guttag, J.V., Horning, J.J., Garland, S.J., Jones, K.D., Modet, A., Wing, J.M.: Larch Languages and Tools for Formal Specification. Springer, New York (1993)
15. HOL (2015). http://hol.sourceforge.net
16. Maude (2015). http://maude.cs.uiuc.edu/
17. Meseguer, J.: Twenty years of rewriting logic. J. Log. Algebr. Program. **81**(7–8), 721–781 (2012)
18. Nipkow, T., Paulson, L.C., Wenzel, M.: Isabelle/HOL. LNCS, vol. 2283, p. 3. Springer, Heidelberg (2002)
19. Ogata, K., Futatsugi, K.: Proof scores in the OTS/CafeOBJ method. In: Najm, E., Nestmann, U., Stevens, P. (eds.) FMOODS 2003. LNCS, vol. 2884, pp. 170–184. Springer, Heidelberg (2003)
20. PVS (2015). http://pvs.csl.sri.com
21. Rocha, C., Meseguer, J.: Proving safety properties of rewrite theories. Technical report, University of Illinois at Urbana-Champaign (2010)
22. Rocha, C., Meseguer, J.: Proving safety properties of rewrite theories. In: Corradini, A., Klin, B., Cîrstea, C. (eds.) CALCO 2011. LNCS, vol. 6859, pp. 314–328. Springer, Heidelberg (2011)

Function Calls at Frozen Positions in Termination of Context-Sensitive Rewriting

Raúl Gutiérrez(✉) and Salvador Lucas

DSIC, Universitat Politècnica de València,
Camino de Vera S/N, 46022 Valencia, Spain
{rgutierrez,slucas}@dsic.upv.es

Abstract. *Context-sensitive rewriting* (CSR) is a variant of rewriting where only selected arguments of function symbols can be rewritten. Consequently, the subterm positions of a term are classified as either *active*, i.e., positions of subterms that *can be rewritten*; or *frozen*, i.e., positions that can*not*. Frozen positions can be used to denote subexpressions whose evaluation is *delayed* or just *forbidden*. A typical example is the *if-then-else* operator whose second and third arguments are not evaluated until the evaluation of the first argument yields either *true* or *false*. Imposing replacement restrictions can *improve* the termination behavior of rewriting-based computational systems. Termination of CSR has been investigated by several authors and a number of automatic tools are able to prove it. In this paper, we analyze how frozen subterms affect termination of CSR. This analysis helps us to improve our *Context-Sensitive Dependency Pair (CS-DP) framework* for automatically proving termination of CSR. We have implemented these improvements in our tool MU-TERM. The experiments show the power of the improvements in practice.

Keywords: Context-sensitive rewriting · Termination · Dependency pairs

1 Introduction

During the *4th International Workshop on Rewriting Logic and its Applications, WRLA 2002*, a tutorial by the second author entitled *Context-Sensitive Rewriting Techniques for Programs With Strategy Annotations* was the starting point of a friendly cooperation with José Meseguer leading to multiple exchanges of students and people from the UIUC and the UPV, and to the development of fruitful joint work on Rewriting Logic, Maude, and, in general, the analysis, verification, and optimization of declarative programming languages.

Partially supported by the EU (FEDER), MINECO project TIN 2013-45732-C4-1-P, and GV project PROMETEO/2011/052. Salvador Lucas' research was developed during a sabbatical year at the CS Dept. of the UIUC and was also partially supported by NSF grant CNS 13-19109. Raúl Gutiérrez is also partially supported by a Juan de la Cierva Fellowship from the Spanish MINECO, ref. JCI-2012-13528.

© Springer International Publishing Switzerland 2015
N. Martí-Oliet et al. (Eds.): Meseguer Festschrift, LNCS 9200, pp. 311–330, 2015.
DOI: 10.1007/978-3-319-23165-5_15

Actually, the idea of *strategy annotation* (where the list of arguments whose evaluation is *allowed* is explicitly given for each function symbol) originally introduced by José and other colleagues as part of the design of OBJ2 [11] anticipated the main ideas underlying the development of *Context-Sensitive Rewriting* for a rather different purpose[1]. On the basis of previous work in [24,25], in the aforementioned tutorial *Context-Sensitive Rewriting* (CSR, [23]) was shown useful to model rewriting-based programming languages like CafeOBJ [12], ELAN [8], OBJ [15], and Maude [9] that are able to use such kind of strategies.

In CSR, we start with a pair $(\mathcal{R}, \mu)$ (often called a CS-TRS) consisting of a *Term Rewriting System* (TRS) $\mathcal{R}$ and a *replacement map* μ, i.e., a mapping from a signature $\mathcal{F}$ into natural numbers that satisfies $\mu(f) \subseteq \{1, \ldots, \mathrm{ar}(f)\}$ for each function symbol f in the signature $\mathcal{F}$, where $\mathrm{ar}(f)$ is the arity of f. Here, μ is used to *discriminate* the argument positions on which the rewrite steps are allowed. In this way, we can avoid undesired computations and (in many cases) obtain a terminating behavior for the TRS (with respect to the context-sensitive rewrite relation). Strategy annotations are still used in CafeOBJ and Maude. In Maude, actually, *frozen arguments* have been recently introduced as a powerful mechanism to avoid undesired reductions. Frozen arguments are even closer to CSR, as they are just the complement of the *replacing arguments* specified by a replacement map μ: the i-th argument of f is frozen iff $i \notin \mu(f)$.

Using CSR, we can easily model the evaluation of expressions which *avoid* or *delay* the evaluation of some of their arguments. Paramount examples are *if-then-else* expressions, some boolean operators (*and/or*) and *lazy cons* operators for list construction.

Example 1. The following TRS $\mathcal{R}$ [29] provides a definition of factorial

$$0+x \to x \quad (1) \qquad \mathsf{zero}(0) \to \mathsf{true} \quad (6)$$
$$\mathsf{s}(x)+y \to \mathsf{s}(x+y) \quad (2) \qquad \mathsf{zero}(\mathsf{s}(x)) \to \mathsf{false} \quad (7)$$
$$\mathsf{p}(\mathsf{s}(x)) \to x \quad (3) \qquad \mathsf{fact}(x) \to \mathsf{if}(\mathsf{zero}(x), \mathsf{s}(0), x*\mathsf{fact}(\mathsf{p}(x))) \quad (8)$$
$$\mathsf{if}(\mathsf{true}, x, y) \to x \quad (4) \qquad 0*x \to 0 \quad (9)$$
$$\mathsf{if}(\mathsf{false}, x, y) \to y \quad (5) \qquad \mathsf{s}(x)*y \to y+(x*y) \quad (10)$$

With $\mu(\mathsf{if}) = \{1\}$ and $\mu(f) = \{1, \ldots, k\}$ for any other k-ary symbol f (i.e., the only function symbol which is restricted by μ is if), we can advantageously use CSR for handling the if-then-else operator: the second and third arguments of an expression $\mathsf{if}(b, s, t)$ are not evaluated until the guard b is evaluated to true or false. Without the replacement map, $\mathcal{R}$ is nonterminating because $\mathsf{fact}(x)$ calls $\mathsf{fact}(\mathsf{p}(x))$, which then calls $\mathsf{fact}(\mathsf{p}(\mathsf{p}(x)))$ and so on. Thanks to the replacement restrictions, though, we can evaluate $\mathsf{fact}(\mathsf{s}^n(0))$ to obtain the factorial $\mathsf{s}^{n!}(0)$ of a number n (encoded as $\mathsf{s}^n(0)$) by using CSR as follows:

$$\underline{\mathsf{fact}(\mathsf{s}^n(0))} \hookrightarrow_{(8),\mu} \mathsf{if}(\underline{\mathsf{zero}(\mathsf{s}^n(0))}, \mathsf{s}(0), \mathsf{s}^n(0)*\mathsf{fact}(\mathsf{p}(\mathsf{s}^n(0)))) \hookrightarrow_{(7),\mu} \cdots$$

---

[1] The notion of context-sensitive rewriting was developed as part of Lucas' Master Thesis (1994) to implement concurrent programming languages that, like the π-calculus, forbid reductions on some arguments of its operations.

This can be formally proved (see [23,27] and also [20] for an account of the algebraic semantics of context-sensitive specifications). Note that zero($s^n(0)$) *is forced to be reduced first to either* true *or* false *before evaluating the 'then' or 'else' expression, thus avoiding undesired reductions until the guard is fully evaluated.*

Direct techniques and frameworks for proving termination of CSR have been developed [1,3,17]. But, in practice, proving termination of some CS-TRSs with certain lazy structures as the *if-then-else* in the example can be difficult. In fact, finding an automatic proof of Example 1, and other examples like [13, Example 1] or [10, Example 3.2.14] are open problems since 1997, 2003 or 2008, respectively. The reason why these problems cannot be proved terminating by existing termination tools lies in the lack of sufficiently precise models of how the evaluation of expressions is *delayed* in context-sensitive computations. In this paper, we revisit this problem to obtain easier and mechanizable proofs of termination.

After some preliminaries in Sects. 2, 3 analyzes the role of frozen subterms in infinite μ-rewrite sequences, Sect. 4 models the activation of delayed subexpressions. Section 5 revises the characterization of the termination of CSR. Section 6 proposes a new notion of CS usable rules, the extended basic CS usables, that allows us to simplify termination proofs if the application conditions are satisfied, Sect. 7 shows the experimental evaluation and Sect. 8 concludes. An extended version of this paper including proofs can be found in [18].

2   Preliminaries

See [7] and [23] for basics on term rewriting and CSR, respectively. Throughout the paper, $\mathcal{X}$ denotes a countable set of variables and $\mathcal{F}$ denotes a signature, i.e., a set of function symbols each having a fixed arity given by a mapping ar : $\mathcal{F} \rightarrow \mathbb{N}$. The set of terms built from $\mathcal{F}$ and $\mathcal{X}$ is $\mathcal{T}(\mathcal{F}, \mathcal{X})$. Terms are viewed as labeled trees in the usual way. The symbol labeling the root of the term s is denoted as root(s). Positions $p, q, \ldots$ are represented by chains of positive natural numbers used to address subterms of s. Given positions p, q, we denote their concatenation as $p.q$. We denote the empty chain by Λ. Positions are ordered by the standard prefix ordering: $p \leq q$ if $\exists q'$ such that $q = p.q'$. The set of positions of a term s is $\mathcal{P}os(s)$. If p is a position, and Q is a set of positions, $p.Q = \{p.q \mid q \in Q\}$. For a replacement map μ, the set of *active positions* $\mathcal{P}os^\mu(s)$ of $s \in \mathcal{T}(\mathcal{F}, \mathcal{X})$ is: $\mathcal{P}os^\mu(s) = \{\Lambda\}$, if $s \in \mathcal{X}$ and $\mathcal{P}os^\mu(s) = \{\Lambda\} \cup \bigcup_{i \in \mu(\text{root}(s))} i.\mathcal{P}os^\mu(s|_i)$, if $s \notin \mathcal{X}$. We write $s \trianglerighteq t$, t is a subterm of s, if there is $p \in \mathcal{P}os(s)$ such that $t = s|_p$ and $s \triangleright t$, t is a proper subterm of s, if $s \trianglerighteq t$ and $s \neq t$. Given a replacement map μ, we write $s \trianglerighteq_\mu t$, t is a μ-replacing subterm of s, if there is $p \in \mathcal{P}os^\mu(s)$ such that $t = s|_p$ and $s \triangleright_\mu t$, t is a proper μ-replacing subterm of s, if $s \trianglerighteq_\mu t$ and $s \neq t$. Moreover, we write $s \triangleright_{\mu\!\!\!/} t$, t is a non-μ-replacing subterm of s, if there is a *frozen position* p, i.e. $p \in \mathcal{P}os^{\mu\!\!\!/}(s)$ where $\mathcal{P}os^{\mu\!\!\!/}(s) = \mathcal{P}os(s) - \mathcal{P}os^\mu(s)$, such that $t = s|_p$. Let $\mathcal{V}ar(s) = \{x \in \mathcal{X} \mid \exists p \in \mathcal{P}os(s), s|_p = x\}$, $\mathcal{V}ar^\mu(s) = \{x \in \mathcal{V}ar(s) \mid \exists p \in \mathcal{P}os^\mu(s), s|_p = x\}$ and $\mathcal{V}ar^{\mu\!\!\!/}(s) = \{x \in \mathcal{V}ar(s) \mid s \triangleright_{\mu\!\!\!/} x\}$. A *context* is a term

$C \in \mathcal{T}(\mathcal{F} \cup \{\Box\}, \mathcal{X})$ with zero or more 'holes' $\Box$ (a fresh constant symbol). We write $C[\ ]_p$ to denote that there is a (usually single) hole $\Box$ at position p of C. Generally, we write $C[\ ]$ to denote an arbitrary context (where the number and location of the holes is clarified 'in situ') and $C[t_1, \ldots, t_n]$ to denote the term obtained by filling the holes of a context $C[\ ]$ with terms $t_1, \ldots, t_n$. $C[\ ] = \Box$ is called the *empty* context.

A rewrite rule is an ordered pair (ℓ, r), written $\ell \rightarrow r$, with $\ell, r \in \mathcal{T}(\mathcal{F}, \mathcal{X})$, $\ell \notin \mathcal{X}$ and $\mathcal{V}ar(r) \subseteq \mathcal{V}ar(\ell)$. A TRS is a pair $\mathcal{R} = (\mathcal{F}, R)$ where R is a set of rewrite rules. Given $\mathcal{R} = (\mathcal{F}, R)$, we consider $\mathcal{F}$ as the disjoint union $\mathcal{F} = \mathcal{C} \uplus \mathcal{D}$ of symbols $c \in \mathcal{C}$, called *constructors* and symbols $f \in \mathcal{D}$, called *defined functions*, where $\mathcal{D} = \{\mathrm{root}(\ell) \mid \ell \rightarrow r \in R\}$ and $\mathcal{C} = \mathcal{F} - \mathcal{D}$. Given a CS-TRS $(\mathcal{R}, \mu)$, we have $s \hookrightarrow_{\mathcal{R},\mu} t$ (alternatively $s \overset{p}{\hookrightarrow}_{\mathcal{R},\mu} t$ if we want to make the position explicit) if there are $\ell \rightarrow r \in \mathcal{R}$, $p \in \mathcal{P}os^\mu(s)$ and a substitution σ with $s|_p = \ell\sigma$ and $t = s[r\sigma]_p$. A CS-TRS $(\mathcal{R}, \mu)$ is *terminating* if $\hookrightarrow_{\mathcal{R},\mu}$ is well-founded.

3   Minimal Non-μ-Terminating Terms at Frozen Positions

In this section we investigate how frozen subterms affect termination of CSR. Our analysis is used in Sect. 4 to obtain a more precise model of termination of CSR using Context-Sensitive Dependency Pairs (CS-DPs, [3]). If a TRS $\mathcal{R}$ is nonterminating, then *terms* are either terminating or nonterminating. The subset $\mathcal{T}_\infty$ of *minimal* nonterminating terms consists of nonterminating terms whose proper subterms are all terminating. And the following observations are in order [21,22]: (1) every nonterminating term s contains a subterm $t \in \mathcal{T}_\infty$, (2) $\mathrm{root}(t)$ is a defined symbol of $\mathcal{R}$, and (3) minimality is preserved under inner rewritings:

Lemma 1. *Let $\mathcal{R}$ be a TRS. For every term $s \in \mathcal{T}_\infty$, if $s \overset{>\Lambda}{\longrightarrow}_{\mathcal{R}} t$ and t is nonterminating then $t \in \mathcal{T}_\infty$.*

In CSR, if a CS-TRS $(\mathcal{R}, \mu)$ is nonterminating, among non-μ-terminating terms we distinguish the subset $\mathcal{T}_{\infty,\mu}$ of *strongly minimal* non-μ-terminating terms, whose proper subterms are *all* μ-terminating. But unlike minimality for rewriting, strong minimality is *not* preserved under inner μ-rewritings.

Example 2. Consider the following TRS $\mathcal{R}$ [3, Example 3]:

$$\mathsf{a} \rightarrow \mathsf{c}(\mathsf{f}(\mathsf{a}))\ (11) \qquad\qquad \mathsf{f}(\mathsf{c}(x)) \rightarrow x\ (12)$$

together with $\mu(\mathsf{c}) = \emptyset$ and $\mu(\mathsf{f}) = \{1\}$, and the term $\mathsf{f}(\mathsf{a}) \in \mathcal{T}_{\infty,\mu}$. If we apply (11) to the proper subterm a, we obtain $\mathsf{f}(\mathsf{c}(\mathsf{f}(\mathsf{a}))) \notin \mathcal{T}_{\infty,\mu}$ because $\mathsf{f}(\mathsf{a})$ is a subterm of $\mathsf{f}(\mathsf{c}(\mathsf{f}(\mathsf{a})))$.

Unfortunately, strong minimality does *not* distinguish active and frozen positions and a result as Lemma 1 is not possible for strongly minimal terms. The set of *minimal* non-μ-terminating terms $\mathcal{M}_{\infty,\mu}$ consists of all non-μ-terminating terms whose proper subterms *at active positions* are all μ-terminating. Minimal non-μ-terminating terms are preserved under inner μ-rewritings, as we show in the following lemma.

Lemma 2 [3, Lemma 4]. *Let $(\mathcal{R}, \mu)$ be a CS-TRS. For all $s \in \mathcal{M}_{\infty,\mu}$, if $s \overset{>\Lambda}{\hookrightarrow}_{\mathcal{R},\mu} t$ and t is non-μ-terminating, then $t \in \mathcal{M}_{\infty,\mu}$.*

Furthermore, $\mathcal{T}_{\infty,\mu} \subseteq \mathcal{M}_{\infty,\mu}$. And now, $\mathsf{f}(\mathsf{c}(\mathsf{f}(\mathsf{a})))$ in Example 2 is minimal: $\mathsf{f}(\mathsf{c}(\mathsf{f}(\mathsf{a}))) \in \mathcal{M}_{\infty,\mu}$. The following result establishes that, given a minimal non-μ-terminating term, there are only two ways for an infinite μ-rewrite sequence to proceed.

Proposition 1 [3, Proposition 5]. *Let $(\mathcal{R}, \mu)$ be a CS-TRS. For all $s \in \mathcal{M}_{\infty,\mu}$, there exist a rewrite rule $\ell \to r \in \mathcal{R}$, a substitution σ and a term $u \in \mathcal{M}_{\infty,\mu}$ such that $s \overset{>\Lambda*}{\hookrightarrow}_{\mathcal{R},\mu} \ell\sigma \overset{\Lambda}{\hookrightarrow}_{\ell\to r,\mu} r\sigma \trianglerighteq_{\mu} t$ and either (1) there is a nonvariable subterm u at an active position of r such that $t = u\sigma$, or (2) there is $x \in \mathcal{V}ar^{\mu}(r) - \mathcal{V}ar^{\mu}(\ell)$ such that $x\sigma \trianglerighteq_{\mu} t$.*

What Proposition 1 says is that minimal non-μ-terminating terms at frozen positions (as $\mathsf{f}(\mathsf{a})$ in $\mathsf{f}(\mathsf{c}(\mathsf{f}(\mathsf{a}))))$ show up at active positions by means of migrating variables (a variable x is migrating in a rule $\ell \to r$ if $x \in \mathcal{V}ar^{\mu}(r) - \mathcal{V}ar^{\mu}(\ell)$, as x in rule (12)). If (1) happens, information about the shape of t is provided because it is partially introduced by an active subterm of r. This information is crucial to efficiently mechanize proofs of termination. But if (1) happens, information about the shape of t is *hidden* below a binding $x\sigma$ of the matching substitution σ. The frozen occurrence of x in the left-hand side ℓ of the rule is responsible for this information showing up later in the sequence. In the following, we analyze how minimal non-μ-terminating terms appear at frozen positions in infinite μ-rewrite sequences and how they evolve until getting activated by a migrating variable. Without loss of generality, in the following all the considered infinite μ-rewrite sequences start from strongly minimal non-μ-terminating terms.

Example 3. Consider the following non-μ-terminating TRS $\mathcal{R}$ [1, modified(I)]:

$$\mathsf{a} \to \mathsf{f}(\overline{\mathsf{g}(\mathsf{b})}) \quad (13) \qquad\qquad \mathsf{h}(\overline{x}) \to x \quad (15)$$

$$\mathsf{f}(\overline{x}) \to \mathsf{h}(\overline{\mathsf{c}(x)}) \quad (14) \qquad\qquad \mathsf{b} \to \mathsf{a} \quad (16)$$

with $\mu(\mathsf{g}) = \mu(\mathsf{c}) = \{1\}$ and $\mu(f) = \emptyset$ for all $f \in \mathcal{F} - \{\mathsf{g},\mathsf{c}\}$. Subexpressions at frozen positions are identified using the overbar. And consider the following infinite μ-rewrite sequence (Fig. 1 shows it graphically, where shaded triangles are minimal non-μ-terminating terms[2]):

$$\underline{\mathsf{a}} \hookrightarrow_{(13),\mu} \mathsf{f}(\overline{\mathsf{g}(\mathsf{b})}) \hookrightarrow_{(14),\mu} \mathsf{h}(\overline{\mathsf{c}(\mathsf{g}(\mathsf{b}))}) \hookrightarrow_{(15),\mu} \mathsf{c}(\mathsf{g}(\underline{\mathsf{b}})) \hookrightarrow_{(16),\mu} \mathsf{c}(\mathsf{g}(\underline{\mathsf{a}})) \hookrightarrow_{(13),\mu} \cdots$$

As we can see in the sequence, $\mathsf{a} \in \mathcal{T}_{\infty,\mu}$, and the first μ-rewriting step introduces the minimal non-μ-terminating term b at a frozen position by using rule (13) which introduces the context $\mathsf{g}(\square)$ where b is located. Afterwards, the context $\mathsf{c}(\square)$ is inserted *above* term $\mathsf{g}(\mathsf{b})$ which is *"pushed down"* by the right-hand side

---

[2] Note that minimal non-μ-terminating terms may contain minimal non-μ-terminating terms (at frozen positions, though). We use darker shades for such nested minimal non-μ-terminating terms.

of rule (14). Finally, the migrating variable x in rule (15) is *instantiated* (in the third step) to $c(g(b))$. The application of rule (15) finally *activates* b, which is now *active* inside $c(g(b))$.

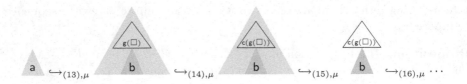

Fig. 1. Infinite μ-rewrite sequence in Example 3

Example 3 shows how minimal non-μ-terminating terms are partially "introduced" in an infinite μ-rewrite sequence: there is a rule $\ell \to r$ (in this case (13)), a subterm u of r at a frozen position (b) and a possible context with a hole at an active position $(g(\square))$.

As discussed above, the context surrounding those "hidden" minimal non-μ-terminating terms t can be "increased", i.e., t can be "pushed down" into a bigger context. Furthermore, the context can be "decreased" as well, as we can see in the following example.

Example 4. Consider the following TRS $\mathcal{R}$ [1, modified (II)]:

$$a \to f(\overline{g(c(g(b)))}) \quad (17) \qquad h(\overline{c(x)}) \to x \quad (19)$$
$$f(\overline{g(x)}) \to h(\overline{x}) \quad (18) \qquad b \to a \quad (20)$$

with $\mu(g) = \mu(c) = \{1\}$ and $\mu(f) = \emptyset$ for all $f \in \mathcal{F} - \{g, c\}$, and:

$$\underline{a} \hookrightarrow_{(17),\mu} f(\overline{g(c(g(b)))}) \hookrightarrow_{(18),\mu} h(\overline{c(g(b))}) \hookrightarrow_{(19),\mu} g(\underline{b}) \hookrightarrow_{(20),\mu} g(\underline{a}) \hookrightarrow_{(17),\mu} \cdots$$

Figure 2 shows it graphically. Once again, the first μ-rewriting step introduces the minimal non-μ-terminating term b at a frozen position by using rule (17) which introduces the context $g(c(g(\square)))$. But, in the second μ-rewriting step, part of the active context $g(c(g(\square)))$ which is frozen at $s_2 = f(g(c(g(b))))$, i.e. $g(\square)$, is *removed* from s_2 due to *pattern matching* with the left-hand side of rule (18). Finally, in the same way, part of the active context $c(g(\square))$ which is frozen at $s_3 = h(c(g(b)))$, i.e. $c(\square)$, is removed from s_3 in the third μ-rewriting step by *pattern matching* with rule (19) and, furthermore, the migrating variable x is instantiated to $g(b)$.

We describe these "incoming" and "outcoming" contexts surrounding frozen subterms. First, we notice that, when examining the rules (14), (18) and (19) (which are responsible for the introduction and removal of contexts discussed in Examples 3 and 4) they all share the following features:

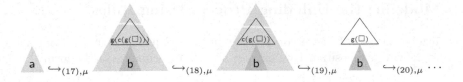

Fig. 2. Infinite μ-rewrite sequence in Example 4

- if a rule $\ell \to r$ adds a context C_i, then there is a term $s = C_i[x]_p$ such that $r = D[s]_q$, being q a frozen position of r and p an active position of s. Furthermore, if $\ell \to r$ is applied in a minimal non-μ-terminating sequence, the variable x cannot occur at active positions, i.e., $x \in (\mathcal{V}ar^{\not\!\!/}(\ell) \cap \mathcal{V}ar^{\not\!\!/}(r)) - (\mathcal{V}ar^{\mu}(\ell) \cup \mathcal{V}ar^{\mu}(r))$ (if not, minimality is violated); and,
- if a rule $\ell \to r$ removes a context C_o, then there is a term $s = C_o[x]_p$ such that $\ell = D[s]_q$, being q a frozen position of ℓ and p an active position of s. Furthermore, if $\ell \to r$ is applied in a minimal non-μ-terminating sequence, the variable x cannot occur at active positions, i.e., $x \in (\mathcal{V}ar^{\not\!\!/}(\ell) \cap \mathcal{V}ar^{\not\!\!/}(r)) - (\mathcal{V}ar^{\mu}(\ell) \cup \mathcal{V}ar^{\mu}(r))$ or $\ell|_{q.p}$ is migrating (in this case, we are in the second case of Proposition 1, where the minimal non-μ-terminating term shows up and is the responsible of continuing the sequence).

Rules involving these incoming and outcoming contexts can be applied several times and in different orders.

Example 5. Consider the following TRS $\mathcal{R}$ [1, modified (III)]:

$$a \to f(\overline{g(b)}) \qquad (21) \qquad\qquad h(\overline{c(x)}) \to x \qquad (24)$$
$$f(\overline{x}) \to h(\overline{g(c(c(x)))}) \qquad (22) \qquad\qquad b \to a \qquad (25)$$
$$h(\overline{g(x)}) \to h(\overline{x}) \qquad (23)$$

with $\mu(g) = \mu(c) = \{1\}$ and $\mu(f) = \emptyset$ for all $f \in \mathcal{F} - \{g, c\}$. And consider the following infinite μ-rewrite sequence (graphically in Fig. 3):

$$\underline{a} \hookrightarrow_{(21),\mu} \underline{f(\overline{g(b)})} \hookrightarrow_{(22),\mu} \underline{h(g(c(c(g(b)))))} \hookrightarrow_{(23),\mu} \underline{h(\overline{c(c(g(b))))}} \hookrightarrow_{(24),\mu} c(g(\underline{b})) \cdots$$

Note that the migrating variable x is instantiated to term $x\sigma = C[u] = c(g(b))$ where $u = b$ is minimal non-μ-terminating and the context $C[\Box] = c(g(\Box))$ with a hole at an active position is a combination of *fragments* of contexts added at frozen positions by rewrite rules.

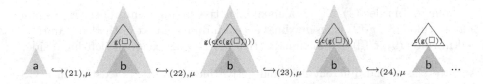

Fig. 3. Infinite μ-rewrite sequence in Example 5

4   Modeling the Unhiding Process Using Rules

Recapitulating Sect. 3, if we consider an infinite sequence starting from $s_1 \in \mathcal{T}_{\infty,\mu}$, following Proposition 1 we extract an infinite sequence of the form:

$$ s_1 \xrightarrow{>\Lambda_*}_{\mathcal{R},\mu} \ell_1\sigma \xrightarrow{\Lambda}_{\ell_1 \to r_1,\mu} r_1\sigma \trianglerighteq_\mu s_2 \xrightarrow{>\Lambda_*}_{\mathcal{R},\mu} \ell_2\sigma \xrightarrow{\Lambda}_{\ell_2 \to r_2,\mu} r_2\sigma \trianglerighteq_\mu \cdots $$

where $s_i \in \mathcal{M}_{\infty,\mu}$, for all $i > 0$. If Proposition 1(2) is applied on step j, $j > 0$, we know that: (a) previously in the chain there is a rule (like (13), (17) and (21)) that introduces the minimal non-μ-terminating term in the sequence together with an active context, (b) there are rules that modify this active context (like (14), (18), (22) and (23)) and, finally, (c) rule $\ell_j \to r_j$ (like (15), (19) and (24)) shows up the minimal non-μ-terminating by means of a migrating variable x together with part of its active context, $x\sigma = C[u]$. In this section, we use the knowledge of the previous section to define a TRS that can be used to extract u from $C[u]$ by using a minimum set of rules. Furthermore, we introduce the new notion of *unhidable*. All this prepares the introduction of a *new notion of minimality* which is the basis of our new characterization of termination of CSR.

Following the observations in the previous section, we can get the patterns which introduce the minimal non-μ-terminating term at a frozen position in a μ-rewrite sequence together with its active context, as $\mathsf{g(b)}$ in rule (13) in Example 3 and in rule (21) in Example 5 and $\mathsf{g(c(b))}$ in rule (17) in Example 4.

Definition 1. *Let* $\mathcal{R} = (\mathcal{F}, R) = (\mathcal{C} \uplus \mathcal{D}, R)$ *be a TRS,* $\ell \to r \in R$ *and* μ *a replacement map on* $\mathcal{F}$. *We say that* $s = C[t]_p$ *is a raw hidden term of* $\ell \to r$ *if* $r = D[C[t]_p]_q$, $q \in \mathcal{P}os^{\not\mu}(r)$, $p \in \mathcal{P}os^\mu(C[t]_p)$, $\mathsf{root}(t) \in \mathcal{D}$ *and* $q.p$ *is minimal in* r *(i.e., there is no* $q'.p'$ *such that* $r = D'[C'[t]_{p'}]_{q'}$, $q' \in \mathcal{P}os^{\not\mu}(r)$, $p' \in \mathcal{P}os^\mu(C'[t]_{p'})$ *and* $p' < p$). *Let* $\mathsf{H_{raw}}(\mathcal{R}, \mu)$ *be the set of all raw hidden terms from rules in* $(\mathcal{R}, \mu)$.

Example 6. In Example 1, we have $\mathsf{H_{raw}}(\mathcal{R}, \mu) = \{x{*}\mathsf{fact(p}(x))\}$; in Example 3, we have $\mathsf{H_{raw}}(\mathcal{R}, \mu) = \{\mathsf{g(b)}\}$; in Example 4 we have $\mathsf{H_{raw}}(\mathcal{R}, \mu) = \{\mathsf{g(c(g(b)))}\}$; and, in Example 5, we have $\mathsf{H_{raw}}(\mathcal{R}, \mu) = \{\mathsf{g(b)}\}$.

We identify the shape of the patterns that increase or decrease the active context attached to delayed subexpressions.

Definition 2. *Let* $u \in \mathcal{T}(\mathcal{F}, \mathcal{X})$ *and* μ *a replacement map on* $\mathcal{F}$. *We say that* $s = C[\Box]_p$ *is a* maximal active hiding context *in* u *if* $u = D[C[x]_p]_q$, $q \in \mathcal{P}os^{\not\mu}(u)$, $p \in \mathcal{P}os^\mu(C[x]_p)$ *and* $q.p$ *is minimal in* u.

Example 7. In rule (14), $\mathsf{c}(\Box)$ is a maximal active hiding contex of the right-hand side, in rule (18), $\mathsf{g}(\Box)$ is an maximal active hiding context of the left-hand side and in rule (19), $\mathsf{c}(\Box)$ is a maximal active hiding context of the left-hand side.

And we clasify the different maximal active hiding contexts existing in a CS-TRS.

Definition 3. *Let* $\mathcal{R} = (\mathcal{F}, R)$ *be a TRS,* $\ell \to r \in R$, μ *a replacement map on* $\mathcal{F}$, $D[\square]_q$ *a context with a hole at a frozen position* q, $C[\square]_p$ *a context with a hole at an active position and* $x \in \mathcal{X}$. *We say that* $s = C[\square]_p$ *is either:*

1. *An* incoming context *of* $\ell \to r$ *if* s *is a maximal active hiding context of* r, $r = D[C[x]_p]_q$, *and* $x \in (Var^{\not\mu}(\ell) \cap Var^{\not\mu}(r)) - (Var^{\mu}(\ell) \cup Var^{\mu}(r))$.
2. *An* outcoming context *of* $\ell \to r$ *if* s *is a maximal active hiding context of* ℓ, $\ell = D[C[x]_p]_q$, *and* $x \in (Var^{\not\mu}(\ell) \cap Var^{\not\mu}(r)) - (Var^{\mu}(\ell) \cup Var^{\mu}(r))$.
3. *A* terminal outcoming context *of* $\ell \to r$ *if* s *is a maximal active hiding context of* ℓ, $\ell = D[C[x]_p]_q$, *and* $x \in Var^{\mu}(r) - Var^{\mu}(\ell)$.

Let $C_i(\mathcal{R}, \mu)/C_o(\mathcal{R}, \mu)/C_t(\mathcal{R}, \mu)$ *be the set of all incoming / outcoming / terminal outcoming contexts from rules in* $(\mathcal{R}, \mu)$.

Example 8. In rule (14), $c(\square)$ is an incoming context of the right-hand side, in rule (18), $g(\square)$ is an outcoming context of the left-hand side and in rule (19), $c(\square)$ is a terminal outcoming context of the left-hand side.

In Example 1, we have $C_i(\mathcal{R}, \mu) = C_o(\mathcal{R}, \mu) = C_t(\mathcal{R}, \mu) = \emptyset$; in Example 3, we have $C_i(\mathcal{R}, \mu) = \{c(\square)\}$, and $C_o(\mathcal{R}, \mu) = C_t(\mathcal{R}, \mu) = \emptyset$; in Example 4, we have $C_i(\mathcal{R}, \mu) = \emptyset$, $C_o(\mathcal{R}, \mu) = \{g(\square)\}$, and $C_t(\mathcal{R}, \mu) = \{c(\square)\}$; and, in Example 5, we have $C_i(\mathcal{R}, \mu) = \{g(c(c(\square)))\}$, $C_o(\mathcal{R}, \mu) = \{g(\square)\}$, and $C_t(\mathcal{R}, \mu) = \{c(\square)\}$.

Outcoming contexts represent the fragments of active contexts which can be *removed* by a rule. Incoming contexts represent the active contexts that can be *added*. The following fixed-point definition obtains any combination of added/removed contexts (this will allow us to model the contexts that appear in the infinite μ-rewrite sequence in Example 5).

Definition 4. *Let* $\mathcal{R} = (\mathcal{F}, R)$ *be a TRS and* $\mu \in M_{\mathcal{F}}$. *The set* $\mathsf{XC}_i(\mathcal{R}, \mu)$ *and* $\mathsf{XC}_o(\mathcal{R}, \mu)$ *are the least sets satisfying:*

1. $C_i(\mathcal{R}, \mu) \subseteq \mathsf{XC}_i(\mathcal{R}, \mu)$, $C_o(\mathcal{R}, \mu) \subseteq \mathsf{XC}_o(\mathcal{R}, \mu)$ *and* $C_t(\mathcal{R}, \mu) \subseteq \mathsf{XC}_t(\mathcal{R}, \mu)$.
2. *If* $C_i[\square] \in \mathsf{XC}_i(\mathcal{R}, \mu)$, $C_o[\square] \in \mathsf{XC}_o(\mathcal{R}, \mu)$, *and there exist* $\theta = mgu(C_i[x],$ $C_o[y])$ *(rename variables if necessary) where* x *and* y *are fresh variables, such that* $y\theta \notin \mathcal{X}$ *and* $y\theta = C_i'[x]$, *then* $C_i'[\square] \in \mathsf{XC}_i(\mathcal{R}, \mu)$.
3. *If* $C_o[\square] \in \mathsf{XC}_o(\mathcal{R}, \mu)$, $C_i[\square] \in \mathsf{XC}_i(\mathcal{R}, \mu)$, *and there exist* $\theta = mgu(C_o[x],$ $C_i[y])$ *(rename variables if necessary) where* x *and* y *are fresh variables, such that* $y\theta \notin \mathcal{X}$ *and* $y\theta = C_o'[x]$, *then* $C_o'[\square] \in \mathsf{XC}_o(\mathcal{R}, \mu)$.
4. *If* $C_t[\square] \in \mathsf{XC}_t(\mathcal{R}, \mu)$, $C_i[\square] \in \mathsf{XC}_i(\mathcal{R}, \mu)$, *and there exist* $\theta = mgu(C_t[x],$ $C_i[y])$ *(rename variables if necessary) where* x *and* y *are fresh variables, such that* $y\theta \notin \mathcal{X}$ *and* $y\theta = C_t'[x]$, *then* $C_t'[\square] \in \mathsf{XC}_t(\mathcal{R}, \mu)$.

Note that, when the most general unifier (mgu) is computed, terms do not share variables, so a variable renaming is applied if necessary. The computation of $\mathsf{XC}_i(\mathcal{R}, \mu)$, $\mathsf{XC}_o(\mathcal{R}, \mu)$ and $\mathsf{XC}_t(\mathcal{R}, \mu)$ terminates (in each step, the resulting context is a instantiated fragment of one of the contexts that are unified).

Example 9. In Examples 1, 3 and 4, we have $\mathsf{XC_i}(\mathcal{R}, \mu) = \mathsf{C_i}(\mathcal{R}, \mu)$, $\mathsf{XC_o}(\mathcal{R}, \mu) = \mathsf{XC_o}(\mathcal{R}, \mu)$ and $\mathsf{XC_t}(\mathcal{R}, \mu) = \mathsf{C_t}(\mathcal{R}, \mu)$. In Example 5, we have $\mathsf{C_o}(\mathcal{R}, \mu) = \mathsf{XC_o}(\mathcal{R}, \mu)$, $\mathsf{C_t}(\mathcal{R}, \mu) = \mathsf{XC_t}(\mathcal{R}, \mu)$, but $\mathsf{XC_i}(\mathcal{R}, \mu) = \{\mathsf{g}(\mathsf{c}(\mathsf{c}(\square))), \mathsf{c}(\mathsf{c}(\square))\}$. The context $\mathsf{c}(\mathsf{c}(\square))$ represents a fragment of the active incoming context that remains after applying rule (22) and rule (23).

Terminal outcoming contexts can only be applied just before the minimal non-μ-terminating term shows up at an active position. Therefore, $\mathsf{FXC_i}(\mathcal{R}, \mu)$ extends $\mathsf{XC_i}(\mathcal{R}, \mu)$ obtaining the fragments of contexts obtained after removing the the terminal outcoming context.

Definition 5. *Let $\mathcal{R} = (\mathcal{F}, R)$ be a TRS and $\mu \in M_{\mathcal{F}}$. The set $\mathsf{FXC_i}(\mathcal{R}, \mu)$ satisfies:*

1. $\mathsf{XC_i}(\mathcal{R}, \mu) \subseteq \mathsf{FXC_i}(\mathcal{R}, \mu)$.
2. *If $C_i[\square] \in \mathsf{XC_i}(\mathcal{R}, \mu)$, $C_t[\square] \in \mathsf{XC_t}(\mathcal{R}, \mu)$, and there exist $\theta = mgu(C_i[x], C_t[y])$ (rename variables if necessary) where x and y are fresh variables, such that $y\theta \notin \mathcal{X}$ and $y\theta = C[x]$, then $C[\square] \in \mathsf{FXC_i}(\mathcal{R}, \mu)$.*

Example 10. In Examples 1, 3 and 4, we have $\mathsf{FXC_i}(\mathcal{R}, \mu) = \mathsf{XC_i}(\mathcal{R}, \mu) = \mathsf{C_i}(\mathcal{R}, \mu)$. In Example 5, $\mathsf{FXC_i}(\mathcal{R}, \mu) = \{\mathsf{g}(\mathsf{c}(\mathsf{c}(\square))), \mathsf{c}(\mathsf{c}(\square)), \mathsf{c}(\square)\}$. The context $\mathsf{c}(\square)$ represents a final fragment of the active incoming context that remains after applying rule (24) (when the minimal non-μ-terminating term shows up at an active position).

In the same way, we apply the outcoming contexts to the raw hidden terms to obtain the possible shape of those terms when they show up by means of migrating variables.

Definition 6. *The set $\mathsf{XH_{raw}}$ is the least set satisfying (1) $\mathsf{H_{raw}} \subseteq \mathsf{XH_{raw}}$, and (2) if $C_i[t] \in \mathsf{XH_{raw}}$, $C_o[\square] \in \mathsf{XC_o}(\mathcal{R}, \mu)$ and there exist $\theta = mgu(C_i[t], C_o[x])$ where x is a fresh variable, such that $x\theta = C[t\theta]$, then $C[t\theta] \in \mathsf{XH_{raw}}$.*

The set $\mathsf{FXH_{raw}}$ satisfies (1) $\mathsf{XH_{raw}} \subseteq \mathsf{FXH_{raw}}$, and (1) if $C_i[t] \in \mathsf{XH_{raw}}$, $C_t[\square] \in \mathsf{XC_t}(\mathcal{R}, \mu)$ and there exist $\theta = mgu(C_i[t], C_t[x])$ where x is a fresh variable, such that $x\theta = C[t\theta]$, then $C[t\theta] \in \mathsf{FXH_{raw}}$.

Example 11. In Examples 1 and 3, $\mathsf{FXH_{raw}}(\mathcal{R}, \mu) = \mathsf{XH_{raw}}(\mathcal{R}, \mu) = \mathsf{XH_{raw}}(\mathcal{R}, \mu)$; in Example 4, we have $\mathsf{XH_{raw}}(\mathcal{R}, \mu) = \{\mathsf{g}(\mathsf{c}(\mathsf{g}(\mathsf{b}))), \mathsf{c}(\mathsf{g}(\mathsf{b}))\}$ and $\mathsf{FXH_{raw}}(\mathcal{R}, \mu) = \{\mathsf{g}(\mathsf{c}(\mathsf{g}(\mathsf{b}))), \mathsf{c}(\mathsf{g}(\mathsf{b})), \mathsf{g}(\mathsf{b})\}$; and, in Example 5, $\mathsf{XH_{raw}}(\mathcal{R}, \mu) = \mathsf{FXH_{raw}}(\mathcal{R}, \mu) = \{\mathsf{g}(\mathsf{b}), \mathsf{b}\}$.

Previous definitions will be helpful in the next section to obtain a notion of minimality that gives us more information about non-μ-terminating terms at frozen positions.

4.1 A New Notion of Minimal Non-μ-terminating Term

The following notion of *unhidable* prepares a notion of *minimality* that provides more information about minimal non-μ-terminating terms at *frozen* positions.

Definition 7. *Let* $\mathcal{R} = (\mathcal{F}, R)$ *and* $\mathcal{S} = (\mathcal{F}, S_0 \uplus S_1)$ *be TRSs, and* $\mu \in M_{\mathcal{F}}$. *Let* $s, t \in \mathcal{T}(\mathcal{F}, \mathcal{X})$. *We say that* s unhides t *using* $\mathcal{S}$ *if* $s \xrightarrow{\Lambda}{}^{*}_{S_0} \circ \xrightarrow{\Lambda}_{S_1} t$. *We say that a term* u *is* unhidable *using* $\mathcal{S}$ *if for every subterm* $v \in M_{\infty, \mu}$ *such that* $u = D[C[v]_p]_q$, $q \in \mathcal{P}os^{\not\models}(u)$, $p \in \mathcal{P}os^{\mu}(C[v]_p)$, $q.p$ *minimal,* $C[v]_p$ *unhides* v *using* $\mathcal{S}$ *and* v *is unhidable using* $\mathcal{S}$.

Setting $S_0 = \mathsf{FXCR_i}(\mathcal{R}, \mu)$ and $S_1 = \mathsf{FXHR_{raw}}(\mathcal{R}, \mu)$ in Definition 7, where we define $\mathsf{FXCR_i}(\mathcal{R}, \mu) = \{C_i[x] \to x \mid C_i[\square] \in \mathsf{FXC_i}(\mathcal{R}, \mu)\}$ and $\mathsf{FXHR_{raw}}(\mathcal{R}, \mu) = \{C_i[s]_p \to s \mid C_i[s]_p \in \mathsf{FXH_{raw}}(\mathcal{R}, \mu), p \in \mathcal{P}os^{\mu}(C_i[s]_p), s \in \mathcal{D}\}$, we obtain the following properties.

Proposition 2. *Let* $\mathcal{R} = (\mathcal{F}, R)$ *be a TRS,* $\mu \in M_{\mathcal{F}}$, $S_0 = \mathsf{FXCR_i}(\mathcal{R}, \mu)$, $S_1 = \mathsf{FXHR_{raw}}(\mathcal{R}, \mu)$, $\mathcal{S} = (\mathcal{F}, S_0 \uplus S_1)$, σ *be a substitution and* u, v *be terms such that* u *unhides* v *using* $\mathcal{S}$. *Then,*

1. $S_0 \cap S_1 = \emptyset$.
2. *If* $C_o[\square] \in \mathsf{XC_o}(\mathcal{R}, \mu) \cup \mathsf{XC_t}(\mathcal{R}, \mu)$, *and* $u = C_o\sigma[C[v]]$ *then* $C[v]$ *unhides* v *using* $\mathcal{S}$.
3. *If* $C_i[\square] \in \mathsf{XC_i}(\mathcal{R}, \mu)$, *then* $C_i\sigma[u]$ *unhides* v *using* $\mathcal{S}$.

We are ready now to introduce our new notion of minimality.

Definition 8 (Unhidable minimal term). *Let* $\mathcal{R} = (\mathcal{F}, R)$ *be a TRS,* $\mu \in M_{\mathcal{F}}$, $S_0 = \mathsf{FXCR_i}(\mathcal{R}, \mu)$, $S_1 = \mathsf{FXHR_{raw}}(\mathcal{R}, \mu)$ *and* $\mathcal{S} = (\mathcal{F}, S_0 \uplus S_1)$. *We define the set of* unhidable minimal non-μ-terminating *terms* $M^{*}_{\infty, \mu}$ *as follows:* $s \in M^{*}_{\infty, \mu}$ *iff* $s \in M_{\infty, \mu}$ *and* s *is unhidable using* $\mathcal{S}$.

The following result improves Proposition 1 by using then new notion of minimal non-μ-terminating term.

Proposition 3. *Let* $\mathcal{R} = (\mathcal{F}, R) = (\mathcal{C} \uplus \mathcal{D}, R)$ *be a TRS,* $\mu \in M_{\mathcal{F}}$, $S_0 = \mathsf{FXCR_i}(\mathcal{R}, \mu)$, $S_1 = \mathsf{FXHR_{raw}}(\mathcal{R}, \mu)$ *and* $\mathcal{S} = (\mathcal{F}, S_0 \uplus S_1)$. *Then for all* $t \in M^{*}_{\infty, \mu}$, *there exist* $\ell \to r \in R$, *a substitution* σ *and a term* $u \in M^{*}_{\infty, \mu}$ *such that* $t \xrightarrow{\geq \Lambda}{}^{*}_{\mathcal{R}, \mu} \ell\sigma \xrightarrow{\Lambda} r\sigma \trianglerighteq_{\mu} u$ *and either:*

1. *There is a nonvariable active subterm* s *of* r, $r \trianglerighteq_{\mu} s$, *such that* $\mathsf{root}(s) \in \mathcal{D}$ *and* $u = s\sigma$, *or*
2. *There is* $x \in \mathcal{V}ar^{\mu}(r) - \mathcal{V}ar^{\mu}(\ell)$ *such that* $x\sigma = C[u]$ *for a possibly empty context* $C[\square]$ *with a hole at an active position and* $C[u]$ *unhides* u *using* $\mathcal{S}$.

5   From Minimal Terms to the CS-DP Framework

Dependency pairs [6] describe the *propagation* of minimal non-μ-terminating terms in non-terminating rewrite sequences. The notion of CS-DP is a consequence of Proposition 1. The notation $f^{\sharp}$ for a given symbol f means that f is *marked*. For $s = f(s_1, \ldots, s_n)$, we write $s^{\sharp}$ to denote the marked term $f^{\sharp}(s_1, \ldots, s_n)$. We often capitalize f and use F instead of $f^{\sharp}$ in our examples.

Definition 9 (Context-Sensitive Dependency Pairs [3]**).** *Given a CS-TRS* $(\mathcal{R}, \mu)$, *let* $\mathsf{DP}(\mathcal{R}, \mu) = \mathsf{DP}_{\mathcal{F}}(\mathcal{R}, \mu) \cup \mathsf{DP}_{\mathcal{X}}(\mathcal{R}, \mu)$ *the set of CS-DPs where* $\mathsf{DP}_{\mathcal{F}}(\mathcal{R}, \mu) = \{\ell^\sharp \to s^\sharp \mid \ell \to r \in \mathcal{R}, r \unrhd_\mu s, \mathrm{root}(s) \in \mathcal{D}, \ell \not\unrhd_\mu s\}$, *and* $\mathsf{DP}_{\mathcal{X}}(\mathcal{R}, \mu) = \{\ell^\sharp \to x \mid \ell \to r \in \mathcal{R}, x \in Var^\mu(r) - Var^\mu(\ell)\}$. *We extend* μ *into* $\mu^\sharp$ *by* $\mu^\sharp(f) = \mu(f)$ *if* $f \in \mathcal{F}$, *and* $\mu^\sharp(f^\sharp) = \mu(f)$ *if* $f \in \mathcal{D}$.

Example 12. For $(\mathcal{R}, \mu)$ in Example 1, we obtain the following CS-DPs:

$$\mathsf{s}(x) +^\sharp y \to x +^\sharp y \qquad (26) \qquad\qquad \mathsf{FACT}(x) \to \mathsf{ZERO}(x) \ (30)$$
$$\mathsf{s}(x) *^\sharp y \to y +^\sharp (x * y) \qquad (27) \qquad \mathsf{IF}(\mathsf{true}, x, y) \to x \qquad (31)$$
$$\mathsf{s}(x) *^\sharp y \to x *^\sharp y \qquad (28) \qquad \mathsf{IF}(\mathsf{false}, x, y) \to y \qquad (32)$$
$$\mathsf{FACT}(x) \to \mathsf{IF}(\mathsf{zero}(x), \mathsf{s}(0), x * \mathsf{fact}(\mathsf{p}(x))) \ (29)$$

DPs (26)-(30) capture the direct function calls and collapsing DPs (31)-(32) capture the activation of delayed function calls.

As usual when dealing with DPs, we *abstract* the notion of chain using generic TRSs $\mathcal{P}$, $\mathcal{R}$ and $\mathcal{S}$. Termination of CS-TRSs is characterized by the absence of infinite chains of CS-DPs [2,3].

Definition 10 (Chain of Pairs [17]**).** *Let* $\mathcal{P}$, $\mathcal{R}$ *and* $\mathcal{S}$ *be TRSs and* μ *a replacement map where* $\mathcal{S} = \mathcal{S}_{\unrhd_\mu} \uplus \mathcal{S}_\sharp$, $\mathcal{S}_{\unrhd_\mu}$ *are rules of the form* $s \to t \in \mathcal{S}$ *such that* $s \unrhd_\mu t$ *and* $\mathcal{S}_\sharp = \mathcal{S} - \mathcal{S}_{\unrhd_\mu}$. *A* $(\mathcal{P}, \mathcal{R}, \mathcal{S}, \mu)$-chain *is a finite or infinite sequence of pairs* $u_i \to v_i \in \mathcal{P}$, *together with a substitution* σ *satisfying that, for all* $i \geq 1$,

1. *If* $v_i \notin Var(u_i) - Var^\mu(u_i)$, *then* $v_i\sigma = w_i \hookrightarrow^*_{\mathcal{R},\mu} u_{i+1}\sigma$, *and*
2. *If* $v_i \in Var(u_i) - Var^\mu(u_i)$, *then* $v_i\sigma \xrightarrow{\Lambda}^*_{\mathcal{S}_{\unrhd_\mu}} \circ \xrightarrow{\Lambda}_{\mathcal{S}_\sharp} w_i \hookrightarrow^*_{\mathcal{R},\mu} u_{i+1}\sigma$.

An infinite $(\mathcal{P}, \mathcal{R}, \mathcal{S}, \mu)$-chain *is called* minimal *if for all* $i \geq 1$, w_i *is* $(\mathcal{R}, \mu)$-*terminating.*

In Definition 10, $\mathcal{P}$ plays the role of $\mathsf{DP}(\mathcal{R}, \mu)$ and $\mathcal{S}$ has two components $\mathcal{S}_{\unrhd_\mu}$ and $\mathcal{S}_\sharp$ which are useful to model the connection between a collapsing pair to another pair. The connection between the results obtained in the previous section and the notion of chain is straightforward, we only have to introduce the marking in our unhiding rules.

Definition 11 (Unhiding TRS). *Let* $\mathcal{R}$ *be a TRS and* $\mu \in M_{\mathcal{R}}$. *We define* $\mathsf{unh}(\mathcal{R}, \mu) = \mathsf{unh}_{\unrhd_\mu}(\mathcal{R}, \mu) \uplus \mathsf{unh}_\sharp(\mathcal{R}, \mu)$, *where* $\mathsf{unh}_{\unrhd_\mu}(\mathcal{R}, \mu) = \mathsf{FXCR}_i(\mathcal{R}, \mu)$ *and* $\mathsf{unh}_\sharp(\mathcal{R}, \mu) = \{s \to t^\sharp \mid s \to t \in \mathsf{FXHR}_{\mathsf{raw}}(\mathcal{R}, \mu)\}$.

Example 13. The unhiding TRS $\mathsf{unh}(\mathcal{R}, \mu)$ in Example 1 consists of the following rules:

$$x * \mathsf{fact}(\mathsf{p}(x)) \to x *^\sharp \mathsf{fact}(\mathsf{p}(x)) \ (33) \qquad\quad x * \mathsf{fact}(\mathsf{p}(x)) \to \mathsf{FACT}(\mathsf{p}(x)) \ (34)$$
$$x * \mathsf{fact}(\mathsf{p}(x)) \to \mathsf{P}(x) \qquad\qquad (35)$$

where $\mathsf{FXC}_i(\mathcal{R}, \mu) = \emptyset$. In [17], the definition of the unhiding TRS is different. We would have the following *bigger* set of rules:

$$x*\mathsf{fact}(\mathsf{p}(x)) \rightarrow x*^{\sharp}\mathsf{fact}(\mathsf{p}(x)) \qquad\qquad x*y \rightarrow y$$
$$\mathsf{fact}(\mathsf{p}(x)) \rightarrow \mathsf{FACT}(\mathsf{p}(x)) \qquad\qquad \mathsf{fact}(x) \rightarrow x$$
$$\mathsf{p}(x) \rightarrow \mathsf{P}(x)$$

The following result provides a new characterization of termination of CSR.

Theorem 1. *Let $\mathcal{R}$ be a TRS and $\mu \in M_{\mathcal{R}}$. $\mathcal{R}$ is μ-terminating if and only if there is no infinite minimal $(\mathsf{DP}(\mathcal{R}, \mu), \mathcal{R}, \mathsf{unh}(\mathcal{R}, \mu), \mu^{\sharp})$-chain.*

Example 14. For $(\mathcal{R}, \mu)$ in Example 3, we obtain the following CS-DPs:

$$\mathsf{A} \rightarrow \mathsf{F}(\overline{\mathsf{g}(\mathsf{b})}) \qquad (36) \qquad\qquad \mathsf{F}(\overline{x}) \rightarrow x \qquad (38)$$
$$\mathsf{F}(\overline{x}) \rightarrow \mathsf{F}(\overline{\mathsf{c}(x)}) \qquad (37) \qquad\qquad \mathsf{B} \rightarrow \mathsf{A} \qquad (39)$$

The infinite sequence in Example 3 is captured by the following $(\mathcal{P}, \mathcal{R}, \mathcal{S}, \mu^{\sharp})$-chain, where $\mathcal{P} = \mathsf{DP}(\mathcal{R}, \mu)$ and $\mathcal{S} = \mathsf{unh}(\mathcal{R}, \mu)$:

$$\underline{\mathsf{A}} \rightarrow_{(36)} \underline{\mathsf{F}(\overline{\mathsf{g}(\mathsf{b})})} \rightarrow_{(37)} \underline{\mathsf{F}(\overline{\mathsf{c}(\mathsf{g}(\mathsf{b}))})} \rightarrow_{(38)} \underline{\mathsf{c}(\mathsf{g}(\mathsf{b}))} \xrightarrow{\Lambda}_{\mathcal{S}_{\rhd_{\mu}}} \mathsf{g}(\mathsf{b}) \xrightarrow{\Lambda}_{\mathcal{S}_{\sharp}} \underline{\mathsf{B}} \rightarrow_{(39)} \underline{\mathsf{A}} \rightarrow_{(36)} \cdots$$

5.1   Context-Sensitive Dependency Pair Framework

In the DP framework [14], the focus is on the so-called *termination problems* involving two TRSs $\mathcal{P}$ and $\mathcal{R}$ instead of just the 'target' TRS $\mathcal{R}$. In our setting we start with the following definition (see also [1,3]).

Definition 12 (CS Problem and Processor). *A CS problem τ is a tuple $\tau = (\mathcal{P}, \mathcal{R}, \mathcal{S}, \mu)$, where $\mathcal{P}$, $\mathcal{R}$ and $\mathcal{S}$ are TRSs, and μ is a replacement map on the signatures of $\mathcal{R}$, $\mathcal{P}$ and $\mathcal{S}$. The CS problem $(\mathcal{P}, \mathcal{R}, \mathcal{S}, \mu)$ is finite if there is no infinite minimal $(\mathcal{P}, \mathcal{R}, \mathcal{S}, \mu)$-chain.*

A CS processor Proc is a mapping from CS problems into sets of CS problems. A CS-processor Proc is sound if for all CS problems τ, τ is finite whenever $\forall \tau' \in \mathsf{Proc}(\tau)$, τ' is finite[3].

In order to prove the μ-termination of a TRS $\mathcal{R}$, we adapt the result from [14] to CSR.

Theorem 2 (CS-DP Framework [3]). *Let $\mathcal{R}$ be a TRS and μ a replacement map on the signature of $\mathcal{R}$. We construct a tree whose nodes are labeled with CS problems or "yes", and whose root is labeled with $(\mathsf{DP}(\mathcal{R}, \mu), \mathcal{R}, \mathsf{unh}(\mathcal{R}, \mu), \mu^{\sharp})$. For every inner node labeled with τ, there is a sound processor Proc satisfying one of the following conditions:*

1. $\mathsf{Proc}(\tau) = \emptyset$ *and the node has just one child, labeled with "yes".*
2. $\mathsf{Proc}(\tau) \neq$ no, $\mathsf{Proc}(\tau) \neq \emptyset$, *and the children of the node are labeled with the CS problems in $\mathsf{Proc}(\tau)$.*

If all leaves of the tree are labeled with "yes", then $\mathcal{R}$ is μ-terminating.

---

[3] In order to keep our presentation simple, we do not introduce here the notions related with completeness of processors, needed for *nontermination* proofs.

6   Usable Rules in the CS-DP Framework

One of the most powerful CS processors to deal with CS problems is the μ-*reduction pair processor*, a processor that discards pairs that can be strictly oriented using orderings. A μ-*reduction pair* $(\gtrsim, \sqsupset)$ consists of a stable and μ-monotonic[4] quasi-ordering $\gtrsim$, and a well-founded stable relation $\sqsupset$ on terms in $\mathcal{T}(\mathcal{F}, \mathcal{X})$ which are compatible, i.e., $\gtrsim \circ \sqsupset \subseteq \sqsupset$ or $\sqsupset \circ \gtrsim \subseteq \sqsupset$ [2]. Given a CS problem $\tau = (\mathcal{P}, \mathcal{R}, \mathcal{S}, \mu)$, if there is a μ-reduction pair such that $\mathcal{P} \cup \mathcal{S} \subseteq \gtrsim \cup \sqsupset$ and $\mathcal{R} \subseteq \gtrsim$ then $(\mathcal{P}, \mathcal{R}, \mathcal{S}, \mu)$ is finite if $(\mathcal{P} - \mathcal{P}_\sqsupset, \mathcal{R}, \mathcal{S} - \mathcal{S}_\sqsupset, \mu)$ is finite, where $\mathcal{P}_\sqsupset$ and $\mathcal{S}_\sqsupset$ represent the set of rules from $\mathcal{P}$ and $\mathcal{S}$ oriented using $\sqsupset$. The μ-reduction pair processor can be improved using the notion of *usable rule* [5]. Usable rules, initially connected to innermost termination, allow us to discard those rules from $\mathcal{R}$ that are not directly involved in (possible) infinite minimal $(\mathcal{P}, \mathcal{R}, \mathcal{S}, \mu)$-chains. In rewriting (and also in CSR), the notion of usable rule is connected with $\mathcal{C}_\varepsilon$-termination [16,28]. A TRS $\mathcal{R} = (\mathcal{F}, R)$ is $\mathcal{C}_\varepsilon$-terminating if $\mathcal{R} \uplus \mathcal{C}_\varepsilon$ is terminating, where $\mathcal{C}_\varepsilon = \{c(x, y) \to x, c(x, y) \to y\}$ (with $c \notin \mathcal{F}$). The idea behind the usable rules is that for every infinite minimal $(\mathcal{P}, \mathcal{R}, \mathcal{S}, \mu)$-chain we can construct an infinite sequence where rewrite steps using $\mathcal{R}$ can be simulated by rewrite steps using $\mathcal{U}_\tau(\mathcal{R})$ and $\mathcal{C}_\varepsilon$, where $\mathcal{U}_\tau(\mathcal{R})$ is the set of usable rules of τ. So, instead of $\mathcal{R} \subseteq \gtrsim$, we only need to satisfy $\mathcal{U}_\tau(\mathcal{R}) \uplus \mathcal{C}_\varepsilon \subseteq \gtrsim$.

In [19], the notion of CS usable rule was given for chains of pairs. This notion is different from the one given in unrestricted rewriting. For example, if we consider the following CS problem $\tau_1 = (\{(29), (31), (32)\}, \mathcal{R}, \{(33), (34)\}, \mu^\sharp)$ obtained from Example Example 13, the set of CS usable rules in τ_1 is $\mathcal{R}$. This is caused by the presence of migrating variables. In the presence of migrating variables, every rule headed by a symbol appeared at a frozen positions in the right-hand side of a rule in $\mathcal{R}$ must be considered usable (in this case $*$, fact and p, and by transitivity $+$, if and zero).

But, if we look closely at the μ-rewrite sequence from Example 1 and its translation into a $(\mathsf{DP}(\mathcal{R}, \mu), \mathcal{R}, \mathsf{unh}(\mathcal{R}, \mu), \mu^\sharp)$-chain:

$$\underline{\mathsf{FACT}(\mathsf{s}^n(x))} \to_{(29)} \mathsf{IF}(\underline{\mathsf{zero}(\mathsf{s}^n(x))}, \overline{\mathsf{s}(0)}, \overline{\mathsf{s}^n(x)*\mathsf{fact}(\mathsf{p}(\mathsf{s}^n(x)))}) \hookrightarrow_{(7),\mu}$$

$$\mathsf{IF}(\mathsf{false}, \overline{\mathsf{s}(0)}, \overline{\mathsf{s}^n(x)*\mathsf{fact}(\mathsf{p}(\mathsf{s}^n(x)))}) \to_{(32)} \underline{\mathsf{s}^n(x)*\mathsf{fact}(\mathsf{p}(\mathsf{s}^n(x)))} \xrightarrow{\Lambda}_{(34)}$$

$$\underline{\mathsf{FACT}(\mathsf{p}(\mathsf{s}^n(x)))} \hookrightarrow_{(3),\mu} \cdots$$

we notice that x in $\mathsf{FACT}(\mathsf{p}(\mathsf{s}^n(x)))$ appears at an *active* position, but x comes from the initial term $\mathsf{FACT}(\mathsf{s}^n(x))$ where it was also at an active position, i.e., x does not behave as a migrating variable in the $(\mathsf{DP}(\mathcal{R}, \mu), \mathcal{R}, \mathsf{unh}(\mathcal{R}, \mu), \mu^\sharp)$-chain. Intuitively, this is equivalent to consider a pair $\mathsf{FACT}(x) \to \mathsf{FACT}(\mathsf{p}(x))$ and remove the intermediate steps. This *"conservative"* behavior allows us to ensure that only the rules defining zero and p are usable and, hence, obtain a

---

[4] A binary relation R on terms is μ-monotonic if for all terms $s, t, t_1, \ldots, t_m$, and m-ary symbols f, whenever $s\,\mathsf{R}\,t$ and $i \in \mu(f)$ we have $f(\ldots, t_{i-1}, s, \ldots)\,\mathsf{R}\,f(\ldots, t_{i-1}, t, \ldots)$.

smaller set of usable rules. Therefore, we have to find the general conditions that allow us to use this suitable set of usable rules in the μ-reduction pair processor.

6.1 Strongly Minimal Terms

The first stumbling rock in our goal comes when we try to control the shape of infinite terms (minimal non-μ-terminating terms in infinite μ-rewrite sequences) that appear at frozen positions. In the analysis of infinite μ-rewrite sequences, we obtain this control by imposing strong minimality on the initial term of the sequence, but this notion is lost in the notion of chain. Therefore, our first step is to introduce the notion of *strongly minimal* $(\mathcal{P}, \mathcal{R}, \mathcal{S}, \mu)$-chain. This notion ensures that the initial term of an infinite $(\mathcal{P}, \mathcal{R}, \mathcal{S}, \mu)$-chain does not contain any subterm that can generate an infinite $(\mathcal{P}, \mathcal{R}, \mathcal{S}, \mu)$-chain.

Definition 13. *An infinite* $(\mathcal{P}, \mathcal{R}, \mathcal{S}, \mu)$-*chain* $u_1 \to v_1, u_2 \to v_2, \ldots$ *is strongly minimal if it is minimal and there is no rule* $s \to t \in \mathcal{S}_\sharp$ *and substitutions* σ, θ *such that* $u_1\sigma \rhd s\theta$ *and* $t\theta$ *starts an infinite minimal* $(\mathcal{P}, \mathcal{R}, \mathcal{S}, \mu)$-*chain.*

But the absence of infinite strongly minimal $(\mathcal{P}, \mathcal{R}, \mathcal{S}, \mu)$-chain do not characterize the finiteness of CS problems. For example, if $\mathcal{S}_\sharp = \{a \to F(a)\}$, $\mathcal{P} = \{F(x) \to x\}$, $\mathcal{R} = \emptyset$ and $\mu(f) = \emptyset$ for all f in the signature, we have the infinite minimal $(\mathcal{P}, \mathcal{R}, \mathcal{S}, \mu)$-chain $F(a) \to_\mathcal{P} a \xrightarrow{\Lambda}_\mathcal{S} F(a) \to_\mathcal{P} \cdots$ which is not strongly minimal. Furthermore, there is no infinite strongly minimal $(\mathcal{P}, \mathcal{R}, \mathcal{S}, \mu)$-chain. The following result allows us to use strongly minimal chains in the CS-DP framework by imposing an structural condition on rules in $\mathcal{S}_\sharp$. Rules in $\mathrm{unh}_\sharp(\mathcal{R}, \mu)$ always satisfy the condition imposed on $\mathcal{S}_\sharp$ in Theorem 3.

Theorem 3. *Let* $\tau = (\mathcal{P}, \mathcal{R}, \mathcal{S}, \mu)$ *be a CS problem such that for every* $s \to t \in \mathcal{S}_\sharp$, $s = f(s_1, \ldots, s_m)$ *and* $t = g(s_1, \ldots, s_m)$. *Then,* τ *is finite if there is no infinite strongly minimal* $(\mathcal{P}, \mathcal{R}, \mathcal{S}, \mu)$-*chain.*

6.2 Left-Linearity and μ-Conservativity

The second stumbling rock in our goal comes when we want to ensure that any term occurring at a frozen position does not show up at an active position by means of a variable instantiation after pair or rule applications. We will make use of left-linearity and conservativity conditions. Left-linearity allow us to discard rules which left-hand side variables are at the same time at frozen and active positions, because we impose its unicity. A rule $\ell \to r$ is μ-conservative if $Var^\mu(r) \subseteq Var^\mu(\ell)$, i.e., there is no migrating variable. Collapsing pairs are not conservative, but if we ensure that when we introduce a possible infinite term at a frozen position in the chain (as $\mathsf{fact}(\mathsf{p}(x))$ in rule (7) or pair rule (29)) it remains unaltered until it shows up by means of a $\mathcal{S}_\sharp$ rule application (in this case, rule (34)), we only need to pay attention to the rule or pair $\ell \to r$ that introduce the possible infinite term u in the chain at a frozen position, $r \rhd_\mu u$, (i.e., rule (7) or pair rule (29)) and check that $\ell \to u$ (i.e. $\mathsf{fact}(x) \to \mathsf{fact}(\mathsf{p}(x))$ and $\mathsf{FACT}(x) \to \mathsf{fact}(\mathsf{p}(x))$) is conservative. If so, we say that the CS problem is *conservative with respect to* $\mathcal{S}$.

Definition 14 (Conditions for S). *Let $\tau = (\mathcal{P}, \mathcal{R}, \mathcal{S}, \mu)$ be a CS problem. We say that τ is conservative with respect to $\mathcal{S}$ if $\mathcal{S}$ is conservative and the following conditions hold:*

- *for all $s \to t \in \mathcal{S}_\sharp$, $s = f(s_1, \ldots, s_m)$ and $t = g(s_1, \ldots, s_m)$; and,*
- *for each $s \to t \in \mathcal{S}_\sharp$ and for each $u \to v \in \mathcal{P} \cup \mathcal{R}$, if there is a nonvariable subterm v' of v at a frozen position such that $\theta = mgu(v', s)$, then $v' = s$ up to renaming of variables and $u \to v'$ must be conservative.*

These conditions always hold if $\mathcal{S}_\sharp \subseteq \mathsf{unh}_\sharp(\mathcal{R}, \mu)$.

6.3 Extended Basic CS Usable Rules

We define our set of usable rule in the usual way. Let $\mathcal{F}un^\mu(s)$ be the set of symbols at active positions in a term $s \in \mathcal{T}(\mathcal{F}, \mathcal{X})$, $\mathcal{F}un^\mu(s) = \{f \mid \exists p \in Pos^\mu(s), f = \mathsf{root}(s|_p)\}$. and $\mathcal{F}un^\#(s)$ the set of symbols at frozen positions in a term $s \in \mathcal{T}(\mathcal{F}, \mathcal{X})$, $\mathcal{F}un^\#(s) = \{f \mid \exists p \in Pos(s) - Pos^\mu(s), f = \mathsf{root}(s|_p)\}$. Let $Rls_\mathcal{R}(f) = \{\ell \to r \in \mathcal{R} \mid \mathsf{root}(\ell) = f\}$.

Definition 15 (Extended Basic μ-Dependency). *Given a TRS $(\mathcal{F}, R)$ and a replacement map μ, we say that $f \in \mathcal{F}$ has an extended basic μ-dependency on $h \in \mathcal{F}$, written $f \rhd_{\mathcal{R}, \mu} h$, if $f = h$ or there is a function symbol g with $g \rhd_{\mathcal{R}, \mu} h$ and a rule $\ell \to r \in Rls_\mathcal{R}(f)$ with $g \in \mathcal{F}un^\#(\ell) \cup \mathcal{F}un^\mu(r)$.*

Definition 16 (Extended Basic CS Usable Rules). *Let $\tau = (\mathcal{P}, \mathcal{R}, \mathcal{S}, \mu)$ be a CS problem. The set $\mathcal{U}_\tau^\rhd(\mathcal{R})$ of extended basic context-sensitive usable rules of τ is*

$$\mathcal{U}_\tau^\rhd(\mathcal{R}) = \bigcup_{u \to v \in \mathcal{P} \cup \mathcal{S}, f \in \mathcal{F}un^\#(u) \cup \mathcal{F}un^\mu(v), f \rhd_{\mathcal{R}, \mu} g} Rls_\mathcal{R}(g)$$

We obtain the processor Proc_{UR}. The pairs $\mathcal{P}$ in a CS problem $(\mathcal{P}, \mathcal{R}, \mathcal{S}, \mu)$, where $\mathcal{P}$ is a TRS over the signature $\mathcal{G}$, are partitioned as follows: $\mathcal{P}_\mathcal{X} = \{u \to v \in \mathcal{P} \mid v \in Var(u) - Var^\mu(u)\}$ and $\mathcal{P}_\mathcal{G} = \mathcal{P} - \mathcal{P}_\mathcal{X}$.

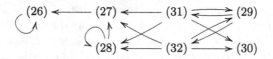

Fig. 4. CS Dependency Graph for Example 1

Theorem 4. *Let $\tau = (\mathcal{P}, \mathcal{R}, \mathcal{S}, \mu)$ be a CS problem such that (a) $\mathcal{P}_\mathcal{G} \cup \mathcal{U}_\tau^\rhd(\mathcal{R}) \cup \mathcal{S}_{\rhd_\mu}$ is left-linear and conservative, and (b) whenever $\mathcal{P}_\mathcal{X} \neq \emptyset$ we have that $\mathcal{P}_\mathcal{X}$ is left-linear and τ is conservative with respect to $\mathcal{S}$. Let $(\gtrsim, \sqsupset)$ be a μ-reduction pair such that (1) $\mathcal{P} \subseteq \gtrsim \cup \sqsupset$, $\mathcal{U}_\tau^\rhd(\mathcal{R}) \uplus \mathcal{C}_\varepsilon \subseteq \gtrsim$, (2) whenever $\mathcal{P}_\mathcal{X} \neq \emptyset$ we have* *that* $S \subseteq \gtrsim \cup \sqsupset$. *Let* $\mathcal{P}_\sqsupset = \{u \to v \in \mathcal{P} \mid u \sqsupset v\}$ *and* $S_\sqsupset = \{s \to t \in S \mid s \sqsupset t\}$. *Then, the processor* Proc_{UR} *given by*

$$\mathsf{Proc}_{UR}(\tau) = \begin{cases} \{(\mathcal{P} - \mathcal{P}_\sqsupset, \mathcal{R}, S - S_\sqsupset, \mu)\} & \text{if (1) and (2) hold} \\ \{(\mathcal{P}, \mathcal{R}, S, \mu)\} & \text{otherwise} \end{cases}$$

is sound.

Example 15. In Example 1, we start with the CS problem $\tau_0 = (\mathsf{DP}(\mathcal{R}, \mu), \mathcal{R},$ $\mathsf{unh}(\mathcal{R}, \mu), \mu^\sharp)$. Applying the well-known SCC processor [17] to τ_0, $\mathsf{Proc}_{SCC}(\tau_0)$, we get the new set of CS problems $\mathsf{Proc}_{SCC}(\tau_0) = \{\tau_1, \tau_2, \tau_3\}$ using the computed CS dependency graph from Fig. 4, where $\tau_1 = (\{(26)\}, \mathcal{R}, \emptyset, \mu)$, $\tau_2 = (\{(28)\}, \mathcal{R}, \emptyset, \mu)$ and $\tau_3 = (\{(29), (31), (32)\}, \mathcal{R}, \{(34)\}, \mu)$. Applying the well-known μ-subterm processor [17] to CS problems τ_1 and τ_2 we get $\mathsf{Proc}_{sub}(\tau_1) = \tau_4$ and $\mathsf{Proc}_{sub}(\tau_2) = \tau_4$, where $\tau_4 = (\emptyset, \mathcal{R}, \emptyset, \mu)$ and, hence, we can conclude that τ_1 and τ_2 are finite.

But, until now, CS problem τ_3 could not be handled by any automatic tool. By Definition 16, the set of extended basic CS usable rules $\mathcal{U}_\tau^\triangleright(\mathcal{R})$ is:

$$\mathsf{zero}(0) \to \mathsf{true} \qquad \mathsf{zero}(\mathsf{s}(x)) \to \mathsf{false} \qquad \mathsf{p}(\mathsf{s}(x)) \to x$$

when in the previous approach all the rules are usable. We can use the extended basic CS usable rules instead of $\mathcal{R}$ because the CS problem satisfies the restrictions in Theorem 4 and the following polynomial interpretation [26] allows us to remove pair (32):

$$
\begin{aligned}
[\mathsf{fact}](x) &= 2x & [*](x, y) &= \tfrac{1}{2}xy + 2 \\
[\mathsf{p}](x) &= \tfrac{1}{2}x & [\mathsf{zero}] &= \tfrac{1}{2}x^2 \\
[0] &= 2 & [\mathsf{s}](x) &= 2x + 1 \\
[\mathsf{false}] &= \tfrac{1}{2} & [\mathsf{true}] &= 2 \\
[\mathsf{FACT}](x) &= 2x^2 + 2 & [\mathsf{IF}](x, y, z) &= \tfrac{1}{2}xy + \tfrac{1}{2}x + z
\end{aligned}
$$

The new CS problem $\tau_5 = (\{(29), (31)\}, \mathcal{R}, \{(34)\}, \mu)$ can be handled again using Theorem 4. The following polynomial interpretation removes pair (31):

$$
\begin{aligned}
[\mathsf{fact}](x) &= 1 & [*](x, y) &= 2x + 2 \\
[0] &= 0 & [\mathsf{s}](x) &= 2x \\
[\mathsf{p}](x) &= 2x + 1 & [\mathsf{zero}] &= 2x + 1 \\
[\mathsf{false}] &= 1 & [\mathsf{true}] &= 1 \\
[\mathsf{FACT}](x) &= 2 & [\mathsf{IF}](x, y, z) &= y + 1
\end{aligned}
$$

and we obtain a finite CS problem by applying Proc_{SCC} to the resulting CS problem.

7   Experimental Evaluation

We have performed an experimental evaluation of the new improvements introduced by these new results presented in the paper in our tool for proving termination properties, MU-TERM [4]. We compared our new version, we call it MU-TERM 5.1, with respect to the previous version, MU-TERM 5.08 [17]. The experiments have been performed on an Intel Core 2 Duo at 2.4GHz with 8GB of RAM, running OS X 10.9.1 using a 120 seconds timeout. We used the last version of the termination problem database, TPDB 8.0.7[5], context-sensitive category. Results are in http://zenon.dsic.upv.es/muterm/ benchmarks/lrc15-csr/benchmarks.html and summarized in Table 1. MU-TERM 5.1 also participated in the CSR category in the 2014 termination competition (http://termination-portal.org/wiki/Termination_Competition_2014) and the same results were confirmed.

Table 1. MU-TERM 5.1 vs. MU-TERM 5.08 comparison

Tool version	Proved	Total time (Av. time)
MU-TERM **5.1**	102/109	1.62s
MU-TERM **5.08**	99/109	2.23s

The practical improvements revealed by the experimental evaluation are twofold. First, we can prove (now) termination of 102 of the 109 examples, 3 more examples than our previous version, including [29, Example 1], [13, Example 1] and [10, Example 3.2.14], whose automatic proofs were open problems since 1997, 2003 and 2008. To our knowledge, there is no other tool that can prove more than those 99 examples from this collection of problems. Second, the new definitions yield a faster implementation; this is witnessed by a speed-up of 1.37 with respect to our previous version.

8   Conclusions

In this paper, we revisit infinite μ-rewrite sequences to obtain a new notion of minimal non-μ-terminating term and a new set of unhiding rules. Since the introduction of the CS-DPs in 2006, the constraints introduced by the unhiding process have been a headache for constraint solvers. For example, in the original approach for each symbol f in the signature and replacing argument $i \in \mu(f)$, a *projection constraint* $f(x_1, \ldots, x_n) \geq x_i$ should be satisfied in order to find a proof. Subsequent works [1,17] reduced these projection constraints to a subset of those for the hidden symbols. Now, in many cases, as in the leading example of the paper, we can avoid these projection/embedding constraints and the unhiding rules are a very small set of rules. In the context of the CS-DP framework, we propose a new notion of chain, the notion of strongly minimal $(\mathcal{P}, \mathcal{R}, \mathcal{S}, \mu)$-chain and a new set of CS usable rules, the extended basic CS usable rules, that allows us to simplify termination proofs on CS problems with respect to the set of unhiding rules. The new processor leads us to a faster and more

---

[5] See http://termcomp.uibk.ac.at/termcomp/.

powerful CS-DP framework. We show an example where the technique is success-fully applied [29, Example1] (included in the TPDB), whose automatic proof was an open problem since 1997. An implementation and an experimental evaluation was performed in our tool for proving termination properties, MU-TERM [4]. With these improvements, MU-TERM won the CSR category in the 2014 termination competition.

References

1. Alarcón, B., Emmes, F., Fuhs, C., Giesl, J., Gutiérrez, R., Lucas, S., Schneider-Kamp, P., Thiemann, R.: Improving context-sensitive dependency pairs. In: Cervesato, I., Veith, H., Voronkov, A. (eds.) LPAR 2008. LNCS (LNAI), vol. 5330, pp. 636–651. Springer, Heidelberg (2008)
2. Alarcón, B., Gutiérrez, R., Lucas, S.: Context-sensitive dependency pairs. In: Arun-Kumar, S., Garg, N. (eds.) FSTTCS 2006. LNCS, vol. 4337, pp. 297–308. Springer, Heidelberg (2006)
3. Alarcón, B., Gutiérrez, R., Lucas, S.: Context-Sensitive dependency pairs. Inf. Comput. **208**, 922–968 (2010)
4. Alarcón, B., Gutiérrez, R., Lucas, S., Navarro-Marset, R.: Proving termination properties with MU-TERM. In: Johnson, M., Pavlovic, D. (eds.) AMAST 2010. LNCS, vol. 6486, pp. 201–208. Springer, Heidelberg (2011)
5. Arts, T., Giesl, J.: Proving innermost normalisation automatically. In: Comon, H. (ed.) RTA 1997. LNCS, vol. 1232, pp. 151–171. Springer, Heidelberg (1997)
6. Arts, T., Giesl, J.: Termination of term rewriting using dependency pairs. Theor. Comput. Sci. **236**(1–2), 133–178 (2000)
7. Baader, F., Nipkow, T.: Term Rewriting and All That. Cambridge University Press, Cambridge (1998)
8. Borovanský, P., Kirchner, C., Kirchner, H., Moreau, P., Ringeissen, C.: An overview of ELAN. Electr. Notes Theor. Comput. Sci. **15**, 55–70 (1998). http://dx.doi.org/10.1016/S15710661(05)825526
9. Clavel, M., Durán, F., Eker, S., Lincoln, P., Martí-Oliet, N., Meseguer, J., Talcott, C. (eds.): All About Maude - A High-Performance Logical Framework. LNCS, vol. 4350. Springer, Heidelberg (2007)
10. Emmes, F.: Automated Termination Analysis of Context-Sensitive Term Rewrite Systems. Master's thesis, Fakultät für Mathematik, Informatik und Naturwissenschaften der Rheinisch-Westfälischen Technischen Hochschule Aachen, Aachen, Germany (2008)
11. Futatsugi, K., Goguen, J.A., Jouannaud, J.P., Meseguer, J.: Principles of OBJ2. In: Proceedings of the 12th ACM SIGACT-SIGPLAN Symposium on Principles of Programming Languages, POPL 1985, pp. 52–66. ACM (1985)
12. Futatsugi, K., Nakagawa, A.: An overview of CAFE specification environment-an algebraic approach for creating, verifying, and maintaining formal specifications over networks. In: Proceedings of the 1st International Conference on Formal Engineering Methods, ICFEM 1997, p. 170. IEEE Computer Society (1997)
13. Giesl, J., Middeldorp, A.: Innermost termination of context-sensitive rewriting. In: Ito, M., Toyama, M. (eds.) DLT 2002. LNCS, vol. 2450, pp. 231–244. Springer, Heidelberg (2003)
14. Giesl, J., Thiemann, R., Schneider-Kamp, P., Falke, S.: Mechanizing and improving dependency pairs. J. Autom. Reasoning **37**(3), 155–203 (2006)

15. Goguen, J.A., Winkler, T., Meseguer, J., Futatsugi, K., Jouannaud, J.P.: Software Engineering with OBJ: Algebraic Specification in Action. Kluwer, Boston (2000). chap. Introducing OBJ
16. Gramlich, B.: Generalized sufficient conditions for modular termination of rewriting. Appl. Algebra Eng. Commun. Comput. **5**, 131–151 (1994)
17. Gutiérrez, R., Lucas, S.: Proving termination in the context-sensitive dependency pair framework. In: Ölveczky, P.C. (ed.) WRLA 2010. LNCS, vol. 6381, pp. 18–34. Springer, Heidelberg (2010)
18. Gutiérrez, R., Lucas, S.: Function calls at frozen positions in termination of context-sensitive rewriting. Technical report, DSIC, Universitat Politècnica de València, May 2015. http://hdl.handle.net/10251/50750
19. Gutiérrez, R., Lucas, S., Urbain, X.: Usable rules for context-sensitive rewrite systems. In: Voronkov, A. (ed.) RTA 2008. LNCS, vol. 5117, pp. 126–141. Springer, Heidelberg (2008)
20. Hendrix, J., Meseguer, J.: On the completeness of context-sensitive order-sorted specifications. In: Baader, F. (ed.) RTA 2007. LNCS, vol. 4533, pp. 229–245. Springer, Heidelberg (2007)
21. Hirokawa, N., Middeldorp, A.: Dependency pairs revisited. In: van Oostrom, V. (ed.) RTA 2004. LNCS, vol. 3091, pp. 249–268. Springer, Heidelberg (2004)
22. Hirokawa, N., Middeldorp, A.: Tyrolean termination tool: techniques and features. Inf. Comput. **205**(4), 474–511 (2007)
23. Lucas, S.: Context-sensitive computations in functional and functional logic programs. J. Funct. Logic Program. **1998**(1), 1–61 (1998)
24. Lucas, S.: Termination of on-demand rewriting and termination of OBJ programs. In: De Nicola, R., Søndergaard, H. (eds.) Proceedings of the 3rd Internatinal Conference on Principles and Practice of Declarative Programming, PPDP 2001, pp. 82–93. ACM Press (2001)
25. Lucas, S.: Termination of rewriting with strategy annotations. In: Nieuwenhuis, R., Voronkov, A. (eds.) LPAR 2001. LNCS (LNAI), vol. 2250, pp. 666–680. Springer, Heidelberg (2001)
26. Lucas, S.: Polynomials over the reals in proofs of termination: from theory to practice. RAIRO Theor. Inf. Appl. **39**(3), 547–586 (2005)
27. Lucas, S.: Completeness of context-sensitive rewriting. Inf. Process. Lett. **115**(2), 87–92 (2015). http://dx.doi.org/10.1016/j.ipl.2014.07.004
28. Ohlebusch, E.: On the modularity of termination of term rewriting systems. Theor. Comput. Sci. **136**, 333–360 (1994)
29. Zantema, H.: Termination of context-sensitive rewriting. In: Comon, H. (ed.) RTA 1997. LNCS, vol. 1232. Springer, Heidelberg (1997)

Model-Checking HELENA Ensembles with Spin

Rolf Hennicker, Annabelle Klarl$^{(\boxtimes)}$, and Martin Wirsing

Ludwig-Maximilians-Universität München, München, Germany
klarl@pst.ifi.lmu.de

Abstract. The HELENA approach allows to specify dynamically evolving ensembles of collaborating components. It is centered around the notion of roles which components can adopt in ensembles. In this paper, we focus on the early verification of HELENA models. We propose to translate HELENA specifications into PROMELA and check satisfaction of LTL properties with Spin [11]. To prove the correctness of the translation, we consider an SOS semantics of (simplified variants of) HELENA and PROMELA and establish stutter trace equivalence between them. Thus, we can guarantee that a HELENA specification and its PROMELA translation satisfy the same LTL formulae (without *next*). Our correctness proof relies on a new, general criterion for stutter trace equivalence.

1 Introduction

The HELENA approach [10] proposes to model large distributed systems of goal-oriented collaborations of components by dynamically evolving ensembles where the participating components play certain roles. By adopting a role, a component executes a role-specific behavior. The introduction of roles allows to focus on the particular tasks which components fulfill in specific collaborations and to structure the implementation of ensemble-based systems [13,14].

Ensembles always collaborate towards some global goals. Such goals are often temporal properties which we specify by linear temporal logic (LTL) formulae [17]. In this paper, we focus on the early (pre-implementation) verification of HELENA models for their intended goals. We propose to translate HELENA specifications into PROMELA and check satisfaction of LTL properties with the model-checker Spin [11]. PROMELA is well-suited as a target language since it supports dynamic creation of concurrent processes and asynchronous communication. Our contribution is as follows: Firstly, we propose a syntactic translation of a simplified variant of HELENA (called HELENALIGHT), which restricts ensemble specifications to the core concepts of our modeling approach, to a subset of PROMELA (called PROMELALIGHT), which is sufficient to express all HELENALIGHT concepts. Secondly, we prove the correctness of the translation. For this purpose, we define a formal SOS semantics for HELENALIGHT and PROMELALIGHT specifications. The latter is based on the semantics for full PROMELA

This work has been partially sponsored by the European Union under the FP7-project ASCENS, 257414.

Dedicated to José Meseguer

N. Martí-Oliet et al. (Eds.): Meseguer Festschrift, LNCS 9200, pp. 331–360, 2015.
DOI: 10.1007/978-3-319-23165-5_16

in [20]. On this semantic basis, we establish stutter trace equivalence between the semantics of a HELENALIGHT specification and its PROMELALIGHT translation. Then, we reuse results from the literature [1] that satisfaction of LTL formulae (without *next*) is preserved by stutter trace equivalence. As a consequence, we can verify LTL properties for a HELENALIGHT specification by model-checking its PROMELALIGHT translation with Spin. To prove stutter trace equivalence between HELENALIGHT and PROMELALIGHT, we investigate a new, general criterion that Kripke structures are stutter trace equivalent if particular stutter simulations (called ≈-stutter simulations) can be established in both directions.

In Sect. 2, we explain foundations on LTL and propose our criterion for stutter trace equivalence which entails $LTL_{\setminus \mathbf{X}}$ preservation. In Sect. 3, we summarize the HELENA modeling approach and present the running example of a peer-2-peer network storing files. Sect. 4 defines syntax and semantics of HELENALIGHT and Sect. 5 for PROMELALIGHT resp. In Sect. 6, we provide the formal translation from HELENALIGHT to PROMELALIGHT and, in Sect. 7, establish the desired correctness results. Sect. 8 discusses model-checking of HELENA specifications.

Personal Note. José, Rolf, and Martin know each other since the eighties where they investigated the foundations of algebraic specifications, José with Joseph Goguen in the initial algebra setting, and Martin and Rolf inspired by the CIP program development methodology from the loose and observational semantics point of view. Behavioral specifications and equivalence of algebras were already studied by José and Joseph in [8] which was a fruitful source for the thesis of Rolf. In the beginning of the nineties, Martin was looking for an appropriate semantical framework for underpinning the typically informal or semi-formal object-oriented methods with a formal framework. Rewriting logic invented by José a few years earlier appeared to be the perfect tool: a simple computational logic that supports concurrent computation, logical deduction, and object-oriented features. Martin is still very grateful that José gave him the opportunity to present his first results on integrating formal specifications into pragmatic object-oriented development at the first WRLA in 1996 [22]. Later, José, Rolf, and Martin became all members of IFIP WG 1.3. and José and Martin worked together on the algebraic foundation and analysis of modern programming paradigms including studies of the semantic foundations of multi-paradigm languages [2], the analysis of denial of service attacks [6], and the specification and correct implementation of the distributed programming language KLAIM [7]. Our current paper follows the same aim; we present the formal foundation of HELENA ensemble specifications and use results related to José's seminal paper on algebraic simulations [19] for proving the correctness of model-checking HELENA ensembles with Spin.

It is a great pleasure to know and work with José since so many years; we admire his broad and deep knowledge in many areas; it is inspiring to discuss with him and he is a very kind and warm hearted colleague and friend. We are looking forward to many further exciting scientific exchanges with José.

2   Foundations on $LTL_{\backslash \mathbf{X}}$ Preservation

In this section, we review Kripke structures, linear temporal logic (LTL), and satisfaction of LTL formulae. We propose how to induce Kripke structures from labeled transitions systems (which will be used for the semantics of HELE-NALIGHT and PROMELALIGHT specifications). Furthermore, we propose a criterion for stutter trace equivalence of Kripke structures which entails $LTL_{\backslash \mathbf{X}}$ [1] preservation according to the literature in [1].

LTL in Kripke Structures: A Kripke structure consists of a set of states connected by (unlabeled) transitions. The states are labeled by sets of atomic propositions which hold in the state and some states are marked as initial states.

Definition 1 (Kripke Structure). *Let AP be a set of atomic propositions. A Kripke structure K over AP is a tuple $(S_K, I_K, \to_K, F_K)$ such that S_K is a set of states, $I_K \subseteq S_K$ is a set of initial states, $\to_K \subseteq S_K \times S_K$ is an (unlabeled) transition relation without terminal states (i.e., $\forall s \in S_K. \exists s' \in S_K \ . \ s \to_K s'$), and $F_K : S_K \to 2^{AP}$ is a labeling function associating to each state the set of atomic propositions that hold in it.*

For a Kripke structure $K = (S_K, I_K, \to_K, F_K)$, we further define: A *path of K* is an infinite sequence $p = s_0 s_1 s_2 \dots$ (with $s_i \in S_K$ for all $i \in \mathbb{N}$) such that $s_0 \in I_K$ and $s_i \to_K s_{i+1}$. A *trace of K* is an infinite sequence $t = t_0 t_1 t_2 \dots$ such that there exists a path $p = s_0 s_1 s_2 \dots$ in K and $t_i = F_K(s_i)$ for all $i \in \mathbb{N}$.

To describe temporal properties, we use linear temporal logic (LTL).

Definition 2 (LTL [1]**).** *Let AP be a set of atomic propositions. LTL formulae over AP are inductively defined by:*

$$
\begin{aligned}
\phi \quad &= \quad p \in AP & \textit{(atomic proposition)} \\
&| \ \neg\phi \mid \phi \wedge \psi & \textit{(proposition logic operators)} \\
&| \ \mathbf{X}\phi \mid \Diamond\phi \mid \Box\phi \mid \phi\,\mathbf{U}\psi & \textit{(linear temporal logic operators)}
\end{aligned}
$$

Disjunction, implication, and equivalence are given by the usual abbreviations. The set of LTL formulae over AP is denoted by $LTL(AP)$.

Satisfaction of LTL formulae is defined by the usual inductive definition [1].

Definition 3 (Satisfaction of LTL in Kripke Structures). *Let $K = (S_K, I_K, \to_K, F_K)$ be a Kripke structure over AP, $t = t_0 t_1 t_2 \dots$ a trace of K, and $\phi \in LTL(AP)$; $t|_i$ denotes the subsequence $t_i t_{i+1} t_{i+2} \dots$ of t.*
The satisfaction of ϕ for trace t, written $t \models \phi$, is inductively defined by

$- \ t \models p$, *if $p \in t_0$,*
$- \ t \models \neg\phi$, *if $t \not\models \phi$,*
$- \ t \models \phi \wedge \psi$, *if $t \models \phi$ and $t \models \psi$,*

---

[1] $LTL_{\backslash \mathbf{X}}$ is the fragment of LTL that does not contain the *next* operator **X**.

- $t \models \boldsymbol{X}\phi$, if $t|_1 \models \phi$,
- $t \models \Diamond\phi$, if there exists $k \geq 0$ such that $t|_k \models \phi$,
- $t \models \Box\phi$, if for all $k \geq 0$ holds $t|_k \models \phi$,
- $t \models \phi \boldsymbol{U}\psi$, if there exists $k \geq 0$
  such that $t|_k \models \psi$ and for all $0 \leq j < k$ holds $t|_j \models \phi$,

The Kripke structure K satisfies an LTL formula ϕ, written $K \models \phi$, if all traces of K satisfy ϕ.

LTL in Labeled Transition Systems: In contrast to Kripke structures, labeled transition systems do not label states with atomic propositions, but transitions with actions.

Definition 4 (Labeled Transition System). A labeled transition system (LTS) T is a tuple $(S_T, I_T, A_T, \to_T)$ such that S_T is a set of states, $I_T \subseteq S_T$ is a set of initial states, A_T is a set of actions such that the silent action $\tau \notin A_T$, and $\to_T \subseteq S_T \times (A_T \cup \tau) \times S_T$ is a labeled transition relation.

For an LTS $T = (S_T, I_T, A_T, \to_T)$, we further define: a^* denotes a (possibly empty) sequence of a actions. If $w = a_1 \ldots a_n$ holds for some $n \in \mathbb{N}$ and $a_1, \ldots, a_n \in (A_T \cup \tau)$, then $s \xrightarrow{w}_T s'$ stands for $s = s'$, if $n = 0$, and $s \xrightarrow{a_1}_T s_1 \ldots s_{n-1} \xrightarrow{a_n}_T s'$ with appropriate $s_1, \ldots, s_{n-1}$ otherwise. The LTS T together with a set of atomic propositions AP and a satisfaction relation $s \models \phi$ (for $s \in S_T$ and $\phi \in LTL(AP)$) induces a Kripke structure $K(T) = (S_T, I_T, \to_T^{\bullet}, F)$. The labeled transition relation $\to_T$ is transformed into an unlabeled, total transition relation $\to_T^{\bullet}$ which forgets the actions and adds a new transition $s \to_T^{\bullet} s$ for each terminal state $s \in S_T$. The labeling function $F : S_T \to 2^{AP}$ is defined by $F(s) = \{p \in AP \mid s \models p\}$.

Definition 5 (Satisfaction of LTL in Labeled Transition Systems). Let $T = (S_T, I_T, A_T, \to_T)$ be a labeled transition system, AP a set of atomic propositions, $s \models \phi$ a satisfaction relation for $s \in S_T$ and $\phi \in LTL(AP)$.
  T satisfies ϕ, written $T \models \phi$, if $K(T) \models \phi$, i.e., the induced Kripke structure $K(T)$ satisfies ϕ.

LTL$_{\backslash \mathbf{X}}$ Preservation: Lastly, we investigate when two Kripke structures satisfy the same set of $LTL_{\backslash \mathbf{X}}$ formulae. Therefore, we introduce the notion of stutter trace equivalence.

- Two paths of Kripke structures over the same set of atomic propositions AP are *stutter trace equivalent* if their traces only differ in the number of their stutter steps, i.e., there exist sets of atomic propositions $P_i \subseteq AP$ (with $i \in \mathbb{N}$) such that the traces of both paths have the form $P_0^+ P_1^+ P_2^+ \ldots$ where P_i^+ denotes a non-empty sequence of the same set P_i.
- Two Kripke structures K_1 and K_2 are *stutter trace equivalent* if for each path of K_1 there exists a stutter trace equivalent path of K_2 and vice versa.

To provide a criterion for stutter trace equivalence of Kripke structures, we propose the notion of a $\approx$-stutter simulation.

Definition 6 ($\approx$-Stutter Simulation). *Let $K_1 = (S_1, I_1, \rightarrow_1, F_1)$ and $K_2 = (S_2, I_2, \rightarrow_2, F_2)$ be two Kripke structures over AP. Let $\approx \subseteq S_1 \times S_2$ be a relation.*

A relation $\sim \subseteq S_1 \times S_2$ is a $\approx$-stutter simulation of K_1 by K_2 if (1) $\sim \subseteq \approx$ and (2) for all $s \in S_1, t \in S_2$ with $s \sim t$, if $s \rightarrow_1 s'$, then $s' \sim t$ or there exists $t \rightarrow_2 t_1 \rightarrow_2 \ldots \rightarrow_2 t_n \rightarrow_2 t'_1 \rightarrow_2 \ldots \rightarrow_2 t'_m \rightarrow_2 t'$ $(n, m \geq 0)$ such that $s \approx t_i$ for all $i \in \{1, \ldots, n\}$, $s' \approx t'_j$ for all $j \in \{1, \ldots, m\}$ and $s' \sim t'$.

K_1 is $\approx$-stutter simulated by K_2 if there exists a $\approx$-stutter simulation $\sim$ of K_1 by K_2 such that $s_0 \sim t_0$ for all $s_0 \in I_1, t_0 \in I_2$.

Stutter trace equivalence does not require preservation of the branching structure of the underlying Kripke structures. Therefore, interestingly, the notion of $\approx$-stutter simulations (compared to stutter bisimulations [19] preserving branching) is sufficient to provide a criterion whether two Kripke structures are stutter trace equivalent.

Theorem 1 (Stutter Trace Equivalence). *Let K_1 and K_2 be two Kripke structures over AP with states S_1, S_2 resp. Let $\approx \subseteq S_1 \times S_2$ be a property-preserving relation, i.e., for all $s \in S_1, t \in S_2$, if $s \approx t$, then $F_1(s) = F_2(t)$, and $\approx^{-1}$ its inverse relation. If K_1 is $\approx$-stutter simulated by K_2 and K_2 is $\approx^{-1}$-stutter simulated by K_1, then K_1 and K_2 are stutter trace equivalent.*

Proof. Since K_1 is $\approx$-stutter simulated by K_2, each path of K_1 is simulated by a corresponding path of K_2 such that on all paths, the states related by $\approx$ have the same properties. The same holds vice versa for $\approx^{-1}$.   □

The question arises which LTL formulae are satisfied by two stutter trace equivalent Kripke structures. It is clear that the *next* operator $\mathbf{X}$ of temporal logic is not preserved since stutter steps are allowed. However, if we restrict our attention to the temporal logic $LTL_{\backslash \mathbf{X}}$, we can use a result of [1] which shows that all formulae of $LTL_{\backslash \mathbf{X}}$ are preserved. In practice, eliminating the *next* operator is not a great loss since interesting properties are not so much concerned with what happens in the next step as to what eventually happens [16].

Theorem 2 ($LTL_{\backslash \mathbf{X}}$ Preservation). *Let K_1 and K_1 be two stutter trace equivalent Kripke structures over AP. For any $LTL_{\backslash \mathbf{X}}$ formula ϕ over AP, we have $K_1 \models \phi \Leftrightarrow K_2 \models \phi$.*

Proof. The proof can be found in [1, pp. 534–535] (Theorem 7.92 and Corollary 7.93).

3   The HELENA Approach

The role-based modeling approach HELENA [10] provides concepts to describe systems where components team up in ensembles to perform global goal-oriented tasks. To participate in an ensemble, a component plays a certain role. This role adds role-specific behavior to the component and allows collaboration with (the roles of) other components. By switching between roles, the component changes its currently executed behavior. By adopting several roles in parallel, a component can concurrently execute different behaviors and participate at the same time in different ensembles.

Components: *Component instances* are classified by *component types*. They are considered as carriers of basic information relevant across many ensembles. They provide basic capabilities to store data in attributes and to perform computations by operations. Additionally, they can be connected to other component instances by storing references to them.

Roles: Whenever a component instance joins an ensemble, the component adopts a role by creating a new *role instance* and assigning it to itself. The kind of roles a component is allowed to adopt is determined by *role types*. A role type defines role-specific attributes and a set of incoming and outgoing message types which are supported for interaction and collaboration between role instances.

P2P Example: We consider a peer-2-peer network supporting the distributed storage of files which can be retrieved upon request. Several peers are connected in a ring structure and work together to request and transfer a file: One peer plays the role of the `Requester` of the file, other peers act as `Routers` and the peer storing the requested file adopts the role of the `Provider`. All these roles can be adopted by components of type `Peer`. Figure 1 shows the component type `Peer` and the role type `Requester` in a graphical representation similar to UML classes. For simplicity, we only consider peers which can store one single file. The attribute `hasFile` of a `Peer` (cf. Fig. 1a) indicates whether the peer has the file; the file's content information is represented by the attribute `content`. A `Peer` is connected to its neighbor depicted by the association in Fig. 1a. The role type `Requester` indicates by the notation `Requester:{Peer}` that any component instance of type `Peer` can adopt that role. It stores whether it already has the file in its attribute `hasFile` and supports two incoming and two outgoing messages.

Ensemble Structures: To define the structural characteristics of a collaboration, an *ensemble structure* specifies the role types whose instances form the

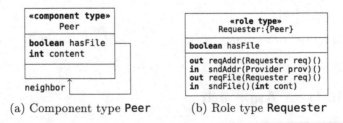

(a) Component type `Peer`   (b) Role type `Requester`

Fig. 1. Types occurring in the p2p example

ensemble, determines how many instances of each role type may contribute by a multiplicity (like 0..1, 1, *, 1..* etc.), and defines the capacity of the input queue for each role type. We assume that between instances of two role types the messages which are output on one side and input on the other side can be exchanged on the input queues of the role instances.

P2P Example: Figure 2 shows a graphical representation of the ensemble structure for the p2p example. It consists of the three role types Requester, Router, and Provider with associated multiplicities and input queue capacities.

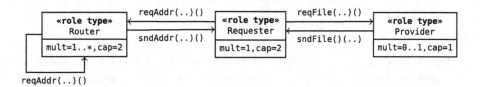

Fig. 2. Ensemble structure $\Sigma_{transfer}$ for the p2p example

Role Behaviors and Ensemble Specifications: An *ensemble specification* adds dynamic behavior to an ensemble structure Σ by equipping each role type occurring in Σ with a *role behavior*. A role behavior is given by a process expression built from the null process **nil**, action prefix $a.P$, conditional selection **if** (*condition1*) **then** $\{P1\}$(**or** (*condition2*) **then** $\{P2\}$)* (with nondeterministic choice if several branches are executable), and process invocation. There are actions for creating (**create**) and retrieving (**get**) role instances, sending (!) or receiving (?) messages, and invoking operations of the owning component. Additionally, state labels can be used to mark a certain progress of execution in the role behavior (we will use these labels in atomic propositions to express goals). We additionally use predefined variables like **self** to refer to the current role instance and **owner** to refer to the owning component instance. The attributes of the current role instance and its owning component instance are accessed in a Java like style and we provide a predefined query plays(rt,ci) to ask whether the component instance ci currently plays the role rt.

P2P Example: Figure 3 shows the behavior specification of a Router. Initially, a router can receive a request for an address. Depending on whether its owner has the file, it either creates a provider role instance and sends it back to the requester in $P_{provide}$ or forwards the request to another router in P_{fwd} if possible.

LTL for Ensemble Specifications: To express goals over HELENA ensemble specifications, we use linear temporal logic (LTL) formulae over a particular set of atomic HELENA propositions AP: A *state label proposition* is of the form $rt[i]@label$. It is satisfied if there exists a role instance i of type rt whose next performed action is the state label *label*. An *attribute proposition* must be boolean and is built from arithmetic and relational operators, data constants, and propositions of the form $rt[i]:attr$ (or $ct[i]:attr$). An attribute proposition $rt[i]:attr$ is

roleBehavior $Router$ = $?reqAddr(Requester\ rq)()$.
                  if $(\mathbf{owner}.hasFile)$ then $\{P_{provide}\}$
                  or $(!\mathbf{owner}.hasFile)$ then $\{P_{fwd}\}$
    $P_{provide}$ = $p\leftarrow$create$(Provider, \mathbf{owner})$. $rq!sndAddr(p)()$. nil
      P_{fwd} = if $(\mathbf{plays}(Router, \mathbf{owner}.neighbor))$ then $\{$nil$\}$
              or $(!\mathbf{plays}(Router, \mathbf{owner}.neighbor))$ then $\{P_{create}\}$
    P_{create} = $r\leftarrow$create$(Router, \mathbf{owner}.neighbor)$. $r!reqAddr(rq)()$. $Router$

Fig. 3. Role behavior of a `Router`

satisfied if there exists a role instance i of type rt such that the value of its attribute $attr$ evaluates to **true** (and analogously for component attributes). LTL formulae over HELENA propositions and their satisfaction are inductively defined as already described in Sect. 2.

For the p2p example, we want to express that the requester will always receive the requested file if the file is available in the network. We assume a network of three peers and formulate the following achieve goal in LTL which refers to the values of the attribute `hasFile` of component type `Peer` and role type `Requester`:

$$(Peer[1]:hasFile \lor Peer[2]:hasFile \lor Peer[3]:hasFile) \Rightarrow \Diamond Requester[1]:hasFile)$$

In the next sections we will consider a simpler variant of HELENA and present a precise formalization of ensemble specifications, their semantics, satisfaction of atomic propositions and model-checking by translation to PROMELA.

4   HELENALIGHT

We restrict full HELENA specifications to some core concepts which leads to the definition of HELENALIGHT. We first formally define the syntax of HELENALIGHT ensemble specifications. Afterwards, we introduce an SOS-style semantics for such specifications and define satisfaction of LTL formulae.

4.1   Syntax of HELENALIGHT Ensemble Specifications

In HELENALIGHT, we abstract from the underlying component types of a full HELENA specification and consider only role types, whose instances can be dynamically created, and their interactions. Additionally, we omit any notion of data such that we do not consider attributes and data parameters anymore.

Role Types: Role types are characterized by their name and a set of outgoing and incoming message types. In contrast to full HELENA, we omit role attributes and consider message types with exactly one role parameter.

Definition 7 (Message Type). *A message type msg is of the form msgnm(rt X) such that msgnm is the name of the message type and X is a formal parameter of role type rt.*

Definition 8 (Role Type). *A role type is a tuple $rt = (rtnm, rtmsgs_{out}, rtmsgs_{in})$ such that rtnm is the name of the role type, and $rtmsgs_{out}$ and $rtmsgs_{in}$ are sets of message types for outgoing and incoming messages supported by rt.[2]*

Ensemble Structures: Ensemble structures specify which role types are needed for a collaboration. In contrast to full HELENA, we omit multiplicities constraining the number of admissible role instances for each role type. We assume asynchronous communication and specify for each role type the (positive) capacity of the input queue of each role instance of that type.

Definition 9 (Ensemble Structure). *An ensemble structure Σ is a tuple $\Sigma = (nm, roletypes, roleconstraints)$ such that nm is the name of the ensemble structure, roletypes is a set of role types, and for each $rt \in roletypes, roleconstraints(rt)$ is a finite capacity $c > 0$ of the input queue of rt.*

In this paper, we consider only closed ensemble structures Σ. This means that any outgoing message of some role type of Σ must occur as an incoming message of at least one role type of Σ and vice versa, and any parameter type occurring in a message type is a role type of Σ.

Role Behavior Declarations: Given an ensemble structure Σ, process expressions (over Σ) will be used to specify role behaviors. They are built from the process constructs and actions in Definition 10. Opposed to full HELENA, we omit component instances on which role instances are created and any data in message exchange. Furthermore, we omit the **get** action, operation calls and any attribute setters since we do not have attributes in HELENALIGHT.

Definition 10 (Process Expression). *A process expression is built from the following grammar, where N is the name of a process, msgnm is the name of a message type, X and Y are names of variables, rt is a role type (more precisely the name of a role type), and label is the name of a state label:*

$P ::= $ **nil**	*(null process)*
$\mid aP$	*(action prefix)*
$\mid P_1 + P_2$	*(nondeterministic choice)*
$\mid N$	*(process invocation)*
$a ::= X \leftarrow$ **create**(rt)	*(role instance creation)*
$\mid Y!msgnm(X)$	*(sending a message)*
$\mid ?msgnm(rt\,X)$	*(receiving a message)*
$\mid label$	*(state label)*

---

[2] In the following, we often write *rt* synonymously for the role type name *rtnm*.

A receive action $?msgnm(rt\,X)$ (and resp. a create action $X \leftarrow \mathbf{create}(rt)$) declares and opens the scope for a local variable X of type rt. We assume that the names of the declared variables are unique within a process expression and different from **self** which is a predefined variable that can always be used.

Definition 11 (Well-Formedness of Process Expressions). *Let* $\Sigma = (nm, roletypes, roleconstraints)$ *be an ensemble structure. A process expression* P *is* well-formed *for a role type* $rt' \in roletypes$ *w.r.t.* Σ, *if all actions occurring in* P *are well-formed for* rt' *w.r.t.* Σ. *This means:*

- *For a role instance creation action* $X \leftarrow \mathbf{create}(rt)$: $rt \in roletypes$.
- *For a send action* $Y!msgnm(X)$,
    - *the role type* rt' *supports the message type* $msgnm(rt''\,X'')$ *as outgoing message and the variable* X *is of type* rt'',[3]
    - *the role type of the variable* Y *supports the message type* $msgnm(rt''\,X'')$ *as incoming message,*
    - *the variables* X *and* Y *have been declared before, with the exception that* X *can be the special, predefined variable* **self** *of type* rt'.
- *For a receive action* $?msgnm(rt\,X)$, *the role type* rt' *supports the message type* $msgnm(rt\,X)$ *as incoming message.*
- *State labels are unique within the process expression* P
- *State labels are not the first action of the process expressions* P_1 *or* P_2 *in the nondeterministic choice* $P_1 + P_2$.

Building on process expressions, we can now define role behavior declarations. Opposed to full HELENA, a role behavior declaration can not invoke other processes, but can invoke itself recursively.

Definition 12 (Role Behavior Declaration). *Let* Σ *be an ensemble structure and* rt *be a role type in* Σ. *A role behavior declaration for* rt *has the form* **roleBehavior** $rt = P$ *where* P *is a process expression which is well-formed for* rt *w.r.t.* Σ *such that* P *contains (recursive) process invocations at most for* rt.[4]

Ensemble Specifications: An ensemble specification consists, as in full HELENA, of two parts: an ensemble structure and a set of role behavior declarations for all role types occurring in the ensemble structure.

Definition 13 (Ensemble specification). *An ensemble specificationis a pair* $EnsSpec = (\Sigma, behaviors)$ *such that* Σ *is an ensemble structure and behaviors is a set of role behavior declarations which contains exactly one declaration* **role-Behavior** $rt = P$ *for each role type* $rt \in \Sigma$.

---

[3] We must distinguish here between the role type rt', whose behavior is going to be defined, and the role type rt'' used for the parameter.

[4] Note that in the above definition we use rt also as a process name for the role behavior of the role type rt.

```
1  roleType Requester {
2    rolemsg out reqAddr(Requester req);
3    rolemsg in sndAddr(Provider prov);
4    rolemsg out reqFile(Requester req);
5    rolemsg in sndFile(Provider prov);
6  }
7
8  roleType Router {
9    rolemsg in/out reqAddr(Requester req);
10   rolemsg out sndAddr(Provider prov);
11 }
12
13 roleType Provider {
14   rolemsg in reqFile(Requester req)();
15   rolemsg out sndFile(Provider prov);
16 }
17
18 ensembleStructure TransferEnsemble {
19   roleTypes = {<Requester, cap = 2>,
20               <Router,    cap = 2>,
21               <Provider,  cap = 1>};
22 }
```

```
1  roleBehavior Requester =
2    router <- create(Router) .
3    router ! reqAddr(self) .
4    ? sndAddr(Provider prov) .
5    prov ! reqFile(self) .
6    ? sndFile(Provider prov2) .
7    stateSndFile . nil
8
9  roleBehavior Provider =
10   ? reqFile(Requester req) .
11   stateReqFile .
12   req ! sndFile(self) . nil
13
14 roleBehavior Router =
15     ? reqAddr(Requester req) .
16     { prov <- create(Provider) .
17       req ! sndAddr(prov) .
18       nil }
19   +
20     { router <- create(Router) .
21       router ! reqAddr(req) .
22       Router }
```

(a) Role types and ensemble structure   (b) Role behavior declarations

Fig. 4. The p2p example in HELENALIGHT

P2P Example: A simplified variant of the p2p example of Sect. 3, written in HELENATEXT [13], is shown in Fig. 4. In contrast to the specification in full HELENA, we omit the underlying component type Peer and all role attributes as well as data parameters. The ensemble structure names the participating role types and their capacity, but no multiplicities. HELENALIGHT also restricts the specification of dynamic behavior. Process expressions can only use nondeterministic choice instead of conditional selection. Thus, in contrast to the router behavior in full HELENA (cf. Fig. 3), the router *nondeterministically* either provides the file or forwards the request (cf. line 16–22 in Fig. 4b).

4.2   Semantics of HELENALIGHT Ensemble Specifications

The semantic domain of ensemble specifications are labeled transition systems describing the evolution of ensembles. Structured operational semantics (SOS) rules define the allowed transitions. We pursue an incremental approach, similar to [9] and [20], by splitting the semantics into two different layers. The first layer describes how a single role behavior evolves according to the constructs for process expressions of the last section. The second layer builds on the first one by defining the evolution of a whole ensemble from the concurrent evolution of its constituent role instances.

Evolution of Roles: On the first level, we do not have any information about the global state of the whole ensemble (involving all active role instances). Therefore, we only formalize the progress of a single role behavior given by a process expression. Figure 5 defines the SOS rules inductively over the structure

$$
\begin{array}{ll}
\text{(action prefix)} & a.P \overset{a}{\hookrightarrow} P \\[2em]
\text{(choice-left)} & \dfrac{P_1 \overset{a}{\hookrightarrow} P_1'}{P_1 + P_2 \overset{a}{\hookrightarrow} P_1'} \\[2em]
\text{(choice-right)} & \dfrac{P_2 \overset{a}{\hookrightarrow} P_2'}{P_1 + P_2 \overset{a}{\hookrightarrow} P_2'} \\[2em]
\text{(process invocation)} & \dfrac{Q \overset{a}{\hookrightarrow} Q'}{rt \overset{a}{\hookrightarrow} Q'} \qquad \text{if } \textbf{roleBehavior } rt = Q
\end{array}
$$

Fig. 5. SOS rules for the evolution of process expressions in HELENALIGHT

of process expressions in Definition 10. Note that the rule for process invocation relies on a given role behavior declaration. We use the symbol $\hookrightarrow$ to describe transitions on this level. Since it does not involve instances and considers just the behavior of single role types, this level concerns behavioral types.

Evolution of Ensembles: On the next level, we consider global states, which we call *ensemble states*, and the concurrent execution of role instances. For the semantics of an ensemble specification $EnsSpec = (\Sigma, behaviors)$, we describe the possible evolutions of ensemble states (for any admissible, initial ensemble state). An ensemble state captures the set of currently existing role instances together with their local states. Transitions between those ensemble states, denoted by the symbol $\rightarrow$, describe the evolution of an ensemble. They are initiated by the actions for sending and receiving messages, role instance creation, and state labels. According to the specified capacity of input queues (for roles in ensemble structures), we use bounded asynchronous communication for message exchange between role instances, i.e., each role instance has exactly one (bounded) input queue which receives the messages issued by (other) role instances and directed to the current one.

Let us now look more closely to the formal definition of an ensemble state. Intuitively, an ensemble state describes the local states of all participating roles. Formally, a local state of a role instance is a tuple (rt, v, q, P) which stores the following information: the (non modifiable) role type rt of the instance, a local environment function v mapping local variables to values (the empty environment is denoted by $\emptyset$), the current content q of the input queue of the instance (the empty queue is denoted by ϵ, the length of q is denoted by $|q|$), and a process expression P representing the current control state of the instance.

We furthermore assume that each role instance has a unique identifier, represented by a positive natural number. Hence, an ensemble state representing the local states of all currently existing role instances is given by a finite function $\sigma : \mathbb{N}^+ \rightarrow \mathcal{L}$, such that $\mathcal{L}$ is the set of local states explained before. Finiteness of σ means that there exists $n \in \mathbb{N}$, denoted by $size(\sigma)$, such that $\sigma(i) = \bot$

for all $i > n$ and $\sigma(i) \neq \bot$ for all $0 < i \leq n$.[5] The definition domain of σ is denoted by $dom(\sigma)$. For instance creation, an ensemble state σ is extended by a new role instance together with its local state by assigning an element $\lambda \in \mathcal{L}$ to the next free identifier, which is $size(\sigma) + 1$ and denoted by $next(\sigma)$ in the following. Such an extension is denoted by $\sigma_{[next(\sigma) \mapsto \lambda]}$. We can also update the value of an identifier $i < next(\sigma)$ with a new value λ which is denoted by $\sigma_{[i \mapsto \lambda]}$. In summary, an ensemble state associates a local state to each currently existing role instance. Thus, a role instance i is characterized by a unique identifier and its associated local state $\lambda \in \mathcal{L}$. In the following, we often write i synonymously for the role instance identifier.

For a given ensemble specification $EnsSpec = (\Sigma, behaviors)$, the allowed transitions between ensemble states, denoted by $\rightarrow$, are described by the SOS rules in Fig. 6. For each rule, the transition between two ensemble states is inferred from a transition of process expressions on the type level, denoted by $\hookrightarrow$ in Fig. 5. The rules concern state changes of existing role instances in accordance to communication actions, the creation of new role instances (which start execution in the initial state of the behavior of their corresponding role type) and state label actions. The labels on the transitions of $\rightarrow$ indicate which role instance i currently executes which action from its role behavior specification.

Initial States: An ensemble state σ is an *admissible initial state* for the ensemble specification $EnsSpec$, if for all $i \in dom(\sigma), \sigma(i) = (rt, \emptyset[\textbf{self} \mapsto i], \varepsilon, P)$ such that P is the process expression used in the declaration of the role behavior for rt, i.e., $EnsSpec$ contains the declaration **roleBehavior** $rt = P$.

Well-Definedness of Ensemble States: A HELENALIGHT ensemble state $\sigma :$ $\mathbb{N}^+ \rightarrow \mathcal{L}$ is well-defined if for all $i \in \mathbb{N}^+$ and $\sigma(i) = (rt, v, q, P)$:

- $\textbf{self} \in dom(v)$,
- for any (local) variable $X \in dom(v)$: $v(X) \in dom(\sigma)$,
- for $q = msgnm_1(k_1) \cdot \ldots \cdot msgnm_m(k_m)$: $k_1, \ldots, k_m \in dom(\sigma)$.

Well-definedness is not a restriction since any admissible initial state is well-defined and the SOS rules of HELENALIGHT preserve well-definedness. This follows from the syntactic restriction for well-formed role behavior declarations.

Semantics: The rules in Fig. 6 generate, for an ensemble specification $EnsSpec$ and any admissible initial ensemble state σ_{init}, a labeled transition system $T_{\text{HEL}} = (S_{\text{HEL}}, I_{\text{HEL}}, A_{\text{HEL}}, \rightarrow_{\text{HEL}})$ with $I_{\text{HEL}} = \{\sigma_{init}\}$.

4.3 LTL for HELENALIGHT

To express goals over HELENALIGHT ensemble specifications, we use a subset of the LTL formulae defined for full HELENA in Sect. 3. We omit atomic propositions involving attributes and only use state label propositions of the form

---

[5] Here and in the following, we assume that the range of a finite function is implicitly extended by the undefined value $\bot$.

$rt[i]@label$ where rt is a role type, $i \in \mathbb{N}^+$ and $label$ is a state label. Therefore, the set $AP(EnsSpec)$ of all atomic propositions for a HELENALIGHT ensemble specification $EnsSpec$ consists of all such state label expressions $rt[i]@label$. LTL formulae are built over these propositions as explained in Sect. 3.

An atomic proposition $p = rt[i]@label$ is satisfied in an ensemble state s, written $s \models p$, if there exists a role instance i of type rt whose next performed action in T_{HEL} is the state label $label$. This is well-defined since, due to well-formedness, labels are not allowed as first actions in branches. The LTS T_{HEL} for a given HELENALIGHT ensemble specification and an admissible initial ensemble state together with the above set $AP(EnsSpec)$ of atomic propositions and the satisfaction relation $s \models p$ induces a Kripke structure which is denoted by $K(T_{\mathrm{HEL}})$ (cf. Sect. 2). Following Definition 5, we define satisfaction of LTL in HELENALIGHT as follows: The LTS T_{HEL} for an ensemble specification $EnsSpec$ and an admissible initial state satisfies an LTL formulae ϕ over the set $AP(EnsSpec)$, written $T_{\mathrm{HEL}} \models \phi$, if $K(T_{\mathrm{HEL}}) \models \phi$.

P2P Example: We reformulate the goal from Sect. 3 to[6]

$$\Box(Provider@stateReqFile \Rightarrow \Diamond Requester@stateSndFile).$$

In HELENALIGHT, we omit component types and cannot refer to attributes. Therefore, we express that the file exists in the network by the provider reaching its state labeled by $stateReqFile$ (note that we have added $\Box$ since this state label expression does not hold in the initial state). Similarly, we express that the file was transferred to the requester by the requester reaching its state labeled by $stateSndFile$.

5   PROMELALIGHT

PROMELA [11] is a language for modeling systems of concurrent processes. Its most important features are the dynamic creation of processes and support for synchronous and asynchronous communication via message channels. PROMELA verification models serve as input for the model-checker Spin [11]. On the one hand, Spin can be used to run a randomized simulation of the model. On the other hand, it can check LTL properties, formulated over a PROMELA specification, and find and display counterexamples.

To verify LTL properties for HELENA specifications, we exploit PROMELA and Spin. We first translate a HELENA specification to PROMELA and then check the specified LTL properties with Spin. Dynamic role creation in HELENA can easily be expressed by dynamic process creation in PROMELA, and asynchronous message exchange between roles in HELENA by asynchronous communication via message channels in PROMELA. For formally proving the correctness of the translation, we use HELENALIGHT and for the target of the translation into PROMELA, we use an appropriate sub-language which we call PROMELALIGHT.

---

[6] $Provider@stateReqFile$ and $Requester@stateSndFile$ are shorthand notations without identifier which can only be used if there exists at most one instance of the role type.

$$
\text{(send)} \quad \frac{P_i \xrightarrow{\;Y!msgnm(X)\;} P_i'}{\sigma \xrightarrow{\;i:Y!msgnm(X)\;} \sigma[i \mapsto (rt_i, v_i, q_i, P_i')][j \mapsto (rt_j, v_j, q_j \cdot msgnm(k), P_j)]}
$$

if $i \in dom(\sigma), \sigma(i) = (rt_i, v_i, q_i, P_i),$
$\quad v_i(Y) = j \in dom(\sigma), \sigma(j) = (rt_j, v_j, q_j, P_j),$
$\quad |q_j| < roleconstraints(rt_j), v_i(X) = k \in dom(\sigma).$

$$
\text{(receive)} \quad \frac{P_i \xrightarrow{\;?msgnm(rt_j\ X)\;} P_i'}{\sigma \xrightarrow{\;i:?msgnm(rt_j\ X)\;} \sigma[i \mapsto (rt_i, v_i[X \mapsto j], q_i, P_i')]}
$$

if $i \in dom(\sigma), \sigma(i) = (rt_i, v_i, msgnm(j) \cdot q_i, P_i),$
$\quad j \in dom(\sigma), \sigma(j) = (rt_j, v_j, q_j, P_j).$

$$
\text{(create)} \quad \frac{P_i \xrightarrow{\;X \leftarrow \mathbf{create}(rt_j)\;} P_i'}{\sigma \xrightarrow{\;i:X \leftarrow \mathbf{create}(rt_j)\;} \sigma'}
$$

if $\sigma' = \sigma[i \mapsto (rt_i, v_i[X \mapsto next(\sigma)], q_i, P_i')][next(\sigma) \mapsto (rt_j, \emptyset[\mathbf{self} \mapsto next(\sigma)], \varepsilon, P_j)],$
$\quad i \in dom(\sigma), \sigma(i) = (rt_i, v_i, q_i, P_i), \mathbf{roleBehavior}\ rt_j = P_j.$

$$
\text{(label)} \quad \frac{P_i \xrightarrow{\;label\;} P_i'}{\sigma \xrightarrow{\;i:label\;} \sigma[i \mapsto (rt_i, v_i, q_i, P_i')]}
$$

if $i \in dom(\sigma), \sigma(i) = (rt_i, v_i, q_i, P_i).$

Fig. 6. SOS rules for the evolution of ensembles in HELENALIGHT

5.1   Syntax of PROMELALIGHT Specifications

The following syntax is a simplified version of the PROMELA syntax defined in [20]. The constructs specify a significant sub-language of the Promela definition which is sufficient as a target for the translation of HELENALIGHT.

PROMELALIGHT **Specifications:** Intuitively, a PROMELALIGHT specification consists of a set of process types whose behavior is specified by process expressions. We first define process expressions in PROMELALIGHT based on [20]. We use the same names for nonterminals as in [20], but sometimes we unfold the original definitions to get a smaller grammar for our purposes. In contrast to [20], we added the PROMELA expression **skip** as an explicit construct (corresponding to **nil** in HELENALIGHT). Furthermore, the conditional statement and the **goto** statement are not treated as process steps, but as a processes itself. Consequently, gotos can only occur at the end of a process expression. We have also removed guards from the conditional statements, thus obtaining nondeterministic choice.

Definition 14 (Process Expressions). *A process expression seq is built from the following grammar, where label is the name of a state label (used for gotos and verification), var, var_1, and var_2 are names of variables, const is a constant, pt is the name of a process type, and typelist is a comma-separated list of types:*

$$
\begin{aligned}
seq:: =\ & \textbf{skip} & \textit{(empty process)} \\
| \ & step;\, seq & \textit{(sequential composition)} \\
| \ & \textbf{if} :: seq_1 :: seq_2\, \textbf{fi} & \textit{(nondeterministic choice)} \\
| \ & \textbf{goto}\ label & \textit{(goto)} \\
step:: =\ & label : \textbf{true} & \textit{(state label)} \\
| \ & var_1!const,\, var_2 & \textit{(send)} \\
| \ & var_1?const,\, var_2 & \textit{(receive)} \\
| \ & \textbf{run}\ pt(var) & \textit{(run)} \\
| \ & \textbf{chan}\ var & \textit{(channel declaration)} \\
| \ & \textbf{chan}\ var = [const]\ \textbf{of}\ \{typelist\} & \textit{(channel declaration with initialization)}
\end{aligned}
$$

Note that send and receive steps always concern data tuples $const, var_2$ consisting of a constant and a variable. A channel declaration **chan** $var = \ldots$ opens the scope for a local channel variable var. We assume that the names of the declared variables are unique within a process expression and different from **self**, which is a predefined variable of type **chan** that can always be used. A variable is *initialized* if either the variable occurs in a receive step as var_2 or in a channel declaration with initialization as var or is the special variable **self**.

Definition 15 (Well-Formedness of Process Expressions). *A process expression is* well-formed *if (1) all variables occurring in a send or run step have been initialized before, (2) the variable var_1 in a receive step has been initialized before and the variable var_2 has been declared before, and (3) label : **true** is not the first statement in seq_1 or seq_2 in $\textbf{if} :: seq_1 :: seq_2\, \textbf{fi}$.*

Process expressions are used to define process types. In PROMELALIGHT, a process type has always one parameter **self** of type **chan** which represents a distinguished input channel for each process instance.

Definition 16 (Process Type Declaration). *A process type declaration has the form* **proctype** $pt(\textbf{chan}\ \textbf{self})\{start_{pt} : \textbf{true};\, seq\}$ *where pt is the name of the process type, seq is a well-formed process expression not containing a state label $start_{pt} : \textbf{true}$, and any **goto** expression occurring in seq has the form* **goto** $start_{pt}$.

The above definition associates a process expression to a process type pt. It allows a restricted version of recursion by introducing the state label $start_{pt} : \textbf{true}$ at the beginning of the process and allowing to jump back to that via **goto** $start_{pt}$. This syntactic restriction simplifies the semantics since the continuation of a **goto** is then uniquely determined. Hence, we do not need to carry the full body of a process type declaration in the semantic states and to search for labels in the body to find the continuation as in [20].

Definition 17 (PROMELALIGHT specification). *A PROMELALIGHT specification consists of a set of process type declarations.*

```
1  mtype { reqAddr, sndAddr,
2         reqFile, sndFile }
3  proctype Requester(chan self) {
4    startRequester: true;
5    chan router = [2] of { mtype, chan };
6    run Router(router);
7    router!reqAddr,self;
8    chan prov;
9    self?sndAddr,prov;
10   prov!reqFile,self;
11   chan prov2;
12   self?sndFile,prov2;
13   stateSndFile: true;
14   skip
15 }
16 proctype Provider(chan self) {
17   startProvider: true;
18   chan req;
19   self?reqFile,req;
20   stateReqFile: true;
21   req!sndFile,self;
22   skip
23 }
```

```
1  proctype Router(chan self) {
2    startRouter: true;
3
4    chan req;
5    self?reqAddr,req;
6
7    if
8    ::
9      chan prov = [1] of { mtype, chan };
10     run Provider(prov);
11     req!sndAddr,prov;
12     skip
13   ::
14     chan router = [2] of { mtype, chan };
15     run Router(router);
16     router!reqAddr,req;
17     goto startRouter
18   fi
19 }
20 init {
21   chan req = [2] of { mtype, chan };
22   run Requester(req);
23 }
```

(a) Message definitions and process type declarations for **Requester** and **Provider**

(b) Process type declaration for **Router**

Fig. 7. The p2p example in PROMELALIGHT

P2P Example: The formal translation from HELENALIGHT to PROMELALIGHT will be discussed in Sect. 7. To illustrate PROMELALIGHT, we already present here, in Fig. 7, the PROMELALIGHT translation of the simplified variant of the p2p example. Let us briefly look at the process type declaration for a router in Fig. 7b in comparison to the role behavior declaration in Fig. 4b. Nondeterministic choice is expressed by the **if** construct of PROMELALIGHT. Role instance creation in HELENALIGHT is translated to starting a new process in PROMELALIGHT (line 10 and 15 in Fig. 4b). Asynchronous message exchange is obtained by passing an asynchronous channel to the newly created process for communication (line 9 and 14 in Fig. 4b).

5.2 Semantics of PROMELALIGHT Specifications

The semantic domain of PROMELALIGHT specifications are again labeled transition systems. We also follow a two-level SOS approach which has been advocated for the formal PROMELA semantics in [20]. On the first level, the SOS rules only deal with the progress of process expressions specified by the nonterminal symbol *seq* in Definition 14. Process instances and their concurrent execution are considered on the second level.

Evolution of Process Expressions: On the first level, we only formalize the progress determined by a single process expression. Figure 8 defines the SOS rules inductively over the structure of PROMELALIGHT process expressions in Definition 14 where the symbol $\hookrightarrow$ describes transitions on this level. In contrast

(sequential composition)	$step; seq \overset{step}{\hookrightarrow} seq$
(choice-left)	$\dfrac{seq_1 \overset{step}{\hookrightarrow} seq_1'}{\text{if} :: seq_1 :: seq_2 \text{ fi} \overset{step}{\hookrightarrow} seq_1'}$
(choice-right)	$\dfrac{seq_2 \overset{step}{\hookrightarrow} seq_2'}{\text{if} :: seq_1 :: seq_2 \text{ fi} \overset{step}{\hookrightarrow} seq_2'}$
(goto)	$\textbf{goto } start_{pt} \overset{\textbf{goto } start_{pt}}{\hookleftarrow} start_{pt} : \textbf{true}; seq$ $\textbf{if proctype } pt(\textbf{chan self})\{start_{pt} : \textbf{true}; seq\}$

Fig. 8. SOS rules for the evolution of a process expression in PROMELALIGHT

to [20], we postpone not only the treatment of process instances, but also the treatment of local environments and the consideration of channel instances to the second level.

Evolution of Concurrent Process Instances: On the next level, we consider global states and the concurrent execution of process instances. Similarly to ensemble states in HELENALIGHT, a global state in PROMELALIGHT captures the currently existing process instances. However, in contrast to input queues in HELENALIGHT, process instances communicate via channels which are not owned by a local process, but belong to the global state. Hence, a global state of a PROMELALIGHT specification captures (1) the set of the currently existing channel instances (together with their states) and (2) the set of the currently existing process instances (together with their local states). Transitions between global states are initiated by the actions for sending and receiving a message, running a new process, channel declarations, **goto**s, and state labels.

Let us now look more closely to the formal definition of a global state in PROMELALIGHT. Intuitively, a global state describes the local states of all currently existing channels and the local states of all currently existing process instances. Each channel instance is uniquely identified by a positive natural number and the currently existing channel instances are represented by a finite function (called *channel function*) $ch : \mathbb{N}^+ \to \mathcal{C}$ such that $\mathcal{C}$ is the set of local channel states. A local state of a channel is a tuple (T, ω, κ) consisting of the (non-modifiable) type T of entries, the content ω which is a word of T-values (we write ε for the empty word), and the (non-modifiable) capacity $\kappa > 0$ of the channel[7]. Similarly, each process instance is uniquely identified by a positive natural number[8], and the currently existing process instances are represented by

---

[7] In PROMELALIGHT, we only consider asynchronous communication ($\kappa > 0$).

[8] For technical reasons, explained in the discussion of initial states below, we deviate from [20] and do not use 0 as an identifier for channels and processes.

a finite function $\mathbf{proc} : \mathbb{N}^+ \to \mathcal{P}$ such that $\mathcal{P}$ is the set of local process (instance) states. A local state of a process instance is a tuple (pt, β, π) where pt is the process type of the instance, β is a local environment function mapping local variables to values (i.e., channel identifiers or **null**) and π is a process expression representing the current control state of the instance. Finally, a global state is a pair $(\mathbf{ch}, \mathbf{proc})$ of a channel function $\mathbf{ch}$ and a function $\mathbf{proc}$ representing the currently existing process instances.[9]

For a given PROMELALIGHT specification, the allowed transitions between global states, denoted by $\to$, are described by the SOS rules in Fig. 9. They evolve a set of process instances which execute in accordance with their process types under the assumption of asynchronous communication. For each rule, the transition between two global states is inferred from a transition of process expressions on the type level, denoted by $\hookrightarrow$ in Fig. 8. The labels on the transitions of $\to$ indicate which process instance i currently executes which step from its process type specification. In the rules, we use the shorthand notations for the extension and update of finite functions from Sect. 4.2.

Initial States: A global state $(\mathbf{ch}, \mathbf{proc})$ is an *admissible initial state* for a PROMELALIGHT specification, if

- for all $c \in dom(\mathbf{ch})$: $\mathbf{ch}(c) = (T, \varepsilon, \kappa)$ for some T and κ,
- for all $i \in dom(\mathbf{proc})$: $\mathbf{proc}(i) = (pt, \emptyset[\mathbf{self} \mapsto c_i], \mathrm{start}_{pt} : \mathbf{true}; seq)$ such that $c_i \in dom(\mathbf{ch})$ with $c_i \neq c_j$ for $i \neq j$ and the PROMELALIGHT specification contains the process type declaration **proctype** $pt(\mathbf{chan}\ \mathbf{self})\{\mathrm{start}_{pt} :$ **true**; $seq\}$.

Concrete initial states in PROMELALIGHT are constructed by running an appropriate initialization as shown in line 20–23 of Fig. 7b where one channel and one requester instance, using that channel as input, are created. The initialization is executed by a root process **init** which implicitly obtains the identifier 0. However, we do not consider this process in a PROMELALIGHT specification and are not interested in the verification of properties for the root process (which anyway does not have any counterpart in a HELENA specification). Thus, we use in our semantic framework and in atomic propositions of LTL formulae only positive natural numbers for process identifiers.

Well-Definedness of Global States: A global PROMELALIGHT state $\gamma = (\mathbf{ch}, \mathbf{proc})$ with $\mathbf{ch} : \mathbb{N}^+ \to \mathcal{C}$ and $\mathbf{proc} : \mathbb{N}^+ \to \mathcal{P}$ is well-defined if for all $i \in \mathbb{N}^+$, $\mathbf{ch}(i) = (T, \omega, \kappa)$ and $\mathbf{proc}(i) = (pt, \beta, \pi)$:

- $\beta(\mathbf{self}) \in dom(\mathbf{ch})$,
- for any (local) variable $X \in dom(\beta)$: $\beta(X) \in dom(\mathbf{ch}) \cup \{\mathbf{null}\}$,
- for $\omega = (msgnm_1, c_1) \cdot \ldots \cdot (msgnm_m, c_m)$: $c_1, \ldots, c_m \in dom(\mathbf{ch})$ and $\kappa \geq m$.

Semantics: As for HELENALIGHT, the rules in Fig. 9 generate, for a PROMELALIGHT specification and any admissible initial state γ_{init}, a labeled transition system $T_{\mathrm{PRM}} = (S_{\mathrm{PRM}}, I_{\mathrm{PRM}}, A_{\mathrm{PRM}}, \to_{\mathrm{PRM}})$ with $I_{\mathrm{PRM}} = \{\gamma_{init}\}$.

---

[9] In [20], ch is denoted by $\mathcal{C}$, proc by **act**, and β by $\mathcal{L}$.

(goto)
$$\dfrac{\pi_i \xrightarrow{\textbf{goto } label} \pi_i'}{(\mathsf{ch}, \mathsf{proc}) \xrightarrow{i:\textbf{goto } label} (\mathsf{ch}, \mathsf{proc}_{[i \mapsto (pt_i, \beta_i, \pi_i')]})}$$

if $i \in dom(\mathsf{proc})$, $\mathsf{proc}(i) = (pt_i, \beta_i, \pi_i)$.

(label)
$$\dfrac{\pi_i \xrightarrow{label:\textbf{true}} \pi_i'}{(\mathsf{ch}, \mathsf{proc}) \xrightarrow{i:label:\textbf{true}} (\mathsf{ch}, \mathsf{proc}_{[i \mapsto (pt_i, \beta_i, \pi_i')]})}$$

if $i \in dom(\mathsf{proc})$, $\mathsf{proc}(i) = (pt_i, \beta_i, \pi_i)$.

(send)
$$\dfrac{\pi_i \xrightarrow{var_1 ! const, var_2} \pi_i'}{(\mathsf{ch}, \mathsf{proc}) \xrightarrow{i:var_1 ! const, var_2} (\mathsf{ch}_{[c \mapsto (T, \omega \cdot (const, v), \kappa)]}, \mathsf{proc}_{[i \mapsto (pt_i, \beta_i, \pi_i')]})}$$

if $i \in dom(\mathsf{proc})$, $\mathsf{proc}(i) = (pt_i, \beta_i, \pi_i)$, $\beta_i(var_2) = v \in dom(\mathsf{ch})$,
$\beta_i(var_1) = c \in dom(\mathsf{ch})$, $\mathsf{ch}(c) = (T, \omega, \kappa)$, $|\omega| < \kappa$.

(receive)
$$\dfrac{\pi_i \xrightarrow{var_1 ? const, var_2} \pi_i'}{(\mathsf{ch}, \mathsf{proc}) \xrightarrow{i:var_1 ? const, var_2} (\mathsf{ch}_{[c \mapsto (T, \omega, \kappa)]}, \mathsf{proc}_{[i \mapsto (pt_i, \beta_i[var_2 \mapsto v], \pi_i')]})}$$

if $i \in dom(\mathsf{proc})$, $\mathsf{proc}(i) = (pt_i, \beta_i, \pi_i)$, $var_2 \in dom(\beta_i)$,
$\beta_i(var_1) = c \in dom(\mathsf{ch})$, $\mathsf{ch}(c) = (T, (const, v) \cdot \omega, \kappa)$, $v \in dom(\mathsf{ch})$.

(run)
$$\dfrac{\pi_i \xrightarrow{\textbf{run } pt_j(var)} \pi_i'}{(\mathsf{ch}, \mathsf{proc}) \xrightarrow{i:\textbf{run } pt_j(var)} \gamma'}$$

if $\gamma' = (\mathsf{ch}, \mathsf{proc}_{[i \mapsto (pt_i, \beta_i, \pi_i')]}{[next(\mathsf{proc}) \mapsto (pt_j, \emptyset[\textbf{self} \mapsto c], start_{pt_j}:\textbf{true};seq)]})$
$i \in dom(\mathsf{proc})$, $\mathsf{proc}(i) = (pt_i, \beta_i, \pi_i)$, $\beta_i(var) = c \in dom(\mathsf{ch})$,
proctype $pt_j(\textbf{chan self})\{start_{pt_j} : \textbf{true}; seq\}$.

(chan-1)
$$\dfrac{\pi_i \xrightarrow{\textbf{chan } var} \pi_i'}{(\mathsf{ch}, \mathsf{proc}) \xrightarrow{i:\textbf{chan } var} (\mathsf{ch}, \mathsf{proc}_{[i \mapsto (pt_i, \beta_i[var \mapsto \textbf{null}], \pi_i')]})}$$

if $i \in dom(\mathsf{proc})$, $\mathsf{proc}(i) = (pt_i, \beta_i, \pi_i)$

(chan-2)
$$\dfrac{\pi_i \xrightarrow{\textbf{chan } var=[const] \textbf{ of } \{typelist\}} \pi_i'}{(\mathsf{ch}, \mathsf{proc}) \xrightarrow{i:\textbf{chan } var=...} \gamma'}$$

if $\gamma' = (\mathsf{ch}_{[next(\mathsf{ch}) \mapsto (typelist, \varepsilon, const)]}, \mathsf{proc}_{[i \mapsto (pt_i, \beta_i[var \mapsto next(\mathsf{ch})], \pi_i')]})$
$i \in dom(\mathsf{proc})$, $\mathsf{proc}(i) = (pt_i, \beta_i, \pi_i)$.

Fig. 9. SOS rules for the evolution of concurrent process instances.

5.3   LTL for PROMELALIGHT

To express goals over PROMELALIGHT specifications, we use LTL formulae. As in HELENALIGHT, we restrict the atomic propositions of LTL formulae to state label expressions of the form $pt[i]@label$ where pt is a process type, $i \in \mathbb{N}^+$ and *label* is a state label. Therefore, the set $AP(PrmSpec)$ of all atomic propositions for a PROMELALIGHT specification $PrmSpec$ consists of all such state label expressions $pt[i]@label$. LTL formulae are built over these propositions as explained in Sect. 3.

An atomic proposition $p = pt[i]@label$ is satisfied in a global state γ, written $\gamma \models p$ if there exists a process instance i of type pt whose next performed action in T_{PRM} is the state label $label$. The LTS T_{PRM} for a given PROMELALIGHT specification and an admissible initial state together with the above set $AP(PrmSpec)$ of atomic propositions and the satisfaction relation $\gamma \models p$ induces a Kripke structure denoted by $K(T_{\mathrm{PRM}})$ (cf. Sect. 2). Following Definition 5, we define satisfaction of LTL in PROMELALIGHT as follows: The LTS T_{PRM} for a PROMELALIGHT specification $PrmSpec$ and an admissible initial state satisfies an LTL formulae ϕ over the set $AP(PrmSpec)$, written $T_{\mathrm{PRM}} \models \phi$, if $K(T_{\mathrm{PRM}}) \models \phi$.

6 Translation of HELENALIGHT to PROMELALIGHT

In this section, we propose a transformation from HELENALIGHT ensemble specifications to PROMELALIGHT verification models. We assume given a HELENALIGHT ensemble specification $EnsSpec = (\Sigma, behaviors)$ with $\Sigma = (nm, roletypes, roleconstraints)$ being an ensemble structure. The translation into the PROMELALIGHT specification $trans(EnsSpec)$ proceeds in two steps: First, we provide all message types from HELENALIGHT in PROMELALIGHT by declaring an enumeration type, called **mtype** (not shown here).

Translation of Role Behavior Declarations: Then, for each role type and its corresponding role behavior in HELENALIGHT, we create a process type in PROMELALIGHT which reflects the execution of the role behavior and is inductively defined over the structure of process expressions and actions; see Fig. 10.

$$
\begin{aligned}
trans_{\mathrm{decl}}(\textbf{roleBehavior } rt{=}P) \;&=\; \textbf{proctype } rt(\textbf{chan self}) \; \{ \\
&\quad\; \text{start}_{rt} : \textbf{true}; \; trans_{\mathrm{proc}}(P) \; \}
\end{aligned}
$$

$$
\begin{aligned}
trans_{\mathrm{proc}}(\textbf{nil}) \;&=\; \textbf{skip} \\
trans_{\mathrm{proc}}(aP) \;&=\; trans_{\mathrm{act}}(a); trans_{\mathrm{proc}}(P) \\
trans_{\mathrm{proc}}(P_1 + P_2) \;&=\; \textbf{if} :: trans_{\mathrm{proc}}(P_1) :: trans_{\mathrm{proc}}(P_2) \, \textbf{fi} \\
trans_{\mathrm{proc}}(N) \;&=\; \textbf{goto } \text{start}_N
\end{aligned}
$$

$$
\begin{aligned}
trans_{\mathrm{act}}(Y!msgnm(X)) \;&=\; Y!msgnm, X \\
trans_{\mathrm{act}}(?msgnm(rt\,X)) \;&=\; \textbf{chan } X; \textbf{self}?msgnm, X \\
trans_{\mathrm{act}}(X \leftarrow \textbf{create}(rt)) \;&=\; \textbf{chan } X = [roleconstraints(rt)] \, \textbf{of } \{\textbf{mtype,chan}\}; \\
&\quad\; \textbf{run } rt(X) \\[4pt]
trans_{\mathrm{act}}(label) \;&=\; label : \textbf{true} \\
\quad \textbf{if } label \neq \text{start}_{rt}
\end{aligned}
$$

352   R. Hennicker et al.

```
\begin{figure}[htb]

\begin{formalfigure}[0.85\textwidth]

\vspace{-1.5em}

\begin{align*}

&\transdecl({\scriptstyle\textbf{roleBehavior}\, \RoleTypeVar = P}) &=~& \textbf{proctype}\,

\RoleTypeVar(\pchan~\self)~\{\\

&&& \quad\pstatelabel{\start{rt}}; \transproc(P)~\}\\\[1em]

&\transproc(\nil) &=~& \pskip\\

&\transproc(a\prefix P) &=~& \transact(a);\transproc(P)\\

&\transproc(P_1+P_2) &=~& \pchoice{\transproc(P_1)}{\transproc(P_2)}\\

&\transproc(N) &=~& \pgoto{\start{N}}\\\[1em]

&\transact(\sndLight{\MsgName}{X}{Y}) &=~& Y ! \MsgName,X\\

&\transact(\rcvLight{\MsgName}{\RoleTypeVar\, X}) &=~& \pchande{X}{}; \self ? \MsgName,X\\

&\transact(\createLight{X}{\RoleTypeVar}) &=~&

\pchandecl{X}{{\scriptstyle\EnsStructRoleConstraints(\RoleTypeVar)}}{{\scriptstyle\mtype

,\pchan}};\\

&&&\prun{\RoleTypeVar}{X}\\

&\transact(\statelabel{\lab}) &=~& \pstatelabel{\lab}\\

& \qquad \textbf{if } \statelabel{\lab} \neq \start{\RoleTypeVar}

\end{align*}

\vspace{-2em}

\end{formalfigure}

\vspace{-1.5em}

\end{figure}
```

Fig. 10. Translation of role behavior declarations

As an example, we consider the HELENALIGHT specification in Fig. 4 which is translated with the above rules to the PROMELALIGHT specification in Fig. 7.

Translation of Initial States: To be able to show semantic equivalence between HELENALIGHT and PROMELALIGHT specifications, we have to translate admissible initial states. We assume given an admissible initial HELENALIGHT ensemble state σ with local states $\sigma(i) = (rt, \emptyset[\textbf{self} \mapsto i], \varepsilon, P)$ for all $i \in dom(\sigma)$. Its translation is the admissible initial PROMELALIGHT state $trans_{\text{init}}(\sigma) = (\text{ch}, \text{proc})$, such that the content of all existing channels in ch is empty, $dom(\text{proc}) = dom(\sigma)$, and for all $i \in dom(\text{proc})$, $\text{proc}(i) = (rt, \emptyset[\textbf{self} \mapsto c_i], \text{start}_{rt} : \textbf{true}; trans_{\text{proc}}(P))$ with $c_i \in dom(\text{ch})$ and $c_i \neq c_j$ for $i \neq j$.

7   ≈-Stutter Equivalence of the Translation

We now prove the correctness of the translation from HELENALIGHT to PROMELALIGHT, i.e., that a HELENALIGHT specification and its PROMELALIGHT translation satisfy the same set of $LTL_{\backslash \textbf{X}}$ formulae. We first define two relations $\sim$ and $\approx$ between the Kripke structures induced from a HELENALIGHT specification and its PROMELALIGHT translation. To be able to apply Theorem 1 from Sect. 2, we show that $\approx$ preserves satisfaction of atomic propositions and any admissible initial state of a HELENALIGHT ensemble specification and its PROMELALIGHT translation are related by $\sim$. Furthermore, we prove that the relation $\sim$ is a $\approx$-stutter simulation of the Kripke structure of the HELENALIGHT specification by the Kripke structure of the PROMELALIGHT translation and the inverse relation $\approx^{-1}$ is a $\approx^{-1}$-stutter simulation in the other direction. Having proven stutter trace equivalence, we can then apply Theorem 2 entailing preservation of $LTL_{\backslash \textbf{X}}$.

Silent Actions: To prove stutter trace equivalence of Kripke structures, we rely on the transitions in the labeled transition systems which induce the Kripke structures. Thereby, some transitions in HELENALIGHT are reflected by several transitions in PROMELALIGHT, e.g., the transition with the action $?msgnm(X)$ is reflected by two transitions with the actions **chan** X and **self**?$msgnm, X$ (cf. definition of $trans_{\text{act}}$ in Sect. 6). These additional transitions do not change satisfaction of atomic propositions. Thus, we consider the following steps and their corresponding actions in PROMELALIGHT as silent and denote them by τ:

- the transition from **chan** X; **self**?$msgnm, X$ to **self**?$msgnm, X$,
- the transition from **chan** $X = [roleconstraints(rt)]$ **of** $\{\text{mtype,chan}\}$; **run** $rt(X)$ to **run** $rt(X)$,
- the transition start_{pt_i} : **true**; π to π since start state labels start_{pt_i} : **true** only exist in PROMELALIGHT, and
- the transition **goto** start_{pt_i}; π to π since in PROMELALIGHT, recursive process invocation is expressed by a jump (i.e., a **goto** step) while in HELENALIGHT, the body of the invoked role behavior is directly applied without any execution step for recursion.

Simulation Relations: We define two relations which both express a correspondence between HELENALIGHT ensemble states and global PROMELALIGHT states, but require a different level of correspondence.

Definition 18 (Relation $\sim$ and $\approx$). *Let $K(T_{\mathrm{HEL}}) = (S_{\mathrm{HEL}}, I_{\mathrm{HEL}}, \rightarrow^{\bullet}_{\mathrm{HEL}}, F_{\mathrm{HEL}})$ be the induced Kripke structure of a HELENALIGHT ensemble specification and $K(T_{\mathrm{PRM}}) = (S_{\mathrm{PRM}}, I_{\mathrm{PRM}}, \rightarrow^{\bullet}_{\mathrm{PRM}}, F_{\mathrm{PRM}})$ be the induced Kripke structure of a PROMELALIGHT specification. The relation $\sim \subseteq S_{\mathrm{HEL}} \times S_{\mathrm{PRM}}$ is defined as follows: $\sigma \sim (\boldsymbol{ch}, \boldsymbol{proc})$ if it holds that*

1. *$dom(\sigma) = dom(\boldsymbol{proc})$ and*
2. *for all $i \in dom(\sigma)$ with $\sigma(i) = (rt_i, v_i, q_i, P_i)$ and $\boldsymbol{proc}(i) = (pt_i, \beta_i, \pi_i)$:*
   (a) *$rt_i = pt_i$,*
   (b) *$dom(v_i) \subseteq dom(\beta_i)$ such that for all $X \in dom(v_i)$:*
      $v_i(X) = j \Leftrightarrow \beta_i(X) = \beta_j(\boldsymbol{self})$ (where $\boldsymbol{proc}(j) = (pt_j, \beta_j, \pi_j)$),
   (c) *$q_i = msgnm_1(k_1) \cdot \ldots \cdot msgnm_m(k_m) \Leftrightarrow$*
      $\boldsymbol{ch}(\beta_i(\boldsymbol{self})) = (T, (msgnm_1, \beta_{k_1}(\boldsymbol{self})) \cdot \ldots \cdot (msgnm_m, \beta_{k_m}(\boldsymbol{self})), \kappa),$
   (d) *$trans_{proc}(P_i) = \pi_i$ or $\pi_i \xrightarrow{\;start_{pt_i}:true\;}_{\mathrm{PRM}} trans_{proc}(P_i).$*

The relation $\approx \subseteq S_{\mathrm{HEL}} \times S_{\mathrm{PRM}}$ is defined just as the relation $\sim$ with the exception of item (2d) where $trans_{proc}(P_i) = \pi_i$ is replaced by $trans_{proc}(P_i) \xrightarrow{\tau^*}_{\mathrm{PRM}} \pi_i$. Obviously, it holds that $\sim \subseteq \approx$.

Firstly, in the defined relations, there must be as many role instances in HELENALIGHT as process instances in PROMELALIGHT. Secondly, the local state of each role instance i must be related to the local state of the process instance with the same identifier i: (a) The role type rt_i must match the process type pt_i. (b) The local variables in v_i must have counterparts in β_i, but note that the value types of HELENALIGHT and PROMELALIGHT are subtly different. A local variable in HELENALIGHT points to a role instance whereas a local variable in PROMELALIGHT points to a channel. Furthermore, note that vice versa, there might be local variables in β_i which do not have any counterparts in v_i. (c) The content of the input queue of the role instance must match the content of the corresponding channel of the process instance. As for local variables, the input queue of the role instance consists of role instance identifiers whereas the related PROMELALIGHT input channel contains the identifiers of the input channels of the process instances (corresponding to these role instances). (d) For the process expression π_i occurring in the local state of the process instance, we either require that it is the same as the translation of the process expression P_i occurring in the local state of the role instance or that it can evolve by the single action $start_{pt_i}$: **true** to the translation of P_i. The latter takes into account that the translation of a role behavior into PROMELALIGHT adds a start label at the beginning of the translated role behavior. For the relation $\approx$, we weaken the first condition such that π_i must only be reachable by evolving the translation of P_i with arbitrary many τ actions.

Properties of the Simulation Relations: Based on the induced Kripke structures, we show some interesting properties of the two relations: the relation $\approx$ preserves satisfaction of atomic propositions and any admissible initial state in HELENALIGHT and its PROMELALIGHT translation are related by the relation $\sim$.

Lemma 1 (Preservation of Atomic Propositions). *Let* $K(T_{\text{HEL}}) = (S_{\text{HEL}}, I_{\text{HEL}}, \rightarrow^{\bullet}_{\text{HEL}}, F_{\text{HEL}})$ *be an induced Kripke structure of a* HELENALIGHT *ensemble specification* $EnsSpec = (\Sigma, behaviors)$ *such that no role behavior in* behaviors *starts with a state label and let* $K(T_{\text{PRM}}) = (S_{\text{PRM}}, I_{\text{PRM}}, \rightarrow^{\bullet}_{\text{PRM}}, F_{\text{PRM}})$ *be the induced Kripke structure of a* PROMELALIGHT *specification.*

For all $\sigma \in S_{\text{HEL}}, \gamma \in S_{\text{PRM}}$*, if* $\sigma \approx \gamma$*, then* $F_{\text{HEL}}(\sigma) = F_{\text{PRM}}(\gamma)$*.*

Proof. A HELENALIGHT state σ satisfies an atomic proposition $p = rt[i]@label$ only if there exists a role instance i of type rt whose next performed action is the state label *label*; similarly, γ satisfies p only if there exists a process instance i of type rt whose next performed action is the state label *label*. In any other case p is not satisfied. For each such proposition p, the proof proceeds by induction on the depth of the derivation of the next action of the process expression for role instance i. The first interesting case in the induction is nondeterministic choice. In both, HELENALIGHT and PROMELALIGHT, labels are not allowed as first actions in a branch such that the nondeterministic choice and each branch separately do not satisfy any atomic proposition p. Therefore, the induction step is trivial. Another interesting case is process invocation. HELENALIGHT executes directly the first action of its role behavior. On the other hand, in PROMELALIGHT process invocation is realized by a **goto** jump followed by the start label and the first action of the (translated) role behavior. These steps (trivially) preserve satisfaction of atomic propositions since start labels are not allowed as atomic propositions and since the first action of a role behavior cannot be a state label. □

Lemma 2 (Relationship between Initial States). *Let* σ *be an admissible initial state of a* HELENALIGHT *ensemble specification, then* $\sigma \sim trans_{init}(\sigma)$*.*

Proof. In HELENALIGHT, an admissible initial state σ consists of local states $\sigma(i) = (rt, \emptyset[\mathbf{self} \mapsto i], \varepsilon, P)$ (cf. definition of admissible initial states in Sect. 4.2). For the PROMELALIGHT translation holds $trans_{init}(\sigma) = (\mathbf{ch}, \mathbf{proc})$; the content of all existing channels in $\mathbf{ch}$ is empty, $dom(\mathbf{proc}) = dom(\sigma)$, and for all $i \in dom(\mathbf{proc})$, we have $\mathbf{proc}(i) = (rt, \emptyset[\mathbf{self} \mapsto c_i], \text{start}_{rt} : \mathbf{true}; trans_{proc}(P))$ with $c_i \in dom(\mathbf{ch})$ (cf. definition of $trans_{init}$ in Sect. 6). Therefore, all conditions for $\sigma \sim trans_{init}(\sigma)$ are satisfied, in particular item (2d) is satisfied since $\text{start}_{rt} : \mathbf{true}; trans_{proc}(P) \xrightarrow{\text{start}_{rt}:\mathbf{true}}_{\text{PRM}} trans_{proc}(P_i)$. □

Stutter Simulations: Based on the previous two lemmata, we move on to show that the relation $\sim$ is a $\approx$-stutter simulation of a HELENALIGHT specification by its PROMELALIGHT translation and that the inverse relation $\approx^{-1}$ itself is a $\approx^{-1}$-stutter simulation in the other direction. Note that $\approx$ itself would not preserve the branching structure of $K(T_{\text{HEL}})$ (due to branching with silent actions in PROMELALIGHT), but the coarser relation $\sim$ does.

Proposition 1 (Stutter Simulation of HELENALIGHT Specifications). *Let $K(T_{\text{HEL}})$ and $K(T_{\text{PRM}})$ be the induced Kripke structures of a HELENALIGHT ensemble specification and of its PROMELALIGHT translation trans(EnsSpec) as in Lemma 1. Then, $\sim$ is a $\approx$-stutter simulation of K_{HEL} by K_{PRM}.*

Proof. With Lemma 2, we proved that any initial state of a HELENALIGHT specification and its PROMELALIGHT translation are in the relation $\sim$. It remains to show that the relation $\sim$ fulfills the property of a $\approx$-stutter simulation described in Definition 6. In the proof, we rely on the underlying labeled transition systems T_{HEL} and T_{PRM} of the Kripke structures $K(T_{\text{HEL}})$ and $K(T_{\text{PRM}})$. To reflect labels of HELENALIGHT in PROMELALIGHT, we introduce a notation which translates a HELENALIGHT label to its corresponding PROMELALIGHT label by omitting silent actions:

$$trans_{\text{label}}(i : Y!msgnm(X)) \qquad\qquad = i : Y!msgnm, X$$
$$trans_{\text{label}}(i :?msgnm(rt_j\, X)) \qquad\quad = i : \mathbf{self}?msgnm, X$$
$$trans_{\text{label}}(i : X \leftarrow \mathbf{create}(rt_j)) \quad\; = i : \mathbf{run}\ rt_j(X)$$
$$trans_{\text{label}}(i : label) \qquad\qquad\qquad\; = i : label : \mathbf{true}$$

By relying on that notation, we show the following property which entails the required property for a $\approx$-stutter simulation:

For all $\sigma \in S_{\text{HEL}}, \gamma \in S_{\text{PRM}}$ with $\sigma \sim \gamma$, if $\sigma \xrightarrow{a}_{\text{HEL}} \sigma'$,
then there exists $\gamma \xrightarrow{\tau}_{\text{PRM}} \gamma_1 \ldots \xrightarrow{\tau}_{\text{PRM}} \gamma_n \xrightarrow{trans_{\text{label}}(a)}_{\text{PRM}} \gamma'\ (n \geq 0)$
such that $\sigma \approx \gamma_k$ for all $k \in \{1, \ldots, n\}$, and $\sigma' \sim \gamma'$.

The proof proceeds by induction on the depth of the derivation of $\sigma \xrightarrow{a}_{\text{HEL}} \sigma'$. The induction relies on the following fact: Each action a in HELENALIGHT is reflected in PROMELALIGHT, but some internal steps might be necessary in PROMELALIGHT before the corresponding action $trans_{\text{label}}(a)$ can actually be executed, e.g., message reception with the action $i : \, ?msgnm(rt_j\, X)$ is translated to two actions $i : \mathbf{chan}\ X$ and $i : \mathbf{self}?msgnm, X$ or process invocation in PROMELALIGHT uses first a $\mathbf{goto}$ step and then a start label step to reach the beginning of the (translated) role behavior. Since the relation $\approx$ just requires that the translation of the HELENALIGHT process expression P for role instance i can evolve by τ actions to the PROMELALIGHT process expression π for the corresponding process instance i, all those intermediate steps result in states remaining in the relation $\approx$. Only the translated action $trans_{\text{label}}(a)$ evolves the PROMELALIGHT translation according to the evolution of the HELENALIGHT specification such that the resulting states are again in the relation $\sim$.   □

In the other direction, the inverse relation $\approx^{-1}$ serves as $\approx^{-1}$-stutter simulation.

Proposition 2 (Stutter Simulation of PROMELALIGHT Translations). *Let $K(T_{\text{HEL}})$ and $K(T_{\text{PRM}})$ be the induced Kripke structures of a HELENALIGHT ensemble specification and of its PROMELALIGHT translation trans(EnsSpec) as in Lemma 1. Then, $\approx^{-1}$ is a $\approx^{-1}$-stutter simulation of K_{PRM} by K_{HEL}.*

Proof. We rely on Lemma 2 as before. The proof that the relation $\approx^{-1}$ satisfies the property for a $\approx^{-1}$-stutter simulation is based, as before, on the underlying labeled transition systems. We show the following property which entails the required property for a $\approx^{-1}$-stutter simulation:

$$\text{For all } \gamma \in S_{\text{PRM}}, \sigma \in S_{\text{HEL}} \text{ with } \gamma \approx^{-1} \sigma, \text{ if } \gamma \xrightarrow{b}_{\text{PRM}} \gamma',$$
$$\text{then } \gamma' \approx^{-1} \sigma \text{ if } b = \tau \text{ or there exists } \sigma \xrightarrow{a}_{\text{HEL}} \sigma' \text{ if } b \neq \tau$$
$$\text{such that } trans_{\text{label}}(a) = b \text{ and } \gamma' \approx^{-1} \sigma'.$$

The proof proceeds by induction on the depth of the derivation of $\gamma \xrightarrow{b}_{\text{PRM}} \gamma'$. The induction relies on the following fact: Silent actions in PROMELALIGHT, denoted by τ might only change the value of local variables which are not yet in relation to HELENALIGHT, i.e., silent steps preserve the relationship according to $\approx^{-1}$. For all non-silent actions, the relation $\approx^{-1}$ is sufficient to transfer executability of a PROMELALIGHT action b to its corresponding HELENALIGHT action a with $trans_{\text{label}}(a) = b$ such that $\approx^{-1}$ is again established by the transition. □

Lemma 1, Proposition 1, and Proposition 2 allow us to infer, by Theorem 1, that the induced Kripke structures of a HELENALIGHT ensemble specification and its PROMELALIGHT translation are stutter trace equivalent. Thus, we can apply Theorem 2 to show that the both labeled transitions systems satisfy the same $LTL_{\backslash \mathbf{X}}$ formulae.

Theorem 3 (HELENALIGHT $LTL_{\backslash \mathbf{X}}$ Preservation). *Let T_{HEL} be the labeled transition system of a HELENALIGHT ensemble specification EnsSpec $= (\Sigma, behaviors)$ together with an admissible initial state such that no role behavior in behaviors starts with a state label. Let T_{PRM} be the labeled transition system of its PROMELALIGHT translation trans(EnsSpec).*
For any $LTL_{\backslash \mathbf{X}}$ formula ϕ over AP(EnsSpec), $T_{\text{HEL}} \models \phi \Leftrightarrow T_{\text{PRM}} \models \phi$.

8 Model-Checking HELENALIGHT with Spin

The results from the previous sections allow us to verify LTL properties for a HELENALIGHT ensemble specification by model-checking its PROMELALIGHT translation in Spin. However, the semantics of HELENALIGHT and therefore satisfaction of LTL formulae is defined relatively to a given initial state σ_{init}. Thus, when model-checking the corresponding PROMELALIGHT translation, we have to establish the corresponding initial state $trans_{\text{init}}(\sigma_{init})$ in PROMELALIGHT and verify properties relatively to this initial state. We setup the initial state in a dedicated `init`-process (cf. Fig. 7). To reflect satisfaction of LTL formulae relatively to this initial state, we further extend the original HELENALIGHT LTL formula ϕ to $\Box(init \Rightarrow \phi)$. The *init* is thereby a property which only holds when the initialization in PROMELALIGHT according to the given initial state in HELENALIGHT was finished.

P2P Example: Respecting the aforementioned adaptations to LTL formulae, the goal for our p2p example in Sect. 4.3 is translated to

$$\Box \ (\ Requester@startRequester \Rightarrow$$
$$\Box(Provider@stateReqFile \Rightarrow \Diamond Requester@stateSndFile)\ ).$$

If we restrict the number of routers, this property holds for the PROME-LALIGHT translation in Fig. 7 and we can therefore conclude that it also holds for the HELENALIGHT ensemble specification in Sect. 4.1 for the initial state where only one requester exists. To model-check the more interesting goal from Sect. 3, we have to extend the translation to full HELENA by supporting two additional features, data and components, which adopt roles. In [12], we report the extended translation and argue that stutter trace equivalence holds for this extension as well. Furthermore, we present an automatic code generator based on the XTEXT workbench of Eclipse which takes a HELENA ensemble specification written in HELENATEXT, our domain specific language [13] for HELENA ensembles, as input and generates the PROMELA translation as output.

9 Conclusion

HELENA specifications provide models for dynamically evolving ensembles. This paper deals with a missing link in the HELENA development methodology concerning the early verification of ensemble specifications against goals described by LTL formulae. For this purpose, we proposed to translate HELENA ensemble specifications into PROMELA which can be checked with Spin. To prove the correctness of the translation, we have (a) defined an SOS semantics for simpler variants of HELENA and PROMELA and (b) shown that both are stutter trace equivalent. Hence, LTL formulae (without *next*) are preserved; cf. [1].

Our approach of verification is in-line with goal-oriented requirements approaches like KAOS [17]. They also specify goals by LTL properties. However, they translate their system specifications into the process algebra FSP [18], which is not sufficient to represent the dynamics of ensembles since dynamic process creation and directed communication are not supported. Techniques for the development of ensembles have been thoroughly studied in the recent ASCENS project [21]: In [5], ensemble-based systems are described by simplified SCEL programs and translated to PROMELA. However, the translation is neither proved semantically correct nor automated. DFINDER [4] implements efficient strategies exploiting compositional verification of invariants to prove safety properties for BIP ensemble models, but does not deal with dynamic creation of components. DEECo ensemble models [3] are implemented with the Java framework jDEECo and verified with Java Pathfinder [4]. Thus, opposed to HELENA, they do not need any translation. However, since DEECo relies on knowledge exchange rather than message passing, they do not verify communication behaviors. Finally, we would like to point out, that our approach has been strongly inspired by the way how the distributed language KLAIM has been transferred to Maude in [7]. There, the correctness of the translation was established by a stutter bisimulation which preserves CTL^* properties (without *next*). The translation of HELENA into PROMELA is, however, not stutter bisimilar but stutter trace equivalent and thus only preserves LTL formulae (without *next*).

For future work, we plan to conduct more experiments to examine the power of our verification approach. For instance, the question arises how big ensembles can get in terms of role instances to still provide results in reasonable time. For model-checking full HELENA, it is also interesting what impact the topology of the underlying component network (e.g., ring structure, graph structure, etc.) has on the verification of goals for ensemble specifications. As a larger case study, we are currently investigating the power of our verification method for the science cloud platform, a voluntary peer-to-peer cloud computing platform, which was modeled in HELENA in [15].

References

1. Baier, C., Katoen, J.: Principles of Model Checking. MIT Press, Cambridge (2008)
2. Boronat, A., Knapp, A., Meseguer, J., Wirsing, M.: What is a multi-modeling language? In: Corradini, A., Montanari, U. (eds.) WADT 2008. LNCS, vol. 5486, pp. 71–87. Springer, Heidelberg (2009)
3. Bures, T., Gerostathopoulos, I., Hnetynka, P., Keznikl, J., Kit, M., Plasil, F.: The invariant refinement method. In: Wirsing, M., Hölzl, M., Koch, N., Mayer, P. (eds.) Software Engineering for Collective Autonomic Systems. LNCS, vol. 8998, pp. 405–428. Springer, Switzerland (2015)
4. Combaz, J., Bensalem, S., Kofron, J.: Correctness of service components and service component ensembles. In: Wirsing, M., Hölzl, M., Koch, N., Mayer, P. (eds.) Software Engineering for Collective Autonomic Systems. LNCS, vol. 8998, pp. 107–159. Springer, Switzerland (2015)
5. De Nicola, R., Lluch Lafuente, A., Loreti, M., Morichetta, A., Pugliese, R., Senni, V., Tiezzi, F.: Programming and verifying component ensembles. In: Bensalem, S., Lakhneck, Y., Legay, A. (eds.) From Programs to Systems. LNCS, vol. 8415, pp. 69–83. Springer, Heidelberg (2014)
6. Eckhardt, J., Mühlbauer, T., AlTurki, M., Meseguer, J., Wirsing, M.: Stable availability under denial of service attacks through formal patterns. In: de Lara, J., Zisman, A. (eds.) Fundamental Approaches to Software Engineering. LNCS, vol. 7212, pp. 78–93. Springer, Heidelberg (2012)
7. Eckhardt, J., Mühlbauer, T., Meseguer, J., Wirsing, M.: Semantics, distributed implementation, and formal analysis of KLAIM models in Maude. Sci. Comput. Program. **99**, 24–74 (2015)
8. Goguen, J.A., Meseguer, J.: Universal realization, persistent interconnection and implementation of abstract modules. In: Nielsen, M., Schmidt, E.M. (eds.) Automata, Languages and Programming. LNCS, vol. 140, pp. 265–281. Springer, Heidelberg (1982)
9. Havelund, K., Larsen, K.G.: The fork calculus. In: Lingas, A., Carlsson, S., Karlsson, R. (eds.) ICALP 1993. LNCS, vol. 700, pp. 544–557. Springer, Heidelberg (1993)

10. Hennicker, R., Klarl, A.: Foundations for ensemble modeling – the HELENA approach. In: Iida, S., Meseguer, J., Ogata, K. (eds.) Specification, Algebra, and Software. LNCS, vol. 8373, pp. 359–381. Springer, Heidelberg (2014)
11. Holzmann, G.: The Spin Model Checker. Addison-Wesley, Reading (2003)
12. Klarl, A.: From Helena ensemble specifications to promela verification models. Technical report, LMU Munich (2015). http://goo.gl/G0sU6U
13. Klarl, A., Cichella, L., Hennicker, R.: From HELENA ensemble specifications to executable code. In: Lanese, I., Madelaine, E. (eds.) FACS 2014. LNCS, vol. 8997, pp. 183–190. Springer, Heidelberg (2015)
14. Klarl, A., Hennicker, R.: Design and implementation of dynamically evolving ensembles with the Helena framework. In: Proceedings of the Australasian Software Engineering Conference, pp. 15–24. IEEE (2014)
15. Klarl, A., Mayer, P., Hennicker, R.: HELENA@Work: modeling the science cloud platform. In: Margaria, T., Steffen, B. (eds.) ISoLA 2014, Part I. LNCS, vol. 8802, pp. 99–116. Springer, Heidelberg (2014)
16. Lamport, L.: What good is temporal logic? In: IFIP 9th World Congress, pp. 657–668 (1983)
17. van Lamsweerde, A.: Requirements Engineering: From System Goals to UML Models to Software Specifications. Wiley, New York (2009)
18. Magee, J., Kramer, J.: Concurrency-State Models and Java Programs. Wiley, New York (2006)
19. Meseguer, J., Palomino, M., Martí-Oliet, N.: Algebraic simulations. J. Logic Algebraic Program. **79**(2), 103–143 (2010)
20. Weise, C.: An incremental formal semantics for PROMELA. In: Third SPIN Workshop (1997)
21. Wirsing, M., Hölzl, M., Koch, N., Mayer, P. (eds.): Software Engineering for Collective Autonomic Systems. LNCS, vol. 8998. Springer, Switzerland (2015)
22. Wirsing, M., Knapp, A.: A formal approach to object-oriented software engineering. Electr. Notes Theoret. Comput. Sci. **4**, 322–360 (1996)

Modularity of Ontologies in an Arbitrary Institution

Yazmin Angelica Ibañez[1], Till Mossakowski[2]([✉]), Donald Sannella[3], and Andrzej Tarlecki[4]

[1] Department of Computer Science, University of Bremen,
Bremen, Germany
[2] Faculty of Computer Science, Otto-von-Guericke University of Magdeburg,
Magdeburg, Germany
till@iws.cs.ovgu.de
[3] Laboratory for Foundations of Computer Science,
University of Edinburgh, Edinburgh, UK
[4] Institute of Informatics,
University of Warsaw, Warsaw, Poland

Abstract. The notion of module extraction has been studied extensively in the ontology community. The idea is to extract, from a large ontology, those axioms that are relevant to certain concepts of interest (formalised as a subsignature). The technical concept used for the definition of module extraction is that of inseparability, which is related to indistinguishability known from observational specifications.

Module extraction has been studied mainly for description logics and the Web Ontology Language OWL. In this work, we generalise previous definitions and results to an arbitrary inclusive institution. We reveal a small inaccuracy in the formal definition of inseparability, and show that some results hold in an arbitrary inclusive institution, while others require the institution to be weakly union-exact.

This work provides the basis for the treatment of module extraction within the institution-independent semantics of the distributed ontology, modeling and specification language (DOL), which is currently under submission to the Object Management Group (OMG).

1 Introduction

Goguen's and Burstall's invention of the concept of institution to formalise the notion of logical system has stimulated a research programme with the general idea that modular structuring of complex specifications can be studied largely

T. Mossakowski This work has been partially supported by the Future and Emerging Technologies (FET) programme within the Seventh Framework Programme for Research of the European Commission, FET-Open Grant number: 611553, project COINVENT.
A. Tarlecki This work has been partially supported by the National Science Centre, grant 2013/11/B/ST6/01381.

N. Martí-Oliet et al. (Eds.): Meseguer Festschrift, LNCS 9200, pp. 361–379, 2015.
DOI: 10.1007/978-3-319-23165-5_17

independently of the details of the underlying logical system. José Meseguer has made important contributions to institutions and their translations [1–4,7, 11–15,19], and to the study of module systems over arbitrary institutions, see especially [7]. His contributions have been inspiring for our work, and some of his papers are among those we cite most frequently. In the present work, we study modularity over an arbitrary institution using a concept of *inseparability*, which has some resemblance to José's notion of *indistinguishability* [22] of protocols, although context and technical details are very different. This paper is dedicated to José on the occasion of his 65th birthday — our congratulations and best wishes, José!

The notion of modularity studied in the specification community is modularity *by construction*: complex specifications are formed from basic specifications (which are simply logical theories in some institution) by means of specification-building operations [7,20,21].

In the ontology community, a different notion of modularity has emerged: while even large ontologies with tens of thousands of axioms are often formalised as flat logical theories, the notion of ontological *module extraction* [27] provides an *a posteriori* extraction of relevant parts of ontologies. Module extraction has been studied mainly for description logics, but first attempts for first-order logic have also been made.

In the present paper, we try to cast module extraction in the institution-independent framework and compare it with module notions from specification theory. This work thereby also provides a semantics for certain modularity concepts and constructs in the distributed ontology, modeling and specification language DOL [16,17], which is currently under submission as a standard to the Object Management Group (OMG).

The problem of module extraction can be phrased as follows: given a subset Σ of the signature of an ontology, find a (minimal) subset of that ontology that is "relevant" for the terms in Σ. For example, the size of well-established ontologies such as SNOWMED CT[1] or GALEN[2] makes it difficult for current tools to navigate through them on a standard computer. Therefore, in an application where only a specific subset of the terms in such huge ontologies is used, it is more practical to reuse only those parts that *cover* all the knowledge about that subset of relevant terms.

The key concept of "relevance" may be formalised in different ways. We will not discuss in any detail here approaches based on syntactic structure of axioms and hierarchy of concepts [6,25,26]. Instead, we will focus on *logic-based modules*, for which relevance amounts to entailment (or model) preservation over a signature Σ. That is, given an ontology $\mathcal{O}$, when we say that a module $\mathcal{M}$ (which is a subset of $\mathcal{O}$) "is relevant for" the terms in Σ, we mean that all consequences of $\mathcal{O}$ that can be expressed over Σ are also consequences of $\mathcal{M}$. Then $\mathcal{O}$ is said to be a *conservative extension* (CE) of $\mathcal{M}$. A stronger property

---

[1] http://ihtsdo.org/snomed-ct/.
[2] http://www.opengalen.org/.

is that every model of $\mathcal{M}$ extends to a model of $\mathcal{O}$ — we refer to this as *model conservative extension*.

One of the reasons why one might be interested in modularity aspects of an ontology is for reusing information about relevant terms it captures. Reusing a module $\mathcal{M} \subseteq \mathcal{O}$ within another ontology $\mathcal{O}'$ is referred to in the literature as the *module importing scenario*. In this scenario, the signature Σ used to extract $\mathcal{M}$ from $\mathcal{O}$ acts as the *interface* signature between $\mathcal{O}'$ and $\mathcal{O}$ in the sense that it contains the set of terms that one is interested in reusing and that might be shared between $\mathcal{O}'$ and $\mathcal{O}$.

Example 1.1. Assume that we have the following OWL-ontology $\mathcal{O}$:

Male $\equiv$ Human $\sqcap \neg$Female,	Father $\sqsubseteq$ Human,
Human $\sqsubseteq \forall$has_child.Human,	Father $\equiv$ Male $\sqcap \exists$has_child.$\top$

For readers not familiar with OWL, we provide the translation to first-order logic:

$\forall x.$Male$(x) \leftrightarrow$ Human$(x) \land \neg$Female(x),
$$\forall x.\text{Father}(x) \rightarrow \text{Human}(x),$$
$\forall x.$Human$(x) \rightarrow \forall y.has_child(x, y) \rightarrow$ Human(y)
$$\forall x.\text{Father}(x) \leftrightarrow \text{Male}(x) \land \exists y.\text{has\_child}(x, y)$$

Now further assume that we are interested in the terms in $\Sigma = \{$Male, Human, Female, has_child$\}$. Then the subset $\mathcal{M}$ containing only the grey shaded axioms is a Σ-module of $\mathcal{O}$. Indeed, one can show that $\mathcal{O}$ has the same Σ-consequences as $\mathcal{M}$. For example, Male $\sqcap \exists$has_child.$\top \sqsubseteq$ Human follows from $\mathcal{O}$, but also from $\mathcal{M}$. □

Ideally, an imported module $\mathcal{M}$ should be as small as possible while still guaranteeing to capture all the relevant knowledge w.r.t. Σ. Importing $\mathcal{M}$ into $\mathcal{O}'$ would have the same observable effect as importing the entire ontology $\mathcal{O}$, e.g., one should get the same answers to a query in both cases.

Observe that the logical view appears to be theoretically sound and elegant and guarantees that by reusing only terms from Σ one is not able to distinguish between importing $\mathcal{M}$ and importing $\mathcal{O}$ into some ontology $\mathcal{O}'$.

This paper contributes the generalization of central notions of ontology module extraction to an arbitrary institution. While doing this, we were also able to correct a small inaccuracy appearing in the definition of inseparability used in the literature. Our paper is organized as follows: in Sect. 2, we recall institutions and inclusion systems (the latter leading to a set-theoretic flavour of signatures, which is generally assumed in the ontology community). Section 3 studies conservative extensions and inseparability as a prequisite for module extraction, which is studied in Sect. 4, together with some robustness properties. Section 5 concludes the paper.

2   Institutions

The large variety of logical languages in use can be captured at an abstract level using the concept of *institutions* [8]. This allows us to develop results independently of the specific features of a logical system. We use the notions of institution and logical language interchangeably throughout the rest of the paper.

The main idea is to collect the non-logical symbols of the language in signatures and to assign to each signature the set of sentences that can be formed using its symbols. Informally, in typical examples, each signature lists the symbols it consists of, together with their kinds. Signature morphisms are mappings between signatures. We do not assume any details except that signature morphisms can be composed and that there are identity morphisms; this amounts to a category of signatures. Readers unfamiliar with category theory may replace this with a partial order; signature morphisms are then just inclusions. See [18] for details of this simplified foundation.

Institutions also provide a model theory, which introduces semantics for the language and gives a satisfaction relation between the models and the sentences of a signature. The main restriction imposed is the satisfaction condition, which captures the idea that truth is invariant under change of notation (and enlargement of context) along signature morphisms. This relies on two further components of institutions: the translation of sentences along signature morphisms, and the reduction of models against signature morphisms (generalising the notion of model reduct known from logic).

Definition 2.1. *An **institution** [8] is a quadruple* $I = (\mathbf{Sign}, \mathbf{Sen}, \mathbf{Mod}, \models)$ *consisting of the following:*

- *a category* $\mathbf{Sign}$ *of* signatures *and* signature morphisms;
- *a functor* $\mathbf{Sen}\colon \mathbf{Sign} \to \mathbb{S}et$[3] *giving, for each signature* Σ*, the set of* sentences $\mathbf{Sen}(\Sigma)$*, and for each signature morphism* $\sigma\colon \Sigma \to \Sigma'$*, the* sentence translation map $\mathbf{Sen}(\sigma)\colon \mathbf{Sen}(\Sigma) \to \mathbf{Sen}(\Sigma')$*, where* $\mathbf{Sen}(\sigma)(\varphi)$ *is often written as* $\sigma(\varphi)$;
- *a functor* $\mathbf{Mod}\colon \mathbf{Sign}^{op} \to \mathbb{C}at$[4] *giving, for each signature* Σ*, the* category of models $\mathbf{Mod}(\Sigma)$*, and for each signature morphism* $\sigma\colon \Sigma \to \Sigma'$*, the* reduct functor $\mathbf{Mod}(\sigma)\colon \mathbf{Mod}(\Sigma') \to \mathbf{Mod}(\Sigma)$*, where* $\mathbf{Mod}(\sigma)(M')$ *is often written as* $M'|_\sigma$*. Then* $M'|_\sigma$ *is called the* σ-reduct *of* M'*, while* M' *is called a* σ-expansion *of* $M'|_\sigma$; *and*
- *a* satisfaction relation $\models_\Sigma \subseteq |\mathbf{Mod}(\Sigma)| \times \mathbf{Sen}(\Sigma)$ *for each* $\Sigma \in |\mathbf{Sign}|$,

such that for each $\sigma\colon \Sigma \to \Sigma'$ *in* $\mathbf{Sign}$ *the following* **satisfaction condition** *holds:*

$$(\star) \qquad M' \models_{\Sigma'} \sigma(\varphi) \text{ iff } M'|_\sigma \models_\Sigma \varphi$$

for each $M' \in |\mathbf{Mod}(\Sigma')|$ *and* $\varphi \in \mathbf{Sen}(\Sigma)$.   □

---

[3] $\mathbb{S}et$ is the category having sets as objects and functions as arrows.

[4] $\mathbb{C}at$ is the category of categories and functors. Strictly speaking, $\mathbb{C}at$ is a quasicategory (which is a category that lives in a higher set-theoretic universe).

As usual, the satisfaction relation between models and sentences determines a semantic notion of consequence: for any signature $\Sigma \in |\mathbf{Sign}|$, a Σ-sentence $\varphi \in \mathbf{Sen}(\Sigma)$ is a (semantic) *consequence* of a set of Σ-sentences $\Phi \subseteq \mathbf{Sen}(\Sigma)$, written $\Phi \models_{\Sigma} \varphi$, if for each model $M \in |\mathbf{Mod}(\Sigma)|$, $M \models_{\Sigma} \varphi$ whenever $M \models_{\Sigma} \Phi$ (i.e., $M \models_{\Sigma} \psi$ for all $\psi \in \Phi$). This is an example of how logical notions can be defined in an arbitrary institution. It is easy to see that semantic consequence is preserved under translation w.r.t. signature morphisms: given $\sigma \colon \Sigma \to \Sigma'$, if $\Phi \models_{\Sigma} \varphi$ then $\sigma(\Phi) \models_{\Sigma'} \sigma(\varphi)$. The opposite implication does not hold in general though.

It is also possible to complement an institution with a proof theory, introducing a derivability or *deductive consequence* relation between sentences, formalised as an *entailment system* [13]. In particular, this can be done for the institutions presented below.

Several institution-independent languages for structured theories have been defined, see e.g. [7,21]. One of them is the distributed ontology, modeling and specification language DOL [17], which also provides language constructs for module extraction.

Example 2.2. In the institution Prop of propositional logic, signatures are sets of propositional variables and signature morphisms are functions. Models are valuations into $\{T, F\}$ and model reduct is just composition. Sentences are formed inductively from propositional variables by the usual logical connectives. Sentence translation means replacement of propositional variables along the signature morphism. Satisfaction is the usual satisfaction of a propositional sentence under a valuation.   □

Example 2.3. OWL signatures consist of sets of atomic classes, individuals, object and data properties. OWL signature morphisms map classes to classes, individuals to individuals, object properties to object properties and data properties to data properties. For an OWL signature Σ, sentences include subsumption relations between classes or properties, membership assertions of individuals in classes and pairs of individuals in properties, and complex role inclusions. Sentence translation along a signature morphism simply replaces non-logical symbols with their image along the morphism. The kinds of symbols are class, individual, object property and data property, respectively, and the set of symbols of a signature is the union of its sets of classes, individuals and properties. Models are (unsorted) first-order structures that interpret concepts as unary and properties as binary predicates, and individuals as elements of the universe of the structure, and satisfaction is the standard satisfaction of description logics. This gives us an institution for OWL.

Strictly speaking, this institution captures OWL 2 DL *without restrictions* in the sense of [23]. The reason is that in an institution, sentences can be used for arbitrary formation of theories. This is related to the presence of union as a specification-building operation, which is also present in DOL. OWL 2 DL's specific restrictions on theory formation can be modelled *inside* this institution, as a constraint on ontologies (theories). This constraint is generally not preserved under unions or extensions. DOL's multi-logic capability allows the clean distinction between ordinary OWL 2 DL and OWL 2 DL without restrictions. □

Example 2.4. In the institution FOL$^=$ of many-sorted first-order logic with equality, signatures are many-sorted first-order signatures, consisting of a set of sort and sorted operation and predicate symbols. Signature morphisms map sorts, operation and predicate symbols in a compatible way. Models are many-sorted first-order structures. Sentences are closed first-order formulae with atomic formulae including equality between terms of the same sort. Sentence translation means replacement of symbols along the signature morphism. A model reduct interprets a symbol by first translating it along the signature morphism and then interpreting it in the model to be reduced. Satisfaction is the usual satisfaction of a first-order sentence in a first-order structure. □

A *presentation* in an institution $I = (\mathbf{Sign}, \mathbf{Sen}, \mathbf{Mod}, \models)$ is a pair $P = (\Sigma, \Phi)$, where $\Sigma \in |\mathbf{Sign}|$ is a signature and $\Phi \subseteq \mathbf{Sen}(\Sigma)$ is a set of Σ-sentences. Σ is also denoted as $\mathsf{Sig}(P)$, Φ as $\mathsf{Ax}(P)$. We extend the model functor to presentations and write $\mathbf{Mod}(\Sigma, \Phi)$ (or sometimes $\mathbf{Mod}(\Phi)$ if the signature is clear) for the full subcategory of $\mathbf{Mod}(\Sigma)$ that consists of the *models* of (Σ, Φ), i.e., $|\mathbf{Mod}(\Sigma, \Phi)| = \{M \in |\mathbf{Mod}(\Sigma)| \mid M \models_\Sigma \Phi\}$.

A *presentation morphism* $\sigma \colon (\Sigma, \Phi) \to (\Sigma', \Phi')$ is a signature morphism $\sigma \colon \Sigma \to \Sigma'$ such that for all models $M' \in |\mathbf{Mod}(\Sigma', \Phi')|$, $M'|_\sigma \in |\mathbf{Mod}(\Sigma, \Phi)|$. This defines the category **Pres** of *presentations* in I. An easy consequence of the satisfaction condition is that presentation morphisms preserve semantic consequence:

Proposition 2.5. $\sigma \colon (\Sigma, \Phi) \to (\Sigma', \Phi')$ *is a presentation morphism iff for all* Σ*-sentences* φ*, if* $\Phi \models_\Sigma \varphi$ *then* $\Phi' \models_{\Sigma'} \sigma(\varphi)$. □

Each presentation (Σ, Φ) *generates* a *theory* $(\Sigma, \mathsf{cl}_\models(\Phi))$, where $\mathsf{cl}_\models(\Phi) = \{\varphi \in \mathbf{Sen}(\Sigma) \mid \Phi \models_\Sigma \varphi\}$ is the closure of Φ under semantic consequence. The category **Th** of *theories* in I is the full subcategory of **Pres** with objects (Σ, Φ) such that Φ is closed under semantic consequence. The closure under semantic consequence extends to the functor $\mathsf{cl}_\models \colon \mathbf{Pres} \to \mathbf{Th}$, which together with the inclusion $\mathbf{Th} \hookrightarrow \mathbf{Pres}$ establishes the equivalence between **Pres** and **Th**.

A presentation morphism $\sigma \colon (\Sigma, \Phi) \to (\Sigma', \Phi')$ is *model-conservative* if for each model $M \in |\mathbf{Mod}(\Sigma, \Phi)|$ there is a model $M' \in |\mathbf{Mod}(\Sigma', \Phi')|$ that is a σ-*expansion* of M, i.e., $M'|_\sigma = M$. A presentation morphism $\sigma \colon (\Sigma, \Phi) \to (\Sigma', \Phi')$ is *consequence-conservative* if for all Σ-sentences $\varphi \in \mathbf{Sen}(\Sigma)$, $\Phi \models_\Sigma \varphi$ whenever $\Phi' \models_{\Sigma'} \sigma(\varphi)$ (the opposite implication always holds).

Proposition 2.6. *If a presentation morphism is model-conservative then it is consequence-conservative as well.* □

The opposite implication does not hold in general: model-conservativity is a strictly stronger notion than consequence-conservativity. However, in some logics, the two notions may coincide:

Example 2.7. In the institution **Prop** of propositional logic (see Example 2.2), a presentation morphism is model-conservative iff it is consequence-conservative. Consider a presentation morphism $\sigma : (V, \Phi) \to (V', \Phi')$ in **Prop**. Assume that σ is not model-conservative, and let $m\colon V \to \{T, F\}$ be such that $m \models \Phi$ and m has no σ-expansion to a model of Φ'. For each propositional variable $p \in V$, let $\varphi_{m,p}$ be p if $m(p) = T$ and $\neg p$ if $m(p) = F$. Consider $\Psi' = \Phi' \cup \{\sigma(\varphi_{m,p}) \mid p \in V\}$. Since there is no model $m'\colon V' \to \{T, F\}$ such that $m' \models \Phi'$ and $m'|_\sigma = m$, Ψ' has no model, and so *false* is a semantic consequence of Ψ'. By compactness of propositional logic, for some finite set of variables $p_1, \ldots, p_n \in V$, the implication $\sigma(\varphi_{m,p_1}) \wedge \ldots \wedge \sigma(\varphi_{m,p_n}) \Rightarrow false$ is a consequence of Φ'. However, the implication $\varphi_{m,p_1} \wedge \ldots \wedge \varphi_{m,p_n} \Rightarrow false$ is not a consequence of Φ, and hence σ is not consequence-conservative.   □

The signatures of the standard institutions presented above come naturally equipped with a notion of subsignature, hence signature inclusion, and a well-defined way of forming a union of signatures. These concepts can be captured in a categorical setting using *inclusion systems* [5,9]. However, we will work with a slightly different version of this notion:

Definition 2.8. *An **inclusive category** is a category with a broad subcategory[5] which is a partially ordered class with a least element (denoted $\emptyset$), non-empty products (denoted $\cap$) and finite coproducts (denoted $\cup$), such that for each pair of objects A, B, the following is a pushout in the category:*

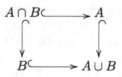

□

For any objects A and B of an inclusive category, we write $A \subseteq B$ if there is an inclusion from A to B; the unique such inclusion will then be denoted by $\iota_{A \subseteq B}\colon A \hookrightarrow B$, or simply $A \hookrightarrow B$.

A functor between two inclusive categories is inclusive if it takes inclusions in the source category to inclusions in the target category.

Definition 2.9. *An institution $I = (\mathbf{Sign}, \mathbf{Sen}, \mathbf{Mod}, \models)$ is **inclusive**[6] if*

- **Sign** *is an inclusive category,*
- **Sen** *is inclusive and preserves intersections,[7] and*
- *each model category is inclusive, and reduct functors are inclusive.[8]*

---

[5] That is, with the same objects as the original category.

[6] Even though we use the same term as in [9], since the overall idea is the same, on one hand, some of our assumptions here are weaker than in [9], and on the other hand, we require a bit more structure on the category of signatures.

[7] That is, for any family of signatures $\mathbb{S} \subseteq |\mathbf{Sign}|$, $\mathbf{Sen}(\cap \mathbb{S}) = \bigcap_{\Sigma \in \mathbb{S}} \mathbf{Sen}(\Sigma)$.

[8] That is, we have a model functor $\mathbf{Mod}\colon \mathbf{Sign}^{op} \to \mathbb{I}\mathcal{C}at$, where $\mathbb{I}\mathcal{C}at$ is the (quasi)category of inclusive categories and inclusive functors.

Moreover, we asume that reducts w.r.t. signature inclusions are surjective on objects.   □

The empty object in the category of signatures will be referred to as the empty signature (indeed, in typical signature categories it is empty) and will be written as $\Sigma_\emptyset$.

Since in any inclusive institution the category of signatures has arbitrary intersections, for any set of sentences $\Phi \subseteq \bigcup_{\Sigma \in |\mathbf{Sign}|} \mathbf{Sen}(\Sigma)$, there exists the least signature $\mathsf{Sig}(\Phi)$ such that $\Phi \subseteq \mathbf{Sen}(\mathsf{Sig}(\Phi))$.

The assumption that reducts are surjective on models is rather mild and ensures that semantic consequence is not only preserved but also reflected under signature extension. Then, given $\Phi \subseteq \mathbf{Sen}(\Sigma)$ and $\varphi \in \mathbf{Sen}(\Sigma)$ (or, equivalently, $\mathsf{Sig}(\Phi) \cup \mathsf{Sig}(\varphi) \subseteq \Sigma$), we have that $\Phi \models_\Sigma \varphi$ if and only if $\Phi \models_{\mathsf{Sig}(\Phi) \cup \mathsf{Sig}(\varphi)} \varphi$. In particular, this justifies use of the notation $\Phi \models \varphi$ without any explicit reference to the signature over which sentences and consequence between them are considered. Moreover, $\Phi \models \varphi$ if and only if $|\mathbf{Mod}(\Sigma, \Phi)| \subseteq |\mathbf{Mod}(\Sigma, \{\varphi\})|$ for every signature $\Sigma \supseteq \mathsf{Sig}(\Phi) \cup \mathsf{Sig}(\varphi)$.

The institutions Prop, OWL and FOL$^=$ sketched above in Examples 2.2, 2.3 and 2.4 can be equipped with the obvious inclusion system on their signatures and models, and then become inclusive institutions.

In inclusive institutions, if $\Sigma_1 \subseteq \Sigma_2$ via an inclusion $\iota\colon \Sigma_1 \hookrightarrow \Sigma_2$ and $M \in \mathbf{Mod}(\Sigma_2)$, we write $M|_{\Sigma_1}$ for $M|_\iota$. Note that $\mathbf{Sen}(\iota)\colon \mathbf{Sen}(\Sigma_1) \to \mathbf{Sen}(\Sigma_2)$ is the usual set-theoretic inclusion, hence its application may be omitted.

For some results, we need an amalgamation property on models. An inclusive institution I is called *(weakly) union-exact* if all intersection-union signature pushouts in **Sign** are (weakly) amalgamable. More specifically, the latter means that for any pushout

$$
\begin{array}{ccc}
\Sigma_1 \cap \Sigma_2 & \longrightarrow & \Sigma_1 \\
\downarrow & & \downarrow \\
\Sigma_2 & \longrightarrow & \Sigma_1 \cup \Sigma_2
\end{array}
$$

in **Sign**, any pair $(M_1, M_2) \in \mathbf{Mod}(\Sigma_1) \times \mathbf{Mod}(\Sigma_1)$ that is *compatible* in the sense that M_1 and M_2 reduce to the same $(\Sigma_1 \cap \Sigma_2)$-model can be *amalgamated* to a unique (or weakly amalgamated to a not necessarily unique) $(\Sigma_1 \cup \Sigma_2)$-model: there exists a (unique) $M \in \mathbf{Mod}(\Sigma_1 \cup \Sigma_2)$ that reduces to M_1 and M_2, respectively.

The institutions Prop, OWL and FOL$^=$ sketched above are all union-exact.

3   Conservative Extensions and Inseparability

An ontology is typically presented as a collection of concepts/objects, relations, properties and axioms — thus a presentation of a theory in some suitable logic, with OWL being a typical example. The goal of this paper is to study some concepts used in the research on ontologies and their modularisation independently of the logic in use. We make this precise by presenting these concepts in the context of an arbitrary (but fixed for now) inclusive institution $I = (\mathbf{Sign}, \mathbf{Sen}, \mathbf{Mod}, \models)$. The presentation below is based on the general concepts and facts conerning inclusive institutions, as spelled out in Sect. 2. To stay in tune with the literature and concerns of the domain we consider, we will adjust the terminology and notation appropriately.

An *ontology* $\mathcal{O}$ in a logic given as the institution I is just a set of sentences $\mathcal{O} \subseteq \bigcup_{\Sigma \in |\mathbf{Sign}|} \mathbf{Sen}(\Sigma)$ in I. As explained in Sect. 2, for each ontology $\mathcal{O}$ we have its signature $\mathsf{Sig}(\mathcal{O})$, which is the least signature over which all the sentences in $\mathcal{O}$ may be considered.

Note that if we want an ontology to be always considered over a larger signature with some extra symbols without changing its intended meaning, we need to add trivially true sentences that involve the additional symbols. In many typical institutions such sentences always exists (for instance, $p \vee \neg p$ in Prop, etc.); if this is not the case, we may want to expand our institution by some trivial sentences to "declare" that some extra symbols are to be considered.

Ontology inclusions give a starting notion to study relationships between ontologies. If $\mathcal{O} \subseteq \mathcal{O}'$ then we say that $\mathcal{O}'$ is an extension of $\mathcal{O}$. As in Sect. 2, conservativity of such an extension may be defined in two variants: based on models and based on semantic consequence (deduction), respectively. However, we are often interested in further nuances, where conservativity is considered up to an indicated signature of current interest.

Consider ontologies $\mathcal{O} \subseteq \mathcal{O}'$ and a signature $\Sigma \in |\mathbf{Sign}|$.

1. $\mathcal{O}'$ is a *model Σ-conservative extension* (Σ-mCE) of $\mathcal{O}$ if for every $(\mathsf{Sig}(\mathcal{O}) \cup \Sigma)$-model $\mathcal{I}$ of $\mathcal{O}$ there exists a $(\mathsf{Sig}(\mathcal{O}') \cup \Sigma)$-model $\mathcal{I}'$ of $\mathcal{O}'$ such that $\mathcal{I}|_\Sigma = \mathcal{I}'|_\Sigma$.
2. $\mathcal{O}'$ is a *consequence Σ-conservative extension* (Σ-cCE) of $\mathcal{O}$ if for every Σ-sentence α, we have $\mathcal{O}' \models \alpha$ iff $\mathcal{O} \models \alpha$.

Proposition 2.6 essentially applies here as well, so that the notion of model Σ-conservative extension is strictly stronger than that of consequence Σ-conservative extension, and it clearly does not depend on the expressiveness of the institution. Thus if $\mathcal{O}'$ is a Σ-mCE of $\mathcal{O}$ then $\mathcal{O}'$ is a Σ-cCE of $\mathcal{O}$ as well, while the converse does not hold. However, for propositional logic, the two concepts are equivalent, see Example 2.7.

We have parameterised above both concepts of conservative extension by a specific signature to indicate the focus of current interest. Further concepts will be developed in a similar fashion, taking the signature of interest into account. As this signature may vary here rather arbitrarily, we will need to adjust any ontology to cover it explicitly, turning the ontology into a presentation in I: for an ontology $\mathcal{O}$ and a signature Σ, we define $\mathcal{O} \uparrow \Sigma = (\mathsf{Sig}(\mathcal{O}) \cup \Sigma, \mathsf{Ax}(\mathcal{O}))$.

Now, when a signature of interest is indicated, the notion of ontology extension may be refined as follows. Again, this comes in two flavours: one based on models, the other on sentences (consequence).

Given ontologies $\mathcal{O}'$ and $\mathcal{O}$ and a signature Σ:

1. $\mathcal{O}'$ is a model Σ-extension of $\mathcal{O}$ if for all models $\mathcal{I}' \in |\mathbf{Mod}(\mathcal{O}' \uparrow \Sigma)|$ there is $\mathcal{I} \in |\mathbf{Mod}(\mathcal{O} \uparrow \Sigma)|$ such that $\mathcal{I}'|_\Sigma = \mathcal{I}|_\Sigma$.

2. $\mathcal{O}'$ is a consequence Σ-extension of $\mathcal{O}$ if for all Σ-sentences α, we have $\mathcal{O}'{\upharpoonright}\Sigma \models \alpha$ if $\mathcal{O}{\upharpoonright}\Sigma \models \alpha$.

Clearly, if $\mathcal{O} \subseteq \mathcal{O}'$ then $\mathcal{O}'$ is a model Σ-extension of $\mathcal{O}$, for any signature Σ. Moreover, essentially by Proposition 2.5, if $\mathcal{O}'$ is a model Σ-extension of $\mathcal{O}$ then it is a consequence Σ-extension of $\mathcal{O}$ as well.

The model Σ-extension condition may be rewritten as follows:

$$\{\mathcal{I}'|_\Sigma \mid \mathcal{I}' \in |\mathbf{Mod}(\mathcal{O}'{\upharpoonright}\Sigma)|\} \subseteq \{\mathcal{I}|_\Sigma \mid \mathcal{I} \in |\mathbf{Mod}(\mathcal{O}{\upharpoonright}\Sigma)|\}$$

One may feel tempted to simplify this and instead write

$$\{\mathcal{I}'|_\Sigma \mid \mathcal{I}' \models \mathcal{O}'\} \subseteq \{\mathcal{I}|_\Sigma \mid \mathcal{I} \models \mathcal{O}\}$$

However, this formally makes little sense unless we assume $\Sigma \subseteq \mathsf{Sig}(\mathcal{O}') \cap \mathsf{Sig}(\mathcal{O})$. This would be a strong assumption concerning the signature of interest (even if $\mathcal{O} \subseteq \mathcal{O}'$), especially when we come to discussing robustness properties below. If this condition does not hold, it is not entirely clear what $\mathcal{I}|_\Sigma$ should mean. One possible interpretation might be "remove all model components whose names do not occur in Σ" (consider reducts to $\Sigma \cap \mathsf{Sig}(\mathcal{O}')$ and $\Sigma \cap \mathsf{Sig}(\mathcal{O})$, respectively). But even then, the two definitions depart: consider (in OWL) $\mathcal{O}' = \{C \sqsubseteq C\}$, $\mathcal{O} = \{C' \sqsubseteq C'\}$, and $\Sigma = \{C, C'\}$. Then according to our definition, $\mathcal{O}'$ is a model Σ-extension of $\mathcal{O}$, but this would not be the case if the apparently simplified condition was used.[9] In fact, the simpler condition cannot be met in a non-trivial way unless $\Sigma \cap \mathsf{Sig}(\mathcal{O}') = \Sigma \cap \mathsf{Sig}(\mathcal{O})$, another strong assumption we rather avoid.

One may now want to define a module in an ontology $\mathcal{O}$ to be another ontology $\mathcal{M}$ such that $\mathcal{M} \subseteq \mathcal{O}$ and the inclusion is conservative (either in the model-based sense, or in the consequence-based sense). However, we want this concept to work for an arbitrary signature of interest. The appropriate requirement is formulated in terms of *inseparability*. One intuition is that inseparability is a proper generalisation of conservative extension to a more symmetric situation.

Let $\mathcal{O}_1$ and $\mathcal{O}_2$ be ontologies and Σ a signature. Then $\mathcal{O}_1$ and $\mathcal{O}_2$ are *model Σ-inseparable*, written $\mathcal{O}_1 \equiv_\Sigma^m \mathcal{O}_2$ if

$$\{\mathcal{I}|_\Sigma \mid \mathcal{I} \in |\mathbf{Mod}(\mathcal{O}_1{\upharpoonright}\Sigma)|\} = \{\mathcal{I}|_\Sigma \mid \mathcal{I} \in |\mathbf{Mod}(\mathcal{O}_2{\upharpoonright}\Sigma)|\}$$

Note that in the literature, a simpler condition is commonly used:

$$\{\mathcal{I}|_\Sigma \mid \mathcal{I} \models \mathcal{O}_1\} = \{\mathcal{I}|_\Sigma \mid \mathcal{I} \models \mathcal{O}_2\}$$

However, this "simplification" is dubious — all the comments concerning the definition of model Σ-extension above apply here as well.

Clearly, $\mathcal{O}_1$ and $\mathcal{O}_2$ are model Σ-inseparable iff $\mathcal{O}_1$ is a model Σ-extension of $\mathcal{O}_2$ and $\mathcal{O}_2$ is a model Σ-extension of $\mathcal{O}_1$. Moreover if $\mathcal{O}_1 \subseteq \mathcal{O}_2$ and $\Sigma \subseteq$

---

[9] This remains true even if $\mathcal{I}$ varies over models of arbitrary signatures, which seems to be a widespread understanding in the ontology modularity community. Note that $\mathcal{I} \models \mathcal{O}$ still entails that $\mathcal{I}$ interprets at least the symbols occurring in $\mathcal{O}$.

$\mathsf{Sig}(\mathcal{O}_1)$ then $\mathcal{O}_1$ and $\mathcal{O}_2$ are model Σ-inseparable iff $\mathcal{O}_2$ is a model Σ-conservative extension of $\mathcal{O}_1$.

Model Σ-inseparability provides a very strong form of equivalence between ontologies considered from the perspective given by Σ: $\mathcal{O}_1 \equiv_{\Sigma}^{m} \mathcal{O}_2$ guarantees that $\mathcal{O}_1$ can be replaced by $\mathcal{O}_2$ in any application that refers only to symbols from Σ. Moreover, since this notion does not depend on the expressibility of the underlying institution, we may arbitrarily strengthen the logic without affecting this equivalence.

Weaker versions of inseparability relations can be defined. To begin with, as usual, we consider a deductive variant: $\mathcal{O}_1$ and $\mathcal{O}_2$ are *consequence Σ-inseparable*, written $\mathcal{O}_1 \equiv_{\Sigma}^{s} \mathcal{O}_2$, if for all Σ-sentences φ

$$\mathcal{O}_1 \models \varphi \text{ iff } \mathcal{O}_2 \models \varphi$$

Let us recall here again that the semantic consequences might be taken over any signatures that encompass all symbols used either in the ontology or in the sentences considered.

Given a signature Σ, in many contexts we are not interested in preservation of all Σ-sentences, but it is sufficient to consider only sentences of some specific form that express the properties we really care about. This extra twist may be captured by considering a set of Σ-sentences $\Lambda \subseteq \mathbf{Sen}(\Sigma)$, and weakening consequence Σ-inseparability as follows: $\mathcal{O}_1$ and $\mathcal{O}_2$ are *Λ-consequence Σ-inseparable*, written $\mathcal{O}_1 \equiv_{\Sigma}^{\Lambda} \mathcal{O}_2$, if for all Σ-sentences $\varphi \in \Lambda$

$$\mathcal{O}_1 \models \varphi \text{ iff } \mathcal{O}_2 \models \varphi$$

For instance, in OWL, one relevant choice of the set Λ is to consider all subsumptions between atomic concepts.

It is easy to see now that indeed, the above three equivalences are gradually coarser:

Proposition 3.1. *For any signature Σ and set of Σ-sentences $\Lambda \subseteq \mathbf{Sen}(\Sigma)$, we have $\equiv_{\Sigma}^{m} \subseteq \equiv_{\Sigma}^{s} \subseteq \equiv_{\Sigma}^{\Lambda}$.* $\square$

In particular, this means that if two ontologies are model Σ-inseparable then they are Σ-inseparable by any set of sentences, even if we strengthen the logic in use. Whatever sentences we add to our institution, no matter how strong they would be, two ontologies that are model Σ-inseparable will have the same consequences among them.

We mentioned above that one may want to consider various signatures Σ, changing the focus of interest through which ontologies are considered. In particular, this means that to use Λ-consequence Σ-inseparability, we have to provide the set of sentences over each such signature Σ. What one wants then is an inclusive functor $\Lambda\colon \mathbf{Sign} \to \mathbb{S}et$ with $\Lambda(\Sigma) \subseteq \mathbf{Sen}(\Sigma)$ for all signatures Σ. This implies that for $\Sigma' \subseteq \Sigma$, $\Lambda(\Sigma') \subseteq \Lambda(\Sigma)$, capturing the intuition that the sentences to be preserved cannot be disregarded when signature is enlarged. For any signature Σ, slightly abusing the notation, we write $\equiv_{\Sigma}^{\Lambda}$ for $\equiv_{\Sigma}^{\Lambda(\Sigma)}$.

Given the above arrangements, the inseparability relations defined are preserved when the signature considered is narrowed:

Proposition 3.2. *Given any signatures* $\Sigma' \subseteq \Sigma$, *we have* $\equiv_\Sigma^m \subseteq \equiv_{\Sigma'}^m$, $\equiv_\Sigma^s \subseteq \equiv_{\Sigma'}^s$, *and* $\equiv_\Sigma^\Lambda \subseteq \equiv_{\Sigma'}^\Lambda$. $\hfill\square$

For a given institution, an *inseparability relation* is a family $\mathcal{S} = \langle\equiv_\Sigma^\mathcal{S}\rangle_{\Sigma \in |\mathbf{Sign}|}$ of equivalence relations on the family of presentations. The informal intuition we want to capture is that for any two ontologies $\mathcal{O}_1$ and $\mathcal{O}_2$, $\mathcal{O}_1 \equiv_\Sigma^\mathcal{S} \mathcal{O}_2$ means that $\mathcal{O}_1$ and $\mathcal{O}_2$ are indistinguishable w.r.t. Σ, i.e., they represent the same knowledge of interest about the topics expressible in the signature Σ. Any specific definition of the inseparability relation determines the exact meaning of the terms "indistinguishable" and "the knowledge of interest". However, since "the knowledge of interest" relevant for a signature should not be disregarded when the signature is enlarged, it is desirable that the inseparability relations are monotone in the following sense:

Definition 3.3 ([10]). *An inseparability relation* $\mathcal{S} = \langle\equiv_\Sigma^\mathcal{S}\rangle_{\Sigma \in |\mathbf{Sign}|}$ *is monotone if*

1. *for any signatures* $\Sigma' \subseteq \Sigma$, $\equiv_\Sigma^\mathcal{S} \subseteq \equiv_{\Sigma'}^\mathcal{S}$ *(the inseparability relation gets finer when the signature gets larger), and*
2. *if* $\mathcal{O}_1 \subseteq \mathcal{O}_2 \subseteq \mathcal{O}_3$ *and* $\mathcal{O}_1 \equiv_\Sigma^\mathcal{S} \mathcal{O}_3$ *then* $\mathcal{O}_1 \equiv_\Sigma^\mathcal{S} \mathcal{O}_2$ *and* $\mathcal{O}_2 \equiv_\Sigma^\mathcal{S} \mathcal{O}_3$ *(the intuition here is: since larger ontologies capture more of "the knowledge of interest", we also require that any ontology squeezed between an ontology and its inseparable extension is inseparable from both of them).* $\hfill\square$

The inseparability relations defined above ($\langle\equiv_\Sigma^m\rangle_{\Sigma\in|\mathbf{Sign}|}$, $\langle\equiv_\Sigma^s\rangle_{\Sigma\in|\mathbf{Sign}|}$, and $\langle\equiv_\Sigma^\Lambda\rangle_{\Sigma\in|\mathbf{Sign}|}$) are typical examples we will use in the following. It is easy to show that all are monotone.

Monotonicity can be reformulated as robustness under signature restrictions. We now recall further robustness properties from the literature [10,27].

Definition 3.4. *An inseparability relation* $\mathcal{S} = \langle\equiv_\Sigma^\mathcal{S}\rangle_{\Sigma\in|\mathbf{Sign}|}$ *is*

- *robust under signature extensions if for all ontologies* $\mathcal{O}_1$ *and* $\mathcal{O}_2$ *and all signatures* Σ, Σ' *with* $\Sigma' \cap (\mathsf{Sig}(\mathcal{O}_1) \cup \mathsf{Sig}(\mathcal{O}_2)) \subseteq \Sigma$

$$\mathcal{O}_1 \equiv_\Sigma \mathcal{O}_2 \text{ implies } \mathcal{O}_1 \equiv_{\Sigma'} \mathcal{O}_2$$

- *robust under replacement if for all ontologies* $\mathcal{O}$, $\mathcal{O}_1$ *and* $\mathcal{O}_2$ *and all signatures* Σ *with* $\mathsf{Sig}(\mathcal{O}) \subseteq \Sigma$, *we have*

$$\mathcal{O}_1 \equiv_\Sigma \mathcal{O}_2 \text{ implies } \mathcal{O}_1 \cup \mathcal{O} \equiv_\Sigma \mathcal{O}_2 \cup \mathcal{O}$$

- *robust under joins if for all ontologies* $\mathcal{O}_1$ *and* $\mathcal{O}_2$ *and all signatures* Σ *with* $\mathsf{Sig}(\mathcal{O}_1) \cap \mathsf{Sig}(\mathcal{O}_2) \subseteq \Sigma$, *we have for* $i = 1, 2$

$$\mathcal{O}_1 \equiv_\Sigma \mathcal{O}_2 \text{ implies } \mathcal{O}_i \equiv_\Sigma \mathcal{O}_1 \cup \mathcal{O}_2$$

$\hfill\square$

We have the following result on robustness:

Theorem 3.5. *Model inseparability is robust under replacement. In a union-exact inclusive institution, model inseparability is also robust under signature extensions and joins.*

Proof. Robustness under replacement: Consider ontologies $\mathcal{O}$, $\mathcal{O}_1$ and $\mathcal{O}_2$ and a signature Σ such that $\mathsf{Sig}(\mathcal{O}) \subseteq \Sigma$ and $\mathcal{O}_1 \equiv^m_\Sigma \mathcal{O}_2$. We need to show that $\mathcal{O}_1 \cup \mathcal{O} \equiv^m_\Sigma \mathcal{O}_2 \cup \mathcal{O}$, which amounts to showing

$$\{\mathcal{I}|_\Sigma \mid \mathcal{I} \in |\mathbf{Mod}((\mathcal{O}_1 \cup \mathcal{O})\!\uparrow\!\Sigma)|\} = \{\mathcal{I}|_\Sigma \mid \mathcal{I} \in |\mathbf{Mod}((\mathcal{O}_2 \cup \mathcal{O})\!\uparrow\!\Sigma)|\}$$

By symmetry, it suffices to prove one inclusion. Let $\mathcal{I} \in |\mathbf{Mod}((\mathcal{O}_1 \cup \mathcal{O})\!\uparrow\!\Sigma)|$. Define $\mathcal{I}' = \mathcal{I}|_{\mathsf{Sig}(\mathcal{O}_1) \cup \Sigma}$. By $\mathcal{O}_1 \equiv^m_\Sigma \mathcal{O}_2$, we know that $\mathcal{I}'|_\Sigma$ has an expansion to an $\mathcal{O}_2\!\uparrow\!\Sigma$-model $\mathcal{I}''$. But since $\mathcal{I} \models \mathcal{O}$ and $\mathsf{Sig}(\mathcal{O}) \subseteq \Sigma$, also $\mathcal{I}'' \models \mathcal{O}$. Hence $\mathcal{I}'' \in |\mathbf{Mod}((\mathcal{O}_2 \cup \mathcal{O})\!\uparrow\!\Sigma)|$, and obviously $\mathcal{I}|_\Sigma = \mathcal{I}''|_\Sigma$.

Robustness under signature extensions: Let $\mathcal{O}_1$ and $\mathcal{O}_2$ be ontologies Σ, Σ' be signatures with $\Sigma' \cap (\mathsf{Sig}(\mathcal{O}_1) \cup \mathsf{Sig}(\mathcal{O}_2)) \subseteq \Sigma$. Assume $\mathcal{O}_1 \equiv^m_\Sigma \mathcal{O}_2$. We need to show that $\mathcal{O}_1 \equiv^m_{\Sigma'} \mathcal{O}_2$, which amounts to showing

$$\{\mathcal{I}|_\Sigma \mid \mathcal{I} \in |\mathbf{Mod}(\mathcal{O}_1\!\uparrow\!\Sigma')|\} = \{\mathcal{I}|_\Sigma \mid \mathcal{I} \in |\mathbf{Mod}(\mathcal{O}_2\!\uparrow\!\Sigma')|\}$$

By symmetry, it suffices to prove one inclusion. Let $\mathcal{I}'_1 \in \mathbf{Mod}(\mathcal{O}_1\!\uparrow\!\Sigma')$. Since $\mathcal{O}_1 \equiv^m_\Sigma \mathcal{O}_2$, $\mathcal{I}|_\Sigma$ has an expansion to an $\mathcal{O}_1\!\uparrow\!\Sigma$-model $\mathcal{I}_2$. From $\Sigma \subseteq \Sigma'$ and $\Sigma' \cap \mathsf{Sig}(\mathcal{O}_2) \subseteq \Sigma$ we get $\Sigma' \cap \mathsf{Sig}(\mathcal{O}_2\!\uparrow\!\Sigma) = \Sigma$. Since also $\Sigma' \cup \mathsf{Sig}(\mathcal{O}_2\!\uparrow\!\Sigma) = \mathsf{Sig}(\mathcal{O}_2\!\uparrow\!\Sigma')$ the following diagram

$$
\begin{array}{ccc}
(\Sigma', \emptyset) & \hookrightarrow & \mathcal{O}_2\!\uparrow\!\Sigma' \\
\uparrow & & \uparrow \\
(\Sigma, \emptyset) & \hookrightarrow & \mathcal{O}_2\!\uparrow\!\Sigma
\end{array}
$$

is an intersection-union-pushout in **Pres**. Hence, by weak union-exactness, we can amalgamate $\mathcal{I}'_1|_{\Sigma'}$ and $\mathcal{I}_2$ to $\mathcal{I}'_2 \in \mathbf{Mod}(\mathcal{O}_2 \uparrow \Sigma')$, which gives us the desired expansion of $\mathcal{I}'_1|_{\Sigma'}$.

Robustness under joins: Consider ontologies $\mathcal{O}_1$ and $\mathcal{O}_2$ and a signature Σ such that $\mathsf{Sig}(\mathcal{O}_1 \cap \mathcal{O}_2) \subseteq \Sigma$ and $\mathcal{O}_1 \equiv^m_\Sigma \mathcal{O}_2$. Then we need to show $\mathcal{O}_1 \equiv^m_\Sigma \mathcal{O}_1 \cup \mathcal{O}_2$ and $\mathcal{O}_2 \equiv^m_\Sigma \mathcal{O}_1 \cup \mathcal{O}_2$. We only prove the former; the latter follows by symmetry. We need to show

$$\{\mathcal{I}|_\Sigma \mid \mathcal{I} \in |\mathbf{Mod}(\mathcal{O}_1\!\uparrow\!\Sigma)|\} = \{\mathcal{I}|_\Sigma \mid \mathcal{I} \in |\mathbf{Mod}((\mathcal{O}_1 \cup \mathcal{O}_2)\!\uparrow\!\Sigma)|\}$$

The inclusion from right to left is obvious. For the converse inclusion, let $\mathcal{I}_1 \in |\mathbf{Mod}(\mathcal{O}_1\!\uparrow\!\Sigma)|$. Since $\mathcal{O}_1 \equiv^m_\Sigma \mathcal{O}_2$, $\mathcal{I}_1|_\Sigma$ has an expansion $\mathcal{I}_1 \in |\mathbf{Mod}(\mathcal{O}_2\!\uparrow\!\Sigma)|$. From $\mathsf{Sig}(\mathcal{O}_1 \cap \mathcal{O}_2) \subseteq \Sigma$ we get $\mathsf{Sig}(\mathcal{O}_1\!\uparrow\!\Sigma) \cap \mathsf{Sig}(\mathcal{O}_2\!\uparrow\!\Sigma) = \Sigma$. Moreover, we have $\mathsf{Sig}(\mathcal{O}_1\!\uparrow\!\Sigma) \cup \mathsf{Sig}(\mathcal{O}_2\!\uparrow\!\Sigma) = (\mathcal{O}_1 \cup \mathcal{O}_2)\!\uparrow\!\Sigma$. This implies that

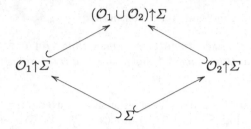

is an intersection-union-pushout in **Pres**. Hence, by weak union-exactness, we can amalgamate $\mathcal{I}_1$ and $\mathcal{I}_2$ to $\mathcal{I}'' \in \mathbf{Mod}((\mathcal{O}_1 \cup \mathcal{O}_2)\!\upharpoonright\!\Sigma)$, which gives us the desired expansion of $\mathcal{I}_1|_\Sigma$.

4   Module Notions

Equipped with the concepts introduced in the previous sections, we are now ready to introduce the notion of an ontology module. In fact, following the literature, we will put forward a number of concepts, and will study their properties and their mutual relationships. As in Sect. 3, we work in the framework of a logical system formalised as an inclusive institution $I = (\mathbf{Sign}, \mathbf{Sen}, \mathbf{Mod}, \models)$ (Sect. 2).

The notions of a module we present below may be parameterised by an arbitrary inseparability relation $\mathcal{S} = \langle \equiv^{\mathcal{S}}_\Sigma \rangle_{\Sigma \in |\mathbf{Sign}|}$.

Definition 4.1 ([10]). *Let $\mathcal{O}$ be an ontology, $\mathcal{M} \subseteq \mathcal{O}$ and Σ a signature. We call $\mathcal{M}$*

- *a (plain) Σ-module of $\mathcal{O}$ induced by $\mathcal{S}$ if $\mathcal{M} \equiv^{\mathcal{S}}_\Sigma \mathcal{O}$;*
- *a self-contained Σ-module of $\mathcal{O}$ induced by $\mathcal{S}$ if $\mathcal{M} \equiv^{\mathcal{S}}_{\Sigma \cup \mathsf{Sig}(\mathcal{M})} \mathcal{O}$;*
- *a depleting Σ-module of $\mathcal{O}$ induced by $\mathcal{S}$ if $\mathcal{O} \setminus \mathcal{M} \equiv^{\mathcal{S}}_{\Sigma \cup \mathsf{Sig}(\mathcal{M})} \emptyset$.*   □

Example 1.1 shows a plain ontology module. The intuition is that the module $\mathcal{M}$ already contains all the relevant information from $\mathcal{O}$ if attention is restricted to the concepts (symbols) in signature Σ. Note however that the module in Example 1.1 is not depleting: this follows from the fact that $\mathcal{O} \setminus \mathcal{M}$ has still some non-trivial consequences relevant w.r.t. Σ, e.g., $\mathcal{O} \setminus \mathcal{M} \models \mathsf{Male} \sqcap \exists \mathsf{has\_child}.\top \sqsubseteq \mathsf{Human}$.

The main advantage of depleting over plain modules is that minimal depleting modules exist, see Theorem 4.6 below. Therefore, DOL uses the minimal depleting module as semantics of the module extraction operator. It is unclear how one could give a definite semantics to this operator in terms of plain modules, because there may be multiple pairwise incomparable minimal plain modules.

The intuition of depleting modules is as follows: In addition to the properties of plain Σ-modules, for a depleting Σ-module $\mathcal{M}$ of $\mathcal{O}$, the difference $\mathcal{O} \setminus \mathcal{M}$ has no knowledge about $\Sigma \cup \mathsf{Sig}(\mathcal{M})$. This means that the difference of $\mathcal{O}$ and its module $\mathcal{M}$ does not entail any axioms over $\Sigma \cup \mathsf{Sig}(\mathcal{M})$ other than tautologies.

A different formulation of this observation involves the notion of safety. We say that $\mathcal{O}$ is *safe* for Σ if, for every ontology $\mathcal{O}'$ with $\mathsf{Sig}(\mathcal{O}) \cap \mathsf{Sig}(\mathcal{O}') \subseteq \Sigma$, we have that $\mathcal{O} \cup \mathcal{O}'$ is a model Σ-conservative extension of $\mathcal{O}'$. Alternatively, this notion can be formulated in terms of inseparability: an ontology $\mathcal{O}$ is safe for a signature Σ if and only if $\mathcal{O} \equiv_{\Sigma}^{m} \emptyset$.

Now if $\mathcal{M}$ is a depleting Σ-module, then $\mathcal{O} \setminus \mathcal{M}$ is safe for $\mathsf{Sig}(\mathcal{M})$ and so the module can be maintained separately outside of $\mathcal{O}$ without the risk of unintended interaction with the rest of $\mathcal{O}$. Also note that checking depleting Σ modules is exactly the same problem as deciding Σ-inseparability from the empty ontology.

In the rest of this paper we will focus on modules induced by model insepa-rability $\langle \equiv_{\Sigma}^{m} \rangle_{\Sigma \in |\mathbf{Sign}|}$, leaving similar developments for other inseparability rela-tions introduced in Sect. 3 for future study. We therefore drop all qualifications "induced by $\mathcal{S}$" in the terminology below.

Modules induced by model inseparability are essentially based on model con-servative extensions:

Proposition 4.2. *For any ontology $\mathcal{O}$, $\mathcal{M} \subseteq \mathcal{O}$ and signature Σ, $\mathcal{M}$ is a Σ-module of $\mathcal{O}$ if and only if $\mathcal{O}$ is a model Σ-conservative extension of $\mathcal{M}$.* □

We say that a subontology $\mathcal{M} \subseteq \mathcal{O}$ *covers* all the knowledge that $\mathcal{O}$ has about Σ if $\mathcal{O}$ is a consequence Σ-conservative extension of $\mathcal{M}$, that is, if for every sentence $\alpha \in \mathbf{Sen}(\Sigma)$, we have that $\mathcal{O} \models \alpha$ if and only if $\mathcal{M} \models \alpha$.

A "plain" Σ-module $\mathcal{M}$ of $\mathcal{O}$ covers all knowledge that $\mathcal{O}$ has about Σ. In fact, this claim holds also when any extension of the institution I with arbitrarily strong sentences (but the same signatures and models) is allowed.

The notion of self-contained module is stronger than the plain Σ-module notion in that it requires the module to preserve entailments that can be for-mulated in the interface signature *plus* the signature of the module. That is, it covers all the knowledge that $\mathcal{O}$ has about $\Sigma \cup \mathsf{Sig}(\mathcal{M})$. Formally, monotonicity of the model inseparability relations, see Proposition 3.2, easily implies:

Proposition 4.3. *If $\mathcal{M}$ is a self-contained Σ-module of $\mathcal{O}$, then $\mathcal{M}$ is a (plain) Σ-module of $\mathcal{O}$ as well.* □

Since $\equiv_{\Sigma}^{m}$ enjoys robustness under replacement (Theorem 3.5), we get as in [10]:

Proposition 4.4. *If $\mathcal{M}$ is a depleting Σ-module of $\mathcal{O}$, then it is a self-contained Σ-module.* □

Comparison of the various module notions can be carried out examining prop-erties relevant for ontology reuse. The robustness properties for inseparability (see Definition 3.4) can be transferred to modules as follows:

Robustness under signature restrictions. This property means that a module of an ontology w.r.t. a signature Σ is also a module of this ontology w.r.t. any subsignature of Σ. This property is important because it means that we do not need to import a different module when we restrict the set of terms that we are interested in.

Robustness under signature extensions. This means that a module of an ontology $\mathcal{O}$ w.r.t. a signature Σ is also a module of $\mathcal{O}$ w.r.t. any $\Sigma' \supseteq \Sigma$ as long as $\Sigma' \cap \mathsf{Sig}(\mathcal{O}) \subseteq \Sigma$. This means that we do not need to import a different module when extending the set of relevant terms with terms not from $\mathcal{O}$.

Robustness under replacement. This property means that if $\mathcal{M}$ is a module of $\mathcal{O}$ w.r.t. Σ, then the result of importing $\mathcal{M}$ into another ontology $\mathcal{O}'$ is a module of the result of importing $\mathcal{O}$ into $\mathcal{O}'$. Formally, for any ontology $\mathcal{O}'$, if $\mathcal{M}$ is a Σ-module of $\mathcal{O}$, then $\mathcal{O}' \cup \mathcal{M}$ is a Σ-module of $\mathcal{O}' \cup \mathcal{O}$. (The precise restrictions to signatures that are needed to ensure this property can vary.) This is called *module coverage* in the literature: importing a module does not affect its property of being a module.

Robustness under joins. It seems that this property of inseparability relations cannot be usefully transferred to ontology modules. However, together with robustness under replacement, it implies that it is not necessary to import two indistinguishable versions of the same ontology. This shows that it is still useful to have the property.

We have summarized the relevant properties of the modules of each kind in Table 1, which follow from the properties of inseparability relations stated in Sect. 3:

Theorem 4.5. *The module notions appearing as column heads in Table 1 have the properties appearing as row head, if marked with a ✓ or some additional assumptions that are needed. If marked with a ✗ , there is a counterexample showing that the property does not hold. It is assumed that all module notions are based on model inseparability.*

Indeed, the condition needed for robustness under replacement is very limited for plain modules, since the importing ontology $\mathcal{O}'$ must have a signature contained in the signature of interest Σ. This seems to be unrealistic in practice. The other module notions have a more liberal condition: $\mathsf{Sig}(\mathcal{O}') \cap \mathsf{Sig}(\mathcal{O}) \subseteq \Sigma \cup \mathsf{Sig}(\mathcal{M})$, which means that the importing ontology $\mathcal{O}'$ may overlap with the imported ontology $\mathcal{O}$ only w.r.t. the signature of interest plus that of the module. This is more realistic.

Given an ontology $\mathcal{O}$ and a signature of interest Σ, the crucial task is to determine a module $\mathcal{M}$ of $\mathcal{O}$ w.r.t. Σ. Clearly, such a module always exists: the entire ontology $\mathcal{O}$ is one example. However, what we are really interested in is *small* modules of $\mathcal{O}$ w.r.t. Σ. The following theorem establishes existence of such modules (see Theorem 72 in [10]), and so is a starting point for various methods of module extraction.

Theorem 4.6. *Let $\mathcal{O}$ be an ontology and Σ be a signature. Then there is a unique minimal depleting Σ-module of $\mathcal{O}$.* $\quad\quad\quad\quad\quad\quad\quad\quad\quad\quad$ □

Table 1. Properties of Σ-modules

Properties	Module Notions		
	plain	self-contained	depleting
inseparability	$\mathcal{O} \equiv^m_\Sigma \mathcal{M}$	$\mathcal{O} \equiv^m_{\Sigma \cup \mathsf{Sig}(\mathcal{M})} \mathcal{M}$	$\mathcal{O} \setminus \mathcal{M} \equiv^m_{\Sigma \cup \mathsf{Sig}(\mathcal{M})} \emptyset$
mCE (cCE)	✓	✓	✓
self-contained	✗	✓	✓
depleting	✗	✗	✓
robustness under signature restrictions	✓	✓	✓
robustness under signature extensions	$\Sigma' \cap \mathsf{Sig}(\mathcal{O}) \subseteq \Sigma$ plus weak union-exactness	$\Sigma' \cap \mathsf{Sig}(\mathcal{O}) \subseteq \Sigma$ plus weak union-exactness	$\Sigma' \cap \mathsf{Sig}(\mathcal{O}) \subseteq \Sigma$ plus weak union-exactness
robustness under replacement	$\mathsf{Sig}(\mathcal{O}') \subseteq \Sigma$	$\mathsf{Sig}(\mathcal{O}') \cap \mathsf{Sig}(\mathcal{O})$ $\subseteq \Sigma \cup \mathsf{Sig}(\mathcal{M})$	$\mathsf{Sig}(\mathcal{O}') \cap \mathsf{Sig}(\mathcal{O})$ $\subseteq \Sigma \cup \mathsf{Sig}(\mathcal{M})$

5 Conclusions

We have generalised the basic notions of ontology module extraction to an arbitrary institution. They can now be applied to logics other than OWL, most notably first-order logic, but also modal logics and more exotic logics. For some nice properties of modules, union-exactness of the institution is needed. While many institutions enjoy this property, some do not, e.g. CASL [24].

We have entirely neglected questions of decidability or efficient computability of modules. While Theorem 4.6 provides a general method for computing the minimum depleting module, it is based on an oracle for inseparability. Future work should hence study computationally interesting approaches to module extraction, like different versions of locality, and generalize these to an arbitrary institution as well.

Acknowledgement. We thank Thomas Schneider for discussions and feedback.

References

1. Basin, D.A., Clavel, M., Meseguer, J.: Reflective metalogical frameworks. ACM Trans. Comput. Log. **5**(3), 528–576 (2004)
2. Boronat, A., Knapp, A., Meseguer, J., Wirsing, M.: What Is a multi-modeling language? In: Corradini, A., Montanari, U. (eds.) WADT 2008. LNCS, vol. 5486, pp. 71–87. Springer, Heidelberg (2009)
3. Cerioli, M., Meseguer, J.: May I borrow your logic? (transporting logical structures along maps). Theor. Comput. Sci. **173**, 311–347 (1997)

4. Clavel, M., Meseguer, J.: Reflection in conditional rewriting logic. Theor. Comput. Sci. **285**(2), 245–288 (2002)
5. Diaconescu, R., Goguen, J.A., Stefaneas, P.: Logical support for modularisation. In: 2nd Workshop on Logical Environments, pp. 83–130. CUP, New York (1993)
6. Paul, D., Valentina, T., Luigi, I.: Ontology module extraction for ontology reuse: An ontology engineering perspective. In: Proceedings of the Sixteenth ACM Conference on Conference on Information and Knowledge Management, CIKM 2007, pp. 61–70, New York, NY, USA. ACM (2007)
7. Durán, F., Meseguer, J.: Structured theories and institutions. Theor. Comput. Sci. **309**(1–3), 357–380 (2003)
8. Goguen, J.A., Burstall, R.M.: Institutions: abstract model theory for specification and programming. J. Assoc. Comput. Mach. **39**, 95–146 (1992). (Predecessor. LNCS **164**, (221–256) (1984))
9. Goguen, J., Roşu, G.: Composing hidden information modules over inclusive institutions. In: Owe, O., Krogdahl, S., Lyche, T. (eds.) From Object-Orientation to Formal Methods. LNCS, vol. 2635, pp. 96–123. Springer, Heidelberg (2004)
10. Kontchakov, R., Wolter, F., Zakharyaschev, M.: Logic-based ontology comparison and module extraction, with an application to DL-Lite. Artif. Intell. **174**(15), 1093–1141 (2010)
11. Lucas, S., Meseguer, J.: Localized operational termination in general logics. In: De Nicola, R., Hennicker, R. (eds.) Wirsing Festschrift. LNCS, vol. 8950, pp. 91–114. Springer, Heidelberg (2015)
12. Martí-Oliet, N., Meseguer, J., Palomino, M.: Theoroidal maps as algebraic simulations. In: Fiadeiro, J.L., Mosses, P.D., Orejas, F. (eds.) WADT 2004. LNCS, vol. 3423, pp. 126–143. Springer, Heidelberg (2005)
13. Meseguer, J.: General logics. In: Logic Colloquium 87, pp. 275–329. North Holland (1989)
14. Meseguer, J.: Membership algebra as a logical framework for equational specification. In: Presicce, F.P. (ed.) WADT 1997. LNCS, vol. 1376, pp. 18–61. Springer, Heidelberg (1998)
15. Meseguer, J., Martí-Oliet, N.: From abstract data types to logical frameworks. In: Astesiano, E., Reggio, G., Tarlecki, A. (eds.) Recent Trends in Data Type Specification. LNCS, vol. 906, pp. 48–80. Springer, Heidelberg (1995)
16. Mossakowski, T., Codescu, M., Neuhaus, F., Kutz, O.: The distributed ontology, modeling, and specification language - DOL. In: Koslow, A., Buchsbaum, A. (eds.) The Road to Universal Logic, volume II of Studies in Universal Logic, pp. 498–520. Springer, Switzerland (2015)
17. Mossakowski, T., Kutz, O., Codescu, M., Lange,C.: The distributed ontology, modeling and specification language. In: Del Vescovo, C.., Hahmann, T., Pearce, D., Walther, D. (eds.) WoMo 2013, CEUR-WS Online Proceedings, vol. 1081 (2013)
18. Mossakowski, T., Kutz, O., Lange, C.: Semantics of the distributed ontology language: institutes and institutions. In: Martí-Oliet, N., Palomino, M. (eds.) WADT 2012. LNCS, vol. 7841, pp. 212–230. Springer, Heidelberg (2013)
19. Palomino, M., Meseguer, J., Martí-Oliet, N.: A categorical approach to simulations. In: Fiadeiro, J.L., Harman, N.A., Roggenbach, M., Rutten, J. (eds.) CALCO 2005. LNCS, vol. 3629, pp. 313–330. Springer, Heidelberg (2005)
20. Sannella, D., Tarlecki, A.: Specifications in an arbitrary institution. Inf. Control **76**, 165–210 (1988). (Earlier version in Proceedings, International Symposium on the Semantics of Data Types, LNCS, vol. 173. Springer (1984))

21. Sannella, D., Tarlecki, A.: Foundations of Algebraic Specification and Formal Software Development. Monographs in Theoretical Computer Science. An EATCS Series. Springer, Berlin (2012)
22. Santiago, S., Escobar, S., Meadows, C., Meseguer, J.: A formal definition of protocol indistinguishability and its verification using maude-NPA. In: Mauw, S., Jensen, C.D. (eds.) STM 2014. LNCS, vol. 8743, pp. 162–177. Springer, Heidelberg (2014)
23. Schneider, M., Rudolph, S., Sutcliffe, G.: Modeling in OWL 2 without restrictions. In: Rodriguez-Muro, M., Jupp, S., Srinivas, K. (eds.) Proceedings of the 10th International Workshop on OWL: Experiences and Directions (OWLED 2013) Co-located with 10th Extended Semantic Web Conference (ESWC 2013), Montpellier, France, 26–27 May 2013, CEUR Workshop Proceedings, vol. 1080. CEUR-WS.org (2013)
24. Schröder, L., Mossakowski, T., Tarlecki, A., Klin, B., Hoffman, P.: Amalgamation in the semantics of CASL. Theor. Comput. Sci. **331**(1), 215–247 (2005)
25. Seidenberg, J., Rector, A.L.: Web ontology segmentation: analysis, classification and use. In: Carr, L., De Roure, D., Iyengar, A., Goble, C.A., Dahlin, M. (eds.) Proceedings of the 15th international conference on World Wide Web, WWW 2006, pp. 13–22, Edinburgh, Scotland, UK, 23–26 May 2006. ACM (2006)
26. Stuckenschmidt, H., Klein, M.: Structure-based partitioning of large concept hierarchies. In: McIlraith, S.A., Plexousakis, D., van Harmelen, F. (eds.) ISWC 2004. LNCS, vol. 3298, pp. 289–303. Springer, Heidelberg (2004)
27. Stuckenschmidt, H., Parent, C., Spaccapietra, S. (eds.): Modular Ontologies: Concepts, Theories and Techniques for Knowledge Modularization. LNCS, vol. 5445. Springer, Heidelberg (2009)

Rewriting Strategies and Strategic Rewrite Programs

Hélene Kirchner[✉]

Inria, Domaine de Voluceau, BP 105,
78153 Rocquencourt, Le Chesnay Cedex, France
helene.kirchner@inria.fr

Abstract. This survey aims at providing unified definitions of strategies, strategic rewriting and strategic programs. It gives examples of main constructs and languages used to write strategies. It also explores some properties of strategic rewriting and operational semantics of strategic programs. Current research topics are identified.

1 Introduction

Since the 80s, many aspects of rewriting have been studied in automated deduction, programming languages, equational theory decidability, program or proof transformation, but also in various domains such as chemical or biological computing, plant growth modelling, security policies, etc. Facing this variety of applications, the question arises to understand rewriting in a more abstract way, especially as a logical framework to encode different logics and semantics. Discovering the universal power of rewriting, in particular through its matching and transformation power, led first to the emergence of Rewriting Logic and Rewriting Calculus.

On the other hand, with the development of rewrite frameworks and languages, more and more reasoning systems have been modeled, for proof search, program transformation, constraint solving, SAT solving. It then appeared that straightforward rule-based computations or deductions are often not sufficient to capture complex computations or proof developments. A formal mechanism is needed, for instance, to sequentialize the search for different solutions, to check context conditions, to request user input to instantiate variables, to process subgoals in a particular order, etc. This is the place where the notion of strategy comes in and this leads to the design and study of strategy constructs and strategy languages also in these contexts.

A common understanding is that rules describe local transformations and strategies describe the control of rule application. Most often, it is useful to distinguish between rules for computations, where a unique normal form (i.e. syntactic expressions which cannot be rewritten anymore) is required and where the strategy is fixed, and rules for deductions, in which case no confluence nor termination is required but an application strategy is necessary. Due to the strong correlation of rules and strategy in many applications, claiming the

© Springer International Publishing Switzerland 2015
N. Martí-Oliet et al. (Eds.): Meseguer Festschrift, LNCS 9200, pp. 380–403, 2015.
DOI: 10.1007/978-3-319-23165-5_18

universal character of rewriting also requires the formalisation of its control. This is achieved through strategic rewriting.

This survey aims at providing unified definitions of strategies, strategic rewriting and strategic programs, with the goal to show the progression of ideas and definitions of the concept, as well as their correlations. It gives examples of main constructs and languages used to write strategies, together with the definition of an operational semantics for strategic programs. Well-studied properties of strategic rewriting are reviewed and current research topics are identified.

Accordingly, following this introduction, the paper is organised as follows: after a brief history of the notion of strategy in rewriting and automated deduction in Sect. 2, we first explain in Sect. 3, what are strategic rewriting and strategic programs. In Sect. 4, several approaches to describe strategies and strategic rewriting are reviewed. In order to catch the higher-order nature of strategies, a strategy is first defined as a proof term expressed in rewriting logic then as a ρ term in rewriting calculus. Looking at a strategy as a set of paths in a derivation tree, the extensional description of strategies, defined as a subset of derivations, is briefly explored. Then a strategy is considered as a partial function that associates to a reduction-in-progress, the possible next steps in the reduction sequence. Last, positional strategies that choose where rules apply are studied. Section 5 presents a few strategy languages, and extracts comon constructs with their variants. We propose an operational semantics for strategic programs in Sect. 6, study properties of their executions together with correctness and completeness results. We then address in Sect. 7 various properties, namely termination, confluence and normalizing properties of strategic rewriting. The conclusion points out further work and possible improvements.

This survey is an extended version of [45], and of a lecture given at ISR2014. Although this research on strategies has been largely influenced by related works on proof search, automated deduction and constraint solvers, this paper does not cover these domains and restricts to the area of rewriting.

2   Historical Considerations

In programming languages, strategies have been primarily provided to describe the operational semantics of functional languages and the related notions of call by value, call by name, call by need. In languages such as Clean [62], OBJ [31], ML [6], and more recently Haskell [39] or Curry [36], strategies are used to improve efficiency of interpreters or compilers, and are not directly accessible to the programmer. In relation with the operational semantics of functional and algebraic languages, strategies have been studied from a long time in λ-calculus [9,47], in the classical setting of first-order term and graph rewriting, or in abstract rewriting systems.

In the context of functional languages, the notions of termination and confluence of reductions to provide normal forms are meaningful to ensure the existence and unicity of results. Significant research was devoted to design computable and efficient strategies that are guaranteed to find a normal form for any input term, whenever it exists. Motivated by the need to avoid useless infinite computations in functional programming languages, local strategies were used in eager languages such as Lisp (with its lazy cons), in the OBJ family of languages (OBJ, CafeOBJ, Maude,...) to guide the evaluation using local strategies for functions, in lazy functional programming, via different kinds of syntactic annotations on the program (strictness annotations, or global and local annotations). For instance, Haskell allows for syntactic annotations on the arguments of datatype constructors.

Besides functional or logic programming, strategies also frequently occur in automated deduction and reasoning systems which have been developed in a different community. Beginning with the ML meta-language of LCF [32], strategies are fundamental in several proof environments, such as Coq [20], TPS [4], PVS [61] but also in automated theorem proving [56], constraint solving [15], SAT or SMT solvers [19]. In these contexts, they are more often called tactics, action plans, search plans or priorities.

From the 1990s, attempts have been made to look at the concept of strategy per se, with the intent to confront point of views and to look at computation and deduction in a logical and uniform approach. Already in [43], the notion of computational system, defined as a rewrite theory and a strategy describing the control for rules application, was applied in a uniform way to straightforward computations with rewrite rules, to constraint solving and to combination of these paradigms in the same logical framework. Since 1997, there has been two series of workshops whose goal was to address the concept of strategy and mix different point of views. The Strategies workshops held by the CADE-IJCAR community[1] and the Workshops on Reduction Strategies held by the RTA-RDP community[2]. More recently in 2013, the International Workshop on Strategic Reasoning is emerging from the game theory community[3].

Once the idea was there, the challenge was to propose good descriptions of the concept of strategy. The approach followed in the rewriting community was to formalize a notion of strategy relying on rewriting logic [53] and rewriting calculus [17] that are powerful formalisms to express and study uniformly computations and deductions in automated deduction and reasoning systems. Briefly speaking, rules describe local transformations and strategies describe the control of rule application. Most often, it is useful to distinguish between rules for computations, where a unique normal form is required and where the strategy is fixed, and rules for deductions, in which case no confluence nor termination is required but an application strategy is necessary. Regarding rewriting as a relation and considering abstract rewrite systems leads to consider derivation tree exploration: derivations are computations and strategies describe selected computations.

With the idea to understand and unify strategies in reduction systems and deduction systems, abstract strategies are defined in [41] and in [14] as a subset

---

[1] See http://www.logic.at/strategies.

[2] http://users.dsic.upv.es/~wrs/.

[3] See http://www.strategicreasoning.net/.

of the set of all derivations (finite or not). Another point of view is to see a strategy as a partial function that, at each step of reduction, gives the possible next steps. Strategies are thus considered as a way of constraining and guiding the steps of a reduction. So at any step in a derivation, it should be possible to say which next step obeys the strategy.

In the 1990s, inspired by tactics in proof systems and by constraint programming, the idea came up to provide a *strategy language* to specify which derivations we are interested in. Various approaches have followed, yielding different strategy languages such as Elan [12,13,44], APS [46], Stratego [70,71], Tom [7] or Maude [18,54,55]. For such languages, rules are the basic strategies and additional constructs are provided to combine them and express the control.

Strategy constructs are also present in graph transformation tools such as PROGRES [67], AGG [23], Fujaba [58], GROOVE [66], GrGen [27], GP [64] and Porgy [2,24,25]. Graph rewriting strategies are especially useful in Porgy, an environment to facilitate the specification, analysis and simulation of complex systems, using port graphs. In Porgy, a complex system is represented by an initial graph, a collection of graph rewriting rules, and a user-defined strategy to control the application of rules. The Porgy strategy language includes constructs to deal with graph traversal and management of rewriting positions in the graph. Indeed in the case of graph rewriting, top-down or bottom-up traversals do not make sense. There is a need for a strategy language which includes operators to select rules and the positions where the rules are applied, and also to change the positions along the derivation.

All these languages share the concern to provide abstract ways to express control of rule applications. In these flexible and expressive strategy languages, elaborated strategies are defined by combining a small number of primitives.

3 What are Strategic Rewriting and Strategic Programs?

Strategic rewrite programs considered in this paper combine the general concept of rewriting applied to syntactic structures (like terms, graphs, propositions, states, etc.) with a strategy to express the control on rule application. In this way, strategic programming follows the separation of concerns principle [21] since different strategies can be designed and experimented with a same rewrite system. Strategic rewrite programs so contribute to improve agility and modularity in programming.

This section reminds notions of rewriting and abstract rewrite systems and introduces related definitions of strategic rewrite programs and strategic rewriting.

3.1 Rewriting

In the various domains where rewrite rules are applied, rewriting definitions have the same basic ingredients. Rewriting transforms syntactic structures that may be words, terms, propositions, dags, graphs, geometric objects like segments, and in general any kind of structured objects. In order to emphasize this fact, we use t, G or a to denote indifferently terms, graphs of any other syntactic structure, used for instance to abstractly model the state of a complex system.

Transformations are expressed with patterns called rules. Rules are built on the same syntax but with an additional set of variables, say $\mathcal{X}$, and with a binder $\Rightarrow$, relating the left-hand side and the right-hand side of the rule, and optionally with a condition or constraint that restricts the set of values allowed for the variables. Performing the transformation of a syntactic structure t is applying the rule labelled ℓ on t, which is basically done in three steps: (1) match to select a redex of t at position p denoted $t_{|p}$ (possibly modulo some axioms, constraints,...); (2) instantiate the rule variables by the result(s) of the matching homomorphism (or substitution) σ; (3) replace the redex by the instantiated right-hand side.

Formally, t rewrites to t' using the rule $\ell : l \Rightarrow r$ if $t_{|p} = \sigma(l)$ and $t' = t[\sigma(r)]_p$. This is denoted $t \longrightarrow_{p,\ell,\sigma} t'$.

The transformation process is similar on graphs (see for instance [34,65]) and many other structured objects can be encoded by terms or graphs.

When $\mathcal{R}$ is a set of rules, this transformation generates a relation $\longrightarrow_{\mathcal{R}}$ on the set of syntactic structures. Its (reflexive) transitive closure is denoted $(\overset{*}{\longrightarrow}_{\mathcal{R}})\ \overset{+}{\longrightarrow}_{\mathcal{R}}$.

Given a set of rewrite rules $\mathcal{R}$, a *derivation*, or computation from G is a sequence of rewriting steps $G \rightarrow_{\mathcal{R}} G' \rightarrow_{\mathcal{R}} G'' \rightarrow_{\mathcal{R}} \ldots$

In this transformation process, there are many possible choices: for the rule itself, the position(s) in the structure, the matching homomorphism(s). For instance, one may choose to apply a rule concurrently at all disjoint positions where it matches, or using matching modulo an equational theory like associativity-commutativity, or also according to some probability. Since in general, there is more than one way of rewriting a structure, the set of rewrite derivations can be organised as a derivation tree. The *derivation tree* of G, written $DT(G, \mathcal{R})$, is a labelled tree whose root is labelled by G, the children of which being all the derivation trees $DT(G_i, \mathcal{R})$ such that $G \rightarrow_{\mathcal{R}} G_i$. The edges of the derivation tree are labelled with the rewrite rule and the morphism used in the corresponding rewrite step. A derivation tree may be infinite, if there is an infinite reduction sequence out of G.

3.2   Strategic Rewrite Programs

Intuitively, a strategic program consists of an initial structure G (or t when it is a term), together with a set of rules $\mathcal{R}$ that will be used to reduce it, according to a given strategy expression S, used to decide which rewrite steps should be performed on G. This amounts to identify the branches in G's derivation tree that satisfy the strategy S and to view strategic rewriting derivations as selected computations.

Formally, a *strategic rewrite program* consists of a finite set of rewrite rules $\mathcal{R}$, a strategy expression S, built from $\mathcal{R}$ using a strategy language, and a given structure G.

We denote it $[S_{\mathcal{R}}, G]$, or simply $[S, G]$ when $\mathcal{R}$ is clear from the context.

Several questions come up with this definition: how to describe a strategy expression S, how to characterize strategic rewriting derivations, how to design a language for strategy expressions, how to define an operational semantics for strategic programs?

3.3   Abstract Reduction System

Dealing with a general notion of rewriting is well addressed in abstract reduction systems. There, rewriting is considered as an abstract relation on structured objects. Even if different variants of the definition of Abstract Reduction System have been given in the literature [41,42,69], they agree on the following basis. An *Abstract Reduction System (ARS)* is a labelled oriented graph $(\mathcal{O}, \mathcal{S})$ with a set of labels $\mathcal{L}$. The nodes in $\mathcal{O}$ are called *objects*. The oriented labelled edges in $\mathcal{S}$ are called *steps*: $a \xrightarrow{\phi} b$ or (a, ϕ, b), with *source* a, *target* b and *label* ϕ. Two steps $a \xrightarrow{\phi} b$ and $c \xrightarrow{\phi'} d$ can be composed if b and c are the same object. Derivations are composition of steps and may be finite or infinite.

For a given ARS $\mathcal{A}$, a *finite derivation* is denoted $\pi : a_0 \xrightarrow{\phi_1} a_1 \ldots \xrightarrow{\phi_{n-1}} a_n$ or $a_0 \xrightarrow{\pi} a_n$, where $n \in \mathbb{N}$ is the length of the derivation. The *source* of π is a_0 and its domain $Dom(\pi) = \{a_0\}$. The *target* of π is a_n and applying π to a_0 gives the singleton set $\{a_n\}$, which is denoted $\pi \bullet \{a_0\} = \{a_n\}$, or $\pi \bullet a_0 = a_n$ by abusively identifying elements and singletons. The *concatenation* of two finite derivations $\pi_1; \pi_2$ is defined as $a \xrightarrow{\pi_1} b \xrightarrow{\pi_2} c$ if $\{a\} = Dom(\pi_1)$ and $\pi_1 \bullet a = Dom(\pi_2) = \{b\}$. Then $(\pi_1; \pi_2) \bullet \{a\} = \pi_2 \bullet (\pi_1 \bullet \{a\}) = \{c\}$, or more simply $(\pi_1; \pi_2) \bullet a = \pi_2 \bullet (\pi_1 \bullet a) = c$.

Termination and confluence properties for ARS are then expressed as follows. For a given ARS $\mathcal{A} = (\mathcal{O}, \mathcal{S})$:

- An object a in $\mathcal{O}$ is *irreducible* if a is the source of no edge.
- A derivation is *normalizing* when its target is irreducible.
- An ARS is *weakly terminating* if every object a is the source of a normalizing derivation.
- $\mathcal{A}$ is *terminating* (or *strongly normalizing*) if all its derivations are of finite length.
- An ARS $\mathcal{A} = (\mathcal{O}, \mathcal{S})$ is *confluent* if

$$\text{for all objects } a, b, c \text{ in } \mathcal{O}, \text{ and all } \mathcal{A}\text{-derivations } \pi_1 \text{ and } \pi_2,$$
$$\text{when } a \xrightarrow{\pi_1} b \text{ and } a \xrightarrow{\pi_2} c,$$
$$\text{there exist } d \text{ in } \mathcal{O} \text{ and two } \mathcal{A}\text{-derivations } \pi_3, \pi_4 \text{ such that}$$
$$c \xrightarrow{\pi_3} d \text{ and } b \xrightarrow{\pi_4} d.$$

3.4   Strategic Rewriting

Given a rewrite system $\mathcal{R}$, defined on a set of objects $\mathcal{O}$ that may be terms, equivalence classes of terms, graphs, or states, we can consider the rewrite relation $\longrightarrow_{\mathcal{R}}$ defined in Sect. 3.1 to get the ARS $\mathcal{A} = (\mathcal{O}, \longrightarrow_{\mathcal{R}})$.

Based on the ARS concept, we can consider strategic rewriting in two dual ways:

– The first one emphasizes the selective purpose of strategies among the set of rewriting derivations. Abstract strategies are defined in [41] and in [14] as follows: for a given ARS $\mathcal{A}$, an *abstract strategy* ζ is a subset of the set of all derivations (finite or not) of $\mathcal{A}$. To relate this definition to the functional aspect of strategies, the notions of domain and application are then defined as follows: $Dom(\zeta) = \bigcup_{\pi \in \zeta} Dom(\pi)$ and $\zeta \bullet a = \{b \mid \exists \pi \in \zeta$ such that $a \xrightarrow{\pi} b\} = \{\pi \bullet a \mid \pi \in \zeta\}$.

– The second way emphasizes the reduction relation itself and relies on a restriction of the rewrite relation. Instead of the rewrite relation $\longrightarrow_{\mathcal{R}}$ we may consider on the same set of objects, another relation (induced by strategic steps) corresponding to strategic rewriting. A *strategic reduction step* is a relation $\xrightarrow{S}$ such that $\xrightarrow{S} \subseteq \xrightarrow{+}_{\mathcal{R}}$.

This leads to consider another ARS $\mathcal{A}' = (\mathcal{O}, \xrightarrow{S})$ and to compare it with the previous one $\mathcal{A} = (\mathcal{O}, \longrightarrow_{\mathcal{R}})$.

The derivation tree defined in Sect 3.1 is a representation of the ARS $\mathcal{A} = (\mathcal{O}, \longrightarrow_{\mathcal{R}})$. The selected branches in the derivation tree is then a representation of the ARS$=\mathcal{A}' = (\mathcal{O}, \xrightarrow{S})$.

Indeed termination, confluence and irreducible objects are in general different for the two ARS. In the term rewriting approach of strategic reduction described in [10], it is required that moreover $NF(\xrightarrow{S}) = NF(\mathcal{R})$ where NF denotes the set of terms which are not reducible any more by the considered relation. We will come back later in Sect. 7 on these properties.

But first, in the following Sect. 4, we consider different ways to describe strategies and strategic rewriting.

4 Strategy Description: Different Points of View

Different definitions of strategy have been given in the rewriting community in the last twenty years, when strategies began to be studied per se. We review them in this section, making clear that they all actually define either selected sets of rewriting derivations, or selected sets of positions where rules should be applied.

4.1 Rewriting Logic

The Rewriting Logic is due to Meseguer [53,57]: *Rewriting logic (RL) is a natural model of computation and an expressive semantic framework for concurrency, parallelism, communication, and interaction. It can be used for specifying a wide range of systems and languages in various application fields. It also has good properties as a metalogical framework for representing logics. In recent years,*

Reflexivity For any $t \in \mathcal{T}(\mathcal{F}, \mathcal{Y})$:
$$\mathbf{t} : t \to t$$

Transitivity
$$\frac{\pi_1 : t_1 \to t_2 \qquad \pi_2 : t_2 \to t_3}{\pi_1 ; \pi_2 \ : \ t_1 \to t_3}$$

Congruence For any $f \in \mathcal{F}$ with $arity(f) = n$:
$$\frac{\pi_1 : t_1 \to t'_1 \quad \cdots \quad \pi_n : t_n \to t'_n}{\mathbf{f}(\pi_1, \ldots, \pi_n) : f(t_1, \ldots, t_n) \to f(t'_1, \ldots, t'_n)}$$

Replacement For any $\ell(x_1, \ldots, x_n) : l \Rightarrow r \in \mathcal{R}$,
$$\frac{\pi_1 : t_1 \to t'_1 \quad \cdots \quad \pi_n : t_n \to t'_n}{\ell(\pi_1, \ldots, \pi_n) : l(t_1, \ldots, t_n) \to r(t'_1, \ldots, t'_n)}$$

Fig. 1. Deduction rules for rewriting logic

several languages based on RL (ASF+SDF, CafeOBJ, ELAN, Maude) have been designed and implemented.[4]

In Rewriting Logic, the syntax is based on a set of terms $\mathcal{T}(\mathcal{F}, \mathcal{Y})$ built with an alphabet $\mathcal{F}$ of function symbols with arities and with variables in $\mathcal{Y}$. A theory is given by a set $\mathcal{R}$ of labeled rewrite rules denoted $\ell(x_1, \ldots, x_n) : l \Rightarrow r$, where labels $\ell(x_1, \ldots, x_n)$ record the set of variables occurring in the rewrite rule. Formulas are sequents of the form $\pi : t \to t'$, where π is a *proof term* recording the proof of the sequent: $\mathcal{R} \vdash \pi : t \to t'$ if $\pi : t \to t'$ can be obtained by finite application of equational deduction rules [57] given in Fig. 1. In this context, a proof term π encodes a sequence of rewriting steps called a derivation.

Let us consider the following example of sorting a list of natural numbers, where natural numbers are a subsort of lists of natural numbers, which is denoted as "Nat < List"; the concatenation operateur "_ _ : List x List -> List" is associatif with the empty list "nil : -> List" as identity; operators profiles are "sort, rec, fin : List -> List"; natural numbers are denoted as "1, 2, 3,..." for simplicity and compared with the usual ordering "<". The rules are expressed as follows:

```
rules for List
   X, Y : Nat ; L L' L'' : List;
   rec :   sort (L X L' Y L'') => sort (L Y L' X L'')
              if Y < X
   fin :   sort (L) => L
end
```

For the derivation:

$$\text{sort}(3\ 1\ 2) \to \text{sort}(1\ 3\ 2) \to \text{sort}(1\ 2\ 3) \to (1\ 2\ 3)$$

---

[4] http://wrla2012.lcc.uma.es/.

the proof term is

```
rec(nil,3,nil,1,(2));rec((1),3,nil,2,nil);fin((1 2 3)).
```

The Elan language, designed in the 1990's, introduced the concept of strategy by giving explicit constructs for expressing control on the rule application [11, 43]. Beyond labeled rules and concatenation denoted ";", other constructs for choice, failure, iteration, were also defined in Elan. A strategy is there defined as a set of proof terms in rewriting logic and can be seen as a higher-order function : if the strategy ζ is a set of proof terms π, applying ζ to the term t means finding all terms t' such that $\pi : t \to t'$ with $\pi \in \zeta$. Since rewriting logic is reflective, strategy semantics can be defined inside the rewriting logic by rewrite rules at the meta-level. This is the approach followed by Maude in [54, 55].

4.2 Rewriting Calculus

The rewriting calculus, also called ρ-calculus, has been introduced in 1998 by Horatiu Cirstea and Claude Kirchner [17]. *The rho-calculus has been introduced as a general means to uniformly integrate rewriting and λ-calculus. This calculus makes explicit and first-class all of its components: matching (possibly modulo given theories), abstraction, application and substitutions.*

The rho-calculus is designed and used for logical and semantical purposes. It could be used with powerful type systems and for expressing the semantics of rule based as well as object oriented paradigms. It allows one to naturally express exceptions and imperative features as well as expressing elaborated rewriting strategies.[5]

Some features of the rewriting calculus are worth emphasizing here: first-order terms and λ-terms are ρ-terms $(\lambda x.t$ is $(x \Rightarrow t))$; a rule is a ρ-term as well as a strategy, so rules and strategies are abstractions of the same nature and "first-class concepts"; application reduction generalizes β−reduction; composition of strategies is function composition and is denoted explicitly here by the operator •; recursion can be for example expressed as in λ calculus with a recursion operator μ.

To illustrate the notion of ρ-term on a simple example, let us come back to the list sorting algorithm. For the derivation:

```
sort (3 1 2) -> sort (1 3 2) -> sort (1 2 3) -> (1 2 3)
```

the corresponding ρ-term can be written :

$$fin \bullet rec_2 \bullet rec_1 \bullet sort(312)$$

with $fin=(sort(L_3) \Rightarrow L_3)$, $rec_2=(sort(L_2 X_2 L'_2 Y_2 L''_2) \Rightarrow sort(L_2 Y_2 L'_2 X_2 L''_2))$ and $rec_1 = (sort(L_1 X_1 L'_1 Y L''_1) \Rightarrow sort(L_1 Y_1 L'_1 X_1 L''_1))$.

In the ρ-calculus, strategies expressed by a well-typed ρ-term of type $term \mapsto term$ evaluates to a set of rewrite derivations [16].

---

[5] http://rho.loria.fr/index.html.

The Abstract Biochemical Calculus (or ρ_{Bio}-calculus) [3] illustrates a useful instance of the ρ-calculus. The ρ_{Bio}-calculus models autonomous systems as *biochemical programs* which consist of the following components: collections of molecules (objects and rewrite rules), higher-order rewrite rules over molecules (that may introduce new rewrite rules in the behaviour of the system) and strategies for modelling the system's evolution. A visual representation via *port graphs* and an implementation are provided by the Porgy environment described in [2]. In this calculus, strategies are abstract molecules, expressed with an arrow constructor ($\Rightarrow$ for rule abstraction), an application operator • and a constant operator stk (for *stuck*) for explicit failure.

4.3   Extensional Strategies

The *extensional* definition of abstract strategies as a set of derivations of an abstract reduction system is given in [14]. The concept is useful to understand and unify reduction systems and deduction systems as explored in [41].

The extensional approach is also useful to address infinite elements. Since abstract reduction systems may involve infinite sets of objects, of reduction steps and of derivations, we can schematize them with constraints at different levels: (i) to describe the objects occurring in a derivation (ii) to describe, via the labels, requirements on the steps of reductions (iii) to describe the structure of the derivation itself (iv) to express requirements on the histories. The framework developed in [42] defines a strategy ζ as all instances $\sigma(D)$ of a derivation schema D such that σ is solution of a constraint C involving derivation variables, object variables and label variables. As a simple example, the infinite set of derivations of length one that transform a into $f(a^n)$ for all $n \in \mathbb{N}$, where $a^n = a * \ldots * a$ (n times), is simply described by: $(a \rightarrow f(X) \mid X * a =_A a * X)$, where $=_A$ indicates that the constraint is solved modulo associativity of the operator $*$.

4.4   Intensional Strategies

Extensional strategies do not capture the idea that a strategy is a partial function that associates to each step in a reduction sequence, the possible next steps. Here, the strategy as a function may depend on the object and the derivation so far. This notion of strategy coincides with the definition of strategy in sequential path-building games, with applications to planning, verification and synthesis of concurrent systems [22]. This remark leads to the following *intensional* definition given in [14]. Again, the essence of the definition is that strategies are considered as a way of constraining and guiding the steps of a reduction. So at any step in a derivation, it should be possible to say which is the next step that obeys the strategy ζ. In order to take into account the past derivation steps to decide the next possible ones, the history of a derivation has to be memorized and available at each step. Through the notion of traced-object $[\alpha]\, a = [(a_0, \phi_0), \ldots, (a_n, \phi_n)]\, a$ in $\mathcal{O}^{[\mathcal{A}]}$, each object a memorizes how it has been reached with the trace α.

An *intensional strategy* for $\mathcal{A} = (\mathcal{O}, \mathcal{S})$ is a partial function λ from $\mathcal{O}^{[\mathcal{A}]}$ to $2^{\mathcal{S}}$ such that for every traced object $[\alpha]\, a$, $\lambda([\alpha]\, a) \subseteq \{\pi \in \mathcal{S} \mid Dom(\pi) = a\}$. If $\lambda([\alpha]\, a)$ is a singleton, then the reduction step under λ is deterministic.

As described in [14], an intensional strategy λ naturally generates an abstract strategy, called its *extension*: this is the abstract strategy ζ_λ consisting of the following set of derivations:

$$\forall n \in \mathbb{N}, \ \pi : a_0 \xrightarrow{\phi_0} a_1 \xrightarrow{\phi_1} a_2 \dots \xrightarrow{\phi_{n-1}} a_n \ \in \zeta_\lambda$$

$$\text{iff } \forall j \in [0, n-1], \quad (a_j \xrightarrow{\phi_j} a_{j+1}) \in \lambda([\alpha]\, a_j).$$

This extension may obviously contain infinite derivations; in such a case it also contains all the finite derivations that are prefixes of the infinite ones, and so is closed under taking prefixes.

A special case are memoryless strategies, where the function λ does not depend on the history of the objects. This is the case of many strategies used in rewriting systems, as shown in the next example. Let us consider an abstract rewrite system $\mathcal{A}$ where objects are terms, reduction is term rewriting and labels are positions where the rewrite rules are applied. Let us consider an order $<$ on the labels which is the prefix order on positions. Then the intensional strategy that corresponds to innermost rewriting is $\lambda_{inn}(t) = \{\pi : t \xrightarrow{p} t' \mid p = max(\{p' \mid t \xrightarrow{p'} t' \in \mathcal{S}\})\}$. When a lexicographic order is used, the classical *rightmost-innermost* strategy is obtained.

Another example, to illustrate the interest of traced objects, is the intensional strategy that restricts the derivations to be of bounded length k. Its definition makes use of the size of the trace α, denoted $|\alpha|$: $\lambda_{ltk}([\alpha]\, a) = \{\pi \mid \pi \in \mathcal{S}, \ Dom(\pi) = a, \ |\alpha| < k - 1\}$. However, as noticed in [14], the fact that intensional strategies generate only prefix closed abstract strategies prevents us from computing abstract strategies that look straightforward: there is no intensional strategy that can generate a set of derivations of length exactly k. Other solutions are provided in [14].

4.5   Positional Strategies

In order to build the function that gives the next possible steps in a reduction sequence, mechanisms to choose the positions in the syntactic structure where a rule or a set of rules can be applied. This can be done in two different ways: either by traversing the syntactic structure, or by using annotations to select a set of positions.

In term rewriting, the first way is illustrated by leftmost-innermost (resp. outermost) reduction strategies on terms that choose the rewriting position according to suffix (resp. prefix) ordering on the set of positions in the term. The second way inspired from OBJ, uses local annotations. Informally, a *strategy annotation* is a list of argument positions and rule names [1,26]. The argument positions indicate the next argument to evaluate and the rule names indicate rules to apply. For instance, the leftmost-innermost strategy for a function symbol C

corresponds to an annotation $strat(C) = [1, 2, .., k, R_1, R_2, ...R_n]$ that indicates that all its arguments should be evaluated from left to right and that the rules R_i should be tried. This is also called *on-demand rewriting*. Note that including (labels of) rules is not allowed in such strategy annotations. It is, however, allowed in the so-called *just-in-time* strategies developed in [68].

Context-sensitive rewriting is a rewriting restriction which can be associated to every term rewriting system [48]. Given a signature $\mathcal{F}$, a mapping $\mu : \mathcal{F} \mapsto \mathcal{P}(N)$, called the *replacement map*, discriminates some argument positions $\mu(f) \subseteq \{1, ..., k\}$ for each k-ary symbol f. Given a function call $f(t_1, ..., t_k)$, the replacements are allowed on arguments t_i such that $i \in \mu(f)$ and are forbidden for the other argument positions. Examples are given in [48,50].

A different approach is followed on graphs. Motivated by the need to apply rules on huge graphs, Porgy [24] introduces annotations to focus on or to avoid part of the graph. A *located graph* G_P^Q consists of a port graph G and two distinguished subgraphs P and Q of G, called respectively the *position subgraph*, or simply *position*, and the *banned subgraph*. In a located graph G_P^Q, P represents the subgraph of G where rewriting steps may take place (i.e., P is the focus of the rewriting) and Q represents the subgraph of G where rewriting steps are forbidden. The intuition is that subgraphs of G that overlap with P may be rewritten, if they are outside Q. The subgraph P generalises the notion of rewrite position in a term: if G is the tree representation of a term t then we recover the usual notion of rewrite position p in t by setting P to be the node at position p in the tree G, and Q to be the part of the tree above P (to force the rewriting step to apply from P downwards). When applying a port graph rewrite rule, not only the underlying graph G but also the position and banned subgraphs may change. A *located rewrite rule* specifies two disjoint subgraphs M and N of the right-hand side r that are used to update the position and banned subgraphs, respectively. If M (resp. N) is not specified, r (resp. the empty graph) is used as default. In general, for a given located rule and located graph G_P^Q, several rewriting steps at P avoiding Q might be possible. Thus, the application of the rule at P avoiding Q produces a *set of located graphs*.

The precise definitions and details are given in [25]. Such definitions of forbidden positions are quite useful to formalize deduction process that for instance prevents rewriting in the parts brought by instantiating rules variables, or needs to always apply at some interface nodes.

5   Strategy Languages

A *strategy language* gives syntactic means to describe strategies. Various *strategy languages* have been proposed by different teams, giving rise to different families. Five of them, representative of these families, are reviewed in this section: Elan [13] puts emphasis on rules and strategies as a paradigm to combine deduction and computation by rewriting[6], and its successor Tom [7] is strongly based

---

[6] http://elan.loria.fr/elan.html.

on the ρ-calculus[7]. Stratego [70] is a successor of ASF+SDF, mainly dedicated to program transformation[8]. Maude [55] inherits from the OBJ family, order-sorted equational rewriting and strategic annotations of operators, and is strongly based on rewriting logic[9]. Porgy [25] took inspiration, partly from the aforementioned languages and also from graph transformation languages, and puts emphasis on strategies which can be useful for modeling and analysing big graphs[10].

Language design is largely a matter of choice and the idea here is not to give a catalogue of constructs present in these languages, but rather extract from them some common features and understand how they address the two main purposes of strategies: on one hand, build derivation steps and derivations; on the other hand, operationaly compute the next strategic reduction steps.

Let us classify the constructs to see which ones are commonly agreed and which ones are specific to one language. Remind that t or G denotes a syntactic expression (term, graph,...) and S is a strategy expression in a strategy language on a rewrite rule system $\mathcal{R}$ with rules $R_1, \ldots, R_n$. Application of S to G is denoted $S \cdot G$.

Elementary strategies are the basis of all languages. The most basic strategy is a labelled rule $\ell : l \Rightarrow r$ ($\ell \triangleq l \Rightarrow r$). id and fail are two strategies that respectively denote success and failure. They can be encoded either as constant or as rules id $\triangleq X \Rightarrow X$ and fail $\triangleq X \Rightarrow$ stk where stk denotes a special constant denoting failure.

However, even for a single rule, rewriting can be performed in various ways, according to redexes or homomorphisms. There are mainly two options there: all(R) denotes all possible applications of the transformation R on the current object, creating a new one for each application. In the derivation tree, this creates as many children as there are possible applications. Instead one(R) chooses only one of the possible applications of the transformation and ignores the others; again there are some variations here, in the way to choose, either by taking the first found application, of by making a random choice between all the possible applications, with equal probabilities.

Note however that the all and one constructs are not available in all strategy languages and are sometimes implicit.

Building derivations is always present under different syntaxes. Composition of two strategies S_1 and S_2 is primarily done by sequential application of S_1 followed by S_2. It is denoted Sequence(S_1, S_2) or seq(S_1, S_2) or S_1 Then S_2 or S_1 ; S_2.

Selection of branches in the derivation tree is obviously needed and present in all languages although with different syntaxes: first(S_1, S_2), (S_1)orelse(S_2)

---

[7] https://gforge.inria.fr/projects/tom/.
[8] http://strategoxt.org/Stratego/WebHome.
[9] http://maude.cs.uiuc.edu/.
[10] http://tulip.labri.fr/TulipDrupal/?q=porgy.

or $S_1 <^+ S_2$ selects the first strategy that does not fail; it fails if both fail. As a variant, $\text{try}(S)$ tries the strategy S but never fails and $\text{try}(S) \triangleq \text{first}(S, \text{id})$.

While first selects the strategy according to the order of its arguments, in the Elan language, the don't care construct $\text{dc}(R_1, \ldots, R_n)$ randomly chooses one of the rules for application. In its implementation however, the first rule that is applicable is chosen and the dc construct is actually a first.

Probabilistic choice is provided in Porgy. When probabilities $p_1, \ldots, p_n \in [0, 1]$ are associated to strategies $S_1, \ldots, S_n$ such that $p_1 + \ldots + p_n = 1$, the construct $\text{ppick}(S_1, p_1, \ldots, S_n, p_n)$ picks one of the strategies for application, according to the given probabilities.

Conditionals and Tests again are present in all languages but with some variations. $\text{if}(S)\text{then}(S')\text{else}(S'')$ checks if application of S is successful (i.e. returns id), in which case S' is applied, otherwise S'' is applied. In Elan, Tom, Stratego and Porgy, in case the application of S to the current object G succeeds, S' is applied to S, while in Maude, S' is applied to $S \cdot G$. Maude also provides the construct $\text{match}(S)$ that matches the term S to G and returns G if success or fail otherwise. As a derived operator, $\text{not}(S) \triangleq \text{if}(S)\text{then}(\text{fail})\text{else}(\text{id})$ fails if S succeeds and succeeds if S fails.

Recursive strategies and iterations are essential due to the functional aspect of strategies. Expressed in Tom with a fixpoint operator $\mu x.S = S[x \leftarrow \mu x.S]$, $\text{repeat}(S)$ keeps on sequentially applying S until it fails and returns the last result: $\text{repeat}(S) = \mu x.\text{first}(\text{Sequence}(S, x), \text{id})$. As a variant, $\text{while}(S)\text{do}(S')$ keeps on sequentially applying S' while the expression S is successful; if S fails, then id is returned.

Stratego [70] instead introduces recursive closure strategies. The recursive closure $recx(S)$ of the strategy S attempts to apply S to the entire subject term and the strategy $recx(S)$ to each occurrence of the variable x in S. Iterators are provided based on this construction.

$$
\begin{aligned}
try(S) &= S <^+ id \\
repeat(S) &= recx(try(S; x)) \\
while(c, S) &= recx(try(c; S; x)) \\
do - while(S, c) &= recx(S; try(c; x)) \\
while - not(c, S) &= recx(c <^+ S; x) \\
for(i, c, S) &= i; while - not(c, S)
\end{aligned}
$$

Exploiting the structure of objects. Traversal strategies are useful to traverse structures, be terms or graphs. They are based on local neighbourhood exploration and iteration.

- On a term $t = f(t_1, ..., t_n)$, $\text{AllSuc}(S)$ applies the strategy S on all immediate subterms: $\text{AllSuc}(S) \cdot f(t_1, ..., t_n) = f(t'_1, ..., t'_n)$ if $S \cdot t_1 = t'_1, ..., S \cdot t_n = t'_n$; it fails if there exists i such that $S \cdot t_i$ fails. $\text{OneSuc}(S)$ applies the strategy S on the first immediate subterm (if it exists) where S does not fail:

$\mathtt{OneSuc}(S) \bullet f(t_1, ..., t_n) = f(t_1, ..., t_i', ..., t_n)$ if for all $1 \leq j < i$, $S \bullet t_j$ fails, and $S \bullet t_i = t_i'$; it fails if f is a constant or if for all i, $S \bullet t_i$ fails.

- On a graph G, $\mathtt{AllNbg}(S)$ applies the strategy S on all immediate successors of the nodes in G, where an immediate successor of a node v is a node connected to v. $\mathtt{OneNbg}(S)$ applies the strategy S on one immediate successor of a node in G, randomly chosen.

Traversal strategies are expressed in Tom with the following fixpoint equations:

$$\begin{aligned}
OnceBottomUp(S) &= \mu x.First(OneSuc(x), S) \\
BottomUp(S) &= \mu x.Sequence(AllSuc(x), S) \\
TopDown(S) &= \mu x.Sequence(S, AllSuc(x)) \\
Innermost(S) &= \mu x.Sequence(AllSuc(x), Try(Sequence(S, x)))
\end{aligned}$$

Focusing strategies. Instead of traversing the structure through a systematic exploration, one may want to focus on or to avoid on sub-structures. Strategy annotations may be seen as precursors of this idea. Porgy allows combining applications of rewrite rules and position updates, using *focusing expressions*. The direct management of positions in strategy expressions, via the distinguished subgraphs P and Q in the target graph and the distinguished graphs M and N in a located port graph rewrite rule are original features of the language. The grammar generates expressions that are used to define positions for rewriting in a graph, or to define positions where rewriting is not allowed. They denote functions used in strategy expressions to change the positions P and Q in the current located graph (e.g. to specify graph traversals). The constructs $\mathtt{CrtGraph}$ (current graph), $\mathtt{CrtPos}$ (current positions) and $\mathtt{CrtBan}$ (current banned positions), applied to a located graph G_P^Q, return respectively the graphs G, P and Q. To generate traversal strategies on graphs, Porgy uses neighbourhood constructs $\mathtt{Nbg}()$ that returns the neighbours of a set of nodes possibly satisfying some user-defined properties.

6   Operational Semantics of Strategic Programs

There are several ways to describe the operational semantics of a programming language. Due to the fact that rewriting logic is reflexive, it is tempting to describe the operational semantics of a strategy language with a set of rewrite rules. This has been done for instance for Elan [11], Maude [18] and Porgy [2] at least. We sketch below another way by defining a transition relation on configurations using semantic rules in the SOS style of [63].

Let us consider a strategic rewrite program consisting of a finite set of rewrite rules $\mathcal{R}$, a strategy expression S (built from $\mathcal{R}$ using a strategy language $\mathcal{L}(\mathcal{R})$) and a given structure G. The intuition behind a strategic program is to use the strategy expression S to decide which rewrite steps should be performed on G. As already said, in general, there may be more than one way of rewriting a structure according to S. In order to keep track of the various rewriting alternatives, we introduce the notion of a *configuration* as a multiset of strategic rewrite programs. A *configuration* C is a multiset $\{O_1, \ldots, O_n\}$ where each O_i is a strategic program $[S_i, G_i]$. The *initial configuration* is $\{[S, G]\}$.

The transition relation $\longmapsto$ is a binary relation on configurations defined as follows:

$$\{O_1, \ldots, O_k, \ldots, O_n\} \longmapsto \{O_1, \ldots, O'_{k_1}, \ldots, O'_{k_m}, \ldots, O_n\}$$

if $O_k \mapsto \{O'_{k_1}, \ldots, O'_{k_m}\}$, for $1 \leq k \leq n$. The transition relation $\mapsto$ is defined through semantic rules. For instance, a few semantic rules are given in Fig. 2 coming from the Porgy operational semantics. More are given in [25].

Given a configuration $\{O_1, \ldots, O_k, \ldots, O_n\}$, there may be several strategic programs O_k where a $\mapsto$-step can be applied, so there is also a $\longmapsto$-derivation tree whose nodes are configurations. Intuitively these configurations provide another view of the derivation tree of the strategic program, or equivalently of the ARS of the relation $\xrightarrow{S}$, with root G. One can recover it by projecting a strategic program $O = [S, G]$ on its second component G and by associating to a $\mapsto$-

$$\frac{G' \in LS_{l \Rightarrow r}(G)}{[\mathtt{one}(l \Rightarrow r), G] \mapsto \{[\mathtt{id}, G']\}}$$

where LS is the legal set of reducts

$$\frac{LS_{l \Rightarrow r}(G) = \emptyset}{[\mathtt{one}(l \Rightarrow r), G] \mapsto \{[\mathtt{fail}, G]\}}$$

$$\frac{\{G_1, \ldots, G_k\} = LS_{l \Rightarrow r}(G)}{[\mathtt{all}(l \Rightarrow r), G] \mapsto \{[\mathtt{id}, G_1], \ldots, [\mathtt{id}, G_k]\}}$$

$$\frac{LS_{l \Rightarrow r}(G) = \emptyset}{[\mathtt{all}(l \Rightarrow r), G] \mapsto \{[\mathtt{fail}, G]\}}$$

$$[\mathtt{id}; S, G] \mapsto \{[S, G]\}$$

$$[\mathtt{fail}; S, G] \mapsto \{[\mathtt{fail}, G]\}$$

$$\frac{[S_1, G] \mapsto \{[S_1^1, G_1], \ldots, [S_1^k, G_k]\}}{[S_1; S_2, G] \mapsto \{[S_1^1; S_2, G_1], \ldots, [S_1^k; S_2, G_k]\}}$$

$$\frac{\exists G', M \text{ s.t. } \{[S_1, G]\} \mapsto^* \{[\mathtt{id}, G'], M\}}{[\mathtt{if}(S_1)\mathtt{then}(S_2)\mathtt{else}(S_3), G] \mapsto \{[S_2, G]\}}$$

$$\frac{\nexists G', M \text{ s.t. } \{[S_1, G]\} \mapsto^* \{[\mathtt{id}, G'], M\}}{[\mathtt{if}(S_1)\mathtt{then}(S_2)\mathtt{else}(S_3), G] \mapsto \{[S_3, G]\}}$$

$$[\mathtt{while}(S_1)\mathtt{do}(S_2), G] \mapsto$$
$$\{[\mathtt{if}(S_1)\mathtt{then}(S_2; \mathtt{while}(S_1)\mathtt{do}(S_2))\mathtt{else}(\mathtt{id}), G]\}$$

Fig. 2. Examples of semantic rules for strategy language

step $O_k \mapsto \{O'_{k_1}, \ldots, O'_{k_m}\}$, for $1 \leq k \leq n$, a set of m strategic reduction steps $G_k \xrightarrow{S} G'_{k_i}$ for $1 \leq i \leq m$.

For a given configuration $C = \{O_1, \ldots, O_k, \ldots, O_n\}$, where each O_i is a strategic program $[S_i, G_i]$, let $Reach(C) = \{G_1, \ldots, G_k, \ldots, G_n\}$ be the set of associated reachable structures. For a derivation $T = C_1 \longmapsto \ldots \longmapsto C_n$ let $Reach(T) = \bigcup_{1 \leq k \leq n} Reach(C_k)$ be the set of associated reachable structures.

As presented in [55], it is expected from a strategy language to satisfy the properties of correctness and completeness w.r.t. rewriting derivations.

Correctness: If T is the derivation $C_0 = \{[S, G]\} \longmapsto \ldots \longmapsto C_k = \{\ldots[S'_k, G'_k]\ldots\}$ and if $G' \in Reach(T)$, then $G \rightarrow^*_{\mathcal{R}} G'$.

Completeness: If $G \rightarrow^*_{\mathcal{R}} G'$, there exists $S \in \mathcal{L}(\mathcal{R})$ and a derivation T of the form $C_0 = \{[S, G]\} \longmapsto \ldots \longmapsto C_k = \{\ldots[S'_k, G'_k]\ldots\}$ such that $G' \in Reach(T)$.

Special strategic programs called *results* in [25], are those of the form $[\mathtt{id}, G']$ or $[\mathtt{fail}, G']$. For a given configuration $C = \{O_1, \ldots, O_k, \ldots, O_n\}$, where each O_i is a strategic program $[S_i, G_i]$, let $Results(C)$ (respectively $Results(T)$) be the subset of $Reach(C)$ (respectively $Reach(T)$) that are results. The result set associated to a configuration or a derivation can be empty, which can be the case for non-terminating programs.

A configuration is *terminal* if no transition can be performed. A meaningful property to prove is that all terminal configurations consist of *results* of the form $[\mathtt{id}, G']$ or $[\mathtt{fail}, G']$. This is expressed through the following *Progress property*:

Characterisation of Terminal Configurations. For every strategic rewrite program $[S, G]$ that is not a result (i.e., $S \neq \mathtt{id}$ and $S \neq \mathtt{fail}$), there exists a configuration C such that $\{[S, G]\} \mapsto C$. In other words, in this case, there are no blocked programs: the transition system ensures that, for any configuration, either there are transitions to perform, or we have reached results.

Strategic programs are not terminating in general, however it may be suitable to identify a terminating sublanguage (i.e. a sublanguage for which the transition relation is terminating). For instance, it is not difficult (but not surprising) to prove that in Porgy, the sublanguage that excludes iterators (such as the while/repeat construct) is strongly terminating.

Last, with respect to the computation power of the language, it is easy to state, as in [35], the Turing completeness property.

Computational Completeness property: The set of all strategic programs $[S_R, G]$ is Turing complete, i.e. can simulate any Turing machine. (Sequential composition and iteration are enough) [35].

7 Properties of Strategic Rewriting

Since strategic rewriting restricts the set of rewriting derivations, it needs careful definitions of termination and confluence under strategies, explored in [41, 42].

These properties of confluence or termination for rewriting under strategies have been largely addressed in the rewriting community for specific term rewriting strategies. Different approaches have been explored, either based on schematization of derivation trees, as in [30], or by tuning proof methods to handle specific strategies (innermost, outermost, lazy strategies) as in [28,29]. Termination of on-demand rewriting in the context of OBJ programs is studied in [1,49,50]. Other approaches as [8] use strategies transformation to equivalent rewrite systems to be able to reuse well-known methods.

When the concept of normal form is important, like in the context of term rewriting systems (TRS for short) where rewriting strategies look for efficient ways to compute normal forms, a relevant question is: which (computable) strategies are guaranteed to find a normal form for any term whenever it exists? Having in mind that, given a set of rules $\mathcal{R}$, a strategic term rewriting reduction *normalizes the term t* if there is no infinite $\xrightarrow{S}$-rewrite sequence starting from t, a strategic rewriting reduction is *normalizing* or *complete* if it normalizes every term that has an $\mathcal{R}$-normal form. Proving completeness of strategic rewriting w.r.t. normal forms is actually a difficult problem and results have been most often obtained in the context of orthogonal systems (i.e. with left-linear non-overlapping left-hand sides). Innermost and outermost reduction are studied for instance in [10,59] where it is shown that the leftmost outermost strategy is normalizing for orthogonal left-normal TRS, but not in general [37,38]. Innermost strategy is complete for terminating TRS and some other restricted class as explored in [60].

Special efforts have been devoted to needed reductions. Needed reduction is interesting for orthogonal term rewriting systems occurring in combinatory logic, λ-calculus, functional programming. Already in 1979, later published in [38], Huet and Lévy defined the notions of needed and strongly needed redexes for orthogonal rewrite systems. The main idea here is to find the optimal way, when it exists, to reach the normal form of a term. A redex is needed when there is no way to avoid reducing it to reach the normal form. Reducing only needed redexes is clearly the optimal reduction strategy, as soon as needed redexes can be decided, which is not the case in general. In an orthogonal TRS, every reducible term contains a needed redex and repeated contraction of needed redexes results in a normal form, if it exists. Unfortunately neededness of a redex is not decidable [69] except for some classes of rewrite systems: in *sequential* TRS [5], every term which is not in normal form contains a needed redex [10]. Strong sequentiality is decidable for left-linear TRS. External redexes (outermost until contracted) are needed. But outermost redexes may fail to be needed if the TRS is not orthogonal. For instance, with $\mathcal{R} = \{f(a) \Rightarrow b,\ a \Rightarrow b\}$, the term $f(a)$ contains two redexes, but the outermost one is not needed: the rewriting step $f(a) \longrightarrow_{\mathcal{R}} f(b)$ normalizes the term without contracting the outermost redex[11]. Again combinatory logic and λ-calculus satisfy these conditions and have motivated their study.

Sufficient conditions to ensure that context-sensitive rewriting is able to compute head-normal forms (terms that do not rewrite into a redex) have been established in [48]. In fact, for a given TRS, it is possible to automatically provide replacement maps supporting such computations. In this setting, the *canonical replacement map* (denoted by μ_{can}) specifies the most restrictive replacement

---

[11] Remark due to an external referee.

map which can be automatically associated to a TRS $\mathcal{R}$ in order to achieve completeness of context-sensitive computations, whenever the TRS is left-linear. So left-linear, confluent, and μ_{can}-terminating TRS admit a computable normalizing strategy to head-normal forms.

8   Conclusion and Further Work

A lot of questions about strategies are yet open, going from the definition of this concept and the interesting properties we may expect to prove, up to the definition of domain specific strategy languages. As further research topics, several directions seem really worth exploring. The first one is the connection with game theory strategies. In the fields of system design and verification, *games* have emerged as a key tool. Such games have been studied since the first half of 20th century in descriptive set theory [40], and they have been adapted and generalized for applications in formal verification; introductions can be found in [33,72]. The coincidence of the term "strategy" in the domains of rewriting and games is more than a pun. It should be fruitful to explore further the connection and to be guided in the study of strategies by some of the insights in the literature of games.

The second research direction is related to proving properties of strategies and strategic reductions. A lot of work has already begun in the rewriting community and have been presented in journals, workshops or conferences of this domain. Properties of confluence, termination, or completeness for rewriting under strategies have been largely addressed. However, as mentioned in Sect. 3.1, the application of rules to the considered objects can optionally be restricted by conditions or constraints, and this generalization has to be carefully studied. When conditional rules are allowed, a number of concepts and computational properties that are mentioned here may crucially depend on the conditional part of the rules. For instance, regarding termination, the notion of operational termination (defined as the absence of infinite proof trees), studied in [51] for conditional term rewriting (CTRS) systems, is different from the notion of termination considered here (the absence of infinite reduction sequences). As another example, a discussion about how irreducible terms and normal forms are also different for CTRSs can be found in [52]. Taking into account these phenomena could provide more insights on strategies.

In addition, other properties of strategies such as fairness or loop-freeness could be worthfully explored, again by making connections between different communities (functional programming, proof theory, verification, game theory,...).

Acknowledgements. The results presented here are based on pioneer work in the Elan language designed in the Protheo team from 1990 to 2002. They rely on joint work with many people, in particular Marian Vittek and Peter Borovanský, Claude Kirchner and Florent Kirchner, Dan Dougherty, Horatiu Cirstea and Tony Bourdier, Oana Andrei, Maribel Fernandez and Olivier Namet. I am grateful to the members of the Protheo and the Porgy teams, for many inspiring discussions on the topics of this paper. I sincerely thank the reviewers for their careful reading and pertinent and constructive remarks. A special tribute is given to José Meseguer, for his inspiring works on rewriting logic and strategies, and this paper is dedicated to him.

References

1. Alpuente, M., Escobar, S., Lucas, S.: Correct and complete (positive) strategy annotations for OBJ. In: Proceedings of the 5th International Workshop on Rewriting Logic and its Applications (WRLA) , vol. 71, Elecronic Notes in Theoretical Computer Science, pp. 70–89 (2004)
2. Andrei, O., Fernandez, M., Kirchner, H., Melançon, G., Namet, O., Pinaud, B.: PORGY: strategy-driven interactive transformation of graphs. In: Echahed, R. (ed.) TERMGRAPH, 6th International Workshop on Computing with Terms and Graphs, vol. 48, Electronic Proceedings in Theoretical Computer Science (EPTCS), pp. 54–68 (2011)
3. Andrei, O., Kirchner, H.: A port graph calculus for autonomic computing and invariant verification. Electron. Notes Theor. Comput. Sci. 253(4), 17–38 (2009)
4. Andrews, P.B., Brown, C.E.: TPS: a hybrid automatic-interactive system for developing proofs. J. Appl. Logic 4(4), 367–395 (2006)
5. Antoy, S., Middeldorp, A.: A sequential reduction strategy. Theor. Comput. Sci. 165(1), 75–95 (1996)
6. Augustsson, L.: A compiler for lazy ML. In: Proceedings of the 1984 ACM Symposium on LISP and Functional Programming, LFP 1984, pp. 218–227, New York, NY, USA. ACM (1984)
7. Balland, E., Brauner, P., Kopetz, R., Moreau, P.-E., Reilles, A.: Tom: piggybacking rewriting on java. In: Baader, F. (ed.) RTA 2007. LNCS, vol. 4533, pp. 36–47. Springer, Heidelberg (2007)
8. Balland, E., Moreau, P.-E., Reilles, A.: Effective strategic programming for java developers. Softw. Pract. Exp. 44(2), 129–162 (2012)
9. Barendregt, H.: The Lambda-calculus, its syntax and semantics. Studies in Logic and the Foundation of Mathematics, Second edition. Elsevier Science Publishers B. V. (North-Holland), Amsterdam (1984)
10. Bezem, M., Klop, J.W., de Vrijer, R. (eds.): Term Rewriting Systems. Cambridge Tracts in Theoretical Computer Science. Cambridge University Press, Cambridge (2003)
11. Borovanský, P., Kirchner, C., Kirchner, H., Moreau, P.-E.: ELAN from a rewriting logic point of view. Theor. Comput. Sci. 285(2), 155–185 (2002)
12. Borovanský, P., Kirchner, C., Kirchner, H., Moreau, P.-E., Ringeissen, C.: An overview of ELAN. Electr. Notes Theor. Comput. Sci. 15, 55–70 (1998)
13. Borovanský, P., Kirchner, C., Kirchner, H., Ringeissen, C.: Rewriting with strategies in ELAN: a functional semantics. Int. J. Found. Comput. Sci. 12(1), 69–98 (2001)
14. Bourdier, T., Cirstea, H., Dougherty, D.J., Kirchner, H.: Extensional and intensional strategies. In: Proceedings Ninth International Workshop on Reduction Strategies in Rewriting and Programming, vol. 15, Electronic Proceedings in Theoretical Computer Science, pp. 1–19 (2009)
15. Castro, C.: Building constraint satisfaction problem solvers using rewrite rules and strategies. Fundamenta Informaticae 34(3), 263–293 (1998)

16. Cirstea, H., Kirchner, C., Liquori, L., Wack, B.: Rewrite strategies in the rewriting calculus. In: Gramlich, B., Lucas, S. (eds.) Electronic Notes in Theoretical Computer Science, vol. 86. Elsevier (2003)

17. Cirstea, H., Kirchner, C.: The rewriting calculus - Part I and II. Logic J. Interest Gr. Pure Appl. Logics **9**(3), 427–498 (2001)

18. Clavel, M., Durán, F., Eker, S., Lincoln, P., Martí-Oliet, N., Meseguer, J., Talcott, C.: All About Maude - A High-Performance Logical Framework: How to Specify, Program, and Verify Systems in Rewriting Logic. Programming and Software Engineering, vol. 4350. Springer, Heidelberg (2007)

19. de Moura, L., Passmore, G.O.: The strategy challenge in SMT solving. In: Bonacina, M.P., Stickel, M.E. (eds.) Automated Reasoning and Mathematics. LNCS, vol. 7788, pp. 15–44. Springer, Heidelberg (2013)

20. Delahaye, D.: A tactic language for the system Coq. In: Parigot, M., Voronkov, A. (eds.) LPAR 2000. LNCS (LNAI), vol. 1955, pp. 85–95. Springer, Heidelberg (2000)

21. Dijkstra, E.W.: Selected Writings on Computing - A Personal Perspective. Texts and Monographs in Computer Science. Springer, New York (1982)

22. Dougherty, D.J.: Rewriting strategies and game strategies. Internal report, August 2008

23. Ermel, C., Rudolf, M., Taentzer, G.: The AGG approach: language and environment. In: Ehrig, H., Engels, G., Kreowski, H.-J., Rozenberg, G. (eds.) Handbook of Graph Grammars and Computing by Graph Transformations: Applications, Languages, and Tools, pp. 551–603. World Scientific, Singapore (1997)

24. Fernández, M., Kirchner, H., Namet, O.: A strategy language for graph rewriting. In: Vidal, G. (ed.) LOPSTR 2011. LNCS, vol. 7225, pp. 173–188. Springer, Heidelberg (2012)

25. Fernández, M., Kirchner, H., Namet, O.: Strategic portgraph rewriting: an interactive modelling and analysis framework. In: Lafuente, A.L., Bosnacki, D., Edelkamp, S., Wij, A. (eds.) Proceedings 3rd Workshop on GRAPH Inspection and Traversal Engineering (GRAPHITE 2014), Grenoble, France, 5th April 2014, vol. 159, Electronic Proceedings in Theoretical Computer Science, pp. 15–29 (2014)

26. Futatsugi, K., Goguen, J.A., Jouannaud, J.-P., Meseguer, J.: Principles of OBJ2. In: Reid, B. (ed.) Proceedings 12th ACM Symposium on Principles of Programming Languages, pp. 52–66. ACM Press (1985)

27. Geiß, R., Batz, G.V., Grund, D., Hack, S., Szalkowski, A.: GrGen: a fast SPO-based graph rewriting tool. In: Corradini, A., Ehrig, H., Montanari, U., Ribeiro, L., Rozenberg, G. (eds.) ICGT 2006. LNCS, vol. 4178, pp. 383–397. Springer, Heidelberg (2006)

28. Giesl, J., Middeldorp, A.: Innermost termination of context-sensitive rewriting. In: Ito, M., Toyama, M. (eds.) DLT 2002. LNCS, vol. 2450, pp. 231–244. Springer, Heidelberg (2003)

29. Giesl, J., Raffelsieper, M., Schneider-Kamp, P., Swiderski, S., Thiemann, R.: Automated termination proofs for haskell by term rewriting. ACM Trans. Program. Lang. Syst. **33**(2), 7:1–7:39 (2011)

30. Gnaedig, I., Kirchner, H.: Termination of rewriting under strategies. ACM Trans. Comput. Logic **10**(2), 1–52 (2009)

31. Goguen, J., Malcolm, G. (eds.): Software Engineering with OBJ: Algebraic Specification in Action. Advances in Formal Methods. Kluwer Academic Publishers, Boston (2000). ISBN 0-7923-7757-5

32. Gordon, M., Milner, R., Morris, L., Newey, M., Wadsworth, C.: A metalanguage for interactive proof in LCF. In: Proceedings of 5th ACM Symposium on Principles of Programming Languages, pp. 119–130. ACM Press, January 1978

33. Grädel, E., Thomas, W., Wilke, T. (eds.): Automata, Logics, and Infinite Games: A Guide to Current Research [outcome of a Dagstuhl seminar, February 2001]. LNCS, vol. 2500. Springer, Heidelberg (2002)

34. Habel, A., Müller, J., Plump, D.: Double-pushout graph transformation revisited. Math. Struct. Comput. Sci. **11**(5), 637–688 (2001)

35. Habel, A., Plump, D.: Computational completeness of programming languages based on graph transformation. In: Honsell, F., Miculan, M. (eds.) FOSSACS 2001. LNCS, vol. 2030, pp. 230–245. Springer, Heidelberg (2001)

36. Hanus, M.: Curry: a multi-paradigm declarative language (system description). In: Twelfth Workshop Logic Programming (WLP 1997), Munich (1997)

37. Huet, G., Lévy, J.-J.: Computations in non-ambiguous linear term rewriting systems. Technical report, INRIA Laboria (1979)

38. Huet, G., Lévy, J.-J.: Computations in orthogonal rewriting systems, I and II. In: Lassez, J.-L., Plotkin, G. (eds.) Computational Logic, chapter 11, 12, pp. 395–414. MIT press (1991)

39. Jones, S.L.P.: Haskell 98 Language and Libraries: The Revised Report. Cambridge University Press, Cambridge (2003)

40. Kechri, A.S.: Classical Descriptive Set Theory. Graduate Texts in Mathematics, vol. 156. Springer, New York (1995)

41. Kirchner, C., Kirchner, F., Kirchner, H.: Strategic computations and deductions. In: Benzmüller, C., Brown, C.E.., Siekmann, J., Statman, R. (eds.) Reasoning in Simple Type Theory. Festchrift in Honour of Peter B. Andrews on His 70th Birthday, vol. 17, Studies in Logic and the Foundations of Mathematics, pp. 339–364. College Publications (2008)

42. Kirchner, C., Kirchner, F., Kirchner, H.: Constraint based strategies. In: Escobar, S. (ed.) WFLP 2009. LNCS, vol. 5979, pp. 13–26. Springer, Heidelberg (2010)

43. Kirchner, C., Kirchner, H., Vittek, M.: Implementing computational systems with constraints. In: Principles and Practice of Constraint Programming, pp. 156–165 (1993)

44. Kirchner, C., Kirchner, H., Vittek, M.: Designing constraint logic programming languages using computational systems. In: Van Hentenryck, P., Saraswat, V. (eds.) Principles and Practice of Constraint Programming. The Newport Papers, chapter 8, pp. 131–158. The MIT Press (1995)

45. Kirchner., H.: A rewriting point of view on strategies. In: Mogavero, F., Murano, A., Vardi, M.Y. (eds.) Proceedings 1st International Workshop on Strategic Reasoning(SR 2013), Rome, Italy, March 16–17, vol. 112, Electronic Proceedings in Theoretical Computer Science (EPTCS), pp. 99–105 (2013)

46. Letichevsky, A.: Development of rewriting strategies. In: Penjam, J., Bruynooghe, M. (eds.) PLILP 1993. LNCS, vol. 714. Springer, Heidelberg (1993)

47. Lévy, J.-J.: Optimal reductions in the lambda-calculus. In: Seldin, J.P., Hindley, J.R. (eds.) To H.B.Curry: Essays on Combinatory Logic, Lambda Calculus and Formalism, pp. 159–191. Academic Press, New York (1980)

48. Lucas, S.: Context-sensitive computations in functional and functional logic programs. J. Func. Logic Program. **1**, 1–61 (1998)

49. Lucas, S.: Termination of on-demand rewriting and termination of OBJ programs. In: Sondergaard, H. (ed.) Proceedings of the 3rd International ACM SIGPLAN Conference on Principles and Practice of Declarative Programming (PPDP 2001), pp. 82–93, Firenze, Italy, September 2001. ACM Press, New York (2001)

50. Lucas, S.: Termination of rewriting with strategy annotations. In: Nieuwenhuis, R., Voronkov, A. (eds.) LPAR 2001. LNCS (LNAI), vol. 2250, pp. 669–684. Springer, Heidelberg (2001)
51. Lucas, S., Marché, C., Meseguer, J.: Operational termination of conditional term rewriting systems. Inf. Process. Lett. **95**(4), 446–453 (2015)
52. Lucas, S., Meseguer, J.: Strong and weak operational termination of order-sorted rewrite theories. In: Escobar, S. (ed.) WRLA 2014. LNCS, vol. 8663, pp. 178–194. Springer, Heidelberg (2014)
53. Martí-Oliet, N., Meseguer, J.: Rewriting logic as a logical and semantic framework. In: Meseguer, J. (ed.) Electronic Notes in Theoretical Computer Science, vol. 4. Elsevier Science Publishers (2000)
54. Martí-Oliet, N., Meseguer, J., Verdejo, A.: Towards a strategy language for maude. In: Martí-Oliet, N. (ed.) Proceedings Fifth International Workshop on Rewriting Logic and its Applications (WRLA 2004), Barcelona, Spain, March 27 - April 4, vol. 117, Electronic Notes in Theoretical Computer Science, pp. 417–441. Elsevier Science Publishers B. V. (North-Holland) (2005)
55. Martí-Oliet, N., Meseguer, J., Verdejo, A.: A rewriting semantics for maude strategies. Electron. Notes Theor. Comput. Sci. **238**(3), 227–247 (2008)
56. McCune, W.: Semantic guidance for saturation provers. In: Calmet, J., Ida, T., Wang, D. (eds.) AISC 2006. LNCS (LNAI), vol. 4120, pp. 18–24. Springer, Heidelberg (2006)
57. Meseguer, J.: Conditional rewriting logic as a unified model of concurrency. Theor. Comput. Sci. **96**(1), 73–155 (1992)
58. Nickel, U., Niere, J., Zündorf, A.: The FUJABA environment. In: ICSE, pp. 742–745 (2000)
59. O'Donnell, M.J. (ed.): Computing in Systems Described by Equations. LNCS. Springer, Heidelberg (1977)
60. Okamoto, K., Sakai, M., Nishida, N., Sakabe, T.: Weakly-innermost strategy and its completeness on terminating right-linear TRSs. In: Proceedings 5th International Workshop on Reduction Strategies in Rewriting and Programming April 22: Nara, p. 2005. ENTCS, Japan (2005)
61. Owre, S., Rushby, J.M., Shankar, N.: PVS: a prototype verification system. In: Kapur, D. (ed.) CADE 1992. LNCS, vol. 607. Springer, Heidelberg (1992)
62. Plasmeijer, M.J., van Eekelen, M.C.J.D.: Functional Programming and Parallel Graph Rewriting. Addison-Wesley, Boston (1993)
63. Plotkin, G.D.: A structural approach to operational semantics. J. Log. Algebr. Program. **60–61**, 17–139 (2004)
64. Plump, D.: The graph programming language GP. In: Bozapalidis, S., Rahonis, G. (eds.) CAI 2009. LNCS, vol. 5725, pp. 99–122. Springer, Heidelberg (2009)
65. Plump, D., Steinert, S.: The semantics of graph programs. In: Mackie, L., Moreira, A.M. (eds.) Proceedings Tenth International Workshop on Rule-Based Programming (RULE 2009), Brasília, Brazil, 28th June 2009, vol. 21, Electronic Proceedings in Theoretical Computer Science (EPTCS), pp. 27–38 (2009)
66. Rensink, A.: The GROOVE simulator: a tool for state space generation. In: Pfaltz, J.L., Nagl, M., Böhlen, B. (eds.) AGTIVE 2003. LNCS, vol. 3062, pp. 479–485. Springer, Heidelberg (2004)
67. Schürr, A., Winter, A.J., Zündorf, A.: The PROGRES approach: language and environment. In: Ehrig, H., Engels, G., Kreowski, H-J., Rozenberg, G. (eds.) Handbook of Graph Grammars and Computing by Graph Transformations, Vol. 2, Applications, Languages, and Tools, pp. 479–546. World Scientific (1997)

68. Van de Pol, J.: Just-in-time: on strategy annotations. In: Proceedings of WRS 2001, 1st International Workshop on Reduction Strategies in Rewriting and Programming, vol. 57, Elecronic Notes in Theoretical Computer Science, pp. 41–63 (2001)
69. van Oostrom, V., de Vrijer, R.: Term Rewriting Systems, vol. 2, Cambridge Tracts in Theoretical Computer Science, chapter 9: Strategies. Cambridge University Press (2003)
70. Visser, E.: Stratego: a language for program transformation based on rewriting strategies system description of stratego 0.5. In: Middeldorp, A. (ed.) RTA 2001. LNCS, vol. 2051, pp. 357–361. Springer, Heidelberg (2001)
71. Visser, E.: A survey of strategies in rule-based program transformation systems. J. Symbolic Comput. **40**(1), 831–873 (2005)
72. Walukiewicz, L.: A landscape with games in the background. In: 19th IEEE Symposium on Logic in Computer Science (LICS 2004), pp. 356–366 (2004)

Network-on-Chip Firewall: Countering Defective and Malicious System-on-Chip Hardware

Michael LeMay$^{(\boxtimes)}$ and Carl A. Gunter

University of Illinois at Urbana-Champaign, Urbana, IL, USA
m@lemays.org

Abstract. Mobile devices are in roles where the integrity and confidentiality of their apps and data are of paramount importance. They usually contain a System-on-Chip (SoC), which integrates microprocessors and peripheral Intellectual Property (IP) connected by a Network-on-Chip (NoC). Malicious IP or software could compromise critical data. Some types of attacks can be blocked by controlling data transfers on the NoC using Memory Management Units (MMUs) and other access control mechanisms. However, commodity processors do not provide strong assurances regarding the correctness of such mechanisms, and it is challenging to verify that all access control mechanisms in the system are correctly configured. We propose a NoC Firewall (NoCF) that provides a single locus of control and is amenable to formal analysis. We demonstrate an initial analysis of its ability to resist malformed NoC commands, which we believe is the first effort to detect vulnerabilities that arise from NoC protocol violations perpetrated by erroneous or malicious IP.

1 Introduction

Personally administered mobile devices are being used or considered for banking, business, military, and healthcare applications where integrity and confidentiality are of paramount importance. The practice of dedicating an entire centrally administered phone to each of these apps is being abandoned in favor of granting access to enterprise data from personal devices as workers demand the sophistication available in the latest consumer mobile devices [6].

Security weaknesses of popular smartphone OSes have motivated isolation mechanisms for devices handling critical data, including hypervisors that operate at a lower level within the system [20]. For example, hypervisors can isolate a personal instance from a sensitive instance of Android, where both instances run simultaneously within Virtual Machines (VMs) on a single physical device. However, virtualized and non-virtualized systems both rely on the correctness of various hardware structures to enforce the memory access control policies that the system software specifies to enforce isolation.

M. LeMay was with the University of Illinois at Urbana-Champaign while performing the work described herein, but he was employed by Intel Corporation at the time of submission. The views expressed are those of the authors only.

© Springer International Publishing Switzerland 2015
N. Martí-Oliet et al. (Eds.): Meseguer Festschrift, LNCS 9200, pp. 404–426, 2015.
DOI: 10.1007/978-3-319-23165-5_19

Mobile devices are usually based on a System-on-Chip (SoC) containing microprocessor cores and peripherals connected by a Network-on-Chip (NoC). Each component on the SoC is referred to as an Intellectual Property (IP) core or block. A single SoC may contain IP originating from many different entities. SoC IP may be malicious intrinsically at the hardware level, or it may be used to perform an attack orchestrated by software, and such IP may lead to compromises of critical data. Such attacks would involve data transfers over the NoC. Memory Management Units (MMUs) and IO-MMUs can potentially prevent such attacks.

Commodity processors do not provide strong assurances that they correctly enforce memory access controls, but recent trends in system design may make it feasible to provide such assurances using enhanced hardware that is amenable to formal analysis. In this paper, we propose the hardware-based *Network-on-Chip Firewall (NoCF)* that we developed using a functional hardware description language, Bluespec. Bluespec is a product of Bluespec, Inc. Although Bluespec has semantics based on term-rewriting systems, those semantics also reflect characteristics of hardware [1]. We developed an embedding of Bluespec into Maude, which is a language and set of tools for analyzing term-rewriting systems. At a high level, term-rewriting systems involve the use of atomic rules to transform the state of a system. We know of no elegant way to directly express the hardware-specific aspects of Bluespec in a Maude term-rewriting theory, so we used Maude strategies to control the sequencing between rules in the theories to match the hardware semantics [16]. We then used our model to detect attacks that violate NoC port specifications, which have previously received little attention.

A lightweight processor core is dedicated to specifying the NoCF policy using a set of *policy configuration interconnects* to interposers, which provides a single locus of control. It also permits NoCF to be applied to NoCs lacking access to memory, avoids the need to reserve system memory for storing policies when that memory is available, and simplifies the internal logic of the interposers. The policy can be pre-installed or specified dynamically by some entity such as a hypervisor within the system. The interposers and associated policies are distributed to accommodate large NoCs.

To demonstrate one type of attack that can be blocked by NoCF, we construct a malicious IP block analogous to a Graphics Processing Unit (GPU) and show how it can be instructed to install a network keylogger by any app that simply has the ability to display graphics. This attack could be used to achieve realistic, malicious objectives. For example, a government seeking to oppress dissidents could convince them to view an image through a web browser or social networking app and subsequently record all of their keystrokes.

Our contributions include:

- An efficient, compact NoCF interposer design that is amenable to formal analysis and provides a single locus of control.
- An embedding of Bluespec into the Maude modeling language.
- Use of formal techniques to discover a new attack.
- A triple-core FPGA prototype that simultaneously runs two completely isolated, off-the-shelf instances of Linux with no hypervisor present on the cores or attached to the NoCs hosting Linux at runtime.

The rest of this paper is organized as follows. Section 2 provides background on SoC technology. Section 3 describes the threat model. Section 4 describes a core-based isolation approach. Section 5 discusses the design of NoCF interposers. Section 6 describes a NoCF prototype system. Section 7 discusses how NoCF can help to mitigate a sample attack. Section 8 formally analyzes the prototype. Section 9 discusses related work. Section 10 concludes the paper. Please refer to our technical report for additional details [14].

2   Background

Each block of IP on an SoC can be provided by an organization within the SoC vendor or by an external organization. SoCs commonly contain IP originating from up to hundreds of people in multiple organizations and spread across multiple countries [27]. Some IP (e.g. a CPU core) may be capable of executing software whereas other IP may only offer a more rudimentary configuration interface, e.g. one based on control registers. It is difficult to ensure that all of the IP is high-quality, let alone trustworthy [5,9]. The general trend is towards large SoC vendors acquiring companies to bring IP development in-house [24]. However, even in-house IP may provide varying levels of assurance depending on the particular development practices and teams involved and the exact nature of the IP in question. For example, a cutting-edge, complex GPU may reasonably be expected to exhibit more errors than a relatively simple Wi-Fi controller that has been in use for several years. Furthermore, malicious hardware can be inserted at many points within the SoC design and manufacturing process and can exhibit a variety of behaviors to undermine the security assurances of the system [4]. Memory Management Units (MMUs) and IO-MMUs are commonly used to restrict the accesses from IP blocks, which can constrain the effects of erroneous or malicious IP. Thus, errors that can permit memory access control policies to be violated are the most concerning. A sample system topology is depicted in Fig. 1. It shows two CPU cores, a GPU, a two-level interconnect, and some examples of peripherals. Note that the GPU is both an interconnect master and a peripheral.

An MMU is a component within a processor core that enforces memory access control policies specified in the form of page tables that are stored in main memory. Some SoCs incorporate IO-MMUs that similarly restrict and redirect peripheral master IP block NoC data transfers. A page table contains entries that are indexed by part of a virtual address and specify a physical address to which the virtual address should be mapped, permissions that restrict the accesses performed using virtual addresses mapped by that entry, whether the processor must be in privileged (supervisor) mode when the access is performed, and auxiliary data. Page tables are often arranged hierarchically in memory, necessitating multiple memory accesses to map a particular virtual address. To reduce the expense incurred by page table lookups, the MMU contains a Translation Lookaside Buffer (TLB) that caches page table entries in very fast memory inside the MMU. In the case of an MMU, each isolated software component (such as

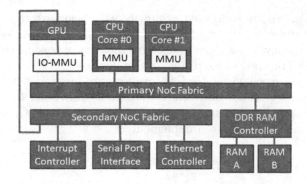

Fig. 1. Example SoC system topology.

a process or VM) is typically assigned a dedicated page table. Correspondingly for an IO-MMU, one or more page tables may be assigned to each device. By only mapping a particular region of physical memory in one of the component's page tables, that memory is protected from accesses by other components. The relatively high complexity of modern MMUs and IO-MMUs increases the likelihood of errors that undermine their access control assurances [10]. Since NoCF does not use page tables nor does it provide address translation support, it is much less complex and can constrain an attack leveraging a vulnerable MMU or IO-MMU. Note that these technologies are not mutually exclusive. In fact, it is useful to provide defense-in-depth by enforcing coarse partitions with NoCF and relying on MMUs and IO-MMUs to implement finer-grained controls within each partition.

It could be preferable to formally verify existing MMUs and IO-MMUs rather than devising new protection mechanisms. However, it is challenging to formally verify MMUs and IO-MMUs. Formal verification techniques can prove the absence of design errors within commercial processor cores, but they currently only provide a good return-on-investment when used instead to detect errors [3]. To the best of our knowledge, MMUs have only been formally verified in experimental processors [8,22]. The policy data for MMUs and IO-MMUs is itself protected by them, so it is likely to be more challenging to verify that the policy is trustworthy compared to the NoCF policy implemented on an isolated core. Finally, the MMU is a central part of each processor core with many interfaces to other parts of the core, complicating analysis. We have not formally verified NoCF either, but we demonstrate how to develop a model of it that is amenable to formal analysis. This is a non-trivial precondition for formal verification.

Individual blocks of IP communicate using one or more NoCs within a single SoC. A NoC is not simply a scaled-down network comparable to, e.g. an Ethernet LAN. Networks for large systems, such as LANs, have traditionally been connection-oriented, predominantly relying on protocols such as TCP/IP. Networks for small systems, such as NoCs, have traditionally lacked support for persistent connections. Older SoC designs relied on buses, which are subtly distinct from NoCs. For our purposes, it is not necessary to distinguish between buses and NoCs. We are concerned primarily with their external ports, which are common between both types of interconnects. Slave devices accessible over a NoC are assigned physical address ranges, so memory access controls like those in NoCF can also be used to control access to devices.

The protection mechanisms that we propose are inserted between the NoC and the IP, and they do not necessitate changes to individual IP blocks. Thus, NoCF could be added quite late in the design process for an SoC, after the main functionality of the SoC has been implemented.

3   Threat Model

Software running on a particular core is assumed to be arbitrarily malicious and must be prevented from compromising the confidentiality, integrity, and availability of software on other cores. The system software that configures NoCF must correctly specify a policy to enforce isolation between the cores. Recent work on minimizing the Trusted Computing Base (TCB) of hypervisors and formally verifying them may be helpful in satisfying this requirement [13, 25].

Our concern in this paper is that isolation between cores that are protected in this manner could potentially be compromised by misbehaving IP. Note that the memory controller in Fig. 1 is connected to two Random Access Memories (RAMs). For the purpose of the threat model discussion, the data in RAM A should only be accessible to the GPU and CPU Core #0 and RAM B should only be accessible to CPU Core #1. We now define the types of compromises we seek to prevent:

1. *Confidentiality:* Some misbehaving IP may construct an unauthorized information flow from some other target IP transferring data that the misbehaving IP or the VM controlling it is not authorized to receive. This flow may be constructed with or without the cooperation of the target IP. The misbehaving IP may have authorization to access a portion of the target IP, but not the portion containing the confidential data. For example, CPU Core #0 or the GPU could potentially transfer data from RAM B to RAM A, since both CPU cores and the GPU have access to the shared memory controller. As another example, a misbehaving memory controller itself could perform that transfer.
2. *Integrity:* Some misbehaving IP may unilaterally construct an unauthorized information flow to other target IP to corrupt data. For example, the GPU or CPU Core #0 may modify executable code or medical sensor data in RAM B.
3. *Availability:* Resource sharing is an intrinsic characteristic of SoCs, so there is the possibility that misbehaving IP may interfere with other IP using those shared resources. For example, the GPU or CPU Core #0 could flood the NoC with requests to monopolize the NoC and interfere with NoC requests from CPU Core #1.

IP can manipulate wires that form its NoC port in an arbitrary manner. The IP might not respect the port clock and can perform intra-clock cycle wire manipulations. The IP might also violate the protocol specification for the port.

Since NoCF performs address-based access control, the NoC fabric is assumed to be trusted to selectively and accurately route requests and responses to and from the appropriate IP to prevent eavesdropping and interference from other IP cores. The integrity core expects each peripherals to be associated with particular ranges of addresses and each master IP core to be associated with particular NoC ports, and it uses that information to configure NoCF policies. Thus, a necessary condition for the correct enforcement of the security policy intended by the integrity core is that the NoC fabric operate in a trustworthy manner. Establishing trust in NoC fabrics is an orthogonal research issue.

We assume that slave devices are trusted to correctly process requests. For example, the memory controller must properly process addresses that it receives to enforce policies that grant different IP access to different regions of a memory accessible through a single shared memory controller. Establishing trust in such devices is an orthogonal research issue.

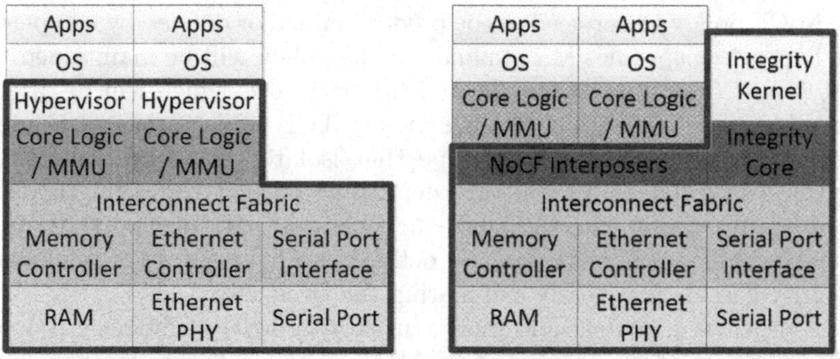

(a) Unaltered, hypervisor-based system.

(b) Similar system protected by NoCF.

Fig. 2. Comparison of TCBs, which are within the thick lines. Colored areas depict layers of hardware (Color figure online).

Covert channels are more prevalent between components that have a high degree of resource sharing, such as between software that shares a processor cache. Thus, NoCF provides tools to limit covert channels by restricting resource sharing. However, we do not attempt to eliminate covert channels in this work.

A mobile device may be affected by radiation and other environmental influences that cause unpredictable modifications of internal system state. A variety of approaches can handle such events and are complementary to our effort to handle misbehaviors that arise from the design of the device [19].

System software could be maliciously altered so that even if the intended NoC access control policy is correctly enforced, some overarching system security objective may still be violated. The operation of each IP core is determined not only by its hardware design and its connectivity to other IP cores, but also by how it is configured at runtime and what software it executes (if applicable).

Trusted computing techniques can defeat attacks that alter system software by ensuring that only specific system software is allowed to execute [2]. We focus on techniques whereby the SoC vendor can constrain untrustworthy IP in its chip designs. Software security and hardware tamper-resistance techniques can further improve assurances of overall system security by checking for the correct operation of trusted IP.

4   Core-Based Isolation

Assigning software components to separate cores eliminates vulnerabilities stemming from shared resources such as registers and L1 caches. Regulating their activities on NoCs with a dynamic policy addresses vulnerabilities from sharing main memory or peripherals. We initially focus on isolating complete OS instances, since the memory access control policies required to accomplish that are straightforward and coarse-grained. However, NoCF could also be used to implement other types of policies.

The NoCF policy either needs to be predetermined or defined by a hypervisor, like the hypervisor specifies MMU policies. The policy will be maintained by an *integrity kernel* that runs on a dedicated *integrity core*, which will be discussed further below. The effect that this has on the TCB of a system with minimal resource sharing, such as our prototype that isolates two Linux instances, is depicted in Fig. 2. The TCB will vary depending on how the policy is defined, since any software that can influence the policy is part of the TCB. In this example, the policy that was originally defined in a hypervisor is now defined in the integrity kernel, completely eliminating the hypervisor.

NoCF provides a coarser and more trustworthy level of memory protection in addition to that of the MMU and IO-MMU. These differing mechanisms can be used together to implement trade-offs between isolation assurances and costs stemming from an increased number of cores and related infrastructure.

The integrity core must be able to install policies in NoCF interposers and must have sufficient connectivity to receive policy information from any other system entities that are permitted to influence policies, such as a hypervisor. It may be possible to place the integrity kernel in firmware with no capability to communicate with the rest of the system, if a fixed resource allocation is desired. On the other end of the spectrum of possible designs, the integrity core may have full access to main memory, so it can arbitrarily inspect and modify system state. Alternately, it may have a narrow communication channel to a hypervisor. Placing the integrity kernel on an isolated integrity core permits the pair of them to be analyzed separately from the rest of the system. However, it is also possible to assign the role of integrity core to a main processor core to reduce hardware resource utilization, even if the core is running other code.

5   NoCF Interposer Design

We now discuss the design decisions underlying NoCF. We base our design on the widely-used AMBA AXI4 NoC port standard. The rule format and storage mechanism of the Policy Decision Point (PDP) are loosely modeled after those of a TLB. The PDP decides which accesses should be permitted so that a Policy Enforcement Point (PEP) can enforce those decisions. Policy rules are inserted directly into the PDP using a policy configuration interconnect to an integrity core. This reduces the TCB of the PDP relative to a possible alternate design that retrieves rules from memory like an MMU. The integrity core is a dedicated, lightweight processor core that is isolated from the rest of the system to help protect it from attack. The decisions from the PDP are enforced for each address request channel by that channel's PEP.

The AXI4 specification defines two matching port types. The master port issues requests and the slave port responds to those requests. Each pair of ports has two distinct channels, one for read requests and one for write requests. This port architecture enables us to easily insert NoCF interposers, each of which contains a PDP, an integrity core interface, and two PEPs, one for each channel. Each interposer provides both a master and slave port so that it can be interposed between each IP master port and the NoC slave port that it connects to. A single interposer is depicted in Fig. 3.

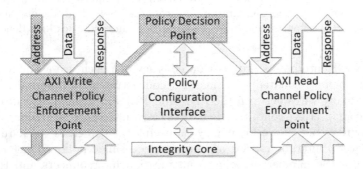

Fig. 3. Internal configuration of NoCF interposer. Each interposer contains all of these components. Hatched regions are formally analyzed in Sect. 8.

We evaluate our design in a prototype system containing two main processor cores in addition to the integrity core, plus a malicious GPU. We now consider it as a sample system arrangement, although many other system arrangements are possible. Each main core has four AXI4 master ports. They connect to two NoCs in the system, one of which solely provides access to the DDR3 main memory controller, while the other connects to the other system peripherals. Each main core has two ports connected to each NoC, one each for instruction and data accesses. The GPU has a single master port connected to the NoC with the main memory, along with a slave port connected to the peripheral NoC (not shown). We depict this topology in Fig. 4. One interposer is assigned to each of the ports between the master IP and the NoCs, with a corresponding policy configuration interconnect to the integrity core. The depicted interconnect topology is slightly simplified compared to the one used in the commercial ARM Cortex-A9 MP processor, which shares an L2 cache between up to four cores. Thus, it would be necessary in that processor to place interposers between the cores and the L2 cache controller and to trust that controller to implement memory addressing correctly.

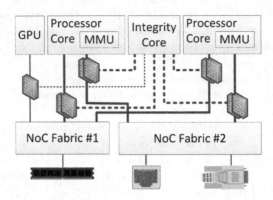

Fig. 4. System topology. A brick wall represents a NoCF interposer on one NoC port. Dashed lines denote policy configuration interconnects. Solid lines denote NoC ports. For interconnects and ports, thin lines denote single items and thick lines denote pairs.

Since NoCF interposers are distributed, they can each use a policy tailored to the port being regulated and this also concentrates the internal interfaces containing many wires between the PDP and PEPs in a small area of the chip while using an interface containing few wires to span the potentially long distance to the integrity core. However, it may be useful in some cases to share a PDP between several interposers that are subject to a single policy. That approach would reduce the number of policy configuration interconnects and the total PDP policy storage. A more complex approach would be to support selectively-shared rules for separate interposers in a single PDP, which would still reduce interconnect logic and could provide some reduction in PDP storage.

Each policy rule specifies a region of memory to which read and/or write access is permitted. A region is defined by a base address and a mask length specifier, which indicates the size of the region as one of a set of possible powers of two. This type of policy can be implemented very efficiently in hardware and corresponds closely to the policies defined by conventional MMU page tables.

Address requests are regulated by PEPs in cooperation with the PDP. The PDP stores a fixed number of rules in its database. The PDP checks the address in the request against all policy rules in parallel. Whenever a request matches some rule that has the appropriate read or write permission bit set, it will be permitted to pass through the PEP. Otherwise, the PDP sends an interrupt to the integrity core and also sends it information about the failing request. It then blocks the request until the integrity core instructs it to resume.

The integrity core can modify the policy rules prior to issuing the resume command. To modify policy rules, the integrity core sends commands over the policy configuration interconnect to insert a policy rule or flush all existing policy rules. Other commands could be defined in the future. When the interposer receives the resume command, it re-checks the request and either forwards it if it now matches some rule, or drops it and returns an error response to the master. It could also do something more drastic, such as blocking the clock signal or power lines feeding the master that issued the bad request.

Addresses other than the one in the request may be accessed during the ensuing data transfer. A variety of addressing modes are supported by AXI4 that permit access to many bytes in a burst of data transfers initiated by a single request. The policy administrator must account for these complexities by ensuring that all bytes that can actually be accessed should be accessible.

It can be useful to physically separate a protection mechanism from the surrounding logic and constrain its interfaces to that logic so that it can be independently analyzed [11]. This is possible in the case of NoCF, since its only interfaces are the controlled NoC ports and the policy configuration interconnect.

The PDP, integrity core interface, and PEP are all implemented in Bluespec to leverage its elegant semantics and concision, with interface logic to the rest of the prototype hardware system written in Verilog and VHDL. For details, please see our technical report [14].

6   Prototype Implementation

We used a Xilinx ML605 evaluation board, which includes a Virtex-6 FPGA, to implement a prototype of NoCF. We use MicroBlaze architecture processor cores implemented by Xilinx, because they are well-supported by Xilinx tools and Linux. The integrity core is very lightweight, with no cache, MMU, or superfluous optional instructions. It is equipped with a 16KiB block of on-chip RAM directly and exclusively connected to the instruction and data memory ports on the integrity core. This RAM is thus inaccessible from the other cores.

The prototype runs Linux 3.1.0-rc2 on both main cores, including support for a serial console from each core and exclusive Ethernet access from the first core. We compiled the Linux kernel using two configurations corresponding to each core so that they use different regions of system memory and different sets of peripherals. This means that no hypervisor beyond the integrity kernel is required, because the instances are completely separated. The system images are loaded directly into RAM using a debugger.

The integrity kernel specifies a policy that constrains each Linux instance to the minimal memory regions that are required to grant access to the memory and peripherals allocated to the instance. Attempts to access addresses outside of an instance's assigned memory regions cause the instance to crash with a bus error, which is the same behavior exhibited by a system with or without NoCF when an instance attempts to access non-existent physical memory.

The interposers each contain two policy rules and replace them in FIFO order, except that the interposers for data loads and stores to the peripherals contain four policy rules each, since they are configured to regulate fine-grained memory regions.

7   Constraining a Malicious GPU

A malicious GPU could perform powerful attacks, since it would have bus-master access and be accessible from all apps on popular mobile OSes. Almost all apps have a legitimate need to display graphics, so software protection mechanisms that analyze app behavior could not be expected to flag communications with the GPU as suspicious, nor could permission-based controls be configured to block such access.

We constructed hardware IP that is analogous to a hypothetical malicious GPU. It has both master and slave AXI4 interfaces. In response to commands received on its slave interface, the IP reads data from a specified location in physical memory. This is analogous to reading a framebuffer. The IP inspects the least significant byte of each pixel at the beginning of the framebuffer. This is a very basic form of steganographic encoding that only affects the value of a single color in the pixel, to reduce the chance of an alert user visually detecting the embedded data. More effective steganographic techniques could easily be devised. If those bytes have a specific "trigger" value, then the IP knows that part of the framebuffer contains a malicious command. The trigger value is selected so that it is unlikely to appear in normal images. The IP then continues reading steganographically-embedded data from the image and interprets it as a command to write arbitrary data embedded in the image to an arbitrary location in physical memory.

We developed a simple network keylogger to be injected using the malicious IP. The target Linux system receives user input via a serial console, so the keylogger modifies the interrupt service routine for the serial port to invoke the main keylogger routine after retrieving each character from the serial port hardware. This 20 byte hook is injected over a piece of error-checking code that is not activated in the absence of errors. The physical address and content of this error-checking code must be known to the attacker, so that the injected code can gracefully seize control and later resume normal execution. The keylogger hook is generated from a short assembly language routine.

The main keylogger routine is 360 bytes long and sends each keystroke as a UDP packet to a hardcoded IP address. It uses the optional netpoll API in the Linux kernel to accomplish this in such a compact payload. This routine is generated from C code that is compiled by the attacker as though it is a part of the target kernel. The attacker must know the addresses of the relevant netpoll routines as well as the address of a region of kernel memory that is unused, so that the keylogger can be injected into that region without interfering with the system's business functions. We chose a region pertaining to NFS functionality. The NFS functionality was compiled into the kernel, but is never used on this particular system.

All of the knowledge that we identified as being necessary to the attacker could reasonably be obtained if the target system is using a standard Linux distribution with a known kernel and if the attacker knows which portion of the kernel is unlikely to be used by the target system based on its purpose. For example, other systems may use NFS, in which case it would be necessary to find a different portion of the kernel that is unused on that system in which to store the keylogger payload.

To constrain the GPU in such a way that this attack fails, it is simply necessary to modify the NoCF policy to only permit accesses from the GPU to its designated framebuffer in main memory, as is depicted in Fig. 5.

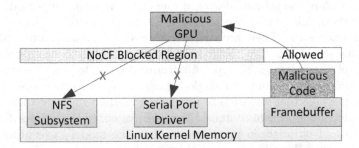

Fig. 5. NoCF can be configured to block attacks that rely on writes by malicious hardware to specific memory locations that it has no legitimate need to access.

This particular attack could also be blocked by a kernel integrity monitor, which ensures that only approved kernel code is permitted to execute [23]. The malware injected by the GPU would not be approved, so it would be unable to execute. However, kernel integrity monitors fail to address attacks on userspace and can be complex, invasive, and high-overhead.

8   Formal Analysis

8.1   Analysis Overview

We developed a shallow embedding of a subset of Bluespec into Maude, a native term rewriting system, and used a Maude model of NoCF to precisely identify a subtle vulnerability in NoCF. A shallow embedding is one where source terms are mapped to target terms whose semantics are natively provided by the target system. We only model the portion of the system that is shown with a hatched background in Fig. 3. This model was sufficient to detect an interesting vulnerability, although a complete model would be necessary to analyze the entire NoCF system in the future.

We manually converted substantial portions of the Bluespec code for NoCF to Maude using a straightforward syntactic translation method that could be automated. The hardware design is described in our technical report [14]. We developed our Bluespec description with no special regard for its amenability to analysis, so the subset of the Bluespec syntax that we modeled has not been artificially restricted. We modeled each variable name and value for Bluespec structures, enumerations, and typedefs as a Maude term. We defined separate sorts for variable names and for data values that can be placed in Bluespec registers or wires. We defined subsorts for specific types of register data, such as the types of data that are transferred through AXI interfaces and the state values for each channel. We defined a separate sort for policy rules.

The model was structured as an object-oriented system. Several distinct message types can be sent between objects. All of them specify a method to be invoked, or that was previously invoked and is now returning a value to its caller. Anonymous and return-addressed messages are both supported. The latter specify the originator of the message. These are used to invoke methods that return some value. There are staged variants of the anonymous and return-addressed message types that include a natural number indicating the stage of processing for the message. This permits multiple rewrite rules to sequentially participate in the processing of a single logical message. Return messages wrap some other message that was used to invoke the method that is returning. They attach an additional piece of content, the return value, to the wrapped message. Special read and write token messages regulate the model's execution, as will be described below. Finally, two special types of messages are defined to model interactions over the FSL interface. An undecided message contains an address request, modeling an interposer notifying the integrity core of a blocked request. An enforce write message models the integrity core instructing the interposer to recheck the blocked request. Those two message types abstract away the details of FSL communication, since those are not relevant to the Critical Security Invariant described below.

We defined equations to construct objects modeling the initial state of each part of the system. We defined Maude object IDs as hierarchical lists of names to associate variables with the specific subsystem in which they are contained and to represent the hierarchical relationships between subsystems. We defined five classes corresponding to the Bluespec types of variables in the model. Registers persistently store some value. The value that was last written into the register in some clock cycle prior to the current one is the value that can be read. A register always contains some value. Wires can optionally store some value within a single clock cycle. Pulse wires behave like ordinary wires, but they can only store a single unary value. OR pulse wires behave like pulse wires, but it is possible for them to be driven multiple times within a single clock cycle. They will only store a unary value if they are driven at least once during the clock cycle.

We modeled Bluespec methods as rewrite rules. The required activation state for the relevant objects is written on the left hand side of the rule, and the transformed state of those objects is written on the right hand side. Either side can contain Maude variables. Simple Bluespec rule conditions can be represented by embedding the required variable values into the left hand side of the corresponding Maude rule. More complex conditions can be handled by defining a conditional Maude rule that evaluates variables from the left hand side of the Maude rule. Updates to register variables require special handling in Maude. We define a wire to store the value to be written to the register prior to the next clock cycle, and include a Maude rewriting rule to copy that value into the register before transitioning to the next cycle.

We modeled Bluespec functions as Maude equations. We also defined Maude functions to model complex portions of Bluespec rules, such as a conditional expression.

The main challenge that we overcame in embedding Bluespec in Maude stems from the fact that Maude by default implements something similar to pure term rewriting system semantics, in which no explicit ordering is defined over the set of rewrite rules. To model the modified term rewriting semantics of Bluespec, we imposed an ordering on the rules in the Maude theory that correspond to Bluespec rules and restricted them to fire at most once per clock cycle. This includes rules to model the implicit Bluespec rules that reset ephemeral state between cycles. We used and extended the Maude strategy framework to control rule execution [16]. The Bluespec compiler output a total ordering of the rules that was logically equivalent to the actual, concurrent schedule it implemented in hardware. We applied that ordering to the corresponding Maude rules.

To model bit vectors, we relied on a theory that had already been developed as part of a project to model the semantics of Verilog in Maude [17].

To search for vulnerabilities in NoCF, we focused on the following Critical Security Invariant:

Invariant 1. *If an address request is forwarded by a NoCF interposer, then it is permitted by the policy within that interposer.*

We modeled some basic attack behaviors to search for ways in which that invariant could be violated. In particular, we specified that during each clock cycle attackers may issue either a permissible or impermissible address request, relative to a predefined policy, or no address request. The AMBA AXI4 specification requires master IP to wait until its requests have been acknowledged by the slave IP before modifying them in any way, but our model considers the possibility that malicious IP could violate that.

We used the Maude "fair rewriting" command to perform a breadth-first search of the model's possible states for violations of the Critical Security Invariant. We extended the Maude strategy framework to trace the rule invocations so that the vulnerabilities underlying detected attacks could be independently verified and remedied.

8.2 Formalization Details

We extended each solution term with an ordered list of quoted identifiers identifying the rules that were invoked to reach the solution. Correspondingly, we extended each task term with a list of quoted identifiers, so that the extended signature of the primary task constructor is as follows:

```
op <_@_via_> : Strat Term QidList -> Task .
```

We extended the equations and rules for evaluating strategies that were necessary for our particular model to also propagate and modify the list of invoked rules.

We defined an object-based model, so we adapted the Maude strategy framework to encapsulate a Maude term of sort `Configuration` directly in each solution term to enhance readability of the output [7]. We marked the solution terms as frozen to prevent any further transformation of the encapsulated configuration term. We used Maude reflection to transform the representation of the configuration term being manipulated by the strategy framework when a solution is recorded. For example, consider our modified definition for the rule for the idle strategy:

```
vars T T' : Term .
var QL : QidList .
rl < idle @ T via QL > => sol-from(T, QL) .
```

For reference, the original definition was:

```
rl < idle @ T > => sol(T) .
```

Analogous modifications were made to other rules in the strategy framework. We defined `sol-from` as follows:

```
var CNF : Configuration .
op sol-from : Term QidList -> Task .
eq sol-from(T, QL) = sol(downTerm(T, err-cnf), QL) .
```

Note that this relies on a term of kind `Configuration` we defined to represent an error, `err-cnf`.

The reverse transformation is needed in the concatenation rules and other rules that need to manipulate the configuration term in the solution, such as:

```
var TASKS : Tasks .
var E : Strat .
eq < sol(CNF, QL) TASKS ; seq(E) > =
  < E @ upTerm(CNF) via QL > < TASKS ; seq(E) > .
```

The rule application equations extend the list of invoked rules, as in the following:

```
var L : Qid .
var Sb : Substitution .
var N : Nat .
var Ty : Type .
ceq apply-top(L, Sb, T, N, QL) =
  sol-from(T', QL L) apply-top(L, Sb, T, N + 1, QL)
    if { T', Ty, Sb' } := metaApply(MOD, T, L, Sb, N) .
```

We now describe the specific strategy used for model checking the Critical Security Invariant. The following defines the initial state for model checking. `mkNoCF` is defined elsewhere to initialize the configuration term for the main NoCF model, and `read-tok` and `write-tok` are the read and write tokens, resp., that were described previously.

```
op init-stt : -> Configuration .
eq init-stt = mkNoCF(nocf) read-tok write-tok .
```

The following represents an abbreviated list of the methods and rules *in the Bluespec model* in the order in which they should be invoked, as specified by the Bluespec compiler:

```
sort FireRule .
ops mandatory-fr fr : NeBluespecIdList Qid -> FireRule .

op bluespec-sched : -> NeFireRuleList .
eq bluespec-sched =
  fr(nocf wrtAddrFilter, 'in_issue)
  fr(nocf wrtAddrFilter, 'ready)
  ...
  fr(nocf, 'wrt_addr_resume)
  fr(nocf, 'upd_wrt_stt)
  ...
  fr(nocf wrtAddrFilter, 'clear)
  mandatory-fr(nocf, 'nocf_clear) .
```

The first parameter for each `FireRule`, a term of sort `NeBluespecIdList`, is a list of identifier terms that identifies an object in a hierarchical Bluespec design relative to which the modeled Bluespec method or rule should be invoked. The second parameter is the name of the Maude rule that models the corresponding Bluespec method or rule. The objects in the configuration term use these same hierarchical identifiers prepended to individual variable identifers. The distinction between `fr` and `mandatory-fr` rules is that `fr` rules are invoked iff they can possibly be applied and strategy execution continues regardless whereas `mandatory-fr` rules are always invoked. We use mandatory invocation only at the end of the schedule list for a special rule that models the Bluespec behavior of clearing certain variables at the end of a hardware clock cycle. It is a conditional rule that is only enabled if a NoC request has not been finally denied. Thus, if a NoC request has been finally denied, then the rule is disabled and strategy execution terminates for that branch of the model state space.

The rules below convert the concise list of ordered Bluespec rules into a Maude strategy to enforce the desired ordering of rule invocation. They also bind the `PFX` variable that is used in the Maude rule identified by `Q` to the specified identifier, so that the Maude rule is applied specifically to the Bluespec object with that identifier.

```
op bluespec-strat : -> Strat .
eq bluespec-strat = sched-to-strat(bluespec-sched) .

op sched-to-strat : FireRuleList -> Strat .
eq sched-to-strat(fr(BI:NeBluespecIdList, Q:Qid)) =
  try(Q:Qid['PFX:NeBluespecIdList <-
```

```
    upTerm(BI:NeBluespecIdList)]) .
eq sched-to-strat(mandatory-fr(BI:NeBluespecIdList, Q:Qid)) =
  Q:Qid['PFX:NeBluespecIdList <- upTerm(BI:NeBluespecIdList)] .
eq sched-to-strat(FR:FireRule FRL:FireRuleList) =
  sched-to-strat(FR:FireRule) ;
  sched-to-strat(FRL:FireRuleList) [owise] .
```

The following strategy fragment may install zero or one out of a set of two policy rules to the NoCF interposer:

```
op add-policy : -> Strat .
eq add-policy =
  try('add_pol_rule_1[none]) | try('add_pol_rule_2[none]) .
```

The amatch test below searches each state term for any subterm that matches the term provided as the first parameter such that the condition in the second parameter is satisfied.

```
op bad-match : -> Strat .
```

```
eq bad-match = amatch('_~>_['_<'{_'}-_['PFX:NeBluespecIdList,
  'PFX1:NeBluespecIdList,'out-read.Method],'RD:RegData],
    '_==_['RD:RegData,upTerm(good-addr-req)] = 'false.Bool) .
```

The first parameter is a term representing a return message in our model. Taking it down a level in reflection makes the syntax more clear:

```
(PFX:NeBluespecIdList <{PFX1:NeBluespecIdList}- out-read)
  ~> RD:RegData
```

PFX is the object to which the original method call was directed, PFX1 is the object that invoked the method and to which this return message is directed, out-read is the identifier of a method that was invoked with no parameters, and RD is the returned term resulting from modeling the method's execution. The out-read method models the interposer passing an approved address request along to the NoC. The condition in the amatch test checks for any such approved request that is not the single request that we defined to be allowed by policy in our Maude model. Thus, this test checks for any address requests that violate policy and yet are still (incorrectly) approved by the NoCF interposer. The Critical Security Invariant specifies that this test should never be satisfied.

The terms below define the initial state and strategy for the model checker.

```
op issue-addr-req : -> Strat .
eq issue-addr-req =
  ('issue_good[none]) | ('issue_bad[none]) | idle .

op init-term : -> Term .
op init-strat : -> Strat .
op init-task : -> Task .
```

```
eq init-term = upTerm(init-stt) .
eq init-strat =
  (((('issue_read[none]) | idle) ; (issue-addr-req) ;
  (add-policy) ; (bluespec-strat)) *) ; bad-match .
eq init-task = < init-strat @ init-term via nil > .
```

The strategy is simply a loop representing zero or more hardware clock cycles of the device being modeled, followed by the test for violations of the Critical Security Invariant. The model can inject stimuli at the beginning of each clock cycle. The issue_read rule consumes the read token and updates the state to cause any currently approved address request from the NoCF interposer to be read out in that clock cycle, or any address request that is approved in a future clock cycle to be read out as soon as possible after that approval. This non-determinism models the fact that the NoC may not be immediately ready to accept an approved address request in the same clock cycle that the NoCF interposer approves it. The issue_good and issue_bad rules likewise consume the write token and issue an address request that either complies with or violates the policy, respectively, to the NoCF interposer. The possibility that the idle strategy may be selected instead of issuing an address request models the fact that address requests may or may not be issued every clock cycle in the hardware design. A rule is in place to reintroduce a new write token after an address request has been issued so that zero or one address requests can be issued during each clock cycle. Finally, fair rewriting is used to check the model represented by init-task. Solutions represent counterexamples to the Critical Security Invariant.

8.3   Analysis Results

We detected a subtle possible attack applicable to a straightforward implementation of the interposer. First, the attacker issues the permissible request, when the slave IP is not yet ready to accept a new request. The attacker then issues the impermissible request after the PEP has approved the first request and is simply waiting for the slave IP to accept the request. The PEP assumes the master adheres to the protocol specification and will wait for the initial request to be acknowledged, so it passes the request through from the master. This attack is depicted in Fig. 6. This type of model is powerful, since it is a simple matter to model basic attacker behaviors which can then be automatically analyzed to detect complex attacks.

We implemented a countermeasure in the Bluespec code to block this attack. It now buffers the request that is subjected to access control checking, and then issues that exact request to the slave if it is allowed, regardless of the current state of the request interface from the master. This countermeasure introduces additional space overhead, so it is not something that a designer would reasonably be expected to include at the outset in a straightforward implementation.

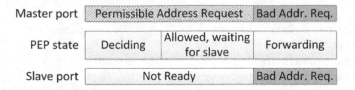

Fig. 6. Timing of address requests at port relative to PEP state for an attack forwarding an unchecked address request.

8.4   Analysis Discussion

We extended the Verilog specification of the NoC ports to interface with the Verilog generated by the Bluespec compiler. This implies that we must trust or verify the Verilog interface code. The interface code consists almost entirely of straightforward wire connections, so it should be amenable to manual or formal analysis. We must also trust the Bluespec compiler to output Verilog code corresponding to the input Bluespec code.

Our threat model allows malicious IP to perform intra-clock cycle manipulations of the wires in the port to the interposer. The effects of such manipulations are difficult to analyze at the level of abstraction considered in this section. If this type of behavior is a concern, it can be easily suppressed by buffering the port using a slice of registers that effectively forces the IP to commit to a single port state for a full clock cycle from the perspective of the interposer. The commercial NoC IP that we used in our prototype supports the creation of such register slices with a simple design parameter. This solution would introduce a single clock cycle delay at the port.

Ultimately, NoCF and other elements of the TCB should be formally verified to be resistant to foreseeable attack types, and the analysis described here suggests that the elegant semantics of Bluespec helps to make such an effort more tractable than it would be if we had used Verilog or VHDL.

As an intermediate goal, it will be important to model more potential attacker behaviors to potentially identify additional vulnerabilities and formally verify the absence of vulnerabilities when possible. The model should be expanded to model all possible sequences of values that misbehaving IP could inject into the NoC ports to which it has access. A challenge is that each master controls a large number of input wires that feed into the NoC. Many of these wires carry 32-bit addresses and data, so inductive proof strategies may permit that number to be substantially reduced by showing that a model using narrower address and data ports is equivalent to the full model in the context of interesting theorems. Similarly, induction may permit long sequences of identical input values or repetitive sequences of input values to be collapsed to shorter sequences, if in fact the NoC logic does not impart significance to the patterns in question.

We considered one theorem for which we detected a counterexample, but there are many other theorems that are foundational to the system's trustworthiness and that should be used as guidance while analyzing NoCF. The process of identifying these theorems should be informed by past vulnerabilities, system requirements, and desirable information-theoretic properties.

Analyzing NoCF and the rest of the TCB with respect to the theorems and the detailed model we have proposed is an important step towards providing strong assurance that the system can be trusted to process sensitive data alongside potentially misbehaving hardware and software components. Our formal analysis of the existing NoCF prototype demonstrates the improved analysis capabilities that are enabled by formal hardware development practices and modern formal analysis tools. This suggests that the broader analysis effort we have proposed is feasible, given sufficient resources.

9   Related Work

In this section, we consider tools and techniques that enable formal reasoning about hardware. The primary novelty of NoCF is that it is a NoC access control mechanism designed to be amenable to formal analysis.

Advances have been made in languages for formally specifying information-flow properties in hardware like Caisson [15]. Tiwari et al. developed and verified an information-flow secure processor and microkernel, but that was not in the context of a mobile-phone SoC and involved radical modifications to the processor compared to those required by NoC-based security mechanisms [26]. Volpano proposed dividing memory accesses in time to limit covert channels [28]. Information-flow techniques could be generally applicable to help verify the security of the trusted components identified in Sect. 3.

Other techniques are complementary to these lines of advancement in that they offer approaches for satisfying the assumptions of our threat model. "Moats and Drawbridges" is the name of a technique for physically isolating components of an FPGA and connecting them through constrained interfaces so that they can be analyzed independently [11,12].

SurfNoC schedules multiple protection domains onto NoC resources in such a way that non-interference between the domains can be verified at the gate level [29]. This could complement NoCF by preventing unauthorized communications channels between domains from being constructed in the NoC fabric.

Richards and Lester defined a shallow, monadic embedding of a subset of Bluespec into PVS and performed demonstrative proofs using the PVS theorem prover on a 50-line Bluespec design [21]. Their techniques may be complementary to our model checking approach for proving properties that are amenable to theorem proving. Katelman defined a deep embedding of BTRS into Maude [18]. BTRS is an intermediate language used by the Bluespec compiler. Our shallow embedding has the potential for higher performance, since we translate Bluespec rules into native Maude rules. Bluespec compilers could potentially output multiple BTRS representations for a single design, complicating verification. Finally, our embedding corresponds more closely to Bluespec code, which could make it easier to understand and respond to output from the verification tools.

10   Conclusion

Mobile devices that became popular for personal use are increasingly being relied upon to process sensitive data, but they are not sufficiently trustworthy to make such reliance prudent. Various software-based techniques are being developed to process data with different levels of sensitivity in a trustworthy manner, but they assume that the underlying hardware memory access control mechanisms are trustworthy. We discuss how to validate this assumption by introducing a NoC Firewall that is amenable to formal analysis. We present a prototype NoCF that is implemented using a hardware description language with elegant semantics. We demonstrate its utility by using it to completely isolate two Linux instances without running any hypervisor code on the cores hosting the instances.

Acknowledgments. This paper is dedicated to José Meseguer, whose work has inspired and formed the basis of so many studies like it.

The work was partially supported by HHS 90TR0003-01 (SHARPS) and NSF CNS 13-30491 (THaW). The views expressed are those of the authors only. We measured lines of code using David A. Wheeler's 'SLOCCount'.

References

1. Bluespec SystemVerilog overview. Technical report, Bluespec, Inc. (2006). http://www.bluespec.com/products/documents/BluespecSystemVerilogOverview.pdf
2. Arbaugh, W.A., Farber, D.J., Smith, J.M.: A secure and reliable bootstrap architecture. In: 18th IEEE Symposium on Security and Privacy, Oakland, CA, USA, pp. 65–71, May 1997
3. Arditi, L.: Formal verification: so many applications. In: Design Automation Conference on Electronic Chips & Systems Design Initiative 2010 Presentation, Anaheim, CA, USA, June 2010
4. Beaumont, M., Hopkins, B., Newby, T.: Hardware trojans - prevention, detection, countermeasures (A literature review). Technical report DSTO-TN-1012, DSTO Defence Science and Technology Organisation, Edinburgh, South Australia, July 2011. http://www.eetimes.com/author.asp?section_id=36&doc_id=1266011
5. Butler, S.: Managing IP quality in the SoC era. Electronic Engineering Times Europe p. 5, October 2011. http://www.wsj.com/articles/SB10001424052748704641604576255223445021138
6. Cheng, R.: So you want to use your iPhone for work? Uh-oh. Wall Street J., April 2011
7. Clavel, M., Durán, F., Eker, S., Lincoln, P., Martí-Oliet, N., Meseguer, J., Talcott, C.: Maude manual (version 2.6). Technical report, January 2011
8. Dalinger, I.: Formal Verification of a Processor with Memory Management Units. Saarland University, Saarbrücken (2006)
9. Goering, R.: Panelists discuss solutions to SoC IP integration challenges. Industry Insights - Cadence Community, May 2011
10. Gotze, K.: A survey of frequently identified vulnerabilities in commercial computing semiconductors. In: 4th IEEE International Symposium on Hardware-Oriented Security and Trust, pp. 122–126. HOST, San Diego (2011)

11. Huffmire, T., Brotherton, B., Wang, G., Sherwood, T., Kastner, R., Levin, T., Nguyen, T., Irvine, C.: Moats and drawbridges: an isolation primitive for reconfigurable hardware based systems. In: 28th IEEE Symposium on Security and Privacy, Oakland, CA, USA, pp. 281–295, May 2007
12. Huffmire, T., Irvine, C., Nguyen, T.D., Levin, T., Kastner, R., Sherwood, T.: Handbook of FPGA Design Security. Springer, The Netherlands (2010)
13. Klein, G., Elphinstone, K., Heiser, G., Andronick, J., Cock, D., Derrin, P., Elkaduwe, D., Engelhardt, K., Kolanski, R., Norrish, M., Sewell, T., Tuch, H., Winwood, S.: seL4: formal verification of an OS kernel. In: 22nd ACM Symposium on Operating Systems Principles, SOSP, Big Sky, MT, USA, pp. 207–220, October 2009
14. LeMay, M., Gunter, C.A.: Network-on-chip firewall: countering defective and malicious system-on-chip hardware, April 2014. http://arxiv.org/abs/1404.3465
15. Li, X., Tiwari, M., Oberg, J.K., Kashyap, V., Chong, F.T., Sherwood, T., Hardekopf, B.: Caisson: a hardware description language for secure information flow. In: 32nd ACM SIGPLAN Conference on Programming Language Design and Implementation, PLDI, San Jose, CA, USA, pp. 109–120, June 2011
16. Martí-Oliet, N., Meseguer, J., Verdejo, A.: A rewriting semantics for Maude strategies. In: 7th International Workshop on Rewriting Logic and its Applications, WRLA, pp. 227–247. Elsevier, Budapest (2008)
17. Meredith, P., Katelman, M., Meseguer, J., Rosu, G.: A formal executable semantics of Verilog. In: 8th ACM/IEEE International Conference on Formal Methods and Models for Codesign, MemoCODE, Grenoble, France, pp. 179–188, July 2010
18. Katelman, M.K.: A Meta-Language for Functional Verification. Ph.D. Dissertation, University of Illinois at Urbana-Champaign, Urbana, Illinois (2011)
19. Mukherjee, S.S., Emer, J., Reinhardt, S.K.: The soft error problem: an architectural perspective. In: 11th International Symposium on High-Performance Computer Architecture, HPCA, pp. 243–247. IEEE, San Francisco (2005)
20. Nachenberg, C.: A window into mobile device security: examining the security approaches employed in Apple's iOS and Google's Android. Technical report, Symantec Security Response, June 2011
21. Richards, D., Lester, D.: A monadic approach to automated reasoning for Bluespec SystemVerilog. Innovations Syst. Softw. Eng. 7(2), 85–95 (2011). Springer
22. Schubert, E.T., Levitt, K., Cohen, G.C.: Formal verification of a set of memory management units. Contractor Report 189566, National Aeronautics and Space Administration, Hampton, VA, USA, March 1992
23. Seshadri, A., Luk, M., Qu, N., Perrig, A.: SecVisor: a tiny hypervisor to provide lifetime kernel code integrity for commodity OSes. In: 21st ACM Symposium on Operating Systems Principles, SOSP, Stevenson, WA, USA, pp. 335–350, October 2007
24. Shimpi, A.L.: NVIDIA to acquire Icera, adds software baseband to its portfolio, May 2011. AnandTech.com
25. Szefer, J., Keller, E., Lee, R.B., Rexford, J.: Eliminating the hypervisor attack surface for a more secure cloud. In: 18th ACM Conference on Computer and Communications Security, CCS, Chicago, IL, USA, October 2011
26. Tiwari, M., Oberg, J.K., Li, X., Valamehr, J., Levin, T., Hardekopf, B., Kastner, R., Chong, F.T., Sherwood, T.: Crafting a usable microkernel, processor, and I/O system with strict and provable information flow security. In: 38th International Symposium on Computer Architecture, ISCA, pp. 189–200. ACM, San Jose (2011)
27. Villasenor, J.: Ensuring hardware cybersecurity. The Brookings Institution, May 2011

28. Volpano, D.: Towards provable security for multilevel reconfigurable hardware. Technical report, Naval Postgraduate School (2008)
29. Wassel, H.M.G., Gao, Y., Oberg, J.K., Huffmire, T., Kastner, R., Chong, F.T., Sherwood, T.: SurfNoC: a low latency and provably non-interfering approach to secure networks-on-chip. In: 40th International Symposium on Computer Architecture, ISCA, pp. 583–594. ACM, Tel-Aviv (2013)

Discretionary Information Flow Control for Interaction-Oriented Specifications

Alberto Lluch Lafuente$^{(\boxtimes)}$, Flemming Nielson,
and Hanne Riis Nielson

DTU Compute, The Technical University of Denmark,
Kongens Lyngby, Denmark
albl@dtu.dk

Abstract. This paper presents an approach to specify and check discretionary information flow properties of concurrent systems. The approach is inspired by the success of the interaction-oriented paradigm to concurrent systems (cf. choreographies, behavioural types, protocols,...) in providing behavioural guarantees of global properties such as deadlock-absence. We show how some information flow properties are easier to formalise and check on a global interaction-oriented description of a concurrent system rather than on a local process-oriented description of the components of the system. We use a simple choreography description language adapted from the literature of choreographies and session types. We provide a generic method to instrument the semantics with information flow annotations. Policies are used to specify the admissible flows of information. The main contribution of the paper is a sound type system for statically checking if a system specification ensures an information flow policy. The approach is illustrated with two archetypal examples of distributed and parallel computing systems: a protocol for an identity-secured data providing service and a parallel MapReduce computation.

Keywords: Information flow control · Discretionary access control · Choreographies · Communication protocols · Interaction-oriented computing · Parallel computing · Service-oriented computing · High-performance computing

1 Introduction

The flow of information within a concurrent system is often expected to satisfy some properties related to which components can access which data and how. Such properties are known as *discretionary* access control policies and provide a fine-grained control over the flow of information, as opposed to other kinds of security policies that often regard the non-interference between security levels of information. Consider for instance, the following concurrent program:

$$\left[\ k!x\ \right]_p \qquad \left[\ k?y\ ;\ k'!\text{``go''}\ ;\ k!y'\ \right]_q \qquad \left[\ k'?z\ ;\ k?z\ \right]_r$$

© Springer International Publishing Switzerland 2015
N. Martí-Oliet et al. (Eds.): Meseguer Festschrift, LNCS 9200, pp. 427–450, 2015.
DOI: 10.1007/978-3-319-23165-5_20

428   A. Lluch Lafuente et al.

where $u!v$ denotes the sending of a message v over a channel u, $u?w$ denotes the reception on variable w of a message on channel u, sequential composition is denoted by ; and the sequential code of the concurrent processes p, q, r composing the system is enclosed between square brackets. Is there a flow of information from variable x to variable z? A simple analysis of all possible executions of the system may provide a negative answer depending on the kind of information flow one is interested in (explicit, implicit and so on). However, such an *a posteriori* verification is undecidable in general and often unfeasible (e.g. due to state space explosion). An *a priori* static analysis, however, should be smart enough to discard the potential flow of information over channel k if one is interested in explicit data flows. Indeed, x is sent over channel k and z is obtained from k as well. It is a matter of synchronisation that z will not receive the value of x.

Figure 1 shows a graphical representation of some flow of information in our example. A detailed explanation of our graphical notation will be provided later, here it suffices to understand that the flow of information is represented with a graph (circles and arrows) equipped with an interface (left and right list of principal and variable names). Of course, the figure depicts only *some* flows of information, in particular it represents explicit data flows between variables, write access from principals to variables and control dependencies due to interactions. The pre-

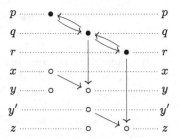

Fig. 1. A flow of information

cise notion of information flow one would like to consider may vary, depending of the properties of interest, the application domain, the context of execution where the system will be deployed and so on. Note, in particular, the absence of channels in the depicted flow, which may be an appropriate choice in case of deployment on a framework with synchronous channels that retain no data after synchronisation. Letting apart this simple illustrative example, a priori verification of discretionary information flow policies in concurrent systems is a challenging task that urges suitable solutions to ease the engineering of trustworthy-by-design systems in the era of big data, data-drivenness and massive parallel/concurrent/high-performance computing.

Contribution. We investigate in this paper an application of the interaction-oriented paradigm to the specification of *trustworthy-by-design* concurrent systems. Our work shares the same constructive attitude towards security promoted in [4] of providing methodologies and techniques to support *security-by-design*. Even if the work was motivated by security concerns, our approach can be applied to other aspects of concurrent and distributed systems where information flows play a fundamental role, like performance issues related to locality of data access, or robustness issues related to control dependencies among processes. Our work is also inspired by the success of the interaction-oriented paradigm in providing deadlock-freedom by design in distributed systems (see e.g. [8]). Many information flow properties are global in nature, which suggests that they should be easier to formalise and check on a global description of the system rather than on the local description of the individuals.

The basis of the approach is the specification of the system by means of a choreography, i.e. a global description of the expected interactions of a system in terms of the messages exchanged between its components. Our choreographies are also enriched with a specification of the information being used by processes in their decision points and about local updates of data. Such descriptions allow one to design and analyse information systems top-down, where the behaviour of the individuals is synthesised from (or specified in) the description of the global behaviour of the system. As an example, consider the following choreography:

$$C1 \triangleq \text{p.x} \rightarrow \text{q.y} : \text{k} ; \quad \text{q. "go"} \rightarrow \text{r.z} : \text{k'} ; \quad \text{q.y'} \rightarrow \text{r.z} : \text{k}$$

where $u.e \rightarrow u'.e' : c$ specifies that component u sends expression e over channel c to component u', that stores the message according to the pattern e'. This choreography specifies the very same concurrent system we saw before. However, statically checking the absence of an explicit data flow from x to z is easier as the choreography-based description resembles a sequential program where traditional information flow analysis techniques may be adapted and applied. For instance, the flow of Fig. 1 can be easily extracted from the static description of C1.

We start our work defining a formal choreography description language for specifying the global behaviour of the system. The language is strongly inspired by existing approaches based on process calculi (e.g. session calculi) and behavioural types (e.g. session types). As usual in those traditions, we consider a notion of well-formedness for choreographies, to rule out systems for which providing information flow guarantees is not trivial. The main contribution of the paper is a sound type system for information flow policies, i.e. one that ensures that if a choreography C is typed with a policy Π, denoted $\mathbf{Ent} \vdash C : \Pi$, we can conclude that C is Π-secure, i.e. the information flows of the behaviours described by C satisfy the policy Π, denoted $C \models \Pi$. In the judgement $\mathbf{Ent} \vdash C : \Pi$, $\mathbf{Ent}$ denotes the set of *entities* (principals, variables, channels) involved in C and Π. A key role in our approach is played by the use of an *instrumented semantics* [24] for our language, where semantic rules are enriched with annotations relevant to the flows of information. The instrumentation of the semantics is parametric with respect to the flows associated to the main events in the choreography (interactions, local updates, choices). This provides a convenient degree of flexibility to the user, who can specify the notion of information flow that better suits his purposes. This is one of the reasons why information flow assurances in our approach are not related to a *non-interference* [10,16] result. In our experience, non-interference cannot be easily conveyed to software, safety or security engineers and often provides a too strong requirement with respect to the kind of information flow properties of interest. Our information flow policies are based on the *Decentralized Label Model* [23]. We use here a graphical notation for information flows and policies based on graphs with interfaces [6,11]. Though not technically different from relational-based notations we think that the use of

```
u.name -> rp.user : a ;          ⎡ a!name ;        ⎤    ⎡ b?id ;                    ⎤
rp.user -> ip.id : b ;           ⎢ a!my_pwd ;      ⎥    ⎢ a?pwd ;                   ⎥
u.my_pwd -> ip.pwd : a ;         ⎢ loop            ⎥    ⎢ if check(id,pwd) then     ⎥
if check(id,pwd)@ip then         ⎢   a?info        ⎥    ⎢     b! "ok"                ⎥
    ip. "ok" -> rp. "ok" : b ;   ⎢   ⊕ (a? "end" ; ⎥    ⎢ else                      ⎥
    rp.class(user) -> s.class : c⎢     break )     ⎥    ⎢     b! "fail" ;            ⎥
else                             ⎣                 ⎦ u  ⎣ b!rep(id) ;               ⎦ ip
    ip. "fail" -> rp. "fail" : b ;
    rp. "na" -> s.class : c ;
( ip.rep(id) -> rp.report : b    ⎡ a?user ;        ⎤    ⎡ c?class ;                 ⎤
| ( data := first(class) @ s;    ⎢ b!user ;        ⎥    ⎢ data := first(class) ;    ⎥
  while data ≠ nil @s do         ⎢ ( (b? "ok" ;    ⎥    ⎢ while data ≠ nil then      ⎥
    s.data -> u.info : a ;       ⎢   c!class(user))⎥    ⎢     a!data ;              ⎥
    data := data.next @ s        ⎢ ⊕ (b? "fail" )  ⎥    ⎢     data := data.next ;   ⎥
  then                           ⎢   c! "na") ;    ⎥    ⎢ then                      ⎥
    s. "end" -> u.info : a  ) )  ⎣ b?report ;      ⎦ rp ⎣     a! "end" ;            ⎦ s
```

Fig. 2. Interaction- (left) and process-oriented (right) specification of a service

graphs provides a formal and visually appealing presentation, suitable for software, safety and security engineers and in line with other successful graphical notations such as message sequence charts, fault trees and attack trees. The type system is as well parametric with respect to the flows associated to the choreography events and its soundness relies on some well-formedness constraints of the flow annotations.

Structure of the Paper. Section 2 presents two paradigmatic case studies aimed at providing some additional motivation and insights, and to serve as running examples. Section 3 presents a simple choreography description language, defining its formal semantics and a notion of well-formedness. Section 4 presents our graphical notation for information flows, the annotation-parametric instrumentation of the semantics and the well-formedness conditions on flow annotations. Section 5 formalises the notion of satisfaction of a policy by a choreography and presents the sound type system that allows us to statically check if a choreography satisfies a security policy. Section 6 discusses related works, concludes the paper and describes our current and future research investigations.

2   Applications: Protocols, Services and HPC

We present in this section two case studies from different domains, to provide additional motivations and insights on the approach, as well as to serve as running examples throughout the paper. The first case study (Sect. 2.1) is a archetypal example of a distributed system, namely a protocol used in an identity-secured data providing service. The second case study (Sect. 2.2) is an archetypal example of a parallel program, namely a parallel MapReduce computation.

2.1   An Identity-Secured Data Providing Service

Our first case study is inspired on the OpenID example of [22], slightly adapted for presentation purposes. The system consists of a user u trying to retrieve some data from a server s. The access to the desired data is subject to authentication by an identity authentication party ip. The interaction between u, s and ip is coordinated by a relying party rp.

Figure 2 provides two descriptions of this case study. The one on the left provides an interaction-oriented description (namely, a choreography), while the one on the right provides a process-oriented description (i.e. a distributed specification). The choreography, P for short, is specified in our language, to be presented later, while the process-oriented description is specified in a language based on standard constructs of concurrent languages (featuring asymmetric, binary, synchronous, pattern-based communication over channels). The four components of the system (u, rp, ip and s) exchange some information through three channels (a, b and c). Some information may be sensitive (e.g. passwords, data, user classes, etc.) so that we may want to impose some policy on the way such information flows.

As in the simple example presented in the Introduction, a first look at the process-oriented specification may suggest some potential flows of information. For instance, we can see that the content of data is sent over channel a, from which both the relying party rp and the authentication party ip read messages. The user password and the user class are involved in similar situations. A security engineer may want to restrict and check those flows of information.

An example of a policy that one may be interested in is depicted in Fig. 3. The policy focuses on explicit data flows only. The policy allows explicit data flows between some of the variables involved in the system but forbids others. For instance, flows from my_pwd to report or data, which could compromise the password of the user, are not allowed. Information contained in data is only allowed to flow to data itself or to info, thus we forbid data to flow to any of the other variables that principals ip and rp may access.

The satisfaction of this policy depends on the kind of information flows one wants to consider. If only explicit flows are considered, the policy is satisfied. This may not be trivial by inspecting the process-oriented specification but a look at the interaction-oriented specification may be more reassuring. For instance, my_pwd flows only directly to pwd through the interaction u.my_pwd -> ip.pwd : a,

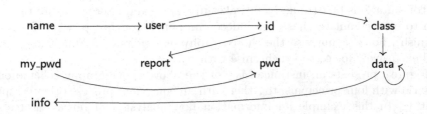

Fig. 3. Policy $\Pi_{s,1}$ for the data providing service

but pwd is not explicity used to update any variable and is never communicated. Moreover, the choreography clearly specifies that data is only used to calculate the next piece of data or to be transferred to u's info, and similarly for the rest of the sensitive informations we have mentioned.

The situation is different, of course, if explicit flows are considered as well. We will see this and examples of other policies in the rest of the paper.

2.2   A Binary Parallel Map-Reduce Computation

Our next case study is an archetypal example of parallel computations, namely MapReduce [12]. MapReduce is very popular pattern for processing large data sets in parallel, which has its origins in functional programming primitives such as Lisp's map and reduce and High-Performance Computing primitives such as MPI's scatter and reduce operations. We consider here a simple case in which the programmer is interested in computing the function $red(map(x_0, \ldots, x_m))$, where $map : T \to T'$ is a function that processes single values of some type T into single values of some other type T' (and that can be piece-wise lifted to vectors as we do above) and $red : T'^* \to T'$ is a function to accumulate a vector of values of type T' into a single value of type T'. We assume, as usual in MapReduce, that red is associative and commutative and amounts to the identity function on vectors of size 1. This allows one to decompose function red as a binary function, which greatly helps the accumulation of values in parallel.

Figure 4 is a general scheme of a possible interaction-oriented MapReduce specification in our choreography description language. The main idea of the scheme is that in a MapReduce choreography $M_{i,n+1}$, a principal p_i will act as the *leader* of 2^{n+1} principals in the computation of $red(map(x_i, \ldots, x_{i+2^{n+1}}))$ to be stored in y_i. The leader principal will decompose that computation into the problem

$$
\begin{aligned}
M_{i,0} &\triangleq y_i := map(x_i)@p_i \\
M_{i,n+1} &\triangleq p_i \text{->} p_{i+2^n} : k_{i,i+2^n} \text{ ;} \\
& \quad (M_{i,n} \mid M_{i+2^n,n}) \text{ ;} \\
& \quad p_{i+2^n} \text{->} p_i : k_{i+2^n,i} \text{ ;} \\
& \quad y_i := red(y_i, y_{i+2^n})@p_i
\end{aligned}
$$

Fig. 4. MapReduce scheme

of computing $red(map(x_i, \ldots, x_{i+2^n}))$ in y_i and $red(map(x_{i+2^n+1}, \ldots, x_{i+2^{n+1}}))$ in y_{i+2^n} first, and accumulating the results afterwards. The first sub-problem will be solved by p_i itself, while the second sub-problem will be delegated to principal p_{i+2^n}. Both sub-problems are solved applying the very same scheme. The interactions do not involve any value passing and are simply used to trigger the computations first (roughly, p_i wakes up p_{i+2^n}) and later to ensure that the data to be accumulated has been indeed computed (roughly, p_i waits for p_{i+2^n} to finish). For this purpose the choreography uses pairs of channels $k_{u,v}$ to be used exclusively for p_u to synchronise with p_v.

Figure 5 presents an instance $M_{0,2}$ of the above choreography scheme on a scenario with four principals, together with an equivalent process-oriented specification. In this example an information flow analysis can reveal interesting information regarding data locality or control dependencies. For instance, one

$p_0 \text{->} p_2 \colon k_{0,2}$;
(

    $p_0 \text{->} p_1 \colon k_{0,1}$;
    ($y_0 := \mathsf{map}(x_0)@p_0 \mid y_1 := \mathsf{map}(x_1)@p_1$) ;
    $p_1 \text{->} p_0 \colon k_{1,0}$;
    $y_0 := \mathsf{red}(y_0, y_1)@p_0$
|
    $p_2 \text{->} p_3 \colon k_{2,3}$;
    ($y_2 := \mathsf{map}(x_2)@p_2 \mid y_3 := \mathsf{map}(x_3)@p_3$) ;
    $p_3 \text{->} p_2 \colon k_{3,2}$;
    $y_2 := \mathsf{red}(y_2, y_3)@p_2$
) ;
$p_2 \text{->} p_0 \colon k_{2,0}$;
$y_0 := \mathsf{red}(y_0, y_2)@p_0$

$$\left[\begin{array}{l} k_{0,2}! \ ; \\ k_{0,1}! \ ; \\ y_0 := \mathsf{map}(x_0) \ ; \\ k_{1,0}? \ ; \\ y_0 := \mathsf{red}(y_0, y_1)@p_0 \ ; \\ k_{2,0}? \ ; \\ y_0 := \mathsf{red}(y_0, y_2)@p_0 \end{array}\right]_{p_0} \quad \left[\begin{array}{l} k_{0,1}? \ ; \\ y_1 := \mathsf{map}(x_1) \ ; \\ k_{1,0}! \end{array}\right]_{p_1}$$

$$\left[\begin{array}{l} k_{0,2}? \ ; \\ k_{2,3}! \ ; \\ y_2 := \mathsf{map}(x_2) \ ; \\ k_{3,2}? \ ; \\ y_2 := \mathsf{red}(y_2, y_3)@p_0 \ ; \\ k_{2,0}! \end{array}\right]_{p_2} \quad \left[\begin{array}{l} k_{2,3}? \ ; \\ y_3 := \mathsf{map}(x_3) \ ; \\ k_{3,2}! \end{array}\right]_{p_3}$$

Fig. 5. Interaction- (left) and process-oriented (right) specification of MapReduce

would be interested in controlling which principals can affect which other principals for the sake of analysing the robustness of the computation in terms of failure dependencies. In addition, one would like to control the access of data by principals for the sake of ensuring performance by maximising access locality.

As an example of a concrete policy, consider Fig. 6. The policy focuses on how principals access data, i.e. which principal in the computation is accessing which data variable. The policy allows each principal p_i to read variable x_i and to read and write on variable y_i. Additional write permissions are granted to allow accumulation: p_0 is allowed to read y_2 and y_1 and p_2 is allowed to read y_3. This policy is actually satisfied by the choreography (and its distributed process-oriented counterpart) and is indeed more permissive than needed. For instance, p_1 is allowed to read variable y_1 but that does not occur. This policy illustrates that sometimes one is not interested in explicit data flows. Indeed, controlling similar policies can be very useful when the system is to be deployed on a parallel architecture and principals and data have to be allocated in computational resources such as processors, machines, memory locations, etc.

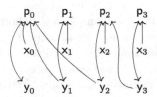

Fig. 6. Policy $\Pi_{m,1}$ for MapReduce

All in all, the case studies we have presented and the examples of information flow policies motivate the need to consider different notions of information flow and that is the reason why our framework is parametric with respect to the notion of flow to be considered.

3   Choreographic Specifications of Concurrent Systems

A choreography describes the expected interactions of a system in terms of the messages exchanged between its components. Choreographies can be used to automatically derive (via *endpoint projections*) distributed code skeletons or local specifications to be checked on existing implementations. The latter case is typical, for instance, of legacy systems (where existing implementations may be available) or open systems (where the principals may be governed by independent parties). Choreographies are usually given a so called *weak* (or *partial* or *constraint*) interpretation [21]: a (distributed) realisation S of a choreography C is admitted if it exhibits a subset of the behaviours specified by the choreography. In a trace-based setting, the question can be rephrased as $Traces(C) \subseteq Trace(S)$. This does not necessarily prescribe the presence of unobservable/hidden interactions aimed at realising the choreography. For instance, in a trace-based setting the above mentioned notion of admissibility of realisations can be relaxed to $Traces(C)_{|\mathcal{O}} \subseteq Trace(S)_{|\mathcal{O}}$, where $_{|\mathcal{O}}$ is the projection on a set of observables $\mathcal{O}$ that would discard hidden interactions. However, most approaches to choreographies are based on the idea that the choreography specifies *all* the interactions that may be observed in the system and assume that no additional interactions will take place in the realisation of the choreography. We believe that this is methodologically more adequate for information flow control: implicit hidden interactions may introduce unexpected flows of informations, whereas requiring all interactions to be explicitly declared should help understanding the actual flows of information and should mitigated the unintentional introduction of undesired flows.

As a consequence, not all choreographies are *realisable* in a distributed way. A typical example is the choreography p.e -> q.x : k ; r.e' -> s.y : k' where, clearly, there is no way of imposing the order of the interactions without introducing additional ones. Typically, well-formedness conditions are given to impose some semantic and syntactic constraints on choreographies that ensure good properties in terms of realizability and soundness of the endpoint projections.

Choreographies: Syntax. We consider a simple choreography description language inspired by process calculi and session types approaches to choreographies. In particular, the syntax of our language is close to the *interaction oriented language* [20], the *choreography calculus* [8,22] and the *global types* used in [3,9].

We use universes of variables **Var**, pattern expressions **Expr** over variables, principals **Prin**, and channels **Chan**. We denote the union of principals, variables and channels the *entities* and denote them with **Ent** = **Prin** ∪ **Var** ∪ **Chan**.

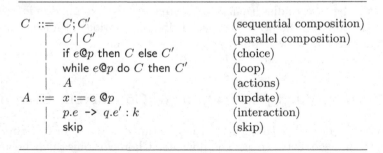

$$
\begin{array}{lll}
C & ::= & C; C' & \text{(sequential composition)} \\
& | & C \mid C' & \text{(parallel composition)} \\
& | & \text{if } e@p \text{ then } C \text{ else } C' & \text{(choice)} \\
& | & \text{while } e@p \text{ do } C \text{ then } C' & \text{(loop)} \\
& | & A & \text{(actions)} \\
A & ::= & x := e \ @p & \text{(update)} \\
& | & p.e \ \text{->} \ q.e' : k & \text{(interaction)} \\
& | & \text{skip} & \text{(skip)}
\end{array}
$$

Fig. 7. A simple choreography description language

Definition 1 (syntax of choreographies). *The syntax of our choreography description language is defined by the grammar of Fig. 7, where* $e, e' \in$ **Expr**, $x \in$ **Var**, $p, q \in$ **Prin** *are distinct principals* $(p \neq q)$, *and* $k \in$ **Chan**.

The set of principals (reps. entities) in a choreography C is denoted by $pn(C)$ (reps. $en(C)$), the set of variables in an expression e is denoted by $v(e)$. These functions are defined as expected and we hence neglect their formal definition. We assume that variable names cannot be used as values in expressions, to forbid mechanisms such as indirect references and name passing in interactions, which may pose additional challenges in our technique.

The syntactic category C corresponds to choreographies. To avoid confusion between individual choreographies and the syntactic category, we sometimes use $\mathcal{C}$ to denote the set of all terms generated by C. The syntactic category A corresponds to actions. This syntactic category is not strictly necessary but simplifies the presentation of our approach. We shall use a set $\mathcal{A}$ of *events* defined as the union of all terms generated by A and all expressions of the form $e@p$.

The language includes classical constructs such as sequential and parallel composition, branching and loops. One feature to be remarked is that decision points are annotated with the name of a principal (cf. $@p$ in loops and choices). By doing so one can specify which principal p is the *selector*, i.e. the principal responsible for taking the control decision. Another remarkable difference with respect to standard languages is that the while construct has a termination code in addition to the body. Similar annotations are not new in choreographies and are used e.g. in [3,9]. The idea is that such code is used by the selector principal to notify termination to all passive processes involved in the body. This feature is not strictly necessary in our work but we prefer to have it in order make our language closer to the ones used in the literature of choreographies.

Actions include local updates of the form $x := e \ @p$ where principal p updates variable x with the result of evaluating expression e, the void action skip and a binary interaction $p.e \ \text{->} \ q.e' : k$ between principals p and q. In such an interaction principal p aims at sending over channel k the result of evaluating expression e to principal q. Expression e is to be matched against the pattern e' whose variables act as binders to collect the result of the interaction. As we have seen, e and e' can be void in which case we use the simplified notation $p \ \text{->} \ q : k$.

Semantics of Choreographies. We present two semantics for our language: an operational semantics, aimed at providing a first insight to the reader, and a denotational semantics, which eases the presentation of the main results.

The operational semantics of our language is the relation $\rightarrow \ \subseteq \ \mathcal{C} \times \mathcal{A} \times \mathcal{C}$ defined by the rules in Fig. 8. The rules are very similar to those of standard parallel programming languages or process calculi. We just remark here that the semantics is abstract with respect to the actual evaluation of expressions: the branching in choices and loops can be seen as non-deterministic choices.

The denotational semantics defines the traces of a choreography as words over the alphabet of events $\mathcal{A}$. As in some approaches to choreographies (e.g. [9]), we restrict ourselves to finite words, and hence finite traces.

$$\frac{C_1 \xrightarrow{\alpha} C_1'}{C_1 ; C_2 \xrightarrow{\alpha} C_1' ; C_2} \qquad \frac{}{\text{skip}; C \xrightarrow{\text{skip}} C} \qquad \frac{A \neq \text{skip}}{A \xrightarrow{A} \text{skip}}$$

$$\frac{i \neq j \in \{1,2\} \qquad C_i \xrightarrow{\alpha} C_i' \qquad C_j' = C_j}{C_1 \mid C_2 \xrightarrow{\alpha} C_1' \mid C_2'}$$

$$\frac{i \in \{1,2\}}{\text{if } e@p \text{ then } C_1 \text{ else } C_2 \xrightarrow{e@p} C_i} \qquad \frac{i \in \{1,2\} \quad C_1 = C \text{ ; while } e@p \text{ do } C \text{ then } C_2}{\text{while } e@p \text{ do } C \text{ then } C_2 \xrightarrow{e@p} C_i}$$

Fig. 8. Operational semantics of choreographies

Definition 2 (trace semantics). *The trace semantics of our language is given by function* $Traces : \mathcal{C} \rightarrow 2^{\mathcal{A}^*}$ *defined by*

$$Traces(C_1 ; C_2) = Traces(C_1)\, Traces(C_2)$$
$$Traces(C_1 \mid C_2) = Traces(C_1) \bowtie Traces(C_2)$$
$$Traces(\text{if } e@p \text{ then } C_1 \text{ else } C_2) = e@p \, (Traces(C_1) \cup Traces(C_2))$$
$$Traces(\text{while } e@p \text{ do } C_1 \text{ then } C_2) = e@p \, (Traces(C_1) \, e@p)^* Traces(C_2)$$
$$Traces(A) = \{A\}$$

Above, juxtaposition denotes the concatenation of traces, the unary operator _* is the usual Kleene star of regular expressions, and the binary operator $\bowtie$ denotes the *shuffling* of trace sets, i.e. $T \bowtie T' = \{\sigma_1 \sigma_1' \ldots \sigma_n \sigma_n' \mid \sigma_1 \ldots \sigma_n \in T \wedge \sigma_1' \ldots \sigma_n' \in T'\}$. The empty trace will be denoted by ϵ. Both semantics can be shown to be equivalent for finite behaviours: the finite traces of the transition system defined by the operational semantics coincide with the traces defined by the denotational semantics (up to occurrences of skip).

Well-Formed Choreographies. As mentioned, choreographies should enjoy a couple of properties to be useful in practice, e.g. to ensure distributed realizability and soundness of endpoint projections. It is common practice to define a notion of well-formed choreography and, possibly, syntactic restrictions to ensure well-formedness. In our work, well-formedness is just needed for the correctness of our type system and we hence provide a simple notion tailored for our purpose.

Definition 3 (well-formed choreography). *Let C be a choreography. We say that C is well-formed if the following conditions hold:*

1. *every occurrence of $C_1 \mid C_2$ in C should be such that $en(C_1) \cap en(C_2) = \emptyset$;*
2. *all traces $\sigma \in Traces(C)$ satisfy the following condition: If $\sigma = \sigma' \alpha \beta \sigma''$, with $\alpha, \beta \in \mathcal{A}$ then $pn(\alpha) \cap pn(\beta) \neq \emptyset$ or $\sigma' \beta \alpha \sigma'' \in Traces(C)$.*

Our notion of well-formedness is reminiscent of the semantic notion of well-formedness used in [9] and some syntactic restrictions taken from [19]. Intuitively, well-formedness requires (1) no entity can be involved in both branches of a parallel composition, and (2) that the set of traces of a choreography is closed under the transposition of actions involving disjoint principals. Note also, that our notion of well-formedness is not strong enough to guarantee realisability. For example, the choreography p . e->q . e' : k ; r . e"->q . e''' : k (adapted from examples of ill-formed choreographies of [19]) is well-formed according to our notion, but cannot be realised due to the race condition on channel k. This is not a problem: it just means that our technique applies to more choreographies than needed in practice.

4   Information Flows in Choreographies

We provide in this section our notion of information flows and the mechanism to instrument the semantics of choreographies with flow annotations.

Information Flows as Graphs with Interfaces. A flow F is essentially a relation among entities, possibly expressing how entities influence or depend on each other. We represent flows in this paper using graphs with interfaces [6,11] as they provide an intuitive visual representation and elegant and well-defined notions of flow composition.

A graph with a discrete interface (cf. Definition 11 in Sect. A) is denoted by $I \xrightarrow{i} G \xleftarrow{o} O$ and is defined by an input interface I (a set of nodes), and output interface O (a set of nodes), a body graph G and a pair i, o of injective mappings from I and O to the nodes of G. A detailed presentation of graphs with interfaces can be found in Sect. A.

Definition 4 (flow graph). *A flow graph (or briefly a flow) is a graph with interface $I \xrightarrow{i} G \xleftarrow{o} O$.*

Examples of flow graphs can be found in Figs. 1 and 9. The visual representation places the input and output interfaces to the left and to the right of the body graph, respectively. The mappings are denoted with dotted lines, while normal arrows are used for the edges. We use two sorts of nodes to distinguish principals (•) from variables (○). We neglect channels in our examples, for the sake of simplicity, so we do not use any specific node sort for them.

As we have seen in the case studies of Sect. 2 we sometimes distinguish different kinds of information flows. A direct flow between two entities is denoted by an arrow, but we sometimes use a specific terminology depending on the sort of the source and the target of an edge. More precisely, we call an edge between variables a *data flow*, an edge between processes *a control flow*, and edge from a variable to a process an *data-to-control flow* and vice versa for *control-to-data flows*. We also call *flow* to a path in a body graph. A path between variables is called *explicit flow* if it does not contain any control point. Otherwise it is called *implicit flow*. The input interface can be understood as the entities being used in the flow, while the output interface can be seen as the entities being provided by the flow. Note that the input and output mappings may not agree

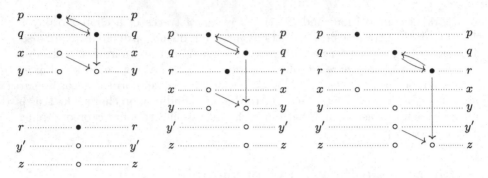

Fig. 9. Flows F_1 (top left), $\mathbf{Id}_{\{r,y',z\}}$ (bottom left), $F_1 \otimes \mathbf{Id}_{\{r,y',z\}}$ (mid) and $F2$ (right).

on a given entity η, see e.g. how the overwrite of variable of z is modelled in the flow of Fig. 1.

In the following we denote the set of entities $I \cup O$ involved in a flow graph $F = I \xrightarrow{i} G \xleftarrow{o} O$ by $en(F)$. Our flow graphs can be related to standard concepts and terminology of information flow control. For instance, given a flow $F = I \xrightarrow{i} G \xleftarrow{o} O$, we can define the set of *influencers* of a set of entities $E \subseteq O$, denoted $\mathcal{I}_F(E)$, as $\{\eta \in I \mid i(\eta) \rightarrow_G^+ \eta' \wedge \eta' \in o(E)\}$, i.e. the set of input entities that have a flow towards some entity in E exposed in the output of F. Conversely, we can define the set of *readers* of a set of entities $E \subseteq I$, denoted $\mathcal{R}_F(E)$, as $\{\eta \in O \mid \exists \eta' \in i(E) \wedge \eta' \rightarrow_G^+ o(\eta)\}$, i.e. the set of output entities that have a flow from some entity in E exposed in the input of F. Here $u \rightarrow_G^+ v$ denotes that v is reachable from u through a path of positive length in graph G.

Flow graphs are equipped with suitable operations such as the empty flow $\mathbf{0}$, a family of identities $\mathbf{Id}_N$ indexed by a set of entities N (see e.g. $\mathbf{Id}_{\{r,y',z\}}$ in the bottom left of Fig. 9), a binary (associative, commutative) parallel composition operation $\otimes$ (see e.g. $F_1 \otimes \mathbf{Id}_{\{r,y',z\}}$ in the middle of Fig. 9) and a binary (associative) sequential composition operation $\circ$ (e.g. the flow graph in Fig. 1 can be obtained as the composition $(F_1 \otimes \mathbf{Id}_{\{r,y',z\}}) \circ F2$ of the flows in Fig. 9). A precise definition of those operations can be found in Sect. A (taken from [15]), which recasts the original presentations of [6,11] in the exact shape we need. The intuitive idea is that the sequential composition of a flow F with a flow G is the result of identifying the outputs of F with the inputs of G and merging their body graphs accordingly. The resulting graph has the input interface of F as input and the output of G as output. Instead, the parallel composition of a flow F with a flow G is the result of identifying inputs of F with inputs of G and outputs F with outputs G and merging their body graphs accordingly. The resulting graph has as input (resp. output) interface the union of the input (resp. output) interfaces of F and G.

Instrumenting the Semantics. We denote the set of all flows by $\mathcal{F}$. The instrumentation of the semantics with flows is obtained by mapping actions in $\mathcal{A}$ into flows via some suitable function $flows : \mathcal{A} \rightarrow \mathcal{F}$.

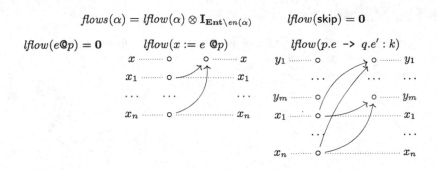

Fig. 10. Mapping of events into flows: explicit data flows

Figure 10 presents an example of such a mapping function, where $v(e) = \{x_1, \ldots, x_n\}$ and $v(e') = \{y_1, \ldots, y_m\}$. In particular, the function represents an annotation based on explicit data flows, where one is not interested in channels or principals, but just in direct transfer of data in interactions and assignments. The example function *flows* relies on a function *lflows* that defines the *local* flow of an event α. Such local flow is composed in parallel with $\mathbf{I}_{\mathbf{Ent}\setminus en(\alpha)}$, i.e. the identity on all entities not involved in α. The local flows for skip are defined to be the empty graph (no flow at all). It is worth remarking that in those flows the variables x or y_i can belong to $v(e)$ and hence coincide with some variable x_i. This kind of flow annotation can be useful, for instance, in the data providing case study of Sect. 2.1. We will refer to this flow annotation by $flows_e$.

Another example is depicted in Fig. 11. This flow annotation considers explicit data flows and some implicit data and control flows. For example, the local flows associated to conditions in decision points record the data-to-control flow from the free variables of e to the principal p and the data flow to variable x. Moreover, the flows for assignments and interactions are similar with the difference being that in interactions we record the mutual control flow between the interacting principals. We call this flow annotation function $flows_e$.

Yet another example can be found in Fig. 12. In this case, that we will refer to as $flows_a$, we are interested in flows related to how processes directly access data by either reading or writing variables. Note that, contrary to the previous cases, we are not interested in observing the fact that a variable has been overwritten. This is the kind of flow annotation that would make sense in the example we saw in Sect. 2.2, related to the MapReduce computation, where one is interested in controlling the locality of data accesses.

As we have seen, our approach provides some flexibility in the definition of flow annotations. However, the soundness of our approach relies on some *well-formedness* restrictions of those annotations.

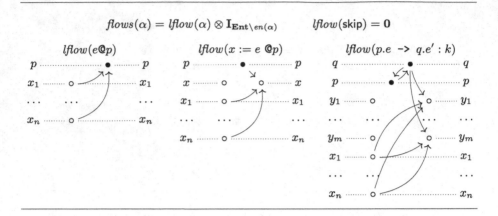

Fig. 11. Mapping of events into flows: explicit and implicit flows

Definition 5 (well-formed flow annotation). *A·flow annotation function* $flows : \mathcal{A} \to \mathcal{F}$ *is well-formed iff* $\forall \alpha \in \mathcal{A} : flows(\alpha) = F \otimes \mathbf{I}_{\mathbf{Ent} \setminus en(F)} \wedge en(F) \subseteq en(\alpha)$.

Intuitively, the idea is that a well-formed flow annotation associates a flow to an event α which is composed by an arbitrary flow F on some entities occurring in α and the identity on all other entities not occurring in F. It is easy to see that the flow annotations of Figs. 10, 11 and 12 are well-formed. Note that well-formedness also forbids the introduction of fresh entities outside **Ent** in the interface of the defined flows. Of course, flow annotations are a doubly-sharped mechanism: it provides a lot of flexibility to the information flow engineer, but it also discharges on him the responsibility of specifying the local flows of interest. As an extreme case consider that a flow annotation could be just defined as the constant $\mathbf{Id_{Ent}}$. In this case, no flow would be observed and all policies would be satisfied.

From now on, we restrict our attention to well-formed annotation functions. Given a well-formed flow annotation function *flows* the flow-instrumentation of the operational semantics of our language can be obtained as the relation $\to \subseteq \mathcal{C} \times \mathcal{F} \times \mathcal{C}$ defined as $\{C \xrightarrow{flows(\alpha)} C' \mid C \xrightarrow{\alpha} C'\}$. Similarly, the traces defined by the denotational semantics can be transformed into flows by sequentially composing the flows associated to the events of a trace.

Definition 6 (trace flows). *The flows of a trace σ, denoted $flows(\sigma)$, is given by function $flows : \mathcal{A}^* \to \mathcal{F}$ defined as*

$$flows(\epsilon) = \mathbf{Id_{Ent}} \qquad flows(\sigma\sigma') = flows(\sigma) \circ flows(\sigma')$$

The above definition does not define $flows(\alpha)$ to provide the afore mentioned flexibility in the observation of flows in events. Function *flows* is lifted to set of traces T and choreographies C in the obvious way, i.e. $flows(T) = \{flows(\sigma) \mid$

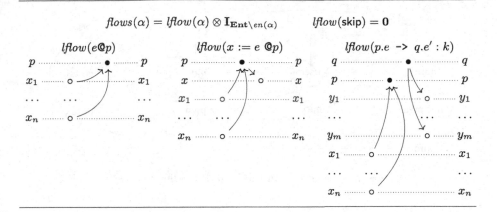

$$flows(\alpha) = lflow(\alpha) \otimes \mathbf{I}_{\mathbf{Ent} \setminus en(\alpha)} \qquad lflow(\mathsf{skip}) = \mathbf{0}$$

Fig. 12. Mapping of events into flows: data access flows

$\sigma \in T\}$ and $flows(C) = flows(Traces(C))$. As a simple example, the flow in Fig. 1 represents the flow of the only trace of the choreography C1 discussed in the Introduction obtained with the mapping of events into flows defined in Fig. 11.

5 Typing Choreographies

We represent information flow policies with flow graphs with the idea that a flow graph denotes *all* flows that are allowed in a system.

Definition 7 (information flow policy). *An information flow policy Π is a graph with interface $\mathbf{Ent} \xrightarrow{i} G \xleftarrow{o} \mathbf{Ent}$. The set of all policies is denoted by $\mathcal{P}$.*

It is worth to note that we require the input and output of a policy to coincide with the set of all entities $\mathbf{Ent}$ of interest. We call a policy *coherent* if $i = o$, i.e. if all entities are mapped to the same node in the body by both input and output mappings. Note that when a policy $F = \mathbf{Ent} \xrightarrow{i} G \xleftarrow{o} \mathbf{Ent}$ agrees in its input and output interfaces, we can rename some nodes in the body G with their (unique) images in the interface, possibly after alpha-renaming some internal nodes to avoid name clashes. This way we can provide the more compact notation that we use some of our figures. An additional simplification that we do in our graphical notation is that if a flow or policy is the identity on some entity η, then we neglect η in the visual notation. For instance, in all our examples we neglect the channels used in the choreographies.

We have already seen some examples of policies in Sect. 2, namely in Figs. 3 and 6. An additional example can be found in Fig. 13. It specifies a policy for the data providing case study which extends the policy of Fig. 3 to implicit flows. Figure 14, instead, provides two additional policies for the MapReduce case study. The one on the left focuses on control flows and may be useful to

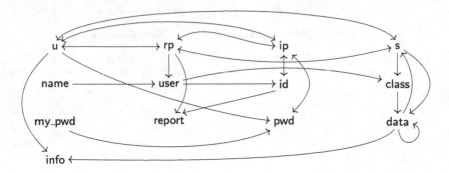

Fig. 13. Policy $\Pi_{s,2}$ in the Data Providing Service

control how principals depend on each other, e.g. in case of failure. The one on the right is oriented to explicit data flows.

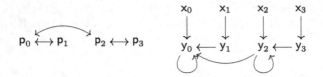

Fig. 14. Policies $\Pi_{m,2}$ (left) and $\Pi_{m,3}$ (right) for MapReduce

The following definition formalizes the notion of satisfaction of an information flow policy by a choreography, resp. a set of traces, a *decomposable* trace, an *atomic* trace, and an information flow. Here, a *decomposable* refers to the fact that we are interested in observing all components (i.e. subtraces) of a trace and *atomic* refers to the fact that we want to observe the trace as a whole.

Definition 8 (policy satisfaction). *The set of policy satisfaction relations* $\_ \models \_ : \mathcal{C} \cup 2^{\mathcal{A}^*} \cup \mathcal{A}^* \cup \mathcal{F} \to \mathcal{P}$, $\_ \vdash \_ : \mathcal{A}^* \to \mathcal{P}$, *is defined by:*

$$
\begin{aligned}
C &\models \Pi \quad \textit{iff} \quad Traces(C) \models \Pi \\
T &\models \Pi \quad \textit{iff} \quad \forall \sigma \in T : \sigma \models \Pi \\
\sigma &\models \Pi \quad \textit{iff} \quad \sigma = \sigma'' \sigma' \sigma''' \Rightarrow \sigma' \vdash \Pi \\
\sigma &\vdash \Pi \quad \textit{iff} \quad flows(\sigma) \models \Pi \\
F &\models \Pi \quad \textit{iff} \quad \forall \eta, \eta' \in en(\Pi) : i_F(\eta) \to^*_{G_F} o_F(\eta') \ \Rightarrow \ i_{\Pi}(\eta) \to^*_{G_{\Pi}} o_{\Pi}(\eta')
\end{aligned}
$$

Intuitively, the idea is that a choreography satisfies a policy Π if all its traces satisfy the policy Π. A trace σ satisfies Π if no subtrace of σ introduces a flow from an entity η to an entity η' that is not allowed in Π.

Our type system statically checks if a choreography C satisfies a policy Π. Our types are thus policies and our type judgements are of the form $\mathbf{Ent} \vdash C : \Pi$.

$$\dfrac{\mathbf{Ent} \vdash C_1 : \Pi \quad \mathbf{Ent} \vdash C_2 : \Pi}{\mathbf{Ent} \vdash C_1 ; C_2 : \Pi} \qquad \dfrac{\mathbf{Ent} \vdash C_1 : \Pi \quad \mathbf{Ent} \vdash C_2 : \Pi}{\mathbf{Ent} \vdash C_1 \mid C_2 : \Pi} \qquad \dfrac{A \models \Pi}{\mathbf{Ent} \vdash A : \Pi}$$

$$\dfrac{e@p \models \Pi \quad \mathbf{Ent} \vdash C_1 : \Pi \quad \mathbf{Ent} \vdash C_2 : \Pi}{\mathbf{Ent} \vdash \text{if } e@p \text{ then } C_1 \text{ else } C_2 : \Pi} \qquad \dfrac{e@p \models \Pi \quad \mathbf{Ent} \vdash C_1 : \Pi \quad \mathbf{Ent} \vdash C_2 : \Pi}{\mathbf{Ent} \vdash \text{while } e@p \text{ do } C_1 \text{ then } C_2 : \Pi}$$

Fig. 15. Type system

The type system is sound for *coherent* policies, i.e. policies whose input and output interfaces agree.

Definition 9 (type system). *The type system for judgements* $\mathbf{Ent} \vdash C : \Pi$ *is defined by the rules of Fig. 15.*

The main result of our work is the soundness of the type system (cf. Theorem 1). In order to prove such result we have first to prove the following lemma that formalises the fact that trace concatenation preserves the satisfaction of policies.

Lemma 1 (concatenation). *Let* $\sigma_1, \sigma_2 \in \mathcal{A}^*$ *be two traces and* $\Pi \in \mathcal{P}$ *a coherent policy. If* $\sigma_1 \models \Pi$ *and* $\sigma_2 \models \Pi$ *then* $\sigma_1 \sigma_2 \models \Pi$.

Proof. The proof is by induction on the length of σ_1 and σ_2.

[$\sigma_1 = \epsilon$ or $\sigma_2 = \epsilon$] These are trivial cases.

[$\sigma_1 = \alpha_1$ and $\sigma_2 = \alpha_2$, $\alpha_i \in \mathcal{A}$]. To prove $\sigma_1 \sigma_2 \models \Pi$ we have to show that $\sigma \vdash \Pi$ for all subtraces of $\alpha_1 \alpha_2$. Those are four: ϵ, α_1, α_2 and $\alpha_1 \alpha_2$. The first three cases are trivial. The interesting case is the latter. The proof is by contradiction. Suppose that $\alpha_1 \alpha_2 \not\models \Pi$. This means that there exist two different entities $\eta, \eta' \in en(\Pi)$ such that $\eta \in \mathcal{R}_{flows(\alpha_1 \alpha_2)}(\eta')$ but $\eta \notin \mathcal{R}_\Pi(\eta')$. Since $\alpha_i \models \Pi, i \in \{1, 2\}$ we know that $\eta \notin \mathcal{R}_{flows(\alpha_i)}(\eta')$. Hence, the flow from η' to η must have been introduced in the composition of the flows of α_1 and α_2. There must exist an entity η'' such that $\eta'' \in \mathcal{R}_{flows(\alpha_1)}(\eta')$ and $\eta'' \in \mathcal{I}_{flows(\alpha_2)}(\eta)$. But since $\eta'' \in \mathbf{Ent}$ (by well-formedness of *flows*) and $\alpha_i \models \Pi, i \in \{1, 2\}$ it must be the case that $\eta'' \in \mathcal{R}_\Pi(\eta')$ and $\eta'' \in \mathcal{I}_\Pi(\eta)$. Moreover, since Π s coherent, we know that $i(\eta'') = o(\eta'')$. Hence Π must be such that $\eta \in \mathcal{R}_\Pi(\eta')$. This is a contradiction. Hence, $\sigma_1 \sigma_2 \vdash \Pi$. Since we have shown that all subtraces of $\alpha_1 \alpha_2$ satisfy Π we can conclude that $\sigma_1 \sigma_2 \models \Pi$.

[$\sigma_1 = \alpha_1 \sigma_1'$, $\alpha_1 \in \mathcal{A}$] We have to show that $\sigma \vdash \Pi$ for all subtraces σ of $\alpha_1 \sigma_1' \sigma_2$. We distinguish two cases (i) traces of the form σ' with $\sigma' \sigma'' = \sigma_1 \sigma_2$ and (ii) traces of the form $\alpha_1 \sigma'$ with $\sigma' \sigma'' = \sigma_1 \sigma_2$. Consider case (i) first. We know that $\sigma_1' \models \Pi$ and $\sigma_2 \models \Pi$. By induction, we can conclude that $\sigma_1' \sigma_2 \models \Pi$. Hence all sub-traces of $\sigma_1 \sigma_2$ not starting with α_1 satisfy Π. Consider now case (ii), i.e. traces of the form $\alpha_1 \sigma'$ with $\sigma' \sigma'' = \sigma_1 \sigma_2$. Again, we can apply induction since $\alpha_1 \models \Pi$ and $\sigma' \models \Pi$ (by case i). Thus we conclude that $\sigma_1 \sigma_2 \models \Pi$. $\qquad\qquad\qquad\qquad\qquad\qquad\qquad\qquad\qquad\qquad\qquad\qquad\qquad\qquad\quad\square$

Theorem 1 (soundness). *Let $C \in \mathcal{C}$ be a well-formed choreography and $\Pi \in \mathcal{P}$ be a coherent policy. If $\mathbf{Ent} \vdash C : \Pi$ then $C \models \Pi$.*

Proof. The proof is by induction on the structure of C.

$[C = A]$ This case is trivial since $A \models \Pi$ is precisely the premise for typing A.

$[C = C_1; C_2]$ By definition, every trace of $C_1; C_2$ is of the form $\sigma_1 \sigma_2$ with $\sigma_i \in \mathit{Traces}(C_i), i \in \{1, 2\}$. We have that $\sigma_i \models \Pi, i \in \{1, 2\}$ by induction since the typing rules for $C_1; C_2$ require $\mathbf{Ent} \vdash C_i : \Pi, i \in \{1, 2\}$. Hence, we can apply Lemma 1 to conclude that $\sigma_1 \sigma_2 \models \Pi$.

$[C = C_1 \mid C_2]$ The proof of this case relies on well-formedness, which allows us to transform every trace $\sigma \in \mathit{Traces}(C_1 \mid C_2)$ into a trace σ' in one of the forms used in the above case, while having $\mathit{flows}(\sigma) = \mathit{flows}(\sigma')$.

We start proving that the transformation is possible and later prove that it indeed preserves the flows. Every trace $\sigma \in \mathit{Traces}(C_1 \mid C_2)$ is of the form $\sigma_{1,1}\sigma_{1,2} \ldots \sigma_{n,1}\sigma_{n,2}$ with $\sigma_{1,i} \ldots \sigma_{n,i} \in \mathit{Traces}(C_i), i \in \{1, 2\}$. It is easy to see if σ is of the form $\sigma' \sigma_{k,2}\sigma_{k+1,1}\sigma''$ with $\sigma_{k,2} = \sigma'_{k,2}\alpha$ and $\sigma_{k+1,1} = \beta\sigma'_{k+1,1}$, with $\alpha, \beta \in \mathcal{A}$, we can transform σ into trace $\sigma''' = \sigma'\sigma'_{k,2}\beta\alpha\sigma'_{k+1,1}\sigma''$ which belongs to $\mathit{Traces}(C_1 \mid C_2)$ by conditions (1) and (2) of well-formedness. Indeed, condition (1) requires α and β two involve disjoint entities ($en(\alpha) \cap en(\beta)$) and (2) ensures that transposing α and β in σ yields a trace that belongs to $\mathit{Traces}(C_1 \mid C_2)$.

It remains to show that σ and σ''' have the same flows. The main idea is that $\mathit{flows}(\alpha\beta) = \mathit{flows}(\beta\alpha)$ since $en(\alpha) \cap en(\beta) = \emptyset$. This can be shown as follows

$$
\begin{aligned}
\mathit{flows}(\alpha\beta) &= \mathit{flows}(\alpha) \circ \mathit{flows}(\beta) \\
&= (F \otimes \mathbf{Id}_{\mathbf{Ent} \setminus en(F)}) \circ (G \otimes \mathbf{Id}_{\mathbf{Ent} \setminus en(G)}) && \text{(well-formed } \mathit{flows}\text{)} \\
&= \begin{pmatrix} F \\ \otimes \\ \mathbf{Id}_{\mathbf{Ent} \setminus (en(F) \cup en(G))} \\ \otimes \\ \mathbf{Id}_{en(G)} \end{pmatrix} \circ \begin{pmatrix} \mathbf{Id}_{en(F)} \\ \otimes \\ \mathbf{Id}_{\mathbf{Ent} \setminus (en(F) \cup en(G))} \\ \otimes \\ G \end{pmatrix} && \text{(since } en(F) \cap en(G) = \emptyset\text{)} \\
&= \begin{pmatrix} (F & \circ \ \mathbf{Id}_{en(F)}) \\ & \otimes \\ (\mathbf{Id}_{\mathbf{Ent} \setminus (en(F) \cup en(G))} & \circ \ \mathbf{Id}_{\mathbf{Ent} \setminus (en(F) \cup en(G))}) \\ & \otimes \\ (\mathbf{Id}_{en(G)} & \circ \ G) \end{pmatrix} && \text{(distribution)} \\
&= F \otimes \mathbf{Id}_{\mathbf{Ent} \setminus (en(F) \cup en(G))} \otimes G && \text{(identity)} \\
&= G \otimes \mathbf{Id}_{\mathbf{Ent} \setminus (en(F) \cup en(G))} \otimes F && \text{(commutativity)}
\end{aligned}
$$

from which we can apply the same equations (upwards, replacing G by F) to obtain $\mathit{flows}(\beta\alpha)$.

Applying the above described transpositions in σ as much as needed results in σ being rewritten into $\sigma_1 \sigma_2$ with $\sigma_i \in \mathit{Traces}(C_i), i \in \{1, 2\}$. Since $\mathbf{Ent} \vdash C_i : \Pi, i \in \{1, 2\}$, the proof schema based on the application of Lemma 1 used in the above case can then be applied to conclude that $C \models \Pi$.

[C = if $e@p$ then C_1 else C_2] By definition, every trace σ in *Traces* (if $e@p$ then C_1 else C_2) is of the form $e@p \, \sigma'$ with $\sigma' \in Traces(C_i), i \in \{1, 2\}$. The typing rule for C requires $e@p \models \Pi$ and $\mathbf{Ent} \vdash C_i : \Pi, i \in \{1, 2\}$. By induction we have $C_i \models \Pi, i \in \{1, 2\}$ so that we can apply the Lemma 1 to conclude that $e@p \, \sigma' \models \Pi$.

[C = while $e@p$ do C_1 then C_2] This case is similar to the above one.   □

Let us now consider the choreographies P and $M_{0,2}$ of our case studies, the flow annotation functions $flows_e$, $flows_i$, and $flows_a$ and the policies $\Pi_{s,1}$, $\Pi_{s,2}$, $\Pi_{m,1}$, $\Pi_{m,2}$ and $\Pi_{s,3}$. For ease of notation, let us use the notation $C \models_f \Pi$ to refer to $C \models \Pi$ when *flows* is f.

One can easily see that the following satisfaction statements regarding the choreography P of the identity-secured data providing service can be concluded from our type system: P $\models_{flows_e} \Pi_{s,1}$, P $\models_{flows_e} \Pi_{s,2}$, and P $\models_{flows_i} \Pi_{s,2}$. An easy way to see this is to note that all events in the choreography (actions and expressions used in the branching statements) satisfy the corresponding policy. Instead the statement P $\models_{flows_i} \Pi_{s,1}$ cannot be concluded. As a matter of fact, policy $\Pi_{s,1}$ is not satisfied by P, since $\Pi_{s,1}$ does not allow implicit flows, that P actually has. For instance, the flow annotation function $flows_i$ would reveal an implicit flow from my_pwd to class that is not allowed by the policy $\Pi_{s,1}$.

$$flows(\alpha) = lflow(\alpha) \otimes \mathbf{I}_{\mathbf{Ent} \setminus en(\alpha)} \qquad lflow(\text{skip}) = \mathbf{0}$$

$$lflow(e@p) = \mathbf{0} \qquad lflow(x := e \,\, @p) = \mathbf{0} \qquad lflow(p.e \rightarrow q.e' : k)$$

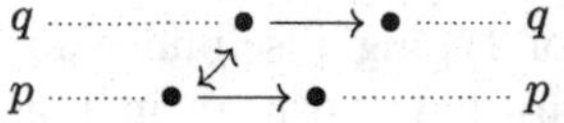

Fig. 16. Mapping of events into flows: control flows with temporal dependencies

Regarding the MapReduce case study, we can conclude, for instance that $M_{0,2} \models_{flows_a} \Pi_{m,1}$, $M_{0,2} \models_{flows_e} \Pi_{m,3}$, and $M_{0,2} \models_{flows_i} \Pi_{m,3}$. However, we cannot conclude that $M_{0,2} \models_{flows_i} \Pi_{m,2}$ since $flows_i$ requires us to observe data flows that $\Pi_{m,2}$ does not allow. A flow annotation function like $flows_e$ but limited to control flows would then allow us to check the policy.

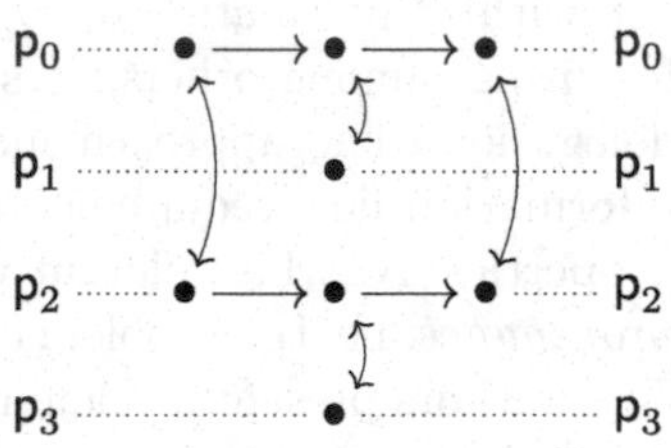

Actually, the policy $\Pi_{m,2}$ is more permissive

Fig. 17. Policy $\Pi_{m,4}$

than it could be. It allows one to have dependencies between all principals, including p_1 and p_3, whose respective controls do not actually depend on each other in P. A more relaxed policy forbidding control flow dependencies between p_1 and p_3 can be found in Fig. 17. Contrary to all policies we have presented so far, policy $\Pi_{m,4}$ is not coherent, i.e. it does not agree in the interface and, thus, our type system cannot be applied. That is, some entities in the interface correspond to different nodes in the body graph. This is essential to capture some information about the temporal order of flows. It is easy to check that the policy allows flows from p_1 to p_0 and p_2 but not to p_3.

To check policies like $\Pi_{m,4}$ we would have to consider the richer control flow annotation $flows_d$ of Fig. 16 to avoid spurious flows. Indeed, flow annotation functions like $flows_e$ abstract away from temporal information in the control flow. Such abstractions introduce spurious flows between, for instance, p_1 and p_3 whose respective controls do not depend on each other. Instead, $flows_d$ records some

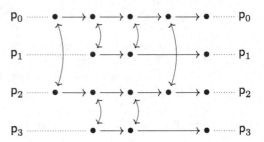

Fig. 18. Control dependencies in MapReduce

basic information about the temporal order of interactions. Figure 18 illustrates the flow of all maximal traces of choreography P. The policy $\Pi_{m,4}$ would be satisfied. For example, the above mentioned spurious flows between p_1 and p_3 are not present now. Indeed, p_1 interacts directly with p_0 and indirectly with p_2 but only *after* p_2 has finished interacting with p_3. Extending our type system to deal with such policies is subject of current work.

6   Related Works and Conclusion

The use of typing disciplines for security systems has a long tradition. The recent years have seen an increasing interest in applying techniques based on behavioural types to security analysis. We refer to [2] for a comprehensive survey and limit our discussion to the most relevant and recent works in that area.

A first work worth mentioning is [7], which presents an approach for dealing with non-interference properties in distributed systems where components interact within multiparty sessions. Systems are described with a session calculus featuring, among others, session creation, inter-session interaction, and session delegation. The approach includes a session type system whose rules include information flow requirements to ensure both behavioural and non-interference properties. Another relevant work is presented in [14]. The work focuses on *data provenance*, i.e. the problem of keeping track of how data flow and are processed. The authors present a calculus to describe how processes use, consume and publish linked data. The approach is equipped with type systems for the calculus to ensure mandatory access control properties based on security levels [14] and role-based access control properties [13].

There are two main differences between the above discussed works and our own work. The first one is our focus on *discretionary* information flow, instead of other forms of information flow control (mandatory, role-based, non-interference, etc.). A second difference lies in the specification languages used; our choreography description language differs from the calculi used in the above mentioned works. It would not be trivial, for instance, to integrate our type system with the type system of [7] as both focus on related but significantly different languages and properties. We would also like to remark that our work is still in a preliminary phase and does not yet consider aspects of choreography realizability, local projections and systems with multiple-sessions.

As a matter of fact, those aspects are part of our research agenda. We are currently developing a suitable notion of local projections and a type system that ensure that projections enjoy suitable semantic relations with choreographies, from which we can conclude that local projections do not introduce flows not specified in the choreography. The main contribution of the paper would then be applied to ensure Π-secure distributed implementations.

We also plan to investigate the development of our approach in several directions, including the possibility to specify and check temporal aspects by extending our type system beyond coherent policies. We would also like to develop an inference system to compute over- and under-approximations of information flows, and to consider *intransitive* information flow properties. On the application side, we plan to investigate the suitability of our approach to popular parallel programming frameworks such as MPI. There is indeed an urgent demand of formal guarantees for systems developed in such frameworks [18] and some flow analysis works already exist (e.g. [1,5]) also based on behavioural types (e.g. [17]).

Acknowledgement. We would like to dedicate this work to José Meseguer, for his inspiring and influencing works in the areas of Concurrency Theory, Algebraic Specifications and Security. We hope that José will understand (and forgive!) the absence of a non-interference result in our work. The first author would like to express his gratitude to José for honouring him with his friendship and giving him the unique experience of scientific collaboration. The first author is also grateful to Fabrizio Montesi, Emilio Tuosto and Marco Carbone for fruitful feedback on early versoins of this work, and to the organizers of the BETTY COST Action for their gentle invitation to present an early version of the work in a meeting.

A   Graphs with Interfaces

We recall and adapt a few definitions concerning graphs, and their extension with *interfaces*, referring to [6,11,15] for a more detailed presentation.

Definition 10 (graphs). *A is a four-tuple $\langle V, E, s, t \rangle$ where V is the set of nodes, E is the set of edges and $s, t : E \to V$ are the source and target functions. A graph morphism is a pair of functions $\langle f_V, f_E \rangle$ preserving the source and target functions, i.e. $f_V \circ s = s \circ f_E$ and $f_V \circ t = t \circ f_E$.*

Definition 11 (graphs with interfaces). *A graph with interfaces is a span of graph morphisms $I \xrightarrow{i} G \xleftarrow{o} O$, where G is a body graph, I and O are the input and output graph interfaces, and $i : I \to G$, $o : O \to G$ are the input and output graph morphisms. An interface graph morphism $f : \mathbb{G} \Rightarrow \mathbb{H}$ is a triple of graph morphisms $\langle f_I, f, f_O \rangle$, preserving the input and output morphisms.*

With an abuse of notation, we sometimes refer to the image of the input and output morphisms as inputs and outputs, respectively, and we use the term *graph* to refer to abbreviate *graphs with interfaces*. We restrict our attention to graphs with *discrete* interfaces, i.e., such that their set of edges is empty.

The following definitions define the sequential and parallel composition of graphs with interfaces, whose informal description can be found in Sect. 4.

Definition 12 (sequential composition of graphs). *Let* $\mathbb{G} = I \xrightarrow{i} G \xleftarrow{j} J$ *and* $\mathbb{G}' = J \xrightarrow{j'} G' \xleftarrow{o} O$ *be graphs with interfaces. Then, their* sequential composition *is the graph* $\mathbb{G} \circ \mathbb{G}' = I \xrightarrow{i'} G'' \xleftarrow{o'} O$, *for* G'' *the disjoint union* $G \uplus G'$, *modulo the equivalence on nodes induced by* $j(x) = j'(x)$ *for all* $x \in N_J$, *and* i', o' *the uniquely induced arrows.*

Definition 13 (parallel composition of graphs). *Let* $\mathbb{G} = I \xrightarrow{i} G \xleftarrow{o} O$ *and* $\mathbb{H} = I' \xrightarrow{i'} H \xleftarrow{o'} O'$ *be two graphs with interfaces. Then, their* parallel composition *is the* GWDI $\mathbb{G} \otimes \mathbb{H} = (I \cup I') \xrightarrow{i''} G' \xleftarrow{o''} (O \cup O')$, *for* G' *the disjoint union* $G \uplus H$, *modulo the equivalence on nodes induced by* $o(y) = o'(y)$ *for all* $y \in N_O \cap N_{O'}$ *and* $i(y) = i'(y)$ *for all* $y \in N_I \cap N_{I'}$, *and* i'', o'' *the uniquely induced arrows.*

With an abuse of notation, the set-theoretic operators are defined component-wise. The operations are concretely defined, modulo the choice of canonical representatives for the set-theoretic operations: the result is independent of such a choice, up-to isomorphism of the body graphs.

A *graph expression* is a term over the syntax containing all graphs with discrete interfaces as constants, and parallel and sequential composition as binary operators. An expression is *well-formed* if all occurrences of those operators are defined for the interfaces of their arguments, according to Definitions 12 and 13; its interfaces are computed inductively from the interfaces of the graphs occurring in it, and its *value* is the graph obtained by evaluating all operators in it. For the axiomatic properties of the operations (e.g. associativity, identity of $\circ$, associativity, commutativity and identity of $\otimes$) we refer to [6,11].

References

1. Aananthakrishnan, S., Bronevetsky, G., Gopalakrishnan, G.: Hybrid approach for data-flow analysis of MPI programs. In: Malony, A.D., Nemirovsky, M., Midkiff, S.P. (eds.) International Conference on Supercomputing, ICS 2013, pp. 455–456. ACM, Eugene, OR, USA, 10–14 June 2013. http://doi.acm.org/10.1145/2464996. 2467286
2. Bartoletti, M., Castellani, I., Denielou, P.M., Dezani-Ciancaglini, M., Ghilezan, S., Pantovic, J., Pérez, J.A., Thiemann, P., Toninho, B., Vieira, H.T.: Combining behavioural types with security analysis, state of The Art Report of WG2 - European Cost Action IC1201 BETTY (Behavioural Types for Reliable Large-Scale Software Systems). http://www.behavioural-types.eu/publications/WG2-State-of-the-Art.pdf/at_download/file

3. Bocchi, L., Melgratti, H., Tuosto, E.: Resolving non-determinism in choreographies. In: Shao, Z. (ed.) ESOP 2014. LNCS, vol. 8410, pp. 493–512. Springer, Heidelberg (2014). http://dx.doi.org/10.1007/978-3-642-54833-8_26

4. Boudol, G.: Secure information flow as a safety property. In: Degano, P., Guttman, J., Martinelli, F. (eds.) FAST 2008. LNCS, vol. 5491, pp. 20–34. Springer, Heidelberg (2009). http://dx.doi.org/10.1007/978-3-642-01465-9_2

5. Bronevetsky, G.: Communication-sensitive static dataflow for parallel message passing applications. In: Proceedings of the CGO 2009, The Seventh International Symposium on Code Generation and Optimization, pp. 1–12, Seattle, Washington, USA, 22–25 March 2009. IEEE Computer Society (2009). http://dx.doi.org/10.1109/CGO.2009.32

6. Bruni, R., Gadducci, F., Montanari, U.: Normal forms for algebras of connection. Theor. Comput. Sci. **286**(2), 247–292 (2002). http://dx.doi.org/10.1016/S0304-3975(01)00318-8

7. Capecchi, S., Castellani, I., Dezani-Ciancaglini, M.: Typing access control and secure information flow in sessions. Inf. Comput. **238**, 68–105 (2014). http://dx.doi.org/10.1016/j.ic.2014.07.005

8. Carbone, M., Montesi, F.: Deadlock-freedom-by-design: multiparty asynchronous global programming. In: Giacobazzi, R., Cousot, R. (eds.) The 40th Annual ACM SIGPLAN-SIGACT Symposium on Principles of Programming Languages (POPL 2013), pp. 263–274. ACM (2013). http://doi.acm.org/10.1145/2429069.2429101

9. Castagna, G., Dezani-Ciancaglini, M., Padovani, L.: On global types and multi-party session. Log. Methods Comput. Sci. **8**(1), 1–45 (2012). http://dx.doi.org/10.2168/LMCS-8(1:24)2012

10. Cohen, E.: Information transmission in computational systems. In: Sixth ACM Symposium on Operating Systems Principles, SOSP 1977, pp. 133–139. ACM, New York, NY, USA (1977). http://doi.acm.org/10.1145/800214.806556

11. Corradini, A., Gadducci, F.: An algebraic presentation of term graphs, via GS-monoidal categories. Appl. Categorical Struct. **7**(4), 299–331 (1999). http://dx.doi.org/10.1023/A:1008647417502

12. Dean, J., Ghemawat, S.: Mapreduce: simplified data processing on large clusters. Commun. ACM **51**(1), 107–113 (2008)

13. Dezani-Ciancaglini, M., Ghilezan, S., Jakśić, S., Pantović, J.: Types for role-based access control of dynamic web data. In: Mariño, J. (ed.) WFLP 2010. LNCS, vol. 6559, pp. 1–29. Springer, Heidelberg (2010)

14. Dezani-Ciancaglini, M., Horne, R., Sassone, V.: Tracing where and who provenance in linked data: a calculus. Theor. Comput. Sci. **464**, 113–129 (2012). http://dx.doi.org/10.1016/j.tcs.2012.06.020

15. Gadducci, F.: Graph rewriting for the pi-calculus. Math. Struct. Comput. Sci. **17**(3), 407–437 (2007). http://dx.doi.org/10.1017/S096012950700610X

16. Goguen, J.A., Meseguer, J.: Security policies and security models. In: IEEE Symposium on Security and Privacy, pp. 11–20 (1982)

17. Gopalakrishnan, G., Kirby, R.M., Siegel, S.F., Thakur, R., Gropp, W., Lusk, E.L., de Supinski, B.R., Schulz, M., Bronevetsky, G.: Formal analysis of MPI-based parallel programs. Commun. ACM **54**(12), 82–91 (2011). http://doi.acm.org/10.1145/2043174.2043194

18. Honda, K., Marques, E.R.B., Martins, F., Ng, N., Vasconcelos, V.T., Yoshida, N.: Verification of MPI programs using session types. In: Träaff, J.L., Benkner, S., Dongarra, J.J. (eds.) EuroMPI 2012. lncs, vol. 7490, pp. 291–293. Springer, Heidelberg (2012). http://dx.doi.org/10.1007/978-3-642-33518-1_37

19. Honda, K., Yoshida, N., Carbone, M.: Multiparty asynchronous session types. In: Necula, G.C., Wadler, P. (eds.) 35th ACM SIGPLAN-SIGACT Symposium on Principles of Programming Languages (POPL 2008), pp. 273–284. ACM (2008). http://doi.acm.org/10.1145/1328438.1328472

20. Lanese, I., Guidi, C., Montesi, F., Zavattaro, G.: Bridging the gap between interaction- and process-oriented choreographies. In: Cerone, A., Gruner, S. (eds.) Sixth IEEE International Conference on Software Engineering and Formal Methods, SEFM 2008, pp. 323–332, Cape Town, South Africa, 10–14 November 2008. IEEE Computer Society (2008). http://dx.doi.org/10.1109/SEFM.2008.11

21. Lohmann, N., Wolf, K.: Decidability results for choreography realization. In: Kappel, G., Maamar, Z., Motahari-Nezhad, H.R. (eds.) ICSOC 2011. LNCS, vol. 7084, pp. 92–107. Springer, Heidelberg (2011). http://dx.doi.org/10.1007/978-3-642-25535-9_7

22. Montesi, F.: Choreographic Programming. Ph.D. thesis, IT University of Copenhagen (2013). http://www.fabriziomontesi.com/files/m13_phdthesis.pdf

23. Myers, A.C., Liskov, B.: A decentralized model for information flow control. In: SOSP. pp. 129–142 (1997). http://doi.acm.org/10.1145/268998.266669

24. Nielson, F., Nielson, H.R., Hankin, C.: Principles of Program Analysis. Springer, Heidelberg (1999)

Verifying Reachability-Logic Properties on Rewriting-Logic Specifications

Dorel Lucanu[1]([⊠]), Vlad Rusu[2], Andrei Arusoaie[2], and David Nowak[3]

[1] Faculty of Computer Science, Alexandru Ioan Cuza University, Iași, Romania
dlucanu@info.uaic.ro
[2] Inria Lille Nord Europe, Villeneuve-d'Ascq, France
{Vlad.Rusu,Andrei.Arusoaie}@inria.fr
[3] CRIStAL, CNRS and University of Lille, Villeneuve-d'Ascq, France
david.nowak@univ-lille1.fr

Abstract. Rewriting Logic is a simply, flexible, and powerful framework for specifying and analysing concurrent systems. Reachability Logic is a recently introduced formalism, which is currently used for defining the operational semantics of programming languages and for stating properties about program executions. Reachability Logic has its roots in a wider-spectrum framework, namely, in Rewriting Logic Semantics. In this paper we show how Reachability Logic can be adapted for stating properties of transition systems described by Rewriting-Logic specifications. We propose a procedure for verifying Rewriting-Logic specifications against Reachability-Logic properties. We prove the soundness of the procedure and illustrate it by verifying a communication protocol specified in Maude.

1 Introduction

Since its original formulation [1] by José Meseguer, Rewriting Logic (RWL) has been defined both as a *semantical framework*, suitable for describing concurrent and distributed systems, and as a *logical framework*, i.e. a meta-logic where other logics can be naturally represented. Both directions have dynamic and strong development. A concurrent system is specified by a rewrite theory (Σ, E, R), where Σ defines the syntax of the system and of its states, E defines the states of the system as an algebraic data type, and R is a set of rewriting rules defining the local transitions of the system. RWL deduction consists of a set of inference rules used to prove sequents $t \to t'$, meaning that the term-pattern t "becomes" t' by concurrently applying the rewrite rules R modulo the equations E. Here the relation "becomes" refers the dynamics of the specified concurrent system.

Maude [2] is the most well-known and most used implementation of RWL. Maude is accompanied by a set of tools for analysing rewrite theories and the systems they describe.

Reachability Logic (hereafter, RL) was introduced in a series of papers [3–6] as means for specifying the operational semantics of programming languages and for

© Springer International Publishing Switzerland 2015
N. Martí-Oliet et al. (Eds.): Meseguer Festschrift, LNCS 9200, pp. 451–474, 2015.
DOI: 10.1007/978-3-319-23165-5_21

stating reachability properties between states of program executions. Reachability Logic is a very promising framework. It has emerged from the Rewriting Logic Semantics project introduced by José Meseguer and Grigore Roşu [7]. Briefly, an RL formula $\varphi \Rightarrow \varphi'$ expresses *reachability relationships* between two sets of states of a system, denoted by the *patterns* φ and φ' respectively. Depending on the interpretation of formulas, the relationships can express either programming-language semantic steps, or safety properties of programs in the languages in question. Several widely-used languages (including C [8], Java [9]) have been completely specified using an RL-based semantics in the $\mathbb{K}$ tool [10], and non-trivial programs have been proved using an implementation of RL [1].

Contributions. In this paper we show that RL can be used beyond its (by now, traditional) domain of programming languages. Specifically, we adapt RL for stating properties of systems described in Rewriting Logic [11] (hereafter, RWL). We propose a procedure for proving RL properties of RWL specifications, we prove the soundness of our procedure, and illustrate its use by verifying RL properties of a communication protocol written in Maude [2].

Our contribution with respect to RL is a proved-sound verification procedure. Previous works [3–6] include sound and relatively complete proof systems for various versions of RL, but these systems lack strategies for rule applications, making them unpractical for verification; our procedure can be seen as such a strategy.

With respect to RWL, our contribution is the adaptation of the above procedure for verifying RWL theories against *reachability* properties $\varphi \Rightarrow \varphi'$, which say that all terminating executions starting from a state in the set φ eventually reach a state in the set φ'. Both φ and φ' denote possibly infinite sets of states. We note that RL properties for RWL theories are different from the reachability properties that can be checked in Maude using the **search** command or the Linear Temporal Logic (LTL) model checker [2]. The difference resides in the possibility of using first-order logic for constraining the initial and the final state terms, and in the interpretation of RL formulas. Specifically, the version of RL that we consider (the *all-paths* interpretation) corresponds to a subset of LTL interpreted on *finite paths*, whereas Maude's LTL model checker uses the standard (infinite-paths) interpretation of LTL; and both the model checker and the **search** command are bound to checking reachability properties starting from finitely many initial states by exploring finitely many execution paths.

Related work. We focus only on related verification approaches for RWL specifications. These fall under the usual classification of verification techniques: *algorithmic* ones, which essentially consist in using an automatic model checker; *deductive* ones, which involve an interaction with a theorem prover; and *abstraction-based* ones, which consist in first reducing the state-space of a system from unmanageable (e.g., infinite/large) to manageable (e.g., finite/small), a step that typically involves human interaction, and then using a model checker on the reduced system.

---

[1] Available at http://www.matching-logic.org/index.php/Special:MatchCOnline.

Algorithmic techniques include Maude's finite-state LTL model checker [12] with its more recent extensions to the *temporal logic of rewriting* [13], and a narrowing-based symbolic model-checker for handling classes of infinite-state systems [14]. Among the deductive techniques, [15,16] propose two different approaches for reducing safety properties of RWL to equational reasoning, and then using equational reasoning tools for proving the resulting encoded properties. We note that the encoding of RWL into equational logic was proposed earlier in [17] for defining the semantics of RWL. Among the abstraction-based techniques, equational abstractions [18], and algebraic simulations [19] are key contributions.

Finally, our verification procedure uses an operation called *derivative* that consists in computing the symbolic successors of a given set of states (represented by a formula). This symbolic computation is inspired from [20,21].

Outline. After preliminaries in Sect. 2, we introduce in Sect. 3 the notion of derivative, which is essential for our approach. We then introduce in Sect. 4 our procedure for verifying RL properties on transition systems also defined by RL formulas, and state its soundness. In Sect. 5, we adapt our approach to transition systems defined by RWL theories. We illustrate in Sect. 6 the usability of our procedure by applying it to a communication protocol described in Maude. Proofs of the technical results are included in an Appendix.

2 Preliminaries

2.1 Matching Logic

We recall the syntax and the semantics of Matching Logic (ml) as presented in [3]. Since ml is based on the many-sorted first-order logic (FOL), we recall first the basic definitions from FOL.

Given S a set of sorts, an S-sorted *first order signature* Φ is a pair (Σ, Π), where Σ is an algebraic S-sorted signature and Π is an indexed set of the form $\{\Pi_w \mid w \in S^*\}$ whose elements are called *predicate symbols*, where $\pi \in \Pi_w$ is said to have *arity* w. A Φ-*model* consists of a Σ-algebra M together with a subset $M_p \subseteq M_{s_1} \times \cdots \times M_{s_n}$ for each predicate $p \in \Pi_w$, where $w = s_1 \ldots s_n$.

Next, we define the syntax of FOL formulas over a first order signature $\Phi = (\Sigma, \Pi)$ and a possible infinite S-indexed set of variables Var. We let S denote the set of sorts in Φ and $T_\Sigma(Var)$ the algebra of Σ-terms with variables in Var.

The set of Φ-*formulas* is defined by

$$\phi ::= \top \mid p(t_1, \ldots, t_n) \mid \neg\phi \mid \phi \wedge \phi \mid (\exists V)\phi$$

where p ranges over predicate symbols Π, each t_i ranges over $T_\Sigma(Var)$ of appropriate sort, and V over finite subsets of Var.

Given a first order Φ-model M, a Φ-formula ϕ, and a valuation $\rho : Var \to M$, the satisfaction relation $\rho \models \phi$ is defined as follows:

1. $\rho \models \top$;
2. $\rho \models p(t_1, \ldots, t_n)$ iff $(\rho(t_1), \ldots, \rho(t_n)) \in M_p$;
3. $\rho \models \neg\phi$ iff $\rho \models \phi$ does not hold;
4. $\rho \models \phi_1 \wedge \phi_2$ iff $\rho \models \phi_1$ and $\rho \models \phi_2$;
5. $\rho \models (\exists V)\phi$ iff there is $\rho' : Var \to M$ with $\rho'(X) = \rho(X)$, for all $X \notin V$, such that $\rho' \models \phi$.

A formula ϕ is *valid in* M, denoted by $M \models \phi$, if it is satisfied by all valuations ρ.

We recall below the ml concepts and results used in this paper. Their presentation is based on [3].

Definition 1 (ML Formulas). *An* ml *signature* $\Phi = (\Sigma, \Pi, State)$ *is a first-order signature* (Σ, Π) *together with a distinguished sort State for states. The set of* ml*-formulas over* Φ *is defined by*

$$\varphi ::= \pi \mid \top \mid p(t_1, \ldots, t_n) \mid \neg\varphi \mid \varphi \wedge \varphi \mid (\exists V)\varphi$$

where the basic pattern π *ranges over* $T_{\Sigma, State}(Var)$, *p ranges over predicate symbols* Π, *each* t_i *ranges over* $T_\Sigma(Var)$ *of appropriate sorts, and V over finite subsets of Var.*

The sort *State* is intended to model system states. The free occurrence of variables in ml formulas is defined as usual (i.e., like in FOL) and we let *FreeVars*(φ) denote the set of variables freely occurring in φ. We often use particular ml formulas $\varphi \triangleq \pi \wedge \phi$, where π represents a state and ϕ is a FOL formula used for constraining this state.

Example 1. Assume that S includes the sorts *Nat, State*, Σ includes a binary operation symbol $\langle \_, \_ \rangle : Nat \times Nat \to State$, and Π the predicate symbols *div* and $\_>\_$, with arguments of sort *Nat*. Then $\varphi \triangleq \langle x, y \rangle \wedge (\exists z)((z > 1) \wedge div(z, x) \wedge div(z, y))$ is an ml formula. We have *FreeVars*$(\varphi) = \{x, y\}$.

Definition 2 (ML satisfaction relation). *Given* $\Phi = (\Sigma, \Pi, State)$ *an* ml *signature,* M *a* (Σ, Π)-*model,* φ *an* ml *formula over* Φ, $\gamma \in M_{State}$ *a state, and* $\rho : Var \to M$ *a valuation, the satisfaction relation* $(\gamma, \rho) \models \varphi$ *is defined as follows:*

1. $(\gamma, \rho) \models \pi$ *iff* $\rho(\pi) = \gamma$;
2. $(\gamma, \rho) \models \top$;
3. $(\gamma, \rho) \models p(t_1, \ldots, t_n)$ *iff* $(\rho(t_1), \ldots, \rho(t_n)) \in M_p$;
4. $(\gamma, \rho) \models \neg\varphi$ *iff* $(\gamma, \rho) \models \varphi$ *does not hold;*
5. $(\gamma, \rho) \models \varphi_1 \wedge \varphi_2$ *iff* $(\gamma, \rho) \models \varphi_1$ *and* $(\gamma, \rho) \models \varphi_2$; *and*
6. $(\gamma, \rho) \models (\exists V)\varphi$ *iff there is* $\rho' : Var \to M$ *with* $\rho'(X) = \rho(X)$, *for all* $X \notin V$, *such that* $(\gamma, \rho') \models \varphi$.

Example 2. Let M be a model defined such that M_{Nat} is the set of natural numbers, $M_<$ is the inequality "greater than" over natural numbers, and $M_{div}(m, n)$ holds iff m divides n. Let φ denote the ml formula $\langle x, y \rangle \wedge ((\exists z)(z > 1) \wedge div(z, x) \wedge div(z, y))$. If we consider $\rho(x) = 4$ and $\rho(y) = 6$ then we have $(\langle 4, 6 \rangle, \rho) \models \varphi$

because $\rho(\langle x, y \rangle) = \langle 4, 6 \rangle$ and $\rho \models ((\exists z)(z > 1) \wedge div(z, x) \wedge div(z, y))$. We do not have $(\langle 3, 5 \rangle, \rho) \models \varphi$ because $\rho(\langle x, y \rangle) \neq \langle 3, 5 \rangle$. Even if we consider $\rho'(x) = 3$ and $\rho'(y) = 5$ we still have $(\langle 3, 5 \rangle, \rho') \not\models \varphi$ because there is no m greater than 1 such that $M_{div}(m, 3)$ and $M_{div}(m, 5)$ hold.

Definition 3 (FOL encoding of ML). *If φ is an ml-formula then $\varphi^{=?}$ is the FOL formula $(\exists z)\varphi'$, where φ' is obtained from φ by replacing each basic pattern occurrence π with $z = \pi$, and z is a variable that does not occur in φ.*

Example 3. Here are a few examples of ml formulas and their FOL encodings:

φ	$\varphi^{=?}$
$(\pi_1 \wedge \phi_1) \vee (\pi_2 \wedge \phi_2)$	$(\exists z)((z = \pi_1 \wedge \phi_1) \vee (z = \pi_2 \wedge \phi_2))$
$\neg \pi$	$(\exists z)\neg(z = \pi)$
$\pi_1 \wedge \neg \pi_2$	$(\exists z)((z = \pi_1) \wedge \neg(z = \pi_2))$
$\pi \vee \neg \pi$	$(\exists z)(z = \pi \vee \neg(z = \pi))$

The relationship between ml formulas and their FOL encodings is given by the following result:

Proposition 1. $\rho \models \varphi^{=?}$ *iff there is γ such that $(\gamma, \rho) \models \varphi$.*

The following proposition is needed later for our soundness result.

Proposition 2. *If ϕ is a FOL (i.e. structureless) formula, then $(\phi \wedge \varphi)^{=?}$ is equivalent to $\phi \wedge \varphi^{=?}$. Moreover, if $\rho \models \phi$ and $(\gamma, \rho) \models \varphi$ then $(\gamma, \rho) \models \phi \wedge \varphi$.*

Example 4. We consider first an ml formula including two state terms. Let φ denote the ml formula $\langle x, y \rangle \wedge x < 5 \wedge \langle u, v \rangle \wedge 8 < v$. Then $\varphi^{=?}$ is equivalent to $(\exists z)z = \langle x, y \rangle \wedge z = \langle u, v \rangle \wedge x < 5 \wedge 8 < v$, which in turn is equivalent to $\langle x, y \rangle \wedge = \langle u, v \rangle \wedge x < 5 \wedge 8 < v$. We have $\rho \models \varphi^{=?}$ iff $\rho(\langle x, y \rangle) = \rho(\langle u, v \rangle) \wedge \rho(x) < 5 \wedge \rho(v) > 8$ iff $(\gamma, \rho) \models \varphi$, where $\gamma = \rho(\langle x, y \rangle) = \rho(\langle u, v \rangle)$.

If φ_1 is an ml formula not including a state term (i.e., it is *structureless* according to ml terminology), then $\varphi_1^{=?}$ is the same with φ_1.

2.2 Reachability Logic

In this section we recall reachability-logic formulas, the transition systems that they induce, and their all-paths interpretation [6]. We consider a fixed ml signature $\Phi = (\Sigma, \Pi, State)$, a set of variables Var, and a fixed Φ-model M.

Definition 4 (RL Formulas). *An RL formula is a pair $\varphi \Rightarrow \varphi'$ of ml-formulas.*

Definition 5 (Transition-System Specification). *An RL transition-system specification is a set S of RL-formulas. The transition system defined by S over M is $(M_{State}, \Rightarrow_S)$, where $\Rightarrow_S = \{(\gamma, \gamma') \mid (\exists \varphi \Rightarrow \varphi' \in S)(\exists \rho)(\gamma, \rho) \models \varphi \wedge (\gamma', \rho) \models \varphi'\}$. We write $\gamma \Rightarrow_S \gamma'$ for $(\gamma, \gamma') \in \Rightarrow_S$.*

Example 5. The following set of rules is meant to compute the gcd of two natural numbers:

$$\mathcal{S} = \{\langle x, y\rangle \wedge x > y \wedge y > 0 \Rightarrow (\exists k)\langle y, x - k * y\rangle \wedge x \geq k * y \wedge k > 0,$$
$$\langle x, y\rangle \wedge y \geq x \Rightarrow \langle y, x\rangle\}$$

We further assume that M interprets $+$ and $*$ as being the usual operations over natural numbers; $m - n$ is the difference between m and n, if $m > n$, or 0, otherwise. Examples of transitions are: $\langle 8, 2\rangle \Rightarrow_{\mathcal{S}} \langle 2, 0\rangle$ as instance of the first rule, and $\langle 8, 10\rangle \Rightarrow_{\mathcal{S}} \langle 10, 8\rangle$ and $\langle 2, 2\rangle \Rightarrow_{\mathcal{S}} \langle 2, 2\rangle$ as instances of the second rule.

In the sequel we consider a fixed transition system $(M_{State}, \Rightarrow_{\mathcal{S}})$.

Definition 6 (Execution Paths). *An* execution path *is a (possibly infinite) sequence of transitions* $\tau \triangleq \gamma_0 \Rightarrow_{\mathcal{S}} \gamma_1 \Rightarrow_{\mathcal{S}} \cdots$.

If $i \geq 0$ *then* $\tau|_{i..}$ *is the execution path consisting of the (possibly infinite) subsequence starting from* γ_i, *if any. An execution path is* complete *iff it is not a strict prefix of an another execution path.*

A pair (τ, ρ), *consisting of an execution path* $\tau \triangleq \gamma_0 \Rightarrow_{\mathcal{S}} \cdots$ *and a valuation* ρ, *starts from an ml formula* φ *if* $(\gamma_0, \rho) \models \varphi$.

Example 6. Examples of executions are $\tau \triangleq \langle 8, 10\rangle \Rightarrow_{\mathcal{S}} \langle 10, 8\rangle \Rightarrow_{\mathcal{S}} \langle 8, 2\rangle \Rightarrow_{\mathcal{S}} \langle 2, 0\rangle$ and $\tau' \triangleq \langle 8, 10\rangle \Rightarrow_{\mathcal{S}} \langle 10, 8\rangle \Rightarrow_{\mathcal{S}} \langle 8, 2\rangle \Rightarrow_{\mathcal{S}} \langle 2, 2\rangle \Rightarrow_{\mathcal{S}} \langle 2, 2\rangle \Rightarrow_{\mathcal{S}} \cdots$. Both executions are complete.

Since an infinite execution path cannot be the prefix of an another one, it follows that infinite execution paths are complete and hence the above definition is slightly different from that given in [6].

Definition 7 (All-Paths Interpretation of rl formulas). *We say that a pair* (τ, ρ) *satisfies an* RL *formula* $\varphi \Rightarrow \varphi'$, *written* $(\tau, \rho) \models \varphi \Rightarrow \varphi'$, *iff* (τ, ρ) *starts from* φ *and one of the following two conditions holds: there exists* $i \geq 0$ *such that* $(\gamma_i, \rho) \models \varphi'$ *or* τ *is infinite. We say that* $\Rightarrow_{\mathcal{S}}$ *satisfies* $\varphi \Rightarrow \varphi'$, *written* $\Rightarrow_{\mathcal{S}} \models \varphi \Rightarrow \varphi'$, *iff* $(\tau, \rho) \models \varphi \Rightarrow \varphi'$ *for all* (τ, ρ) *starting from* φ *with* τ *complete.*

We let $\llbracket \varphi \Rightarrow \varphi' \rrbracket \triangleq \{\tau \mid (\exists \rho)(\tau, \rho) \models \varphi \Rightarrow \varphi'\}$. *If* F *is a set of* RL *formulas, then* $\llbracket F \rrbracket = \bigcup_{\varphi \Rightarrow \varphi' \in F} \llbracket \varphi \Rightarrow \varphi' \rrbracket$.

Example 7. An RL formula specifying that any execution path satisfying it computes the greatest common divisor (gcd) is $\langle x, y\rangle \Rightarrow (\exists x', y')\langle x', y'\rangle \wedge gcd(x', x, y)$, where gcd is a predicate symbol with the interpretation: $M_{gcd}(d, m, n)$ holds iff d is the gcd of m and n. If $\rho(x) = 10, \rho(y) = 8$, τ and τ' are the execution paths defined in Example 6, then both (τ, ρ) and (τ', ρ) satisfy the given formula.

The definition of all-paths interpretation of ml formulas given above is slightly different from that given in [6], where $\Rightarrow_{\mathcal{S}} \models \varphi \Rightarrow \varphi'$ iff $(\tau, \rho) \models \varphi \Rightarrow \varphi'$ for all (τ, ρ) starting from φ with τ finite and complete. By contrast, our definition lets infinite paths satisfy formulas vacuously. The infinite paths are introduced from technical reasons,e.g., in order to prove $(\tau, \rho) \models \varphi \Rightarrow \varphi'$ we do not need to prove or know that τ is finite and complete.

2.3   Rewrite Theories

In this paper we propose a new approach for analysing rewrite theories. Our approach is dedicated to rewrite theories that model systems having at least *some* terminating executions, since nonterminating executions satisfy RL formulas vacuously. Communication protocols, such as the one we illustrate our approach on later in the paper, are an example of such systems. A counterexample are reactive systems, which (in the ideal case) execute forever in interaction with their environment.

Here we briefly recall the definition for (a particular case of) rewrite theories and their rewriting relation. A *rewrite theory* $\mathcal{R} = (\Sigma, E \cup A, R)$ consists of a signature Σ, a set of equations E, a set of axioms A, e.g., associativity, commutativity, identity or combinations of these, a set of rewrite rules R of the form $l \rightarrow r$ **if** b, where l and r are terms with variables and b is a term of a distinguished sort *Bool*. We further assume that there is a special constant *true* of sort *Bool*.

In this paper we shall consider rewrite theories $\mathcal{R} = (\Sigma, E \cup A, R)$ with a distinguished sort *State* such that R is topmost w.r.t. *State*. Moreover, the actual theories $\mathcal{R}$ that we can analyse impose some additional technical restrictions on their components, which are briefly discussed in Sect. 5.

We use the standard notation for RWL artefacts: $=_{E \cup A}$ denotes the equality modulo the equations given by E and A, $T_{\Sigma, E \cup A}(X)$ denotes the set of $=_{E \cup A}$-equivalences classes of Σ-terms with variables in X, $T_{\Sigma, E \cup A} \triangleq T_{\Sigma, E \cup A}(\emptyset)$ is the set of $=_{E \cup A}$-equivalences classes of ground Σ-terms, and $FreeVars(t)$ denotes the set of variables occurring in the term t^2. The relation $\rightarrow_{\mathcal{R}}$ denotes the one-step rewriting relation defined by applying a rule from R modulo axioms $E \cup A$ over ground terms of sort *State*: $[u] \rightarrow_{\mathcal{R}} [v]$ iff there are a rule $l \rightarrow r$ **if** b in R and a (ground) substitution $\sigma : FreeVars(l, r, b) \rightarrow T_{\Sigma, E \cup A}$ such that $\sigma(l) =_{E \cup A} \sigma(u)$, $\sigma(r) =_{E \cup A} \sigma(v)$, and $\sigma(b) =_{E \cup A} true$.

3   Derivatives of ML and RL Formulas

The notion of derivative is essential for our approach. Roughly speaking, the derivative of a formula specifies states/execution paths obtained from those satisfying the initial formula after executing one step. For the remaining part of this section we consider a fixed transition system specification $\mathcal{S}$ and its associated transition system $(M_{State}, \Rightarrow_{\mathcal{S}})$ over a fixed model M.

Assumption 1. *In what follows we consider only ml formulas φ with the following property: if φ does not occur as a member of a rule in $\mathcal{S}$ and $\varphi_l \Rightarrow \varphi_r \in \mathcal{S}$ then $FreeVars(\varphi) \cap FreeVars(\varphi_l, \varphi_r) = \emptyset$. This is not a real restriction since the free variable in rules can always be renamed.*

---

[2] For the sake of uniformity, we keep the notation $FreeVars(t)$ for the set of variables occurring in the term t. This is a consistent notation since all occurrences of variables in term are considered as being free. $FreeVars(t_1, t_2)$ is $FreeVars(t_1) \cup FreeVars(t_2)$.

Definition 8 (Semantic Definition of Derivatives for rl Formulas). *We say that $\varphi_1 \Rightarrow \varphi'$ is a $\mathcal{S}$-derivative of $\varphi \Rightarrow \varphi'$ if for all $(\tau_1, \rho) \models \varphi_1 \Rightarrow \varphi'$ there is $(\tau, \rho) \models \varphi \Rightarrow \varphi'$ such that $\tau_1 = \tau|_{1\ldots}$.*

Example 8. An $\mathcal{S}$-derivative for $\langle x, y \rangle \Rightarrow (\exists x', y') \langle x', y' \rangle \wedge gcd(x', x, y)$ is the following formula: $\langle y, x - y \rangle \wedge x > y \wedge y > 0 \Rightarrow (\exists x', y') \langle x', y' \rangle \wedge gcd(x', x, y)$.

Definition 9 (Complete Sets of Derivatives). *A set D of $\mathcal{S}$-derivatives of $\varphi \Rightarrow \varphi'$ is complete iff $[\![\varphi_1 \Rightarrow \varphi']\!] \subseteq [\![D]\!]$ for each $\mathcal{S}$-derivative $\varphi_1 \Rightarrow \varphi'$ of $\varphi \Rightarrow \varphi'$.*

Example 9. The set

$$\{(\exists k)\langle y, x - k * y \rangle \wedge y > 0 \wedge x \geq k * y \wedge k > 0 \Rightarrow (\exists x', y')\langle x', y' \rangle \wedge gcd(x', x, y),$$
$$\langle y, x \rangle \wedge y \geq x \Rightarrow (\exists x', y')\langle x', y' \rangle \wedge gcd(x', x, y)\}$$

is a complete set of $\mathcal{S}$-derivatives for $\langle x, y \rangle \Rightarrow (\exists x', y')\langle x', y' \rangle \wedge gcd(x', x, y)$.

The next definition and lemma provide us with syntactical means of computing complete sets of derivates for RL formulas.

Definition 10 (Syntactic Definition of Derivative for RL Formulas). *If φ is an ml formula then*

$$\Delta_\mathcal{S}(\varphi) \triangleq \{(\exists Free\,Vars(\varphi_l, \varphi_r))(\varphi_l \wedge \varphi)^{=?} \wedge \varphi_r \mid \varphi_l \Rightarrow \varphi_r \in \mathcal{S}\}.$$

If $\varphi \Rightarrow \varphi'$ is an RL*-formula then*

$$\Delta_\mathcal{S}(\varphi \Rightarrow \varphi') \triangleq \{\varphi_1 \Rightarrow \varphi' \mid \varphi_1 \in \Delta_\mathcal{S}(\varphi)\}.$$

Lemma 1. *If $\varphi_1 \in \Delta_\mathcal{S}(\varphi)$ then $\varphi_1 \Rightarrow \varphi'$ is an $\mathcal{S}$-derivative of $\varphi \Rightarrow \varphi'$.*

Lemma 2. *Let $\varphi_1 \Rightarrow \varphi'$ be an $\mathcal{S}$-derivative of $\varphi \Rightarrow \varphi'$, τ_1 be an execution path and ρ a valuation. If $(\tau_1, \rho) \models \varphi_1 \Rightarrow \varphi'$ then there is $\varphi'_1 \in \Delta_\mathcal{S}(\varphi)$ such that $(\tau_1, \rho) \models \varphi'_1 \Rightarrow \varphi'$.*

From Lemmas 1 and 2 we directly obtain:

Proposition 3. *$\Delta_\mathcal{S}(\varphi \Rightarrow \varphi')$ is a complete set of $\mathcal{S}$-derivatives for $\varphi \Rightarrow \varphi'$.*

Example 10.

$$\Delta_\mathcal{S}(\langle x, y \rangle \wedge y \geq 0) = \{(\exists x', y', k)\langle y', x' - k * y' \rangle \wedge \langle x', y' \rangle = \langle x, y \rangle \wedge y' > 0 \wedge x' > y'$$
$$\wedge \, x' \geq k * y' \wedge k > 0 \wedge y \geq 0,$$
$$(\exists x', y')\langle y', x' \rangle \wedge \langle x', y' \rangle = \langle x, y \rangle \wedge y' \geq x' \wedge y \geq 0\},$$

which can be simplified to

$$\Delta_\mathcal{S}(\langle x, y \rangle \wedge y \geq 0) = \{(\exists k)\langle y, x - k * y \rangle \wedge y > 0 \wedge x \geq k * y \wedge k > 0,$$
$$\langle y, x \rangle \wedge y \geq x \wedge y \geq 0\},$$

using the implications $M \models \langle x', y' \rangle = \langle x, y \rangle \longrightarrow (x = x' \wedge y = y')$, $M \models (x \geq k * y \wedge k > 0) \longrightarrow x > y$ and $M \models y > 0 \longrightarrow y \geq 0$, where M is the model defined in the previous examples, and $\longrightarrow$ is the usual implication in FOL.

$\Delta_\mathcal{S}(\langle x, y \rangle \Rightarrow (\exists x', y')\langle x', y' \rangle \wedge gcd(x', x, y))$ is the set given in Example 9.

The following definition of S-*derivability* is used in our verification procedure for RL formulas. The lemma following it gives an equivalent characterisation in terms of FOL, which enables the checking of S-derivability using SMT solvers.

Definition 11 (S-derivability of ML-formulas). *An ml formula φ is S-derivable iff there is at least a transition starting from it, i.e., there exist a model $(\gamma, \rho) \models \varphi$ and a transition $\gamma \Rightarrow_S \gamma_1$.*

Lemma 3. *φ is S-derivable iff $\bigvee_{\varphi_1 \in \Delta_S(\varphi)} \varphi_1^{=?}$ is satisfiable.*

Lemma 3 also shows the strong relationship between the S-derivability of a ml formula φ and the S-derivatives of RL-formulas $\varphi \Rightarrow \varphi'$. Hence it does make sense to name the elements of the set $\Delta_S(\varphi)$ as being S-*derivatives* of φ.

The notion of totality, defined below, is essential for the soundness of our verification procedure. Intuitively, a transition-system specification S is total if its rules cover all models (γ, ρ) of any S-derivable formula φ. For instance, if $\langle x, y \rangle \wedge y \neq 0 \Rightarrow \langle y, x \% y \rangle \in S$ then in order to be total S must also include a rule for the case $y = 0$.

Definition 12. *S is total iff for for each S-derivable φ and each pair (γ, ρ) such that $(\gamma, \rho) \models \varphi$, there is γ_1 such that $\gamma \Rightarrow_S \gamma_1$.*

Note the difference between S-derivability and totality: S-derivability requires to have at least one transition starting from φ and the totality requires to have at least one transition starting from γ for any model (γ, ρ) of φ.

The next result enables the use of SIT solvers for checking totality.

Proposition 4. *S is total iff for each S-derivable φ,*

$$M \models \varphi^{=?} \longrightarrow \bigvee_{\varphi_1 \in \Delta_S(\varphi)} \varphi_1^{=?}.$$

We note that in general SMT solvers do not support theories for high-level algebraic structures. However, in practice, one can either introduce the theories in the solver, or use simplifications rules before sending the formulas to the solver.

4   A Procedure for Verifying RL Properties

We now introduce our procedure for verifying RL properties on transition systems also defined by RL formulas. We assume given an ml signature Φ and Φ-model M.

The soundness result, stated below, says that if the procedure returns **success** when presented with a given input (consisting of a transition system RL specification S, a set of goals G_0, and the S-derivatives of G_0, then the transition system $\Rightarrow_S$ satisfies all the goals. We note that this result is not a trivial consequence of the soundness of the RL proof system [6]; our initial attempts at proving soundness by reducing it to the soundness of the RL proof system showed that one step of our procedure corresponds to several (many) steps of the RL proof system. Thus, the soundness of several nontrivial derived rules of the RL proof system would have to be proved first before attempting to prove the soundness of our procedure. We thus opted for a direct proof.

460   D. Lucanu et al.

procedure prove(S, G_0, G)

1: **if** $G = \emptyset$ **then return success**
2:     **else choose** $\varphi \Rightarrow \varphi' \in G$
3:         **if** $M \models \varphi \longrightarrow \varphi'$ **then return prove**$(S, G_0, G \setminus \{\varphi \Rightarrow \varphi'\})$
4:         **else if** there is $\varphi_c \Rightarrow \varphi'_c \in G_0$ s. t. $M \models \varphi \longrightarrow (\exists FreeVars(\varphi_c))\varphi_c$ **then**
5:             **return prove**$(S, G_0, G \setminus \{\varphi \Rightarrow \varphi'\} \cup \Delta_{\varphi_c \Rightarrow \varphi'_c}(\varphi \Rightarrow \varphi'))$
6:         **else if** φ is S-derivable **then**
7:             **return prove**$(S, G_0, G \setminus \{\varphi \Rightarrow \varphi'\} \cup \Delta_S(\varphi \Rightarrow \varphi'))$
8:         **else return failure.**

Fig. 1. RL verification procedure.

Theorem 1 (Soundness). *Let* prove *be the procedure given in Fig. 1. Assume that S is total. Let G_0 be such that for each $\varphi_c \Rightarrow \varphi'_c \in G_0$, φ_c is S-derivable and satisfies FreeVars$(\varphi'_c) \subseteq$ FreeVars(φ_c). If* prove$(S, G_0, \Delta_S(G_0))$ *returns* success *then* $\Rightarrow_S \models G_0$.

Example 11. Let S be the RL specification defined in Example 5 and $G_0 \triangleq \{\varphi_0 \Rightarrow \varphi'_0\}$, where $\varphi_0 \Rightarrow \varphi'_0$ is

$$\langle x, y \rangle \wedge y \geq 0 \Rightarrow (\exists x', y')\langle x', y' \rangle \wedge gcd(x', x, y).$$

We illustrate in a step-by-step manner the procedure call prove(S, G_0, G), where G is initially $\Delta_S(G_0) = \{\varphi_1 \Rightarrow \varphi'_0, \varphi_2 \Rightarrow \varphi'_0\}$, and

$$\varphi_1 \triangleq (\exists k)\langle y, x - k * y \rangle \wedge y > 0 \wedge x \geq k * y \wedge k > 0 \Rightarrow \varphi'_0,$$
$$\varphi_2 \triangleq \quad \langle y, x \rangle \wedge y \geq x \wedge y \geq 0 \Rightarrow \varphi'_0\}.$$

Let us consider that $\varphi_1 \Rightarrow \varphi'_0$ is the current chosen goal from G. Obviously, $M \models \varphi_1 \longrightarrow \varphi'_0$ does not hold. Since $M \models x \geq k * y \longrightarrow x - k * y \geq 0$, we obtain $M \models \varphi_1 \longrightarrow (\exists x', y')\langle x', y' \rangle \wedge y' \geq 0$ (i.e. the condition of the if statement at line 4 holds) and hence the goal $\varphi_1 \Rightarrow \varphi'_0$ is replaced with $\varphi_{11} \Rightarrow \varphi'_0$ by the statement in line 5, where

$$\varphi_{11} \triangleq (\exists x_1, y_1, x'_1, y'_1, k)\langle x'_1, y'_1 \rangle \wedge gcd(x'_1, x_1, y_1) \wedge \langle y, x - k * y \rangle = \langle x_1, y_1 \rangle$$
$$\wedge y > 0 \wedge x \geq k * y \wedge k > 0$$

Since $M \models (gcd(x', y, x - k * y) \wedge k > 0 \wedge x \geq k * y) \longrightarrow gcd(x', x, y)$, it follows that $M \models \varphi_{11} \longrightarrow \varphi'_0$ and the goal $\varphi_{11} \Rightarrow \varphi'_0$ is removed from G by the if statement in line 2.

Now the only goal in G is $\varphi_2 \Rightarrow \varphi'_0$. It is easy to see that $M \not\models \varphi_2 \longrightarrow \varphi'_0$ and $M \not\models \varphi_2 \longrightarrow (\exists FreeVars(\varphi'_0))\varphi'_0$, i.e. the conditions on then lines 3 and 4 do not hold. We have $\Delta_S(\varphi_2 \Rightarrow \varphi'_0) = \{\varphi_{21} \Rightarrow \varphi'_0, \varphi_{22} \Rightarrow \varphi'_0\}$, where

$$\varphi_{21} \triangleq (\exists k)\langle x, y - k * x \rangle \wedge x > 0 \wedge y \geq k * x \wedge k > 0 \wedge y > x \Rightarrow \varphi'_0,$$
$$\varphi_{22} \triangleq \quad \langle x, y \rangle \wedge x \geq y \wedge y \geq x \wedge y \geq 0 \Rightarrow \varphi'_0\}.$$

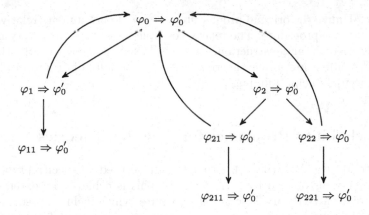

Fig. 2. The graph $\mathcal{G}$ corresponding to the **prove** procedure call for Example 11

The $\varphi_{21} \Rightarrow \varphi'_0$ is processed in the same way like $\varphi_1 \Rightarrow \varphi'_0$. Since $M \models \varphi_{22} \longrightarrow$ $(\exists x', y')\langle x', y'\rangle \wedge y' \geq 0$ and hence the goal $\varphi_{22} \Rightarrow \varphi'_0$ is replaced with $\varphi_{221} \Rightarrow \varphi'_0$ by the **if** statement in line 4, where

$$\varphi_{221} \triangleq (\exists x_1, y_1, x'_1, y'_1, k)\langle x'_1, y'_1\rangle \wedge gcd(x'_1, x_1, y_1) \wedge \langle x, y\rangle = \langle x_1, y_1\rangle$$
$$\wedge\, x \geq y \wedge y \geq x \wedge y \geq 0$$

It is easy to see that $M \models \varphi_{221} \longrightarrow \varphi'_0$ and hence the goal $\varphi_{221} \Rightarrow \varphi'_0$ is removed from G by the **if** statement in line 2.

Now the set of current goals G is empty and the execution of the procedure call **prove**$(\mathcal{S}, G_0, \Delta_{\mathcal{S}}(G_0))$ returns **success**. The execution of the procedure **prove** corresponding to this call is graphically represented in Fig. 2: the sinks correspond to the implications on line 3, the forward arrows correspond to the calls on line 7, and the backward arrows correspond to the calls on line 5 in the procedure. This graph covers the symbolic executions starting from φ_0.

Remark 1. A *sound approximated check* for validity statements $M \models \varphi$ is a procedure that, when presented with M and φ, if it answers *true* then $M \models \varphi$. We conjecture that Theorem 1 still holds when one uses sound approximated checks for the various validity statements occurring in the procedure (including $\mathcal{S}$-derivability and totality of $\mathcal{S}$, which amount to validity, cf. the previous section). Approximated checks, such as those implemented in SMT solvers, are required here since exact checks do not exist due to undecidability issues.

Remark 2. Theorem 1 says nothing about executions of the procedure that return **failure** or that do not terminate. Such outcomes may mean either $\Rightarrow_{\mathcal{S}} \not\models G_0$, or $\Rightarrow_{\mathcal{S}} \models G_0$ but the information contained in the goals G_0 is not sufficient for proving them, or, again, that the approximation induced by the validity checkers was too coarse. It is the user's burden to come up with a set of goals containing enough information such that the procedure terminates successfully.

This is similar to proving loop invariants in imperative programs, which requires users to provide a strong-enough invariant (i.e., one that can be proved).

Remark 3. Unlike the original proof system [6] we do not aim at (relative) completeness for our procedure. The relative completeness result is a very nice but essentially theoretical construction, which is based on strong assumptions (an oracle for deciding first-order theories) and is essentially of no practical use (it does not actually help in finding proofs).

5   Reachability Properties for Rewrite Theories

In this section we show that RL formulas can be used to specify properties of transition systems defined by RWL theories. This is achieved by extending the signature of an RWL theory $\mathcal{R}$ to an ml-signature, which includes predicates that can be used to define RL properties of the transition system defined by $\mathcal{R}$.

We also show how the verification procedure in Sect. 4 can be adapted in order to take advantage of RWL-specific operations such as matching. More precisely, we prove that, under reasonable assumptions, a complete set of derivatives of an RL formula can be computed using standard matching-modulo algorithms.

We note that RL properties for rewrite theories are different from the reachability properties that can be checked in Maude using the search command. The difference is given by the possibility of using FOL for constraining the initial and the final state terms, and by the interpretation of RL formulas.

Definition 13 (ML Extension of a Rewrite Theory). *Consider a rewrite theory $\mathcal{R} = (\Sigma, E \cup A, R)$ with a distinguished sort State such that $\mathcal{R}$ is topmost w.r.t. State. An ml extension of $\mathcal{R}$ consists of an ml signature $(\Sigma, \Pi, State)$ together with an interpretation $(T_{\Sigma, E \cup A})_p \subseteq T_{\Sigma, E \cup A, s_1} \times \cdots \times T_{\Sigma, E \cup A, s_n}$ for each predicate symbol $p \in \Pi_{s_1, \ldots, s_n}$. In this way, $T_{\Sigma, E \cup A}$ is a model of the ml extension of $\mathcal{R}$.*

The above definition allows one to see the operations $bop \in \Sigma_{s_1 \ldots s_n, Bool}$ as predicates $bop \in \Pi_{s_1 \ldots s_n}$ with the interpretation $([t_1], \ldots, [t_n]) \in (T_{\Sigma, E \cup A})_{bop}$ iff $bop(t_1, \ldots, t_n) =_{E \cup A} true$. Consequently, each term b of sort $Bool$ defines a FOL formula such that, if $\rho : Var \rightarrow T_{\Sigma, E \cup A}$, then $\rho \models b$ iff $\rho(b) =_{E \cup A} true$.

Any rewrite rule $l \rightarrow r$ if b can be viewed as an RL formula $l \wedge b \Rightarrow r$ and the transition relation $\Rightarrow_R$ is exactly the same with the one-step topmost rewriting relation $\rightarrow_{\mathcal{R}}$. This allows one to naturally define when RL formulas specify properties for RWL theories.

Definition 14 (RL Properties for RWL Theories). *An RL property for $\mathcal{R} = (\Sigma, E \cup A, R)$ is an RL formula $\varphi \Rightarrow \varphi'$, where φ and φ' are ml formulas defined over an ml extension of $\mathcal{R}$ (cf. Definition 13). We say that $\mathcal{R}$ satisfies $\varphi \Rightarrow \varphi'$ iff $\rightarrow_{\mathcal{R}} \models \varphi \Rightarrow \varphi'$ (cf. Definition 7).*

We next focus on adapting the prove procedure for verifying RL properties of RWL theories. Specifically, we show the derivatives can be computed using matching algorithms (under certain assumptions). We give first a technical definition.

Definition 15 (Ground $(E \cup A)$-unifier). *Consider a rewrite theory $\mathcal{R} = (\Sigma, E \cup A, R)$. Two Σ-terms t and t' are* ground $(E \cup A)$-unifiable *if there is $\sigma : FreeVars(t, t') \to T_\Sigma$ such that $\sigma(t) =_{E \cup A} \sigma(t')$. The substitution σ is called $(E \cup A)$-unifier of t and t'.*

The following assumptions are required for computing derivatives using matching. The first assumption restricts the class of RL formulas that can be used as properties to those that are useful in practice (i.e., having more than one pattern, or having negations of patterns, in either side of formulas is not really useful).

Assumption 2. *We assume that $FreeVars(b) \subseteq FreeVars(l, r)$ for each rule $l \to r$ **if** b in R. In this paper we also assume that the RL properties of RWL theories $\mathcal{R}$ are of particular form $(\exists X)(\pi \wedge \phi) \Rightarrow (\exists X')(\pi' \wedge \phi')$ with $FreeVars(\phi) \subseteq FreeVars(\pi)$ (and hence $X \subseteq FreeVars(\pi)$).*

The second assumption relates ground unifiers to matching substitutions.

Assumption 3. *We assume that for each $l \to r$ **if** b in R with l and π ground $(E \cup A)$-unifiable, there is a set of matching substitutions $match(l, \pi)$ such that*

- *$\sigma_0(l) = \pi$ for each $\sigma_0 \in match(l, \pi)$, and*
- *for each ground $E \cup A$-unifier σ of l and π there are $\sigma_0 \in match(l, \pi)$ and $\sigma' : FreeVars(\pi) \to T_\Sigma$ satisfying $\sigma =_{E \cup A} \sigma' \circ \sigma_0$.[3]*

Assumption 3 holds under reasonable constraints, cf. the *Matching Lemma* [21]. In a nutshell, the constraints distinguish a *builtin subtheory* for the equational subtheory of the rewrite theory $\mathcal{R}$, with corresponding builtin equations and axioms, assumed to be manageable by an SMT solver; there are no non-builtin equations, while non-builtin axioms may include the usual combinations of associativity, commutativity, and unity. The next results show that the matching substitutions can be used to compute the derivatives of ml formulas and, consequently, the derivatives of RL formulas.

Lemma 4 (Computing ML Derivatives by Matching). *Let $\varphi \Rightarrow \varphi'$ be an RL property for $\mathcal{R} = (\Sigma, E \cup A, R)$, where $\varphi \triangleq (\exists X)\pi \wedge \phi$. Then, for each derivative $\varphi_1 \in \Delta_\mathcal{R}(\varphi)$ there exists a rewrite rule $l \to r$ **if** b in R such that*

$$T_{\Sigma, E \cup A} \models \varphi_1 \longleftrightarrow \bigvee_{\sigma_0 \in match(l, \pi)} (\exists X \cup FreeVars(r) \setminus FreeVars(l))\sigma_0(r) \wedge \sigma_0(b) \wedge \phi.$$

The following theorem directly follows from Lemma 4:

Theorem 2 (Computing RL Derivatives by Matching). *For each $\varphi_1 \Rightarrow \varphi' \in \Delta_\mathcal{R}(\varphi \Rightarrow \varphi')$ with $\varphi \triangleq (\exists X)\pi \wedge \phi$, there is $l \to r$ **if** $b \in R$ such that*

$$[\![\varphi_1 \Rightarrow \varphi']\!] = [\![\bigvee_{\sigma_0 \in match(l, \pi)} (\exists X \cup FreeVars(r) \setminus FreeVars(l))\sigma_0(r) \wedge \sigma_0(b) \wedge \phi \Rightarrow \varphi']\!].$$

---

[3] $\sigma_1 =_{E \cup A} \sigma_2$ iff $dom(\sigma_1) = dom(\sigma_2)$ and $(\forall x \in dom(\sigma_1))\sigma_1(x) =_{E \cup A} \sigma_2(x)$.

Symbolic Rewrite Rules. We now show that Theorem 2 enables the use of *symbolic rewrite rules* to efficiently compute derivatives and hence to implement the procedure **prove** in RWL. For $\varphi \triangleq (\exists X)(\pi \wedge \phi)$, let $\Delta_R^{match}(\varphi \Rightarrow \varphi')$ be the set

$$\{(\exists X \cup \mathit{FreeVars}(r) \setminus \mathit{FreeVars}(l))\, \sigma_0(r) \wedge \sigma_0(b) \wedge \phi \Rightarrow \varphi'$$
$$| \ l \rightarrow r \ \mathbf{if} \ b \in R, \sigma_0 \in match(l, \pi)\} \qquad (1)$$

We have $[\![\Delta_R^{match}(\varphi \Rightarrow \varphi')]\!] = [\![\Delta_R(\varphi \Rightarrow \varphi')]\!]$ by Theorem 2, which implies that $\Delta_R^{match}(\varphi \Rightarrow \varphi')$ is a complete set of $\mathcal{R}$-derivatives. This allows us to use of Δ_R^{match} in the procedure **prove** instead of Δ_R. Next, we note that formulas $\varphi_1 \Rightarrow \varphi'$ in $\Delta_R^{match}(\varphi \Rightarrow \varphi')$ (where $\varphi \triangleq (\exists X)(\pi \wedge \phi)$) can be computed by applying a *symbolic rewrite rule* of the form $l \wedge \psi \Rightarrow r \wedge b \wedge \psi$ to the left-hand side of $\varphi \Rightarrow \varphi'$, where ψ is a fresh variable of sort *Bool*, with a matching substitution σ_0 such that $\sigma_0(\psi) = \phi$. Moreover, φ is R-derivable iff there are a rule $l \rightarrow r$ **if** b in R and $\sigma_0 \in match(l, \pi)$ such that $(\exists X \cup \mathit{FreeVars}(r) \setminus \mathit{FreeVars}(l))\sigma_0(b) \wedge \phi$ is satisfiable, by Lemma 4. This is equivalent to saying that φ is R-derivable iff the symbolic rewrite rule $l \wedge \psi \Rightarrow r \wedge b \wedge \psi$ is applicable to φ.

Overall, R-derivatives of RL formulas $\varphi \Rightarrow \varphi'$ (where $\varphi \triangleq (\exists X)(\pi \wedge \phi)$) can be computed by transforming each rule $l \rightarrow r$ **if** $b \in R$ into a symbolic rewrite rule $l \wedge \psi \Rightarrow r \wedge b \wedge \psi$ and by applying the symbolic rewrite rule. This is how derivatives are computed in our RWL adaptation of the **prove** procedure.

6   Verifying RL Properties of a Communication Protocol

We illustrate the theory on a simple communication protocol described in Maude.

Protocol Description. The protocol transmits a file between a sender and a receiver. The file is a sequence of records. The sender and receiver communicate through unidirectional, lossy channels, one of which carries messages (a record in the file, together with a sequence number) from send to receiver, while the other one carries retransmission requests (natural numbers) from receiver to sender.

Both the sender and the receiver maintain a counter in order to keep track of the next record to be sent, respectively received. The sender transmits the next record in the file together with the current value of its counter, which then is incremented by one. The receiver accepts a message only if the sequence number of the message is equal to the receiver's counter; if this is the case, the counter is incremented and the record from the message is saved. The receiver discards all other messages (i.e., whose sequence number is not the expected one). It may also (nondeterministically) request a retransmission, by sending the current value of its counter over the retransmission request channel. This nondeterminism can be seen as an abstraction of a timeout mechanism, not modeled here for simplicity.

Upon reception of a retransmission request, the sender ignores it if it is greater than or equal to its counter (indicating a wrong retransmission request). Otherwise the sender updates its counter to the number it received on retransmission request channel, in order to start resending messages from that number on.

```
--- sender
crl [send_to_R] :
< n, R, L, m, recv > => < s(n),(< fileToSend(n),n >: R), L, m, recv >
if n <= Max .

crl [update_request_from_L] :
< n, R, (resend : L), m, recv > => < resend, R, L, m, recv >
if resend < n .

crl [ignore_request_from_L] :
< n, R, (resend : L), m, recv > => < n, R, L, m, recv >
if resend >= n .

--- receiver
crl [accept_element_from_R] :
  < n, (R : < e, nb >), L, m, recv > =>  < n, R, L,  (s m), (recv : e) >
if m == nb .

crl [ignore_element_from_R] :
  < n, (R : < e, nb >), L, m, recv >  => < n, R, L, m, recv >
if m =/= nb .

crl [send_request_to_L] :
< n, R, L, m, recv > => < n, R, L : m, m, recv>
if m <= Max .
```

Fig. 3. Communication Protocol (excerpt).

The protocol's State structure is given as a constructor with five arguments:
<_,_,_,_,_> : Nat List{Pair} List{Nat} Nat List{Element} -> State,
where Pair is a sort of pairs consisting of an Element and a natural number, and
List{} are parameterised lists. They respectively denote: the index of the next
record to be sent, the sender-to-receiver channel, the receiver-to-sender channel,
the next expected record on the receiver side, and the list of records currently
accepted and stored by the receiver.

The file to be sent is modelled by a function fileToSend : Nat -> Elements,
of size Max. The sender and receiver's rules are shown in Fig. 3. There are also rules
for the channels losing elements, not shown here due to lack of space.

Reachability Properties. The protocol's initial state is <0,nil,nil,0,nil>. Its
expected reachability property states that all terminating executions starting
from the initial state should end up in a state of the form
    < (s Max),nil,nil,(s Max),file:List{Element}>
where file should satisfy $(\forall j)\, 0 \le j \le$ Max $\longrightarrow$ fileToSend(j) = file[j],
where _[_] is a function returning an element at a given position in a list. (That
is, the file received is the same as the file sent.) In order to specify the constraints
on the final states we defined in Maude a subset of RL, so that the reachability
property specifying the protocol is written as the following Maude rewrite rule:

```
< 0,nil,nil,0,nil > //\\ True  =>
Exists file : < (s Max),nil,nil,(s Max),file >
//\\ Forall j : ((0 <=? j  And (j <=? Max)
            Implies (fileToSend(j) === file[j]))
```

The operation _//_ is the constructor for our defined subset of ml, which
takes a term of sort State and term of sort FOL and builds a term of sort ML.

Note that the Maude `search` command cannot be used to prove this RL formula: to do so it would have to explore all terminating executions starting from the initial state, which are infinitely many (and can be arbitrarily long).

Thus, we use our verification procedure. Unsurprisingly, the above RL formula is not enough by itself for our procedure to succeed. For this to happen, a "helper" RL formula is required, whose right-hand side is the same as for the one above, but whose left-hand side describes an invariant (to hold for all states reachable from the initial states):

```
< n, R, L, m, file > //\\
  (Forall j : ((0 <=? j And j <? m)
            Implies (fileToSend(j) === file[j]))) And
  (Forall e :
    Forall nb : (< e, nb > In? R
            Implies e === fileToSend(nb))) And
  size(file) + 1 === m
=>
Exists file : < (s Max),nil,nil,(s Max),file > //\\
  Forall j : ((0 <=? j And (j <=? Max)
            Implies (fileToSend(j) === file[j]))
```

This formula says that the currently received file (whose size is m -1) equals the portion of the file being sent (up to that size); and that all messages currently in transition from sender to receiver are records in the `fileToSend` file. It was obtained by trial-and-error, while applying the following verification technique.

Verification. We have implemented key functionality from our verification procedure at Maude's metalevel. A first transformation is applied to rewrite rules as described in the *Symbolic rewrite rules* paragraph of Sect. 5. This reduces the application of rules with unification to application with matching. Derivatives are computed based on matching and rewriting as described in Sect. 5. We also use Maude's metalevel to achieve the following executions of our verification procedure:

- for the first RL formula (the protocol's specification): deriving it with respect to the protocol's rules $\mathcal{S}$, then applying the second formula as a circularity;
- for the second RL formula (designated above as the "helper" formula): deriving it with respect to the protocol's rules, then applying itself as a circularity.

By requiring that these two executions return `success`, Maude generates several proof obligations: essentially, that the condition for applying circularities holds (at line 4 in our verification procedure), and that the condition for returning `success` (line 3) also holds. Several of those proof obligations are discharged automatically by simplification rules we included in Maude (e.g., that FOL disjunction is commutative). The remaining ones are axioms satisfied by the assumed model for the various elements in our problem domain (e.g., lossy channels, files consisting on records, and natural numbers). For example, one proof obligation says that if all messages in a channel contain records from the `fileToSend` file, then by losing a message all the remaining messages satisfy the same property.

There are four such proof obligations left after automatic simplification. We have (manually) checked that they hold (in the assumed model for our problem domain). The trial-and-error process for finding the helper RL formula consisted in examining the generated proof obligations, and noting that some do not hold unless more information is added about the problem domain.

7 Conclusion and Future Work

In this paper we propose a procedure for verifying reachability properties on symbolic transition systems. While the reachability properties are stated as RL formulas, we allow symbolic transition systems to be described by either RL specifications or by RWL specifications. We prove that our procedure is sound. We show with a concrete example that our procedure works in practice.

The paper also contributes to establishing connections between RL and RWL. In [22] it is shown how RL specifications can be encoded as RWL theories. Here, we take an alternative approach, which consists in using RL as a property language for RWL. The proposed procedure adapted for RWL can be implemented in Maude, using reflection and the recently added support for sat checking and one-step rewriting modulo SMT using the CVC4 library (http://cvc4.cs.nyu.edu/).

In terms of future work there are several directions to follow. First, starting from our prototype, we shall develop a tool in the Maude environment that will efficiently implement our procedure; we envision that the tool will generate proof obligations to be discharged by Maude's inductive theorem prover ITP [23]. We also intend to formalise in the Coq proof assistant our procedure and its soundness proof in order to be able not only to verify properties but also to generate certificates. Finally, we will use the extraction mechanism of Coq to obtain certified OCaml code for our procedure and use it as a reference implementation.

Acknowledgments. This paper is to celebrate the 65th birthday of Professor José Meseguer. His seminal achievements, together with his warm and professional advices often guided and inspired the research activity of the first author.

The second author has spent his postdoc a couple of offices away from José's. At the time he was working on another topic and did not really understand what rewriting logic and Maude were about. He became aware of both of them several years later, and has been inspired by them and enjoying them ever since.

Proofs of Technical Results

Proof of **Proposition** 1, Page 5. We use the notation convention from Definition 3.

$\Leftarrow$. If $(\gamma, \rho) \models \varphi$ then we consider ρ' such that $\rho'(z) = \gamma$ and $\rho'(x) = \rho(x)$ for all $x \neq z$. We obtain $\rho' \models \varphi^=$, which implies $\rho \models (\exists z)\varphi^=$.

$\Rightarrow$. Let $\square$ be a fresh variable ($\square \notin Var$) of sort $State$ and $\varphi^\square$ defined in the same way like $\varphi^=$, but using $\square$ instead of z. Note that $\varphi^\square$ is defined on a extended signature. If $\rho : Var \to M$ and $\gamma \in M_{State}$, then let $\rho^\gamma : Var \cup \{\square\} \to M$ denote the extension of ρ with $\rho(\square) = \gamma$. By Proposition 1 in [3] we have $\rho^\gamma \models \varphi^\square$ iff $(\gamma, \rho) \models \varphi$. Assume that $\rho \models \varphi^{=?}$. It follows that for any extension ρ' of ρ to $Var \cup \{\square\}$ we have $\rho' \models \varphi^{=?}$. Since $\varphi^{=?}$ can be obtained from $(\exists\square)\varphi^\square$

by alpha conversion, we obtain $\rho' \models (\exists\Box)\varphi^\Box$. It follows that there exists ρ'' : $Var \cup \{\Box\} \to M$ such that $\rho''(x) = \rho'(x)$ for $x \neq \Box$ and $\rho'' \models \varphi^\Box$. Since ρ' extends ρ, we obtain $\rho''(x) = \rho(x)$ for $x \in Var$. If we take $\gamma = \rho''(\Box)$, then $\rho'' = \rho^\gamma$ and hence $(\gamma, \rho) \models \varphi$. □

Proof of **Proposition** 2, Page 2. The FOL formula $\phi^{=?}$ is the same as ϕ because ϕ has no basic patterns, and hence $\phi^{=?}$ is equivalent to ϕ because the existential variable z does not occur in ϕ. It follows that $(\phi \wedge \varphi)^{=?} \triangleq (\exists z)(\phi \wedge \varphi)^=$ is equivalent $(\exists z)(\phi^= \wedge \varphi^=)$, which in turn is equivalent to $\phi \wedge (\exists z)\varphi^=$, which is the same as $\phi \wedge \varphi^{=?}$.

We now prove the second part of the proposition. By Proposition 1 and $(\gamma, \rho) \models \varphi$ we obtain $\rho \models \varphi^{=?}$. By applying the definition of the FOL satisfaction relation to $\rho \models \phi$ and $\rho \models \varphi^{=?}$ we obtain $\rho \models \phi \wedge \varphi^{=?}$, which, by the first part of this proposition, is equivalent to $\rho \models (\phi \wedge \varphi)^{=?}$. Then, using Proposition 1 and its proof we obtain that $(\gamma, \rho) \models \phi \wedge \varphi$, which concludes the proof. □

Proof of **Lemma** 1, Page 8. Assume that φ_1 is $(\exists FreeVars(\varphi_l, \varphi_r))(\varphi_l \wedge \varphi)^{=?} \wedge \varphi_r$ for some $\varphi_l \Rightarrow \varphi_r \in \mathcal{S}$. If $[\![\varphi_1]\!] = \emptyset$ then $\varphi_1 \Rightarrow \varphi'$ is an $\mathcal{S}$-derivative of $\varphi \Rightarrow \varphi'$ by Definition 8. Assume $[\![\varphi_1]\!] \neq \emptyset$, i.e. there exists (τ_1, ρ_1) starting from φ_1, with $\tau_1 \triangleq \gamma_1 \Rightarrow_{\mathcal{S}} \cdots$. Then

$$(\gamma_1, \rho_1) \models \varphi_1 \qquad\qquad \longleftrightarrow$$
$$(\exists\rho)(\gamma_1, \rho) \models ((\varphi_l \wedge \varphi)^{=?} \wedge \varphi_r) \qquad\qquad \longleftrightarrow$$
$$(\exists\rho)(\rho \models (\varphi_l \wedge \varphi)^{=?} \wedge (\gamma_1, \rho) \models \varphi_r) \qquad\qquad \longleftrightarrow$$
$$(\exists\rho)((\exists\gamma_0)(\gamma_0, \rho) \models (\varphi_l \wedge \varphi) \wedge (\gamma_1, \rho) \models \varphi_r) \qquad\qquad \longleftrightarrow$$
$$(\exists\rho)((\exists\gamma_0)((\gamma_0, \rho_1) \models \varphi \wedge (\gamma_0, \rho) \models \varphi_l) \wedge (\gamma_1, \rho) \models \varphi_r) \qquad\qquad \longleftrightarrow$$
$$(\exists\gamma_0)(\gamma_0, \rho_1) \models \varphi \wedge (\exists\rho)((\gamma_0, \rho) \models \varphi_l \wedge (\gamma_1, \rho) \models \varphi_r) \qquad\qquad \longrightarrow$$
$$(\exists\gamma_0)(\gamma_0, \rho_1) \models \varphi \wedge \gamma_0 \Rightarrow_{\mathcal{S}} \gamma_1$$

where, by Assumption 1, we may w.l.o.g. choose ρ such that $\rho(x) = \rho_1(x)$ for all $x \notin FreeVars(\varphi_l, \varphi_r)$. Hence there is $\tau = \gamma_0 \Rightarrow_{\mathcal{S}} \gamma_1 \Rightarrow_{\mathcal{S}} \cdots$ such that $\tau|_{1..} = \tau_1$ and $(\gamma_0, \rho_1) \models \varphi$. If τ_1 is infinite then τ is infinite. If $(\exists i \geq 1)(\gamma_i, \rho_1) \models \varphi'$ then $(\exists i \geq 0)(\gamma_i, \rho_1) \models \varphi'$. So, finally, we obtain that $(\tau_1, \rho_1) \models \varphi_1 \Rightarrow \varphi'$ implies $(\tau, \rho_1) \models \varphi \Rightarrow \varphi'$. Since γ_1 and ρ_1 have been chosen arbitrarily we conclude that $\varphi_1 \Rightarrow \varphi'$ is an $\mathcal{S}$-derivative of $\varphi \Rightarrow \varphi'$. □

Proof of **Lemma** 2, Page 8. Suppose that $(\tau_1, \rho) \models \varphi_1 \Rightarrow \varphi'$ and $\tau_1 \triangleq \gamma_1 \Rightarrow_{\mathcal{S}} \cdots$. Then (τ_1, ρ) starts from φ and one of the following two claims holds: a) there exists $i \geq 1$ such that $(\gamma_i, \rho) \models \varphi'$ or b) τ is infinite. So, to prove that $(\tau_1, \rho) \models \varphi'_1 \Rightarrow \varphi'$ it is enough to prove that (τ_1, ρ) starts from some $\varphi'_1 \in \Delta_{\mathcal{S}}(\varphi)$. The pair (τ_1, ρ) can be extended to (τ, ρ) such that $\tau|_{1..} = \tau_1$ and $(\tau, \rho) \models \varphi \Rightarrow \varphi'$ by the definition of the $\mathcal{S}$-derivative. It follows that there is γ_0 such that $\tau \triangleq \gamma_0 \Rightarrow_{\mathcal{S}} \gamma_1 \Rightarrow_{\mathcal{S}} \cdots$, $(\gamma_0, \rho) \models \varphi$, and $(\gamma_1, \rho) \models \varphi_1$. There is $\varphi_l \Rightarrow \varphi_r \in \mathcal{S}$ and ρ' such that $(\gamma_0, \rho') \models \varphi_l$ and $(\gamma_1, \rho') \models \varphi_r$ by the definition of $\Rightarrow_{\mathcal{S}}$. By Assumption 1, we may w.l.o.g. choose ρ' such that $\rho'(x) = \rho(x)$ for all $x \notin FreeVars(\varphi_l, \varphi_r)$. Hence $(\gamma_0, \rho') \models \varphi \wedge \varphi_l$. We take $\varphi'_1 \triangleq (\exists FreeVars(\varphi_l, \varphi_r))(\varphi \wedge \varphi_l)^{=?} \wedge \varphi_r$. We obviously have $\varphi_1' \in \Delta_S(\varphi)$ and $(\gamma_1, \rho) \models \varphi_1'$ because there exists ρ' (defined above) such that $(\gamma_1, \rho') \models (\varphi \wedge \varphi_l)^{-?} \wedge \varphi_r$. Since $\tau_1 = \gamma_1 \Rightarrow_S \cdots$, it follows that (τ_1, ρ) starts from $\varphi_1' \in \Delta_S(\varphi)$, which implies $(\tau_1, \rho) \models \varphi_1' \Rightarrow \varphi'$. Since (τ_1, ρ) has been chosen arbitrary, the conclusion of the lemma follows.   □

Proof of **Lemma** 3, Page 9. The following equivalences hold:

$$\bigvee_{\varphi_1 \in \Delta_S(\varphi)} \varphi_1^{=?} \text{ is satisfiable} \qquad\qquad \longleftrightarrow$$

$$(\exists \rho_1)\rho_1 \models \bigvee_{\varphi_1 \in \Delta_S(\varphi)} \varphi_1^{=?} \qquad\qquad \longleftrightarrow$$

$$(\exists \rho_1)(\exists \varphi_1 \in \Delta_S(\varphi))\rho_1 \models \varphi_1^{=?} \qquad\qquad \longleftrightarrow$$

$$(\exists \rho_1)(\exists \varphi_l \Rightarrow \varphi_r \in S)\rho_1 \models ((\exists \mathit{FreeVars}(\varphi_l, \varphi_r))(\varphi_l \wedge \varphi)^{=?} \wedge \varphi_r)^{=?} \qquad \longleftrightarrow$$

$$(\exists \rho_1)(\exists \varphi_l \Rightarrow \varphi_r \in S)(\exists \gamma_1)(\gamma_1, \rho_1) \models (\exists \mathit{FreeVars}(\varphi_l, \varphi_r))(\varphi_l \wedge \varphi)^{=?} \wedge \varphi_r$$

Then, we have:

$$(\gamma_1, \rho_1) \models (\exists \mathit{FreeVars}(\varphi_l, \varphi_r))(\varphi_l \wedge \varphi)^{=?} \wedge \varphi_r \qquad\qquad \longleftrightarrow$$

$$(\exists \rho)(\gamma_1, \rho) \models (\varphi_l \wedge \varphi)^{=?} \wedge \varphi_r \qquad\qquad \longleftrightarrow$$

$$(\exists \rho)\rho \models (\varphi_l \wedge \varphi)^{=?} \wedge (\gamma_1, \rho) \models \varphi_r \qquad\qquad \longleftrightarrow$$

$$(\exists \rho)(\exists \gamma)(\gamma, \rho) \models (\varphi_l \wedge \varphi) \wedge (\gamma_1, \rho) \models \varphi_r \qquad\qquad \longleftrightarrow$$

$$(\exists \rho)(\exists \gamma)(\gamma, \rho) \models \varphi \wedge (\gamma, \rho) \models \varphi_l \wedge (\gamma_1, \rho) \models \varphi_r \qquad\qquad \longleftrightarrow$$

$$(\exists (\gamma, \rho))(\gamma, \rho) \models \varphi \wedge \gamma \Rightarrow_S \gamma_1$$

where, by the definition of $\models$, ρ satisfies $\rho(x) = \rho_1(x)$ for all $x \notin \mathit{FreeVars}(\varphi_l, \varphi_r)$. We obtained that $\bigvee_{\varphi_1 \in \Delta_S(\varphi)} \varphi_1^{=?}$ is satisfiable iff there exists (γ, ρ) such that $(\gamma, \rho) \models \varphi$ and there exists γ_1 such that $\gamma \Rightarrow_S \gamma_1$, i.e., iff φ is S-derivable.   □

Proof of **Proposition** 4, Page 9. We use the notation convention in Definition 3.

$$M \models \varphi^{=?} \longrightarrow \bigvee_{\varphi_1 \in \Delta_S(\varphi)} \varphi_1^{=?} \qquad\qquad \longleftrightarrow$$

$$(\forall \rho)\rho \models \varphi^{=?} \longrightarrow \rho \models \bigvee_{\varphi_1 \in \Delta_S(\varphi)} \varphi_1^{=?} \qquad\qquad \longleftrightarrow$$

$$(\forall \rho)(\exists \gamma)(\gamma, \rho) \models \varphi \longrightarrow$$
$$(\exists \varphi_l \Rightarrow \varphi_r \in S)\rho \models (\exists \mathit{FreeVars}(\varphi_l, \varphi_r))((\varphi_l \wedge \varphi)^{=?} \wedge \varphi_r)^{=?} \qquad \longleftrightarrow$$

$$(\forall \rho)(\exists \gamma)(\gamma, \rho) \models \varphi \longrightarrow$$
$$(\exists \varphi_l \Rightarrow \varphi_r \in S)\rho \models (\exists \mathit{FreeVars}(\varphi_l, \varphi_r))(\varphi_l \wedge \varphi)^{=?} \wedge \varphi_r^{=?} \qquad \longleftrightarrow \qquad (2)$$

$$(\forall \rho)(\exists \gamma)(\gamma, \rho) \models \varphi \longrightarrow$$

$$(\exists \varphi_l \Rightarrow \varphi_r \in S)(\exists \rho')\rho' \models (\varphi_l \wedge \varphi)^{=?} \wedge (\exists \gamma_1)(\gamma_1, \rho') \models \varphi_r \qquad \longleftrightarrow \qquad (3)$$
$$(\forall \rho)(\exists \gamma)(\gamma, \rho) \models \varphi \longrightarrow$$
$$(\exists \varphi_l \Rightarrow \varphi_r \in S)(\exists \rho')(\gamma, \rho') \models (\varphi_l \wedge \varphi) \wedge (\exists \gamma_1)(\gamma_1, \rho') \models \varphi_r \qquad \longleftrightarrow \qquad (4)$$
$$(\forall \rho)(\exists \gamma)(\gamma, \rho) \models \varphi \longrightarrow (\exists \gamma_1)\gamma \Rightarrow_S \gamma_1$$

where, by Assumption 1, we may w.l.o.g. choose ρ' such that $\rho'(x) = \rho(x)$ for all $x \notin \text{FreeVars}(\varphi_l, \varphi_r)$, which implies $(\gamma', \rho') \models \varphi$ iff $\gamma' = \gamma$ and $(\gamma, \rho) \models \varphi$. Therefore in the equivalence (3) $\longleftrightarrow$ (4) we could take $(\gamma, \rho') \models (\varphi_l \wedge \varphi)$. The equivalence (2) follows by Proposition 2. $\qquad \square$

Proof of **Theorem** 1, Page 10. The following lemmas are needed in the proof.

Lemma 5 (Coverage Step). *Let* γ, γ', ρ, φ, *and* $\alpha \triangleq \varphi_l \Rightarrow \varphi_r \in S$ *such that* $\gamma \Rightarrow_{\{\alpha\}} \gamma'$ *and* $(\gamma, \rho) \models \varphi$. *Then,* $(\gamma', \rho) \models \Delta_{\{\alpha\}}(\varphi)$.

Proof. From $\gamma \Rightarrow_{\{\alpha\}} \gamma'$ we obtain a valuation ρ' such that $(\gamma, \rho') \models \varphi_l$ and $(\gamma', \rho') \models \varphi_r$. By Assumption 1, $\text{FreeVars}(\varphi_l, \varphi_r) \cap \text{FreeVars}(\varphi) = \emptyset$. Hence we can choose ρ' such that $\rho'(x) = \rho(x)$ for all $x \in \text{FreeVars}(\varphi)$. Thus, $(\gamma, \rho') \models \varphi$. From the latter and $(\gamma, \rho') \models \varphi_l$ we obtain $(\gamma, \rho') \models \varphi \wedge \varphi_l$, and using Proposition 1 we have $\rho' \models (\varphi \wedge \varphi_l)^{=?}$. Using Proposition 2 we obtain $(\gamma', \rho') \models (\varphi \wedge \varphi_l)^{=?} \wedge \varphi_r$ which implies $(\gamma', \rho) \models (\exists \text{FreeVars}(\varphi_l, \varphi_r))(\varphi \wedge \varphi_l)^{=?} \wedge \varphi_r$ (using Assumption 1). By Definition 10 $(\exists \text{FreeVars}(\varphi_l, \varphi_r))(\varphi \wedge \varphi_l)^{=?} \wedge \varphi_r$ is just $\Delta_{\{\alpha\}}(\varphi)$, which ends the proof. $\qquad \square$

Lemma 6 (Coverage by Derivatives). *Any computation* $\tau \triangleq \gamma_0 \Rightarrow_S \gamma_1 \Rightarrow_S \cdots$ *with* (τ, ρ) *starting from* φ *is "covered" by derivatives, i.e., there exists a sequence* $\varphi_0, \varphi_1, \ldots$ *of ml formulas such that*

1. $\varphi_0 = \varphi$
2. $\varphi_{i+1} \in \Delta_S(\varphi_i), i = 0, 1, \ldots$
3. $(\gamma_i, \rho) \models \varphi_i, i = 0, 1, \ldots$

Proof. By induction on i using Lemma 5 in the induction step. $\qquad \square$

A successful execution of $\mathbf{prove}(S, G_0, \Delta_S(G_0))$ consists of a sequence of calls

$$\mathbf{prove}(S, G_0, \mathcal{G}_1), \ldots, \mathbf{prove}(S, G_0, \mathcal{G}_n)$$

such that

- $\mathcal{G}_0 = G_0$,
- $\mathcal{G}_1 = \Delta_S(G_0)$,
- $\mathcal{G}_n = \emptyset$,
- for all $i \in 0 \ldots n-1$, $\mathcal{G}_{i+1} = \mathcal{G}_i \setminus \{\varphi \Rightarrow \varphi'\} \cup \mathcal{G}_{\varphi \Rightarrow \varphi'}$, for some $\varphi \Rightarrow \varphi' \in \mathcal{G}_i$, where

$$\mathcal{G}_{\varphi \Rightarrow \varphi'} = \begin{cases} \emptyset & , \text{if } M \models \varphi \longrightarrow \varphi' \\ \Delta_{\varphi_c \Rightarrow \varphi'_c}(\varphi \Rightarrow \varphi') & , \text{if there is } \varphi_c \Rightarrow \varphi'_c \in G_0, \text{s.t.} M \models \varphi \longrightarrow \overline{\varphi}_c, \\ \Delta_S(\varphi \Rightarrow \varphi'_c) & , \text{if } \varphi \text{ } S\text{-derivable} \end{cases}$$

In the following we let $\mathcal{F} = \bigcup_{i=0}^{n} \mathcal{G}_i$.

Lemma 7. *Let* $\varphi \Rightarrow \varphi' \in \mathcal{F}$. *Then* $M \models \varphi \longrightarrow \varphi'$ *or* φ *is* $\mathcal{S}$-*derivabile.*

Proof. Let $\varphi \Rightarrow \varphi' \in \mathcal{F}$. If $\varphi \Rightarrow \varphi' \in \mathcal{G}_0 = G_0$ then φ is $\mathcal{S}$-derivabile (any formula in G_0 has the lhs $\mathcal{S}$-derivable). Otherwise, there is i such that $\varphi \Rightarrow \varphi' \in \mathcal{G}_i \setminus \mathcal{G}_{i+1}$. By the definition of $\mathcal{G}_{i+1}$ the formula $\varphi \Rightarrow \varphi'$ was eliminated from $\mathcal{G}_i$ in one of the three situations:

1. $M \models \varphi \longrightarrow \varphi'$
2. $M \models \varphi \longrightarrow \overline{\varphi}_c$ for some $\varphi_c \Rightarrow \varphi_c' \in G_0$
3. φ is $\mathcal{S}$-derivable.

In the first and the third cases we obtain directly the conclusion of our lemma. The only case we have to discuss is the second one. Note that there are γ and ρ such that $(\gamma, \rho) \models \varphi$. Otherwise, we have $M \models \varphi \longrightarrow \varphi'$ which corresponds to the first case. From $(\gamma, \rho) \models \varphi$ and $M \models \varphi \longrightarrow \overline{\varphi}_c$ we have $(\gamma, \rho) \models \overline{\varphi}_c$. By Definition 2, there is ρ' such that $(\gamma, \rho') \models \varphi_c$. Since φ_c is derivable (because $\varphi_c \Rightarrow \varphi_c' \in G_0$ and $\mathcal{S}$ is total, there exists a transition $\gamma \Rightarrow_\mathcal{S} \gamma_1$, which, by Definition 11, implies that φ is $\mathcal{S}$-derivable. $\qquad\square$

Lemma 8. *For all* τ, *for all* ρ, *for all* $\varphi \Rightarrow \varphi' \in \mathcal{F}$, *if* τ *is finite and complete, and* (τ, ρ) *starts from* φ *then* $(\tau, \rho) \models \varphi \Rightarrow \varphi'$.

Proof. We proceed by induction on the length of τ. We an consider arbitrary $\varphi \Rightarrow \varphi' \in \mathcal{F}$ and ρ satisfying the hypotheses of the lemma.

Base case. Assume that $\tau = \gamma_0$ and that $(\gamma_0, \rho) \models \varphi$. Since τ is complete then there is no γ_1 such that $\gamma_0 \Rightarrow_\mathcal{S} \gamma_1$. Therefore, any φ' such that $(\gamma_0, \rho) \models \varphi'$, is not $\mathcal{S}$-derivable (otherwise, it contradicts Definition 11). Thus, φ is not $\mathcal{S}$-derivable.

By Lemma 7 we have $M \models \varphi \longrightarrow \varphi'$, and using the fact that $(\gamma_0, \rho) \models \varphi$ we obtain $(\gamma_0, \rho) \models \varphi'$, i.e. $(\tau, \rho) \models \varphi \Rightarrow \varphi'$.

Induction step. Assume that $\tau = \gamma_0 \Rightarrow_\mathcal{S} \gamma_1 \cdots$, and $(\gamma_0, \rho) \models \varphi$. In this case φ is $\mathcal{S}$-derivable (by Definition 11). Since $\varphi \Rightarrow \varphi' \in \mathcal{F}$ then $\varphi \Rightarrow \varphi'$ has been eliminated at some point, so there is i such that $\varphi \Rightarrow \varphi' \in \mathcal{G}_i \setminus \mathcal{G}_{i+1}$. Again, by the definition of $\mathcal{G}_{i+1}$, we have three possible cases:

1. $M \models \varphi \longrightarrow \varphi'$. Since $(\gamma_0, \rho) \models \varphi$ we obtain $(\gamma_0, \rho) \models \varphi'$, which implies $(\tau, \rho) \models \varphi \Rightarrow \varphi'$.
2. $M \models \varphi \longrightarrow \overline{\varphi}_c$. From $(\gamma_0, \rho) \models \varphi$ we obtain $(\gamma_0, \rho) \models \overline{\varphi}_c$, and, by Definition 2, there is ρ' with $\rho'(x) = \rho(x)$ for all $x \notin FreeVars(\varphi_c)$ such that $(\gamma_0, \rho') \models \varphi_c$. If $\gamma_0 \Rightarrow_\mathcal{S} \gamma_1$ then there is a rule $\alpha \triangleq \varphi_l \Rightarrow \varphi_r \in \mathcal{S}$ such that $\gamma_0 \Rightarrow_{\{\alpha\}} \gamma_1$ (Definition 5). From $(\gamma_0, \rho') \models \varphi_c$ and Lemma 5 we obtain $\varphi_1 \in \Delta_{\{\alpha\}}(\varphi_c) \subseteq \Delta_\mathcal{S}(\varphi_c)$ such that $(\gamma_1, \rho') \models \varphi_1$. Since $\varphi_1 \in \Delta_\mathcal{S}(\varphi_c)$ and $\varphi_c \Rightarrow \varphi_c' \in G_0 = \mathcal{G}_0$ then $\varphi_1 \Rightarrow \varphi_c' \in \Delta_\mathcal{S}(\varphi_c \Rightarrow \varphi_c') \subseteq \mathcal{G}_1 \subseteq \mathcal{F}$. Now, the inductive hypothesis holds for $\varphi_1 \Rightarrow \varphi_c'$, and we have $(\tau|_{1..}, \rho') \models \varphi_1 \Rightarrow \varphi_c'$. Since τ is finite, there exists $j \geq 1$ such that $(\gamma_j, \rho') \models \varphi_c'$. Next, we want show that $(\gamma_j, \rho) \models (\exists FreeVars(\varphi_c, \varphi_c'))((\varphi_c \wedge \varphi)^{=?} \wedge \varphi_c')$. This is equivalent (by Definition 2) to showing that there is a valuation ρ'' with $\rho''(x) = \rho(x)$ for all $x \notin FreeVars(\varphi_{c'})$ such that $(\gamma_j, \rho'') \models (\varphi_c \wedge$

$\varphi)^{=?} \wedge \varphi'_c$. Let us consider $\rho'' = \rho'$. From the hypothesis of Theorem 1 we have $FreeVars(\varphi'_c) \subseteq FreeVars(\varphi_c)$, which implies that $FreeVars(\varphi_c, \varphi'_c) \subseteq FreeVars(\varphi_c)$. Also, note that $\rho'(x) = \rho(x)$, for all $x \notin FreeVars(\varphi_c)$. Using Assumption 1, i.e. $FreeVars(\varphi) \cap FreeVars(\varphi_c) = \emptyset$, and $(\gamma_0, \rho) \models \varphi$ we obtain $(\gamma_0, \rho') \models \varphi$. Given the fact that $(\gamma_0, \rho') \models \varphi_c$, by Definition 2, $(\gamma_0, \rho') \models \varphi_c \wedge \varphi$. By Proposition 1, from $(\gamma_0, \rho') \models \varphi_c \wedge \varphi$ we obtain $\rho' \models (\varphi_c \wedge \varphi)^{=?}$. Moreover, by Proposition 2 and the fact that $(\gamma_j, \rho') \models \varphi'_c$ we obtain $(\gamma_j, \rho') \models (\varphi_c \wedge \varphi)^{=?} \wedge \varphi'_c$. Therefore, there is $\rho'' = \rho'$, such that $(\gamma_j, \rho'') \models (\varphi_c \wedge \varphi)^{=?} \wedge \varphi'_c$, and we can conclude that $(\gamma_j, \rho) \models (\exists FreeVars(\varphi_c, \varphi'_c))((\varphi_c \wedge \varphi)^{=?} \wedge \varphi'_c)$.

Note that the set $\mathcal{G}_{i+1}$ includes $\Delta_{\varphi_c \Rightarrow \varphi'_c}(\varphi \Rightarrow \varphi'_c)$, and we can apply again the inductive hypothesis: $(\tau|_{j..}, \rho) \models (\exists FreeVars(\varphi_c, \varphi'_c))((\varphi_c \wedge \varphi)^{=?} \wedge \varphi'_c) \Rightarrow \varphi'$, i.e. there is $k \geq j$ such that $(\gamma_k, \rho) \models \varphi'$, which implies $(\tau, \rho) \models \varphi \Rightarrow \varphi'$.

3. φ is $\mathcal{S}$-derivable. Then $\Delta_\mathcal{S}(\varphi \Rightarrow \varphi') \subseteq \mathcal{G}_{i+1} \subseteq \mathcal{F}$. If $\gamma_0 \Rightarrow_\mathcal{S} \gamma_1$ then there is a rule $\alpha \triangleq \varphi_l \Rightarrow \varphi_r \in \mathcal{S}$ s. t. $\gamma_0 \Rightarrow_{\{\alpha\}} \gamma_1$ (Definition 5). Since $(\gamma_1, \rho) \models \varphi_1$, then, by Lemma 5, there is $\varphi_1 \in \Delta_{\{\alpha\}}(\varphi) \subseteq \Delta_\mathcal{S}(\varphi)$ such that $(\gamma_1, \rho) \models \varphi_1$. We obtain $(\tau|_{1..}, \rho) \models \varphi_1 \Rightarrow \varphi'$ by the inductive hypothesis, which implies that there is $j \geq 1$ s. t. $(\gamma_j, \rho) \models \varphi'$. Hence $(\tau, \rho) \models \varphi \Rightarrow \varphi'$. □

Proof (of Theorem 1). Let $\tau \triangleq \gamma_0 \Rightarrow_\mathcal{S} \gamma_1 \Rightarrow_\mathcal{S} \cdots$ be a complete execution path, and let the valuation ρ be such that (τ, ρ) starts from φ_0 with $\varphi_0 \Rightarrow \varphi'_0 \in G_0$. If τ is finite then $(\tau, \rho) \models \varphi \Rightarrow \varphi'_c$ by Lemma 8. If τ is infinite then $(\tau, \rho) \models \varphi \Rightarrow \varphi'_c$ by Definition 7. □

Proof of **Lemma** 4, Page 13. By definition of $\Delta_R(\varphi)$, φ_1 is $(\exists FreeVars(l, r))(l \wedge b \wedge (\exists X)(\pi \wedge \phi))^{=?} \wedge r$ for some rewrite rule $l \rightarrow r$ **if** $b \in R$. By Assumption 1, $FreeVars(l, r) \cap X = \emptyset$ and hence φ_1 is equivalent to $(\exists X \cup FreeVars(l, r))(l = \pi) \wedge b \wedge \phi \wedge r$. We have:

$(\gamma_1, \rho_1) \models \varphi_1$ ⟷

$(\gamma_1, \rho_1) \models (\exists X \cup FreeVars(l, r))(l = \pi) \wedge b \wedge \phi \wedge r$ ⟷

$(\exists \rho)(\rho(l) = \rho(\pi)) \wedge \rho \models (b \wedge \phi) \wedge (\rho(r) = \gamma_1)$ ⟷

$(\exists \sigma)(\sigma(l) =_{E \cup A} \sigma(\pi)) \wedge \rho \models b \wedge \rho_1 \models \phi \wedge (\sigma(r) \in \gamma_1)$ ⟷

$(\exists \sigma_0)(\exists \sigma'')(\sigma_0(l) =_{E \cup A} \pi) \wedge \rho \models b \wedge \rho_1 \models \phi \wedge (\sigma''(\sigma'(\sigma_0(r))) \in \gamma_1)$ ⟷

$(\exists \sigma_0 \in match(l, \pi))(\exists \sigma'')\rho \models b \wedge \rho_1 \models \phi \wedge (\sigma'' \uplus \sigma'(\sigma_0(r)) \in \gamma_1)$ ⟷

$\bigvee_{\sigma_0 \in match(l, \pi)} (\exists \sigma'')\rho \models b \wedge \rho_1 \models \phi \wedge (\sigma'' \uplus \sigma'(\sigma_0(r)) \in \gamma_1)$ ⟷

$\bigvee_{\sigma_0 \in match(l, \pi)} (\exists \rho_0)\rho_0 \models \sigma_0(b) \wedge \rho_0 \models \phi \wedge \rho_0(\sigma_0(r)) = \gamma_1$ ⟷

$\bigvee_{\sigma_0 \in match(l, \pi)} (\exists \rho_0)(\gamma_1, \rho_0) \models (\sigma_0(b) \wedge \phi \wedge \sigma_0(r))$ ⟷

$\bigvee_{\sigma_0 \in match(l, \pi)} (\gamma_1, \rho_1) \models (\exists X \cup FreeVars(r) \setminus FreeVars(l))(\sigma_0(b) \wedge \phi \wedge \sigma_0(r))$ ⟷

$$(\gamma_1, \rho_1) \models \bigvee_{\sigma_0 \in match(l,\pi)} (\exists X \cup FreeVars(r) \setminus FreeVars(l))(\sigma_0(b) \wedge \phi \wedge \sigma_0(r))$$

where

- $\gamma_1 \in T_{\Sigma,E\cup A}$ of sort *State*, i.e., γ_1 is an $(E \cup A)$-equivalence class $[t]$ with $t \in T_{\Sigma,State}$;
- by Assumption 1, we may assume w.l.o.g. that $\rho(x) = \rho_1(x)$ for all $x \notin X \cup FreeVars(l,r)$;
- $\sigma : FreeVars(l,r,\phi) \rightarrow T_{\Sigma}$ with $[\sigma(x)] = \rho(x)$;
- the substitutions $\sigma_0 : FreeVars(l) \rightarrow FreeVars(\pi)$ and $\sigma' : FreeVars(\pi) \rightarrow T_{\Sigma}$ are given by Assumption 3, i.e., $\sigma|_{FreeVars(l,\pi)} = \sigma' \circ \sigma_0$; note that σ' is uniquely determined by σ and σ_0;
- $\sigma'' = \sigma|_{FreeVars(r) \setminus FreeVars(l)}$;
- $\sigma'' \uplus \sigma'(x) = \sigma''(x)$ if $x \in FreeVars(r) \setminus FreeVars(l)$, and $\sigma'' \uplus \sigma'(x) = \sigma'(x)$ if $x \in FreeVars(\sigma_0(l))$;
- $\rho_0(x) = [\sigma'(x)]$, for $x \in FreeVars(\pi)$, $\rho_0(x) = [\sigma''(x)]$, for $x \in FreeVars(r) \setminus FreeVars(l)$, and $\rho_0(x) = \rho(x)$ in the rest (hence $\rho_0(x) = \rho_1(x)$ for $x \notin X \cup FreeVars(r) \setminus FreeVars(l)$); and
- $\rho \models b$ iff $\sigma''(\sigma_0(b)) =_{E\cup A} true$ iff $\rho_0 \models \sigma_0(b)$. $\qquad\square$

References

1. Meseguer, J.: Conditional rewriting logic as a unified model of concurrency. Theor. Comput. Sci. **96**(1), 73–155 (1992). Selected Papers of the 2nd Workshop on Concurrency and Compositionality
2. Clavel, M., Durán, F., Eker, S., Lincoln, P., Martí-Oliet, N., Meseguer, J., Talcott, C.: All about Maude - A High-performance Logical Framework: How to Specify, Program and Verify Systems in Rewriting Logic. Springer, Heidelberg (2007)
3. Roşu, G., Ştefănescu, A.: Checking reachability using matching logic. In: Leavens, G.T., Dwyer, M.B. (eds) OOPSLA, pp. 555–574. ACM (2012). also available as technical report http://hdl.handle.net/2142/33771
4. Roşu, G., Ştefănescu, A.: Towards a unified theory of operational and Axiomatic semantics. In: Czumaj, A., Mehlhorn, K., Pitts, A., Wattenhofer, R. (eds.) ICALP 2012, Part II. LNCS, vol. 7392, pp. 351–363. Springer, Heidelberg (2012)
5. Roşu, G., Ştefănescu, A., Ciobâcă, Ş., Moore, B.M.: One-path reachability logic. In: Proceedings of the 28th Symposium on Logic in Computer Science (LICS 2013), pp. 358–367. IEEE, June 2013
6. Ştefănescu, A., Ciobâcă, Ş., Mereuta, R., Moore, B.M., Şerbănută, T.F., Roşu, G.: All-path reachability logic. In: Dowek, G. (ed.) RTA-TLCA 2014. LNCS, vol. 8560, pp. 425–440. Springer, Heidelberg (2014)
7. Meseguer, J., Roşu, G.: The rewriting logic semantics project. Theor. Comput. Sci. **373**(3), 213–237 (2007)
8. Ellison, C., Roşu, G.: An executable formal semantics of C with applications. In: Proceedings of the 39th Symposium on Principles of Programming Languages (POPL 2012), pp. 533–544. ACM (2012)
9. Bogdănaş, D., Roşu, G.: K-Java: a complete semantics of Java. In Proceedings of the 42nd Symposium on Principles of Programming Languages (POPL 2015), pp. 445–456. ACM, January 2015

10. Roşu, G., Şerbănuţă, T.F.: An overview of the K semantic framework. J. Logic Algebraic Program. **79**(6), 397–434 (2010)
11. Meseguer, J.: Twenty years of rewriting logic. J. Logic Algebraic Program. **81**(7), 721–781 (2012)
12. Eker, S., Meseguer, J., Sridharanarayanan, A.: The Maude LTL model checker. Electron. Notes Theor. Comput. Sci. **71**, 162–187 (2004)
13. Bae, K., Meseguer, J.: Model checking linear temporal logic of rewriting formulas under localized fairness. Sci. Comput. Program. **99**, 193–234 (2015)
14. Bae, K., Escobar, S., Meseguer, J.: Abstract logical model checking of infinite-state systems using narrowing. In: 24th International Conference on Rewriting Techniques and Applications, RTA 2013, 24–26 June 2013, pp. 81–96, Eindhoven, The Netherlands (2013)
15. Rocha, C., Meseguer, J.: Proving safety properties of rewrite theories. In: Corradini, A., Klin, B., Cîrstea, C. (eds.) CALCO 2011. LNCS, vol. 6859, pp. 314–328. Springer, Heidelberg (2011)
16. Rusu, V.: Combining theorem proving and narrowing for rewriting-logic specifications. In: Fraser, G., Gargantini, A. (eds.) TAP 2010. LNCS, vol. 6143, pp. 135–150. Springer, Heidelberg (2010)
17. Bruni, R., Meseguer, J.: Semantic foundations for generalized rewrite theories. Theor. Comput. Sci. **360**(1), 386–414 (2006)
18. Meseguer, J., Palomino, M., Martí-Oliet, N.: Equational abstractions. Theor. Comput. Sci. **403**(2), 239–264 (2008)
19. Meseguer, J., Palomino, M., Martí-Oliet, N.: Algebraic simulations. J. Logic Algebraic Program. **79**(2), 103–143 (2009)
20. Arusoaie, A., Lucanu, D., Rusu, V.: A generic framework for symbolic execution. In: Erwig, M., Paige, R.F., Van Wyk, E. (eds.) SLE 2013. LNCS, vol. 8225, pp. 281–301. Springer, Heidelberg (2013). http://hal.inria.fr/hal-00853588
21. Rocha, C., Meseguer, J., Muñoz, C.: Rewriting modulo SMT and open system analysis. In: Escobar, S. (ed.) WRLA 2014. LNCS, vol. 8663, pp. 247–262. Springer, Heidelberg (2014)
22. Arusoaie, A., Lucanu, D., Rusu, V., Şerbănuţă, T.-F., Ştefănescu, A., Roşu, G.: Language definitions as rewrite theories. In: Escobar, S. (ed.) WRLA 2014. LNCS, vol. 8663, pp. 97–112. Springer, Heidelberg (2014)
23. Hendrix, J.: Decision Procedures for Equationally Based Reasoning. PhD thesis, University of Illinois at Urbana Champaign (2008)

Emerging Issues and Trends in Formal Methods in Cryptographic Protocol Analysis: Twelve Years Later

Catherine Meadows[✉]

Naval Research Laboratory, Washington, D.C., WA, USA
meadows@itd.nrl.navy.mil

Abstract. In 2003 I published a paper "Formal Methods in Cryptographic Protocol Analysis: Emerging Issues and Trends", in which I identified the various open problems related to applying formal methods to the analysis of cryptographic protocols as we saw them then, and discussed the state of the art and the problems that still needed to be solved. Twelve years later, it is time for a an update and a reassessment. In this paper I revisit the open problems that I addressed in the original paper, discussing the progress that has been made in the intervening years, and the problems that still remain to be solved. I also discuss some new open problems that have arisen since then.

1 Introduction

In 2003 we published a paper "Formal Methods in Cryptographic Protocol Analysis: Emerging Issues and Trends" [63], in which we identified the various open problems as we saw them then, and discussed the state of the art and the problems that still needed to be solved. Twelve years later, we believe it is time for an update and a reassessment. Formal analysis of cryptographic protocols has come a long way since then, both in power and range. Fifteen years ago, most work in formal verification relied on what is known as the "Dolev-Yao" model, in which the network is assumed to be under the complete control of an attacker and the cryptographic algorithm is treated like a black box. The protocols verified were generally authentication and key exchange protocols, and the properties proved were either various forms of authentication (if X happens, did Y happen before it?) and simple secrecy (can the intruder learn a term in the clear)?.

Now the classes of protocols examined include such varied applications as voting protocols, routing protocols, security APIs, and zero knowledge protocols. The classes of properties proved have also been greatly extended. Most notable is the development of theorem-provers that provide automated assistance in the generation and verification of game transformation proofs used by

© Springer International Publishing Switzerland 2015 (outside the US)
N. Martí-Oliet et al. (Eds.): Meseguer Festschrift, LNCS 9200, pp. 475–492, 2015.
DOI: 10.1007/978-3-319-23165-5_22

cryptographers. There are also tools that can be used to prove a symbolic form of indistinguishability. Finally, the black box itself has been pried upon to allow the treatment of various algebraic properties on the symbolic level.

This raises a number of questions. First of all, how thoroughly have the problems discussed in the 2003 paper been addressed? Which ones are still relevant, and which ones have turned out to be dead ends? What new problems have arisen that we were not considering then, and how are they being addressed?

In this paper we revisit the open problems that we addressed in the original paper, discussing the progress that has been made in the intervening years, and the problems that still remain to be solved. We also discuss some new open problems that have arisen since then.

The rest of this paper is organized as follows. In Sect. 2 we give an overview of the Dolev-Yao model that provides the basis for much of the work in this area. In Sect. 3 we revisit our 2003 paper and discuss the progress that has been made and the open problems that remain. In Sect. 4 we discuss some new areas of research that have arisen in the last twelve years. In Sect. 5 we conclude the paper and summarize.

2   Overview of the Dolev-Yao Model

Much symbolic protocol analysis is based on the simple but powerful paradigm developed by Dolev and Yao in the late 70s and early 80s [40]. In this paradigm messages are represented by symbolic terms constructed using constants, function symbols, and variables. Furthermore, the network is controlled by an intruder who can intercept, destroy, and redirect traffic, and create and send messages of its own. Thus, we can think of the protocol as a distributed program for generating elements of a term algebra, defined by a set of rules $\mathcal{I}$ that defines actions executed by the intruder, and a set of rules $\mathcal{P}$ describing actions executed by the honest principals.

This structure makes it possible to develop decision procedures for evaluating the security of protocols, e.g., whether or not the intruder is able to learn a secret or violate authentication requirements. Indeed, such problems have been shown to be NP-complete in the bounded session model [74], that is, when the honest principals are restricted to a bounded number of executions of the rules in $\mathcal{P}$. But even in the unbounded session model there are a number of tools that offer semi-decision procedures.

Another approach to symbolic protocol analysis is to develop logics that can be used to derive what the receiver of a message can conclude about it, given certain assumptions about the communication channel and the cryptosystems involved. The most prominent early example is the Burrows, Abadi, and Needham logic [27], but there have been a number of later logics that use approach, e.g., [70,73]. They have the disadvantage over state exploration tools in that they are at a higher level of abstraction and cannot be used to find attacks on insecure protocols. On the positive side, it is more straightforward to specify different properties of communication channels, and a failed proof can help the user understand hidden assumptions. CPSA [39] is an example of a logic-based tool.

3   Revisiting the Old Questions

In this section we consider the problems that were addressed in the original 2003 paper. They are presented in the original order.

3.1   Open-Ended Protocols

In the 2003 paper the term "open-ended protocol" was applied to protocols in which some feature of a given protocol execution may be parametrized, such as the number of principals involved, or the size of a data structure used. If the size of the parameter is unbounded, then the protocol is open-ended. We note that even protocols in the traditional "Dolev-Yao" model can be considered open-ended in a sense, since they can involve an arbitrarily large number of sessions. The approach used with respect to unbounded sessions has been to rely on semi-decision procedures. These may not be practical if one has multiple unbounded parameters, so in these cases researchers have generally taken a different approach, e.g., by putting additional conditions on the protocol.

The most general class of open-ended protocol is the class that makes use of *recursive tests*, where the depth of the recursion is the parameter. The approach here, first used by Küsters and Wilke in [56], and later followed in [6,82] and others, is to develop a restricted language for expressing the recursive tests, and to show that size of any possible attack is bounded when the number of sessions is bounded. As work has progressed in this area, the class of protocols that can be handled this way has grown, but there are also some negative results. For example, Kurtz et al. have shown [53] that, for a class of languages expressible as *selecting theories*, if certain terms are allowed to be of unlimited complexity, then the security problem becomes undecidable.

For some classes of open-ended protocols it has been possible to come up, not only with decidability results for the bounded session model, but decidable approximations for the unbounded session model. In [69] Paiola and Blanchet show that under certain conditions, one use approximations to translate a protocol that uses unbounded lists into a protocol that uses lists of length one, whose security implies the security of the original protocol. Progress of this kind has also been made on protocols parametrized by the number of principals in a protocol (aka group protocols). For example, Cortier et al. show in [35] that for a certain class of routing protocols and properties, it is sufficient to analyze only four node topologies. Both these results make it possible to analyze some open-ended protocols using general-purpose cryptographic protocol analysis tools.

What perhaps is most needed in this area is consolidation of these results and methods for using them together, since more than one of these open-ended properties may appear in the same protocol. We discuss this issue further in Sect. 4.4.

3.2   Applications and Threats from the Early Twenty-First Century

Denial of Service. In 2003 denial of service (DoS) was a relatively new problem, and it was not well understood how to manage it. However, one common way that denial of service is implemented is to initiate multiple sessions of a protocol with a responder but not carry them through. The responder would have to devote resources to responding to the protocols and maintaining state, ultimately leading to it being overwhelmed. This suggested that one way of making protocols more secure against denial of service was first, to make it harder to initiate protocols than to respond to them, and secondly, to perform weak (and inexpensive) authentication early on in the protocol, and stronger authentication later on. This approach was formalized in my denial of service attack model [62].

Although this attack model made sense for defense a against a single attacker, it needed further elaboration to be effective against distributed denial of service attacks, in which multiple attackers (e.g., a botnet) would be unleashed by a single controller. What was needed was, not only for protocols to be harder to initiate, but to remain harder to initiate even when the opponent was operating on a massive scale. This observation has led to the invention of *client puzzles* [50], whereby clients are requested to solve a problem of moderate difficulty before the server responds. These puzzles are designed to be resistant to large-scale efforts to break them by the incorporation of computational properties such as their solution being difficult to parallelize.

Client puzzles also have the advantage that different puzzle strategies can be used depending on the threat level perceived by the server. Moreover, they have other applications besides denial of service, such as in Bitcoin [68]. However, they must be carefully designed and deployed. This has lead to an substantial amount of work in formalizing desired properties of both puzzles and the protocols that use them, although mostly on the computational level (see [48] for a survey of recent literature). However, perhaps at this point client puzzles are well enough understood at the computational level that symbolic analysis of the protocols that use them would also be possible and useful.

Anonymous Communication. In 2003 a number of anonymous communication systems had been deployed, including systems such as Onion Routing [46], The Anonymizer, and Crowds [72], although mostly on a relatively small scale. Now the Onion Router has developed into the Tor Network [81], which is used worldwide to protect privacy of communications. In particular, experience with Tor has given a much greater degree of insight into how anonymous communication works in a real environment, and the kinds of threats it faces.

However, it turned out to be difficult to develop formal models and analyses of large-scale anonymous communication. The main stumbling block is the threat model. It is possible to design key distribution and authentication protocols that are secure against a very strong and simple model in which the attacker controls the entire network. However, even in 2003 it was clear that it was not possible to design practical anonymous communication protocols that are secure against such a strong attacker. Instead, one must assume that an attacker has access to only part of the network at the time, and even then the probability of breaking anonymity is usually non-negligible. These features are harder to capture in a formal model, and even harder to analyze; thus research in this area has depended heavily on experimentation and simulation.

That does not mean, however, that mathematical modeling and analysis of anonymous communication is impossible. However, work has concentrated on modeling and proofs of security instead of automated analysis, and on identifying the exact types of guarantees that an anonymous communication system provides. Indeed several researchers, e.g., [8,44] have applied mathematical systems for reasoning about cryptography such as Canetti's Universal Composability Framework [30] to prove results about the types of security properties guaranteed by different types of anonymous communication systems against different types of attackers. These analyses are not formal in the strict sense, since they are not based on logical systems, but they do use precise definitions and rigorous proofs, and increase our understanding of the systems. In the future, as the hardware and software used for the analysis of formal systems becomes more powerful, it may become practical and desirable to produce logical formalizations of these models that can be analyzed with machine assistance.

Electronic Commerce. The 1990s were the years when electronic commerce became a reality, and it was not surprising that there was an explosion of research in electronic commerce protocols, even leading to the founding of a conference in that area, Financial Cryptography. Many of these protocols supported complex security policies (SET [78] was the most famous example), and a substantial number supported various sorts of anonymity as well. Thus the application of formal methods to electronic commerce acted as a spur on the development of cryptographic protocol analysis methods. For example, the formalization of the SET requirements by myself and my colleagues [64] had a substantial impact on our development of the NPATRL protocol requirements language. SET was also formally analyzed by Bella et al. [17,18], making it probably the most complex protocol analyzed with machine assistance at the time.

Twelve years later, electronic commerce is more mature, and although research still continues, it is no longer such a hot topic. However, other new applications come forward. One of the most prominent is electronic voting. Like electronic commerce in the 90's, it is on the point of becoming generally used, but security considerations are holding it back, and indeed are seen as even more of an impediment than they were for electronic commerce. This has lead to a sizable amount of work in formal analysis as well. The challenges faced in the design and verification of voting protocols, which involve assuring both accountability and privacy, are similar to those of electronic commerce. However, they also pose additional problems, most importantly the assurance of the correctness of a function computed over the voters' input. This has provided impetus to areas such as verification of group protocols (discussed in Sect. 3.1) and of symbolic indistinguishability (discussed in Sect. 4.1).

3.3 High Fidelity

By "high fidelity" I mean approaching the computational model of cryptographic protocols as closely as possible. From the very beginning, the gap between symbolic and computational models has been a concern. By 2003 we were starting to see more work in this area, and it has increased dramatically over the years. In this section I consider three ways of approaching high fidelity that have been investigated the most thoroughly. The first is by adding more detail to the symbolic model. The second is by proving computational soundness of a symbolic model. The third is by automating computational proofs directly.

Adding More Information to the Symbolic Model. One can add more detail to the symbolic model by adding *equational theories* to the term algebras used to construct messages sent in a protocol. For example, instead of using a rule that says that if an intruder knows a key K and the encryption of a message M with K, then it can derive M, one can express this directly vie the equation $d(K, e(K, M)) = M$, where e and d stand for encryption and decryption, respectively. This not only allows us to potentially find attacks that are not possible to find in the free algebra model (empty equational theory), but also enables the specification of protocols (e.g., Diffie-Hellman key exchange) with would be difficult to describe without the inclusion of algebraic properties.

In 2003 we were already seeing work in this area, although for the most part it concerned itself with specific equational theories, e.g., exclusive-or or cancellation of encryption and decryption. Twelve years later this work has been substantially expanded to include *classes* of theories that can be handled with the same general-purpose unification or matching algorithm. This is important, because it allows us to mix and match different cryptographic algorithms that obey different equational theories. The biggest divide at this point is between *rewrite theories*, which consist of equations that can be given an orientation, and *AC theories*, which include associative-commutative axioms which are left without an orientation. For example, ProVerif supports *subterm convergent* rewrite theories [22], while Maude-NPA supports AC theories with *finite variant decompositions* [34], using a technique called *variant narrowing* [43], and TAMARIN [65] concentrates on Diffie-Hellman and related theories.

There are still problems that remain. One of these is how to handle important classes of theories, such as associativity without commutativity and homomorphic encryption over AC operators, that do not fall into either of the above categories, and cannot be handled by other means. If we can find workable solutions for these theories, this will greatly expand the types of problems we can handle. Associativity without commutativity will allow us to represent the associative properties of concatenation, which will make realistic analysis of type confusion attacks a possibility. Homomorphic encryption over AC operators will allow us to analyze the many privacy-preserving protocols that use this feature.

Another, potentially more wide-reaching problem, is managing complexity. As theories become more complex, the task of reasoning about them in a sound, complete, and efficient manner becomes harder. General-purpose approaches, although they guarantee soundness and completeness, may not scale well for complex theories. One possible solution is to develop hybrid algorithms that have some but not all of the features of the general purpose algorithms, for example *asymmetric unification* [42], which is being explored as an alternative to the general purpose method of variant unification [43].

Computational Soundness. Another approach to achieving higher fidelity to the computational model is to prove soundness of the symbolic model with respect to a computational model. That is, security in the symbolic model should imply security in the computational model. In 2003, work was just starting in this area, with Abadi and Rogaway's initial result relating symbolic and computational and models of encryption [2], followed by Backes, Pfitzmann, and Waidner's Universally Composable Cryptographic Library [11] based on Canetti's Universal Composability Framework (UC) [30], which was used to produce what may have been the first computationally sound symbolic analysis of a key distribution protocol [10]. Work has continued on refining and expanding this UC-based symbolic model. In particular, it was applied by Backes et al. [7] to proving computational security properties of the Kerberos protocol. Research has also focused on different models and on problems such as indistinguishability properties (e.g., [13]), and symbolic models equipped with equational theories (e.g., [16]). We are also seeing new applications, in particular to the symbolic verification of automatically generated cryptographic algorithms [14, 58].

However, there is a problem that has been noticed as this work has progressed: the difficulty of finding proofs of computational soundness, and the complexity of the proofs once they are found. This is a problem that seems to be inherent, and it results from the gap between the two different models. For example, the symbolic model has two different notions of secrecy: one defined in terms of the attacker not seeing the secret in the clear, the other being a symbolic notion of indistinguishability. The computational model however has *many* different notions of secrecy, based on the way in the which the attacker may interact with the algorithm. Furthermore, there are situations such as key cycles (e.g., expressions of the form $e(k, k)$, $e(k_1, e(k_2, k_1))$, etc., where $e(k, m)$ denotes encryption of m with key k) that the symbolic model treats as innocuous, but for which there are no computational proofs of security. Techniques for avoiding these must either be built into the symbolic model, or states that contain them must be treated as insecure. All this complicates both the model and the soundness proofs.

Comon-Lundh et al. [33] have pointed out that many of these types of problems arise because in the symbolic model proving security depends on showing the *absence* of attacks, while in the computational model proving security depends on showing the *presence* of indistinguishability proofs, and suggest similar approach to symbolic verification. Indeed protocol verification logics have already applied an approach like this to producing computationally sound symbolic systems for proving security protocols correct, e.g., by Roy et al. [73], although at a higher level of abstraction then what Comon-Lundh et al. had in mind. The idea is carried out further by Bana and Comon-Lundh in [12], in which a system is constructed in which the symbolic attacker can make any symbolic deduction that is not explicitly ruled out by an axiom. These axioms themselves are derived from cryptographic properties, e.g. indistinguishability against chosen ciphertext attacks. One then attempts to prove the protocol is secure by proving that any attack is inconsistent with the protocol's axioms. This work suggests that the question we should be asking is, not only how to prove computational soundness of symbolic models, but what is the symbolic model that provides the *best* interface to the computational model, in terms of both concreteness and the ease of proving computational soundness?

Automation of Computational Proofs of Security. One way of avoiding the disconnect between symbolic and computational models is to automate computational proofs directly. This involves *game transformation* proofs [19,49,79] that were introduced to add rigor to computational proofs. In this style of proof one tries to prove that, given certain computational assumptions, the probability of an attacker's winning a certain type of game is a negligible function of the security parameters. The game is transformed, step by step, to a series of equivalent games until finally we wind up with a game in which the attacker's advantage is clearly negligible. For example, one might want to prove security of a cryptoalgorithm by showing that an attacker has only a negligible advantage of winning a game in which it attempts to distinguish between the encryptions of two different messages, even when it chooses the messages. This could be proven by transforming the original game into one in which the attacker attempts to distinguish between two random numbers.

The clearly defined format and the fact that many of the transformation steps can be standardized make game transformation proofs a natural candidate for automation, and indeed the goal of automation was proposed by Halevi [49] as early as 2005. More recently, researchers have been taking up the challenge, with CryptoVerif [21] and EasyCrypt [15] being probably the best known tools. Generally, the tools require substantial interaction with the user, who provides the tool input on the sequence of game transformations, and, in the case of EasyCrypt, guidance for proving equivalence.[1]

Some questions of interest here are: how wide is the range of cryptographic algorithms and protocols that can be verified with these methods, and to what degree do the methods scale up? In particular, if these methods are successful, they could be used for the verification of complex cryptographic systems for which human-generated and human-checkable proofs may not be practical.

3.4 Composability

Protocols do not operate by themselves. Generally they function as a layer in a protocol stack, making use of channel guarantees provided by a lower level protocol. The same is true for cryptographic protocols.

There has been a considerable amount of work on the problem of verifying protocols assuming abstract properties of the channel that they are relying on.

---

[1] CryptoVerif also provides a completely automated option, but it is the interactive one that seems to be usually applied.

Probably the earliest work in this direction was Boyd and Mao's [25] and Mauer and Schmidt's [59] calculi that view cryptographic protocols as a means for transforming communication channels into channels that satisfy progressively stricter properties. This work, and much that followed upon it, relies on protocol logics rather than state exploration (see Sect. 4.3 for more examples). This is not surprising; it is more straightforward to use abstract security properties to derive other abstract security properties than to translate them into conditions that can be used by a search-based tool. However, we are also seeing work in both proving and using abstract channel properties in search-based tools, e.g., the vertical protocol integration of Gross, Mödersheim, and Vigano [47,67].

It appears that reasoning about communication channels is fairly well understood for protocol logics. The question now is: what is the best way to incorporate this reasoning in search-based tools, and how do we compose the guarantees?

3.5   Getting it into the Real World

Back in 2003, we were starting to see formal analyses of real-life protocols used in standards, e.g., SSL/TLS [66], Kerberos [29], SET [17], and IKE Version 1 [61]. In a number of cases this had a positive impact on the standard. However, most of these analyses were done after the protocol had been standardized, which made it harder to address any problems that were found during the analysis. Moreover, most of the analyses were one-offs; a single version of the standard was analyzed. However, a standard is a living, evolving thing that is constantly being changed, and one must verify that these changes do not negatively affect the security functionality. Thus it was clear that in order to be really useful formal analysis needed to be more tightly integrated into the standards process.

In 2015, we still see very little of this integration, for a number of reasons. One is that formal analysis of cryptographic protocols still requires a high degree of skill to specify the protocol and the security properties that it needs to satisfy. Moreover, it is not always practical to have the formal verification researchers perform the analyses, as was done in the instances we cited above. Researchers are interested in research challenges, and not all protocols provide such challenges. Moreover, researchers are often restricted by the scope of their research projects, which will generally last about three to five years, while maintaining a standard can be a long-term commitment.

Although a number of obstacles exist to incorporating formal methods more closely into the standardization process, there are other paths by which formal analysis is moving closer to being applied to real world protocols. Skill and interest in protocol analysis has been gradually spreading through the research community, both via researchers in secure protocol design applying formal methods tools, and researchers in formal protocol analysis moving into protocol design. An example in point is the development of secure voting protocols. The consequences could be severe if a breach of security occurred, so their correctness is important. This has lead to a significant level of activity on the part of the research community in the development and verification of such protocols, including both cryptographic proofs of correctness (e.g., [4,20,32,52,55,75]) and formal symbolic verification (e.g., [36,54,77]). Although we are still some way from adopting any of these as a standard, the need in particular for *voter-verifiable* protocols in which voters can verify that their votes were correctly counted, has already lead to their use on a small scale, e.g., Helios in university elections [3], Scantegrity II in the 2009 and 2011 Takoma Park elections [31], and a Prêt à Voter-based system in the 2014 Victoria state elections [28].

We note that there are still some limitations on these analyses. One is that they often focus on a subset of the security properties, e.g., coercion resistance. Another is that it is still not clear how easy it would be to update these proofs as the protocols evolve. But the fact that the systems being introduced have already been subjected to intense mathematical and formal analysis, in many cases from more than one party, increases the trust in which they can be held, and bodes well for the positive effect of formal analysis on future standards.

Finally, we remark that another way of getting formal analysis of cryptographic protocols into the real world is to concentrate on a particular problem area that the tools can handle really well with minimal input from human beings, and to build optimized tools for these tasks that can be made available commercially. This is the approach that is being followed by a group of researchers on security APIs, which is discussed later in Sect. 4.2.

4   Some New Issues and Trends

In this section I discuss some new problems that have emerged in the last twelve years.

4.1   Privacy and Symbolic Indistinguishability

By *privacy* we mean the protection of private information from disclosure. However, there is an important difference between privacy and protection of secret information such as cryptographic keys. Personal information tends in general to be low entropy, so even if an attacker cannot get secret information in the clear, the ability to distinguish between two encryptions of different data could lead to its guessing it. Inability of an attacker to distinguish in this way is foundation of the standard cryptographic formulations of security, but it was not originally supported in the symbolic model.

Work was already beginning in this area by the early 2000s. In [1] Abadi and Fournet presented a calculus for cryptographic protocol analysis, the applied pi-calculus, for which they formulated an indistinguishability-based security property they called *observational equivalence*. At about the same time Lowe [57] developed automated techniques for the analysis of protocols using weak passwords, in which the attacker should not be able to tell the difference between its choice of a correct password and its choice of an incorrect password. Later, a method for model-checking a property somewhat stronger that observational equivalence, called *uniformity* was implemented in the ProVerif tool [22], and it has been applied to a number of different protocols. More recently, a similar indistinguishability property was formulated and implemented in Maude-NPA [76], thus opening up the possibility of model-checking with respect to AC theories.

One challenge in model-checking indistinguishability is that it requires the solving of disunification problems (find all substitutions to the variables in two terms f and g guaranteeing nonequality) as well as unification problems. One reason for ProVerif's success in this area is that efficient disunification algorithms are known for the class of theories it handles, subterm convergent rewrite rules. On the other hand, much less is known about disunification for theories that involve AC. Thus more understanding of disunification for AC theories is needed to further extend symbolic reasoning about indistinguishability.

4.2 Security APIs

A security application programmer interface, or security API, is an API that performs various cryptographic operations upon data input into it by a program using it. APIs need to enforce various security properties: for example keys managed by the API should not appear in the clear outside the security module.

Some early work on formal methods for cryptographic protocol analysis focused on security APIs, e.g., [51,60], but increased interest followed upon Bond's discovery of an attack on the IBM CCA protocol in 2000 [24].

Security of APIs have much in common with the key exchange protocols verified using symbolic cryptographic protocol analysis tools. We assume a single attacker who interacts with a single principal, the API. The attacker can modify messages it receives from the API and use them to construct new messages, and is assumed to be able to perform any operation available to the API. Thus, in theory it should be possible to apply general-purpose symbolic cryptographic analysis tools to security APIs. But in practice this has proved difficult. For one thing, there is the question of scale. In a key exchange protocol principals may have the opportunity to play two or three different roles. A security API, however, may execute dozens of different types of commands. Furthermore, security APIs may make use of global state information. Finally, a number of operations that are used in certain security APIs are simply difficult to model in the standard symbolic setting, e.g., the format-dependent PIN recovery attacks discussed in [45].

Given the above, the approach taken to research on security APIs has typically been to develop special purpose models and tools that deal with classes of APIs instead of developing or adapting general-purpose tools. This is made more cost-effective by the fact that the same or similar tools can be applied to *installations* of APIs as well as the API specification. This makes it possible to develop a tool that is optimized for a single API such as PKCS#11 that can be applied to multiple installations. This is, for example, the idea behind the business model of Cryptosense [37], a recent start-up that is seeking to commercialize security API analysis techniques.

The formal analysis of security APIs is an example of a field that has come to a high level of maturity in a relatively short time, partly because concentrated focus of the problem allows totally automated methods.

4.3   Multi-channel Protocols, Ceremonies, and Procedures

In Sect. 3.4 we discussed research on modeling communication channels. In recent years work in this area has expanded as new protocols have been developed that take into account the special properties of certain communication channels. These include e.g., *distance bounding protocols* [26], that make use of timed channels to verify a principal's location, and *short authenticated string (SAS)* [83] protocols, that make use of the fact that a human being observing a device can easily verify the source of a message, but cannot remember a long string, to provide low-bandwidth authenticated channels. As mentioned in Sect. 3.4 the logic-based approach to cryptographic protocol analysis seem to lend themselves well to reasoning about channels with special properties. In particular, the *Protocol Derivation Logic* of Pavlovic and myself [70] is a logic that allows the specification of the properties of a channel in terms of what the receiver of a message can conclude about it.

More recently, researchers have begun to take note of the fact that channels and protocols not only appear in computer networks but pervade human life in general, and have started to look at ways for formalizing then. In [41] Ellison notes that cryptographic protocols are embedded in larger *ceremonies* that involve the humans who make use of the protocols. In [23] Blaze notes that many interactions between human beings follow rules and procedures similar to protocols and could be analyzed the same way protocols are.

We note that the concept of ceremony has been of particular interest to people analyzing the security of web browsers, whose security depends in part upon correctly modeling the way human beings interact with the browser interface. However, we believe that a much wider application is possible. In [71] Pavlovic and I consider a model that applies to the class of *procedures* that govern the interactions between groups of humans and machines. Consider, for example, a protocol that involves a customer performing a bank transaction. This involves a customer interacting with her keyboard and monitor, which in turn interact with her computer, which in turn communicates across the Internet with a machine at the bank. Each of these principals and devices uses different communication channels with different properties. We model these groups as *actor networks* inspired by the idea of actor networks in sociology, and communication between principals in an actor network is modeled using a successor to the Protocol Derivation Logic: the Procedure Derivation Logic [71].

Although this area is still in its infancy, we believe it has great potential for application as the Internet of Things evolves.

4.4   Combining Models and Proof Methods

In any area of formal methods, there is always one problem to deal with: the large number of different models and formal systems that need to be dealt with, and that stand in the way of making tools and techniques work together. To some degree this is unavoidable, as we have seen for example in the disconnect between symbolic and computational security; different problems require different techniques. But often different formal systems will be developed to handle the same general problems. The problem comes when one wants to combine two tools a apply a result proved for one system to another. How can we be sure that what is valid for one system remains valid for another? Symbolic protocol analysis is perhaps lucky in that a fairly small number of formal systems appear to dominate: e.g., strand spaces [80], and the applied pi calculus [1]. However, even when different tools use the same formal system, they often modify it. So even though the systems used by two tools may be close, it may still be hard to tell if they are equivalent.

One way of dealing with this problem is to develop interfaces between different tools and formal systems. For example, the AVISPA system [5] provides a single front-end to a number of different tools, and the CoSP framework [9], gives a computationally sound back-end to protocol verification tools. The idea behind CoSP is that, instead of re-proving soundness results for one's own system, one can prove soundness of the system with respect to the CoSP, and then inherit the computational soundness results of CoSP for free.

Building these kinds of interfaces can be tedious, and for that reason it is perhaps not surprising that they are still relatively rare. But as the field matures, and more people want to make use of results without having to reprove them for different systems, we expect that interfaces will become more prevalent.

5   Conclusion

I have given an update of the overview of formal cryptographic protocol analysis that I presented twelve years ago in [63]. As can be expected, although some research directions appear to have led to a (perhaps temporary) dead end, and others still have yet to fulfill their promise, others have made significant advances and are having positive effects on the security of cryptographic protocols. But what is most heartening is the sheer variety of research in this area. Although early research started by concentrated on the very simple Dolev-Yao model, researchers are now developing models and techniques for all types of extensions and variations, as well as totally different models such as game-based cryptography. We can conclude that formal cryptographic protocol analysis is a healthy field that we expect to keep on growing and contributing.

Acknowlegements. I would like to thank José Meseguer for our long and fruitful collaboration, much of which has resulted in research described in this paper. Furthermore, I would like to thank all the other researchers whose efforts have contributed to the work described in this paper, especially those that I have had the good fortune to collaborate with. Finally, I would like to thank Anupam Datta for suggesting that an update of my 2003 paper might be of interest to readers.

References

1. Abadi, M., Fournet, C.: Mobile values, new names, and secure communication. In: Hankin, C., Schmidt, D., (eds.) Conference Record of POPL 2001: The 28th ACM SIGPLAN-SIGACT Symposium on Principles of Programming Languages, 17-19 January 2001, pp. 104–115, London, UK. ACM (2001)

2. Abadi, M., Rogaway, P.: Reconciling two views of cryptography (the computational soundness of formal encryption). J. Cryptol. **15**(2), 103–127 (2002)
3. Adida, B., De Marneffe, O., Pereira, O., Quisquater, J.-J.: Electing a university president using open-audit voting: analysis of real-world use of helios. In: Electrionic Voting Technology/Workshop on Trustworthy Elections: EVT/WOTE (2009). https://www.usenix.org/legacy/event/evtwote09/tech/
4. Adida, B.: Helios: web-based open-audit voting. In: van Oorschot, P.C. (ed.) Proceedings of the 17th USENIX Security Symposium, 28 July - 1 August 2008, San Jose, CA, USA, pp. 335–348. USENIX Association (2008)
5. Armando, A., Basin, D., Boichut, Y., Chevalier, Y., Compagna, L., Cuellar, J., Drielsma, P.H., Heám, P.C., Kouchnarenko, O., Mantovani, J., Mödersheim, S., von Oheimb, D., Rusinowitch, M., Santiago, J., Turuani, M., Viganò, L., Vigneron, L.: The AVISPA tool for the automated validation of internet security protocols and applications. In: Etessami, K., Rajamani, S.K. (eds.) CAV 2005. LNCS, vol. 3576, pp. 281–285. Springer, Heidelberg (2005)
6. Arnaud, M., Cortier, V., Delaune, S.: Deciding security for protocols with recursive tests. In: Bjørner, N., Sofronie-Stokkermans, V. (eds.) CADE 2011. LNCS, vol. 6803, pp. 49–63. Springer, Heidelberg (2011)
7. Backes, M., Cervesato, I., Jaggard, A.D., Scedrov, A., Tsay, J.-K.: Cryptographically sound security proofs for basic and public-key kerberos. Int. J. Inf. Sec. **10**(2), 107–134 (2011)
8. Backes, M., Goldberg, I., Kate, A., Mohammadi, E.: Provably secure and practical onion routing. In: Chong, S. (ed.) 25th IEEE Computer Security Foundations Symposium, CSF 2012, 25–27 June 2012, Cambridge, MA, USA, pp. 369–385. IEEE (2012)
9. Backes, M., Hofheinz, D., Unruh, D.: Cosp: a general framework for computational soundness proofs. In: Al-Shaer, E., Jha, S., Keromytis, A.D. (eds.) Proceedings of the 2009 ACM Conference on Computer and Communications Security, CCS 2009, Chicago, Illinois, USA, 9-13 November2009, pp. 66–78. ACM (2009)
10. Backes, M., Jacobi, C.: Cryptographically sound and machine-assisted verification of security protocols. In: Alt, H., Habib, M. (eds.) STACS 2003. Lecture Notes in Computer Science, vol. 2607, pp. 675–686. Springer, Heidelberg (2003)
11. Backes, M., Pfitzmann, B., Waidner, M.: A universally composable cryptographic library. IACR Cryptology ePrint Archive 2003:15 (2003)
12. Bana, G., Comon-Lundh, H.: Towards unconditional soundness: computationally complete symbolic attacker. In: Degano, P., Guttman, J.D. (eds.) [38], pp.189–208
13. Bana, G., Comon-Lundh, H.: A computationally complete symbolic attacker for equivalence properties. In: Ahn, G-J., Yung, M., Li, N. (eds.) Proceedings of the 2014 ACM SIGSAC Conference on Computer and Communications Security, Scottsdale, AZ, USA, 3–7 November 2014, pp. 609–620. ACM (2014)
14. Barthe, G., Crespo, J.M., Grégoire, B., Kunz, C., Lakhnech, Y., Schmidt, B., Béguelin, S.Z.: Fully automated analysis of padding-based encryption in the computational model. In: Sadeghi, A.R., Gligor, V.D., Yung, M. (eds.) 2013 ACM SIGSAC Conference on Computer and Communications Security, CCS 2013, Berlin, Germany, 4–8 November 2013 pp. 1247–1260. ACM (2013)
15. Barthe, G., Grégoire, B., Heraud, S., Béguelin, S.Z.: Computer-aided security proofs for the working cryptographer. In: Rogaway, P. (ed.) CRYPTO 2011. LNCS, vol. 6841, pp. 71–90. Springer, Heidelberg (2011)

16. Baudet, M., Cortier, V., Kremer, S.: Computationally sound implementations of equational theories against passive adversaries. In: Caires, L., Italiano, G.F., Monteiro, L., Palamidessi, C., Yung, M. (eds.) ICALP 2005. LNCS, vol. 3580, pp. 652–663. Springer, Heidelberg (2005)
17. Bella, G., Massacci, F., Paulson, L.C.: Verifying the SET registration protocols. IEEE J. Sel. Areas Commun. **21**(1), 77–87 (2003)
18. Bella, G., Massacci, F., Paulson, L.C.: Verifying the SET purchase protocols. J. Autom. Reasoning **36**(1–2), 5–37 (2006)
19. Bellare, M.: New proofs for NMAC and HMAC: security without collision-resistance. In: Dwork, C. (ed.) CRYPTO 2006. LNCS, vol. 4117, pp. 602–619. Springer, Heidelberg (2006)
20. Bernhard, D., Cortier, V., Pereira, O., Smyth, B., Warinschi, B.: Adapting helios for provable ballot privacy. In: Atluri, V., Diaz, C. (eds.) ESORICS 2011. LNCS, vol. 6879, pp. 335–354. Springer, Heidelberg (2011)
21. Blanchet, B.: A computationally sound mechanized prover for security protocols. IEEE Trans. Dependable Sec. Comput. **5**(4), 193–207 (2008)
22. Blanchet, B., Abadi, M., Fournet, C.: Automated verification of selected equivalences for security protocols. In: Proceedings of 20th IEEE Symposium on Logic in Computer Science (LICS 2005), 26–29 June 2005, Chicago, IL, USA, pp. 331–340. IEEE Computer Society, Chicago (2005)
23. Blaze, M.: Toward a broader view of security protocols. In: Christianson, B., Crispo, B., Malcolm, J.A., Roe, M. (eds.) Security Protocols 2004. LNCS, vol. 3957, pp. 106–120. Springer, Heidelberg (2006)
24. Bond, M.: A chosen key difference attack on control vectors (2000). http://www.cl.cam.ac.uk/mkb23/research.html
25. Boyd, C., Mao, W.: Design and analysis of key exchange protocols via secure channel identification. In: Safavi-Naini, R., Pieprzyk, J.P. (eds.) ASIACRYPT 1994. LNCS, vol. 917, pp. 171–181. Springer, Heidelberg (1995)
26. Brands, S., Chaum, D.: Distance bounding protocols. In: Helleseth, T. (ed.) EUROCRYPT 1993. LNCS, vol. 765, pp. 344–359. Springer, Heidelberg (1994)
27. Burrows, M., Abadi, M., Needham, R.M.: A logic of authentication. In: SOSP, pp. 1–13 (1989)
28. Burton, C., Culnane, C., Heather, J., Peacock, T., Ryan, P.Y.A., Schneider, S., Teague, V., Wen, R., Xia, Z., Srinivasan, S.: Using prêt à voter in victoria state elections. In: Halderman, J.A., Pereira, O. (eds.) 2012 Electronic Voting Technology Workshop/Workshop on Trustworthy Elections, EVT/WOTE 2012, Bellevue, WA, USA, 6–7 August 2012. USENIX Association (2012)
29. Butler, F., Cervesato, I., Jaggard, A.D., Scedrov, A., Walstad, C.: Formal analysis of kerberos 5. Theor. Comput. Sci. **367**(1–2), 57–87 (2006)
30. Canetti, R.: Universally composable security: a new paradigm for cryptographic protocols. In: 42nd Annual Symposium on Foundations of Computer Science, FOCS 2001, Las Vegas, Nevada, USA, 14–17 October 2001, pp. 136–145. IEEE Computer Society (2001)
31. Carback, R., Chaum, D., Clark, J., Conway, J., Essex, A., Herrnson, P.S., Mayberry, T., Popoveniuc, S., Rivest, R.L., Shen, E., Sherman, A.T., Vora, P.L.: Scantegrity II municipal election at takoma park: the first E2E binding governmental election with ballot privacy. In: Proceedings of 19th USENIX Security Symposium, Washington, DC, USA, 11–13 August 2010, pp. 291–306. USENIX Association (2010)
32. Clarkson, M.R., Chong, S., Myers, A.C.: Civitas: A secure voting system. Technical report, Cornell University (2007)

33. Comon-Lundh, H., Cortier, V.: How to prove security of communication protocols? a discussion on the soundness of formal models w.r.t. computational ones. In: Schwentick, T., Dürr, C. (eds.) 28th International Symposium on Theoretical Aspects of Computer Science, STACS 2011, 10–12 March 2011, Dortmund, Germany, vol. 9, LIPIcs, pp. 29–44. Schloss Dagstuhl - Leibniz-Zentrum fuer Informatik (2011)

34. Comon-Lundh, H., Delaune, S.: The finite variant property: how to get rid of some algebraic properties. In: Giesl, J. (ed.) RTA 2005. LNCS, vol. 3467, pp. 294–307. Springer, Heidelberg (2005)

35. Cortier, V., Degrieck, J., Delaune, S.: Analysing routing protocols: four nodes topologies are sufficient. In: Degano, P., Guttman, J.D. (eds.) [38], pp. 30–50

36. Cortier, V., Smyth, B.: Attacking and fixing helios: an analysis of ballot secrecy. J. Comput. Secur. **21**(1), 89–148 (2013)

37. Cryptosense. Cryptosense web page. cryptosense.com

38. Degano, P., Guttman, J.D. (eds.): POST 2012. LNCS, vol. 7215. Springer, Heidelberg (2012)

39. Doghmi, S.F., Guttman, J.D., Thayer, F.J.: Searching for shapes in cryptographic protocols. In: Grumberg, O., Huth, M. (eds.) TACAS 2007. LNCS, vol. 4424, pp. 523–537. Springer, Heidelberg (2007)

40. Dolev, D., Yao, A, C-C.: On the security of public key protocols (extended abstract). In: 22nd Annual Symposium on Foundations of Computer Science, Nashville, Tennessee, USA, 28–30 October 1981, pp. 350–357. IEEE Computer Society (1981)

41. Ellison, C.M.: Ceremony design and analysis. IACR Cryptology ePrint Archive 2007:399 (2007)

42. Erbatur, S., Escobar, S., Kapur, D., Liu, Z., Lynch, C.A., Meadows, C., Meseguer, J., Narendran, P., Santiago, S., Sasse, R.: Asymmetric unification: a new unification paradigm for cryptographic protocol analysis. In: Bonacina, M.P. (ed.) CADE 2013. LNCS, vol. 7898, pp. 231–248. Springer, Heidelberg (2013)

43. Escobar, S., Sasse, R., Meseguer, J.: Folding variant narrowing and optimal variant termination. J. Log. Algebr. Program. **81**(7–8), 898–928 (2012)

44. Feigenbaum, J., Johnson, A., Syverson, P.F.: Probabilistic analysis of onion routing in a black-box model. ACM Trans. Inf. Syst. Secur. **15**(3), 14 (2012)

45. Focardi, R., Luccio, F.L., Steel, G.: An introduction to security API analysis. In: Aldini, A., Gorrieri, R. (eds.) FOSAD 2011. LNCS, vol. 6858, pp. 35–65. Springer, Heidelberg (2011)

46. Goldschlag, D.M., Reed, M.G., Syverson, P.F.: Onion routing. Commun. ACM **42**(2), 39–41 (1999)

47. Groß, T., Mödersheim, s.: Vertical protocol composition. In: Proceedings of the 24th IEEE Computer Security Foundations Symposium, CSF 2011, Cernay-la-Ville, France, 27–29 June 2011, pp. 235–250. IEEE Computer Society (2011)

48. Groza, B., Warinschi, B.: Cryptographic puzzles and dos resilience, revisited. Des. Codes Crypt. **73**(1), 177–207 (2014)

49. Halevi, S.: A plausible approach to computer-aided cryptographic proofs. IACR Cryptology ePrint Archive 2005:181 (2005)

50. Juels, A., Brainard, J.G.: Client puzzles: a cryptographic countermeasure against connection depletion attacks. In: Proceedings of the Network and Distributed System Security Symposium, NDSS 1999, San Diego, California, USA. The Internet Society (1999)

51. Kemmerer, R.A.: Using formal verification techniques to analyze encryption protocols. In: Proceedings of the 1987 IEEE Symposium on Security and Privacy, Oakland, California, USA, 27–29 April 1987, pp. 134–139. IEEE Computer Society (1987)
52. Khader, D., Tang, Q., Ryan, P.Y.A.: Proving prêt à voter receipt free using computational security models. In: 2013 Electronic Voting Technology Workshop/Workshop on Trustworthy Elections, EVT/WOTE 2013, Washington, D.C., USA, 12–13 August 2013. USENIX Association (2013)
53. Kürtz, K.O., Küsters, R., Wilke, T.: Selecting theories and nonce generation for recursive protocols. In: Ning, P., Atluri, V., Gligor, V.D., Mantel, H. (eds.) Proceedings of the 2007 ACM workshop on Formal methods in security engineering, FMSE 2007, Fairfax, VA, USA, 2 November 2007, pp. 61–70. ACM (2007)
54. Küsters, R., Truderung, T.: An epistemic approach to coercion-resistance for electronic voting protocols. In: 30th IEEE Symposium on Security and Privacy (S&P 2009), 17–20 May 2009, Oakland, California, USA, pp. 251–266. IEEE Computer Society (2009)
55. Küsters, R., Truderung, T., Vogt, A.: Proving coercion-resistance of scantegrity II. In: Soriano, M., Qing, S., López, J. (eds.) ICICS 2010. LNCS, vol. 6476, pp. 281–295. Springer, Heidelberg (2010)
56. Küsters, R., Wilke, T.: Automata-based analysis of recursive cryptographic protocols. In: Diekert, V., Habib, M. (eds.) STACS 2004. LNCS, vol. 2996, pp. 382–393. Springer, Heidelberg (2004)
57. Lowe, G.: Analysing protocols subject to guessing attacks. In: WITS 2002 (2002)
58. Malozemoff, A.J., Katz, J., Green, M.D.: Automated analysis and synthesis of block-cipher modes of operation. In: IEEE 27th Computer Security Foundations Symposium, CSF 2014, Vienna, Austria, 19–22 July 2014, pp. 140–152. IEEE (2014)
59. Maurer, U.M., Schmid, P.E.: A calculus for security bootstrapping in distributed systems. J. Comput. Secur. 4(1), 55–80 (1996)
60. Meadows, C.: Applying formal methods to the analysis of a key management protocol. J. Comput. Secur. 1(1), 5–36 (1992)
61. Meadows, C.: Analysis of the internet key exchange protocol using the NRL protocol analyzer. In: 1999 IEEE Symposium on Security and Privacy, Oakland, California, USA, 9–12 May 1999, pp. 216–231. IEEE Computer Society (1999)
62. Meadows, C.: A cost-based framework for analysis of denial of service networks. J. Comput. Secur. 9(1/2), 143–164 (2001)
63. Meadows, C.: Formal methods for cryptographic protocol analysis: emerging issues and trends. IEEE J. Sel. Areas Commun. 21(1), 44–54 (2003)
64. Meadows, C., Syverson, P.F.: A formal specification of requirements for payment transactions in the SET protocol. In: Hirschfeld, R. (ed.) FC 1998. LNCS, vol. 1465, pp. 122–140. Springer, Heidelberg (1998)
65. Meier, S., Schmidt, B., Cremers, C., Basin, D.: The TAMARIN prover for the symbolic analysis of security protocols. In: Sharygina, N., Veith, H. (eds.) CAV 2013. LNCS, vol. 8044, pp. 696–701. Springer, Heidelberg (2013)
66. Mitchell, J.C., Shmatikov, V., Stern, U.: Finite-state analysis of SSL 3.0. In: Rubin, A.D.: (ed.) Proceedings of the 7th USENIX Security Symposium, San Antonio, TX, USA, 26–29 January 1998. USENIX Association (1998)
67. Mödersheim, S., Viganò, L.: Sufficient conditions for vertical composition of security protocols. In: Moriai, S., Jaeger, T., Sakurai, K. (eds.) 9th ACM Symposium on Information, Computer and Communications Security, ASIA CCS 2014, Kyoto, Japan, 03–06 June 2014, pp. 435–446. ACM (2014)

68. Nakamoto, S.: Bitcoin: a peer-to-peer electronic cash system (2008). https:// bitcoin.org
69. Paiola, M., Blanchet, B.: Verification of security protocols with lists: From length one to unbounded length. J. Comput. Secur. **21**(6), 781–816 (2013)
70. Pavlovic, D., Meadows, C.: Deriving ephemeral authentication using channel axioms. In: Christianson, B., Malcolm, J.A., Matyáš, V., Roe, M. (eds.) Security Protocols 2009. LNCS, vol. 7028, pp. 240–261. Springer, Heidelberg (2013)
71. Pavlovic, D., Meadows, C.: Actor-network procedures. In: Ramanujam, R., Ramaswamy, S. (eds.) ICDCIT 2012. LNCS, vol. 7154, pp. 7–26. Springer, Heidelberg (2012)
72. Reiter, M.K., Rubin, A.D.: Crowds: anonymity for web transactions. ACM Trans. Inf. Syst. Secur. **1**(1), 66–92 (1998)
73. Roy, A., Datta, A., Mitchell, J.C.: Formal proofs of cryptographic security of diffie-hellman-based protocols. In: Barthe, G., Fournet, C. (eds.) TGC 2007. LNCS, vol. 4912, pp. 312–329. Springer, Heidelberg (2008)
74. Rusinowitch, M., Turuani, M.: Protocol insecurity with finite number of sessions is np-complete. In: 14th IEEE Computer Security Foundations Workshop (CSFW-14 2001), 11–13 June 2001, Cape Breton, Nova Scotia, Canada, p. 174. IEEE Computer Society (2001)
75. Ryan, P.Y.A., Bismark, D., Heather, J., Schneider, S., Xia, Z.: Prêt à voter: a voter-verifiable voting system. IEEE Trans. Inf. Forensics Secur. **4**(4), 662–673 (2009)
76. Santiago, S., Escobar, S., Meadows, C., Meseguer, J.: A formal definition of protocol indistinguishability and its verification using Maude-NPA. In: Mauw, S., Jensen, C.D. (eds.) STM 2014. LNCS, vol. 8743, pp. 162–177. Springer, Heidelberg (2014)
77. Schneider, S., Teague, V., Culnane, C., Heather, J.: Special section on vote-id 2013. J. Inf. Sec. Appl. **19**(2), 103–104 (2014)
78. SET Secure Electronic Transactions LLC. SET Secure Electronic Transactions, Version 1.0 (2002). http://www.exelana.com/set/
79. Shoup, V.: Sequences of games: a tool for taming complexity in security proofs. IACR Cryptology ePrint Archive 2004:332 (2004)
80. Thayer, J.F., Herzog, J.C., Guttman, J.D.: Strand spaces: proving security protocols correct. J. Comput. Secur. **7**(1), 191–230 (1999)
81. Tor Project. The Tor Project: Anonymity Online. https://www.torproject.org/
82. Truderung, T.: Selecting theories and recursive protocols. In: Abadi, M., de Alfaro, L. (eds.) CONCUR 2005. LNCS, vol. 3653, pp. 217–232. Springer, Heidelberg (2005)
83. Vaudenay, S.: Secure communications over insecure channels based on short authenticated strings. In: Shoup, V. (ed.) CRYPTO 2005. LNCS, vol. 3621, pp. 309–326. Springer, Heidelberg (2005)

A Denotational Semantic Theory
of Concurrent Systems

Jayadev Misra$^{(\boxtimes)}$

Department of Computer Science, University of Texas, Austin 78712, USA
misra@utexas.edu

Abstract. This paper proposes a general denotational semantic theory suitable for most concurrent systems. It is based on well-known concepts of events, traces and specifications of systems as sets of traces. Each programming language *combinator* is modeled by a *transformer* that combines the specifications of the components to yield the specification of a system. We introduce *smooth* and *bismooth* transformers that correspond to monotonic and continuous functions in traditional denotational theory. We show how fairness under recursion can be treated within this theory.

Keywords: Denotational semantics · Specification transformer · Smooth transformer · Bismooth transformer · Specifications of recursive programs · Fairness and recursion

1 Introduction

This paper proposes a general denotational semantic theory suitable for most concurrent systems. It is based on well-known concepts of events, traces and specifications of systems as sets of traces.

A concurrent system consists of a number of components that are combined using the *combinators* of a specific programming language. A specification of a component is a prefix-closed set of traces. A *transformer* combines the specifications of the components to yield the specification of a system; thus, each combinator of a programming language is modeled by a transformer. The two most significant concepts in this paper are *smooth* and *bismooth* transformers. A smooth transformer is a monotonic function on traces, ordered by prefixes. A bismooth transformer is smooth, and, analogous to continuous functions in traditional denotational theory [13], preserves the upward-closures of specifications. These transformers can model various features of concurrent systems such as, concurrent interactions with memory and objects, independent as well as causally dependent threads, unbounded non-determinism, shared resource, deadlock, fairness, divergence and recursion.

Note on Proofs: A complete paper, that includes proofs of all propositions in an Appendix, is at http://www.cs.utexas.edu/users/misra/DenotationalSemantics.pdf.

© Springer International Publishing Switzerland 2015
N. Martí-Oliet et al. (Eds.): Meseguer Festschrift, LNCS 9200, pp. 493–518, 2015.
DOI: 10.1007/978-3-319-23165-5_23

Treatment of recursion, Sect. 4, requires us to introduce *bismooth* transformer, the counterpart of a continuous function that preserves the limits of chains (upward-closure) as well as the prefixes of traces (downward-closure). We develop a version of the well-known fixed point theorem [7,13] that shows that first computing a simple fixed point and then taking its limit is appropriate for bismooth transformers, see Sect. 4.4. Transformers that encode fairness are smooth (monotonic) but not bismooth (continuous); so this theorem does not apply. We generalize the least fixed point theorem, to *min-max* fixed point theorem, for smooth transformers that allows treatment of certain forms of fairness; see Sects. 4.5 and 4.6.

Monotonic and continuous functions in denotational semantics operate on elements of any complete partial order without any pre-assumed structure. Even though smooth and bismooth transformers are the counterparts of monotonic and continuous functions, they operate on specifications which have structure as sets of traces. We exploit this structural information to obtain strong results about various classes of transformers and fixed points.

We do not to develop the semantics of a specific programming language but of transformers that are of general applicability in all conceivable concurrent systems. Features of specific programming languages can be treated by combining a few elementary transformers, as we demonstrate in the example below. The long-term goal of the research is to suggest a framework for analysis of concurrent programming language constructs.

A Motivating Example. Let $\oplus$ be a 3-way combinator so that in $\oplus(A, B, C)$ synchronization of the executions A and B initiates the execution of C. Operationally, A and B are parent threads that execute concurrently at start. Child thread C starts executing only when the parents synchronize, by A engaging in event e and B in $\bar{e}$. In case both events occur, they have completed a "rendezvous", C is started and A and B resume execution. Neither e nor $\bar{e}$ is shown explicitly as occurring in the execution in case of a rendezvous.

It is possible that a synchronization may never be completed even though one of A and B, say A, has engaged in its synchronization event e. In that case, A remains waiting to synchronize and C is never started, though B may continue to execute forever or halt without synchronization.

We define a transformer $\oplus'$, corresponding to the combinator $\oplus$, that transforms the *specifications* of A, B and C to yield the specification of $\oplus(A, B, C)$. The definition of $\oplus'$ uses a few transformers described in this paper.

Let the specifications of A, B and C be p, q and r, respectively. Introduce C' that behaves as C but indicates the start of its execution by a specific event a; event a does not occur in p, q or r. The specification of C' is $cons(a, r)$ that appends a as the first event to every trace in r; see the definition of *cons* in Sect. 3.3.5. The execution of $\oplus(A, B, C')$ interleaves their individual executions arbitrarily, subject to the constraint that the events e, $\bar{e}$ and a be synchronized. The interleaved executions of A, B and C' is given by their unfair merge, written as $p \mid q \mid cons(a, r)$; see Sect. 3.3.11. The synchronization of e, $\bar{e}$ and a is written using a transformer, called *rendezvous*, that introduces a new event τ to indicate the simultaneous occurrences of e, $\overline{e}$ and a; see Sect. 3.3.14. Finally, event τ is removed from the specification, using transformer *drop*; see Sect. 3.3.4. Thus,

$$\oplus'(p, q, r) = drop(\{\tau\}, rendezvous(\{e, \overline{e}, a\}, \tau, (p \mid q \mid cons(a, r))))$$

We can now assert certain properties of $\oplus'$. For example, that it is bismooth, because all the transformers in its definition are bismooth and composition of bismooth transformers is bismooth.

2   Basic Concepts

A *trace* represents one possible execution of a component. The *specification* of a component is a set of prefix-closed *traces*. We define these concepts and explore their properties in this section.

2.1   Event and Trace

2.1.1   Event

Event types are uninterpreted symbols drawn from an event alphabet. The choice of event types for a component constitutes a design decision about the granularity at which we may wish to examine the component. For the systems that we consider in this paper, event types could be many including, for instance: input and output, binding of a parameter to a value, calling a shared resource for read/write access, receiving a response from a resource, locking and unlocking of a resource, allocation and disposal of storage, or publishing a value as a result of a computation. Each *event* is an instance of some event type in an execution. Instances of the same event type are distinguished, say by subscripts, so that all events in an execution are distinct. The semantic theory makes no assumption about the meanings of events.

2.1.2   Trace

A *trace* is the formal counterpart of partial or complete execution of a program. A trace includes a sequence of events, the events occurring in the execution, and the state at the end of the execution if it is finite; the state is called *status* in this paper[1]. An execution that has halted, i.e., one that can engage in no further event, has status H. A finite execution that is waiting for an event to happen or waiting to halt has status W. An infinite execution has status D, representing divergence. A finite execution may also have status D; this represents an infinite execution that has only a finite prefix of visible events; see Sect. 2.4. A trace is written in the form $y[m]$ where y is the *status* from $\{H, W, D\}$, and m is a finite or infinite sequence of distinct events. If y is H or W then m is finite.

For notational simplicity, different instances of the same event are sometimes written as the same symbol in examples; thus, given that tl is an event type, we may abbreviate the trace $W[tl_1 \; tl_2]$ by $W[tl \; tl]$.

---

[1] We use the term *status* to distinguish the state of execution from the states of other mutable objects in the system.

2.1.3 Example

Consider a component that has the following behavior. It tosses a coin repeatedly until the coin lands heads. Then it halts. Let hd and tl denote the events of coin landing heads and tails, respectively, and tl^i a sequence of length i of tl events. Then any finite execution is represented by, for some $i \geq 0$, either (1) $W[tl^i]$, (2) $W[tl^i\ hd]$, or (3) $H[tl^i\ hd]$. If the coin is fair we expect it to land heads eventually; so, these are the only traces of the component. If the coin is unfair, it is possible to have an infinite sequence of tails, and the corresponding trace is $D[tl^\omega]$. If the coin toss events are invisible, then the only traces in an external spec for fair coin are $W[\ ]$ and $H[\ ]$, and for unfair coin are $W[\ ]$, $H[\ ]$ and $D[\ ]$. Thus, with an unfair coin an external observer can assert only that this component may eventually halt or may compute forever.

The component described in this example does not interact with any other component. To see interaction, suppose the component does not actually toss the coin but requests another component to do so and communicate the result to it. Let $toss$ be a request for a toss, and $rcvhd$ and $rcvtl$ are the events corresponding to the responses received when the toss lands heads and tails, respectively; assume that a response is guaranteed. A trace of the component with a fair coin is, for some $i \geq 0$, either (1) $W[(toss\ rcvtl)^i]$, (2) $W[(toss\ rcvtl)^i\ toss]$, (3) $W[(toss\ rcvtl)^i\ toss\ rcvhd]$, or (4) $H[(toss\ rcvtl)^i\ toss\ rcvhd]$. An external observer can assert eventual termination, because there is no external event for which the program may wait forever. With an unfair coin there is an additional trace $D[(toss\ rcvtl)^\omega]$, and termination can not be asserted.

Tuples of traces. In this paper, transformers are functions that map a set of traces to a set of traces. In dealing with programs that contain several components, a transformer, such as merge, maps each *tuple* of traces, with one trace from each component, to a set of possible traces of the program. In most contexts the distinction between a trace and a tuple of traces is immaterial, so we use the term "trace" to denote a single trace or a finite tuple of traces, the tuple size depending on the context. A tuple of traces is finite if each component trace is finite.

Traceset. A *traceset* is a *non-empty* set of traces. A finitary traceset is one in which each trace is finite. Tracesets are partially ordered by subset order.

2.2 Prefix Order Over Traces

Informally, trace s is a prefix of t when the execution corresponding to s can possibly be extended to that for t. For sequences m and n, let $m \sqsubseteq n$ denote that m is a prefix of n. Impose a partial order $\leq$ over the status values as follows: $W \leq H$ and $W \leq D$. The partial order mimics the causal order in an execution so that $W[m]$ may evolve to $H[m]$ by changing state silently, and $W \leq D$ because any finite computation precedes an extension of it to an infinite computation.

Trace $y[m]$ is a *prefix* of $z[n]$ ($z[n]$ an *extension* of $y[m]$) if $y \leq z$ and $m \sqsubseteq n$. And, $y[m]$ is a *proper prefix* of $z[n]$, if $y[m] \leq z[n]$ and $y[m] \neq z[n]$. So, a trace with status H or D has no extension. An infinite trace is a prefix only of itself. And $W[\,] \leq y[m]$ for every trace $y[m]$.

For tuples of traces define one tuple as a prefix of another if each entry in the former tuple is a prefix of the corresponding entry in the latter. And $(s_0, s_1, \cdots, s_k) < (t_0, t_1, \cdots, t_k)$ if $s_i \leq t_i$ for each i and $s_j < t_j$ for some j.

Properties of prefix order. The following properties are easy to prove.

1. Prefix order, $\leq$, is a partial order over traces.
2. The inverse of proper prefix order, $>$, is a well-founded order over traces.
3. The set of prefixes of a trace are totally ordered.

An Induction Principle over traces. The inverse of proper prefix order, $>$, is a well-founded order even in the presence of infinite traces. This allows us to formulate the following induction principle. Let P be a predicate over traces, both finite and infinite.

> If for all t, $(\forall s : s < t : P(s)) \Rightarrow P(t)$,
> then $P(t)$ holds for all traces t.

2.3   Prefix Closure

The *prefix-closure*, also called *downward-closure*, of trace t is denoted by t_*; it is the set of all prefixes of t. For a traceset p, p_* is the set of prefixes of all traces of p. That is,

$$t_* = \{s \mid s \leq t\} \text{ and } p_* = \cup\{t_* \mid t \in p\}.$$

It follows that for traces s and t, $(s, t)_* = s_* \times t_*$.

Finite Prefix-Closure. Denote the set of *finite* prefixes of trace t by $t_{*'}$. Define $p_{*'}$ for traceset p analogously. Note that an infinite trace t is not in $t_{*'}$, though $t \in t_*$.

Notational Conventions.

1. Prefix-closure and finite prefix-closure operators have the highest binding power among all operators.
2. Prefix closure and finite prefix-closure apply to event sequences, not just traces and tracesets.
3. Write $C_*(p)$ for $(C(p))_*$ for any p in any context C.
4. (singletons and sets) A singleton trace may appear wherever a traceset is expected to appear. That is, if $C(p)$ is a valid expression for any traceset p, so is $C(t)$ for a trace t, and it denotes $C(\{t\})$.

    Conversely, if $C(t)$ is a valid expression for any trace t, so is $C(p)$ for any traceset p, and it denotes $\cup_{t \in p} C(t)$.

Thus, $W[m_*]$ is a shorthand for $\{W[k] \mid k \in m_*\}$. And, $W_*[m] = (W[m])_* = \{s \mid s \leq W[m]\}$.

Elementary Properties of Prefix-Closure. Below p and q are tracesets, and t any trace. The following properties are easy to show. Closure expansion, item (4), is used extensively in subsequent proofs.

1. Prefix-closure is an algebraic closure, i.e., for tracesets p and q,
   (a) (extensive) $p \subseteq p_*$
   (b) (monotonic) $p \subseteq q \Rightarrow p_* \subseteq q_*$
   (c) (idempotent) $(p_*)_* = p_*$.
2. Finite prefix-closure of tracesets is monotonic and idempotent. Extensive property does not hold for the traceset $\{t\}$ where t is an infinite trace.
3. $(t_*)_{*'} = (t_{*'})_* = t_{*'}$.
4. (Closure expansion) For any trace $z[m]$, $z_*[m] = \{z[m]\} \cup W[m_{*'}]$.
5. (closure distributes over set union) For a family F of tracesets, F possibly infinite, $(\cup_{p \in F}(p))_* = (\cup_{p \in F}(p_*))$.
6. (closure distributes over Cartesian product) $(p \times q)_* = p_* \times q_*$.

2.4   Specification

Informally, a specification of a component, henceforth abbreviated as *spec*, is a set of traces, where each trace corresponds to an execution of the component in *some* environment. Different traces may correspond to executions in different environments. Properties of a component may be deduced from its spec, such as that its publications are monotonic in value (a safety property), every execution eventually halts (a progress property), or that the component's execution may deadlock (the spec includes a trace $W[m]$ that has no extension).

Spec. A *spec* is a prefix-closed traceset. A *finitary spec* is a spec consisting of finite traces.

Note that a spec of n-tuples includes the bottom trace, $(W[\,], W[\,], \cdots, W[\,])$, consisting of n individual empty traces.

2.4.1   Properties of Specs

The proofs of the following properties are elementary.

1. For any traceset p, p_* and $p_{*'}$ are both specs.
2. Union of a finite or infinite family of specs is a spec.
3. Intersection of a finite or infinite family of specs is a spec.
4. Cartesian product of a pair of specs is a spec.

Consider the coin toss example of Sect. 2.1.3. With a fair coin we expect the spec to be $H_*[tl^i\ hd]$, and for an unfair coin to be $H_*[tl^i\ hd] \cup \{D[tl^\omega]\}$.

2.4.2   Chains and Their Limits

A *chain* is a finitary spec whose elements are totally ordered under $\leq$. A chain may be finite or infinite. For any trace t the set of its finite prefixes, $t_{*'}$, is a chain.

The limit of chain c, written as $lim(c)$, is the least upper bound of the traces in c with respect to the $\leq$ ordering. For a finite chain c, $lim(c)$ is the longest trace in c. For an infinite chain c, $lim(c)$ is the unique infinite trace such that every trace in c is its prefix. Note that $lim(c)$ does not belong to c for infinite c because c consists of finite traces only. Notationally, use $lim(c)$ as a trace and also as a singleton traceset.

Define the limit of a finite tuple of chains as the tuple of limits of the corresponding chains. That is,

$$lim(c_0, c_1, \cdots c_n) = (lim(c_0), lim(c_1), \cdots lim(c_n))$$

2.4.3   Complete Lattice of Specs

The least upper bound of a set of specs is their union, and the greatest lower bound is the intersection. Thus, specs form a complete lattice under subset order, where $\perp = W[\,]$ and $\top$ is the union of all specs.

3   Transformer

Any component of a system is either a *primitive* component or a *structured* component. A primitive component is defined by its spec. A structured component consists of one or more subcomponents that are combined using the *combinators* of the language. A spec transformer, or simply a *transformer*, corresponding to each combinator is a function mapping the Cartesian product of the specs of the subcomponents to the spec of the structured component. The number of subcomponents, therefore the length of the tuples in the argument of the transformer, is the *arity* of the transformer. A language semantic thus consists of the specs of the primitive components and the transformers corresponding to each combinator. For the moment assume that the domain of a transformer is the set of all traces. We show how to restrict the domain of a transformer in Sect. 3.2.2.

Convention. We develop the theory for transformers of arity 1, a transformer that maps a spec to a spec. Generalizations for other arities are straightforward. Examples of transformers of higher arity appear in Sects. 3.3.2, 3.3.11, 3.3.12 and 3.3.15. For a transformer of arity 2 we adopt infix notation, as in $p \oplus q$.

We restrict ourselves to a class of transformers, called *smooth*. Smooth transformers correspond to monotonic functions in denotational semantic theory. A subset of smooth transformers, called *bismooth*, correspond to continuous functions. We develop the theory of smooth transformers in this section and bismooth transformers in Sect. 4.3.

3.1   Trace-Wise Transformer

A trace-wise transformer is.a total function from traces to tracesets. Using the notational convention introduced earlier, a trace-wise transformer f applied to a traceset p is defined to be: $f(p) = \cup\{f(t) \mid t \in p\}$. For trace-wise combinator $\oplus$ over a pair of specs, $p \oplus q = \cup\{s \oplus t \mid s \in p,\ t \in q\}$.

Any transformer maps a spec to a spec. We restrict ourselves to trace-wise transformers in this paper because a language combinator can combine only individual executions of its components. Non-determinism is represented by mapping a trace to a traceset, every trace of the latter corresponds to a possible execution. The size of the result traceset is arbitrary, thus allowing unbounded non-determinism.

3.1.1   Properties of Trace-Wise Transformers
The following properties follow from the definition of trace-wise transformers.

1. A trace-wise transformer distributes over union (possibly infinite union) of tracesets. That is, given a family F of tracesets, $(\cup_{p \in F} f(p)) = f(\cup_{p \in F}(p))$.
2. Composition of trace-wise transformers is a trace-wise transformer.
3. (Monotonicity) For trace-wise f and tracesets p and q,
   $p \subseteq q \;\Rightarrow\; f(p) \subseteq f(q)$.

A trace-wise transformer may not transform a spec to a spec, i.e., the resulting traceset may not be prefix-closed (consider a transformer that maps every trace to $W[a]$ where a is some event; the resulting traceset does not include $W[\,]$, hence, is not a spec). The smoothness condition, described below, guarantees this property.

3.2   Smooth Transformer

A transformer f is *smooth* if and only if for any traceset p

$$f_*(p) = f(p_*), \text{ where } f_*(p) \text{ stands for } (f(p))_*.$$

And, f is *finitely smooth* if for finitary p, $f_*(p) = f(p_*)$.

3.2.1   Properties of Smooth Transformers
The following properties are proved in Propositions 1–3.

1. A transformer f is smooth if and only if it preserves prefix-closure over individual traces, i.e., $f_*(t) = f(t_*)$, for every trace t.
2. A transformer is smooth if and only if it maps specs to specs.
3. Composition of smooth transformers is smooth.

Terminology and Notation. Henceforth, "transformer" stands for "trace-wise transformer" in this paper. For a binary smooth transformer $\oplus$ written in infix style, $(p \oplus q)_* = p_* \oplus q_*$, for tracesets p and q.

3.2.2   Domain of a Transformer

We have so far assumed that every transformer is defined for all traces. In many cases a transformer f can meaningfully be defined only over some domain $dom(f)$; we assume that $dom(f)$ is a spec. We show how to extend the domain of a transformer while retaining its essential properties. Specifically, we define transformer g over all traces that induces the same mapping over $dom(f)$ as f and retains smoothness and bismoothness.

For any t in $dom(f)$ let $g(t) = f(t)$. For $t \notin dom(f)$ and finite t, let $g(t) = \cup\{f(s) \mid s \leq t \text{ and } s \in dom(f)\}$. For $t \notin dom(f)$ and infinite t, let $g(t) = lim(g(t_{*'}))$, where lim is defined in Sect. 2.4.2). It can be shown that if f is smooth over the traces in $dom(f)$ then so is g over all traces, and if f is bismooth over any spec in $dom(f)$ then so is g over all specs.

Note: For $t \notin dom(f)$ and finite t, alternately let $g(t) = f(s)$ where s is the longest prefix of t in $dom(f)$.

3.3   Some Elementary Smooth Transformers

In this section, we show a number of smooth transformers that are of general utility. Transformer g with arguments is written as $g(args, t)$ where $args$ is a set of parameters and t a trace. Here, g represents a family of transformers, one transformer for each value of $args$. For a specific value of $args$ we abbreviate $g(args, t)$ to $f(t)$, and then prove the smoothness of f. Note that the identity transformer, $id(t) = t$ for all traces t, is smooth.

3.3.1   Status Map

This is a family of transformers each member of which may change the status of a trace but not its event sequence. Applying $statusmap(y[m])$, a generic member of the family, yields $y'[m]$ where y' may differ from y only if $y = H$, or $y = D$ and m is finite; thus, $statusmap(y[m]) = y[m]$, if $y = W$ or m is infinite. Every transformer in $statusmap$ is smooth; see Proposition 4.

3.3.2   Choice

The choice transformer, or, corresponds to a non-deterministic choice between two components to execute. For components f and g with specs p and q, f or g has the spec $p \cup q$. As a trace-wise transformer: s or $t = \{s, t\}$. We show that or is smooth, in Proposition 5.

3.3.3   Hide

Transformer $hide$ is parametrized by a set of events E, which may be finite or infinite; $hide(E, t)$ is the trace obtained after removing all events from t that also occur in E. Application of $hide$ may remove an unbounded, and possibly infinite, number of events from a trace. For example, $hide(\{a\}, D[a^\omega])$ results in $D[\,]$. To see $hide$ is smooth see Proposition 6.

3.3.4   Drop

Transformer *drop* is same as *hide* except that in $drop(E,t)$ (1) the event set E is finite, and (2) only the first occurrence, if any, of an event from E is removed from t, but subsequent occurrences are retained. The proof that *drop* is smooth is similar to the proof for *hide*. The reason we treat *drop* separately is that *drop* is bismooth —see Sect. 4.3.4— whereas *hide* is not. This property permits *drop*, but not *hide*, to be freely used in recursive equations.

3.3.5   Cons

Append a specific event a as the first event of every trace. To ensure that a spec is transformed to a spec, $cons(a, W[])$ includes $W[]$.

$$cons(a, W[]) = \{W[], W[a]\}, \qquad cons(a, y[m]) = \{y[am]\}$$

We show that *cons* is smooth in Proposition 7.

3.3.6   Filter

A class of transformers, called *filter*, is essential for most applications of this theory. A filter can be used to model interactions among components by rejecting the traces that do not implement acceptable interactions, as in accesses to shared resources. A filter can also model rendezvous-style interactions and fairness constraints.

Associated with each filter is a predicate b over traces such that:

F1. $b(W[])$ holds, and
F2. If $b(t)$ holds then $b(s)$ holds for all proper prefixes s of t, i.e., writing $b(t_{*'})$ for the conjunction of $b(s)$ over all finite prefixes s of t: $b(t) \Rightarrow b(t_{*'})$.

Filter f corresponding to predicate b *accepts* t iff $b(t)$ holds and *rejects* it otherwise. Thus, if a filter accepts a trace it accepts all prefixes of that trace; equivalently, if it rejects a trace, it rejects all extensions of that trace. Since $b(W[])$ holds, not all traces are rejected. A filter applied to a spec retains only its acceptable traces.

The natural definition of transformer f corresponding to filter predicate b is $f(t) = \{t\}$ if $b(t)$ and $\{\}$ otherwise. This definition violates the requirement that $f(t)$ be a traceset, a non-empty set of traces, for all t. So, we propose:

$$f(t) = \{s \mid s \leq t \text{ and } b(s)\}$$

Transformer f is smooth; see Proposition 17.

Observe that for filter predicates b and b', $b \wedge b'$ and $b \vee b'$ are also filter predicates. If transformers g and g' implement b and b' respectively, then $g \circ g'$ implements $b \wedge b'$ and $g(t) \cup g'(t)$, for any trace t, implements the disjunction of the filters. Any filter transformer is idempotent, and it distributes over union and intersection of specs. The following identity is used in the min-max fixed point theorem, Sect. 4.5. For filter g and specs p and q,

$$g(p \cap q) = g(p) \cap g(q) = g(p) \cap q$$

Continuous vs. Discontinuous Filter. We distinguish between two kinds of filters, *continuous* and *discontinuous*, depending on the value of $b(t)$ for infinite t. A discontinuous filter models fairness wherein an infinite trace may be rejected even though all its finite prefixes are accepted. A continuous filter rejects an infinite trace only if some finite prefix of it is also rejected. Conversely, a continuous filter accepts an infinite trace if all its finite prefixes are accepted.

Both types of filter predicates obey the conditions (F1) and (F2) given earlier. Additionally, a continuous filter predicate b satisfies the stronger condition (F2') below in place of (F2):

F2'. $b(t) \equiv b(t_{*'})$, for every trace t.

Note that (F2) and (F2') are equivalent for finite t. It is only for infinite t that (F2') imposes the additional constraint: $b(t_{*'}) \Rightarrow b(t)$.

For a continuous filter f, we can let $f(t)$ be the longest prefix of t for which b holds. This is defined for finite t because $b(W[])$ holds, and for infinite t because the longest prefix is t if $b(t)$ holds and some finite prefix of t if $\neg b(t)$ holds. Continuous filters are always bismooth, discontinuous filters are not; see Sect. 4.3.4.

Filters are some of the most useful transformers. The following sections list special cases of filters that arise in concurrent programming.

Partitioning a filter. Any filter can be written as a composition of two filters one of which is continuous and the other rejects only infinite traces. That is, a filter f can be written as $f_{inf} \circ f_{fin}$ where (1) f_{inf} rejects trace t only if t is infinite, and f rejects t though it accepts all finite prefixes of t, and (2) f_{fin} rejects all other traces, finite and infinite, that f rejects. It is possible that neither f_{inf} nor f_{fin} rejects any trace. Clearly, $f = f_{inf} \circ f_{fin}$. Further, f_{fin} is a continuous filter because whenever it rejects an infinite trace it also rejects a finite prefix of it. And, if f_{inf} rejects any trace, it is a discontinuous filter.

3.3.7   Restrict by Inclusion of Events
Reject a trace if it contains a specific event a, or, more generally, an event from a specified set E. This is a filter because (1) it accepts $W[]$, and (2) if it accepts a trace, it accepts all its prefixes. The filter is continuous.

The converse of this rejection criterion is *not* smooth: accept a trace only if it is $W[]$ or contains a specific event a. Then any trace that has a as its last event is accepted but all its prefixes except $W[]$ are rejected. Therefore, it may transform a spec to a traceset that is not prefix-closed.

3.3.8   Restrict by Exclusion of Events
Accept a trace only if it is $W[]$ or its *first* event is drawn from a specified set of events. This condition defines a filter predicate b because: (1) $b(W[])$ holds, and (2) if $b(t)$ holds, it holds for all prefixes of t. The filter is continuous. The requirement that the specified event be the first one in the event sequence is crucial; without this requirement the transformer is not smooth.

504   J. Misra

The acceptance criterion here is stronger than a typical filter: whenever a trace is accepted, all its extensions are also accepted.

3.3.9   Restrict by Precedence Relation

Let R be a binary relation over events. Define a transformer that accepts trace t iff for every (e, e') in R, if e' is in t then e is also in t and e precedes e'. Thus, an acceptable trace is one that either includes (1) none of e and e', (2) just e, or (3) both e and e' with e preceding e'. It is easy to see that $W[\,]$ is accepted and the prefix of an acceptable trace is acceptable. Further, this transformer is a continuous filter.

3.3.10   Atom

Atomicity is a fundamental notion in concurrent programming, particularly in the theory of transactions. Roughly, trace t is atomic with respect to a specified set of events if all the specified events occur contiguously in some order in t. We propose a more general definition that is useful in defining other transformers.

A *pattern alphabet* is a finite subset of the event alphabet. A *pattern* is a finite string over the pattern alphabet. Let P be a finite set of patterns. Trace t is *atomic* with respect to P if the event sequence in t can be written uniquely as a sequence of patterns from P interspersed with events outside the pattern alphabet, optionally followed by a prefix of some pattern if t is finite. Predicate $atom(P, t)$, where P is a finite set of patterns and t a trace, holds iff t is atomic with respect to P.

It is easy to see that *atom* is a filter predicate, because $W[\,]$ is accepted and if t is accepted then so are all its prefixes. Additionally, *atom* defines a continuous filter because if an infinite trace t is rejected then some finite prefix of it is not atomic with respect to P.

3.3.11   Unfair Merge

One of the most important transformers, that models concurrent executions of components, is *merge*. It interleaves the events of two traces arbitrarily yielding a traceset from a pair of traces. Besides interleaving the events, merge also computes the status of the interleaved trace based on those of the given traces. Assume that the events in the traces to be merged are distinct.

There are two forms of interleavings, *unfair* and *fair*, of event sequences m and n. The distinction is significant only when one or both of m and n are infinite. If each interleaving includes all elements of m and n then it is *fair*; we treat fair merge in Sect. 3.3.12. An unfair interleaving may include only a finite prefix of n for infinite m, and analogously for infinite n.

Properties of unfair interleaving. Define unfair interleaving of m and n, $m \otimes n$, by the following program (using pattern matching style):

$$[\,] \otimes n = n$$

$$m \otimes [\,] = m$$
$$(a : m) \otimes (b : n) = (a : (m \otimes (b : n))) \cup (b : ((a : m) \otimes n))$$

Using fixed point induction it can be shown that $\otimes$ is symmetric. It can also be shown that it is monotonic in both arguments, so $m \otimes n \subseteq m' \otimes n$ and $m \otimes n \subseteq m \otimes n'$, where $m \subseteq m'$ and $n \subseteq n'$. Further,

$$(m \otimes n)_* = m_* \otimes n_* \hspace{2cm} (\otimes \text{ distributes over prefixes})$$

Transformer for unfair merge. Unfair merge of two traces applies unfair interleaving to their event sequences. Also, it applies a symmetric binary operation $\cap$ over their status values: $H \cap y = y$ and $W \cap W = W$. Define unfair merge transformer, $|$, as follows, where both y and z are from $\{H, W\}$.

$$
\begin{aligned}
y[m] \,|\, z[n] &= (y \cap z)(m \otimes n) \\
D[m] \,|\, z[n] &= D[m \otimes n_*] \\
D[m] \,|\, D[n] &= D[m \otimes n_*] \cup D[m_* \otimes n]
\end{aligned}
$$

Observe that m and n may be finite or infinite in $D[m]$ and $D[n]$ above. Note that $z[n] \,|\, D[m] = D[m] \,|\, z[n]$, and it is not shown explicitly below.

The intuition behind this definition is as follows. Expression $y[m] \,|\, z[n]$ denotes the concurrent execution of two executions, one corresponding to $y[m]$ and the other to $z[n]$. Both executions are finite and if either fails to halt then the concurrent execution does not halt either, as given by the status $(y \cap z)$. The event sequence in $y[m] \,|\, z[n]$ is an interleaving of m and n, which justifies the result expression $(y \cap z)(m \otimes n)$.

Next, consider infinite executions defined by the next two cases. In $D[m] \,|\, z[n]$, $D[m]$ denotes an infinite execution $D[m']$ where m is the sequence of visible events in m'; thus, m may be finite. The resulting concurrent execution is infinite, so its status is D. Any concurrent execution executes a prefix of $z[n]$ with all of $D[m']$; so the event sequences in all such executions are given by $m' \otimes n$. Since only the events of m are retained from m', the resulting expression is $D[m \otimes n_*]$. Similar remarks apply for $D[m] \,|\, D[n]$, because any execution may use a prefix of the event sequence of one of $D[m]$ or $D[n]$ and all events of the other.

It can be shown from the above definition that unfair merge is commutative, associative and $H[\,]$ is its zero. We show that unfair merge is smooth in Proposition 8.

3.3.12 Fair Merge

Fair merge is based on fair interleaving, which we denote by $\otimes'$:

$$m \otimes' n = \{x \mid x \in m \otimes n, \ x \text{ contains } m \text{ and } n \text{ as subsequences}\}$$

Note that if m is infinite and n non-empty, then $m \in m \otimes n$ and $m \notin m \otimes' n$.

Extend the definition of $\cap$ to apply to all status values $\{H, W, D\}$ as follows. Recall that $\cap$ is symmetric. For any status value y

$H \cap y = y$, $W \cap W = W$ and $D \cap y = D$

Define fair merge transformer, $|'$, of two argument traces $y[m]$ and $z[n]$ for y and z from $\{H, W, D\}$, and finite or infinite m and n.

$$y[m] \, |' \, z[n] = (y \cap z)(m \otimes' n)$$

The proof of smoothness of fair merge can be developed in a manner similar to unfair merge. There is a much simpler alternative proof. Observe that fair merge of $y[m]$ and $z[n]$ is same as their unfair merge followed by application of a filter that removes every infinite trace $D[k]$ from $y[m] \, | \, z[n]$ where $k \notin m \otimes' n$. Both unfair merge and filter are smooth; so, their composition, fair merge, is also smooth.

3.3.13   Replace

We consider a general version of substitution of a sequence of events by a single event. A *source alphabet* and a *target alphabet* are disjoint finite subsets of the event alphabet. A *replacement pair* is of the form (σ, τ) where σ, called the *source*, is a finite string over the source alphabet, and τ, called the *target*, is a single symbol from the target alphabet.

Let R be a finite set of replacement pairs. A source may occur multiple times in R with different targets, and similarly, a target may have multiple occurrences in R with different sources. Transformer *replace* substitutes occurrences of a source by all corresponding targets in an event sequence. The effect of $replace(R, t)$ is to (1) accept t if t is atomic with respect to the sources, see Sect. 3.3.10 and t contain no symbol from the target alphabet, and (2) if t is accepted, replace occurrence of every source by *all* corresponding targets to obtain a set of traces, and (3) then replace occurrence of any proper prefix of a source by the empty string. The situation in (3) arises because the prefix of an atomic trace may contain a prefix of a source as its suffix. The domain of this transformer can be extended to all traces using domain extension described in Sect. 3.2.2.

Henceforth, let $f(t)$ denote $replace(R, t)$ for a specific R. The definition of f for finite t is given in clausal form in a functional style, where the clauses are attempted in the given order from top to bottom.

$$
\begin{aligned}
f(y[\sigma']) &= \quad y[\,], \text{ where } \sigma' \text{ is a proper prefix of a source} \\
f(y[\sigma m]) &= \quad \cup\{cons(\tau, f(y[m])) \mid (\sigma, \tau) \in R\} \\
f(y[am]) &= \quad cons(a, f(y[m])), \, a \notin \text{ source alphabet}
\end{aligned}
$$

Thus the source σ is replaced by every target associated with it in $f(y[\sigma m])$.

For infinite t, it is easier to specify the transformer using limits from Sect. 2.4.2. Such a definition permits simpler proofs of smoothness and bismoothness: $f(t) = lim(f(t_{*'}))$.

We prove that *replace* is smooth in Proposition 9. It can be shown that the "substitution" transformer, that replaces each event e in a trace by event $h(e)$ where h is a function over the event alphabet, is smooth.

3.3.14   Rendezvous

The unfair and fair merge transformers of Sects. 3.3.11 and 3.3.12 implement independent concurrent processes whose executions can be arbitrarily interleaved. We consider more refined versions of concurrent executions in Sect. 3.5 in which the processes call upon shared resources, and hence, their executions can not be arbitrarily interleaved. Here, we introduce a form of synchronization, called rendezvous in CSP [5] and CCS [10], that ensures that a pair of complementary events $\{e, \overline{e}\}$ from the two processes occur simultaneously. Their simultaneous occurrence is shown by an event τ in the combined trace that belongs to neither process.

We define *rendezvous* by composing the transformers *atom* and *replace*. First, perform an appropriate merge, fair or unfair, of the specs of the two processes. Then apply transformer *atom* of Sect. 3.3.10 to eliminate the traces in which $\{e, \overline{e}\}$ do not occur contiguously. Next, using transformer *replace* of Sect. 3.3.13, replace all (contiguous) occurrences of $\{e, \overline{e}\}$ by τ, and remove any e or $\overline{e}$ event that occurs by itself. We generalize this scheme slightly by allowing rendezvous to occur with any finite set of events E instead of just two events $\{e, \overline{e}\}$, as follows.

Let E' be the set of strings obtained by permuting the events of E in all possible order. Henceforth, write $rendezvous(E, \tau, t)$, $\tau \notin E$, for the transformer that (1) accepts t provided t is atomic with respect to E', (2) replaces every pattern of E' in trace t by event τ, and then (3) removes any non-empty proper prefix of a pattern of E'. Here, t would likely be a trace arising out of the concurrent executions of processes. If required, τ can be eliminated by applying transformer *drop* of Sect. 3.3.4. Define *rendezvous* as follows and note that it is smooth since *atom* and *replace* are smooth.

$$rendezvous(E, \tau, t) = replace(\{(\sigma, \tau) \mid \sigma \in E'\}, atom(E', t))$$

A flaw in this definition, as noted by a referee, is that contiguous events of a *single* component may perform rendezvous, because there is no distinction among events from different components. To overcome this flaw distinguish events from different components so that every pattern in E' consists of events from different components.

3.3.15   Sequential Composition

Consider a simple form of sequential composition of f and g in which g starts executing only when f halts. The corresponding transformer; is:

$H[m]\ ;\ z[n] = z[mn],$
$s\ ;\ z[n] = s,$ otherwise

It can be shown that sequential composition is associative. We show that sequential composition is smooth in Proposition 10.

3.4   Fairness

Fairness is a filter that eliminates only certain infinite traces from a spec. For example, a fairness constraint for the coin toss example of Sect. 2.1.3 may specify that the coin is fair so that an infinite sequence of tails is impossible; then, trace $D[tl^\omega]$ is inadmissible. A fairness constraint about a strong semaphore may specify that any execution in which a P event on the semaphore remains waiting forever while V events happen infinitely often is inadmissible. In a real-time computation a fairness constraint may specify that an infinite number of events may not occur within a bounded time interval.

A fairness constraint can be defined by a filter predicate b, where b holds for all finite traces and, possibly, some infinite traces. Therefore, $b(W[\,])$ holds and if b holds for any trace it holds for all its prefixes. For the coin toss example, the filter predicate b holds for every finite trace and every infinite trace that does not have an infinite suffix of either heads or tails (for the example shown, it does not matter if the coin lands heads infinitely often, because the game is terminated after the first landing of a head). Being a filter, fairness is a smooth transformer. Fairness is not bismooth; see Sect. 4.3.4.

Fairness can be composed with other transformers. In particular, different forms of fairness may apply to different parts of a program; a fair and an unfair version of the coin toss program may run concurrently, for example, and our theory would yield their combined spec.

3.5   Shared Resource

Merge transformers, Sects. 3.3.11 and 3.3.12, model independent concurrent executions of processes by interleaving the traces of the individual processes. Concurrent executions are rarely independent. For example, trace s of one process includes the event $read(3)$ that reads value 3 from a read/write shared store, trace t of another process includes $write(3)$ for the same store, and the store is local to these two processes. Then $write(3)$ precedes $read(3)$ in the traces for their concurrent execution; any trace in which the events occur in a different order has to be rejected. Further, if t includes $write(5)$ instead of $write(3)$, no trace for the concurrent execution can include $read(3)$.

Shared resource is a filter. Each resource instance is a filter over an alphabet that denotes the available operations on the resource. Alphabets of different instances of the same resource and of different resources are disjoint. Applied to the merge of traces of individual processes, the filter rejects the traces that violate the semantics of the shared store. For example, for a read/write store that is local to a pair of concurrently executing processes, first the appropriate merge of their traces is constructed, and then a filter applied to ensure that: (1) a value is written to the store before any value is read, and (2) any value that is read is equal to the value last written. Any trace that violates these constraints is rejected. Independent resources are independent filters that may be applied in arbitrary order on a trace.

Local vs. global resource. Consider concurrently executing processes A and B that include traces s and t in their specs, respectively. Suppose s includes $read(3)$ and t includes $write(5)$, as the only events on a shared read/write store. As we have seen earlier, if the store is local to A and B, no trace for the concurrent execution of A and B can include $read(3)$. However, if the store is global, so that other processes may access it, another process may perform $write(3)$. So, a trace for the concurrent execution of A and B may include $read(3)$ for a global store.

It follows from this discussion that each resource has *two* filters corresponding to its local and global behaviors. Suppose processes A, B and C, whose specs are p, q and r, respectively, have a local resource. Let fl be the local filter and fg the global filter for the resource. Then the spec for the concurrent execution of A, B and C (assuming unfair merge for their concurrent execution) is $fl(fg(p\,|\,q)\,|\,r)$. It is easy to see that the global filter for a read/write store accepts all traces, because for any given trace there is a sequence of accesses to the store that validates that trace.

It is possible to develop a more elaborate set of filters for a resource based on access rights that allows different processes to perform different operations on the resource.

Blocking operations on shared resource. Both filters, local and global, for a read/write store are continuous. In fact, a resource for which all operations are non-blocking induces continuous filters. (Note, however, that for processes that share a read/write store, their concurrent execution is modeled by the fair merge of their specs. A fair merge introduces discontinuity; see Proposition 31).

For a resource with blocking operations, the filter may be continuous or discontinuous. Consider a semaphore that has operations P and V on it, where P is blocking and V non-blocking. It is customary to consider P as consisting of two events, a *request* event, which we denote by $\langle P$ and a *response* event $P\rangle$, where $P\rangle$ is always preceded by the corresponding $\langle P$, though a $\langle P$ may never be followed by a corresponding $P\rangle$.

First, consider a weak semaphore that merely ensures that a request is granted (response sent), whenever the semaphore is available, to *some* waiting process (i.e., any that has an outstanding request for it), though any specific waiting process may never be granted the semaphore. A weak semaphore filter, both local and global, has to reject an infinite trace in which the semaphore is continuously available in an infinite suffix, the suffix contains $\langle P$, but contains no subsequent $P\rangle$. The weak semaphore filter is continuous.

Next, consider a strong semaphore that ensures that each process that requests the semaphore is eventually granted it, provided the semaphore is available infinitely often in an infinite execution. The specification of each process identifies the request and response events by the process identity. The corresponding filter rejects an infinite trace that contains an infinite number of occurrences of V, some occurrence of $\langle P_1$ for a specific process numbered 1, but no subsequent $P_1\rangle$. This is a discontinuous filter.

4   Treatment of Recursion

The theory developed so far is adequate for programs that include no recursive definition; now, we enhance the theory to treat recursive definitions. Guarded recursion is usually easier to handle. We treat the general case of unguarded recursion, as in solving an equation of the form $x = f(x)$ in spec x, for a given transformer f. Thus, we will compute the spec of a definition such as

   def loop() = *loop*()

where *loop*(), with no arguments, is defined recursively. As we will see, the spec of this program will not be the bottom spec $\{W[]\}$ but $\{D[]\}_*$ denoting a divergent computation. This is because we expect each recursive call to engage in an internal event in making the call, so the call entails an infinite computation in which the internal events are invisible.

4.1   Classical Treatment of Recursion

The least fixed-point theorem due to Kleene [7], and also in Scott [13], applies for any continuous function f on a complete partial order (cpo).

Theorem 1 (Least Fixed-point Theorem). Let f be a continuous function on a cpo whose bottom element is $\perp$. The least fixed-point of f, $lfp(f)$, is $lub\{f^i(\perp) \mid i \geq 0\}$ where
$f^0(x) = x$, $f^{i+1}(x) = f(f^i(x))$ and lub is the least upper bound of a chain.   □

   In applying this theorem in our context, the set of specs form a complete lattice, hence a cpo. Any trace-wise transformer is continuous over specs because given a chain of specs p_i, $0 \leq i$, where the least upper bound is union:

$$f(\cup\{p_i \mid i \geq 0\}) = \cup\{f(p_i) \mid i \geq 0)\}.$$

Corollary 1. *The least fixed point of a smooth transformer is a spec.*

Proof: It is easily shown by induction on i that for any i, $i \geq 0$, $f^i(W[])$ is a spec. The union of specs is a spec. So, $lfp(f)$ is a spec, from Theorem 1.

4.1.1   Revisiting the Coin Toss Example

As an example of the application of the least fixed-point theorem consider the coin toss example of Sect. 2.1.3. Call the toss program *stutter*. A step of *stutter* either halts the computation, or engages in event *tl* and then calls *stutter*, the choice being non-deterministic and unfair in that an infinite number of calls may be made to *stutter*.

   There are two component computations, *halt* and the recursive call on *stutter*, that are combined through non-deterministic choice. As we have shown in Sect. 3.3.2, the transformer corresponding to choice is set union. The spec of halt is $\{H[]\}_*$. Let x stand for the spec of *stutter*. The recursive call preceded by event *tl* is $cons(tl, x)$; see Sect. 3.3.5 for a definition of *cons*. Thus, we have:

$$x = \{H[\,]\}_* \cup cons(tl, x)$$

Observe that each of the transformers, $\cup$ and *cons* are smooth. So, their composition given above is smooth.

The steps in the application of the least fixed-point theorem successively yield, $\{W[\,]\}$, $\{W[\,], H[\,], W[tl]\}$, $\{W[\,], H[\,], W[tl], H[tl], W[tl^2]\}$, $\cdots$, and for any i, $\{H[tl^j] \mid 0 \le j \le i\}_* \cup \{W[tl^{i+1}]\}$. Then $lfp(stutter)$, the lub of this sequence, is $\{H[tl^i] \mid 0 \le i\}_*$.

4.1.2 The Need for Upward-Closure of Specs

From $lfp(stutter)$ we may deduce that every execution of *stutter* is finite, though unbounded, in length. But this is not what happens in reality. It is possible for an unfair coin to land tails forever, so the trace $D[tl^\omega]$ ought to be included in the spec. And, $\{H[tl^i] \mid 0 \le i\}_*$ is actually the spec of *stutter* where a *fair* coin is used in the toss so that there is no infinite computation.

The present difficulty arises because subset ordering over specs implies that the lub of a chain of specs is simply their union. We overcome this difficulty by introducing the notion of upward-closed specs that include the limits of countable chains of traces in the spec. The lub of a chain of upward-closed specs is not simply their union, but the upward-closure of their union. Thus, the lub of the specs $\{H[tl^j] \mid 0 \le j \le i\}_* \cup \{W[tl^{i+1}]\}$, for all i, $0 \le i$, is $\{H[tl^i] \mid 0 \le i\}_* \cup \{D[tl^\omega]\}$.

This discussion suggests that in solving $x = f(x)$, the transformer f needs to transform an upward-closed spec to an upward-closed spec. Not all smooth transformers have this property. So, we introduce *bismooth transformers*, a subclass of smooth transformers, that have this property. We develop the appropriate concepts of upward-closure, and revisit the least fixed-point theorem.

4.2 Upward-Closure

4.2.1 Definitions

The following definitions use chains and limits from Sect. 2.4.2.

Definition of upward-closure. The upward-closure of spec p is:

$$p^* = \{lim(c) \mid c \text{ a chain in } p\}.$$

It follows that c^*, the *upward-closure* of chain c, is $c \cup lim(c)$. In particular, for finite c, $c^* = c$. A spec is *upward-closed* if $p^* = p$, i.e., if chain c is in p, then so is $lim(c)$.

Trace s in p is *maximal* if there is no t in p such that $s < t$. An arbitrary spec may not have a maximal trace, for example the spec $\{W[a^i] \mid i \ge 0\}$. But p^* always has a maximal trace. Limit of a spec is given by:

$$lim(p) = \{s \mid s \text{ a maximal trace in } p^*\}$$

Chain Continuity. Transformer f is *chain continuous* if $f(c^*) = f^*(c)$ for any chain c ($f^*(c)$ is $(f(c))^*$). Each of the following conditions imply chain continuity: (1) $f(lim(c)) = lim(f(c))$, for any chain c, (2) $f(t) = lim(f(t_{*'}))$ for any infinite trace t.

4.2.2   Properties of Upward-Closure

Proofs of the following properties are in Propositions 11–19.

1. (Proposition 11) Upward-closure is algebraic closure, i.e., for specs p and q,
   (a) (extensive) $p \subseteq p^*$.
   (b) (monotonic) $p \subseteq q \Rightarrow p^* \subseteq q^*$.
   (c) (idempotent) $(p^*)^* = p^*$.
2. Alternate characterizations of upward-closure: For spec p,
   (a) $p^* = p \cup \{lim(c) \mid c$ an infinite chain in $p\}$.
       This is because the limit of every finite chain of p is in p.
   (b) (Proposition 12) $p^* = p \cup lim(p)$.
   (c) (Proposition 13) $p^* = lim_*(p)$.
       It follows that given spec p, p^* is a spec because $p^* = lim_*(p)$, and $lim_*(p)$ is prefix-closed.
3. (Galois Adjoints) Finite prefix closure and upward-closure are Galois Adjoints, i.e., for traceset p and spec q: $p_{*'} \subseteq q \equiv p \subseteq q^*$.
   The following identities are then easily derived for specs p and q.
   (a) $p_{*'} = (p^*)_{*'}$
   (b) $(p_{*'})^* = p^*$
   (c) $p_{*'} \subseteq q_{*'} \equiv p^* \subseteq q^*$
   (d) $p_{*'} = q_{*'} \equiv p^* = q^*$
4. (Distribution over union and intersection)
   (a) (Union) (Proposition 14) Let F be a family of upward-closed specs and $P = \cup_{p \in F}(p)$. Then $P^* = \cup_{p \in F}(p^*)$ iff every chain in P belongs to some spec in F. For finite F, $P^* = \cup_{p \in F}(p^*)$.
   (b) (Intersection) (Proposition 15) Let F be a family of specs and $P = \cap_{p \in F}(p)$. Then $P^* = \cap_{p \in F}(p^*)$.
   (c) (Proposition 16) For any spec q, $q^* = \cup\{c^* \mid c$ a chain in $q\}$.
5. (upward-closure of tuples) For specs p and q, $(p \times q)^* = p^* \times q^*$.
6. (Proposition 17) Let f be a *finitely* smooth transformer, and $f(t) = lim(f(t_{*'}))$ for every infinite trace t. Then, f is smooth.
7. (Proposition 18) Let f be chain continuous. If a finite trace s is in $f(t)$, for some trace t, then $s \in f(t_{*'})$. Equivalently, $f_{*'}(p) = f_{*'}(p^*) = f_{*'}(p_{*'})$, for any spec p.
8. (Proposition 19) For spec p and filter g, $g(p^*) \subseteq g^*(p)$.

Note on Distribution over union. The result in Item(4a) is of interest only if the chain is infinite and F is infinite, because any finite chain in P belongs to some spec in F, and for finite F the result holds unconditionally. To see that $P^* = \cup_{p \in F}(p^*)$ does not hold unconditionally for infinite families, let $p_i = W_*[a^i]$, for every natural number i. Each p_i is a spec, and $p_i^* = W_*[a^i]$. Therefore, $(\cup_i(p_i^*)) = \{W[a^i] \mid 0 \le i\}$. But, $(\cup_i p_i)^* = \{W[a^i] \mid 0 \le i\}^* = \{W[a^i] \mid 0 \le i\} \cup D[a^\omega]$.

4.3   Bismooth Transformer

A smooth transformer does not necessarily preserve upward-closure. To see this, consider transformer f where $f(t) = t_{*\prime}$, for all traces t. It is easy to see that f is smooth. For an infinite chain of traces c, $f(c) = c$, so $f^*(c) = c^*$ whereas $f(c^*) = f(c \cup \{lim(c)\}) = f(c) \cup f(lim(c)) = c$.

Call a transformer *bismooth* if it preserves both upward and downward-closures. That is, for bismooth f:

(smooth; preserves downward-closure) $f(p_*) = f_*(p)$, for any traceset p, and
(preserves upward-closure) $f(p^*) = f^*(p)$, for any spec p.

A *finitely* bismooth transformer is smooth and it preserves upward-closure over *finitary* specs.

4.3.1   Example of a Bismooth Transformer

Consider transformer *or* from Sect. 3.3.2 where *or* maps a tuple of specs (p, q) to $p \cup q$. We have shown in that section that *or* is smooth. To prove that *or* is bismooth show that $or^*(p, q) = or((p, q)^*)$.

$$or((p, q)^*)$$
$=$ $\{or((p, q)^*) = or(p^* \times q^*)$ from item (5); rewriting$\}$
    $or\{(x, y) \mid x \in p^*, y \in q^*\}$
$=$ $\{or\{(x, y) \mid x \in p^*, y \in q^*\} = \{x \mid x \in p^*\} \cup \{y \mid y \in q^*\}$; set theory$\}$
    $p^* \cup q^*$
$=$ $\{$upward-closure distributes over finite union, item (4)$\}$
    $(p \cup q)^*$
$=$ $\{$definition of *or*$\}$
    $or^*(p, q)$

4.3.2   Chain Continuity $\neq$ Bismoothness

From its definition every bismooth transformer, even a finitely bismooth transformer, is chain continuous. In analogy with the definition of smooth transformers based on traces it may seem that we can give a similar characterization of bismooth transformers based on chains, namely, that every smooth and chain continuous transformer is bismooth. The following counterexample is due to Ernie Cohen.

Consider transformer *hide* from Sect. 3.3.3 that was shown to be smooth. Let *hidea* be its instance that removes every a event from a trace. It is not hard to see that $hidea(c^*) = hidea^*(c)$ for any chain c. Yet *hidea* is not bismooth, as shown below.

Let spec p be $\{W[a^i b^i] \mid i \geq 0\}_*$, where a and b are different symbols from the event alphabet. Now $hidea^*(p) \neq hidea(p^*)$:

$p^* = \{W[a^i b^i] \mid i \geq 0\}_* \cup D[a^\omega]$ $hidea(p^*) = \{W[b^i] \mid i \geq 0\}$
$hidea(p) = \{W[b^i] \mid i \geq 0\}$ $hidea^*(p) = \{W[b^i] \mid i \geq 0\} \cup D[b^\omega]$

4.3.3   Properties of Bismooth Transformers

As the counterexample in the previous subsection shows, chain continuity is insufficient for bismoothness. Typically, proving that a transformer is bismooth is considerably more difficult than proving that it is smooth. The properties given below simplify such proofs.

Properties of Bismooth Transformers.

1. The identity transformer, $id(p) = p$, is bismooth. Easy to show.
2. (Bismooth composition) (Proposition 20) Composition of bismooth transformers is bismooth.
3. (Proposition 21) A transformer is bismooth if and only if it is finitely bismooth.
4. (Propositions 22 and 23) Smooth transformer f is bismooth if and only if (1) f is chain continuous, and (2) corresponding to any chain d in $f(p)$, where p is a spec, there is a chain c in p such that $d \subseteq f(c)$.
5. (Sufficient condition for bismoothness) (Proposition 24) Define a transformer to be *co-finite* if it maps only a finite number of finite traces to any finite trace. A transformer that is smooth, co-finite and chain continuous is bismooth.

Property (2), bismooth composition, permits definition of new bismooth transformers using the existing ones. Property (3) simplifies many proofs regarding bismooth transformers by eliminating considerations of infinite traces in a spec. Even though chain continuity by itself is insufficient to guarantee bismoothness, Property (4) shows that an additional condition on chains in p and $f(p)$ is both necessary and sufficient for bismoothness. Property (5), a sufficient condition for bismoothness, is immensely helpful in proofs when a transformer is defined without using any known bismooth transformer. Almost all proofs in Sect. 4.3.4 about the elementary transformers use this sufficient condition. Its proof uses a recent result, from Misra [11], on a variation of Kőnig's infinity lemma [6]. The co-finiteness condition in property (5) is not a necessary condition for bismoothness; for example if $f(t) = \{W[]\}$ for all t then f is bismooth though not co-finite.

4.3.4   Bismoothness of Transformers from Sect. 3.3

We showed a number of useful transformers in Sect. 3.3. All transformers of that section except *hide* of Sect. 3.3.3, discontinuous *filter* of Sect. 3.3.6 and fair merge of Sect. 3.3.12 are bismooth; see Table 1.

4.4   Least Upward-Closed Fixed Point Theorem

An *upward-closed fixed point* of f is a spec that is both a fixed-point and upward-closed. The following theorem shows that the least upward-closed fixed-point of bismooth f, $lufp(f)$, is $lfp^*(f)$. Since $lfp(f)$ is a spec so is $lufp(f)$.

Theorem 2. [Least Upward-Closed Fixed Point Theorem] (Proposition 34)
For bismooth f, $lufp(f) = lfp^*(f)$ □

Table 1. Summary of bismoothness of elementary transformers

Transformer	Bismooth?	Proof
Status map	Yes	Proposition 25
Choice	Yes	Section 4.3
Hide	No	Section 4.3
Drop	Yes	Proposition 26
Cons	Yes	Proposition 27
Discontinuous filter	No	Proposition 28
Continuous filter	Yes	Proposition 29
Restrict by inclusion	Yes	Special case of continuous filter
Restrict by exclusion	Yes	Special case of continuous filter
Restrict by precedence	Yes	Special case of continuous filter
Atom	Yes	Special case of continuous filter
Unfair merge	Yes	Proposition 30
Fair merge	No	Proposition 31
Replace	Yes	Proposition 32
Rendezvous	Yes	Composition of bismooth transformers
Sequential Composition	Yes	Proposition 33

Consider the coin toss example of Sect. 4.1.1 whose least fixed point is $\{H[tl^i] \mid 0 \leq i\}_*$. The least upward-closed fixed point corresponding to this fixed point is $\{H[tl^i] \mid 0 \leq i\}_* \cup \{D[tl^\omega]\}$, which faithfully describes the finite and infinite behaviors with an unfair coin.

4.5   Min-Max Fixed Points of Smooth Transformers

A smooth transformer that includes some aspect of fairness, say, a discontinuous filter, is not bismooth. We develop a theorem that gives a precise characterization of the appropriate least fixed points of smooth transformers. A similar result for a model of actors using event diagrams is given in Clinger's thesis [4] (Chap. 4).

A smooth transformer is monotonic; hence, using the Knaster-Tarski theorem [14], it has a least fixed point. However, this fixed point may not be upward-closed. Consider the coin toss example of Sect. 4.1.1 that uses a *fair* coin so that an infinite run of tails is inadmissible. The recursive equation describing this component is $x = fc(\{H[\ ]\}_* \cup cons(tl, x))$ where transformer fc implements a fair coin and, hence, is a discontinuous filter. There is no upward-closed fixed point of this equation. The desired fixed-point is $\{H[tl^i] \mid 0 \leq i\}_*$, but it is not upward-closed. So, instead of upward-closed fixed point, we look for a least fixed-point that *includes as many limit traces as possible under the fairness constraint*.

For any smooth transformer f define p to be a *maximal* fixed point of f if p is the greatest fixed point of f in p^*; i.e., p includes as many traces as possible from p^*. Observe that the greatest fixed point in any traceset q is the union of all fixed points in q, because union of fixed points is a fixed point for any trace-wise transformer. The least maximal fixed point of f, $mmfp(f)$, also called the min-max fixed point, is: (1) a maximal fixed point of f, and (2) the least among all maximal fixed points of f. Theorem 3 shows that min-max fixed point exists for any smooth transformer.

The following equation $E(X)$, for a given X and unknown r, is important in the study of min-max fixed point:

$$r = X \cap f(r). \qquad\qquad [E(X)]$$

Theorem 3. [Min-Max Fixed Point Theorem] Let f be a smooth transformer and $p = lfp(f)$. Then (1) $mmfp(f)$ is the greatest fixed point of f in p^*. Further, (2) if $f(p^*) \subseteq p^*$, $mmfp(f)$ is the greatest solution of $E(p^*)$.   □

Proof of (1) is in Proposition 35 and of (2) in Proposition 36. This theorem shows that the min-max fixed point can be "computed" by first computing a least fixed point and then a greatest fixed point, but there is no need for nested fixed point computations. The computation of the least fixed point of f is "semi-constructive" for all smooth transformers using the least fixed point theorem. Unfortunately, a smooth transformer is not necessarily continuous with respect to the greatest lower bound. So, the greatest solution of $E(p^*)$ can not be computed in the same manner. The greatest fixed point of f in p^*, given $f(p^*) \subseteq p^*$, is $\cup \{r \mid r \subseteq f(r) \wedge r \subseteq p^*\}$, using a proof similar to that of the Knaster-Tarski theorem [14].

The min-max fixed point theorem is a generalization of the least upward-closed fixed point Theorem 2. To see this, let f be bismooth. Given $p = lfp(f)$, $f(p^*) = f^*(p) = p^*$. So, p^* is a fixed point, therefore, the greatest fixed point in p^*. Hence, $mmfp(f) = p^*$, from the min-max fixed point theorem.

The condition $f(p^*) \subseteq p^*$ in (2) holds if f is chain continuous, see Proposition 37. We consider a class of "fair" transformers in the next section for which the condition in (2) holds, and we give stronger characterizations of min-max fixed points for such transformers.

4.6   Fixed Point Under Fairness

A common form of a smooth transformer is $g \circ h$ where g is a filter, typically modeling fairness, and h a bismooth transformer. It can then be shown that $f(p^*) \subseteq p^*$ where $p = lfp(f)$, so the following stronger version of Theorem 3 applies.

Theorem 4. [Min-Max Fixed Point Theorem under Fairness] (Proposition 38) Let $f = g \circ h$ where g is a filter, h is bismooth and $p = lfp(f)$. Then $mmfp(f)$ is the greatest solution of the equation $E(p^*)$, as well as of $E'(p^*)$, where $E'(X)$ is the equation $r = g(X) \cap h(r)$.   □

A special case of this theorem often arises in practice: for any infinite trace t, $t \in h(t)$. This holds for the coin-toss example shown previously in this section. In this case a simpler characterization exists for the min-max fixed point.

Theorem 5. (Proposition 39) Let $f = g \circ h$ where g is a filter, h is bismooth, and for any infinite trace t, $t \in h(t)$. Then $mmfp(f) = g(lfp^*(f))$.   □

5   Concluding Remarks

This paper grew out of an effort to develop a proof theory for Orc [12,15], a concurrent programming language designed by the author and his collaborators. The concepts developed during that work, such as smooth and bismooth transformers, were found to be applicable for concurrent systems in general. We have constructed the transformers for Orc constructs by combining some of the elementary transformers described here. We have also extended the theory to real time systems.

We are currently developing a proof theory for concurrent systems, based on the theory developed here. A spec is a predicate over traces. Each elementary transformer corresponds to some operation on one or more predicates; for example, choice is simply disjunction over predicates and a filter is a conjunction of the filter predicate to eliminate unacceptable traces. Other transformers, such as merge and rendezvous, have no simple counterpart in predicate calculus though they can be specified using quantification.

Related Work. Applying denotational semantics to a concurrency calculus was pioneered by Hoare and his collaborators for CSP [1]. In a series of papers, they have developed a number of models culminating in a failure-divergence model [2]. They have defined all the relevant features of CSP, including rendezvous-based synchronized communication as well as both internal and external non-determinism. Fairness is not relevant for CSP.

The theory proposed in this paper is inherently asynchronous. Concurrent execution is modeled via interleaving of actions. Yet, it is possible to simulate rendezvous, as we show in Sect. 3.3.14. There is no special treatment for failure in our theory because it can be included as part of the spec of a component.

The distinction between internal and external non-determinism is exemplified by the expressions $ab+ac$ and $a(b+c)$, where a is an internal event of a component X, b and c are events on which X synchronizes with another component Y, and $+$ denotes non-deterministic choice. In $ab+ac$ the choice is made internally by X to synchronize on either the b (if it has chosen the ab alternative) or the c event (with ac alternative). If X has chosen to synchronize on b and Y offers c, there is a deadlock. This distinction is modeled in our theory by X executing an internal decision event, say a coin toss, that decides between b and c in $ab + ac$. The internal specification of X includes the decision event as a visible event though it is invisible in the external spec. In $a(b + c)$, the choice of the synchronizing event is determined externally, by Y offering either b or c.

Broy and Nelson [3] includes a number of important results concerning the existence and non-existence of fixed-points in the presence of fair choice. Their paper develops the theory for the "dovetail" operator that combines fair choice with angelic non-determinism, so that a terminating computation causes competing non-terminating computations to be discarded and rolled-back.

Meseguer, in personal communication, has observed that the theory presented here is an instance of more general constructions in ω-posets [8,9,16].

Acknowledgments. I am truly grateful to José Meseguer whose thorough reading of an earlier draft, and substantive technical comments, especially about connections to category theory and ω-cpo are highly relevant. Tony Hoare has been a constant source of encouragement and inspiration. I am grateful to Ernie Cohen who spent considerable amount of time helping me with several conceptual issues. Manfred Broy had pointed out some deficiencies in this theory. Vladimir Lifschitz has been a sounding board and adviser on many algebraic questions. Members of IFIP WG 2.3, as always, have given many helpful suggestions. Perceptive comments by two anonymous referees have improved the presentation.

References

1. Brookes, S.D., Hoare, C.A.R., Roscoe, A.W.: A theory of communicating sequential processes. J. ACM **31**(3), 560–599 (1984)
2. Brookes, S.D., Roscoe, A.W., Walker, D.J.: An operational semantics for CSP. Technical report, Carnegie Mellon University
3. Broy, M., Nelson, G.: Adding fair choice to Dijkstra's calculus. TOPLAS **16**(3), 924–938 (1994)
4. Clinger, W.D.: Foundations of actor semantics. Technical report, Massachusetts Institute of Technology, Cambridge, MA, USA (1981)
5. Hoare, C.A.R.: Communicating Sequential Processes. Prentice Hall International, Englewood Cliffs (1984)
6. Kőnig, D.: Theorie der Endlichen und Unendlichen Graphen: Kombinatorische Topologie der Streckenkomplexe. Akad. Verlag, Leipzig (1936)
7. Kleene, S.C.: Introduction to Metamathematics. North-Holland, Amsterdam (1952)
8. Meseguer, J.: Completions, factorizations and colimits for omega-posets. Reports, University of California, Los Angeles (1978)
9. Meseguer, J.: Order completion monads. Algebra Univers. **16**(1), 63–82 (1983)
10. Milner, R.: Communication and Concurrency. In: Hoare, C.A.R. (ed.) International Series in Computer Science. Prentice-Hall, London (1989)
11. Misra, J.: Mapping among the nodes of infinite trees: a variation of Kőnig's infinity lemma. Inf. Process. Lett. **15**(5), 548–549 (2015). doi:10.1016/j.ipl.2015.01.005
12. Misra, J., et al.: Orc language project. http://orc.csres.utexas.edu
13. Scott, D.: Outline of a mathematical theory of computation. In: 4th Annual Princeton Conference on Information Sciences and Systems, pp. 169–176 (1970)
14. Tarski, A.: A lattice-theoretical fixpoint theorem and its application. Pac. J. Math. **5**, 285–309 (1955). See http://www.cs.utexas.edu/users/misra/Notes.dir/KnasterTarski.pdf for a detailed proof
15. Wehrman, I., Kitchin, D., Cook, W., Misra, J.: A timed semantics of Orc. Theoret. Comput. Sci. **402**(2–3), 234–248 (2008)
16. Wright, J., Wagner, E., Thatcher, J.: A uniform approach to inductive posets and inductive closure. Theoret. Comput. Sci. **7**, 57–77 (1978)

Weak Bisimulation as a Congruence in MSOS

Peter D. Mosses[✉] and Ferdinand Vesely

Swansea University, Swansea SA2 8PP, UK
{p.d.mosses,csfvesely}@swansea.ac.uk

Abstract. MSOS is a variant of structural operational semantics with a natural representation of unobservable transitions. To prove various desirable laws for programming constructs specified in MSOS, bisimulation should disregard unobservable transitions, and it should be a congruence. One approach, following Van Glabbeek, is to add abstraction rules and use strong bisimulation with MSOS specifications in an existing congruence format. Another approach is to use weak bisimulation with specifications in an adaptation of Bloom's WB Cool congruence format to MSOS. We compare the two approaches, and relate unobservable transitions in MSOS to equations in Rewriting Logic.

1 Introduction

Rewriting logic comes with a built-in "abstraction dial" [26]. The least abstract position of the dial is when the specification consists entirely of rules, with no equations (apart from structural axioms such as associativity, etc.). Turning rules into equations increases the degree of abstraction: the remaining rules specify rewrites on equivalence classes of terms. The set of equations is required to be confluent, so semantic rules for nondeterministic or concurrent constructs cannot be replaced by equations.

In structural operational semantics (SOS) [1,36,37], specifications usually consist entirely of transition rules. Abstraction can be introduced by regarding particular transitions as *unobservable* (conventionally using the label τ) and ignoring unobservable transitions when defining behavioural equivalences. Confluence of unobservable transitions is not required: nondeterministic choice can itself be unobservable. So-called 'structural congruence' equations such as associativity and commutativity are used in the SOS of the π-calculus [28], but their introduction has a non-trivial impact on the meta-theory of SOS [34]; moreover, they are not as general as equations in Rewriting Logic, nor do they subsume the use of unobservable transitions in SOS.

Modular structural operational semantics (MSOS) [29–31] is a variant of SOS with a particularly natural representation of unobservable transitions: labels are the morphisms of a category, and unobservable transitions are those labelled by identity morphisms. This lets us specify unobservable transitions in MSOS without introducing extra labels such as τ.

For compositional reasoning about behavioural equivalence in SOS and MSOS, bisimulation needs to be a congruence. Various rule formats ensuring

N. Martí-Oliet et al. (Eds.): Meseguer Festschrift, LNCS 9200, pp. 519–538, 2015.
DOI: 10.1007/978-3-319-23165-5_24

congruence have been established for SOS [35]. Recently, a rule format was given for MSOS such that strong bisimulation is a congruence [7]. MSOS was there extended with an unlabelled rewriting relation, which was required to be a precongruence, and used to represent unobservable evaluation steps in the definition of strong bisimulation. That is sufficient for proving some simple context-independent absorption laws, such as the null command being a left and right unit for command sequencing. However, we want to be able to prove also that various program optimisations preserve observable behaviour, and the precongruence is too restrictive for that purpose.

This led us to investigate how to ignore unobservable transitions in connection with bisimulation in MSOS, while ensuring that it remains a congruence. After recalling established concepts, notation and results (Sect. 2), we adopt a technique proposed by Van Glabbeek [10,12] (Sect. 3): we add abstraction rules for the unobservable transitions, and use strong bisimulation with MSOS specifications in an existing congruence format [7]. However, that approach requires the introduction of stuttering rules, which might be seen as a drawback (e.g., in connection with executability) and motivates an alternative approach (Sect. 4): to use weak bisimulation with MSOS specifications in an adaptation of Bloom's WB Cool congruence format. We conclude (Sect. 5) by comparing the two approaches. Our main contribution is establishing adequate techniques for disregarding unobservable transitions in MSOS.

This paper was written for the Festschrift in honor of José Meseguer. PDM would like to express the following personal appreciation of José and his research.

As recalled in [23], José and I first met almost 40 years ago, in 1976, in Oxford. I soon became keenly interested in the initial algebra approach to specification of abstract data types, and followed the development of the seminal OBJ system for executing algebraic specifications, initiated by Joseph Goguen [14] and continued through the 1980s in a fruitful collaboration between Joseph, José and their colleagues [9,13,19]. Order-sorted algebra [15–17,25] was a further topic of intense common interest. José is also a founding member of IFIP WG 1.3 (Foundations of System Specification); in that connection he chaired an expert panel that assessed the design of CASL [2,32].

The breadth and depth of José's research is clearly reflected by the hundreds of carefully written journal and conference papers that he has published. A particularly prominent topic since the beginning of the 1990s is Rewriting Logic [21,22] and its implementation in Maude [8]. When I took a sabbatical in 1998–99, José invited me to visit him at SRI International, and obtained funding (with Carolyn Talcott) for a joint research project during the visit. José's group was a stimulating environment for the initial development of MSOS [29–31]. In particular, I recall that José made crucial observations concerning the possibility of treating labels in MSOS as morphisms of categories, and we subsequently co-authored two papers (together with Christiano Braga and Hermann Haeusler) [5,6] on an embedding of MSOS in Rewriting Logic. The recent progress report by José (with Grigore Roşu) on the Rewriting Logic Semantics Project [26] and his review of 20 years of Rewriting Logic [24] reflect his inspiration to a multitude of colleagues, as well as his own contributions. I look forward to following his ongoing research, and to many future meetings.

2   Background

This section serves as an overview of preliminaries regarding SOS, rule formats and MSOS. Everywhere in this paper we assume a set of variables $\mathcal{X}$ with typical elements x, y and z, optionally subscripted.

2.1   SOS

Structural operational semantics (SOS) [1,36,37] is a framework for giving operational semantics of specification and programming languages.

In the most general case, SOS computations are modelled by *labelled terminal transition systems*.

Definition 1 (Labelled Terminal Transition System). *A labelled terminal transition system (LTTS) is a tuple $\langle \Gamma, L, \longrightarrow, T \rangle$, where $\gamma \in \Gamma$ are configurations (or states), $l \in L$ are labels, $\longrightarrow \subseteq \Gamma \times L \times \Gamma$ is a transition relation, and $T \subseteq \Gamma$ a set of final configurations. An assertion of an l-transition $\langle \gamma, l, \gamma' \rangle \in \longrightarrow$ is written $\gamma \xrightarrow{l} \gamma'$ and we say that γ is the* source *and γ' the* target, *or that γ' is an l-derivative of γ. If $\gamma \in T$ then there is no $\gamma' \in \Gamma$ and $l \in L$ such that $\gamma \xrightarrow{l} \gamma'$.*

A computation *in an LTTS is a finite or infinite sequence of transitions $\gamma_0 \xrightarrow{l_0} \gamma_1 \xrightarrow{l_1} \cdots$ such that if the sequence terminates with γ_n, then $\gamma_n \in T$.*

In SOS definitions of process calculi, states are usually just closed terms generated over an algebraic signature Σ, which is a collection of function symbols (operators) and their arities. For programming languages, the terms may contain computed values, and states usually contain additional *auxiliary entities* such as stores or environments.

The transition relation of an LTTS is generated inductively by a set of SOS rules of the form $\frac{H}{c}$, where H is a (possibly empty) set of premises and c is the conclusion. If H is empty, the rule is an *axiom*, otherwise it is a *conditional rule*. Premises and the conclusion are formulas $t \xrightarrow{l} t'$, where t and t' are configuration terms optionally containing variables. The intuitive meaning of a rule is that if all premises in H are satisfied by some closing substitution then so is the conclusion. Usually, premises will assert transitions for subterms of t, giving the rules a structural character.

2.2   Bisimulation and Rule Formats

For labelled transition systems, the finest useful notion of equivalence between states [38] is *strong bisimulation*. Briefly, a strong bisimulation relates two states if they both can make transitions with the same label to states that are again related; final states must be equal.

Definition 2 (Strong Bisimulation on LTTS). *Given an LTTS* $\langle \Gamma, L, \longrightarrow, T \rangle$, *a symmetric relation* $\mathcal{R} \subseteq \Gamma \times \Gamma$ *is a strong bisimulation if whenever* $\gamma_1 \mathcal{R} \gamma_2$ *then* $\gamma_1 \xrightarrow{l} \gamma_1'$ *implies that there is a* γ_2' *such that* $\gamma_2 \xrightarrow{l} \gamma_2'$ *and* $\gamma_1' \mathcal{R} \gamma_2'$; *and* $\gamma_1 \in T$ *implies* $\gamma_1 = \gamma_2$.

Two configurations γ_1 *and* γ_2 *are strongly bisimilar, written* $\gamma_1 \sim \gamma_2$ *iff* $\gamma_1 \mathcal{R} \gamma_2$ *for some strong bisimulation* $\mathcal{R}$.

For equational reasoning, an important property of bisimulation is *congruence*. A bisimulation is a congruence if it is preserved by all operators in the language. In general, strong bisimulation is not necessarily a congruence: terms whose subterms are bisimilar might not be bisimilar themselves. While for small languages one can check the congruence property by checking all constructs in the language, a more principled way is to make all rules adhere to a particular *congruence format*. The *positive GSOS* format [4] is a congruence format for SOS; we adapt it to MSOS in Sect. 4. A survey of congruence formats and associated meta-theory can be found in [35].

Definition 3 (Positive GSOS). *A rule for a construct* f *with arity* n *is in the positive GSOS format if it has the form*

$$\frac{\{x_i \xrightarrow{l_j} y_j \mid i \in I, j \in J_i\}}{f(x_1, \ldots, x_n) \xrightarrow{l} t}$$

where $I \subseteq \{1, \ldots, n\}$, J_i *is a finite index set for each* i, *and all* x_i *and* y_j *are distinct variables. Only variables* x_i *and* y_j *may appear in* t. *A language definition is in the positive GSOS format if all its rules are.*

The full GSOS format allows also rules with negative premises. A congruence theorem for GSOS states that if all rules in a specification are in the GSOS format, then strong bisimulation is guaranteed to be a congruence. This theorem was first proved by Bloom et al. [4]; a more succinct proof was subsequently provided by Van Glabbeek [11].

Traditionally, a designated *silent* label 'τ' is used for transitions, such as selecting a branch in a conditional statement, that correspond to internal housekeeping actions, and are of no inherent interest.

Strong bisimulation does not treat a τ-label differently from other labels: a τ-transition of one state has to be matched with a τ-transition of the other state. This results in a notion of equivalence which is too strong for many purposes.

One approach is to modify the transition relation by adding *abstraction rules* to a specification. This approach is explored by Van Glabbeek in [10,12] by adding special τ-rules. These rules allow hiding of silent transitions, or allow an arbitrary number of silent transitions *after* an observable transition has been made. We adapt this approach to our needs in Sect. 3.

The more common approach is to weaken bisimulation itself: *weak bisimulation* is a notion of bisimulation that permits ignoring τ-transitions.

Definition 4 (Weak Bisimulation on LTTS). *Assume an LTTS* $\langle \Gamma, L, \longrightarrow, T \rangle$. *A symmetric relation* $\mathcal{R} \subseteq \Gamma \times \Gamma$ *is a weak bisimulation if whenever* $\gamma_1 \mathcal{R} \gamma_2$ *then* $\gamma_1 \xrightarrow{l} \gamma_1'$ *implies that there is a* γ_2' *such that* $\gamma_2 \longrightarrow^* \xrightarrow{(l)} \longrightarrow^* \gamma_2'$ *and* $\gamma_1' \mathcal{R} \gamma_2'$; *and* $\gamma_1 \in T$ *implies* $\gamma_1 = \gamma_2$. *Here* $\gamma \xrightarrow{(l)} \gamma'$ *iff* $l = \tau$ *and* $\gamma = \gamma'$, *or* $l \neq \tau$ *and* $\gamma \xrightarrow{l} \gamma'$; *and* $\longrightarrow^*$ *is the reflexive transitive closure of* $\xrightarrow{\tau}$.

Two configurations γ_1 *and* γ_2 *are weakly bisimilar, written* $\gamma_1 \approx \gamma_2$ *iff* $\gamma_1 \mathcal{R} \gamma_2$ *for some weak bisimulation* $\mathcal{R}$.

Weak bisimulation has less desirable congruence properties than the strong variant. Even when a language is in the positive GSOS format, weak bisimulation is not guaranteed to be a congruence. The Cool formats [3] impose further restrictions on positive GSOS to guarantee that weak bisimulation is a congruence. These restrictions prevent constructs in a language from distinguishing between states based on the number of silent transitions they can make. We spell out the restrictions, adjusted for MSOS, in Sect. 4.

2.3   Modular SOS

Modular SOS (MSOS) [31] improves the modularity of SOS definitions. This is achieved by moving auxiliary entities from configurations to labels of transitions, and letting labels be morphisms of a category. Label components, each corresponding to an auxiliary entity such as environments and stores, are combined using an indexed product, which can also be regarded as a finite map or record. Labels on adjacent transitions are required to be composable. This, together with a record pattern notation akin to that of Standard ML, allows rules to only mention entities that are required in specifying a particular transition, while other entities are left unmentioned.

An MSOS specification consists of an algebraic signature, a label profile and a set of rules. The signature specifies how terms can be formed from a set of constructors, the label profile specifies the label entities used in the specification, and rules define the behaviour of terms formed from constructors. We provide an example set of rules in Table 1. Since information about auxiliary entities is moved to the labels, configurations in MSOS are limited to program terms and computed values (value-added syntax). The rules define a transition system with closed terms as configurations and composites of entity morphisms as labels. We give more details on the facets of MSOS specifications, how they specify computations, and how equalities between terms can be established, in the following subsections.

2.4   MSOS Terms

A basic single-sorted signature is a collection of function symbols (operators, constructors) together with a function assigning an arity to each of these symbols. The terms in the language form a freely generated algebra over the signature. In this paper we adopt an extension of the basic signature to a *value-computation*

Table 1. An example language defined in MSOS. Meta-variables v range over values and s over arbitrary terms. Environments ρ and stores σ are partial maps.

$$\frac{s_1 \xrightarrow{\{\ldots\}} s_1'}{\mathbf{if}(s_1, s_2, s_3) \xrightarrow{\{\ldots\}} \mathbf{if}(s_1', s_2, s_3)} \quad (1)$$

$$\frac{}{\mathbf{if}(\mathbf{true}, s_2, s_3) \xrightarrow{\{-\}} s_2} \quad (2)$$

$$\frac{}{\mathbf{if}(\mathbf{false}, s_2, s_3) \xrightarrow{\{-\}} s_3} \quad (3)$$

$$\frac{s_1 \xrightarrow{\{\ldots\}} s_1'}{\mathbf{let}(i, s_1, s_2) \xrightarrow{\{\ldots\}} \mathbf{let}(i, s_1', s_2)} \quad (4)$$

$$\frac{s_2 \xrightarrow{\{\mathbf{env}=\rho[i\mapsto v_1],\ldots\}} s_2'}{\mathbf{let}(i, v_1, s_2) \xrightarrow{\{\mathbf{env}=\rho,\ldots\}} \mathbf{let}(i, v_1, s_2')} \quad (5)$$

$$\frac{}{\mathbf{let}(i, v_1, v_2) \xrightarrow{\{-\}} v_2} \quad (6)$$

$$\frac{\rho(i) = v}{\mathbf{bound}(i) \xrightarrow{\{\mathbf{env}=\rho,-\}} v} \quad (7)$$

$$\frac{s \xrightarrow{\{\ldots\}} s'}{\mathbf{print}(s) \xrightarrow{\{\ldots\}} \mathbf{print}(s')} \quad (8)$$

$$\frac{}{\mathbf{print}(v) \xrightarrow{\{\mathbf{out}'=v,-\}} \mathbf{skip}} \quad (9)$$

$$\frac{s_1 \xrightarrow{\{\ldots\}} s_1'}{\mathbf{assign}(s_1, s_2) \xrightarrow{\{\ldots\}} \mathbf{assign}(s_1', s_2)} \quad (10)$$

$$\frac{s_2 \xrightarrow{\{\ldots\}} s_2'}{\mathbf{assign}(s_1, s_2) \xrightarrow{\{\ldots\}} \mathbf{assign}(s_1, s_2')} \quad (11)$$

$$\frac{}{\mathbf{assign}(v_1, v_2) \xrightarrow{\{\mathbf{sto}=\sigma,\mathbf{sto}'=\sigma[v_1\mapsto v_2],-\}} \mathbf{skip}} \quad (12)$$

$$\frac{s \xrightarrow{\{\ldots\}} s'}{\mathbf{stored}(s) \xrightarrow{\{\ldots\}} \mathbf{stored}(s')} \quad (13)$$

$$\frac{\sigma(v_1) = v_2}{\mathbf{stored}(v_1) \xrightarrow{\{\mathbf{sto}=\sigma,\mathbf{sto}'=\sigma,-\}} v_2} \quad (14)$$

$$\frac{s_1 \xrightarrow{\{\ldots\}} s_1'}{\mathbf{seq}(s_1, s_2) \xrightarrow{\{\ldots\}} \mathbf{seq}(s_1', s_2)} \quad (15)$$

$$\frac{}{\mathbf{seq}(\mathbf{skip}, s_2) \xrightarrow{\{-\}} s_2} \quad (16)$$

signature [7], which distinguishes a set of value constructors. This provides an alternative to specifying a set of final states of a terminal transition system – the set of final states (values) is the set of terms where the outermost constructor is a value constructor. In the example of Table 1, the boolean values **true** and **false** are nullary value constructors and so is **skip**, which represents the successful termination of a command. Further values assumed by the rules are identifiers and locations, which are used as indices for environments and stores, respectively, and which we have left unspecified in this example.

Definition 5 (Value-computation Signature). *A* value-computation signature (vc-signature) Σ *consists of a set of constructors* C_Σ, *each with an arity* $\mathsf{ar}_\Sigma : C_\Sigma \to \mathbb{N}$, *and a set of value constructors* $VC_\Sigma \subseteq C_\Sigma$.

For a signature Σ, $\mathbb{T}_\Sigma$ *is the set of open terms generated by the set* $\mathcal{X}$ *of variables*, $T_\Sigma \subseteq \mathbb{T}_\Sigma$ *the set of closed terms*, $\mathsf{vars}(t) \subseteq \mathcal{X}$ *is the set of variables in a term* t, *and* $V_\Sigma \subseteq T_\Sigma$ *is the set of* value terms *whose outermost constructor is in* VC_Σ.

A Σ*-substitution is a partial function* $\sigma : \mathcal{X} \to \mathbb{T}_\Sigma$ *mapping variables to terms. The domain of a substitution* σ *is written* $\mathsf{dom}(\sigma)$. *A substitution is closing for a term* $t \in \mathbb{T}_\Sigma$ *if* $\mathsf{vars}(t) \subseteq \mathsf{dom}(\sigma)$ *and its image is a subset of the closed terms* T_Σ.

2.5 MSOS Labels

MSOS transition labels are aggregates of label components which are morphisms of their respective categories. The aggregate is formed as an indexed product or finite map from label indices (for example, **sto** for stores or **env** for environments) to morphisms. The full label is composable only if it is composable component-wise. The morphisms-as-labels aspect of MSOS means that it has a very natural representation of *unobservable transitions*: instead of having a special label, conventionally designated as silent, unobservable transitions in MSOS are simply those labelled by *identity morphisms*.

The usual auxiliary entities used in programming language semantics are covered by three kinds of label categories:

- A category for *read-only* entities, such as environments ρ, will have just an identity morphism for each object; then two morphisms are composable only if they are identical.
- Store, being a *read-write* entity, is represented by a category with individual stores σ as objects and pairs of stores as morphisms. The morphism $\langle \sigma, \sigma' \rangle$ represents an update of store σ (readable component) into σ' (writeable component). The morphisms $\langle \sigma, \sigma' \rangle$ and $\langle \sigma'', \sigma''' \rangle$ are composable only if $\sigma' = \sigma''$; the result of their composition is $\langle \sigma, \sigma''' \rangle$.
- *Write-only* entities, such as emitted signals, are represented by a category with a single object. Morphisms are finite (possibly empty) sequences generated over a set of actions (signals); their composition is concatenation. Individual signals are singleton sequences and there is a single identity morphism: the empty sequence.

Label indices are determined by a *label profile* $\mathcal{L}$, which divides them according to what kind of entity they represent. In the example of Table 1 we only have one label index for each kind of entity: **env** is a read-only label index for environments, **sto** is a read-write label index for stores, and **out** is a write-only label index of the output stream. By convention, indices for write-only entities and writeable components of read-write entities are *always* primed (**out′**, **sto′**), while read-only and readable components of read-write entities are always unprimed (**env**, **sto**). The label profile is instantiated by a set of terms T to $\mathcal{L}(T)$, a set of finite maps from label indices to label terms. For instance, for any transition label $L \in \mathcal{L}(T)$ that contains a store entity as its component at index **sto**, $L(\mathbf{sto})$ is the store at the beginning of the transition and $L(\mathbf{sto'})$ is the store at the end of the transition.

Definition 6 (MSOS Labels [7]**).** *A label profile is a triple of disjoint sets* $\mathcal{L} = \langle \mathcal{L}_{RO}, \mathcal{L}_{RW}, \mathcal{L}_{WO} \rangle$. *The set* reads($\mathcal{L}$) *consists of the unprimed indices* $1 \in \mathcal{L}_{RO} \uplus \mathcal{L}_{RW}$. *The set* writes($\mathcal{L}$) *consists of the primed indices* $\{1' : 1 \in \mathcal{L}_{WO} \uplus \mathcal{L}_{RW}\}$. *For any set* T, *the label set* $\mathcal{L}(T)$ *is the set of maps* (reads($\mathcal{L}$)⊎writes($\mathcal{L}$)) $\rightarrow T$. *For a label* $L \in \mathcal{L}(T)$, *we write* reads(L) *and* writes(L) *for the restriction of* L *to* reads($\mathcal{L}$) *and* writes($\mathcal{L}$) *respectively.*

2.6 MSOS Rules and Specifications

The form of MSOS rules is much the same as for SOS rules, except that configurations are now simply terms (possibly with computed values as sub-terms) and labels contain multiple entities. Although a rule might specify entity morphisms explicitly via side-conditions as in [29], a more readable approach is using the *record pattern* notation introduced in [31]. Record patterns allow to specify labels of transitions directly above the arrows, as illustrated in Table 1. Label patterns usually only mention entities directly used by the rule, while the remaining components of the label are propagated throughout the rule using special meta-variables '...' matching any label components, and '−' matching only label components that are identity morphisms. For example, in Rule 8 the label pattern '{**out′** = v, −}' matches any transition label L where $L(\mathbf{out'})$ is the value v (representing the morphism v) and any other component is an identity morphism. The pattern '{−}' matches labels that are completely unobservable, while '{...}' matches any label.

Definition 7 (MSOS Label Patterns). *Given a label profile* $\mathcal{L}$ *and a signature* Σ, *a label pattern is a sequence of equations '*$1 = t$*', enclosed in '*{}*' and (optionally) terminated by label meta-variables '...' or '−'. In '*$1 = t$*', 1 is a label index from* reads($\mathcal{L}$) ⊎ writes($\mathcal{L}$) *and* t *is an open term over* Σ. *The meta-variable '...' matches any label components, except those already mentioned in the sequence, and '−' matches all unmentioned components of the label only if they are identity morphisms. At most one of '...' and '−' may appear.*

An explicit label pattern *is a label pattern that contains no '...' or '−'.*

In practice, label patterns in rules *always* end with the meta-variables '...' or '−' as this ensures that the rules are modular. In the remainder of this paper we will use the meta-variable P for label patterns as defined above, and L for labels (maps), as specified in Definition 6.

Definition 8 (MSOS Rules). *Let Σ be a signature and $\mathcal{L}$ a label profile. An MSOS rule is a pair $\langle H, c \rangle$, usually written $\frac{H}{c}$, where H is the set of premises and c is the conclusion. Premises and the conclusion are formulas of the form $t \xrightarrow{P} t'$, where t and t' are open terms over Σ, and P is an MSOS label pattern over $\mathcal{L}$ and Σ.*

Definition 9 (MSOS Specifications). *An MSOS specification is a tuple $\langle \Sigma, \mathcal{L}, D \rangle$, where Σ is a vc-signature (Definition 5), $\mathcal{L}$ a label profile (Definition 6) and D a set of MSOS rules (Definition 8).*

2.7 Models of MSOS Specifications

The most general model of computation in MSOS is a *generalised transition system (GTS)* $\langle \Gamma, \mathbb{L}, \longrightarrow, T \rangle$, where $\mathbb{L}$ is a category with morphisms L, such that $\langle \Gamma, L, \longrightarrow, T \rangle$ is an LTTS (Definition 1). Computations in a GTS are computations in the underlying LTTS, with the restriction that labels on adjacent transitions have to be composable.

Since a value-computation signature distinguishes value terms syntactically, we can have a transition system without a distinguished set of final states. We simply disallow transitions from value terms. The following notion of a value-computation transition system is adapted from [7].

Definition 10 (Value-computation Transition Systems). *A value-computation transition system (VCTS) is a tuple $\langle \Sigma, L, \longrightarrow \rangle$, where Σ is a vc-signature, L a set of labels, and $\longrightarrow \subseteq T_{\Sigma} \times L \times T_{\Sigma}$ a transition relation. If $s \xrightarrow{l} s'$, then $s \notin V_{\Sigma}$.*

An MSOS specification $\langle \Sigma, \mathcal{L}, D \rangle$ generates a VCTS $\langle \Sigma, \mathcal{L}(T), \longrightarrow \rangle$ for some T (usually T_{Σ}) by letting $\longrightarrow$ be the least relation closed under rules in D. Since we do not allow negative premises in our specifications, there is always a model and it is unique.

The provability of a transition assertion is established by finding a rule in D and a closing substitution for the rule, such that the conclusion matches the transition assertion and the premises are also provable in this way.

Definition 11 (Provable Transitions). *A transition assertion $s \xrightarrow{L} s'$ is provable in an MSOS specification $\langle \Sigma, \mathcal{L}, D \rangle$ iff there is a rule $\langle H, t \xrightarrow{P} t' \rangle \in D$ and a closing substitution σ, such that, $s = t[\sigma]$, $s' = t'[\sigma]$, $L = P[\sigma]$ and for all $t_i \xrightarrow{P_i} t_i' \in H$, $t_i[\sigma] \xrightarrow{P_i[\sigma]} t_i'[\sigma]$ is also a provable transition in the specification.*

In this definition and in the remainder of the paper we lift substitutions to terms and label patterns in the obvious way by letting $s[\sigma]$ denote the term obtained by replacing all occurrences of variables $x \in \mathsf{dom}(\sigma)$ in s by $\sigma(x)$. $P[\sigma]$ for label patterns P is defined similarly.

2.8 Equivalence in MSOS

Just like with ordinary SOS, bisimulation can be used to prove equivalence between program terms. As indicated in [31], the usual notion of strong bisimulation can be used on MSOS directly and weak bisimulation just needs to be modified by replacing τ-transitions with the notion of unobservable transitions in MSOS. A higher-order *MSOS bisimulation* theory was developed in [7] for value-computation transition systems extended with rewriting. In addition to the usual transition relation (as in Definition 10), such a transition system also has a *rewrite* relation '$\Rightarrow$' used for value-computations that are entirely independent of any auxiliary entities. The relation is a *precongruence*, that is, a reflexive and transitive relation which is preserved by all constructors. MSOS bisimulation allows bisimilarities between label components and uses the rewriting relation to achieve some of the advantages of weak bisimulation regarding transitions that are independent of label components.

A reformulation of the *tyft/tyxt* format [18] for MSOS guarantees that this notion of MSOS bisimulation is a congruence. In [7] only the tyft part of the format is considered but extension to tyxt is straightforward as long as value terms are not allowed to make any transitions in accordance with Definition 10.

It is simple to identify context-independent transitions in a specification: they are transitions labelled with '$\{-\}$' (e.g. Rules 2, 3, 6 and 16 in Table 1). Replacing those transition with rewrites (e.g. $\mathbf{seq}(\mathbf{skip}, s) \Rightarrow s$) allows us, for example, to prove unit laws for sequencing, $\mathbf{seq}(s, \mathbf{skip}) \equiv s \equiv \mathbf{seq}(\mathbf{skip}, s)$, which wouldn't hold using strong bisimulation with MSOS. In this paper we are, however, also interested in equivalences between terms that involve *unobservable* but *context-dependent* transitions, such as $\mathbf{let}(i, v, \mathbf{bound}(i)) \equiv v$. These terms are not MSOS bisimilar due to $\mathbf{let}(i, v, \mathbf{bound}(i))$ requiring a transition for looking up the bound value in the environment according to Rule 5. The rest of the paper discusses two approaches to proving such equivalences.

3   Absorbing Unobservable Transitions

We would like to prove equivalences involving context-dependent unobservable transitions. These can justify simple optimisations which replace a program term with a term that takes fewer computation steps but still has the same observable behaviour and results in the same value. For example, given the specification in Table 1, we want the following equivalences to hold.

Example 12. $\mathbf{let}(i, v, \mathbf{bound}(i)) \equiv v$

Example 13. $\mathbf{let}(i, v, \mathbf{assign}(l, \mathbf{bound}(i))) \equiv \mathbf{assign}(l, v)$

In this section we explore an approach where we allow the unobservable transitions in a VCTS to be ignored by adding *abstraction* rules. These rules are inspired by Van Glabbeek [10] and are similar to built-in rules for the rewriting relation in [7].

Absorption rules permit hiding of unobservable transitions occurring before or after a transition:

$$\text{AbsL}\dfrac{s \xrightarrow{\{-\}} s' \quad s' \xrightarrow{L} s''}{s \xrightarrow{L} s''} \qquad\qquad \text{AbsR}\dfrac{s \xrightarrow{L} s' \quad s' \xrightarrow{\{-\}} s''}{s \xrightarrow{L} s''}$$

Either of these rules also gives us transitivity of unobservable transitions.

These rules allow us to prove strong MSOS *simulations* (as bisimulations but not symmetric) between some terms that were excluded before. For example, we now have that $\mathbf{let}(i, v, \mathbf{assign}(l, \mathbf{bound}(i)))$ strongly simulates $\mathbf{assign}(l, v)$. They are still not bisimilar: the former term can make an unobservable transition that cannot be matched by '$\mathbf{assign}(l, v)$'. We do not get any (bi)similarity between $\mathbf{let}(i, v, \mathbf{bound}(i))$ and v.

To achieve (strong) bisimilarity in such cases, we add a *stuttering* (or *waiting*) rule, allowing a term to make an unobservable transition to itself.

$$\dfrac{}{s \xrightarrow{\{-\}} s}$$

A stuttering transition under this rule is similar to stuttering steps in TLA [20]. In bisimulation proofs, this rule will allow terms to stutter and thus to match an unobservable transition of the challenger with a transition that has no observable effect at all.

The rule is less natural than absorption rules since now any term can make an unobservable transition. It also prevents specification of constructs whose behaviour may depend on the number of unobservable transitions their components make.

In the setting of programming languages we are mainly interested in computations resulting in a value. From this perspective, a general stuttering rule still has two issues. The first is that now value terms have transitions. As noted in [7], values should only be inspected. To address this, we argue that stuttering transitions are not proper computation steps. The term that is the target of the transition is equal to the source and the label is an identity morphism, thus there is nothing that can be observed about the transition. Additionally, a value-computation signature distinguishes value terms (final states), so we can always formulate a condition for when a computation is finished.

The second issue is that now any computation term can perform an infinite sequence of transitions without making any progress. All computations are potentially non-terminating. Furthermore, computations which do not terminate in the original system (without abstraction rules) are now identified with stuck terms ('deadlock = livelock'). We can address this by limiting stuttering to terms that either can progress by making an observable transition, or are value terms. We express this as two stuttering rules. We assume that v ranges over value terms and the predicate unobs holds if its argument is an unobservable label.

$$\text{Wait}\dfrac{s \xrightarrow{L} s' \quad \neg\, \mathsf{unobs}(L)}{s \xrightarrow{\{-\}} s} \qquad\qquad \text{Stutter}\dfrac{}{v \xrightarrow{\{-\}} v}$$

These two rules, together with the two absorption rules introduced before, can be used to abstract from unobservable transitions and prove the desired equivalences in Examples 12 and 13 using MSOS bisimulation as illustrated by Fig. 1.

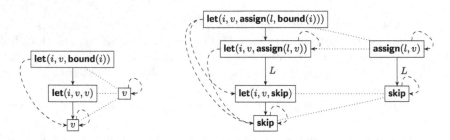

Fig. 1. Transitions of Examples 12 and 13 with abstraction rules added to the system. Unlabelled arrows represent unobservable transitions. Dashed arrows are new transitions due to abstraction rules. Dotted lines connect states related by the bisimulation relation. ($L = \{\mathbf{sto} = \sigma, \mathbf{sto'} = \sigma[l \mapsto v], -\}$).

For our stuttering rule to be admissible, we need to modify the condition in Definition 10 which prohibits transitions from value terms. The condition becomes the following: *If there is a transition* $s \xrightarrow{L} s'$, *then* $s \notin V_\Sigma$, *unless* $s = s'$ *and* $\mathsf{unobs}(L)$. With this restriction relaxed, all our rules are in the MSOS tyft/tyxt format and thus bisimilarity stays a congruence in the 'abstracted' system. The bisimilarity achieved in this way also avoids the usual 'deadlock = livelock' problem.

We stress that the proposed absorption and stuttering rules are to be added to a MSOS specification as *abstraction* rules for reasoning about weak equivalences between terms. They are not useful as *computation* rules (e.g., when animating the semantics), since the two stuttering rules may add many potentially infinite computations.

4   A WB Cool MSOS Format

This section presents a version of the Simply WB Cool format [3] for systems specified in MSOS. To this end we first present an MSOS version of the positive GSOS format (GSOS for MSOS). Then we give a definition of *weak MSOS bisimulation* and place further restrictions on the GSOS for MSOS format to guarantee that weak MSOS bisimulation is a congruence.

4.1   Positive GSOS for MSOS

Our notion of *strong MSOS bisimulation* allows bisimilarities between labels of transitions as well as configurations. To this end we implicitly lift the relation

$\mathcal{R}$ to finite maps: $m_1 \mathcal{R} m_2$ holds for two finite maps m_1 and m_2 with the same domain, whenever $m_1(i) \mathcal{R} m_2(i)$ for all i in the domain of the two maps.

In the following definition the readable label components of the two transitions are required to be the same. This is due to the composability condition of categories for read-only and read-write entities, which requires readable components to be the same. The possibility of transitions with bisimilar readable components is a consequence of the format, as shown by Lemma 18 below.

Definition 14 (Strong MSOS Bisimulation). *A strong MSOS bisimulation over a given VCTS* $\langle \Sigma, \mathcal{L}(T_\Sigma), \longrightarrow \rangle$ *is a symmetric relation* $\mathcal{R} \subseteq T_\Sigma \times T_\Sigma$ *such that, whenever* $s \mathcal{R} t$,

1. *if* $s \xrightarrow{L} s'$, *then* $\exists t', L'.t \xrightarrow{L'} t'$, $\mathsf{reads}(L) = \mathsf{reads}(L')$, $\mathsf{writes}(L) \mathcal{R} \mathsf{writes}(L')$, *and* $s' \mathcal{R} t'$; *and*
2. *if* $s = v(s_1, \ldots, s_n)$ *with* $v \in VC$, *then* $t = v(t_1, \ldots, t_n)$ *and* $s_i \mathcal{R} t_i$ *for* $1 \leq i \leq n$.

Two terms s and t are strongly MSOS bisimilar, written $s \sim_{msos} t$, *iff* $s \mathcal{R} t$ *for some strong MSOS bisimulation* $\mathcal{R}$.

This version allows us to prove associativity laws for constructs. For example, we have $\mathbf{seq}(p, \mathbf{seq}(r, q)) \sim_{msos} \mathbf{seq}(\mathbf{seq}(p, r), q)$, since both terms have the same transitions (observable or unobservable). But unit laws for sequencing, provable by MSOS bisimulation with rewrites, do not hold under strong MSOS bisimulation.

Next, we define a positive GSOS format for MSOS. Following [7] we extend the original GSOS format with value patterns and cater for the data flow discipline between the conclusion and premises of an MSOS rule.

Definition 15 (Patterns [7]). *A pattern u is an open term constructed from value constructors and variables.*

The format definitions use new auxiliary functions ins and outs, in addition to reads and writes lifted to patterns and a selection operation to denote terms in label patterns. These are defined for a label pattern P (over a profile $\mathcal{L}$) as follows:

- $\mathsf{reads}(P) = \mathsf{reads}(\mathcal{L})$ and $\mathsf{writes}(P) = \mathsf{writes}(\mathcal{L})$;
- $P(\mathsf{l})$ denotes the term at label l; and
- $\mathsf{ins}(P)$ denotes the set of all $P(\mathsf{l})$ for $\mathsf{l} \in \mathsf{reads}(P)$ and $\mathsf{outs}(P)$ the set of all $P(\mathsf{l})$ for $\mathsf{l} \in \mathsf{writes}(P)$.

Definition 16 (Positive GSOS for MSOS). *A rule for a construct* $f \in (C_\Sigma \setminus VC_\Sigma)$ *with* $n = \mathsf{ar}(f)$ *is in the positive GSOS for MSOS format if it has the following form:*

$$\frac{\{x \xrightarrow{P_j} y_j \mid x \in X_i, j \in J_x\}}{f(u_1, \ldots, u_n) \xrightarrow{P} t}$$

where for $1 \le i \le n$, $X_i \subseteq \mathsf{vars}(u_i)$ and each u_i is a pattern; J_x is a finite set indexing premises for each x; P and each P_j is a label pattern; and all $\mathsf{ins}(P)$ and $\mathsf{outs}(P_j)$ are patterns. Variables in all u_i, $\{y_j \mid j \in J\}$ (where J is the union of all J_x), $\mathsf{ins}(P)$ and $\mathsf{outs}(P_j)$ are all distinct. No variables from u_i or $\mathsf{ins}(P)$ may appear in any $\mathsf{outs}(P_j)$. Variables in $\mathsf{ins}(P_j)$ must appear in the source of the conclusion or in $\mathsf{ins}(P)$. Finally, variables in t and in each $\mathsf{outs}(P)$ must appear in the source of the conclusion, as targets of premises, in $\mathsf{ins}(P)$, or in $\mathsf{outs}(P_j)$.

An MSOS specification is in the positive GSOS for MSOS format if all its rules are.

The specification in Table 1 is in the positive GSOS for MSOS format, if we consider the mathematical notation for maps and sequences as syntactic sugar for value terms, and we allow side conditions such as $\rho(i) = v$ in Rule 7. In general, side conditions that restrict instantiations of variables with value terms (such as the environment ρ in Rule 7) do not affect the format. They can be understood as generating sets of positive GSOS for MSOS rules. We could also express data operations (such as lookup in maps) using further constructs in the language which would only have unobservable transitions. This approach is used in [7] with rewriting instead of transitions.

We can use the following lemma to obtain bisimilar substitutions for bisimilar value terms.

Lemma 17. *Let $\mathcal{R}$ be the congruence closure of $\sim_{msos}$, u a pattern, and σ a substitution with $\mathsf{dom}(\sigma) = \mathsf{vars}(u)$. Assume $u[\sigma] \mathcal{R} t$. Then there is a substitution π such that $\mathsf{dom}(\pi) = \mathsf{vars}(u)$, $t = u[\pi]$, and for each $x \in \mathsf{vars}(u)$, $\sigma(x) \mathcal{R} \pi(x)$.*

Proof. By induction on the structure of u.   □

The following lemma shows that if a term makes a transition with the label L, the same term can make a transition with a label that is *point-wise bisimilar* to L. The target terms of the transitions will also be bisimilar. Note that as a consequence of the positive GSOS for MSOS format, the lemma holds for any congruence relation, not just bisimulation.

Lemma 18. *Given an MSOS specification in the positive GSOS for MSOS format, let $\mathcal{R}$ be a congruence relation and P an explicit label pattern. Assume $s \xrightarrow{L} s'$ and let σ be a closing substitution such that $\mathsf{reads}(P[\sigma]) \mathcal{R} \mathsf{reads}(L)$. Then there is a term s'' and a substitution σ' such that $s \xrightarrow{P[\sigma']} s''$, $\mathsf{reads}(P[\sigma']) = \mathsf{reads}(P[\sigma])$, $\mathsf{writes}(P[\sigma']) \mathcal{R} \mathsf{writes}(L)$ and $s' \mathcal{R} s''$.*

Proof. By induction on the height of the derivation tree of $s \xrightarrow{L} s'$.   □

Theorem 19. *Given an MSOS specification, if all rules are in the positive GSOS for MSOS format, then $\sim_{msos}$ is a congruence for the specified language.*

Proof. We show that the congruence closure of $\sim_{msos}$ is also a strong MSOS bisimulation. The congruence closure $\mathcal{R}$ is formed by the following rules:

1. $s \sim_{msos} t$ implies $s \mathcal{R} t$
2. $s_i \mathcal{R} t_i$ implies $s = f(s_1, \ldots, s_n) \mathcal{R} f(t_1, \ldots, t_n) = t$, where $n = \mathsf{ar}(f)$ and $1 \le i \le n$.

The proof proceeds by induction on the number of applications of clause 2. The base case is immediate since $\sim_{msos}$ is a strong MSOS bisimulation.

In the inductive step we assume that the bisimulation property holds for all s_i and t_i, derived by fewer applications of clause 2, to show that it also holds for s and t. For condition 1 of Definition 14, we assume $s \xrightarrow{L} s'$ and show that there exists a transition $t \xrightarrow{L'} t'$ that satisfies the condition. We use the known format of the rule used to derive $s \xrightarrow{L} s'$ together with Lemmas 17 and 18 to find a substitution for the same rule so that it can be used to prove $t \xrightarrow{L'} t'$, as required by Definition 11. Condition 2 of Definition 14 follows immediately from the induction hypothesis.   □

For a detailed proof see [33].

4.2   Simply WB Cool MSOS

Now we adapt the WB Cool format as presented in [11] to MSOS. The format is a further restriction of positive GSOS for MSOS. We want the following notion of weak bisimilarity to be a congruence.

Definition 20 (Weak MSOS Bisimulation). *A weak MSOS bisimulation over a given VCTS $\langle \Sigma, \mathcal{L}(T_\Sigma), \longrightarrow \rangle$ is a symmetric relation $\mathcal{R} \subseteq T_\Sigma \times T_\Sigma$ such that, whenever $s \mathcal{R} t$,*

1. *if $s \xrightarrow{L} s'$, then $\exists t', L'.t \longrightarrow^* \xrightarrow{(L')} \longrightarrow^* t'$, $\mathsf{reads}(L) = \mathsf{reads}(L')$, $\mathsf{writes}(L) \mathcal{R} \mathsf{writes}(L')$, and $s' \mathcal{R} t'$; and*
2. *if $s = v(s_1, \ldots, s_n)$ with $v \in VC$, then $\exists t', t \longrightarrow^* t'$ and $t' = v(t_1, \ldots, t_n)$ with $s_i \mathcal{R} t_i$ for $1 \le i \le n$;*

where $t \xrightarrow{(L')} t'$ is $t \xrightarrow{L'} t'$ if L' is observable, and $t \xrightarrow{L'} t'$ or $t = t'$ if it is unobservable.

Two terms s and t are weakly MSOS bisimilar, written $s \approx_{msos} t$, iff $s \mathcal{R} t$ for some weak MSOS bisimulation $\mathcal{R}$.

Under this definition, we now have for example, $\mathsf{seq}(s, \mathsf{skip}) \approx_{msos} s \approx_{msos} \mathsf{seq}(\mathsf{skip}, s)$ and we can also justify the equivalences in Examples 12 and 13.

In Cool formats, *patience rules* allow an argument to perform all its τ-transitions before an observable transition can be performed. Since a VCTS usually has many meaningful final states and arguments may be inspected even after performing all their (observable) transitions, we generalise patience rules to *congruence* rules.

Definition 21 (Congruence Rules). *A congruence rule for a construct f with $n = \mathsf{ar}(f)$ is a rule of the following form:*

$$\frac{x_i \xrightarrow{P_i} y_i}{f(u_1,\ldots,u_{i-1},x_i,u_{i+1},\ldots u_n) \xrightarrow{P} f(u_1,\ldots,u_{i-1},y_i,u_{i+1},\ldots,u_n)}$$

where u_j $(1 \le j \le n)$ are patterns for value terms (Definition 15); x_i and y_i are term variables and for all $\mathsf{l} \in \mathsf{writes}(P) \cup \mathsf{writes}(P_i)$, $P(\mathsf{l}) = P_i(\mathsf{l})$.

In MSOS, a patience rule is a congruence rule with only unobservable labels.

Similarly, our notion of an *active argument* has to cater for patterns in rules.

Definition 22 (Active and Receiving Arguments). *The ith argument of a construct f, with $1 \le i \le \mathsf{ar}(f)$, is active if f has a GSOS rule, where x_i is a variable on the left-hand side of a premise or it appears in $\mathsf{ins}(P_j)$ of a premise; or u_i is a pattern $v(w_1,\ldots,w_{\mathsf{ar}(v)})$, for $v \in VC_\Sigma$, in the source of the conclusion.*

A variable y is receiving in t if t is the target of a rule in which y appears in the right-hand side of a premise. The ith argument of f is receiving if a variable y is receiving in a term t that has a subterm $f(t_1,\ldots,t_{\mathsf{ar}(f)})$ and y appears in t_i.

Definition 23 (Simply WB Cool MSOS). *An MSOS specification is in the Simply WB Cool MSOS format if it is in the positive GSOS for MSOS format and*

1. *no rule has a variable x occurring more than once among the sources of premises;*
2. *in any rule, no variable appears both in the target of the conclusion and in the source of a premise;*
3. *only congruence rules can have premises with unobservable transitions;*
4. *there is a congruence rule for every active argument;*
5. *there is a congruence rule for every receiving argument;*
6. *patterns in the source of the conclusion have the form $v(z_1,\ldots,z_{\mathsf{ar}(v)})$, where all z_i are all variables;*
7. *if a variable x appears in the premises, it must be an argument $u_i = x$ in the source of the conclusion; and*
8. *for all $\mathsf{l} \in \mathsf{writes}(P_j)$, $P_j(\mathsf{l})$ must a variable.*

Conditions 1–5 correspond to the original restrictions of the WB Cool format. Condition 6 forbids nested patterns. (Otherwise one would have to specify 'nested congruence rules' for all patterns with nested value constructors. This generalisation is straightforward, however it complicates the definition of the format while rarely being used in practice. Moreover, nested pattern matching can be split among auxiliary constructs.) Condition 7 ensures that congruence rules can be applied to all terms going into readable components of premise labels. Due to condition 8, a writeable component of a premise can only be pattern matched via an auxiliary construct. This again ensures that congruence rules can be applied to a term obtained this way.

Note that as a consequence of condition 8 the technique used in [31] for speci-fying failure and exceptions using write-only entities might become problematic. A rule cannot check for an exception flag directly, but has to use an auxiliary construct with intermediate unobservable transitions. This could be alleviated by restricting (writeable) label components to value terms and allowing patterns (in accordance with condition 6) for matching on writeable components in the premises.

All rules of Table 1 are in Simply WB Cool MSOS. Rules 1, 4, 5, 8, 10, 11, 13 and 15 are congruence rules.

Theorem 24. *Given an MSOS specification, if all rules are in the Simply WB Cool MSOS format, then $\approx_{msos}$ is a congruence for the specified language.*

Proof. We need to show that the congruence closure of $\approx_{msos}$ is also a weak MSOS bisimulation. The congruence closure $\mathcal{R}$ is formed by the following rules:

1. $s \approx_{msos} t$ implies $s \mathcal{R} t$
2. $s_i \mathcal{R} t_i$ implies $s = f(s_1, \ldots, s_n) \mathcal{R} f(t_1, \ldots, t_n) = t$, where $n = \mathsf{ar}(f)$ and $1 \leq i \leq n$.

The proof proceeds by induction on the number of applications of clause 2. The base case is immediate since $\approx_{msos}$ is a weak MSOS bisimulation.

In the induction step we assume that the bisimulation property holds for all s_i and t_i, derived by fewer applications of clause 2, and show that it also holds for s and t. For condition 1 of Definition 20, we assume that $s \xrightarrow{L} s'$. We use the known format of the rule d used to derive that transition, the conditions of WB Cool MSOS, and Lemma 18 to derive the (possibly empty) transition sequence

$$t = f(t_1, \ldots, t_n) \longrightarrow^* \xrightarrow{(L')} \longrightarrow^* t'.$$

Congruence rules for active arguments of f can be used to derive the first sequence of unobservable transitions $\longrightarrow^*$. Then rule d, used to derive $s \xrightarrow{L} s'$, can be used to derive the transition $\xrightarrow{(L')}$ if L is observable, otherwise there might be no transition. The last unobservable transition sequence $\longrightarrow^*$ can be derived using congruence rules for receiving variables in rule d. Because we find in each step a substitution related by $\mathcal{R}$ to the one used with rule d to derive $s \xrightarrow{L} s'$, we also have that $s' \mathcal{R} t'$, $\mathsf{reads}(L) = \mathsf{reads}(L')$ and $\mathsf{writes}(L) \mathcal{R} \mathsf{writes}(L')$ as required. Condition 2 of Definition 20 follows imme-diately from the assumption in this case. $\qquad\square$

For a detailed proof see [33].

5   Conclusion

By abstracting away from context-free transitions, MSOS bisimulation with pre-congruence from [7] allows us to prove many useful laws in an MSOS setting that do not hold under the usual notions of strong bisimulation. It retains the pleasant properties of strong bisimulation and comes with a liberal tyft congruence format. However, for many useful equivalences we need to ignore also context-dependent unobservable transitions.

We have investigated two approaches of abstracting away from unobservable transitions while ensuring the equivalence is also a congruence. One approach is based on weakening the transition system generated from a specification by adding absorption rules. The main advantage of this approach is that it allows us to use the existing MSOS-tyft congruence format of [7]. However, the cost of this is that computations become potentially diverging. Also, abstraction rules modify the semantics of some constructs in unintended ways. A typical example would be the sum operator from CCS [27].

The other approach we investigated was to define a notion of weak bisimulation for MSOS and equip it with a congruence format. We have chosen the Simply WB Cool format based on the well-studied (positive) GSOS. Although checking whether rules in a specification are in that format could be tedious, it should be straightforward to implement such checks in tool support for MSOS (this is left as future work). While not as powerful as MSOS tyft/tyxt, the Simply WB Cool MSOS format seems to be sufficient for specifying most common programming constructs. The approach based on absorption currently has the advantage of allowing more general premises of rules. This also includes a more straightforward specification of exception handling as mentioned in the previous section.

Finally, let us compare our approach with abstraction in Rewriting Logic. Turning up the "abstraction dial" in a Rewriting Logic semantics for a programming language involves changing some of its rules into equations, and checking that these equations are confluent (modulo structural axioms such as associativity and/or commutativity). This can dramatically reduce the state space. In an MSOS, in contrast, there is no such dial to turn: unobservable transitions arise naturally as a special case of general labelled transitions, and they are not subject to any confluence conditions – although in practice, they are usually deterministic. We conjecture that by identifying the sources of unobservable deterministic transitions with their respective targets, we would obtain state space reductions in MSOS comparable to using the abstraction dial in Rewriting Logic. Unfortunately, we cannot simply add equations to (M)SOS specifications, as that could easily undermine bisimulation being a congruence: even associativity is dangerous [34, Example 9]. However, it should be possible to extend our bisimulation results to specifications in the MSOS tyft format [7], which would allow the use of a precongruence relation that enjoys similar properties to equations in Rewriting Logic.

References

1. Aceto, L., Fokkink, W., Verhoef, C.: Structural operational semantics. In: Smolka, J.B.P. (ed.) Handbook of Process Algebra, pp. 197–292. Elsevier, Amsterdam (2001)

2. Bidoit, M., Mosses, P.D.: Casl User Manual - Introduction to Using the Common Algebraic Specification Language. LNCS, vol. 2900. Springer, Heidelberg (2004)
3. Bloom, B.: Structural operational semantics for weak bisimulations. Theor. Comput. Sci. 146(1–2), 25–68 (1995)
4. Bloom, B., Istrail, S., Meyer, A.R.: Bisimulation can't be traced. J. ACM 42(1), 232–268 (1995)
5. Braga, C.O., Haeusler, E.H., Meseguer, J., Mosses, P.D.: Maude action tool: using reflection to map action semantics to rewriting logic. In: Rus, T. (ed.) AMAST 2000. LNCS, vol. 1816, pp. 407–421. Springer, Heidelberg (2000)
6. de Braga, C.O., Haeusler, E.H., Meseguer, J., Mosses, P.D.: Mapping modular SOS to rewriting logic. In: Leuschel, M. (ed.) LOPSTR 2002. LNCS, vol. 2664, pp. 262–277. Springer, Heidelberg (2003)
7. Churchill, M., Mosses, P.D.: Modular bisimulation theory for computations and values. In: Pfenning, F. (ed.) FOSSACS 2013 (ETAPS 2013). LNCS, vol. 7794, pp. 97–112. Springer, Heidelberg (2013)
8. Clavel, M., Durán, F., Eker, S., Lincoln, P., Martí-Oliet, N., Meseguer, J., Talcott, C.: All About Maude - A High-Performance Logical Framework. LNCS, vol. 4350. Springer, Berlin Heidelberg (2007)
9. Futatsugi, K., Goguen, J.A., Jouannaud, J.P., Meseguer, J.: Principles of OBJ2. In: POPL 1985, pp. 52–66. ACM, New York (1985)
10. van Glabbeek, R.J.: Bounded nondeterminism and the approximation induction principle in process algebra. Technical report CS-R8634, CWI (1986)
11. van Glabbeek, R.J.: On cool congruence formats for weak bisimulations. Theor. Comput. Sci. 412(28), 3283–3302 (2011)
12. van Glabbeek, R.J.: Bounded nondeterminism and the approximation induction principle in process algebra. In: Brandenburg, F.J., Vidal-Naquet, G., Wirsing, M. (eds.) STACS 1987. LNCS, vol. 247, pp. 336–347. Springer, Heidelberg (1987)
13. Goguen, J., Kirchner, C., Kirchner, H., Mégrelis, A., Meseguer, J., Winkler, T.: An introduction to OBJ 3. In: Kaplan, S., Jouannaud, J.-P. (eds.) Conditional Term Rewriting Systems. LNCS, vol. 308, pp. 258–263. Springer, Heidelberg (1988)
14. Goguen, J.A.: Some design principles and theory for OBJ-O, a language to express and execute algebraic specification for programs. In: Blum, E.K., Paul, M., Takasu, S. (eds.) Mathematical Studies of Information Processing. LNCS, vol. 75, pp. 425–473. Springer, Heidelberg (1979)
15. Goguen, J.A., Jouannaud, J.-P., Meseguer, J.: Operational semantics for order-sorted algebra. In: Brauer, W. (ed.) Automata, Languages and Programming. LNCS, vol. 194, pp. 221–231. Springer, Heidelberg (1985)
16. Goguen, J.A., Meseguer, J.: Order-sorted algebra solves the constructor-selector, multiple representation and coercion problems. In: LICS 1987, pp. 18–29. IEEE (1987)
17. Goguen, J.A., Meseguer, J.: Order-sorted algebra I: equational deduction for multiple inheritance, overloading, exceptions and partial operations. Theor. Comput. Sci. 105(2), 217–273 (1992)
18. Groote, J.F., Vaandrager, F.: Structured operational semantics and bisimulation as a congruence. Inf. Comput. 100(2), 202–260 (1992)
19. Kirchner, C., Kirchner, H., Meseguer, J.: Operational semantics of OBJ-3. In: Lepistö, T., Salomaa, A. (eds.) Automata, Languages and Programming. LNCS, vol. 317, pp. 287–301. Springer, Heidelberg (1988)
20. Lamport, L.: The temporal logic of actions. ACM Trans. Program. Lang. Syst. 16(3), 872–923 (1994)

21. Martí-Oliet, N., Meseguer, J.: Rewriting logic as a logical and semantic framework. Electr. Notes Theor. Comput. Sci. **4**, 190–225 (1996)
22. Meseguer, J.: Conditional rewriting logic as a unified model of concurrency. Theor. Comput. Sci. **96**(1), 73–155 (1992)
23. Meseguer, J.: Order-sorted parameterization and induction. In: Palsberg, J. (ed.) Semantics and Algebraic Specification. LNCS, vol. 5700, pp. 43–80. Springer, Heidelberg (2009)
24. Meseguer, J.: Twenty years of rewriting logic. J. Log. Algebr. Program. **81**(7–8), 721–781 (2012)
25. Meseguer, J., Goguen, J.A.: Order-sorted algebra solves the constructor-selector, multiple representation, and coercion problems. Inf. Comput. **103**(1), 114–158 (1993)
26. Meseguer, J., Roşu, G.: The rewriting logic semantics project: A progress report. Inf. Comput. **231**, 38–69 (2013)
27. Milner, R.: Communication and Concurrency. Prentice-Hall Inc., New York (1989)
28. Milner, R.: Communicating and Mobile Systems: The π-Calculus. Cambridge University Press, New York (1999)
29. Mosses, P.D.: Foundations of modular SOS. In: Kutyłowski, M., Wierzbicki, T.M., Pacholski, L. (eds.) MFCS 1999. LNCS, vol. 1672, pp. 70–80. Springer, Heidelberg (1999)
30. Mosses, P.D.: Pragmatics of modular SOS. In: Kirchner, H., Ringeissen, C. (eds.) AMAST 2002. LNCS, vol. 2422, pp. 21–40. Springer, Heidelberg (2002)
31. Mosses, P.D.: Modular structural operational semantics. J. Log. Algebr. Program. **60–61**, 195–228 (2004)
32. Mosses, P.D. (ed.): Casl Reference Manual. LNCS, vol. 2960. Springer, Heidelberg (2004)
33. Mosses, P.D., Vesely, F.: Weak bisimulation as a congruence in MSOS (extended version). Technical report, PLanCompS (2015). http://www.plancomps.org/wbmsos2015
34. Mousavi, M.R.R., Reniers, M.A.: Congruence for structural congruences. In: Sassone, V. (ed.) FOSSACS 2005. LNCS, vol. 3441, pp. 47–62. Springer, Heidelberg (2005)
35. Mousavi, M., Reniers, M.A., Groote, J.F.: SOS formats and meta-theory: 20 years after. Theor. Comput. Sci. **373**(3), 238–272 (2007)
36. Plotkin, G.D.: A structural approach to operational semantics. Technical report DAIMI FN-19, University of Aarhus (1981)
37. Plotkin, G.D.: A structural approach to operational semantics. J. Log. Algebr. Program. **60–61**, 17–139 (2004)
38. Sangiorgi, D.: Introduction to Bisimulation and Coinduction. Cambridge University Press, New York (2011)

Satisfiability of Constraint Specifications on XML Documents

Marisa Navarro[1]([✉]), Fernando Orejas[2], and Elvira Pino[2]

[1] Universidad del País Vasco (UPV/EHU), San Sebastián, Spain
marisa.navarro@ehu.es
[2] Universitat Politècnica de Catalunya, Barcelona, Spain
{orejas,pino}@cs.upc.edu

Abstract. Jose Meseguer is one of the earliest contributors in the area of Algebraic Specification. In this paper, which we are happy to dedicate to him on the occasion of his 65th birthday, we use ideas and methods coming from that area with the aim of presenting an approach for the specification of the structure of classes of XML documents and for reasoning about them. More precisely, we specify the structure of documents using sets of constraints that are based on XPath and we present inference rules that are shown to define a sound and complete refutation procedure for checking satisfiability of a given specification using tableaux.

1 Introduction

The aim of our work is to study how we can specify classes of XML documents and how we can reason about them. Currently, the standard specification of classes of XML documents is done by means of DTDs or XML Schemas. In both cases, we essentially describe the abstract syntax of the class of documents and the types of its attributes. This is quite limited. In particular, we may want to state more complex conditions about the structure of documents in a given class or about their contents. For example, with respect to the structure of documents, we may want to state that if an element includes an attribute with a given content, then these documents should not include some other element. Or, with respect to the contents of documents, we may want to express that the value of some numeric attribute of a certain element is smaller than the value of another attribute of a different element.

In this paper, we concentrate on the specification of the structure of documents, not paying much attention to their contents. In this sense, we present an abstract approach for the specification of (the structure of) classes of XML documents using sets of constraints that are based on XPath [17,21] queries, as

This work has been partially supported by funds from the Spanish Ministry for Economy and Competitiveness (MINECO) and the European Union (FEDER funds) under grant COMMAS (ref. TIN2013-46181-C2-1-R, TIN2013-46181-C2-2-R) and from the Basque Project GIU12/26, and grant UFI11/45.

© Springer International Publishing Switzerland 2015
N. Martí-Oliet et al. (Eds.): Meseguer Festschrift, LNCS 9200, pp. 539–561, 2015.
DOI: 10.1007/978-3-319-23165-5_25

given in [11], using the concept of *tree patterns*. Roughly, a tree pattern describes a basic property on the structure of documents. Its root represents the root of documents. Nodes represent elements that must be present on the given documents and their labels represent their contents, i.e. the names of elements and their value, if any. A wildcard (the symbol ∗), means that we don't know or we don't care about the contents of that element. Finally, single edges represent parent/child relations between elements, while double edges represent a descendant relationship between elements. Again, if any of these two relations is included in a tree pattern, then it should also be included in the documents satisfying that property. For instance, on the left of Fig. 1 we show a tree pattern p describing documents $\mathcal{D}$ whose root node is labelled with a, some child node of the root node in $\mathcal{D}$ is labelled b, and some descendant node of the root node in $\mathcal{D}$ has two child nodes labelled c and d, respectively.

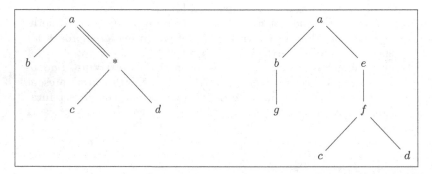

Fig. 1. A tree pattern and a document satisfying the pattern

Similarly, we represent, in an abstract way, XML documents using the same kind of trees. The difference between a document and a tree pattern is that a document does not include double edges or wildcards. For example, on the right of Fig. 1 we show a document that satisfies the pattern on the left. In particular, we may see that the root of the document is labelled by a. Moreover, that root has a child node labelled b and a descendant node (the element labelled f) that has two child nodes labelled c and d, respectively.

We consider three kinds of (atomic) constraints. The first one, called *positive constraints*, are tree patterns. The second one are *negative constraints*, $\neg p$, where p is a tree pattern, expressing that documents should not satisfy p. Finally, the third sort of constraint are *conditional constraints*, written $\forall(c : p \rightarrow q)$, where both p and q are tree patterns. Roughly speaking, these constraints express that if a document satisfies p then it must also satisfy q. Moreover, these constraints can be combined using the connectives $\wedge$ and $\vee$. These kinds of constraints are similar to the graph constraints studied in [15,16] in the context of graph transformation. Nevertheless, the application of the ideas in [15,16] to our setting is not trivial, as discussed in Sect. 3.

Obviously, there are conditions on the structure of XML documents that are not expressible using the kind of constraints studied in this paper. However, our experience in the area of graph transformation [15,16] shows that, in practice, these constraints are sufficient in most cases. Nevertheless, we believe that the ideas presented here can be extended to a class of XML constraints, similar to the class of nested graph conditions that has been shown equivalent to first-order logic of graphs [6]. However, we also believe that this extension is not straightforward.

Since our aim is to be able to reason about these specifications, we present inference rules that are shown, by means of tableaux, to define a sound and complete refutation procedure for checking satisfiability of a given specification.

The paper is organized as follows. Section 2 contains some basic notions and notational conventions we are going to use along the paper. Section 3 introduces the three kinds of constraints that we use as literals of the clauses in a specification. In Sect. 4 we present our tableau method for reasoning about our constraints, and in Sect. 5 we show its soundness and completeness. Finally, in Sects. 6 and 7 we discuss related work and provide some conclusions.

2 Basic Definitions and Notation

In this section we introduce some basic concepts and notations, as well as some definitions and properties on patterns that will be required in the paper.

2.1 Documents and Patterns

As we have seen in the introduction, we consider a *document* as a kind of unordered and unranked tree with nodes labelled from an infinite alphabet Σ and whose edges represent a parent/child relation between nodes. The symbols in Σ represent the element labels, attribute labels, and values that can occur in documents. By considering that the trees are unordered and unranked, the subtrees can commute (the "sibling ordering" is irrelevant), and there are no restrictions on the number of children a node can have.

As also seen, patterns describe properties on the structure of documents and are also represented by trees. However, there is the special label $*$, representing the *wildcard*, and there are two kinds of edges: single and double edges. Patterns (and documents) can be represented textually using the following format: A pattern p with root labelled a and subtrees $p_1, \ldots, p_n$ will be textually written $p = a(!p_1) \ldots (!p_n)$ where each p_i is recursively written in the same format, and ! being / or // to indicate the edge from the root to each subtree p_i. Some parenthesis can be omitted in the case of having only one subtree. For instance, the pattern given in Fig. 1 can be textually written $a(/b)(// * (/c)(/d))$. Similarly, the document in the same Fig. 1 is textually written $a(/b/g)(/e/f(/c)(/d))$.

However, even if the documents and the patterns that we would write would always be finite, in our paper we need to deal with infinite documents and patterns. The reason, is that (as often done), given a specification for a class of documents, we will consider that the specification is consistent if there exist documents that satisfy it, even if these documents are infinite. In this sense, one

might consider that the results shown in Sect. 5 are not fully adequate, in the
sense that we would not have proved the completeness of our proof rules with
respect to the class of finite documents.

For this reason, we need a more precise definition of what documents and
patterns are. In particular, we define them as follows:

Definition 1 (Patterns and Documents). *Given a signature* Σ, *a pattern*
p *on* Σ *is a 5-tuple* $p = (Nodes_p, root_p, Label_p, Edges_p, Paths_p)$ *where,* $Nodes_p$
is a set of nodes, $root_p \in Nodes_p$ *is the root node of* p, $Label_p : Nodes_p \rightarrow$
$\Sigma \cup \{*\}$ *is the labeling function, and* $Edges_p, Paths_p \subseteq Nodes_p \times Nodes_p$ *are two*
relations representing edges and paths between nodes in p, *such that the following*
conditions are satisfied:

1. $Edges_p$ *and* $Paths_p$ *are irreflexive and acyclic (i.e. there are no nodes* n, n'
   such that $\langle n, n' \rangle$ *and* $\langle n', n \rangle$ *are both in* $Edges_p$ *or* $Paths_p$).
2. $Paths_p$ *is transitive and includes* $Edges_p$.
3. *There is no node* n *such that* $\langle n, root_p \rangle$ *is in* $Edges_p$ *or* $Paths_p$.
4. *For any other node* $n \neq root_p$ *in* $Nodes_p$, $\langle root_p, n \rangle \in Paths_p$. *Moreover, if*
   $\langle n', n \rangle$ *and* $\langle n'', n \rangle$ *are both in* $Paths_p$, *then either* $\langle n'', n' \rangle$ *or* $\langle n', n'' \rangle$ *are in*
   $Paths_p$.

Then, *a document* $\mathcal{D}$ *on* Σ *can be defined as a special kind of pattern without*
nodes labelled with $*$, *that is,* $Label_{\mathcal{D}} : Nodes_{\mathcal{D}} \rightarrow \Sigma$, *and, such that, in addition*
it satisfies the following condition:

5. *For every pair of nodes* $n, n' \in Nodes_{\mathcal{D}}$, *if* $\langle n, n' \rangle \in Paths_{\mathcal{D}}$, *then*
   - $\langle n, n' \rangle \in Edges_{\mathcal{D}}$, *or* .
   - *there exist* $n_1, n_2 \in Nodes_{\mathcal{D}}$ *such that* $\langle n, n_1 \rangle, \langle n_2, n' \rangle \in Edges_{\mathcal{D}}$ *and, either*
     $n_1 = n_2$ *or* $\langle n_1, n_2 \rangle \in Paths_{\mathcal{D}}$.

P_Σ *and* D_Σ *will denote, respectively, the set of all patterns and the set of all*
documents on Σ.

Intuitively, the above definition can be easily explained. The relation $\langle n, n' \rangle \in$
$Edges_p$ represents the existence of an edge / between n and n' in the given pattern or document, and $\langle n, n' \rangle \in Paths_p$ represents that there is a path consisting
of edges / or // (in the case of patterns) or just / (in the case of documents)
between n and n'. Conditions 1–4 ensure that our patterns and documents are
trees. Finally, Condition 5 ensures that if $\langle n, n' \rangle \in Paths_{\mathcal{D}}$ then there is a finite
or infinite path, consisting only of edges /, between n and n'. It is easy to see
that, in the case where the given set of nodes is finite, our definition of patterns
and documents would be equivalent to other notions of (finite) trees. In particular, for finite documents, Condition 5 is equivalent to saying that *Paths* is the
transitive closure of *Edges*.

One could think that the second part of Condition 5 could be simplified as
follows:

– there exists $n_1 \in Nodes_{\mathcal{D}}$ such that $\langle n, n_1 \rangle \in Edges_p$ and $\langle n_1, n' \rangle \in Paths_p$.

However, both conditions are not equivalent. In particular, our Condition 5 would exclude infinite paths like $n/n_1/n_2/\ldots/n_k/\ldots$, where for every i, $\langle n_i, n' \rangle \in Paths_p$, which would be allowed by the simpler condition. That is, an infinite path for $\langle n, n' \rangle \in Paths_p$ cannot consist of an infinite sequence $n/n_1/n_2/\ldots/n_k/\ldots$ approaching n'. Instead, our infinite paths must consist of two infinite sequences $n/n_1/n_2/\ldots/n_k/\ldots$ and $\ldots/n'_j/\ldots/n'_2/n'_1/n'$, where for every i, i', $\langle n_i, n'_{i'} \rangle \in Paths_p$.

For example, consider again the pattern and the document in Fig. 1. Abusing of notation, let us identify nodes with labels. Then, for the pattern $p = a(/b)(//*(/c)(/d))$ and the document $\mathcal{D} = a(/b/g)(/e/f(/c)(/d))$, we have:

$Edges_p = \{\langle a,b \rangle, \langle *,c \rangle, \langle *,d \rangle\}$

$Paths_p = \{\langle a,b \rangle, \langle a,* \rangle, \langle a,c \rangle, \langle a,d \rangle, \langle *,c \rangle, \langle *,d \rangle\}$

$Edges_{\mathcal{D}} = \{\langle a,b \rangle, \langle b,g \rangle, \langle a,e \rangle, \langle e,f \rangle, \langle f,c \rangle, \langle f,d \rangle\}$

$Paths_{\mathcal{D}} = \{\langle a,b \rangle, \langle a,e \rangle, \langle a,g \rangle, \langle a,f \rangle, \langle a,c \rangle, \langle a,d \rangle, \langle b,g \rangle, \langle e,f \rangle, \langle e,c \rangle, \langle e,d \rangle, \langle f,c \rangle, \langle f,d \rangle\}$

For the sake of readability, from now on we will omit the signature Σ. Moreover, we will write n/n' instead of $\langle n,n' \rangle \in Edges_p$, and $n//n'$ instead of $\langle n,n' \rangle \in Paths_p$. Notice that, in our simplified notation, the symbol $//$ is overloaded to mean a kind of edge in patterns but also, the relation defining paths in patterns and documents. However, it is easy to distinguish both uses from the context since, in the first case, we will usually refer to "an edge $//$". If some ambiguity could persist then, we will use $//^d$ to denote direct relation between nodes. That is, given $n_1, n_2 \in Nodes_p$ $n_1//^d n_2$ if not n_1/n_2 but, $n_1//n_2$ such that there does not exist $n \in Nodes_p$ with $n_1//n$ and $n//n_2$. Nevertheless, for simplicity, in the examples in the rest of the paper we will use the textual writing for patterns and documents, so that, in those expressions the symbol $//$ always will stand for edges, that is, the direct relation $//^d$.

2.2   Pattern Morphisms and Pattern Models

Morphisms are very important in our work. A document satisfies a pattern if we can identify the structure of the pattern in the document. Formally, we do this by means of morphisms. In addition, we also use a special kind of morphisms to relate the premise and the conclusion in conditional constraints. We define the notion of morphism between two patterns, since documents are a special case of patterns. Then, the same definition applies to morphisms between documents or between patterns and documents. As said, the latter case will be used to define which documents are the models of a pattern. That is, from a logical point of view, we can see patterns as formulae, documents as structures and morphisms defining a notion of pattern satisfaction.

Definition 2 (Morphisms). *Given two patterns $p, q \in P$, a morphism $h : p \to q$, from p to q, is a function $h : Nodes_p \to Nodes_q$ satisfying the following conditions:*

- *Root-preserving: $h(root_p) = root_q$;*

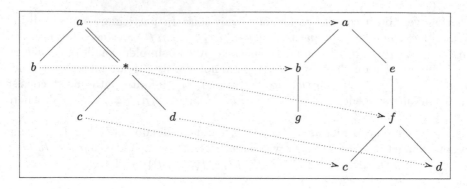

Fig. 2. A pattern p, a document $\mathcal{D}$ and a monomorphism $h : p \to \mathcal{D}$

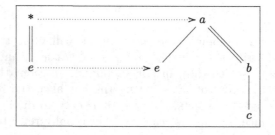

Fig. 3. A monomorphism $h : p \to q$ between two patterns

- *Label-preserving: For each $n \in Nodes_p$, $Label_p(n){=}*$ or $Label_p(n){=}Label_q(h(n))$;*
- *Edge-preserving: For each $n_1, n_2 \in Nodes_p$, if n_1/n_2 then, $h(n_1)/h(n_2)$;*
- *Path-preserving: For each $n_1, n_2 \in Nodes_p$, if $n_1//n_2$ then, $h(n_1)//h(n_2)$;*

As usual, a monomorphism is an injective morphism. $\mathbf{P_\Sigma}$ *and* $\mathbf{D_\Sigma}$ *denote, respectively, the category of patterns and its subcategory of documents on Σ.*

Definition 3 (Models). *Given a pattern $p \in P$ and a document $\mathcal{D} \in D$, we say that $\mathcal{D}$ satisfies p, denoted $\mathcal{D} \models p$, if there exists a monomorphism from p to $\mathcal{D}$. The set of models of a pattern p is the set of documents satisfying p, that is, $Mod(p) = \{\mathcal{D} \in D \mid \mathcal{D} \models p\}$.*

In Fig. 2 there is an example of a monomorphism $h : p \to \mathcal{D}$ from the pattern $p = a(/b)(// * (/c)(/d))$ to the document $\mathcal{D} = a(/e/f(/c)(/d))(/b/g)$. The morphism h is drawn with dotted arrows. We can see that $\mathcal{D}$ satisfies p because its root is labelled with a, it has a child node labelled b, and it has a descendant node (in the example labelled with f) with two child nodes labelled with c and d respectively. In Fig. 3 there is an example of a monomorphism $h : p \to q$ from the pattern $p = *//e$ to the pattern $q = a(/e)(//b/c)$. The monomorphism h is drawn with dotted arrows. The existence of such monomorphism implies that all models of q are also models of p.

The following proposition relates monomorphisms and models for two patterns.

Proposition 1. *Given two patterns $p, q \in P$:*

- *If there exists a monomorphism $h : p \to q$ then $Mod(q) \subseteq Mod(p)$.*
- *$Mod(q) \subseteq Mod(p)$ does not imply that there is a monomorphism $h : p \to q$.*

Proof. For the first claim, let $\mathcal{D}$ be a document in $Mod(q)$, then there exists a monomorphism f from q to $\mathcal{D}$. Then the composition $f \circ h$ is a monomorphism from p to $\mathcal{D}$ and therefore the document $\mathcal{D}$ is also a model for p. The second claim can be shown with an example in [11].

3   Constraints, Clauses and Specifications

As said in the Introduction, following [15,16] we consider three kinds of constraints: positive, negative and, conditional constraints. The underlying idea of our constraints is that they should specify that certain patterns must occur (or must not occur) in a given document. For instance, the simplest kind of constraint, p, specifies that a given document $\mathcal{D}$ should satisfy the pattern p. Obviously, $\neg p$ specifies that a given document $\mathcal{D}$ should not satisfy p. A more complex kind of constraint is of the form $\forall (c : p \to q)$ where c is a prefix morphism, which means that q is a pattern that extends p. Roughly speaking, this constraint specifies that whenever a document $\mathcal{D}$ satisfies the pattern p it should also satisfy the extended pattern q (see Definition 6 below).

However, translating the ideas in [15,16] to our setting is not trivial, mainly for two reasons. On the one hand, in [15,16] models and formulas are both graphs, while in our setting models are documents and formulas are patterns. This difference adds some complication to our setting. In particular, we have solved the problem by defining documents and patterns in such a way that documents are a special case of patterns. This has implied to include explicitly the *paths* relation in the definition of documents. On the other hand, and most importantly, we deal with patterns that are trees having edges of type $//$, but the related notion of "path" is not considered for graph constraints in [15,16]. Actually, in the logic defined in [15,16] or in the more general one defined in [6], the existence of paths is a second order notion [7].

3.1   Constraints and Clauses

Before defining our three kinds of constraints, we must define prefix morphisms.

Definition 4. *Given two patterns p and q, a prefix morphism from p to q is a monomorphism $c : Nodes_p \to Nodes_q$ that satisfies the following conditions:*

- *Root-preserving: $c(root_p) = root_q$;*
- *Label-identity: For each $n \in Nodes_p$, $Label_p(n) = Label_q(c(n))$;*
- *Edge-identity: For each $n_1, n_2 \in Nodes_p$, n_1/n_2 if, and only if, $c(n_1)/c(n_2)$;*
- *Path-identity: For each $n_1, n_2 \in Nodes_p$, $n_1//^d n_2$ if, and only if, $c(n_1)//^d c(n_2)$;*

Recall that $//^d$ stands for direct $//$-relation in patterns, that is, $//$-edges. We will simply write $c : p \to q$ and we will say that p is a prefix of q. Not every monomorphism is a prefix morphism. See for instance that the monomorphism in Fig. 3 is not a prefix morphism since it violates "Label-identity", "Edge-identity", and "Path-identity".

Definition 5. *Given a pattern p, p denotes a* positive constraint *and $\neg p$ denotes a* negative constraint. *A* conditional constraint *is denoted $\forall(c : p \to q)$ where p and q are patterns and $c : p \to q$ is a prefix morphism.*

A clause *α is a finite disjunction of literals $\ell_1 \vee \ell_2 \vee \cdots \vee \ell_n$, where, for each $i \in \{1, \ldots, n\}$, the literal ℓ_i is a (positive, negative or conditional) constraint. The empty disjunction is called the* empty clause *and it can be represented by FALSE.*

Satisfaction of clauses is inductively defined as follows.

Definition 6. *A document $\mathcal{D} \in D$ satisfies a clause α, denoted $\mathcal{D} \models \alpha$, if it holds:*

- *$\mathcal{D} \models p$ if there exists a monomorphism $h : p \to \mathcal{D}$;*
- *$\mathcal{D} \models \neg p$ if $\mathcal{D} \not\models p$ (that is, if there does not exist a monomorphism $h : p \to \mathcal{D}$);*
- *$\mathcal{D} \models \forall(c : p \to q)$ if for every monomorphism $h : p \to \mathcal{D}$ there is a monomorphism $f : q \to \mathcal{D}$ such that $h = f \circ c$.*
- *$\mathcal{D} \models \ell_1 \vee \ell_2 \vee \ldots \vee \ell_n$ if $\mathcal{D} \models \ell_i$ for some $i \in \{1, \ldots, n\}$.*

Let us see what satisfaction of a conditional constraint means. First recall that in a conditional constraint $\forall(c : p \to q)$ the pattern q is an extension of the pattern p so a document $\mathcal{D}$ will be a model of the conditional constraint if whenever $\mathcal{D}$ satisfies p, it also satisfies q. To be more precise, each part in the document $\mathcal{D}$ satisfying p must satisfy q. Consider, for instance, the conditional constraint $\forall(c : p \to q)$ with $p = *//a$, $q = *//a/b$ and c being the obvious prefix morphism from p to q. By Definition 6, a document satisfies this constraint if each node (descendant of the root) labelled a has a child node labelled b. Then the document $\mathcal{D} = g(/a/b)(/a/h)$ does not satisfy the constraint. In fact, for the monomorphism $h : p \to \mathcal{D}$ that applies the node a in p into the second node a in $\mathcal{D}$, there does not exist a monomorphism $f : q \to \mathcal{D}$ such that $h = f \circ c$. However, note that $\mathcal{D} \models q$. Therefore, from a logical point of view, we may notice that, in this framework, $\forall(c : p \to q)$ is not equivalent to $C = \neg p \vee q$.

3.2 Specifications

We assume that a specification $\mathcal{S}$ consists of a set of clauses. As said in the Introduction, our aim is to find a sound and complete refutation procedure for checking satisfiability of specifications consisting of clauses as defined above. Here we give an example of an unsatisfiable specification.

Example 1. Consider the specification $\mathcal{S} = \{C_1, C_2, C_3, C_4\}$ where $C_1 = (*//b) \vee (*//e)$, $C_2 = \forall(c_2 : *//b \rightarrow *(//b)(/e))$, $C_3 = \forall(c_3 : *//e \rightarrow *(//e)(/b))$, and $C_4 = \neg(*(/b)(/e))$.

Clause C_1 specifies that the document(s) must have a node labelled b or e; C_2 says that if the document has some node labelled b then its root must have a child node labelled e; similarly, C_3 says that if the document has some node labelled e then the root must have a child node labelled b; and finally, C_4 says that the root cannot have two children nodes labelled b and e. It is easy to test, for instance, that the document $\mathcal{D}_1 = a(/b)(/f/e)$ satisfies C_1, C_3 and C_4 but $\mathcal{D}_1 \not\models C_2$. Similarly, the document $\mathcal{D}_2 = a/e$ satisfies C_1, C_2 and C_4 but $\mathcal{D}_2 \not\models C_3$. There is no document satisfying all clauses in $\mathcal{S}$.

3.3 Superposition of Patterns

In this section we introduce two operations on patterns that can be seen as a way of pattern deduction, which will be used for obtaining new clauses from a specification. Note, for instance, that if a document $\mathcal{D}$ satisfies both the patterns a/b and a/c then its root, labelled a, must have two children nodes labelled b and c, therefore we can trivially deduce that $\mathcal{D}$ must also satisfy the pattern $a(/b)(/c)$. But not always a single pattern can express the conditions of two patterns: If $\mathcal{D}$ satisfies both the patterns $a/b/e$ and $a/b/c$, then it must be deduced that $\mathcal{D}$ satisfies one of the patterns obtained by superposing both patterns, what means, in this example, that $\mathcal{D}$ must satisfy either the pattern $a/b(/e)(/c)$ or the pattern $a(/b/e)(/b/c)$.

The two superposition operations on patterns will be denoted by the symbols $\otimes$ and $\otimes_{c,m}$ and they are formally introduced in the following definitions.

Given two patterns p_1 and p_2, the operation $p_1 \otimes p_2$ denotes the set of patterns that can be obtained by "combining" p_1 and p_2 in all possible ways.

Definition 7. *Given two patterns p_1 and p_2, $p_1 \otimes p_2$ is defined as the following set of patterns: $p_1 \otimes p_2 = \{s \in P \mid$ there exist jointly surjective monomorphisms $inc_1 : p_1 \rightarrow s$ and $inc_2 : p_2 \rightarrow s\}$ where "jointly surjective" means that $Nodes_s = inc_1(Nodes_{p_1}) \cup inc_2(Nodes_{p_2})$.*

For instance, given the patterns $p_1 = a(/b/e)(//c)$ and $p_2 = a//b/x$, the set $p_1 \otimes p_2$ contains the two patterns: $s_1 = a(/b(/e)(/x))(//c)$ and $s_2 = a(/b/e)(//b/x)(//c)$. Each one corresponds with a way of combining p_1 and p_2; the nodes labelled b are shared in s_1 while there are two different nodes b in $s2$.

The underlying idea is that all patterns s in $p_1 \otimes p_2$ must verify that every document that is a model of s must be a model of p_1 and a model of p_2. Conversely, every document that is a model of both p_1 and p_2 must be a model of some s in $p_1 \otimes p_2$. In some cases the set can be empty.

Proposition 2. *Given two patterns p_1 and p_2, the set of patterns $p_1 \otimes p_2$ is the empty set if and only if $root_{p_1}$ and $root_{p_2}$ have different labels in Σ.*

Proof. If the roots of p_1 and p_2 have different labels in Σ (for instance, a and b) then no combination s is possible since $inc_1 : p_1 \rightarrow s$ implies that the root of s must be labelled a and $inc_2 : p_2 \rightarrow s$ implies that the root of s must be labelled b.

Conversely, if the root s of p_1 and p_2 have the same label a (or if one of them is a and the other one is $*$) then the document s with root labelled a and whose set of subtrees is the union of the subtrees of p_1 and p_2 is an element in $p_1 \otimes p_2$.   ∎

Proposition 3 (Pair Factorization Property). *Given three patterns p_1, p_2, r, and two monomorphisms $f_1 : p_1 \rightarrow r$ and $f_2 : p_2 \rightarrow r$, there exists a pattern $s \in p_1 \otimes p_2$, with monomorphisms $inc_1 : p_1 \rightarrow s$ and $inc_2 : p_2 \rightarrow s$, and there exists a monomorphism $h : s \rightarrow r$ such that $h \circ inc_1 = f_1$ and $h \circ inc_2 = f_2$. In the particular case when r is a document, this property means that r is a model of s. Graphically:*

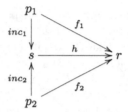

Proof. Since f_1, f_2 are monomorphisms, the root s of p_1 and p_2 cannot have different labels in Σ. Moreover, some pattern $s \in p_1 \otimes p_2$ holds this property. Then, we will have a well-defined morphism h if choose a pattern s such that, for every $m \in Nodes_{p_1}$ and $n \in Nodes_{p_2}$: if $f_1(m) = f_2(n)$ then $inc_1(m) = inc_2(n)$ and if $f_1(m)$ is an ancestor (respectively descendant) of $f_2(n)$, $inc_1(m)$ must not be a descendant (respectively ancestor) of $inc_2(n)$.   ∎

Given a pattern p_1, a prefix morphism $c : p_2 \rightarrow q$ and a monomorphism $m : p_2 \rightarrow p_1$, the operation $p_1 \otimes_{c,m} q$ denotes the set of patterns that can be obtained by combining p_1 and q in all possible ways, but sharing p_2.

Definition 8. *Given a pattern p_1, a prefix morphism $c : p_2 \rightarrow q$, and a monomorphism $m : p_2 \rightarrow p_1$, $p_1 \otimes_{c,m} q$ is defined as the following set of patterns: $p_1 \otimes_{c,m} q = \{s \in P \mid there\ exist\ jointly\ surjective\ monomorphisms\ inc_1 : p_1 \rightarrow s\ and\ inc_2 : q \rightarrow s\ such\ that\ inc_1 \circ m = inc_2 \circ c\}.*

For instance, given the patterns: $p_1 = a(/b/e)(//c/i)$, $p_2 = *//b$, and $q = *(//b//a)(//c/d)$, with the unique possible monomorphism $m : p_2 \rightarrow p_1$ and the unique possible prefix morphism $c : p_2 \rightarrow q$, the set $p_1 \otimes_{c,m} q$ contains the patterns $s_1 = a(/b(/e)(//a))(//c/i)(//c/d)$ and $s_2 = a(/b(/e)(//a))(//c(/i)(/d))$. Note that s_2 is similar to s_1 but with only one node labelled c.

The underlying idea is that all patterns s in $p_1 \otimes_{c,m} q$ must verify that every document $\mathcal{D}$ that is a model of s must be a model of p_1 and a model of q. However, such a document $\mathcal{D}$ is not necessarily a model of the conditional constraint $\forall(c : p_2 \rightarrow q)$. Conversely, every document that is a model of both p_1 and $\forall(c : p_2 \rightarrow q)$ must be a model of some s in $p_1 \otimes_{c,m} q$.

Notice that the set $p_1 \otimes_{c,m} q$ is always non-empty, since given a prefix morphism $c : p_2 \to q$ and a monomorphism $m : p_2 \to p_1$ we can always obtain a pattern s by extending the nodes in $m(p_2)$ as indicated by the function c. However, if c would be a monomorphism instead of a prefix morphism (i.e. if in the definition of conditional constraints we would have used arbitrary monomorphisms), the resulting set could be empty. Take, for instance, $c : p_2 \to q$, with $p_2 = a//b$ and $q = a/b$ (which is not a prefix morphism), and take $p_1 = a/e/b$. Although there is a monomorphism $m : p_2 \to p_1$, there is no pattern s obtained by combining p_1 and q sharing p_2.

4   Tableau-Based Reasoning for XML-patterns

Analogously to tableaux for plain first-order logic reasoning [8], we introduce tableaux for dedicated automated reasoning for XML-document properties. In general, tableaux are trees whose nodes are literals. In our case, these literals are constraints of the form p, $\neg p$ or $\forall (c : p \to q)$.

Definition 9 (Tableau, branch). *A* tableau *is a finitely branching tree whose nodes are constraints. A branch B in a tableau T is a maximal path in T.*

The construction of a tableau for a specification S can be informally explained as follows. We start with a tableau consisting of the single node *true*. For every clause $\alpha = \ell_1 \vee \ldots \vee \ell_n$ in S we extend all the leaves in the tableau with n branches, one for each literal ℓ_i, as depicted in Fig. 5 (see the four first steps) for the specification S in Example 1. Then we continue the extension of each leaf by applying other tableau rules (on two literals in its branch). In Fig. 5 we show the tableau in the left hand side, and the rules applied in its construction in the right hand side.

Before defining the rules that build the tableaux associated to our specifications, we need to introduce some notation. Let p be a positive constraint in B, such that p contains an edge $//$ (that is, $n_1//^d n_2$ for some $n_1, n_2 \in Nodes_p$). Let $prefix(n_1)$ denote the path from $root_p$ to the node n_1, and $hang(n_2)$ the subtree of p hanging from the node n_2; we write $p[n_1//hang(n_2)]$ to highlight the edge $//$ from node n_1 to node n_2 in p. Then $p[n_1/hang(n_2)]$ denotes the pattern obtained by replacing $//$ by $/$. In addition, $p[n_1 \leftarrow]$ denotes that the subtree $hang(n_2)$ has been pruned from p, and $p[n_1 \leftarrow /A]$ (equivalently $p[n_1 \leftarrow //A]$) denotes that the pattern A is hanged as a subtree of node n_1 in p, where $/$ (equivalently $//$) is the edge from n_1 to $root_A$.

For instance, given the pattern $p = e(/i)(/a(/b)(//c(/d)(/j)))$ (see Fig. 4) with an edge $//$ from the node n_1 labelled a to the node n_2 labelled c, we have that $prefix(n_1) = e/a$; $hang(n_2) = c(/d)(/j)$; $p[n_1 \leftarrow] = e(/i)(/a/b)$; and $p[n_1 \leftarrow /A] = e(/i)(/a(/b)(/s/k))$ when hanging, for instance, the pattern $A = s/k$.

Now, the tableau rules that are specific for our logic are the following ones:

Definition 10 (Tableau rules). *Given a specification S, a tableau for S is either a tree consisting of the single node* true, *or for any node x in the tableau that is not a leaf, one of the following conditions hold:*

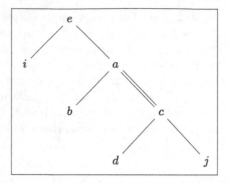

Fig. 4. The pattern $p = e(/i)(/a(/b)(//c(/d)(/j)))$

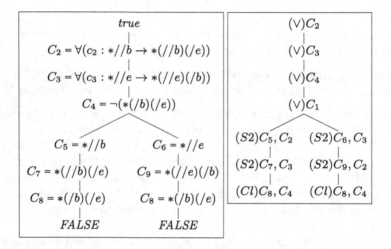

Fig. 5. A closed tableau for the specification S in Example 1 (left) and their applied rules (right)

- **∨-rule (∨):** *There is a clause $\ell_1 \vee \ldots \vee \ell_n$ in S and the children of x are $\ell_1, \ldots \ell_n$.*
- **Superposition rule (S1):** *The constraints p_1 and p_2 are either x or ancestors of x and $p_1 \otimes p_2$ is not empty, and the children of x are the constraints s, for each pattern s in $p_1 \otimes p_2$. Otherwise, if $p_1 \otimes p_2$ is empty, x has the only child FALSE.*
- **Superposition rule (S2):** *The constraints p_1 and $\forall(c : p_2 \to q)$ are either x or ancestors of x such that there is a monomorphism $m : p_2 \to p_1$, and the children of x are the constraints s, for each s in $p_1 \otimes_{c,m} q$.*
- **Unfolding rule (U1):** *The constraint p where $p = p[n_1//hang(n_2)]$ is either x or an ancestor of x and the children of x are the constraint $p[n_1 \leftarrow /hang(n_2)]$, and the constraints $s[inc_2(n) \leftarrow //hang(n_2)]$ for each s in $p[n_1 \leftarrow] \otimes_{c,m} q$, where $q = prefix(n_1)[n_1 \leftarrow /n]$ with $label(n) = *$, and monomorphisms are depicted in the diagram below.*

- **Unfolding rule (U2):** *The constraint p where $p = p[n_1//hang(n_2)]$ is either x or an ancestor of x and the children of x are the constraint $p[n_1 \leftarrow /hang(n_2)]$, and the constraints $s[inc_2(n) \leftarrow /hang(n_2)]$ for each s in $p[n_1 \leftarrow]\otimes_{c,m}q$, where $q = prefix(n_1)[n_1 \leftarrow //n]$ with $label(n) = *$, and monomorphisms are depicted in the diagram below.*

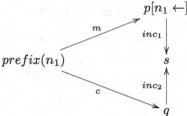

- **Closing rule (Cl):** *The constraints p and $\neg q$ are either x or ancestors of x, such that there is a monomorphism $m : q \to p$, and x has the only child FALSE.*

Obviously the above rules not only describe if we can associate a given tableau to a specification $\mathcal{S}$, but they can also be used in the construction of a tableau for $\mathcal{S}$. The $\vee$-rule is the standard tableaux rule for creating the initial tableau from the clauses of the given specification. The superposition rules state that if we have two non-negative literals in a given branch, then we can extend that branch with the immediate consequences of these literals, as we have seen in Sect. 3.3. We may notice that the superposition rules define a finite number of children for a given node x, because the set of patterns resulting from a superposition of finite patterns is a finite set. The closing rule states that if, in a branch B, we have a positive and a negative literal, $p, \neg q$, that are contradictory, because p embeds q, then we can close B with *FALSE*.

Finally, the unfolding rules extend a branch with all the possible ways of unfolding an edge $n_1//n_2$. In principle, we considered two different ways of doing this unfolding: replacing $n_1//n_2$ by n_1/n_2 and $n_1/n//n_2$ (using rule U1) or by n_1/n_2 and $n_1//n/n_2$ (using rule U2), where $label(n) = *$ in both cases. However, it is necessary to take into account all possible identifications of such new node n with other nodes in the obtained literal. Otherwise, the rule may be unsound as explained in the following example.

Example 2. Consider the specification $\mathcal{S} = \{C_1, C_2, C_3, C_4\}$ with $C_1 = a(//b)(//c)$, $C_2 = \neg(a/b)$, $C_3 = \neg(a/c)$, and $C_4 = \neg(a(/*//b)(/*//c))$.

If rule U1 would just unfold an edge $//$ only in $/$ and $/*//$, then we could easily find a refutation for this specification as follows: By applying such U1 to the edge $a//b$ in C_1 we could get the constraints $C_5 = a(/b)(//c)$ and $C_6 = a(/*//b)(//c)$. The first literal, C_5, is directly contradictory with C_2. If we now unfold C_6 (on the edge $a//c$) by applying the rule U1, we could get the literals $C_7 = a(/*//b)(/c)$ and $C_8 = a(/*//b)(/*//c)$, which could be directly refuted with C_3 and C_4 respectively. But the specification $\mathcal{S}$ is not inconsistent because, for instance, the document $D = a/d(/b)(/c)$ satisfies all its clauses.

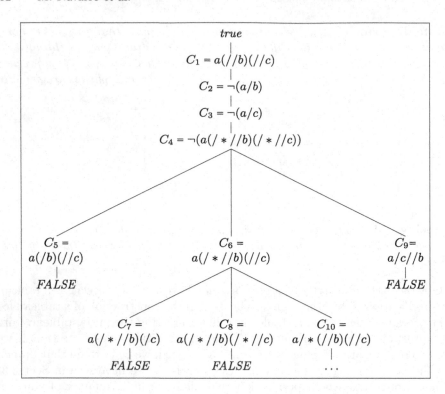

Fig. 6. An open tableau for the specification S in Example 2

However, the situation with the defined unfolding rules is the following. The application of rule U1 on C_1 yields to the literals C_5, C_6, and the literal $C_9 = a/c//b$ obtained by joining the new node $*$ with the old node c in C_6; and the application of the rule U1 on C_6 yields to the literals C_7, C_8, and the literal $C_{10} = a/*(//b)(//c)$ obtained by joining the new node $*$ with the old node $*$ in C_8. While C_9 can be refuted with C_3, the literal C_{10} yields to an open branch as depicted in Fig. 6. See also Fig. 7 for the rules applied to built the tableau in Fig. 6.

Definition 11 (Open/closed branch, tableau proof). *In a tableau T a branch B is* closed *if B contains FALSE; otherwise, it is* open. *A tableau is* closed *if all its branches are closed. A* tableau proof *for (the unsatisfiability of) a specification S is a closed tableau T for S according to the rules given in Definition 10.*

Finally, it will be useful to define tableau satisfiability.

Definition 12 (Branch and tableau satisfiability). *A* branch B *in a tableau T is* satisfiable *if there exists a document $\mathcal{D}$ satisfying all the constraints in B. In this case, we say that $\mathcal{D}$ is a* model *for B, written $\mathcal{D} \models B$. A tableau T is* satisfiable *if there is a satisfiable branch B in T. If $\mathcal{D} \models B$ for a branch B in T, we also say that $\mathcal{D}$ is a model for T and also write $\mathcal{D} \models T$.*

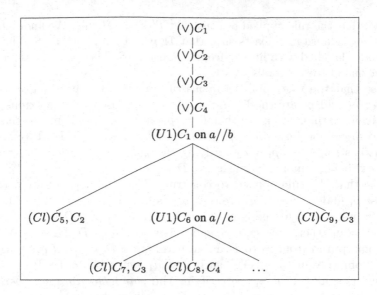

Fig. 7. The applied rules to build the tableau in Fig. 6

5   Soundness and Completeness of the Tableau Method

In this section we prove that our tableau method is sound and complete. In particular, soundness means that if we are able to construct a tableau where all its branches are closed then our original specification S is unsatisfiable. Completeness means that if a *saturated* tableau includes an open branch, where the notion of saturation is defined below, then the original specification is satisfiable. Actually, the open branch provides a model that satisfies the specification.

Theorem 1 (Soundness). *If there is a tableau proof for the specification S, then S is unsatisfiable.*

Proof. We prove that if a specification S is satisfiable, then any associated tableau cannot have all its branches closed. The proof is by induction on the structure of the tableau. Specifically, we show by induction on the construction of T that if $\mathcal{D} \models S$ then $\mathcal{D} \not\models T$.

The base case is trivial since T consists only of the node *true*.

For the general case, assume that $\mathcal{D} \models T$ as inductive hypothesis. We have to show that if T' is constructed by applying a tableau rule to T, then $\mathcal{D} \models T'$. By induction, we know that there exists a branch B in T such that $\mathcal{D} \models B$. If this branch is not extended when constructing T', then it trivially holds that $\mathcal{D} \models T'$ since B is still a branch in T'. Otherwise, if B is extended, then we show that in T' there exists an extended branch B' from B such that $\mathcal{D} \models B'$ and therefore also $\mathcal{D} \models T'$. Now, we proceed by cases depending on what rule is applied in the extension:

– Suppose that the rule applied to construct T' is the $\vee$-rule. We know that one literal ℓ per clause in $\mathcal{S}$ exists such that $\mathcal{D} \models \ell$ because $\mathcal{D} \models \mathcal{S}$. The $\vee$-rule adds nodes labelled with literals from a clause in $\mathcal{S}$. Therefore, $\mathcal{D}$ must satisfy at least one of these literals.

– Suppose that the rule applied to construct T' is the superposition rule S1. Suppose p_1 and p_2 are the literals in B that are used for the extension. By inductive hypothesis we know that $\mathcal{D} \models p_1$ and $\mathcal{D} \models p_2$. It means that there are two monomorphisms $h_1 : p_1 \to \mathcal{D}$ and $h_2 : p_2 \to \mathcal{D}$. By Proposition 3, there exists some $s \in p_1 \otimes p_2$ verifying the *pair factorization property* with $h : s \to \mathcal{D}$ being a monomorphism, so, $\mathcal{D} \models s$.

– Suppose that the rule applied to construct T' is the superposition rule S2. Suppose p_1 and $\forall(c : p_2 \to q)$ are the literals in B that are used for the extension. By inductive hypothesis we know that $\mathcal{D} \models p_1$ and $\mathcal{D} \models \forall(c : p_2 \to q)$. Since $\mathcal{D} \models p_1$, there exists a monomorphism $h_1 : p_1 \to \mathcal{D}$. Then $h_1 \circ m$ is also a monomorphism from p_2 to $\mathcal{D}$. From here, since $\mathcal{D} \models \forall(c : p_2 \to q)$, there is a monomorphism $h_2 : q \to \mathcal{D}$ such that $h_1 \circ m = h_2 \circ c$. By Proposition 3, there exists some $s \in p_1 \otimes_{c,m} q$ verifying the *pair factorization property* with $h : s \to \mathcal{D}$ being a monomorphism, so, $\mathcal{D} \models s$. Graphically:

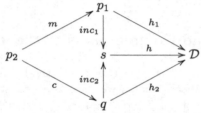

– Suppose that the rule applied to construct T' is one of the unfolding rules. Suppose $p[n_1//hang(n_2)]$ is the literal in B that is used and that (U1) is the rule used for the extension. By inductive hypothesis we know that $\mathcal{D} \models p$. It means that there is a monomorphism $h_1 : p \to \mathcal{D}$, that is, it holds $h_1(n_1)//h_1(n_2)$. Then, accordingly to Condition 5 in Definition 1 we have three cases:

1. If $h_1(n_1)/h_1(n_2)$ holds, it is clear that there exists a monomorphism $h_2 : p[n_1 \leftarrow /hang(n_2)] \to \mathcal{D}$, so we have that $\mathcal{D} \models p[n_1 \leftarrow /hang(n_2)]$.

2. If $h_1(n_1)/m_1//m_2/h_1(n_2)$ holds, for some nodes m_1 and m_2 in $\mathcal{D}$, then the following f_1 and f_2 are monomorphisms:
   - $f_1 : p[n_1 \leftarrow] \to \mathcal{D}$ such that $f_1(p[n_1 \leftarrow]) = h_1(p[n_1 \leftarrow])$,
   - $f_2 : q \to \mathcal{D}$ for $q = prefix(n_1)[n_1 \leftarrow /n]$ with $label(n) = *$, such that $f_2(prefix(n_1)) = h_1(prefix(n_1))$ and $f_2(n) = m_1$.

   Then, by Proposition 3, there exists some $s \in p[n_1 \leftarrow] \otimes_{c,m} q$ verifying the *pair factorization property* with $f : s \to \mathcal{D}$ being a monomorphism.

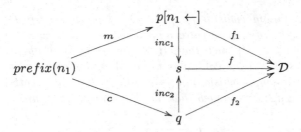

Therefore, $\mathcal{D} \models s[inc_2(n) \leftarrow //hang(n_2)]$ since we can define a monomorphism $h_2 : s[inc_2(n) \leftarrow //hang(n_2)] \rightarrow \mathcal{D}$ such that $h_2 = h_1$ except for:
- $h_2(inc_2(n)) = f(inc_2(n)) = f_2(n) = m_1$

3. Otherwise, $h_1(n_1)/m'/h_1(n_2)$ holds and it is enough to consider morphisms such that:
- $f_2(n) = m'$,
- $h_2(inc_2(n)) = f(inc_2(n)) = f_2(n) = m'$

Similar arguments serve if the rule used is (U2).

Consequently, in all these cases, there exists an extended branch B' in T' such that $\mathcal{D} \models B'$ and therefore $\mathcal{D} \models T'$. ∎

In order to prove completeness, the following notion of saturation of tableaux is required. Saturation describes some kind of fairness that ensures that we do not postpone indefinitely some inference step.

Definition 13 (Saturated Tableau). *Given a tableau T for a specification $\mathcal{S}$, we say that T is saturated if the following conditions hold:*

- *No new literals can be added to any branch B in T using the $\vee$-rule.*
- *For each branch B in T, one of the following conditions is satisfied:*
  - *either it is closed, or*
  - *it is open and all rules have been applied in B.*

It should be clear that it is always possible to build a (possibly infinite) saturated tableau. It is enough to keep, for every branch, a queue of the pending inferences.

To prove completeness we will show that we can associate a canonical model $\mathcal{D}_B$ to any open branch in a given tableau T so that, if T is saturated then $\mathcal{D}_B$ can be proven to be a model for T. In particular, this model is obtained by, first, computing r_B, which is the colimit of the diagram consisting of the patterns in the positive literals in the branch and the monomorphisms induced by rule applications. And, second, by replacing in r_B every $*$ label by a fresh label from Σ that is not present in any literal in the given specification. The existence of these colimits (satisfying an additional minimality property) is described in the Definition 14 and Proposition 4.

Definition 14 (Infinite colimits). *We say that P be a (possibly infinite) directed diagram of patterns, if it is a collection of patterns and monomorphisms between the patterns, such that for every pair of patterns p_1 and p_2 there exists a* *pattern r and monomorphisms $f_1 : p_1 \to r$ and $f_2 : p_2 \to r$ in P. We say that P has a* colimit *if there exists a pattern r_P together with a collection of morphisms $\{h_p : p \to r_P \mid p \in P\}$ such that, if $f : p_1 \to p_2$ in P then $h_{p_1} = h_{p_2} \circ f$. Moreover, we say that the colimit is* minimal, *if for every finite pattern q such that there is a monomorphism $g : q \to r_P$, then, there is a pattern p in P and a monomorphism $g_p : q \to p$ such that the diagram below commutes:*

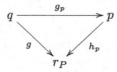

Now, we show that every open branch defines a directed diagram, so that, if the tableau is saturated, it has a minimal colimit.

Proposition 4 (Colimit of open branches in saturated tableaux). *Given an open branch B in a saturated tableau T then, the set of patterns in positive literals in B is a directed diagram P_B that has a minimal colimit r_B.*

Proof. First of all, we define the diagram P_B associated to a branch B as follows:

– If p is a positive literal in a node of B then p is in P_B.
– If s, p_1 and p_2 are literals on nodes of B such that s is one of the child literals obtained from p_1 and p_2 after applying rule (S1), then the corresponding monomorphisms $inc_1 : p_1 \to s$ and $inc_2 : p_2 \to s$ are in P_B.
– If s, p_1 and $\forall(c : p_2 \to q)$ are literals on nodes of B, such that s is one of the child literals obtained from p_1 and $\forall(c : p_2 \to q)$ after applying rule (S2), then the corresponding monomorphism $inc_1 : p_1 \to s$ is in P_B.
– If s and p are literals on nodes of B such that s is one of the child literals obtained from p after applying rule (U1) or (U2), then the associated monomorphism $f : p \to s$ is in P_B.

Then, since T is saturated, if p and q are literals in B, the branch includes the application of the corresponding superposition rule (S1) to these literals. Moreover, since B is open, the superposition $p \otimes q$ is not empty and defines the morphisms $inc_1 : p_1 \to s$ and $inc_2 : p_2 \to s$, so P_B is a directed diagram.

For the colimit construction, it is enough to define r_B as the quotient of the union of the patterns in the diagram modulo the equivalence relation defined by the morphisms of the diagram[1]. For the minimality property, if q is a finite pattern such that $g : q \to r_P$, then there should exist a finite subset of patterns $P_0 \subseteq P_B$ such that the union of the nodes of the patterns in P_0 includes the set of nodes $g(n)$, where n is in $Nodes_q$. Then, using that P_B is directed we can prove the existence of a pattern p in P_B, obtained by doing superpositions on patterns in P_0, such that there is a monomorphism $g_p : q \to p$ verifying $g = h_p \circ g_p$.   ∎

---

[1] The least equivalence relation satisfying that if f is a morphism in the diagram and $f(n) = n'$, then $n \equiv n'$.

Lemma 1 (Canonical models of saturated tableaux). *If B is an open branch of a saturated tableau T, r_B is its colimit, and $\mathcal{D}_B$ is the result of replacing in r_B each $*$ by a label a not present in the given specification $\mathcal{S}$, then $\mathcal{D}_B$ is a document such that $\mathcal{D}_B \models B$ and, hence, $\mathcal{D}_B \models \mathcal{S}$.*

Proof. First, we have to prove that $\mathcal{D}_B$ is indeed a document. This means proving that $\mathcal{D}_B$ satisfies Condition 5 in Definition 1. Let n_1, n_2 be two nodes in $\mathcal{D}_B$ (and, hence, in r_B) such that $n_1//n_2$ holds. By Proposition 4, there must exist a pattern p in P_B containing such nodes n_1 and n_2 such that $n_1//n_2$ holds in p^2. There are several possibilities:

1. If n_1/n_2 is in p then n_1/n_2 is in $\mathcal{D}_B$.
2. If $n_1//^d n_2$ is in p, since the tableau is saturated, at some point in the branch B we would have applied the unfolding rule (U1) to the literal p. As a consequence, B would also include the literal p_1 where either $p_1 = p[n_1 \leftarrow /hang(n_2)]$ or $p_1 = s[inc_2(n) \leftarrow //hang(n_2)]$ for one of the patterns s in $p[n_1 \leftarrow] \otimes_{c,m} q$, where $q = prefix(n_1)[n_1 \leftarrow /n]$ with $label(n) = *$. In the former case, we would know that in $\mathcal{D}_B$ we have n_1/n_2. In the latter case, B has the literal p_1 containing a node m_1 such that $n_1/m_1//^d n_2$ is in p_1. But because the tableau is saturated, at some later point in the branch B we would have applied the unfolding rule (U2) to the literal p_1 on the edge $m_1//^d n_2$. As a consequence, B would also include the literal p_2 where either $p_2 = p_1[m_1 \leftarrow /hang(n_2)]$ or $p_2 = s'[inc_2(n') \leftarrow /hang(n_2)]$ for one of the patterns s' in $p_1[m_1 \leftarrow] \otimes_{c,m} q'$, where $q' = prefix(m_1)[m_1 \leftarrow //n']$ with $label(n') = *$. Now, in the former case, we would know that in $\mathcal{D}_B$ we have $n_1/m_1/n_2$. In the latter case, B has the literal p_2 containing a node m_2 such that $n_1/m_1//^d m_2/n_2$ is in p_2. Therefore we know that in $\mathcal{D}_B$ we have two nodes m_1 and m_2 such that $n_1/m_1//m_2/n_2$ holds.
3. Otherwise, there must be at least two edges between the nodes n_1 and n_2 in p. That is, there must be nodes m_1, m_2 in p with n_1/m_1 or $n_1//^d m_1$, and m_2/n_2 or $m_2//^d n_2$, such that $m_1 = m_2$ or $m_1//m_2$ holds in p. Now:
   (a) If n_1/m_1 and m_2/n_2 then trivially $n_1/m_1//m_2/n_2$ holds in $\mathcal{D}_B$.
   (b) If $n_1//^d m_1$ or $m_2//^d n_2$, then at some point we would apply the first or the second unfolding rule and, as in case 2, we would also prove that n_1, n_2 satisfy Condition 5 in Definition 1.

Now, we prove that $\mathcal{D}_B$ satisfies each literal ℓ in B. Let $rename_B : r_B \to \mathcal{D}_B$ be the isomorphism that renames all the labels $*$ in $Nodes_{r_B}$ by a label a. Then, we have that $\mathcal{D}_B \models \ell$ if, and only if, $r_B \models \ell$ because, for every pattern p, the existence of a monomorphism $h : p \to \mathcal{D}_B$ implies the existence of a monomorphism $rename_B^{-1} \circ h : p \to r_B$; and, conversely, the existence of a monomorphism $f : p \to r_B$ implies the existence of a monomorphism $rename_B \circ f : p \to \mathcal{D}_B$. Therefore, it will be enough to prove that r_B satisfies each literal ℓ in B. We proceed by cases:

---

[2] To be more precise there are nodes m_1, m_2, such that $m_1//m_2$ holds in p, $h_p(m_1) = n_1$ and $h_p(m_2) = n_2$.

- If $\ell = p$ then, as a direct consequence of the colimit construction, we know that there exists a monomorphism $h_p : p \to r_B$, so $r_B \models p$.
- Assume that $\ell = \forall(c : p \to q)$. We will prove that, if there exists $h : p \to r_B$ then there is a monomorphism $f : q \to r_B$ such that $h = f \circ c$. First, we know as a consequence of the colimit construction that since p is finite, there is a pattern $r \in P_B$ (with the corresponding monomorphism $h_r : r \to r_B$) and a monomorphism $g_r : p \to r$, such that the following diagram commutes:

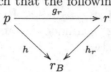

Moreover, since T is saturated, we know that the superposing rule (S2) has been applied between $\forall(c : p \to q)$ and r. Suppose that $s \in r \otimes_{c,g_r} q$ is the one such that s is the literal that was hanged in the branch B. Then, by colimit definition, we know there is a monomorphism $h_s : s \to r_B$ and the following diagram defines the required monomorphism $f = h_s \circ inc_2 : q \to r_B$ such that $h = h_r \circ g_r = f \circ c$.

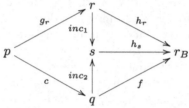

- Let $\ell = \neg p$ and let us see that assuming that there exists $h : p \to r_B$ leads to a contradiction. If it was the case, we know that, since p is finite, there is a pattern $r \in P_B$ (with $h_r : r \to r_B$) and a monomorphism $g_r : p \to r$, such that the following diagram commutes:

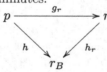

Then, since T is saturated the closing rule (C) must be applied between $\neg p$ and r so the branch should be closed, contradicting the premise.   ∎

Finally, we prove completeness of the tableau method.

Theorem 2 (Completeness). *If the specification S is unsatisfiable, then there is a tableau proof for S.*

Proof. If there is no tableau proof for S, then every tableau for S has an open branch. Hence, if T is a saturated tableau for S, it should have an open branch B and, by Lemma 1, $\mathcal{D}_B \models S$.   ∎

6   Related Work

XPath [17,21] is a well-known language for navigating an XML document (or XML tree) and returning a set of answer nodes. Since XPath is used in many

XML query languages as XQuery, XSLT or XML Schema among others [18–20], a great amount of papers deal with different aspects on different fragments of XPath. For instance, in [4] an overview of formal results on XPath is presented concerning the expressiveness of several fragments, complexity bounds for evaluation of XPath queries, as well as static analysis of XPath queries. In [3] they study the problem of determining, given a query p (in a given XPath fragment) and a DTD D, whether there exists an XML document conforming to D and satisfying p. They show that the complexity ranges from PTIME to undecidable, depending on the XPath fragment and the DTD chosen. The work presented in [5] deals with the same problem (in a particular case) and it uses Hybrid Modal Logic to model the documents and some class of schemas and constraints. They provide a tableau proof technique for constraint satisfiability testing in the presence of schemas.

Our approach is different than the previous ones in two aspects. On the one hand, we do not consider any DTD or schema, and we use a simple fragment of XPath. In this sense our approach is simpler than previous ones. But, on the other hand, our aim is to define specifications of classes of XML documents as sets of constraints on these documents, and to provide a form of reasoning about these specifications. In this sense, our main question is satisfiability, that is, given a set of constraints S, whether there exists an XML document satisfying all constraints in S.

Some other work, which shares part of our aims, is the approach for the specification and verification of semi-structured documents based on extending a fragment of first-order logic [2,12]. They present specification languages that allow us to specify classes of documents, and tools that allow us to check whether a given document (or a set of documents) follows a given specification. However, they do not consider the problem of defining deductive tools to analyze specifications, for instance to look for inconsistencies. Schematron [9] has a more practical nature. It is a language and a tool that is part of an ISO standard (DSDL: Document Schema Description Languages). The language allows us to specify constraints on XML documents by describing directly XML patterns (using XML) and expressing properties about these patterns. Then, the tool allows us to check if a given XML document satisfies these constraints. However, as in the previous approach, Schematron provides no deductive capabilities. Finally, the approach presented in this paper is very related to the work presented in [15,16], showing how to use graph constraints as a specification formalism, and how to reason about these specifications. However, as discussed in Sect. 3, the descendant relation in our constraints makes non-trivial the application of the techniques in [15,16]. In particular, the descendent relation would be second-order in the logic of graph constraints defined in [15,16].

In [1,13] we presented some preliminary work directly related to the work in this paper. In particular, in [13], we introduced the three kinds of constraints considered here and three main inference rules called R1, R2 and R3 (similar to the rules Cl, S1 and S2 in this paper). We proved that these rules were sound, but some counter-examples showed that they were not complete. So, we introduced two new rules, called Unfold1 and Unfold2 (a preliminary version of U1 and U2), that solved these counter-examples, so we conjectured that this was enough to prove completeness. Unfortunately, Unfold1 and Unfold2 are unsound as we explain in Example 2. Moreover, we presented some subsumption and simplification rules in order to produce a more efficient procedure. In parallel, we implemented a prototype tool for reasoning with these rules that is described in [1] by means of examples and screenshots.

7   Conclusion and Further Work

In this paper, we have presented an approach for specifying the structure of XML documents using three kinds of constraints based on XPath, together with a sound and complete method for reasoning about them.

We strongly believe that satisfiability problem for this class of constraints is only semidecidable, since we believe that it would be similar to the (un)decidability of the satisfiability problem for the Horn clause fragment of first-order logic. As a consequence, if a given specification is inconsistent, we can be sure that our procedure will terminate showing that unsatisfiability. However, our procedure may not terminate if the given specification is satisfiable. In this context, we may consider that studying the complexity of a procedure that may not terminate is not very useful. Nevertheless, we may like to have an idea about the performance of our approach when the procedure terminates. One could think, that this performance would be quite poor, since checking if there is a monomorphism between two trees (a basic operation in our deduction procedure) is an NP-complete problem [10]. Actually, this is not our experience with the tool that we have implemented [1]. We think that the situation is similar to what happens with graph transformation tools. In these tools, applying a graph transformation rule means finding a subgraph isomorphism, which is also a well-known NP-complete problem. However, the fact that the graphs are typed (in our case, the trees are labelled), in practice, reduces considerably the search.

In the future, we plan to extend our approach to consider also cross-references and properties about the contents of documents. The former problem means, in fact, to extend our approach to graphs and graph patterns. For the latter case, we plan to follow the same approach that we used to extend our results for graphs in [15,16] to the case of attributed graphs in [14].

References

1. Albors, J., Navarro, M.: SpecSatisfiabilityTool: a tool for testing the satisfiability of specifications on XML documents. In: Proceedings of PROLE 2014, EPTCS, vol. 173, pp. 27–40 (2015)
2. Alpuente, M., Ballis, D., Falaschi, M.: Automated verification of web sites using partial rewriting. Softw. Tools Technol. Transf. **8**, 565–585 (2006)
3. Benedikt, M., Fan, W., Geerts, F.: XPath satisfiability in the presence of DTDs. JACM **55**, 2 (2008)

4. Benedikt, M., Koch, C.: XPath leashed. ACM Comput. Surv. **41**, 1 (2008)
5. Bidoit, N., Colazzo D.: Testing XML constraint satisfiability. In: Proceedings of the International Workshop on Hybrid Logic (HyLo 2006), ENTCS, vol. 174(6), pp. 45–61 (2007)
6. Habel, A., Pennemann, K.H.: Correctness of high-level transformation systems relative to nested conditions. Math. Struct. Comput. Sci. **19**(2), 245–296 (2009)
7. Habel A., Radke H.: Expressiveness of graph conditions with variables. In: International Colloquium on Graph and Model Transformation GraMoT 2010, ECEASST, vol. 30 (2010)
8. Hähnle, R.: Tableaux and related methods. In: Robinson, J.A., Voronkov, A. (eds.) Handbook of Automated Reasoning, pp. 100–178. Elsevier, Amsterdam (2001)
9. Jelliffe, R.: Schematron, Internet Document. http://xml.ascc.net/resource/schematron/
10. Kilpelainen, P., Mannila, H.: Ordered and unordered tree inclusion. SIAM J. Comput. Arch. **24**(2), 340–356 (1995)
11. Miklau, G., Suciu, D.: Containment and equivalence for a fragment of XPath. JACM **51**(1), 2–45 (2004)
12. Nentwich, C., Emmerich, W., Finkelstein, A., Ellmer, E.: Flexible consistency checking. ACM Trans. Softw. Eng. Methodol. **12**(1), 28–63 (2003)
13. Navarro, M., Orejas, F.: A refutation procedure for proving satisfiability of constraint specifications on XML documents. In: SCSS 2014, EasyChair EPiC series, vol. 30, pp. 47–61 (2014)
14. Orejas, F.: Symbolic graphs for attributed graph constraints. J. Symb. Comput. **46**(3), 294–315 (2011)
15. Orejas, F., Ehrig, H., Prange, U.: A logic of graph constraints. In: Fiadeiro, J.L., Inverardi, P. (eds.) FASE 2008. LNCS, vol. 4961, pp. 179–198. Springer, Heidelberg (2008)
16. Orejas, F., Ehrig, H., Prange, U.: Reasoning with graph constraints. Formal Asp. Comput. **22**(3–4), 385–422 (2010)
17. World WIDE WEB CONSORTIUM: XML path language (XPath) recommendation (1999). http://www.w3c.org/TR/XPath/
18. World WIDE WEB CONSORTIUM: XSL transformations (XSLT). W3C recommendation version 1.0 (1999). http://www.w3.org/TR/xslt
19. World WIDE WEB CONSORTIUM: XML schema part 0: Primer. W3C recommendation (2001). http://www.w3c.org/XML/Schema
20. World WIDE WEB CONSORTIUM: XQuery 1.0 and XPath 2.0 formal semantics. W3C working draft (2002). http://www.w3.org/TR/query-algebra/
21. WORLD WIDE WEB CONSORTIUM: XML path language (XPath) 2.0 (2007)

Algebraic Reinforcement Learning
Hypothesis Induction for Relational Reinforcement Learning Using Term Generalization

Stefanie Neubert, Lenz Belzner$^{(\boxtimes)}$, and Martin Wirsing

Ludwig-Maximilians-Universität München, Munich, Germany
belzner@pst.ifi.lmu.de

Dedicated to José Meseguer.

Abstract. The TG relational reinforcement learning algorithm builds
first-order decision trees from perception samples. To this end, it statistically checks the significance of hypotheses about state properties possibly
relevant for decision making. The generation of hypotheses is restricted
by constraints manually specified a priori. In this paper we propose Algebraic Reinforcement Learning (ARL) for eliminating this condition by
employing rewrite theories for state representation, enabling induction
of hypotheses from perception samples directly via term generalization
with the ACUOS system. We compare experimental results for ARL
with and without generalization, and show that generalization positively
influences convergence rates and reduces complexity of learned trees in
comparison to trees learned without generalization.

1 Introduction

A reinforcement learning agent performs actions, observes the effect and eventually perceives a reward when reaching a system goal. Based on this information, reinforcement learning allows the agent to build a data structure that
enables lookup of optimal actions with respect to an agent's situation, probabilistic action effects and system goals [1]. Classical reinforcement learning algorithms build this structure based on propositional state representations, thereby
ignoring any relational or logical structure underlying the domain. This results
in large and unmanageable data structures, and learned results are not easily
transferable to similar situations.

To overcome these problems, relational reinforcement learning (RRL) has been
proposed [2–4]. RRL seeks to incorporate relational domain structure into the
learning process, yielding more compact and transferable results. In particular,
the TG algorithm [5,6] builds first order regression trees as lookup data structure
(i.e. a relational decision tree) by testing whether a state property represented
as first order logic formula has a statistically significant impact on the quality of
executing an action in a given state. This process is repeated recursively, revealing

---

This work has been partially funded by the EU project ASCENS, 257414.

N. Martí-Oliet et al. (Eds.): Meseguer Festschrift, LNCS 9200, pp. 562–579, 2015.
DOI: 10.1007/978-3-319-23165-5_26

relational domain structure that is relevant for decision making. One downside of the TG algorithm is that the state properties to be tested for relevancy have to be *manually* specified by system designers *a priori*. This so called *language bias* constrains the structure of the resulting decision tree. Moreover, this approach requires domain knowledge before starting the learning process and may lead to inadequate tree structure.

We propose Algebraic Reinforcement Learning (ARL) as a novel alternative approach for inducing state properties to be tested for relevancy: Representation of states as algebraic terms allows us to use term generalization to *automatically* construct relational hypotheses about possibly relevant state properties *at runtime*. For generalization we use the ACUOS system [7,8] that allows to generalize rewriting logic terms [9,10]. Ax-matching is used to decide whether such a test holds in a particular state, thus providing data for statistical evaluation of test significance. This approach alleviates specification requirements for TG learning by dropping the need for a predefined language bias. We show experimentally that the benefits of relational over propositional representation as faster learning convergence rates and smaller lookup data structures are maintained.

The paper is structured as follows. Section 2 reviews the TG algorithm and the ACUOS generalization framework. Section 3 introduces our approach of Algebraic Reinforcement Learning, and in particular the integration of term generalization into the TG algorithm to allow for automated hypothesis generation. In Sect. 4 we present experimental results, comparing algorithm performance with and without ACUOS generalization. Section 5 discusses related work, and Sect. 6 concludes this paper and gives a brief outlook on further research directions in the field.

Personal Note. The authors of this paper are "addicts" of rewriting logic and Maude: they like to apply rewriting logic in their research and show that the unique combination of equational reasoning, rewriting and logic leads to new scientific results and insights. E.g. Lenz has used Maude for reasoning about runtime decisions of autonomic components [11,12] and for symbolically solving relational Markov decision processes with rewriting techniques [13]; Martin has applied rewriting logic to give formal foundations to object-oriented software development [14], to study an algebraic approach to soft constraints [15], and Lenz and Martin together to reason about autonomic ensembles [16]. Martin had the opportunity to work with José on the foundations of multi-modeling languages [17], on languages for distributed systems [18] and on cloud computing security [19].

We admire José not only for his creativity, his deep insights, and his broad knowledge in mathematics and computer science but also because of his warm-hearted and kind personality. We are looking forward to many further stimulating exchanges with José.

2   Preliminaries

This section reviews the TG relational reinforcement learning algorithm and the ACUOS framework for term generalization. For a broader view on the relational reinforcement framework the reader is referred to e.g. [2–4]. For in-depth information about term rewriting and rewriting logic, we refer to e.g. [9,10].

2.1   TG Learning

We consider the TG relational reinforcement learning algorithm [5,6]. We start with some definitions.

Definition 1. *Let Φ be the set of conjunctions of first-order predicates with implicit universal quantification. A state formula is a conjunction of first-order predicates $s \in \Phi$.*

Definition 2. *Let $\mathcal{A}$ be the set of possible actions (of an agent), let $\mathcal{R} \subset \mathbb{R}$ be the set of possible rewards, and let $\Xi \subset \Phi \times \mathcal{A} \times \mathcal{R}$ be the set of perception samples. A perception sample $\xi \in \Xi$ is a triple consisting of an observed state, an executed action and an observed reward.*

Notation: We denote the state formula of a perception sample $\xi \in \Xi$ by $state(\xi)$, the action by $action(\xi)$ and the reward by $reward(\xi)$.

Definition 3. *A sample $\xi \in \Xi$ satisfies a state formula $s \in \Phi$ if s is a consequence of $state(\xi)$; i.e. ξ satisfies $s \Leftrightarrow state(\xi) \to s$.*[1]

Definition 4. *A split test is a state formula.*

Definition 5. *A split test hypothesis consists of a state formula $s \in \Phi$ and a set of statistics consisting of the following elements.*

- *A number $n_+ \in \mathbb{N}$, the sum of q-values $\sum_0^{n+} q_i$ and the sum of squared q-values $\sum_0^{n+} q_i^2$ of samples satisfying s.*
- *A number $n_- \in \mathbb{N}$, the sum of q-values $\sum_0^{n-} q_i$ and the sum of squared q-values $\sum_0^{n-} q_i^2$ of samples not satisfying s.*

With these definitions we are able to define first order regression trees. This is the data structure build incrementally by the TG algorithm to encode a reinforcement learning q-function. It can be used by an agent to determine the quality of executing an action in a particular state according to system goal specification. See Fig. 1 for an example of a first order regression tree.

Notation: Variables are denoted by capital letters and constants by small letters.

Definition 6. *A first order regression tree is a tree with the following properties.*

- *Every node has either exactly two children or is a leaf.*
- *Each non-leaf node contains a split test.*
- *Each leaf node contains a q-value and a set of split test hypotheses.*

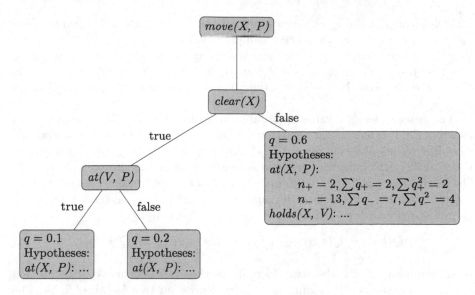

Fig. 1. A first order regression tree. A sample for action $move(x,p)$ with state $at(x,p) \wedge$ $clear(x) \wedge at(v,p)$ (and arbitrary reward) is sorted into the leaf on the left: It passes the tests $clear(X)$ and $at(V,P)$. The q-value of executing this action in the particular state is 0.1. In other words, the leaf on the bottom left characterizes all states that satisfy $clear(X) \wedge at(V,P)$.

The learning agent is provided with the set of actions $\mathcal{A}$ it can perform. Environmental perceptions are represented as first order logic conjunctions. The agent is able to observe any reward gained from action execution. The TG algorithm is shown in Algorithm 1. It is initialized with an empty first order regression tree for each action. The root node of each tree consists of a q-value of zero and a set of tests with empty statistics created according to a *language bias* specified manually (lines 1–2 of Algorithm 1).

The language bias is given in form of *rmode* constraints [20]. These define conjunctions of predicates that may appear as split tests, as well as how often a split test may appear in a path from root to a leaf in a tree to be learned. For any variable occurring in an *rmode* constraint, it is also defined whether it corresponds to a variable already bound in tests further up the tree, or whether they are unbound. For example, the constraint $rmode(3, f(X+, Y-))$ states that the specified test $f(X,Y)$ may occur three times in a path from the node to a leaf. The variable X is bound to a variable already occurring in a test further up the tree; variable Y is unbound (i.e. fresh). Thus, the language bias can be regarded as a function from a given first-order regression tree to a set of split test candidates.

---

[1] In the original TG algorithm implementation, entailment is decided by Prolog's SLD-resolution.

When learning, the agent performs an action in a particular state and observes any gained reward. The perception sample is sorted into the tree of the corresponding action from the root to a leaf according to the tests in the nodes (lines 4–5 of Algorithm 1). A sample satisfying a split test in a node gets subsequently sorted into the true child branch; otherwise it is inserted into the false branch (see Fig. 1).

The reached leaf's q-value is updated according to standard q-learning [1], shown in Eq. 1. It states that an action's q-value in a particular state s is updated according to any observed reward $R(s)$ and the currently best value that can be achieved by performing any possible action $a' \in \mathcal{A}$ in the subsequent state s'. The learning rate $\alpha \in [0; 1]$ determines the impact of new information on already learned structure, and γ is a discount factor defining an agent's preferences towards long-term expected rewards over immediate ones.

$$Q(s,a) \leftarrow Q(s,a) + \alpha[R(s) + \gamma \max_{a' \in \mathcal{A}} Q(s',a') - Q(s,a)] \qquad (1)$$

To determine $Q(s',a')$, the state formula of the state reached by executing a is sorted into the corresponding first order regression tree for all $a' \in \mathcal{A}$. The q-value found in the reached leaf is the actions' current q-value (see Fig. 1). The same mechanism can be used by an agent to determine an optimal action.

For each hypothesis stored in the reached leaf it is tested whether the sample satisfies the hypothesis' state formula. The results are stored in the hypothesis' statistics (lines 6–8 of Algorithm 1). If the updated statistics imply a significant impact of the hypothesis on the expected average q-value of executing the sampled action in a state characterized by the leaf[2], the leaf is replaced by a node containing the significant test. Two new child leaves are attached to the node: The leaf at the true-branch contains the average q-value of the positive samples of the significant hypothesis' statistics, the other leaf at the false-branch the average q-value of the negative samples. For both leaves a new hypothesis set is created according to the language bias. To avoid split decisions without sufficient data, a minimum number of gathered samples (called *minimal sample size*) has to be exceeded before a split is performed (lines 9–17 of Algorithm 1).

The TG algorithm converges towards an optimal q-function yielding agent behavior that maximizes gathered reward. It converges faster than propositional approaches that do not exploit relational domain structure in their representations of the q-function [6].

2.2   ACUOS

The ACUOS system [7,8] allows for term generalization for order-sorted terms from a rewrite theory $(\Sigma, E \cup A)$. Σ contains sorts, sort order and operation symbols to allow for term construction; $E \cup A$ define equivalence classes for terms through specification of directed equations and axioms for operators (e.g. associativity, commutativity, idempotency, identity).

---

[2] This is done with a standard statistical F-Test for variance comparison using the statistics stored with the hypothesis, for details see [6].

input : state and action representation signatures Φ and $\mathcal{A}$
   list *Samples* of samples $\Phi \times \mathcal{A} \times \mathcal{R}$
   integer minimal sample size mss
   language bias LB
output: a first-order regression tree t_a for each action $a \in \mathcal{A}$

1 initialize each t_a (for each $a \in \mathcal{A}$) with a single empty leaf l_{init}^a
2 for each t_α add $LB(t_a)$ to hypothesis set of l_{init}^a

3 **foreach** *sample* $\xi \in$ *Samples* **do**

4   sort ξ into $t_{action(\xi)}$ according to inner node tests, reaching leaf l
5   update q-value in leaf l according to *reward*(ξ)

6   **foreach** *split test hypothesis* h *in* l **do**
7 | update statistics for h
8   **end**

9   **if** *number of samples sorted to l exceeds mss* **then**
10   **if** *f-Test indicates a split for a hypothesis h* **then**
11   create new inner node l_{new} with h as split test
12   add two empty leaves as children to l_{new}
13   set q-value of children according to h
14   add $LB(t_a)$ to hypothesis set of each child of l_{new}
15   replace current leaf l with l_{new}
16   **end**
17   **end**

18 **end**

Algorithm 1. The TG algorithm [5,6].

Definition 7. *A term t is a generalization of terms u and u' if they both are substitution instances of t.*

Example 1. The term $f(X)$ is a generalization of the terms $f(a)$ and $f(b)$ because there are substitutions $\theta = \{X \mapsto a\}$ and $\theta' = \{X \mapsto b\}$ for which $\theta(f(X)) = f(a)$ and $\theta'(f(X)) = f(b)$.

ACUOS is able to find least general generalizations for given expressions. For example, the term $at(X, P)$ is a generalization of $at(x, p_0)$ and $at(x, p_1)$. The *least general* generalization is in fact $at(x, P)$, meaning there is no other generalization of the given terms that is less general. ACUOS generalizes terms from a given rewrite theory taking into account specified sort orders and equivalence classes. In contrast to unsorted generalization algorithms, ACUOS generates a minimal set of least general generalizers. The algorithm has been implemented in the Maude language [9] that allows to execute formally specified rewrite theories.

Example 2. Let x be of sort *Agent*, v of sort *Victim*, p_0, p_1 of sort *Position* and let sorts *Agent* and *Victim* be a subsort of sort *Spatial*. Let $\wedge$ be an associative and commutative operation. Consider the following two terms.

$$at(v, p_0) \wedge at(x, p_1) \wedge clear(x)$$

$$at(v, p_1) \land clear(x) \land at(x, p_0)$$

Generalizing these two terms with the ACUOS system yields a set of least general generalizations containing the following terms.

$$at(S{:}Spatial, p_0) \land at(S'{:}Spatial, p_1) \land clear(x)$$
$$at(v, P{:}Position) \land at(x, P'{:}Position) \land clear(x)$$

Note the induction of sort *Spatial* due to the given sort order. Also note the impact of associativity and commutativity of the $\land$ operation.

3   Algebraic Reinforcement Learning

This section presents Algebraic Reinforcement Learning, an approach which integrates rewriting logic and algebraic state representations with the TG algorithm. This enables automated induction of split test hypotheses from perception samples by term generalization, eliminating the need for the manual specification of a language bias for split test construction.

3.1   Algebraic Regression Trees

We change the state representation of first order regression trees (see Sect. 2.1) to use rewriting logic terms for state representation. Definition 1 is changed to allow for state representation as rewriting logic terms.

Definition 8. *Given a rewrite theory* $(\Sigma, E \cup A)$ *with a sort State* $\in \Sigma$, *a state formula* $s \in \Phi$ *is a term of sort State.*

Also, we adjust the satisfaction relation of perception samples and states (Definition 3) to employ Ax-matching to decide whether a state satisfies a certain property. With regards to the TG algorithm, first order logic entailment is replaced by equational membership logic matching modulo axioms with extension (denoted $\leq_{Ax,\theta}$ for a matching substitution θ, see e.g. [9] for details) to decide whether a property holds in a state or not.

Definition 9. *A sample* $\xi \in \Xi$ *satisfies a state formula* $s \in \Phi$ *iff there exists a matching substitution* θ *of variables (modulo axioms with extension) for their corresponding state terms; i.e.* ξ *satisfies* $s \Leftrightarrow \exists \theta : s \leq_{Ax,\theta} state(\xi)$.

These changes allow to define algebraic regression trees, which are similar to first order regression trees (see Definition 6) but use rewriting logic terms for state representation in nodes and tests.

3.2   Example Domain Specification

As an example of algebraic state representation consider a search and rescue scenario. An agent has to find, pick up, transport and deploy victims of an incident to an ambulance. Agent, ambulance and victims are located at arbitrary positions. The agent can perform actions for moving to a particular position,

```
1    fmod STATE is
2
3        *** State term structure.
4        sort State .
5        op _and_ : State State -> State [assoc comm] .
6
7        *** Domain entities.
8        sort Position Agent Victim Ambulance Spatial .
9        subsort Agent Victim Ambulance < Spatial .
10
11       *** Entity relations.
12       op at : Spatial Position -> State .
13       op clear : Agent -> State .
14       op holds : Agent Victim -> State .
15
16       *** Actions.
17       sort Action .
18       op move : Agent Position -> Action .
19       op grab : Agent Victim -> Action .
20       op drop : Agent Victim -> Action .
21
22   endfm
```

Listing 1.1. A Maude module for state and action representation.

picking up a victim (if the agent is at the same position not already carrying another victim), and deploying a victim at the agent's current position.

The encoding of this domain as a Maude module[3] is shown in Listing 1.1. Domain objects (agents, positions, victims and ambulances) are encoded as sorted variables; relations between objects are defined as operations parametrized by domain objects. Note the polymorphic specification of the relation *at*, making use of the specified sort order. Test hypotheses induced by term generalization in the learning process respect this sort order as well as associativity and commutativity of state terms.

3.3   Hypothesis Induction with ACUOS

With the definitions from Sect. 3.1, it is possible to extend the TG algorithm to employ term generalization to induce split tests (i.e. hypotheses about relevant state properties) for the tree leaves from perception samples gathered in the learning process. This effectively eliminates the requirement of the original TG algorithm to provide a language bias for test construction manually before runtime. We will discuss the extraction of split test candidates from perception

---

[3] Note that no domain dynamics (e.g. rewrite rules) are encoded, as these are implicitly learned by the ARL algorithm w.r.t. observed rewards.

samples and then illustrate the generalization of these candidates. Finally, we discuss the use of multi-predicate feature lookahead for finding relevant split criteria consisting of multiple state features.

Feature Extraction. Each sample may contain features (i.e. state properties) that have impact on the value of an action, as well as features that are not relevant for decision making. For example, when an agent should deploy victims to an ambulance site, a relevant feature is that the agent is at the same position as the ambulance and that the agent carries the victim. On the other hand, it is not necessary to take into account other victims that are located at the ambulance site. To allow for such a distinction, perception samples are subdivided into possibly relevant features. This is done by treating sub-terms of the sample as individual features. They are extracted from a state term by performing a breadth-first search of transitive rewrites of a state according to the rewrite rule in Listing 1.2. Thus, sub-terms are built modulo axioms and equations specified for state terms. Extracted features are added to the split test hypotheses of the leaf the sample was sorted into. Their impact on action values is evaluated statistically as discussed in Sect. 2.1.

```
1  rl S:State and S':State => S:State .
```

Listing 1.2. State feature extraction.

Example 3. Consider the following situation: The agent currently holds v_0 and is co-located with an ambulance a at position p. There is another victim v_1 in the same location. The agent executes action $drop(x, v_0)$, dropping the hold victim. A reward of 1.0 is observed when dropping the victim at the ambulance location. The corresponding perception sample yields the following.

$$[at(x, p_0) \wedge at(v_1, p_0) \wedge at(a, p_0) \wedge holds(x, v_0)] \times drop(x, v_0) \times 1.0$$

Features are extracted from the perception sample by considering all sub-terms of the state formula (i.e. a term of sort *State*). Here, four single predicate features are extracted: $at(x, p_0)$, $at(v_1, p_0)$, $at(a, p_0)$ and $holds(x, v_0)$. Also, subsets containing multiple predicates (up to a user-defined number) are extracted (modulo axioms for sort state), for example $at(x, p_0) \wedge at(a, p_0)$.

Feature Generalization. To exploit relational domain structure, extracted features exposing equal predicates (that differ in arguments) are generalized with the ACUOS algorithm. This yields new, more general features. They are added to the current leaf's hypothesis set. Generalization allows to identify relevant state features faster: As relational structure and variables are used instead of propositional feature representation, more samples yield statistical evidence about the relevancy of that feature. I.e., generalization introduces *abstract states* for classifying samples. Also, if there is a general feature that is relevant for decision making, the learned tree will be more compact.

Variables introduced by generalization are either treated as fresh variables, or substituted by variables occurring in (transitive) parent nodes' split tests of the algebraic regression tree. Note that variables are only substituted if the replacing variable is of equal or lower sort (w.r.t. any given sort order) than the the replaced variable. In other words, the substituted variable has to be of equal or more general sort than the substituting variable.

Example 4. Let $at(x, p_0) \wedge at(a, p_0)$ be a state feature extracted from a perception sample as described above. Let $at(x, p_1) \wedge at(a, p_1)$ be a split test hypothesis of the leaf that the corresponding sample was sorted into. Generalization yields the term $at(x, P) \wedge at(a, P)$, where P is a variable of sort *Position*. It is added to the leaf's hypothesis set for statistical evaluation. Assuming that there is a *Position* variable P_{up} in a split test higher up the tree, an additional hypothesis $at(x, P_{up}) \wedge at(a, P_{up})$ is added to the current hypothesis set.

Note that generalizing $at(x, p_0) \wedge at(a, p_1)$ (with different positions) would yield $at(x, P) \wedge at(a, P')$, not imposing any equality constraints on the introduced variables. Besides this new hypothesis, the introduced variables are substituted combinatorially with free position variables higher up the tree to form new hypotheses incorporated in the the current nodes hypothesis set. Assuming that there is a *Position* variable P_{up} in a split test higher up the tree, two additional hypotheses $at(x, P_{up}) \wedge at(a, P')$ and $at(x, P) \wedge at(a, P_{up})$ are added to the current hypothesis set.

Multi-predicate Feature Lookahead. As illustrated in Example 3, not only single predicate features are extracted from a sample, but also features consisting of multiple predicates. This is an approach to allow for so-called *lookahead*, where a single predicate would not impose any impact on the prediction of the q-value, but a combination of predicates does (see e.g. [21, 22]). This becomes especially important for invariants (e.g. the position of the ambulance $at(a, p_0)$ in a particular setting) and when variables are introduced to features by generalization.

Example 5. To decide about the quality of dropping a victim, neither the predicate $at(x, P)$ nor the predicate $at(a, P)$ alone are relevant criteria (both are invariants), but using their combination $at(x, P) \wedge at(a, P)$ to split samples greatly reduces observed q-values variances. Note that, as is the case in this example, multiple predicates may also impose constraints on variables by equality: $at(x, P) \wedge at(a, P)$ as a split test would require that the positions are equal in order to classify a sample to the true-branch of the corresponding node.

Algorithm 2 shows the ARL algorithm in pseudocode. Note the change to parameters when compared to the TG algorithm (Algorithm 1). State and action representations are now required to be a rewrite theory. As split test hypotheses are extracted from the samples according to the given rewrite theory, there is no need for a language bias to drive hypothesis generation. Thus, the calls to the language bias for hypothesis generation in the TG algorithm (lines 2 and 14 of Algorithm 1) are no longer part of the ARL algorithm. Instead, split test hypotheses are extracted from perception samples and generalized with the ACUOS system as described in this section (lines 5–7 of Algorithm 2).

The other parts of the ARL algorithm correspond (up to the change of the underlying datastructure) exactly to the TG algorithm; in particular, lines 3–4 of Algorithm 2 correspond to lines 4–5 of Algorithm 1, lines 8–9 to lines 6–7, lines 11–15 to lines 9–13, and line 16 corresponds to line 15.

---

input : rewrite theory $(\Sigma, E \cup A)$ with sorts $State$ and $Action$ in Σ
list $Samples$ of samples $State \times Action \times \mathbb{R}$
integer minimal sample size mss
output: a first-order regression tree t_a for each action a specified in R

1 initialize each t_a with a single empty leaf l_{init}^a

2 **foreach** $sample\ \xi \in Samples$ **do**

3   sort ξ into $t_{action(\xi)}$ according to inner node tests, reaching leaf l
4   update q-value in leaf l according to $reward(\xi)$

5   extract features (single- and multi-predicate) from ξ
6   generalize extracted features with ACUOS
7   add generalized features to hypothesis set of l

8   **foreach** $split\ test\ hypothesis\ h\ in\ l$ **do**
9 | update statistics for h
10   **end**

11   **if** $number\ of\ samples\ sorted\ to\ l\ exceeds\ mss$ **then**
12   **if** $f\text{-}Test\ indicates\ a\ split\ for\ a\ hypothesis\ h$ **then**
13   create new inner node l_{new} with h as split test
14   add two empty leaves as children to l_{new}
15   set q-value of children according to h
16   replace current leaf l with l_{new}
17   **end**
18   **end**

19 **end**

Algorithm 2. The ARL algorithm.

---

4 Experimental Results

This section deals with the experimental evaluation of algebraic reinforcement learning performance. It also provides a comparison to learning performance without automated induction of test hypothesis through term generalization.

4.1 Experimental Setup

For experimental evaluation of the ARL algorithm with generalization we used the search and rescue scenario introduced in Sect. 3.2. The agent observes a reward if it deploys a victim at an ambulance. We will refer to the basic experimental setup as *rescue victim* scenario in the following.

As an extension to the *rescue victim* scenario consider fires burning at some positions. Here, the agent only observes reward if all fires are extinguished before deploying a victim at an ambulance. To this end, the agent is additionally equipped with an action to extinguish a fire. This setup will be referred to as *extinguish and rescue* scenario.

We provided a rewrite theory to allow for term representation of states and actions in form of a Maude module for the *rescue victim* scenario (Listing 1.1). The module was extended by a sort *Fire* as a subsort of *Spatial* for the *extinguish and rescue* scenario. An action *extinguish* : *Agent* × *Fire* → *Action* was added as well in this scenario.

Perception samples were generated by simulation. Initial simulation states were randomized; name constants for domain objects were also randomly generated. The learning agent chose its actions according to a Boltzmann exploration strategy [23]: Usually, it performed the action that yielded the maximum value according to the q-function (i.e. the algebraic regression trees) learned so far. However, with some probability a random action was chosen to avoid convergence to local behavior optima. A reward was given when the agent deployed a victim at the position of an ambulance. A learning episode ended when a reward was observed, and a new episode was started with random initial state. The ARL algorithm was used to learn agent behavior to maximize gathered reward and minimize the required number of actions to reach a goal state.

4.2 Results and Discussion

Figures 2 and 3 show the performance of the ARL algorithm with and without using generalization for hypothesis induction for the two scenarios presented in Sect. 4.1. The minimal sample size parameter was set to 2000 samples. The results indicate that automated hypothesis induction with ARL does not affect the benefits of the original TG algorithm when compared to propositional learners. The generalizing learner improves its behavior quickly and converges after ca. 4500 episodes in the *rescue victim* scenario. Its non-generalizing counterpart's behavior improves notably slower. Also, its learned behavior neither converged within 10000 episodes nor reached the level of the generalizing learner (Fig. 2). In the *extinguish and rescue* scenario, the difference is even more explicit: While the generalizing learner converges after ca. 7000 episodes, the non-generalizing algorithm does not converge within 10000 episodes (Fig. 3). In both scenarios, the generalizing variant exploits relations inherent to the domain to structure the representation of the q-function in terms of algebraic regression trees, thus reducing the problem space effectively.

Not only does generalization positively influence convergence rates, also the complexity of the learned trees is reduced in comparison to trees learned without generalization. By introducing equivalence classes of states through term generalization, samples are classified more compactly, using relational information where possible instead of explicitly having to store information about each

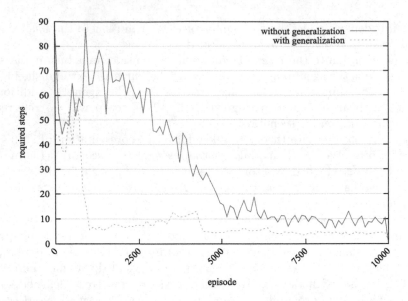

Fig. 2. Required actions to goal per episode in the *rescue victim* scenario.

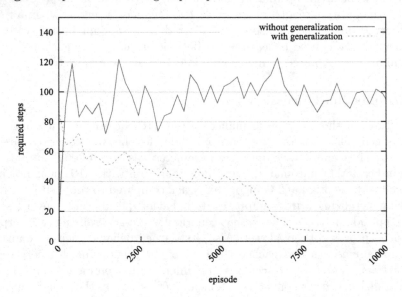

Fig. 3. Required actions to goal per episode in the *extinguish and rescue* scenario.

domain object encountered in the learning process. While learned trees are split 27 times by the non-generalizing learner in the *rescue victim* scenario, ARL only introduces 7 splits that are relevant for decision making. In the *extinguish and rescue* scenario, the former learns trees with 44 splits; ARL requires 23 splits in this case. This reduced complexity of the learned algebraic regression trees (i.e.

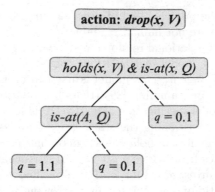

Fig. 4. Tree learned for the *drop* action in the *rescue victim* scenario.

the q-function) allows the ARL learner to classify gathered perception samples faster, thus speeding up the learning process as well as the decision process.

Figure 4 shows the algebraic regression tree learned for the *drop* action in the *rescue victim* scenario. The tree indicates that dropping a victim yields the highest value if the agent currently holds the victim and is located at an ambulance. The tests in the nodes were extracted without the need for an explicit language bias. Only a very general rewrite theory for state and action encoding was provided to the learning agent (see Listing 1.1). The variables for victim V, ambulance A and position Q have been introduced by generalization of extracted state features. Variables V and Q have been unified with (i.e. reference to) variables in parent nodes (see Sect. 3.3). As there was only a single learning agent x, it has not been generalized to a variable by ARL. For a discussion of more complex trees that have been learned, we refer the reader to [24].

5   Related Work

ARL is based on TG learning (see Sect. 2.1), which is in turn based on a first-order decision tree classification algorithm for relational datasets: The TILDE algorithm [20], which uses a similar approach to building relational decision trees by heuristically evaluating split test hypotheses. Other than TG learning, that classifies data w.r.t. continuous categories of q-value variances, TILDE classifies data according to nominal categories. TILDE employs an information theoretic metric instead of a statistical one to decide on the appropriateness of a candidate. Information theoretic induction of decision trees as e.g. TILDE or Quinlan's C4.5 [25] are based on Shannon's notions of information and entropy [26]. The basic idea is to reduce entropy (i.e. uncertainty) in induced partitions of datasets.

Besides TG learning there exist other approaches to relational reinforcement learning. Model-free techniques (as TG learning) use sampled data to learn classifiers that allows to predict action quality. The RIB-RRL algorithm [27] employs a k-nearest-neighbour approach for sample classification, using a user-defined distance metric for first-order formulas. KBR-RRL [28] uses Gaussian kernels for regression. While both approaches yield faster learning rates than TG learning, the learned q-functions are not human-readable.

Another approach to relational reinforcement learning are model-based techniques. Here, an explicit model of transitions triggered by action execution is provided to the learning agent. This model is used to derive goal-based behavior. Symbolic dynamic programming [29–31] uses a relational Markov decision process (RMDP) for model specification. The RMDP is solved by relational value iteration, dynamically partitioning the state space and assigning values to the computed partitions. In effect, a policy is learned that allows to decide on optimal actions for all states from where a given goal is reachable. Relational value iteration has also been investigated for RMDPs specified as rewrite theories [13]. Here, state space partitioning is achieved by narrowing to compute predecessor states and Ax-matching is employed to decide on abstract state subsumption.

The induction of general features from perception samples can also be used to learn explicit action models, for example in the form of action rules [32] or Markov logic networks [33]. In this case, no q-function for decision making is learned, but rather the preconditions and effects of action execution on the environment are identified.

6   Conclusion and Further Work

This paper introduced Algebraic Reinforcement Learning (ARL). We have shown how to integrate term generalization into the TG relational reinforcement learning algorithm. This alleviates the need for user-specified language biases in the learning process. Instead, state properties relevant for learning and decision making are automatically induced from perception samples by term generalization with the ACUOS system. Feature induction respects sort orders and axioms specified for sample representation. Experimental results show that ARL maintains the benefits of using state abstraction in the learning process.

Using a rewrite theory for the specification of perception symbols allows to incorporate sorts and equivalence classes on a very general level. No assumptions have to be made about problem related structure of perception samples gathered at runtime. ARL is able to identify automatically which sampled features are relevant. Feature candidates extracted from perception samples by generalization are only concerned with features the learning agent *actually encounters*, not with features that the specifier thought *could* be relevant. This is especially important in situations where the task to be learned is assigned to an agent dynamically at runtime, and each task only makes use of a small subset of the overall specified knowledge. In this case, only features relevant for the currently assigned task are considered when learning. Also, there may be systems where generalized or refined predicates are learned dynamically from low-level data or given high-level predicates at runtime (for example, high-level predicates used to specify the reward function). It is not trivial to specify a language bias a-priori for dynamically learned predicates. When learning predicates as rewrite theory operations, ARL provides a way to still employ relational reinforcement learning in such a scenario.

Order-sorted term generalization as done by the ACUOS system is a computationally complex task. On the other hand, lookahead by multi-predicate feature generalization is a valuable part of the ARL generalization, making it necessary to generalize more complex terms in particular situations. The current implementation of ARL uses predicate permutation, thus performing a lot of useless or unnecessary generalizations. One approach to systematically reduce generalization effort is to test ground features for information about q-variance before attempting to generalize them: The amount ground features add to classification may eventually indicate classification performance of a correspondingly lifted (i.e. generalized) feature. Thus, generalization of features that clearly do not improve sample classification on a ground level could be avoided in a systematic and mathematically well-defined manner. A problem with this approach is that statistical hypothesis sampling tests as the F-test used with ARL currently do not provide an explicit and comparable metric for independence of split test candidates and q-values. A possible solution could be the combination of information theoretic and statistical approaches.

ARL uses generalization to exploit relational domain structure, but does not build an explicit model of environmental dynamics. Term generalization could as well be used to construct an explicit relational model of the environment, e.g. STRIPS or PDDL models (see for example [23], Chap. 10) or relational MDPs as rewrite theories [13]. Action pre- and postconditions could be generalized explicitly from perception samples, leading to models that are human readable, and which can directly serve as knowledge bases for reasoning methods as e.g. action programming or relational value iteration in rewriting logic [11–13].

Another important direction for future research is to extend ARL to multi-agent scenarios. In its current form, ARL only supports single agent learning. It would be interesting to combine algebraic representational abstraction and feature extraction from samples as performed by ARL with state-of-the-art multi-agent learning algorithms and their corresponding representational models, e.g. H(L)FAQ-Learning and extended behaviour trees [34].

In this paper we discussed the application of generalization to a *model-free* relational reinforcement learner. There are other recent approaches that tackle the problem of learning in a *model-based* manner by exploiting explicit, declarative knowledge about domain dynamics [30,31]. Also, there are techniques that learn explicit relational declarative models of domain dynamics from examples rather than directly learning action values [32,33]. It would be an interesting direction for future research to explore the applicability of algebraic term generalization to these settings.

References

1. Sutton, R.S., Barto, A.G.: Reinforcement learning: an introduction. IEEE Trans. Neural Netw. **9**(5), 1054–1054 (1998)
2. Džeroski, S., De Raedt, L., Driessens, K.: Relational reinforcement learning. Mach. Learn. **43**(1–2), 7–52 (2001)

3. Tadepalli, P., Givan, R., Driessens, K.: Relational reinforcement learning: an overview. In: Proceedings of the ICML 2004 Workshop on Relational Reinforcement Learning (2004)
4. Van Otterlo, M.: A survey of reinforcement learning in relational domains (2005)
5. Driessens, K., Ramon, J., Blockeel, H.: Speeding up relational reinforcement learning through the use of an incremental first order decision tree learner. In: Flach, P.A., De Raedt, L. (eds.) ECML 2001. LNCS (LNAI), vol. 2167, pp. 97–108. Springer, Heidelberg (2001)
6. Driessens, K.: Relational reinforcement learning. In: Sammut, C., Webb, G.I. (eds.) Encyclopedia of Machine Learning, pp. 857–862. Springer, New York (2010)
7. Alpuente, M., Escobar, S., Meseguer, J., Ojeda, P.: A modular equational generalization algorithm. In: Hanus, M. (ed.) LOPSTR 2008. LNCS, vol. 5438, pp. 24–39. Springer, Heidelberg (2009)
8. Alpuente, M., Escobar, S., Espert, J., Meseguer, J.: ACUOS: a system for modular acu generalization with subtyping and inheritance. In: Fermé, E., Leite, J. (eds.) JELIA 2014. LNCS, vol. 8761, pp. 573–581. Springer, Heidelberg (2014)
9. Clavel, M., Durán, F., Eker, S., Lincoln, P., Martí-Oliet, N., Meseguer, J., Talcott, C. (eds.): All About Maude - A High-Performance Logical Framework. LNCS, vol. 4350, pp. 119–129. Springer, Heidelberg (2007)
10. Meseguer, J.: Twenty years of rewriting logic. J. Log. Algebr. Program. **81**(7–8), 721–781 (2012)
11. Belzner, L.: Action programming in rewriting logic. TPLP 13(4-5-Online-Supplement) (2013)
12. Belzner, L.: Verifiable decisions in autonomous concurrent systems. In: Kühn, E., Pugliese, R. (eds.) COORDINATION 2014. LNCS, vol. 8459, pp. 17–32. Springer, Heidelberg (2014)
13. Belzner, L.: Value iteration for relational MDPs in rewriting logic. In: Endriss, U., Leite, J. (eds.) STAIRS 2014 - Proceedings of the 7th European Starting AI Researcher Symposium, Prague, Czech Republic, 18–22 August, 2014. Frontiers in Artificial Intelligence and Applications, vol. 264, pp. 61–70. IOS Press, The Netherlands (2014)
14. Wirsing, M., Knapp, A.: A formal approach to object-oriented software engineering. Theor. Comput. Sci. **285**(2), 519–560 (2002)
15. Wirsing, M., Denker, G., Talcott, C.L., Poggio, A., Briesemeister, L.: A rewriting logic framework for soft constraints. Electr. Notes Theor. Comput. Sci. **176**(4), 181–197 (2007)
16. Belzner, L., De Nicola, R., Vandin, A., Wirsing, M.: Reasoning (on) service component ensembles in rewriting logic. In: Iida, S., Meseguer, J., Ogata, K. (eds.) Specification, Algebra, and Software. LNCS, vol. 8373, pp. 188–211. Springer, Heidelberg (2014)
17. Boronat, A., Knapp, A., Meseguer, J., Wirsing, M.: What is a multi-modeling language? In: Corradini, A., Montanari, U. (eds.) WADT 2008. LNCS, vol. 5486, pp. 71–87. Springer, Heidelberg (2009)
18. Eckhardt, J., Mühlbauer, T., Meseguer, J., Wirsing, M.: Semantics, distributed implementation, and formal analysis of KLAIM models in maude. Sci. Comput. Program. **99**, 24–74 (2015)
19. Eckhardt, J., Mühlbauer, T., AlTurki, M., Meseguer, J., Wirsing, M.: Stable availability under denial of service attacks through formal patterns. In: de Lara, J., Zisman, A. (eds.) FASE 2012. LNCS, vol. 7212, pp. 78–93. Springer, Heidelberg (2012)

20. Blockeel, H., Raedt, L.D.: Top-down induction of first-order logical decision trees. Artif. Intell. **101**(12), 285–297 (1998)
21. Blockeel, H., De Raedt, L.: Lookahead and discretization in ILP. In: Džeroski, S., Lavrač, N. (eds.) ILP 1997. LNCS, vol. 1297, pp. 77–84. Springer, Heidelberg (1997)
22. Castillo, L.P., Wrobel, S.: A comparative study on methods for reducing myopia of hill-climbing search in multirelational learning. In: Proceedings of the Twenty-First International Conference on Machine Learning, p. 19. ACM (2004)
23. Russell, S.J., Norvig, P.: Artificial Intelligence - A Modern Approach, 3rd edn. Pearson Education, New York (2010)
24. Neubert, S.: Solving relational reinforcement learning problems with a combination of incremental decision trees and generalization. Master's thesis, Ludwig-Maximilians-Universität München, Germany (2014)
25. Quinlan, J.R.: C 4.5: Programs for Machine Learning, vol. 1. Morgan Kaufmann, San Mateo (1993)
26. Shannon, C.E.: A mathematical theory of communication. ACM SIGMOBILE Mobile Comput. Commun. Rev. **5**(1), 3–55 (2001)
27. Driessens, K., Ramon, J.: Relational instance based regression for relational reinforcement learning. In: ICML, pp. 123–130 (2003)
28. Gärtner, T., Driessens, K., Ramon, J.: Graph kernels and gaussian processes for relational reinforcement learning. In: Horváth, T., Yamamoto, A. (eds.) ILP 2003. LNCS (LNAI), vol. 2835, pp. 146–163. Springer, Heidelberg (2003)
29. Boutilier, C., Reiter, R., Price, B.: Symbolic dynamic programming for first-order MDPs. In: Nebel, B. (ed.) IJCAI, pp. 690–700. Morgan Kaufmann, Seattle (2001)
30. Wang, C., Joshi, S., Khardon, R.: First order decision diagrams for relational mdps. J. Artif. Intell. Res. **31**, 431–472 (2008)
31. Sanner, S., Kersting, K.: Symbolic dynamic programming for first-order pomdps (2010)
32. Rodrigues, C., Gérard, P., Rouveirol, C., Soldano, H.: Incremental learning of relational action rules. In: 2010 Ninth International Conference on Machine Learning and Applications (ICMLA), pp. 451–458. IEEE (2010)
33. Khot, T., Natarajan, S., Kersting, K., Shavlik, J.: Learning markov logic networks via functional gradient boosting. In: 2011 IEEE 11th International Conference on Data Mining (ICDM), pp. 320–329. IEEE (2011)
34. Hölzl, M., Gabor, T.: Reasoning and learning for awareness and adaptation. In: Wirsing, M., Hölzl, M., Koch, N., Mayer, P. (eds.) Software Engineering for Collective Autonomic Systems: Results of the ASCENS Project. LNCS, vol. 8998, pp. 249–290. Springer, Heidelberg (2015)

The Formal System of Dijkstra and Scholten

Camilo Rocha[✉]

Escuela Colombiana de Ingeniería, AK 45 No. 205-59 (Autopista Norte),
Bogotá, Colombia
camilo.rocha@escuelaing.edu.co

Abstract. The logic of E.W. Dijkstra and C.S. Scholten has been shown to be useful in program correctness proofs and has attracted a substantial following in research, teaching, and programming. However, there is confusion regarding this logic to the point in which, for some time, it was not considered a logic, as logicians use the word. The main objections arise from the fact that: (i) symbolic manipulations seem to be based on the meaning of the terms involved, and (ii) some notation and the proof style of the logic are different, to some extent, from those found in the traditional use of logic. This paper presents the Dijkstra-Scholten logic as a formal system, and explains its proof-theoretic foundations as a formal system, thus avoiding any confusion regarding term manipulation, notation, and proof style. The formal system is shown to be sound and complete, mainly, by using rewriting and narrowing based decision and semi-decision procedures for, respectively, propositional and first-order logic previously developed by C. Rocha and J. Meseguer.

1 Introduction

I feel deeply grateful to José Meseguer for our friendship and his tutelage, encouragement, and support during our collaborative research. In honoring him for his 65th birthday, I have chosen a topic close to our common interest in logic, algebraic specification, and mechanical theorem proving. Mechanizing the logic of E.W. Dijkstra and C.S. Scholten was the first topic of research during my Ph.D. studies under José's guidance starting in the autumn of 2006; it was also a fruitful 'excuse' to learn and use Maude as a formal tool for the first time.

Program derivation is a formal style of program construction originated by the work of E.W. Dijkstra [6] and C.A.R. Hoare [12]. Its formal foundation was consolidated in the work of E.W. Dijkstra and C.S. Scholten [7], under the name of Dijkstra-Scholten logic. This logic can be conceived as a logically sound and complete system, having as its main feature the symbolic manipulation of formulas under the principle known as Leibniz's rule: the substitution of 'equals for equals' does not change the meaning of an entity. However, since its conception, the Dijkstra-Scholten logic was controversial because some of its features are not usual in the traditional study of logic. Also, unsatisfactory presentation has led to confusion, errors, and fallacies in the literature. The purpose of this paper is to present the Dijkstra-Scholten logic as a formal system and explain the proof-theoretic foundations of the logic within this simple setting. This paper relates

© Springer International Publishing Switzerland 2015
N. Martí-Oliet et al. (Eds.): Meseguer Festschrift, LNCS 9200, pp. 580–597, 2015.
DOI: 10.1007/978-3-319-23165-5_27

the novel features of the logic to notions found in the traditional study of logic, thus contributing in clarifying the overall confusion about the foundations, proof style, and notation of the Dijkstra-Scholten logic.

Finding proofs à la Hilbert of predicate formulas can be an unpleasant task even in the presence of results such as the deduction theorem because, for instance, there is no practical mechanism for discharging overwhelming quantities of assumptions that can appear in formal proofs. What can be learned from the Dijkstra-Scholten logic is that the importance of assumptions in formal proofs is overrated; with the right choice of connectives, axioms, and inference rules, proving logical formulas can be easy even without introducing assumptions. This paper presents an axiomatization of the Dijkstra-Scholten logic as a formal system, both at the propositional and first-order logic levels. The main logical connective in the Dijkstra-Scholten logic is equivalence (i.e., 'if and only if') which helps, for instance, in the development of formal proofs by minimizing the use of assumptions.

The Dijkstra-Scholten logic is based on the idea of accumulating a supply of equivalences proved, mainly, by using Leibniz principle: if two formulas are provably equivalent, then substituting one for the other in a formula does not alter the meaning of such a formula. However, Leibniz principle has been traditionally used in the study of traditional logic as a semantic replacement metatheorem [13]. This conception has caused trouble for accepting the Dijkstra-Scholten logic as such because symbolic manipulation in the system does not seem to be a purely syntactic task. As is shown in this paper, Leibniz principle can be formally defined within the setting of a formal system and thus lacks any semantic 'taint'.

The notion of 'proof' in the Dijkstra-Scholten logic is a contribution that has not been fully appreciated by the teaching and research community, in both mathematics and computer science. It is true that the notion of proof proposed by E.W. Dijkstra and C.S. Scholten is not a formal proof in the strict sense of a formal system. For the skeptical, proofs in this logic can be regarded as informal proofs written in a relatively strict format. This paper presents the formalization of such proof style in the form of a *derivation* and relate their meaning to that of a proof in a formal system. In a nutshell, a derivation can come in different flavors and it can always be 'compiled' into a formal proof.

Some of the important proofs about the formal system of E.W. Dijkstra and C.S. Scholten presented in this paper heavily rely on previous results of C. Rocha and J. Meseguer [19–21]. In particular, their equational decision procedure modulo has been used to mechanize part of the soundness proof for the propositional fragment of the system. Similarly, their sequent-based semi-decision procedure modulo for first-order logic has been used in the soundness proof of the system. Surprisingly, up to today, to the best of the author's knowledge the implementation of these procedures in Maude is the only mechanization of a logical system based on the Dijkstra -Scholten logic.

This paper is organized as follows. Section 2 covers background material on formal systems and rewriting logic. Section 3 presents the propositional fragment of the formal system of E.W. Dijkstra and C.S. Scholten; its first-order formulation appears in Sect. 4. Section 5 presents the notion of derivation, explains quantifier notation commonly used in the Dijkstra-Scholten style, and presents an extension of the first-order system with equality. Section 6 presents some related work and a conclusion.

2   Preliminaries

A *formal system* [13] consists of a set of symbols and *rules* for classifying as *formulas* certain expressions over the symbols, a set of formulas called *axioms*, and a set of *inference rules*. The sets of formulas of the formal systems introduced in this paper all have a decidable membership problem. Given a formal system F, a set of F-formulas Γ, and a formula ϕ_n, a *proof of ϕ_n from Γ in* F is a sequence of F-formulas $\phi_0, \phi_1, \ldots, \phi_n$ such that for any $0 \leq i \leq n$: (i) ϕ_i is an axiom, (ii) $\phi_i \in \Gamma$, or (iii) ϕ_i is the conclusion of an inference rule with premises appearing in $\phi_0, \ldots, \phi_{i-1}$. An F-formula ϕ *is a theorem from Γ in* F, written $\Gamma \vdash_F \phi$, if and only if there is a proof of ϕ from Γ in F; in the case when $\Gamma = \emptyset$, ϕ is called a *theorem of* F and it is written $\vdash_F \phi$.

This paper follows notation and terminology from [17] for order-sorted equational logic and from [5] for rewriting logic. An *order sorted signature* Σ is a tuple $\Sigma = (S, \leq, F)$ with finite poset of sorts $(S, \leq)$ and a finite index set of function symbols $F = \{F_{w,s}\}_{(w,s) \in S^* \times S}$. It is assumed that: (i) each connected component of a sort $s \in S$ in the poset ordering has a top sort, denoted by k_s, and (ii) for each operator declaration $f \in F_{s_1 \ldots s_n, s}$ there is also a declaration $f \in F_{k_{s_1} \ldots k_{s_n}, k_s}$. The collection $X = \{X_s\}_{s \in S}$ is an S-sorted family of disjoint sets of variables with each X_s countably infinite. The set of terms of sort s is denoted by $T_\Sigma(X)_s$ and the set of ground terms of sort s is denoted by $T_{\Sigma,s}$, which are assumed nonempty for each s. The expressions $T_\Sigma(X)$ and T_Σ denote the respective term algebras.

A Σ-*equation* is a Horn clause $t = u$ **if** γ, where $t = u$ is a Σ-*equality* with $t, u \in T_\Sigma(X)_s$ for some sort $s \in S$, and the *condition* γ is a finite conjunction of Σ-equalities $\bigwedge_{i \in I} t_i = u_i$. An *equational theory* is a tuple (Σ, E) with order-sorted signature Σ and finite set of Σ-equations E. For φ a Σ-equation, $(\Sigma, E) \vdash \varphi$ iff φ can be proved from (Σ, E) by the deduction rules in [17] iff φ is valid in all models of (Σ, E); assuming $T_{\Sigma,s} \neq \emptyset$ for each $s \in S$, (Σ, E) induces the congruence relation $=_E$ on $T_\Sigma(X)$ defined for any $t, u \in T_\Sigma(X)$ by $t =_E u$ iff $(\Sigma, E) \vdash t = u$. For an equational theory (Σ, E) and a term $t \in T_\Sigma(X)$, the expression $[t]_E$ denotes the equivalence class of t modulo E, i.e., $[t]_E = \{u \mid u \in T_\Sigma(X) \wedge t =_E u\}$.

A Σ-*rule* is a sentence $t \to u$ **if** γ, where $t \to u$ is a Σ-*sequent* with $t, u \in T_\Sigma(X)_s$ for some sort $s \in S$ and the *condition* γ is a finite conjunction of Σ-equalities. A *rewrite theory* is a tuple $\mathcal{R} = (\Sigma, E, R)$ with equational theory $\mathcal{E}_\mathcal{R} = (\Sigma, E)$ and a finite set of Σ-rules R. A *topmost rewrite theory* is a rewrite theory $\mathcal{R} = (\Sigma, E, R)$ such that for some top sort $\mathfrak{s}$ and for each $(t \to u \text{ if } \gamma) \in R$, the terms t, u satisfy $t, u \in T_\Sigma(X)_\mathfrak{s}$ and $t \notin X$, and no operator in Σ has $\mathfrak{s}$ as argument sort. For $\mathcal{R} = (\Sigma, E, R)$ and φ a Σ-rule, $\mathcal{R} \vdash \varphi$ iff φ can be obtained from $\mathcal{R}$ by the deduction rules in [5] iff φ is valid in all models of $\mathcal{R}$. For φ a Σ-equation, $\mathcal{R} \vdash \varphi$ iff $\mathcal{E}_\mathcal{R} \vdash \varphi$. A rewrite theory $\mathcal{R} = (\Sigma, E, R)$ induces the rewrite relation $\to_\mathcal{R}$ on $T_{\Sigma/E}(X)$ defined for every $t, u \in T_\Sigma(X)$ by $[t]_E \to_\mathcal{R} [u]_E$ iff there is a *one-step* rewrite proof $\mathcal{R} \vdash t \to u$. The expressions $\mathcal{R} \vdash t \to u$ and $\mathcal{R} \vdash t \xrightarrow{*} u$ respectively denote a one-step rewrite proof and an arbitrary length (but finite) rewrite proof in $\mathcal{R}$ from t to u.

It is assumed that the set of equations of a rewrite theory $\mathcal{R}$ can be decomposed into a disjoint union $E \uplus B$, with B a collection of axioms (such as associativity, and/or commutativity, and/or identity) for which there exists a *matching algorithm modulo B* producing a finite number of B-matching substitutions, or failing otherwise. It is also assumed that the equations E can be oriented into a set of (possibly conditional) sort-decreasing [8], operationally terminating [16], and confluent [8]. It is finally assumed that the rewrite rules R are weakly coherent relative to the rewrite rules $\vec{E}$ modulo B [25].

3   The Propositional Fragment

This section presents the formal system DS, a formulation of the propositional fragment of the Dijkstra-Scholten logic. Soundness and completeness of DS are proved, mainly, via the rewrite-based decision procedure for propositional logic found by C. Rocha and J. Meseguer [19,21].

Definition 1. *The following are the* symbols *of the formal system* DS:

- *An infinite set of* propositional variables $\{p_0, p_1, p_2, \ldots\}$.
- *Left parenthesis '('* and *right parenthesis ')'*.
- *The set of* logical connectives $\{true, false, \neg, \equiv, \not\equiv, \vee, \wedge, \to, \leftarrow\}$.

Connectives *true* and *false* are constants, and $\neg$ is the only unary connective. Symbols $\equiv$, $\not\equiv$, $\vee$, $\wedge$, $\to$, and $\leftarrow$ represent binary connectives. Table 1 lists the logical connectives with their respective names and intuitive meaning.

Definition 2. *The* formulas *(or* propositions*) of the formal system* DS *are given by the following BNF definition, where p is a propositional variable and ϕ is a formula:*

$$\phi ::= true \mid false \mid p \mid (\neg\phi) \mid (\phi \equiv \phi) \mid (\phi \not\equiv \phi) \mid (\phi \vee \phi) \mid (\phi \wedge \phi)$$
$$\mid (\phi \to \phi) \mid (\phi \leftarrow \phi).$$

The formal system DS includes the propositional constants *true* and *false* as propositions. They, together with the propositional variables, are the *atomic propositions* of the system. The equivalence and disjunction connectives play a major role in DS. In particular, equivalence is important because it represents

Table 1. Name and intuitive meaning of propositional connectives.

Connective	Name	Meaning
true	true	true
false	false	false
¬	negation	not $\cdots$
≡	equivalence	$\cdots$ if and only if $\cdots$
≢	discrepancy	$\cdots$ xor $\cdots$
∨	disjunction	$\cdots$ or $\cdots$
∧	conjunction	$\cdots$ and $\cdots$
→	implication	if $\cdots$, then $\cdots$
←	consequence	$\cdots$ if $\cdots$

Boolean equality and symbolic manipulation in the Dijkstra-Scholten logic is mainly based on the substitution of 'equals for equals'. The remaining connectives play a secondary role because, as it will become apparent from the axioms, they all can be defined in terms of the Boolean constants, equivalence, and disjunction.

Definition 3. *Let* ϕ, ψ, *and* τ *be formulas of* DS. *The* set of axioms *of* DS *is given by the following axiom schemata:*

$(Ax1)$ $((\phi \equiv (\psi \equiv \tau)) \equiv ((\phi \equiv \psi) \equiv \tau))$.
$(Ax2)$ $((\phi \equiv \psi) \equiv (\psi \equiv \phi))$.
$(Ax3)$ $((\phi \equiv true) \equiv \phi)$.
$(Ax4)$ $((\phi \vee (\psi \vee \tau)) \equiv ((\phi \vee \psi) \vee \tau))$.
$(Ax5)$ $((\phi \vee \psi) \equiv (\psi \vee \phi))$.
$(Ax6)$ $((\phi \vee false) \equiv \phi)$.
$(Ax7)$ $((\phi \vee \phi) \equiv \phi)$.
$(Ax8)$ $((\phi \vee (\psi \equiv \tau)) \equiv ((\phi \vee \psi) \equiv (\phi \vee \tau)))$.
$(Ax9)$ $((\neg\phi) \equiv (\phi \equiv false))$.
$(Ax10)$ $((\phi \not\equiv \psi) \equiv ((\neg\phi) \equiv \psi))$.
$(Ax11)$ $((\phi \wedge \psi) \equiv (\phi \equiv (\psi \equiv (\phi \vee \psi))))$.
$(Ax12)$ $((\phi \rightarrow \psi) \equiv ((\phi \vee \psi) \equiv \psi))$.
$(Ax13)$ $((\phi \leftarrow \psi) \equiv (\psi \rightarrow \phi))$.

Axioms $(Ax1)$, $(Ax2)$, and $(Ax3)$ define, respectively, that equivalence is associative, commutative, and has identity element *true*. Similarly, axioms $(Ax4)$, $(Ax5)$, and $(Ax6)$ define, respectively, that disjunction is associative, commutative, and has identity element *false*. Disjunction is idempotent by Axiom $(Ax7)$ and distributes over equivalence by Axiom $(Ax8)$. The remaining axioms $(Ax9)$-$(Ax13)$ present axiomatic definitions for the remaining connectives in DS. In this sense, $\{true, false, \equiv, \vee\}$ is a complete set of connectives for propositional logic.

For formulas ϕ and ψ, and a propositional variable p, the *textual substitution of* p *by* ψ *in* ϕ, denoted $\phi[p := \psi]$, is the proposition obtained from ϕ by substituting ψ for every occurrence of p.

Definition 4. *Let p be a propositional variable, and ϕ, ψ, and τ propositions. The* set of inference rules *of* DS *consists of two rules:*

$$\frac{\psi \quad (\psi \equiv \phi)}{\phi} \text{ EQUANIMITY} \qquad \frac{(\psi \equiv \tau)}{(\phi[p := \psi] \equiv \phi[p := \tau])} \text{ LEIBNIZ.}$$

Rule EQUANIMITY is the stronger version of the traditional MODUS PONENS but it is based on equivalence. Rule LEIBNIZ enables the above-mentioned substitution of 'equals for equals' in DS. Additional inference rules such as MODUS PONENS and MODUS TOLENS for implication, and ASSOCIATIVITY, SYMMETRY, and TRANSITIVITY for equivalence can all be added to DS without changing its provability relation $\vdash_{\mathsf{DS}}$.

Example 1. Consider the following proof of $\vdash_{\mathsf{DS}} true$:

1. $(((true \equiv true) \equiv true) \equiv (true \equiv true))$ $(Ax3)$
2. $((true \equiv true) \equiv true)$ $(Ax3)$
3. $(true \equiv true)$ (Equanimity 2 y 1)
4. $true$ (Equanimity 3 y 2).

The next goal is to prove the soundness and completeness of DS. This is done by using the equational theory T_{DS} in Fig. 1, which corresponds to the Boolean decision procedure of C. Rocha and J. Meseguer [19,21]. Constants TRUE and FALSE denote the propositional constants *true* and *false*. Function symbols equ, or, not, xor, imp, and con respectively stand for equivalence, disjunction, negation, discrepancy, implication, and consequence.

A *valuation* is a function from the set of propositional variables to the Boolean set $\mathbb{B} = \{\mathsf{T}, \mathsf{F}\}$; valuations are extended homomorphically to propositions in the natural way assigning T to *true* and F to *false*. A valuation $\mathbf{v}$ *satisfies* a set of propositions Γ if and only if $\mathbf{v}(\gamma) = \mathsf{T}$ for each $\gamma \in \Gamma$. Given a set of propositions Γ and a proposition ϕ, the *tautological consequence* $\Gamma \models \phi$ is *valid* if and only if $\mathbf{v}(\phi) = \mathsf{T}$ for every valuation $\mathbf{v}$ that satisfies Γ.

Proposition 1 ([21], Theorem 2). *The equational theory T_{DS} is a decision procedure for propositional logic. In particular, for any set of propositions Γ and a proposition ϕ the following equivalence holds:*

$$\Gamma \models \phi \quad \Longleftrightarrow \quad T_{\mathsf{DS}} \vdash \left(\left(\bigwedge_{\gamma \in \Gamma} \gamma \right) \to \phi \right).$$

The theory T_{DS} is the axiomatization of C. Rocha and J. Meseguer [19,21], as a set of confluent and terminating equations modulo AC, of the Dijkstra-Scholten propositional logic [7]. The theory T_{DS} is isomorphic to the theory of Boolean rings T_{BR}, which is based on the isomorphism between Boolean algebras and Boolean rings discovered by M.H. Stone [23]. As a rewrite system, T_{BR} was proposed by J. Hsiang in the 1980s as a decision procedure for propositional logic [14]. See [19,21] for a more detailed account of these facts and for two more equational decision procedures for propositional logic found by C. Rocha and J. Meseguer.

Lemma 1. *The formal system* DS *is sound, namely, for any set Γ of propositions and for a proposition ϕ, the following implication is logically valid:*

$$\Gamma \vdash_{\mathsf{DS}} \phi \quad \Longrightarrow \quad \Gamma \models \phi.$$

Proof (Sketch). The proof of the axioms' soundness is obtained by using Proposition 1 as intermediate result. In particular, since T_{DS} is a decision procedure for propositional logic, the axioms of DS are shown to be sound by mechanical reduction to TRUE in the functional module DS in Maude. Soundness of Rule EQUANIMITY is proved by using simple facts about valuations and soundness of Rule LEIBNIZ can be shown by induction on the structure of ϕ.

Lemma 2 *The formal system* DS *is complete, namely, for any set Γ of propositions and for a proposition ϕ, the following implication is logically valid:*

$$\Gamma \models \phi \quad \Longrightarrow \quad \Gamma \vdash_{\mathsf{DS}} \phi.$$

Proof (Sketch). The proof is obtained by using Proposition 1 as intermediate result. Given a proof of $\Gamma \vdash_{T_{\mathsf{DS}}} \phi$, it is shown by induction that there exists a proof witnessing $\Gamma \vdash_{\mathsf{DS}} \phi$.

Proposition 2. DS *is a sound and a complete propositional logic system.*

4   The First-Order System

This section presents a formulation of the first-order logic of E.W. Dijkstra and C.S. Scholten as the formal system $\mathsf{DS}(\mathcal{L})$. This system is parametric on a first-order language $\mathcal{L}$ and extends the propositional formal system DS. Soundness and completeness are proved, mainly, via the rewrite- and narrowing-based semi-decision procedure for first-order logic found by C. Rocha and J. Meseguer [20,21].

Definition 5. *The symbols of* $\mathsf{DS}(\mathcal{L})$ *are:*

- *An infinite collection $\mathcal{X}$ of variables $\{x_0, x_1, x_2, \ldots\}$.*
- *A set $\mathcal{F}$ of function symbols.*
- *A set $\mathcal{P}$ of predicate symbols, which includes an infinite collection of constants $\{P_0, P_1, P_2, \ldots\}$.*
- *An arity function $ar : \mathcal{F} \cup \mathcal{P} \to \mathbb{N}$ for function and predicate symbols.*
- *Left parenthesis '(', right parenthesis ')', and comma ','.*
- *The set of logical connectives $\{true, false, \neg, \equiv, \not\equiv, \vee, \wedge, \to, \leftarrow, \forall, \exists\}$.*

```
fmod DS is
  sort BoolDS .
  ops TRUE FALSE : -> BoolDS .
  op _equ_ : BoolDS BoolDS -> BoolDS [assoc comm prec 80] .
  op _or_ : BoolDS BoolDS -> BoolDS [assoc comm prec 50] .
  op _and_ : BoolDS BoolDS -> BoolDS [prec 50] .
  op not_ : BoolDS -> BoolDS [prec 10] .
  op _xor_ : BoolDS BoolDS -> BoolDS [assoc comm prec 80] .
  op _imp_ : BoolDS BoolDS -> BoolDS [assoc comm prec 70] .
  op _con_ : BoolDS BoolDS -> BoolDS [assoc comm prec 70] .
  vars P Q R : BoolDS .
  eq P equ P = TRUE .          eq P equ TRUE = P .
  eq P or TRUE = TRUE .        eq P or FALSE = P .
  eq P or P = P .
  eq P or ( Q equ R ) = P or Q equ P or R .
  eq not P = P equ FALSE .
  eq P xor Q = P equ Q equ FALSE .
  eq P and Q = P equ Q equ P or Q .
  eq P imp Q = P or Q equ Q .
  eq P con Q = Q imp P .
endfm
```

Fig. 1. An equational decision procedure for propositional logic in the syntax of Maude.

The symbols of $DS(\mathcal{L})$ are parametric on the sets of function symbols $\mathcal{F}$ and predicate symbols $\mathcal{P}$ defined by the first-order language $\mathcal{L}$. In $DS(\mathcal{L})$ it is assumed that the set of predicate symbols $\mathcal{P}$ contains an infinite collection of constant predicate symbols; they are key for symbol manipulation in the formal system as will be shown by the definition of the inference rules. The logical connectives of $DS(\mathcal{L})$ extend those of DS with the usual symbols '∀' for universal quantification and '∃' for existential quantification.

Definition 6 introduces the sets of terms and formulas of $DS(\mathcal{L})$.

Definition 6. *The set of* terms *and the set of* formulas *of the formal system* $DS(\mathcal{L})$ *are given by the following BNF definitions, where* $x \in \mathcal{X}$, $c \in \mathcal{F}$ *with* $ar(c) = 0$, $f \in \mathcal{F}$ *with* $ar = m > 0$, $P \in \mathcal{P}$ *with* $ar(P) = 0$, $Q \in \mathcal{P}$ *with* $ar(Q) = n > 0$, t *a term, and* ϕ *is a formula:*

$$t ::= x \mid c \mid f(t, \ldots, t).$$
$$\phi ::= true \mid false \mid P \mid Q(t, \ldots, t) \mid (\neg \phi) \mid (\phi \equiv \phi) \mid (\phi \not\equiv \phi) \mid (\phi \vee \phi)$$
$$\mid (\phi \wedge \phi) \mid (\phi \rightarrow \phi) \mid (\phi \leftarrow \phi) \mid (\forall x \, \phi) \mid (\exists x \, \phi).$$

The expressions $\mathcal{T}(\mathcal{X}, \mathcal{F})$ *and* $\mathcal{T}(\mathcal{X}, \mathcal{F}, \mathcal{P})$ *denote, respectively, the set of terms and the set of formulas over* $\mathcal{X}$, $\mathcal{F}$, *and* $\mathcal{P}$.

The set of terms is built from variables and the application of a function symbol to a sequence of terms. The set of formulas is built from the Boolean constants and Boolean combination of formulas, together with the application of a predicate symbol to a sequence of terms and universal/existential quantified formulas. The *atomic formulas* of $DS(\mathcal{L})$ are the Boolean constants *true* and *false*, and the formulas obtained by applying a predicate symbol to zero or more terms. Note that by assuming the infinite collection of constant predicate symbols in $DS(\mathcal{L})$, any propositional variable of DS can be seen as an atomic formula in $DS(\mathcal{L})$ because of the map from propositional variables to predicate symbols given by $p_i \mapsto P_i$; in this way, any DS proposition is a $DS(\mathcal{L})$ formula.

In the Dijkstra-Scholten first-order logic, the textual substitution operator $\_\{\_ := \_\}$ is overloaded both for replacing variables for terms and for replacing constant predicate symbols for formulas. The concept of a free occurrence of a variable in a formula in the Dijkstra-Scholten logic is the traditional one, i.e., an occurrence of a variable x in a formula ϕ is *free* iff such an occurrence of x is not under the scope of a $\forall x$ or $\exists x$. Similarly, a term t is *free* for x in a formula ϕ iff every free occurrence of x in ϕ is such that if it is under the scope of a $\forall y$ or $\exists y$, then y is not a variable in t.

Definition 7. *Let $x \in \mathcal{X}$, $t \in \mathcal{T}(\mathcal{X}, \mathcal{F})$, and $\phi, \psi \in \mathcal{T}(\mathcal{X}, \mathcal{F}, \mathcal{P})$. The set of axioms of $DS(\mathcal{L})$ is given by the following axiom schemata:*

$(Ax\cdot)$ *Any axiom of* DS.
$(Bx1)$ $((\forall x\, \phi) \equiv \phi)$, *if x is not free in ϕ.*
$(Bx2)$ $((\phi \vee (\forall x\, \psi)) \equiv (\forall x\, (\phi \vee \psi)))$, *if x is not free in ϕ.*
$(Bx3)$ $(((\forall x\, \phi) \wedge (\forall x\, \psi)) \equiv (\forall x\, (\phi \wedge \psi)))$.
$(Bx4)$ $((\forall x\, \phi) \rightarrow \phi[x := t])$, *if t is free for x in ϕ.*
$(Bx5)$ $((\exists x\, \phi) \equiv (\neg(\forall x\, (\neg\phi))))$.

Any axiom of DS is an axiom of $DS(\mathcal{L})$, taking into account that any propositional variable p_i in DS is represented in $DS(\mathcal{L})$ by the constant predicate symbols P_i in $\mathcal{P}$. Axiom $(Bx1)$ states that a universal quantifier on variable x can be omitted whenever the formula it quantifies has no free occurrences of x. Axiom $(Bx2)$ states that disjunction distributes over universal quantification whenever there is no variable capture, while Axiom $(Bx3)$ states that conjunction and universal quantification commute. By Axiom $(Bx4)$, it is possible to particularize any universal quantification with a term t whenever the variables in t are not captured by the substitution. Finally, Axiom $(Bx5)$ is an axiomatic definition for existential quantification. In $DS(\mathcal{L})$ the set $\{true, false, \equiv, \vee, \forall\}$ is a complete set of connectives for first-order logic.

Definition 8. *Let $x \in \mathcal{X}$, $P \in \mathcal{P}$ with $ar(P) = 0$, and $\phi, \psi, \tau \in \mathcal{T}(\mathcal{X}, \mathcal{F}, \mathcal{P})$. The set of inference rules of $DS(\mathcal{L})$ consists of three rules:*

$$\frac{\psi \qquad (\psi \equiv \phi)}{\phi} \text{ EQUANIMITY}$$

$$\frac{(\psi \equiv \tau)}{(\phi[P := \psi] \equiv \phi[P := \tau])} \text{ LEIBNIZ}$$

$$\frac{\phi}{(\forall x\, \phi)} \text{ GENERALIZATION.}$$

Rules EQUANIMITY and LEIBNIZ, as in DS, allow symbolic manipulation based on equality and the substitution of 'equals for equals'. Rule GENERALIZA- TION is the usual first-order rule stating that universally quantifying any theorem results in a theorem. The assumption about the *infinite* collection of constant predicate symbols in DS($\mathcal{L}$) is key for the Rule LEIBNIZ to work for substituting formulas in any given formula. Other important fact regarding Rule LEIBNIZ is that from this rule some interesting meta-properties can be proved with almost no effort: (i) any substitution instance of a tautology (i.e., of a DS theorem) is a theorem of DS($\mathcal{L}$) and (ii) the set of theorems of DS($\mathcal{L}$) is closed under formula substitution.

It is fair to say that LEIBNIZ is ultimately at the heart of the Dijkstra- Scholten logic. When it was initially introduced, LEIBNIZ gave the impression of being a semantic tool instead of a deductive one. The reason for this situation may be attributed to an unsatisfactory presentation [2] and also to the fact that traditionally in logic, and except for some logics such as equational logic, the substitution of 'equals for equals' has been more a semantic tool in the form of replacement metatheorems rather than part of a proof system.

The rewrite- and narrowing-based semi-decision procedure $\mathcal{R}_{DS(\mathcal{L})}$ for first- order logic of C. Rocha and J. Meseguer [20, 21] is used to prove soundness and completeness of DS($\mathcal{L}$). The rewrite theory $\mathcal{R}_{DS(\mathcal{L})}$ is a mechanization of the Dijkstra-Scholten logic in a sequent calculus and it satisfies all essential requirements to be executable by rewriting. Also, $\mathcal{R}_{DS(\mathcal{L})}$ is parametric on the underlying equational theory for solving Boolean theorems; in this paper, T_{DS} is used as the Boolean decision procedure of choice.

Figure 2 depicts the order-sorted signature of $\mathcal{R}_{DS(\mathcal{L})}$ in the form of a dia- gram, where dashed lines represent sort inclusion. Sort *Formula* corresponds to first-order formulas built from the Boolean constants, the binary operators for equivalence and disjunction, and universal and existential quantification. The remaining Boolean connectives are added as definitional extensions of the for- mer Boolean connectives. The atomic building blocks for formulas are predicates of sort *Pred* ranging over first-order terms of sort *Term*, constructed from predi- cate symbols $P, Q, \ldots$ of different arities. Sort *Var* represents names of variables and the operator _[_/_] stands for textual substitution of a variable for a term in a formula. Sorts *FSet* represents finite sets of formulas with the constant symbol $\square$ denoting the empty set of formulas. Sorts *Seq* and *SSet* represent first-order sequents and finite sets of first-order sequents, respectively. The trivial sequent is denoted with the constant symbol $\lozenge$. Union of sets of first-order sequents is denoted with •.

Figure 3 presents part of the specification of the rewrite theory $\mathcal{R}_{DS(\mathcal{L})}$ as Maude's system module DSL-SEQ, in which some auxiliary functions have been added to deal with variables and substitutions. Universal quantification is rep- resented with square brackets. Constant mts is used to represent $\lozenge$ and * for •. The first rule specifies the situation in which a proof has been found. The second and third rules deal with equivalence by splitting cases based on mutual implication. The fourth and fifth rules specify how to deal with disjunction,

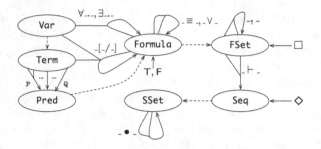

Fig. 2. Order-sorted signature of the rewrite theory $\mathcal{R}_{\mathsf{DS}(\mathcal{L})}$.

while the remaining two rules specify how to handle universally quantified formulas. It is important to note that the underlying Boolean decision procedure $\mathcal{T}_{\mathsf{DS}}$ automatically takes care of the remaining cases of Boolean combinations. The situation for existential quantifiers is dealt equationally with by the usual definitional extension using negation and universal quantification. The last two rules deserve special attention. The next-to-last rule is declared non-executable (attribute **nonexec**) because there is an extra variable in its right-hand side and thus the derivation tree may have infinite branching. The key observation is that the presence of extra variables in a rule's right-hand side is not problematic for *narrowing* with the rules of a coherent rewrite theory modulo its axioms, under the assumption that its rewrite rules are *topmost*. In this scenario, narrowing with rules is a sound and complete deduction process for solving existential queries of the form $\exists x(t(x) \xrightarrow{*} t'(x))$ [18]. The next-to-last rule in Fig. 3 introduces new variables, which are then incrementally instantiated as new terms to narrow the set of sequents at each step. The last rule makes explicit the need for auxiliary function **newVar** to generate fresh variables not occurring in the given formulas.

```
mod DSL-SEQ is
   ...
   vars FSB FSC : FSet . vars B C : Formula var S : Seq .
   rl FSB,B |- B,FSC => mts .
   rl FSB,B equ C |- FSC => FSB,B,C |- FSC * FSB |- B,C,FSC .
   rl FSB |- B equ C,FSC => FSB,B |- C,FSC * FSB,C |- B,FSC .
   rl FSB,B or C |- FSC => FSB,B |- FSC * FSB,C |- FSC .
   rl FSB |- B or C,FSC => FSB |- B,C,FSC .
   rl FSB,[x : B] |- FSC => FSB,B[t/x] |- FSC [nonexec] .
   rl FSB |- [x : B],FSC => FSB |- B[newVar(FSB,B,FSC)/x],FSC .
endm
```

Fig. 3. Rewrite rules of $\mathcal{R}_{\mathsf{DS}(\mathcal{L})}$.

Proposition 3 ([21], Theorem 3). *The rewrite theory $\mathcal{R}_{DS(\mathcal{L})}$ is sound and complete for first-order logic. In particular, a sequent S is provable in the sequent calculus if and only if $\mathcal{R}_{DS(\mathcal{L})} \vdash S \xrightarrow{*} \Diamond$.*

The rewrite rules in $DS(\mathcal{L})$ correspond to a deductively complete subset of the sequent calculus rules presented in [22]. These rules have been implemented in Maude and are executable thanks to the narrowing modulo features of the language/system, as explained before.

This paper adopts Tarski's definition of truth for first-order logic [13]: given a set of formulas Γ and a formula ϕ, the *semantic entailment* $\Gamma \models \phi$ is valid if and only if ϕ is true in any $\mathcal{L}$-model of Γ.

Lemma 3. *The formal system $DS(\mathcal{L})$ is sound, namely, for any set Γ of formulas and for a formula ϕ, the following implication is logically valid:*

$$\Gamma \vdash_{DS(\mathcal{L})} \phi \quad \Longrightarrow \quad \Gamma \models \phi.$$

Proof (Sketch). The proof of the axioms' soundness is obtained by using Proposition 3 as an intermediate result. The proof follows the same ideas in the proof of Lemma 1.

Lemma 4. *The formal system $DS(\mathcal{L})$ is complete, i.e., for any set Γ of formulas and for a formula ϕ, the following implication is logically valid:*

$$\Gamma \models \phi \quad \Longrightarrow \quad \Gamma \vdash_{DS(\mathcal{L})} \phi.$$

Proof (Sketch). The proof is obtained by using Proposition 3 as an intermediate result and it follows the same ideas of the proof of Lemma 2.

Proposition 4. $DS(\mathcal{L})$ *is a sound and a complete first-order logic system.*

5   Derivations, Notation, and Equality

This section presents the notion of derivation and quantifier notation that are specific to the Dijkstra-Scholten logic. This section also presents axioms for extending $DS(\mathcal{L})$ with equality. Examples can be found at the end of the section.

5.1   Derivations

It is fair to say that one of the main reasons the Dijkstra-Scholten approach to logic has gained a significant base of users is because this logic comes with a powerful and agile proof calculus. This situation contrasts with most logical systems for which, traditionally, semantic reasoning is the preferred inference mechanism. 'Proofs' in the Dijkstra-Scholten style of proof are not proofs in the strict sense of a formal system, but instead are sequences of formulas related by equivalence or implication. This approach takes advantage of the transitive properties of these connectives and allows for deriving proofs compactly, mainly, by facilitating the substitution of 'equals for equals'.

Definition 9. *Let Γ be a set of formulas of* $\mathsf{DS}(\mathcal{L})$. *A derivation in* $\mathsf{DS}(\mathcal{L})$ *from Γ is a non-empty finite sequence of formulas $\phi_0, \phi_1, \ldots, \phi_n$ of* $\mathsf{DS}(\mathcal{L})$ *such that $\Gamma \vdash_{\mathsf{DS}(\mathcal{L})} (\phi_{k-1} \equiv \phi_k)$ for $0 < k \leq n$.*

Derivations are sequences of formulas that are pairwise equivalent. The connection between a derivation and a proof is made precise in Proposition 5.

Proposition 5. *Let Γ be a set of formulas of* $\mathsf{DS}(\mathcal{L})$ *and $\phi_0, \phi_1, \ldots, \phi_n$ be a derivation in* $\mathsf{DS}(\mathcal{L})$ *from Γ. It holds that $\Gamma \vdash_{\mathsf{DS}(\mathcal{L})} (\phi_0 \equiv \phi_n)$.*

Proof. The proof follows by induction on $n \in \mathbb{N}$.

As stated in Proposition 5, any derivation yields a proof in the formal system. It is important to note that any proof in the formal system $\mathsf{DS}(\mathcal{L})$ is a derivation in $\mathsf{DS}(\mathcal{L})$, but a derivation is *not* necessarily a proof. Consider for instance the sequence *"false, false"* which is a derivation because Boolean equivalence is reflexive, but this sequence is not a proof because *false* is not a theorem. The key fact about proofs in a formal system is that every formula in a proof is a theorem, while this is not necessarily the case in a derivation.

Derivations are not written directly as a sequence of formulas but instead as a bi-dimensional arrangement of formulas and text explaining the derivation steps.

Remark 1. A derivation $\phi_0, \phi_1, \ldots, \phi_n$ in $\mathsf{DS}(\mathcal{L})$ from Γ is usually written as follows:

$$\phi_0$$
$$\equiv \quad \langle \text{ "explanation}_0\text{" } \rangle$$
$$\phi_1$$
$$\vdots \quad \langle \ldots \rangle$$
$$\phi_{n-1}$$
$$\equiv \quad \langle \text{ "explanation}_{n-1}\text{" } \rangle$$
$$\phi_n$$

in which "explanation$_i$" is a text describing why $\Gamma \vdash_{\mathsf{DS}} (\phi_i \equiv \phi_{i+1})$.

There are other types of derivations based on implication or consequence and not on equivalence. They are called, respectively, *weakening* and *strengthening* derivations.

Definition 10. *Let Γ be a set of formulas. A sequence $\phi_0, \ldots, \phi_n$ of formulas in* $\mathsf{DS}(\mathcal{L})$ *is a:*

1. *A weakening derivation from Γ iff $\Gamma \vdash_{\mathsf{DS}(\mathcal{L})} (\phi_{k-1} \rightarrow \phi_k)$ for each $0 < k \leq n$.*
2. *A strengthening derivation from Γ iff $\Gamma \vdash_{\mathsf{DS}} (\phi_{k-1} \leftarrow \phi_k)$ for each $0 < k \leq n$.*

Note that because equivalence and implication commute, any derivation is a weakening and strengthening derivation.

Proposition 6. *Let Γ be a set of formulas of $\mathsf{DS}(\mathcal{L})$ and $\Phi = \phi_0, \phi_1, \ldots, \phi_n$ be a sequence of formulas in $\mathsf{DS}(\mathcal{L})$.*

- *If Φ is a weakening derivation, then $\Gamma \vdash_{\mathsf{DS}(\mathcal{L})} (\phi_0 \to \phi_n)$.*
- *If Φ is a strengthening derivation, then $\Gamma \vdash_{\mathsf{DS}(\mathcal{L})} (\phi_0 \leftarrow \phi_n)$.*

5.2   Quantifier Notation

The Dijkstra-Scholten logic proposes an alternative notation for writing quantified formulas. Formally, this notation is syntactic sugar that can help in specifying and reasoning about quantifiers and it is specially suited for symbolic manipulation.

Remark 2. Let $x \in \mathcal{X}$ and ϕ, ψ be formulas of $\mathsf{DS}(\mathcal{L})$.

- The expression $(\forall x \mid \psi : \phi)$ is syntactic sugar for $(\forall x \, (\psi \to \phi))$; in particular, $(\forall x \mid true : \phi)$ can be written as $(\forall x \mid: \phi)$.
- The expression $(\exists x \mid \psi : \phi)$ is syntactic sugar for $(\exists x \, (\psi \wedge \phi))$; in particular, $(\exists x \mid true : \phi)$ can be written as $(\exists x \mid: \phi)$.

In the formulas $(\forall x \mid \psi : \phi)$ and $(\exists x \mid \psi : \phi)$, ψ is called the *range* and ϕ the *subject* of the quantification.

This notation has proved in practice to be useful for specifying properties of, for example, indexed data types. For instance, consider the following specification about an 0-based array a of integers that stores only perfect squares:

$$\left(\forall i \mid 0 \leq i < len(a) : \left(\exists n \mid: a[i] = n^2\right)\right).$$

The range of the universal quantification states that i is a valid index of the array a. In general, verification and derivation of code based on specifications of this sort becomes simpler thanks to the resemblance between the notation and the semantics, for example, of an imperative programming language.

5.3   Equality

Definition 11 introduces the axioms for equality.

Definition 11. *Let $x \in \mathcal{X}$, $t \in \mathcal{T}(\mathcal{X}, \mathcal{F})$, and $\phi \in \mathcal{T}(\mathcal{X}, \mathcal{F}, \mathcal{P})$. The following axioms define the equality '$=$' over $\mathcal{T}(\mathcal{X}, \mathcal{F})$:*

$(Bx6)$ $(x = x)$.
$(Bx7)$ $((x = t) \to (\phi \equiv \phi[x := t]))$, *if t is free for x in ϕ.*

Equality is reflexive (Axiom $(Bx6)$) and performing a substitution of 'equal for equal' terms does not affect the truth value of a formula, under some assumptions about variable capture (Axiom $(Bx7)$). Together, axioms $(Bx6)$ and $(Bx7)$, allow for proving in $\mathsf{DS}(\mathcal{L})$ that equality is an equivalence relation on the set of terms $\mathcal{T}(\mathcal{X}, \mathcal{F})$.

Theorem 1. *Let $x \in \mathcal{X}$, $t \in \mathcal{T}(\mathcal{X}, \mathcal{F})$, and $\phi \in \mathcal{T}(\mathcal{X}, \mathcal{F}, \mathcal{P})$. If t is free for x in ϕ and x does not occur in t, then:*

1. $\vdash_{\mathsf{DS}(\mathcal{L})} ((\forall x \mid x = t : \phi) \equiv \phi[x := t])$.
2. $\vdash_{\mathsf{DS}(\mathcal{L})} ((\exists x \mid x = t : \phi) \equiv \phi[x := t])$.

The formulas in Theorem 1 are known as 'one-point' rules. The following proof of the theorem is presented also as an excuse for performing derivations and for using the syntactic sugar for quantified formulas, within the formal system resulting from the extension of $\mathsf{DS}(\mathcal{L})$ with the equality axioms.

Proof. In what follows, a proof of (1) is presented; a proof of (1) can be obtained in a similar way. The proof follows by double implication.

$$(\forall x \mid x = t : \phi)$$
$$\equiv \quad \langle \text{ notation } \rangle$$
$$(\forall x \, ((x = t) \to \phi))$$
$$\to \quad \langle \, (Bx4)\colon t \text{ is free for } x \text{ in } \phi \, \rangle$$
$$((x = t) \to \phi)[x := t]$$
$$\equiv \quad \langle \text{ textual substitution; } x \text{ does not occur in } t \, \rangle$$
$$((t = t) \to \phi[x := t])$$
$$\equiv \quad \langle \, (Bx6) \, \rangle$$
$$(true \to \phi[x := t])$$
$$\equiv \quad \langle \text{ properties of } \to \, \rangle$$
$$\phi[x := t] \, .$$

For the other direction:

$$\phi[x := t]$$
$$\to \quad \langle \text{ weakening } \rangle$$
$$((\neg(x = t)) \vee \phi[x := t])$$
$$\equiv \quad \langle \text{ propositional logic } \rangle$$
$$((x = t) \to \phi[x := t])$$
$$\equiv \quad \langle \, (Bx7); \to \text{ distributes over } \equiv \, \rangle$$
$$((x = t) \to \phi)$$

$$\rightarrow \quad \langle \text{ GENERALIZATION } \rangle$$
$$(\forall x \, ((x = t) \rightarrow \phi))$$
$$\equiv \quad \langle \text{ syntactic sugar } \rangle$$
$$(\forall x \mid x = t : \phi).$$

Finally, consider a short example regarding a simple property of integer arithmetic.

Example 2. Consider the following definition of the 'divides' relation:

$$a \cdot | \, b \equiv \exists x \, (ax = b).$$

The goal is to prove that '$\cdot|$' is reflexive, namely:

$$\vdash_{\mathsf{DS}(\mathcal{L})} \forall a \, (a \cdot | \, a).$$

Consider the following derivation:

$$a \cdot | \, a$$
$$\equiv \quad \langle \text{ definition } \rangle$$
$$\exists c \, (ac = a)$$
$$\leftarrow \quad \langle \text{ Theorem 1.1; strengthening with witness } c = 1 \, \rangle$$
$$(a1 = a)$$
$$\equiv \quad \langle \text{ arithmetic } \rangle$$
$$(a = a)$$
$$\equiv \quad \langle \, (Bx6) \, \rangle$$
$$true.$$

Therefore, $\vdash_{\mathsf{DS}(\mathcal{L})} (a \cdot | \, a)$ and by GENERALIZATION $\vdash_{\mathsf{DS}(\mathcal{L})} \forall a \, (a \cdot | \, a)$.

6 Concluding Remarks

The Dijkstra-Scholten logic was first introduced in [7] by E. Dijkstra and C. Scholten as an alternative and effective notation, method, and calculus for imperative program verification and derivation. In that work, the logic is extended to first-order theories such as integers, sets, and arrays, and is the test-bed for calculating with second-order predicate transformers. It was first introduced as a college textbook by D. Gries and F. Schneider in [9]. R. Backhouse proposed a program construction methodology by calculating implementations from specifications following the style of Dijkstra and Scholten in [1]. More recently, J. Bohórquez has proposed a version of intuitionistic first-order logic based on the substitution of 'equals for equals' [3] and a unified approach to program correctness, unifying Hoare and Dijkstra-Scholten's iterative style of programming with Hehner's recursive predicative programming theory [4].

The foundations of the propositional fragment of the Dijkstra-Scholten logic were first published by D. Gries and F. Schneider in [10]. L. Bijlsma and R. Nederpelt in [2] focus on the logical backgrounds of the Dijkstra-Scholten program development style for correct programs. They also present a number of examples showing that unsatisfactory presentation of DS predicate calculus and some of its features has led to errors and fallacies in the literature. V. Lifschitz defines the concept of calculation (or derivation) in the Dijkstra-Scholten style in terms of proof trees [15]. G. Torulakis presents a metatheory of equational predicate logic that uses Leibniz's substitution of 'equals for equals' as a primary rule of inference, and obtains a complete first-order logic [24]. The main difference between this work and the previous references is that here the foundations of the Dijkstra-Scholten logic are given in terms of the simple notion of a proof system, and some of the main proofs have been obtained mechanically by rewriting and narrowing in Maude.

The main contributions of this paper include a formalization of the Dijkstra-Scholten logic as a formal system, avoiding any confusion regarding term manipulation. Soundness and completeness of the system are proved mainly by induction, and by means of the decision and semi-decision procedures implemented in Maude, respectively, in [19,21] and in [20,21]. This paper also makes clear the distinction between the notions of proof in a formal system and that of derivation, emblematic of the Dijkstra-Scholten style. Explanation regarding the notation for quantified formulas has been given, together with a number of examples. To the best of the author's knowledge, the executable rewriting logic specifications for propositional and first-order logic of C. Rocha and J. Meseguer in [21] are the only theorem proving mechanization developed up to now for the Dijkstra-Scholten logic.

As future work, it seems to be worth pursuing further results in the mechanization of the Dijkstra-Scholten logic, and its adoption in the area of theorem proving and mechanical symbolic verification of programs. The window inference style of reasoning proposed by J. Grundy [11] for the HOL system seems like a promising start point.

References

1. Backhouse, R.: Program Construction: Calculating Implementations from Specifications. Wiley, New York (2003)
2. Bijlsma, L., Nederpelt, R.: Dijkstra-Scholten predicate calculus: concepts and misconceptions. Acta Informatica **35**(12), 1007–1036 (1998)
3. Bohórquez, J.: Intuitionistic logic according to Dijkstra's calculus of equational deduction. Notre Dame J. Formal Log. **49**(4), 361–384 (2008)
4. Bohórquez, J.: An elementary and unified approach to program correctness. Formal Aspects Comput. **22**(5), 611–627 (2010)
5. Bruni, R., Meseguer, J.: Semantic foundations for generalized rewrite theories. Theoret. Comput. Sci. **360**(1–3), 386–414 (2006)

6. Dijkstra, E.: A Discipline of Programming. Prentice-Hall Series in Automatic Computation. Prentice-Hall, Upper Saddle River (1976)
7. Dijkstra, E., Scholten, C.: Predicate Calculus and Program Semantics. Texts and Monographs in Computer Science. Springer-Verlag, New York (1990)
8. Meseguer, J., Durán, F.: On the and coherence properties of conditional rewrite theories. J. Log. Algebraic Program. **81**(7), 816–850 (2011)
9. Gries, D., Schneider, F.B.: A Logical Approach to Discrete Math. Texts and Monographs in Computer Science. Springer, New Yok (1993)
10. Gries, D., Schneider, F.B.: Equational propositional logic. Inf. Process. Lett. **53**(3), 145–152 (1995)
11. Grundy, J.: Window inference in the HOL system. In: Archer, M., Joyce, J.J., Levitt, K.N., Windley, P.J. (eds.) Proceedings of the 1991 International Workshop on the HOL Theorem Proving System and its Applications, August 1991, Davis, California, USA, pp. 177–189. IEEE Computer Society (1992)
12. Hoare, C.A.R.: An axiomatic basis for computer programming. Commun. ACM **12**(10), 576–580 (1969)
13. Hodel, R.: An Introduction to Mathematical Logic. Dover Books on Mathematics Series. Dover Publications Incorporated, New York (2013)
14. Hsiang, J.: Topics in automated theorem proving and program generation. Ph.D. thesis, University of Illinois at Urbana-Champaign, Computer Science Department (1982)
15. Lifschitz, V.: On calculational proofs. Annals Pure Appl. Log. **113**(1–3), 207–224 (2001)
16. Lucas, S., Meseguer, J.: Operational termination of membership equational programs: the order-sorted way. Electron. Notes Theo. Comput. Sci. **238**(3), 207–225 (2009)
17. Meseguer, J.: Membership algebra as a logical framework for equational specification. In: Parisi-Presicce, Francesco (ed.) WADT 1997. LNCS, vol. 1376, pp. 18–61. Springer, Heidelberg (1998)
18. Meseguer, J., Thati, P.: Symbolic reachability analysis using narrowing and its application to verification of cryptographic protocols. Higher-Order Symbolic Comput. **20**(1–2), 123–160 (2007)
19. Rocha, C., Meseguer, J.: Five isomorphic Boolean theories and four equational decision procedures. Technical report 2007–2818, University of Illinois at Urbana-Champaign (2007)
20. Rocha, C., Meseguer, J.: Theorem proving modulo based on Boolean equational procedures. Technical report 2007–2922, University of Illinois at Urbana-Champaign (2007)
21. Rocha, C., Meseguer, J.: Theorem proving modulo based on boolean equational procedures. In: Berghammer, R., Möller, B., Struth, G. (eds.) RelMiCS/AKA 2008. LNCS, vol. 4988, pp. 337–351. Springer, Heidelberg (2008)
22. Socher-Ambrosius, R., Johann, P.: Deduction Systems (Texts in Computer Science). Springer-Verlag, New York (1996)
23. Simmons, G.: Introduction to Topology and Modern Analysis. International Series in Pure and Applied Mathematics. Krieger Publishing Company, Malabar (1963)
24. Tourlakis, G.: On the soundness and completeness of equational predicate logics. J. Log. Comput. **11**(4), 623–653 (2001)
25. Viry, P.: Equational rules for rewriting logic. Theoret. Comput. Sci. **285**, 487–517 (2002)

From Rewriting Logic, to Programming Language Semantics, to Program Verification

Grigore Roşu[(⊠)]

University of Illinois at Urbana-Champaign, Champaign, USA
grosu@illinois.edu

Abstract. Rewriting logic has proven to be an excellent formalism to define executable semantics of programming languages, concurrent or not, and then to derive formal analysis tools for the defined languages with very little effort, such as model checkers. In this paper we give an overview of recent results obtained in the context of the rewriting logic semantics framework $\mathbb{K}$, such as complete semantics of large programming languages like C, Java, JavaScript, Python, and deductive program verification techniques that allow us to verify programs in these languages using a common verification infrastructure.

1 Introduction

Programming language semantics and program analysis and verification are well developed research areas with a long history. In fact, one might think that all problems would have been solved by now: we would hope that any formal semantics for a language should give rise to a proof system and that a verifier for such a system would simply extend the proof system with a proof strategy; or looked at from the other side, we would assume that any verification system for a particular programming language would be grounded in that language's formal semantics. However, reality tells us that most program verifiers are not directly based on a formal semantics, but rather on complex and adhoc hardwired models of their target programming languages. This has at least two negative consequences. First, it makes the development and maintenance of program verifiers hard and uneconomical, particularly for new programming languages or languages which evolve fast. Second, it allows room for subtle bugs in program verifiers themselves. Consider, for example, the following three-line C program:

```
void main() {
  int x = 0;
  return (x = 1) + (x = 2);
}
```

When compiled with some compilers, e.g., GCC3, ICC, Clang, this program evaluates to 3, while when compiled with others, e.g., GCC4, MSVC, it evaluates to 4. This is correct behavior for the compilers, because according to the ISO C11 standard [1] this function is *undefined* (it writes x twice within the same sequence-point interval), and language implementations or compilers are

© Springer International Publishing Switzerland 2015
N. Martí-Oliet et al. (Eds.): Meseguer Festschrift, LNCS 9200, pp. 598–616, 2015.
DOI: 10.1007/978-3-319-23165-5_28

free to treat undefined programs however they find fit, in particular to apply aggressive optimizations, like above. The overall design of the C language has been conceived in the spirit of improved performance, the price to pay being that it becomes programmer's responsibility to ensure that their programs are well-defined. However, what is scary, and in our view unacceptable, is that state-of-the-art program verifiers for C code like VCC [2] and Frama-C [3] *prove* that this program evaluates to 4! That is because in order to simplify implementation, these verifiers use code simplification modules similar to those used by compilers, which work as expected only when the source program is well-defined according to the C semantics.

The unfortunate consequence is that, in spite of more than 40 years of world-wide research and in spite of being hard to use, state-of-the-art software verification tools cannot be trusted. They encode or implement rather adhoc models of programming languages, without any guarantee that their model is faithful to the actual language. The root problem is that most languages *do not even have a formal semantics*, their designers wrongly thinking that a formal semantics is not worth the effort. Hence, the tool developers must rely on informal manuals and on their subjective understanding of the language. Inspired by countless hours of discussions with my mentor, colleague and dear friend José Meseguer, we firmly believe that this can and should change, that *programming languages must have formal semantics!* Moreover, that formal analysis tools and language implementations can and should be derived from such semantics, as shown in Fig. 1, so they are correct by construction. This is not a dream. Not anymore.

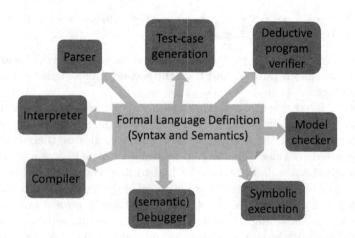

Fig. 1. Rewriting logic semantics can achieve this.

We present a snapshot of recent and current research on using rewriting logic semantics in the field of programming languages through the $\mathbb{K}$ framework, from giving semantics to real programming languages to using such semantics to verify programs.

2   From Rewriting Logic to Language Semantics

Starting with Meseguer's seminal rewriting logic paper [4], which demonstrated how naturally rewriting logic can capture the various computational paradigms, a series of papers have been published on using rewriting logic to give semantics to programming languages. Verdejo and Martì-Oliet [5, 6] show how to use rewriting logic and Maude [7] to define and implement executable semantics for several languages following both big-step [8] and small-step SOS [9, 10] approaches, and Şerbănuţă *et al.* [11] show how several other semantic approaches can be represented in rewriting logic, including Berry and Boudol's chemical abstract machine (CHAM) [12, 13] and Felleisen's reduction semantics with evaluation contests [14, 15].

When representing any of these semantic approaches in rewriting logic, the idea is to define the desired program configurations as an algebraic specification, that is as a signature representing the syntax of configurations and equations defining the underlying mathematical domains, and then to define the various types of transitions uniformly as rewrite rules. For example, in a simple C-like imperative language a program configuration can be a pair < code, state >, where code is a fragment of program and state is a finite-domain map from program variables to values. Fragments of programs are nothing but well-formed terms over an appropriate algebraic signature, and finite-domain maps can be defined as an algebraic data type. In a small-step SOS style rewrite logic semantics, for example, the semantics of the assignment construct can then be defined with rewrite rules as follows (we use the Maude notation):

```
crl < X = E,Sigma > => < X = E',Sigma' > if < E,Sigma > => < E',Sigma' > .
 rl < X = V,Sigma > => < V,Sigma[V / X] > .
```

The first rule reduces the expression E assigned to program variable X step by step until it becomes a value V, and then the second rule assigns that value to X in the state Sigma.

A problem faced when attempting the above for real languages, like C or Java, is that the program configuration tends to be huge, comprising dozens of semantic cells, most of them being unused in most rules or their changes just being propagated by rules. The lack of modularity of SOS was visible even in Plotkin's original notes [9, 10], where he had to modify the definition of simple arithmetic expressions several times as his initial language evolved. Hennessy also makes it even more visible in his book [16]. Each time he adds a new feature, he also has to change the configurations and the entire existing semantics. However, the lack of modularity of language definitional frameworks was not perceived as a major problem until late 1990s, partly because there were few attempts to give complete and rigorous semantics to real programming languages. Hennessy actually used each language extension as a pedagogical opportunity to teach what new semantic components the feature needs and how and where those are located in each sequent.

The first to pinpoint the limitations of plain SOS when defining non-trivial languages were the inventors of alternative semantic frameworks, such as Berry and Boudol [12,13] who proposed the chemical abstract machine model, Felleisen and his collaborators [14,15] who proposed reduction semantics with evaluation contexts, and Mosses and his collaborators [17–19] who proposed the modular SOS (MSOS) approach. Among these, Mosses is perhaps the one who most vehemently criticized the lack of modularity of plain SOS. Meseguer and Braga [20,21] were the first to observe that rewriting logic, through its powerful support for multiset matching, can seamlessly support modular semantic frameworks, by giving a first rewrite logic representation of MSOS. The representation in [20] also led to the development of the Maude MSOS tool [22].

Learning from previous uses of rewriting logic to define programming language semantics, we proposed the $\mathbb{K}$ framework [23] (http://kframework.org) as a formalism and notation inspired from rewrite logic but specialized to the domain of programming languages. In $\mathbb{K}$, programming languages can be defined using configurations, computations and rules. Configurations organize the state in units called cells, which are labeled and can be nested. Computations carry computational meaning as special nested list structures sequentializing computational tasks, such as fragments of program. Computations extend the original language abstract syntax. $\mathbb{K}$ (rewrite) rules make it explicit which parts of the term they read-only, write-only, read-write, or do not care about. This makes $\mathbb{K}$ suitable for defining truly concurrent languages even in the presence of sharing. Computations are like any other terms in a rewriting environment: they can be matched, moved from one place to another, modified, or deleted. This makes $\mathbb{K}$ suitable for defining control-intensive features such as abrupt termination, exceptions or call/cc.

In $\mathbb{K}$, a language is defined in one or more files with extension ".k". A language definition consists roughly of three parts: annotated syntax, configuration, and semantic rules. For syntax, $\mathbb{K}$ uses conventional BNF annotated with $\mathbb{K}$-specific attributes. For example, the syntax of assignment in a language like above can be defined as

```
syntax Stmt ::= Id "=" Exp    [strict(2)]
```

The attribute `strict(2)` states the evaluation strategy of the assignment construct: first evaluate the second argument, and then apply the semantic rule(s) for assignment.

To allow arbitrarily complex and nested program configurations, $\mathbb{K}$ proposes a cell-based approach. Each cell encapsulates relevant information for the semantics, including other cells that can "float" inside it. For our simple language, a "top" cell <T>...</T> containing a code cell <k>...</k> and a state <state>...</state> suffices:

```
configuration <T>
              <k> $PGM </k>
              <state> .Map </state>
              </T>
```

The given cell contents tell $\mathbb{K}$ how to initialize the configuration: $PGM says where to put the input program once parsed, and .Map is the empty map.

Once the syntax and configuration are defined, we can start adding semantic rules. 𝕂 rules are contextual: they mention a configuration context in which they apply, together with local changes they make to that context. The user typically only mentions the absolutely necessary context in their rules; the remaining details are filled in automatically by the tool. For example, here is the 𝕂 rule for assignment:

```
rule <k> X:Id = V:Val => V ...</k>
     <state>... X |-> (_ => V) ...</state>
```

The ellipses are part of the 𝕂 syntax. Recall that assignment was strict(2), so we can assume that its second argument is a value, say V. The context of this rule involves two cells, the k cell which holds the current code and the state cell which holds the current state. Moreover, from each cell, we only need certain pieces of information: from the k cell we only need the first task, which is the assignment "X = V", and from the state cell we only need the binding "X |-> _". The underscore stands for an anonymous variable, the intuition here being that value is discarded anyway, so there is no need to bother naming it. The irrelevant parts of the cells are replaced with ellipses. Then, once the local context is established, we identify the parts of the context which need to change, and we apply the changes using local rewrite rules with the arrow =>, noting that it has a greedy scoping, grabbing everything to the left and everything to the right until a cell boundary (open or closed) or an unbalanced parenthesis is encountered. In our case, we rewrite both the assignment expression and the value of X in the state to the assigned value V. Everything else stays unchanged. The concurrent semantics of 𝕂 regards each rule as a transaction: all changes in a rule happen concurrently; moreover, rules themselves apply concurrently, provided their changes do not overlap.

Once the definition is complete and saved in a .k file, say imp.k, the next step is to generate the desired language model. This is done with the kompile command:

```
kompile imp.k
```

By default, the fastest possible executable model is generated. To generate models which are amenable for symbolic execution, test-case generation, search, model checking, or deductive verification, one needs to provide kompile with appropriate options.

The generated language model is employed on a given program for the various types of analyses using the krun command. By default, with the default language model, krun simply runs the program. For example, if sum.imp contains

```
n=100; s=0;
while(n>0) {
    s=s+n; n=n-1;
}
```

then the command

```
krun sum.imp
```

yields the final configuration

```
<T>
  <k> . </k>
  <state>
    n |-> 0, s |-> 5050
  </state>
</T>
```

Using the appropriate options to the `kompile` and `krun` commands, we can enable all the above-mentioned tools and analyses on the defined programming language and the given program. Many languages are provided with the $\mathbb{K}$ tool distribution, and several others are available from http://kframework.org (start with the $\mathbb{K}$ tutorial). Some of these languages have dozens of cells in their configurations and hundreds of rules.

Besides didactic and prototypical languages, $\mathbb{K}$ has been used to formalize several existing real-life languages and to design and develop analysis and verification tools for them. The most notable are complete $\mathbb{K}$ definitions for the following languages: C11 (POPL'12 [24], PLDI'15 [25]), Java 1.4 (POPL'15 [26]), JavaScript ES5 (PLDI'15 [27]). Each of these language semantics has more than 1,000 semantic rules and has been tested on benchmarks and test suites that implementations of these languages also use to test their conformance, where available. The C semantics, when executed, catches undefinedness; for example, the program discussed in the introduction reports the following error when executed with the C semantics:

```
==============================================================
ERROR! KCC encountered an error while executing this program.
==============================================================
Error: EI01
Description: Unsequenced side effect on scalar object with side effect
  of same object.
Type: Undefined behavior.
See also: C11 sec. 6.5
==============================================================
```

The Java semantics effort has also produced a test suite of several hundreds of programs that thoroughly and systematically test all the Java language constructs, because no such test suite was available.[1] The JavaScript semantics passes all the 2,782 core language tests part of the ECMAScript 5 conformance

---

[1] The official test suite for Java implementations is the Java Compatibility Kit (JCK) from Oracle, which is not publicly available. Oracle offers free access for non-profit organizations willing to implement the whole JDK (http://openjdk.java.net/groups/conformance/JckAccess/), i.e., both the language and the class library. We applied and Oracle rejected our request, because we did not "implement" the complete Java library. Also, note that the NIST Juliet testsuite (http://samate.nist.gov/SARD/testsuite.php) is meant to asses the capability of Java static analysis tools, and not the completeness of Java implementations.

testsuite. To put this in perspective, among the existing implementations of JavaScript, only Chrome's passes all the tests, and no other existing semantics attempt of JavaScript passes more than 90 %. In addition to a reference implementation for a language, a $\mathbb{K}$ executable semantics also yields a simple coverage metric for a test suite: the set of semantic rules it exercises. The semantics of JavaScript revealed that the ECMAScript 5 conformance test suite fails to cover several semantic rules. Guided by the semantics, we wrote tests to exercise those rules and those tests revealed bugs both in production JavaScript engines (Chrome, Safari, Firefox) and in other semantics.

3   From Language Semantics to Program Verification

An operational semantics of a programming language, be it defined in $\mathbb{K}$ or not, defines an execution model the language typically in terms of a transition relation $cfg \Rightarrow cfg'$ between program configurations, and can serve as a formal basis for language understanding, design, implementation, and so on. On the other hand, an axiomatic semantics defines a proof system typically in terms of Hoare triples $\{\psi\}\,\mathsf{code}\,\{\psi'\}$, and can serve as a basis for program verification. To increase confidence in program verifiers and thus avoid problems like the one discussed in the introduction, ideally the axiomatic semantics should be proved sound w.r.t. the operational semantics (see, e.g., [28]). Needless to say that defining an axiomatic semantics for a real language like C is no simpler than defining an operational semantics, and that proving their equivalence is a burden that few can take. Consequently, most program verifiers are actually based on no formal semantics of their target language at all and, as discussed above, end up sometimes proving wrong programs correct.

We have recently proposed a different approach to program verification, *reachability logic* [29–32], which introduces the notion of a reachability rule to express dynamic properties of programs, as a generalization of both a rewrite rule and a Hoare triple. Thus, reachability logic unifies operational and axiomatic semantics. Reachability logic builds upon *matching logic* [33–36] as a formalism to express static state properties. The overall idea of our verification approach is to start with a rewriting-based semantics of a programming language (which is operational), and to derive program properties with the same semantics, without giving the language any other (axiomatic) semantics.

Matching logic allows us to state and reason about structural properties over arbitrary program configurations. The main intuition underlying matching logic is that "terms are predicates and their satisfaction is matching". Syntactically, it introduces a new formula construct, called a *basic pattern*, which is a term possibly containing variables, e.g., a configuration term expressing a desired structure of the program configuration. The formulae, which can be built using basic patterns and arbitrary other conventional logical constructs, are called *patterns*. This way, we can compose structural requirements and add logical constraints over the variables appearing in basic patterns. Semantically, the models of basic patterns are concrete elements, e.g., concrete configurations regarded as ground terms, where such an element satisfies a basic pattern iff it *matches* it. Considering some configuration structure with a top-level cell $\langle ...\rangle_{\text{cfg}}$ holding, in any order, other cells with semantic data such as the code $\langle ...\rangle_k$,[2] an environment $\langle ...\rangle_{\text{env}}$, a heap $\langle ...\rangle_{\text{heap}}$, an input buffer $\langle ...\rangle_{\text{in}}$, an output buffer $\langle ...\rangle_{\text{out}}$, etc., configurations then have the structure:

$$\langle \langle ...\rangle_k \ \langle ...\rangle_{\text{env}} \ \langle ...\rangle_{\text{heap}} \ \langle ...\rangle_{\text{in}} \ \langle ...\rangle_{\text{out}} \ ...\rangle_{\text{cfg}}$$

The contents of the cells can be various algebraic data types, such as trees, lists, sets, maps, etc. Here are two particular concrete configurations (note that x and y are program variables, which unlike in Hoare logics are not logical variables; in matching logic they are constants used to build programs or fragments of programs):

$$\langle \langle \texttt{x=*y; y=x; } ...\rangle_k \ \langle \texttt{x} \mapsto 7, \ \texttt{y} \mapsto 3, \ ...\rangle_{\text{env}} \ \langle 3 \mapsto 5\rangle_{\text{heap}} \ ...\rangle_{\text{cfg}}$$
$$\langle \langle \texttt{x} \mapsto 3\rangle_{\text{env}} \ \langle 3 \mapsto 5, \ 2 \mapsto 7\rangle_{\text{heap}} \ \langle 1, 2, 3, ...\rangle_{\text{in}} \ \langle ..., 7, 8, 9\rangle_{\text{out}} \ ...\rangle_{\text{cfg}}$$

Different languages may have different configuration structures. For example, languages whose semantics are intended to be purely syntactic and based on substitution, such as λ-calculi, may contain only one cell, holding the program itself. Other languages may contain dozens of cells in their configurations. For example, the C semantics has more than 70 nested cells. However, no matter how complex a language is, its configurations can be defined as ground terms over an algebraic signature, using conventional algebraic techniques. Matching logic takes an arbitrary algebraic definition of configurations as parameter and, as mentioned, allows configuration patterns (i.e., terms with variables) as particular formulae. To simplify terminology, all matching logic's formulae are called patterns. As a purposely artificial example, consider the pattern

$$\exists c : Cells, \ e : Env, \ p : Nat, \ i : Int, \ \sigma : Heap$$
$$\langle \langle \texttt{x} \mapsto p, \ e\rangle_{\text{env}} \ \langle p \mapsto i, \ \sigma\rangle_{\text{heap}} \ c\rangle_{\text{cfg}} \ \wedge \ i > 0 \ \wedge \ p \neq i$$

This is satisfied by all configurations where program variable x points to a location p holding a positive integer i different from p. Variables matching the irrelevant parts of a cell, such as the variables e, σ, and c above, are called *structural frames*; when reasoning about languages defined using $\mathbb{K}$, the structural frames typically result from ellipses in rules, that is, from the parts of the configuration which do not change. They are needed in order for the pattern to properly match the expected structure of the desired configurations. For example, if we want to additionally state that p is the only location allocated in the heap, then we can just remove σ from the pattern above and obtain:

$$\exists c : Cells, \ e : Env, \ p : Nat, \ i : Int \ \ \langle \langle \texttt{x} \mapsto p, \ e\rangle_{\text{env}} \ \langle p \mapsto i\rangle_{\text{heap}} \ c\rangle_{\text{cfg}} \ \wedge \ i > 0 \ \wedge \ p \neq i$$

Matching logic allows us to reason about configurations, e.g., to prove:

$$\models \forall c : Cells, \ e : Env, \ p : Nat$$
$$\langle \langle \texttt{x} \mapsto p, \ e\rangle_{\text{env}} \ \langle p \mapsto 9\rangle_{\text{heap}} \ c\rangle_{\text{cfg}} \ \wedge \ p > 10$$
$$\rightarrow \exists i : Int, \ \sigma : Heap \ \langle \langle \texttt{x} \mapsto p, \ e\rangle_{\text{env}} \ \langle p \mapsto i, \ \sigma\rangle_{\text{heap}} \ c\rangle_{\text{cfg}} \ \wedge \ i > 0 \ \wedge \ p \neq i$$

---

[2] In mathematical mode, we prefer the notation $\langle ...\rangle_k$ for cells instead of the XML-like notation `<k>...</k>` preferred in ASCII.

To specify more complex properties, one can use abstractions (e.g., singly-linked lists matched in the heap, etc.), which can be axiomatized and proved sound using conventional means. In fact, as shown in [35–37], like separation logic, matching logic can also be used as a program logic in the context of conventional axiomatic (Hoare) semantics, allowing us to more easily specify structural properties about the program state. However, this way of using matching logic comes with a big disadvantage, shared with Hoare logics in general: the formal semantics of the target language needs to be redefined axiomatically and tedious soundness proofs need to be done. Instead, we prefer to use reachability logic, which allows us to use the operational semantics of the language for program verification as well.

An unconditional *reachability rule* is a pair $\varphi \Rightarrow \varphi'$, where φ and φ' are matching logic patterns (not necessarily closed). The semantics of a reachability rule captures the intuition of *partial correctness* in axiomatic semantics: any configuration satisfying φ either rewrites/transits forever or otherwise reaches through (zero or more) successive transitions a configuration satisfying φ'. In $\mathbb{K}$, programming languages can be given operational semantics based on rewrite rules of the form "$l \Rightarrow r$ if b", where l and r are configuration terms with variables constrained by boolean condition b. Such rules can be expressed as reachability rules $l \wedge b \Rightarrow r$. On the other hand, a Hoare triple of the form $\{\psi\}$ code $\{\psi'\}$ can be regarded as a reachability rule $\langle \cdots \langle \text{code} \rangle_k \cdots \rangle_{\text{cfg}} \wedge \overline{\psi} \Rightarrow \langle \cdots \langle \rangle_k \cdots \rangle_{\text{cfg}} \wedge \overline{\psi'}$ between patterns over minimal configurations holding only the code. Here the ellipses represent appropriate structural frames, $\langle \rangle_k$ is the configuration holding the empty code, and $\overline{\psi}$ and $\overline{\psi'}$ are variants of the original pre/post conditions replacing program variables with appropriate logical variables (an example will be shown shortly). Therefore, reachability rules smoothly capture the basic ingredients of both operational and axiomatic semantics, in that both operational semantics rules and axiomatic semantics Hoare triples are instances of reachability rules.

Figure 2 shows the reachability logic proof system for unconditional reachability rules. This is a simplification of a more general proof system in [30], where conditional reachability rules were also considered, for the particular rewrite logic theories supported by $\mathbb{K}$ (recall that in $\mathbb{K}$ the reachability rules are unconditional, because the side conditions can be moved into the LHS of the rule). We here only discuss the one-path variant of reachability logic, where $\varphi \Rightarrow \varphi'$ means that φ' is matched by some configuration reached after some sequence of transitions from a configuration matching φ. The all-path variant is more complex and can be found in [29]. The one-path and all-path reachability logic variants are equally expressive when the target programming language is deterministic. The target language is given as a reachability system $\mathcal{S}$ (from "semantics"). The soundness result in [30] guarantees that $\varphi \Rightarrow \varphi'$ holds semantically in the transition system generated by $\mathcal{S}$ if $\mathcal{S} \vdash \varphi \Rightarrow \varphi'$ is derivable. Note that the proof system derives more general sequents of the form $\mathcal{A} \vdash_{\mathcal{C}} \varphi \Rightarrow \varphi'$, where $\mathcal{A}$ and $\mathcal{C}$ are sets of reachability rules. Rules in $\mathcal{A}$ are called *axioms* and rules in $\mathcal{C}$ are called *circularities*. If $\mathcal{C}$ does not appear in a sequent, it means it is empty: $\mathcal{A} \vdash \varphi \Rightarrow \varphi'$ is a shorthand for $\mathcal{A} \vdash_{\emptyset} \varphi \Rightarrow \varphi'$. Initially, $\mathcal{C}$ is empty and $\mathcal{A}$ is $\mathcal{S}$.

$$\text{AXIOM}: \frac{\varphi \Rightarrow \varphi' \in \mathcal{A} \qquad \psi \text{ is a FOL formula (the } logical\ frame)}{\mathcal{A} \vdash_C \varphi \wedge \psi \Rightarrow \varphi' \wedge \psi}$$

$$\text{REFLEXIVITY}: \mathcal{A} \vdash_\emptyset \varphi \Rightarrow \varphi$$

$$\text{TRANSITIVITY}: \frac{\mathcal{A} \vdash_C \varphi_1 \Rightarrow \varphi_2 \qquad \mathcal{A} \cup C \vdash_\emptyset \varphi_2 \Rightarrow \varphi_3}{\mathcal{A} \vdash_C \varphi_1 \Rightarrow \varphi_3}$$

$$\text{CONSEQUENCE}: \frac{\models \varphi_1 \rightarrow \varphi_1' \qquad \mathcal{A} \vdash_C \varphi_1' \Rightarrow \varphi_2' \qquad \models \varphi_2' \rightarrow \varphi_2}{\mathcal{A} \vdash_C \varphi_1 \Rightarrow \varphi_2}$$

$$\text{CASE ANALYSIS}: \frac{\mathcal{A} \vdash_C \varphi_1 \Rightarrow \varphi \qquad \mathcal{A} \vdash_C \varphi_2 \Rightarrow \varphi}{\mathcal{A} \vdash_C \varphi_1 \vee \varphi_2 \Rightarrow \varphi}$$

$$\text{ABSTRACTION}: \frac{\mathcal{A} \vdash_C \varphi \Rightarrow \varphi' \text{ where } X \cap FV(\varphi') = \emptyset}{\mathcal{A} \vdash_C \exists X\ \varphi \Rightarrow \varphi'}$$

$$\text{CIRCULARITY}: \frac{\mathcal{A} \vdash_{C \cup \{\varphi \Rightarrow \varphi'\}} \varphi \Rightarrow \varphi'}{\mathcal{A} \vdash_C \varphi \Rightarrow \varphi'}$$

Fig. 2. Proof system for (one-path) reachability using unconditional rules.

During the proof, circularities can be added to C via CIRCULARITY and flushed into $\mathcal{A}$ by TRANSITIVITY or AXIOM.

The intuition is that rules in $\mathcal{A}$ can be assumed valid, while those in C have been postulated but not yet justified. After making progress it becomes (coinductively) valid to rely on them. The intuition for sequent $\mathcal{A} \vdash_C \varphi \Rightarrow \varphi'$, read "$\mathcal{A}$ with circularities C proves $\varphi \Rightarrow \varphi'$", is: $\varphi \Rightarrow \varphi'$ is true if the rules in $\mathcal{A}$ are true and those in C are true after making progress, and if C is nonempty then φ reaches φ' (or diverges) after at least one transition. Let us now discuss the proof rules.

AXIOM states that a trusted rule can be used in any *logical frame* ψ. The logical frame is formalized as a patternless formula, as it is meant to only add logical but no structural constraints. Incorporating framing into the axiom rule is necessary to make logical constraints available while proving the conditions of the axiom hold. Since reachability logic keeps a clear separation between program variables and logical variables the logical constraints are persistent, that is, they do not interfere with the dynamic nature of the operational rules and can therefore be safely used for framing.

REFLEXIVITY and TRANSITIVITY correspond to homonymous closure properties of the reachability relation. REFLEXIVITY requires C to be empty to meet the requirement above, that a reachability property derived with nonempty C takes one or more steps. TRANSITIVITY releases the circularities as axioms for the second premise, because if there are any circularities to release the first premise is guaranteed to make progress.

CONSEQUENCE and CASE ANALYSIS are adapted from Hoare logic. In Hoare logic CASE ANALYSIS is typically a derived rule, but there does not seem to be any way to derive it language-independently. Ignoring circularities, we can think of these five rules discussed so far as a formal infrastructure for symbolic execution.

ABSTRACTION allows us to hide irrelevant details of φ behind an existential quantifier, which is particularly useful in combination with the next proof rule.

CIRCULARITY has a coinductive nature and allows us to make a new circularity claim at any moment. We typically make such claims for code with repetitive behaviors, such as loops, recursive functions, jumps, etc. If we succeed in proving the claim using itself as a circularity, then the claim holds. This would obviously be unsound if the new assumption was available immediately, but requiring progress before circularities can be used ensures that only diverging executions can correspond to endless invocation of a circularity.

Consider, for example, the simple imperative language together with the sum program fragment, SUM, discussed in Sect. 2, but without the initial assignment n=100 , that is:

```
s = 0;
while(n>0) {
  s = s + n;
  n = n - 1;
}
```

In conventional Hoare logic, the property to verify here is

$$\{n = \text{oldn} \wedge n > 0\} \text{ SUM } \{n = 0 \wedge s = \text{oldn} *_{Int} (\text{oldn} +_{Int} 1) /_{Int} 2\}$$

where $\{n = \text{oldn} \wedge n > 0\}$ is the precondition and $\{n = 0 \wedge s = \text{oldn} *_{Int} (\text{oldn} +_{Int} 1) /_{Int} 2\}$ is the postcondition, with Int-subscripted operations being the corresponding mathematical domain operations (note that in Hoare logic no distinction is made between program variables and logical variables). In order to prove it, an axiomatic semantics of the language is needed, which is given as a proof system in terms of such Hoare triples. When correctness is paramount, a proof of correctness of the Hoare logic proof system wrt a trusted semantics, typically an executable one which acts as a reference model of the language, is also needed. This may look easy and even desirable, but the reality is that semantics of real languages are not trivial to define and they continuously evolve as the language itself evolves. Therefore, giving two different semantics and maintaining proofs of correctness between them are highly non-trivial and demotivating tasks. Finally, if one wants the Hoare logic to also be as powerful as it can be for the target language, that is, to allow us to derive any semantically valid Hoare triples, then one has to also prove its relative completeness, yet another highly non-trivial task. Moreover, and even worse from an engineering perspective, each of the above needs to be done for each programming language separately.

On the other hand, reachability logic requires no additional semantics for the target language and no correctness or completeness proofs specific to each language. It takes the executable semantics of the programming language, which is regarded as reference model, as input axioms, and then the language-independent proof rules (e.g., those in Fig. 2) are proved correct and relatively complete once and for all languages [29–32]. Without any syntactic sugar, the reachability logic specifications may be more verbose than the Hoare logic ones. However, we have found that in practice one spends a lot more time on coming up with the conceptually right properties to verify than on writing them in a particular notation. Moreover, in both Hoare logic and reachability logic we typically develop syntactic sugar notations that reduce the user burden (e.g., writing a loop invariant as a formal comment in the code). Without any sugar, the reachability rule below captures the same specification of SUM as the Hoare triple above:

$$\langle \langle \mathrm{SUM} \rangle_k \ \langle \mathtt{s} \mapsto s, \ \mathtt{n} \mapsto n \rangle_{\text{state}} \rangle_{\text{cfg}} \ \wedge \ n \geq_{Int} 0$$
$$\Rightarrow \ \langle \langle \rangle_k \ \langle \mathtt{s} \mapsto n *_{Int} (n +_{Int} 1) /_{Int} 2, \ \mathtt{n} \mapsto 0 \rangle_{\text{state}} \rangle_{\text{cfg}}$$

We encourage the reader to derive the reachability rule above on her own, using the proof system in Fig. 2. Complete details can be found in [34]. We here only give the high-level structure of the proof. By AXIOM with the semantic rule of assignment shown in Sect. 2 and by TRANSITIVITY, we reduce the above reachability rule to one where the s in the left-hand-side pattern is replaced with 0. Let LOOP be the while loop of the SUM program. Like in Hoare logic proofs, we have to derive an invariant for LOOP. In reachability logic, we formalize invariants also as reachability rules, in our case as

$$\langle \langle \mathrm{LOOP} \rangle_k \ \langle \mathtt{s} \mapsto (n -_{Int} n') *_{Int} (n +_{Int} n' +_{Int} 1) /_{Int} 2, \ \mathtt{n} \mapsto n' \rangle_{\text{state}} \rangle_{\text{cfg}} \ \wedge \ n' \geq_{Int} 0$$
$$\Rightarrow \ \langle \langle \rangle_k \ \langle \mathtt{s} \mapsto n *_{Int} (n +_{Int} 1) /_{Int} 2, \ \mathtt{n} \mapsto 0 \rangle_{\text{state}} \rangle_{\text{cfg}}$$

If this "invariant" reachability rule holds, then our original reachability rule can be derived using ABSTRACTION, CONSEQUENCE and TRANSITIVITY. To derive this invariant rule, we first use CIRCULARITY to claim it as a circularity, and then unroll the LOOP using the executable semantic rule for while, then do a CASE ANALYSIS for the resulting conditional, and then run the executable semantics of the corresponding assignments via AXIOM, intermingled with applications of CONSEQUENCE and TRANSITIVITY. Note that we can use the circularity claim in the proof of the positive branch of the conditional, because the TRANSITIVITY added it to the set of axioms once the unrolling of the while loop took place. We leave the rest of the details to the reader, as an exercise.

K implements reachability logic, the same way Maude implements rewriting logic. Since patterns allow variables and constraints on them in configurations, K rewriting becomes symbolic execution with the semantic rules of the language. Its symbolic execution engine is connected to the Z3 SMT solver [38]. We next show an example C program verified with our current implementation of reachability logic in K, mentioning that we have similarly verified various programs manipulating lists and trees, performing arithmetic and I/O operations, and implementing sorting algorithms, binary search trees, AVL trees, and the Schorr-Waite graph marking algorithm. The Matching Logic web page, http://matching-logic.org, contains an online interface to run MatchC, an instance of our verifier for C, where users can try more than 50 existing examples (or upload their own). To simplify writing properties, MatchC allows users to write reachability rules and invariant patterns as comments in the C program.

```
struct listNode { int val; struct listNode *next; };
struct listNode* reverseList(struct listNode *x)
```
rule $\langle \$ \Rightarrow \text{return ?p; } \cdots\rangle_k \langle\cdots \text{ list}(x)(A) \Rightarrow \text{list}(?p)(\text{rev}(A)) \cdots\rangle_{heap}$
```
{
  struct listNode *p, *y;
  p = NULL;
```
inv $\langle\cdots \text{ list}(p)(?B), \text{list}(x)(?C) \cdots\rangle_{heap} \wedge A = \text{rev}(?B)@?C$
```
  while(x != NULL) {
    y = x->next;
    x->next = p;
    p = x;
    x = y;
  }
  return p;
}
```

Fig. 3. C function reversing a singly-linked list.

Figure 3 shows the classic list reverse program, together with all the specifications that the user of MatchC has to provide (grayed areas, given as code annotations). MatchC verifies this program for full correctness, not only memory safety, in 0.06 s. The user-provided specifications are translated into reachability rule proof obligations by MatchC and then attempted to be proved automatically. The "$\$$" stands for the function body, the "$\cdots$" for structural frame variables, the variables starting with "?" are existentially quantified over the current formula, etc. We do not mean to explain the MatchC notation in detail here; we only show this example to highlight the fact that reachability logic verification, in spite of being based on "low-level" operational semantics, still allows a comfortable level of abstraction.

Let $\mathcal{A}$ be the rewrite system giving the semantics of the C language, and let $\mathcal{C}$ be the set of reachability rules corresponding to user-provided specifications (properties that one wants to verify, like the grayed ones above). MatchC derives the rules in $\mathcal{C}$ using the proof system in Fig. 2. It begins by applying CIRCULARITY for each rule in $\mathcal{C}$ and reduces the task to deriving individual sequents of the form $\mathcal{A} \vdash_{\mathcal{C}} \varphi \Rightarrow \varphi'$. To prove them, MatchC rewrites φ using rules in $\mathcal{A} \cup \mathcal{C}$ searching for a formula that implies φ'. Whenever the semantics rule for if in $\mathcal{A}$ cannot apply because its condition is symbolic, a CASE ANALYSIS is applied and formula split into a disjunction. When no rule can be applied, abstraction axioms are attempted. If application of an abstraction axiom would result into a more concrete formula, the verifier applies the respective axiom (for instance, knowing the head of a linked list is not null results in an automatic list unrolling).

Regarding from reachability logic's perspective, $\mathbb{K}$ consists of a collection of heuristics and optimizations to perform *proof search*, that is, to derive proof derivations using the reachability logic proof system. For example, suppose that the initial configuration pattern is all concrete/ground (i.e., when it contains no variables) and that $\mathbb{K}$ is requested to use its rewrite engine to simply execute the program in its initial state. In terms of proof derivation with the one-path reachability proof system in Fig. 2 and its more general variant in [30], this corresponds to a derivation of a one-path reachability rule. If $\mathbb{K}$ is requested to search the entire configuration-space to find all the configurations that can be reached as a consequence of a non-deterministic semantics, that corresponds to a derivation of an all-path reachability rule using the generalized proof system in [29]. Similarly, when using $\mathbb{K}$'s model-checking capabilities or when deductively verifying programs like above, all we do can be framed in terms of deriving proofs using a rigorously defined sound and relatively complete proof system.

4 Additional Related Work

The idea of developing $\mathbb{K}$ as a programming language semantic framework has been first suggested in 2003, in a programming language class taught at the University of Illinois by the author, whose lecture notes were published as a technical report [39]. One year later, the main idea was published in [40]. At that time, we used Maude as an execution language and hand-translated $\mathbb{K}$ definitions to Maude, one language at a time, so $\mathbb{K}$ was simply a Maude methodology for defining rewriting logic semantics to programming languages. $\mathbb{K}$ had several implementations in the meanwhile, most of them consisting of parsers and translators to Maude using Perl or Haskell. The current implementation, reachable from http://kframework.org, is implemented fully in Java except for the mathematical domain solver needed for program verification, which currently is Z3 [38].

In addition to the complete language semantics already mentioned in Sect. 2, there are several incomplete or yet unfinished language semantics, such as Python [41], Scheme [42], as well as various aspects of features of Haskell [43], X10 [44], a RISC assembly [45,46], LLVM [47], Verilog [48], as well as a static policy checker for C [49] and a framework for domain specific languages [50,51].

$\mathbb{K}$'s ability to express truly concurrent computations has been used in researching safe models for concurrency [52], synchronization of agent systems [53], models for P-Systems [54,55], and for the x86-TSO relaxed memory model [56]. $\mathbb{K}$ has been used for designing type checkers/inferencers [57], for model checking executions with predicate abstraction [58,59] and heap awareness [60], for symbolic execution [61–64], computing worst case execution times [65–67], studying program equivalence [68,69], programming language aggregation [70], and runtime verification [56,71]. Additionally, the C definition mentioned above has been used as a program undefinedness checker to analyze C programs [72]. The theoretical relationships between $\mathbb{K}$ language definitions and their Maude counterparts, both at the concrete and at the symbolic level, are studied in [73]. $\mathbb{K}$, through its underlying reachability logic foundation, is organized as an institution in [74]. Finally, some techniques for automatic inference of matching logic specifications are investigated in [75,76].

5   Conclusion

This paper presented a glimpse of the $\mathbb{K}$ framework for formally defining programming languages, which was inspired from using rewrite logic as a semantic framework. $\mathbb{K}$ aims at bringing formal semantics mainstream, by providing an intuitive notation and an attractive set of language-independent tools that can be used with any language once a semantics is given to that language. $\mathbb{K}$ builds upon our firm belief that all programming languages must have formal semantics, and that their semantics is not only a mathematical artifact that helps us better understand the language in question, but that it can in fact be incredibly useful in practice. Virtually all program analysis tools can be generated automatically from formal semantics. $\mathbb{K}$ may not be the final answer to this quest, but we believe that it has proven the concept, that it is indeed possible to build an effective collection of language tools based entirely and only on the language semantics.

References

1. ISO/IEC: Programming languages–C, ISO/IEC WG14, ISO 9899:2011, December 2011. http://www.open-std.org/JTC1/SC22/WG14/www/standards
2. Cohen, E., Dahlweid, M., Hillebrand, M., Leinenbach, D., Moskal, M., Santen, T., Schulte, W., Tobies, S.: VCC: a practical system for verifying concurrent C. In: Berghofer, S., Nipkow, T., Urban, C., Wenzel, M. (eds.) TPHOLs 2009. LNCS, vol. 5674, pp. 23–42. Springer, Heidelberg (2009)
3. Cuoq, P., Kirchner, F., Kosmatov, N., Prevosto, V., Signoles, J., Yakobowski, B.: Frama-C. In: Eleftherakis, G., Hinchey, M., Holcombe, M. (eds.) SEFM 2012. LNCS, vol. 7504, pp. 233–247. Springer, Heidelberg (2012)
4. Meseguer, J.: Conditioned rewriting logic as a united model of concurrency. Theoret. Comput. Sci. **96**(1), 73–155 (1992)
5. Verdejo, A., Martí-Oliet, N.: Executable structural operational semantics in Maude. Departamento de Sistemas Informàticos y Programaciòn, Universidad Complutense de Madrid. Technical report 134-03 (2003)
6. Verdejo, A., Martí-Oliet, N.: Executable structural operational semantics in Maude. J. Logic Algebraic Program. **67**(1–2), 226–293 (2006)
7. Clavel, M., Durán, F., Eker, S., Lincoln, P., Martí-Oliet, N., Meseguer, J., Talcott, C. (eds.): All About Maude. LNCS, vol. 4350. Springer, Heidelberg (2007)
8. Kahn, G.: Natural semantics. In: Brandenburg, F.J., Wirsing, M., Vidal-Naquet, G. (eds.) STACS 1987. LNCS, vol. 247, pp. 22–39. Springer, Heidelberg (1987)
9. Plotkin, G.D.: A structural approach to operational semantics. University of Aarhus, Technical report. DAIMI FN-19 (1981) republished in Journal of Logic and Algebraic Programming, **60–61** (2004)
10. Plotkin, G.D.: A structural approach to operational semantics. J. Logic Algebraic Program. **60–61**, 17–139 (2004)
11. Șerbănuță, T.-F., Roșu, G., Meseguer, J.: A rewriting logic approach to operational semantics. Inf. Comput. **207**(2), 305–340 (2009)
12. Berry, G., Boudol, G.: The chemical abstract machine. In: POPL, pp. 81–94 (1990)
13. Berry, G., Boudol, G.: The chemical abstract machine. Theoret. Comput. Sci. **96**(1), 217–248 (1992)

14. Felleisen, M., Hieb, R.: The revised report on the syntactic theories of sequential control and state. Theoret. Comput. Sci. **103**(2), 235–271 (1992)
15. Wright, A.K., Felleisen, M.: A syntactic approach to type soundness. Inf. Comput. **115**(1), 38–94 (1994)
16. Hennessy, M.: The Semantics of Programming Languages: An Elementary Introduction using Structural Operational Semantics. Wiley, New York (1990)
17. Mosses, P.D.: Foundations of modular SOS. In: Kutyłowski, M., Wierzbicki, T.M., Pacholski, L. (eds.) MFCS 1999. LNCS, vol. 1672, pp. 70–80. Springer, Heidelberg (1999)
18. Mosses, P.D.: Pragmatics of modular SOS. In: Kirchner, H., Ringeissen, C. (eds.) AMAST 2002. LNCS, vol. 2422, pp. 21–40. Springer, Heidelberg (2002)
19. Mosses, P.D.: Modular structural operational semantics. J. Logic Algebraic Program. **60–61**, 195–228 (2004)
20. Meseguer, J., Braga, C.O.: Modular rewriting semantics of programming languages. In: Rattray, C., Maharaj, S., Shankland, C. (eds.) AMAST 2004. LNCS, vol. 3116, pp. 364–378. Springer, Heidelberg (2004)
21. Braga, C., Meseguer, J.: Modular rewriting semantics in practice. Electron. Notes Theor. Comput. Sci. **117**, 393–416 (2005)
22. Chalub, F., Braga, C.: Maude MSOS tool. In: Denker, G., Talcott, C. (eds.) Proceedings of the Sixth International Workshop on Rewriting Logic and its Applications (WRLA 2006). Electronic Notes in Theoretical Computer Science, vol. 176(4), pp. 133–146 (2007)
23. Roşu, G., Şerbănuţă, T.-F.: An overview of the K semantic framework. J. Logic Algebraic Program. **79**(6), 397–434 (2010)
24. Ellison, C., Roşu, G.: An executable formal semantics of C with applications. In: POPL, pp. 533–544 (2012)
25. Hathhorn, C., Ellison, C., Roşu, G.: Defining the undefinedness of C. In: Proceedings of the 36th ACM SIGPLAN Conference on Programming Language Design and Implementation (PLDI 2015). ACM (2015)
26. Bogdănaş, D., Roşu, G.: K-Java: a complete semantics of Java. In: Proceedings of the 42nd Symposium on Principles of Programming Languages (POPL 2015), pp. 445–456. ACM, January 2015
27. Park, D., Ştefănescu, A., Roşu, G.: KJS: a complete formal semantics of JavaScript. In: Proceedings of the 36th ACM SIGPLAN Conference on Programming Language Design and Implementation (PLDI 2015). ACM (2015)
28. Appel, A.W.: Verified software toolchain. In: Barthe, G. (ed.) ESOP 2011. LNCS, vol. 6602, pp. 1–17. Springer, Heidelberg (2011)
29. Ştefănescu, A., Ciobâcă, Ş., Mereuta, R., Moore, B.M., Şerbănuţă, T.F., Roşu, G.: All-path reachability logic. In: Dowek, G. (ed.) RTA-TLCA 2014. LNCS, vol. 8560, pp. 425–440. Springer, Heidelberg (2014)
30. Roşu, G., Ştefănescu, A., Ciobâcă, S., Moore, B.M.: One-path reachability logic. In: LICS 2013. IEEE (2013)
31. Roşu, G., Ştefănescu, A.: Checking reachability using matching logic. In: Proceedings of the 27th Conference on Object-Oriented Programming, Systems, Languages, and Applications (OOPSLA 2012), pp. 555–574. ACM (2012)
32. Roşu, G., Ştefănescu, A.: Towards a unified theory of operational and axiomatic semantics. In: Czumaj, A., Mehlhorn, K., Pitts, A., Wattenhofer, R. (eds.) ICALP 2012, Part II. LNCS, vol. 7392, pp. 351–363. Springer, Heidelberg (2012)
33. Roşu, G.: Matching logic – extended abstract. In: Proceedings of the 26th International Conference on Rewriting Techniques and Applications (RTA 2015). LNCS. Springer, Heidelberg (2015, to appear)

34. Roşu, G., Ştefănescu, A.: From Hoare logic to matching logic reachability. In: Giannakopoulou, D., Méry, D. (eds.) FM 2012. LNCS, vol. 7436, pp. 387–402. Springer, Heidelberg (2012)

35. Roşu, G., Ştefănescu, A.: Matching logic: a new program verification approach. In: ICSE (NIER track), pp. 868–871 (2011)

36. Roşu, G., Ellison, C., Schulte, W.: Matching logic: an alternative to Hoare/Floyd logic. In: Johnson, M., Pavlovic, D. (eds.) AMAST 2010. LNCS, vol. 6486, pp. 142–162. Springer, Heidelberg (2011)

37. Roşu, G., Ştefănescu, A.: Matching logic rewriting: unifying operational and axiomatic semantics in a practical and generic framework. University of Illinois, Technical report, November 2011. http://hdl.handle.net/2142/28357

38. de Moura, L., Bjørner, N.: Z3: an efficient SMT solver. In: Ramakrishnan, C.R., Rehof, J. (eds.) TACAS 2008. LNCS, vol. 4963, pp. 337–340. Springer, Heidelberg (2008)

39. Roşu, G.: CS322, Fall 2003 - Programming language design: Lecture notes. University of Illinois at Urbana-Champaign, Department of Computer Science, Technical report. UIUCDCS-R-2003-2897, December 2003, lecture notes of a course taught at UIUC

40. Meseguer, J., Roşu, G.: Rewriting logic semantics: from language specifications to formal analysis tools. In: Basin, D., Rusinowitch, M. (eds.) IJCAR 2004. LNCS (LNAI), vol. 3097, pp. 1–44. Springer, Heidelberg (2004)

41. Guth, D.: A formal semantics of Python 3.3. Master's thesis, University of Illinois at Urbana-Champaign, July 2013. https://github.com/kframework/python-semantics

42. Meredith, P., Hills, M., Roşu, G.: An executable rewriting logic semantics of K-scheme. In: Dube, D. (ed.) Proceedings of the 2007 Workshop on Scheme and Functional Programming (SCHEME 2007), Technical report DIUL-RT-0701. Laval University, pp. 91–103 (2007)

43. Lazar, D.: K definition of Haskell'98 (2012). https://github.com/davidlazar/haskell-semantics

44. Gligoric, M., Marinov, D., Kamin, S.: CoDeSe: fast deserialization via code generation. In: Dwyer, M.B., Tip, F. (eds.) ISSTA, pp. 298–308. ACM (2011)

45. Asăvoae, M.: K semantics for assembly languages: a case study. In: Hills, M. (ed.) K'11. Electronic Notes in Theoretical Computer Science, vol. 304, pp. 111–125 (2014)

46. Asăvoae, M.: A K-based methodology for modular design of embedded systems. In: WADT (preliminary proceedings). TR-08/12, Universidad Complutense de Madrid, p. 16 (2012). http://maude.sip.ucm.es/wadt2012/docs/WADT2012-preproceedings.pdf

47. Ellison, C., Lazar, D.: K definition of the LLVM assembly language (2012). https://github.com/davidlazar/llvm-semantics

48. Meredith, P.O., Katelman, M., Meseguer, J., Roşu, G.: A formal executable semantics of Verilog. In: Eighth ACM/IEEE International Conference on Formal Methods and Models for Codesign (MEMOCODE 2010), pp. 179–188. IEEE (2010)

49. Hills, M., Chen, F., Roşu, G.: A rewriting logic approach to static checking of units of measurement in C. In: Kniesel, G., Pinto, J.S. (eds.) RULE 2008. Electronic Notes in Theoretical Computer Science, vol. 290, pp. 51–67. Elsevier (2012)

50. Rusu, V., Lucanu, D.: A K-based formal framework for domain-specific modelling languages. In: Beckert, B., Damiani, F., Gurov, D. (eds.) FoVeOOS 2011. LNCS, vol. 7421, pp. 214–231. Springer, Heidelberg (2012)

51. Arusoaie, A., Lucanu, D., Rusu, V.: Towards a K semantics for OCL. In: Hills, M. (ed.) K'11. Electronic Notes in Theoretical Computer Science, vol. 304, pp. 81–96 (2014)
52. Heumann, S., Adve, V.S., Wang, S.: The tasks with effects model for safe concurrency. In: Nicolau, A., Shen, X., Amarasinghe, S.P., Vuduc, R. (eds.) PPOPP, pp. 239–250. ACM (2013)
53. Dinges, P., Agha, G.: Scoped synchronization constraints for large scale actor systems. In: Sirjani, M. (ed.) COORDINATION 2012. LNCS, vol. 7274, pp. 89–103. Springer, Heidelberg (2012)
54. Şerbănuţă, T.-F., Ştefănescu, G., Roşu, G.: Defining and executing P systems with structured data in K. In: Corne, D.W., Frisco, P., Păun, G., Rozenberg, G., Salomaa, A. (eds.) WMC 2008. LNCS, vol. 5391, pp. 374–393. Springer, Heidelberg (2009)
55. Chira, C., Şerbănuţă, T.-F., Ştefănescu, G.: P systems with control nuclei: the concept. J. Logic Algebraic Program. **79**(6), 326–333 (2010)
56. Şerbănuţă, T.-F.: A rewriting approach to concurrent programming language design and semantics. Ph.D. dissertation, University of Illinois at Urbana-Champaign, December 2010. https://www.ideals.illinois.edu/handle/2142/18252
57. Ellison, C., Şerbănuţă, T.-F., Roşu, G.: A rewriting logic approach to type inference. In: Corradini, A., Montanari, U. (eds.) WADT 2008. LNCS, vol. 5486, pp. 135–151. Springer, Heidelberg (2009)
58. Asăvoae, I.M., Asăvoae, M.: Collecting semantics under predicate abstraction in the K framework. In: Ölveczky, P.C. (ed.) WRLA 2010. LNCS, vol. 6381, pp. 123–139. Springer, Heidelberg (2010)
59. Asăvoae, I.M.: Systematic design of abstractions in K. In: WADT (preliminary proceedings), TR-08/12, Universidad Complutense de Madrid, p. 9 (2012). http://maude.sip.ucm.es/wadt2012/docs/WADT2012-preproceedings.pdf
60. Rot, J., Asăvoae, I.M., de Boer, F.S., Bonsangue, M.M., Lucanu, D.: Interacting via the heap in the presence of recursion. In: Carbone, M., Lanese, I., Silva, A., Sokolova, A. (eds.) ICE. Electronic Proceedings in Theoretical Computer Science, vol. 104, pp. 99–113 (2012)
61. Asăvoae, I.M., Asăvoae, M., Lucanu, D.: Path directed symbolic execution in the K framework. In: Ida, T., Negru, V., Jebelean, T., Petcu, D., Watt, S.M., Zaharie, D. (eds.) SYNASC, pp. 133–141. IEEE Computer Society (2010)
62. Asăvoae, I.M.: Abstract semantics for alias analysis in K. In: Hills, M. (ed.) K'11. Electronic Notes in Theoretical Computer Science, vol. 304, pp. 97–110 (2014)
63. Arusoaie, A., Lucanu, D., Rusu, V.: A generic framework for symbolic execution. In: Erwig, M., Paige, R.F., Van Wyk, E. (eds.) SLE 2013. LNCS, vol. 8225, pp. 281–301. Springer, Heidelberg (2013)
64. Arusoaie, A.: A generic framework for symbolic execution: theory and applications. Ph.D. dissertation, Faculty of Computer Science, Alexandru I. Cuza, University of Iasi, September 2014. https://fmse.info.uaic.ro/publications/193/
65. Asăvoae, M., Lucanu, D., Roşu, G.: Towards semantics-based WCET analysis. In: Healy, C. (ed.) WCET. Austian Computer Society (OCG) (2011)
66. Asăvoae, M., Asăvoae, I.M., Lucanu, D.: On abstractions for timing analysis in the K framework. In: Peña, R., van Eekelen, M., Shkaravska, O. (eds.) FOPARA 2011. LNCS, vol. 7177, pp. 90–107. Springer, Heidelberg (2012). http://dx.doi.org/10.1007/978-3-642-32495-6_6

67. Asăvoae, M., Asăvoae, I.M.: On the modular integration of abstract semantics for WCET analysis. In: Lago, U.D., Peña, R. (eds.) FOPARA 2013. LNCS, vol. 8552, pp. 19–37. Springer, Heidelberg (2014). http://dx.doi.org/10.1007/978-3-319-12466-7_2

68. Lucanu, D., Rusu, V.: Program equivalence by circular reasoning. In: Johnsen, E.B., Petre, L. (eds.) IFM 2013. LNCS, vol. 7940, pp. 362–377. Springer, Heidelberg (2013). https://hal.inria.fr/hal-01065830

69. Ciobâcă, Ş., Lucanu, D., Rusu, V., Roşu, G.: A language-independent proof system for mutual program equivalence. In: Merz, S., Pang, J. (eds.) ICFEM 2014. LNCS, vol. 8829, pp. 75–90. Springer, Heidelberg (2014). https://hal.inria.fr/hal-01030754

70. Ciobâcă, S., Lucanu, D., Rusu, V., Roşu, G.: A theoretical foundation for programming languages aggregation. In: 22nd International Workshop on Algebraic Development Techniques. LNCS, Sinaia, Romania. Spriger, Heidelberg, September 2014 (to appear). https://hal.inria.fr/hal-01076641

71. Roşu, G., Schulte, W., Şerbănuţă, T.F.: Runtime verification of C memory safety. In: Bensalem, S., Peled, D.A. (eds.) RV 2009. LNCS, vol. 5779, pp. 132–151. Springer, Heidelberg (2009)

72. Regehr, J., Chen, Y., Cuoq, P., Eide, E., Ellison, C., Yang, X.: Test-case reduction for C compiler bugs. In: Vitek, J., Lin, H., Tip, F. (eds.) PLDI, pp. 335–346. ACM (2012)

73. Arusoaie, A., Lucanu, D., Rusu, V., Şerbănuţă, T.-F., Ştefănescu, A., Roşu, G.: Language definitions as rewrite theories. In: Escobar, S. (ed.) WRLA 2014. LNCS, vol. 8663, pp. 97–112. Springer, Heidelberg (2014). https://hal.inria.fr/hal-00950775

74. Chiriţă, C.E., Şerbănuţă, T.-F.: An institutional foundation for the K semantic framework. In: 22nd International Workshop on Algebraic Development Techniques (WADT 2014). LNCS (2014, to appear)

75. Feliú, M.A.: Logic-based techniques for program analysis and specification synthesis. Ph.D. dissertation, Universitat Politècnica de València, Departamento de Sistemas Informáticos y Computación, September 2013

76. Alpuente, M., Feliú, M.A., Villanueva, A.: Automatic inference of specifications using matching logic. In: Proceedings of the ACM SIGPLAN 2013 Workshop on Partial Evaluation and Program Manipulation, PEPM 2013, Rome, Italy, 21–22 January 2013, pp. 127–136. ACM (2013)

ICEMAN: A Practical Architecture for Situational Awareness at the Network Edge

Samuel Wood[1,2], James Mathewson[1,2], Joshua Joy[3], Mark-Oliver Stehr[4], Minyoung Kim[4(✉)], Ashish Gehani[4], Mario Gerla[3], Hamid Sadjadpour[1,2], and J.J. Garcia-Luna-Aceves[1]

[1] UC Santa Cruz, Santa Cruz, USA
[2] SUNS-tech, Inc., Santa Cruz, USA
{sam,james,hamid}@suns-tech.com, jj@soe.ucsc.edu
[3] UC Los Angeles, Los Angeles, USA
{jjoy,gerla}@cs.ucla.edu
[4] SRI International, Menlo Park, USA
{stehr,mkim,gehani}@csl.sri.com

Abstract. Situational awareness applications used in disaster response and tactical scenarios require efficient communication without support from a fixed infrastructure. As commercial off-the-shelf mobile phones and tablets become cheaper, they are increasingly deployed in volatile ad-hoc environments. Despite wide use, networking in an efficient and distributed way remains as an active research area, and few implementation results on mobile devices exist. In these scenarios, where users both produce and consume sensed content, the network should efficiently match content to user interests without making any fixed infrastructure assumptions. We propose the ICEMAN (Information CEntric Mobile Adhoc Networking) architecture which is designed to support distributed situational awareness applications in tactical scenarios. We describe the motivation, features, and implementation of our architecture and briefly summarize the performance of this novel architecture.

Dedication: As a Principal Investigator and Technical Lead of the ENCODERS project that funded this work as part of the DARPA Content-Based Mobile Edge Networking (CBMEN) program, Dr. Stehr is dedicating this paper to his doctoral adviser Prof. Dr. Meseguer, who introduced him to so many exciting research topics over the years. In fact, this work began in 2005 at the University of Illinois at Urbana-Champain (UIUC) when Dr. Stehr worked with Prof. Meseguer as a Research Associate. At that time Dr. Stehr assembled and led

This work was supported in part by SRI International and by the Defense Advanced Research Projects Agency (DARPA) and SPAWAR Systems Center Pacific (SSC Pacific) under Contract N66001-12-C-4051. The views expressed are those of the author and do not reflect the official policy or position of the Department of Defense or the U.S. Government. Approved for Public Release, Distribution Unlimited, Case 21098 and 23317.

© Springer International Publishing Switzerland 2015
N. Martí-Oliet et al. (Eds.): Meseguer Festschrift, LNCS 9200, pp. 617–631, 2015.
DOI: 10.1007/978-3-319-23165-5_29

618   S. Wood et al.

a team involving UIUC, SRI, and UCSC in the scope of the DARPA Disruption-
Tolerant Networking (DTN) Program, which also funded a subsequent collaboration between SRI and PARC Palo Alto Research Center. The underlying idea
that delay-/disruption-tolerant and content-based networking should be treated
on an equal footing was already advocated by our team, especially J.J. Garcia-
Luna-Aceves, at that time, but in our view not satisfactorily realized until now.
From a theoretical perspective, we are concerned with an extreme and hence
interesting case of a loosely coupled distributed system that requires new decentralized approaches to content-access, dissemination, reliability, and security.
Hence, the ENCODERS project also benefits from a good amount of theoretical research that is mostly hidden from the user. In addition to network coding
and attribute-based encryption, it implements a version of the partially-ordered
knowledge sharing model at the content level that we already used as the basis of
our DTN work at UIUC. In this paper, we focus on ICEMAN, which constitutes
the core of the ENCODERS architecture that was first demonstrated in the field
in May 2013 at Ft. AP Hill, VA.

1   Introduction

The immense global adoption of commercial off-the-shelf mobile phones and
tablets has lead to inexpensive devices with sufficient performance, size, weight,
and power (SWAP) characteristics for deployment at the network edge of tactical
and disaster response scenarios [15,16]. The predominate reason for deploying
these devices is to support applications that increase situational awareness for the
user (the warfighter or emergency responder). Increased situational awareness is
paramount in scenarios where fixed infrastructure is limited to non-existent. In
this case, the network must communicate opportunistically by using whatever
resources are available, to provide the most recent information as early as possible to situational awareness applications. For example, a blue force tracking
application provides the most recent GPS coordinates of each squad member's
position, to each squad member, to avoid friendly fire. Applications must support an assortment of sensing and communication hardware to efficiently produce and consume content to increase situational awareness. Despite the increase
in sensing and communication hardware capabilities in mobile devices, efficient
communication to and from applications on the volatile network edge remains
as a challenging research problem, and large engineering effort.[1]

This paper highlights problems with applying existing network architectures
to moderate sized networks running situational awareness applications at the tactical edge, and introduces a new architecture, ICEMAN (Information-CEntric
Mobile Ad-hoc Networking) aimed at supporting such networks. Our approach is
practical: we evaluate our implementation (which builds on Haggle [32]) on hardware with similar SWAP characteristics as those found on the tactical edge. ICE-
MAN adopts an Information Centric Networking (ICN) philosophy [18] where
the network provides a data object publish/subscribe abstraction to applications.

---

[1] It is sometimes referred to as the "last tactical mile" problem.

We use attribute-based naming, where users express queries as a set of attribute-value pairs and a matching threshold. Among other mechanisms, ICEMAN supports UDP broadcast, network coding, and utility-based content caching to increase data object delivery and reduce delivery latency. Due to the modest size of our target networks, ICEMAN pushes more intelligence into the network layer to increase performance.

The main contributions of this paper are: (1) a description of a complete ICEMAN architecture that integrates multiple content-dissemination, utility-based caching, and transport mechanisms to provide a publish/subscribe API with attribute-based content naming; (2) a description of content and context-based policies which utilize these mechanisms to achieve efficient communication at the tactical edge.

The paper is organized as follows. Section 2 describes related architectures, and discusses their differences in assumptions and design. Section 3 describes ICEMAN in detail. A brief summary of our evaluation can be found in Sect. 4 followed by the conclusion in Sect. 5.

2   Related Work

We summarize a few representative architectures related to ICEMAN and highlight their different assumptions and approaches.

Information centric networking (ICN) encompasses several approaches that share the same content-centric philosophy. The paradigm that distinguishes ICN from other approaches is the principle that the network should provide a host-to-content abstraction, as opposed to the traditional host-to-host abstraction. Indeed, in most ICN proposals there is not an explicit mechanism to communicate with a specific host. ICN architectures share three key design principles [18] that are also used in ICEMAN: (i) publish/subscribe-type primitives, (ii) universal caching, and (iii) content-oriented security model. Unlike other ICN architectures, subscribers in ICEMAN specify constraints on how to match the data object to the interests. Using these constraints, each node can construct a ranked list of the best data objects that match the subscriber's interest.

As in most ICN architectures, an ICEMAN node can cache any data object that it receives, and can forward this data object to any interested node on behalf of the publisher. ICEMAN does not establish secure tunnels for host-to-host content transport nor shared group keys. It secures the content directly by cryptographically enforcing access policies using attribute-based encryption [10], which, by scoping content, plays a role mathematically dual to attribute-based naming.

CCN [21] is a well-known example of an ICN architecture based on hierarchical names and prefix matching. It uses an interest-driven paradigm with the characteristic that an interest is consumed by the first piece of matching content and needs to be refreshed for each successive piece of content to maintain a TCP-like flow-balance property. A generalization of CCN that supports push-and-pull paradigms to make it more suitable for tactical mobile ad-hoc networks (MANETs) has been developed [33].

Pocket switching is similar to delay/disruption-tolerant networking (DTN), and focuses on exploiting contacts between wearable wireless devices. Initial work started at Intel Research Laboratory in Cambridge, and led to the first prototype of the Haggle architecture [19]. This line of research has been advanced in the European Framework Program, which has developed a wide range of routing algorithms that can naturally deal with mobility and exploit social relationships [20,30,31]. Service discovery [13] and opportunistic routing [17] for low duty-cycle lossy wireless sensor networks have been recently studied. A range of innovative approaches to security and metadata privacy has also been developed in this project [5,34,35]. The European project has led to a second generation of the Haggle architecture [32]. Due to its inherent content-based foundation, we have identified Haggle as a suitable basis for ICEMAN.

While most routing algorithms for DTNs are — like IP — based on endpoint identifiers, previous work on interest-driven routing [36,37] in the context of the DARPA DTN program allows persistent subscriptions to content under a name that will be syntactically matched (using simple prefixes or arbitrary patterns) against content stored in the network caches. Matched content travels to the subscribers on the reverse path of the interest. DTN approaches are based on semantically meaningful units of information (content is packaged in so-called bundles, defined in RFC 5050), which has been extended to include metadata in so-called extension blocks. Despite this extension, the interface to applications is based on end-point identifiers or (hierarchical) names (with syntactic matching) and is not sufficiently general to convey the common needs of applications. Descriptive destinations [9] are a noteworthy generalization that add the capability to declaratively constrain the *scope* of destinations, but does not provide a means for the dual goal of content-based access, e.g., by declaratively expressing *interest*. Finally, the notion of single-node custody (and custody transfer) developed in the context of DTN point-to-point links does not match well with the capabilities of today's wireless networks that can utilize broadcast, opportunistically overhear, and assume collective custody of content.

3   Architecture

By using a publish/subscribe paradigm, we attempt to unify two different common views of a network, namely that of a communication medium (most MANET research falls into this category) with that of a distributed data store (which is the focus of most research in peer-to-peer networking). This unification naturally leads to an architecture for integrated multi-party communication and search with in-network caching, temporal decoupling, and late binding, as exemplified by Haggle, which serves as our starting point.

As an extension and partial refactoring of Haggle, the ICEMAN architecture is an event-based architecture (see Fig. 1), in which multiple managers cooperate in a layer-less fashion to provide content-based services. It is a highly multi-threaded architecture where managers coordinate with each other asynchronously through events and manage a set of dynamically instantiated modules to perform computationally expensive operations in their own threads. For

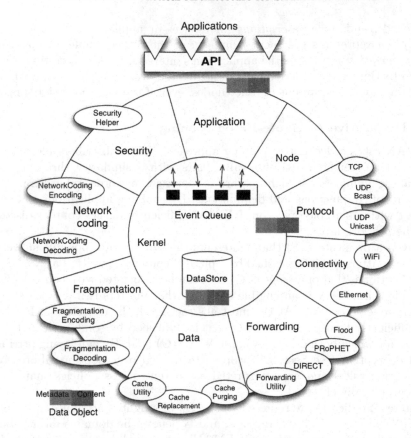

Fig. 1. ICEMAN Architecture and API. The circle depicts the layer-less interaction. Each wedge of the circle is a Manager, and they interact through the Event Queue and DataStore in the Kernel (center). The managers may use the options provided by the modules in the small circles.

instance, data objects are managed by Haggle's data manager, which uses SQLite to store metadata and serves as a matching engine running in its own background thread with a separate task queue.

The fundamental unit of abstraction is a *data object* C associated with *metadata* $\mathcal{M}(C)$, represented as a set of attribute/value pairs, and a *payload* $\mathcal{P}(C)$, which is represented by a file. Each data object has a creation timestamp attribute, so that its creation time $\mathcal{TS}(C)$ is well defined. A *data object identifier* $\mathcal{ID}(C)$ is defined as the SHA1 hash over all this information, which is globally unique with high probability.

To provide content-based network services, two classes of data objects are disseminated: (1) *Exogenous data objects* that are directly or indirectly (e.g., using coding or encryption) used to transport *content*, i.e., application payload and associated attributes. (2) *Endogenous data objects* that support coordination and awareness between network nodes, such as *routing information objects* (if needed), and *node descriptions* for devices or applications. A *device node description* represents the cache summary of the device, while an *application node description* represents the application's interests. Node descriptions have a limited lifetime and are periodically disseminated over multiple hops. Each node maintains node descriptions of other nodes, even if they are not neighbors.

3.1 Declarative Attribute-Based Naming

ICEMAN takes a declarative naming approach where subscribers identify content through weighted attribute-value pairs with a similarity threshold. This generalization makes it straightforward to represent keywords and arbitrary combinations of conjunctions and disjunctions. By enabling applications to express interest with a suitable precision, ICEMAN efficiently pushes content discovery into the network layer.

Given a predicate $\mathcal{I}(S)$ that represents the *interest* of a possible subscriber N, what matters is if it is satisfied by a piece of content C, written as $C \models \mathcal{I}(N)$. A dual notion is that of a *scope* $\mathcal{S}(C)$ that can be associated with content C, and to decide if a node N is an eligible receiver, what matters is if it is satisfied by a given node $N \models \mathcal{S}(C)$. At the most abstract level, the objective of ICEMAN is to efficiently transport content C that is published by some node P to each node N for which both $C \models \mathcal{I}(N)$ and $N \models \mathcal{S}(C)$ hold. By employing attribute-based encryption (see Sect. 3.5) scopes are framed over node attributes (e.g., representing roles) and are interpreted as content access policies that can be cryptographically enforced.

Interest predicates are represented as a set of weighted attribute/value pairs, i.e., $\mathcal{I}(N) \subseteq \mathbb{A} \times \mathbb{V} \times \mathbb{N}$, where $\mathbb{A}$, $\mathbb{V}$, and $\mathbb{N}$ denote the domains for attributes, values, and weights, respectively. ICEMAN's naming allows applications to logically specify how to quantify (and refine) the satisfaction of an interest predicate for a given piece of content C, by the degree of similarity metric. Specifically, we say that C satisfies $\mathcal{I}(N)$ with a threshold t, written as $C \models_t \mathcal{I}(N)$, iff

$$\frac{\sum\{w_i \mid (a_i, v_i, w_i) \in \mathcal{I}(N) \cap (\mathcal{M}(C) \times \mathbb{N})\}}{\sum\{w_i \mid (a_i, v_i, w_i) \in \mathcal{I}(N)\}} \geq t$$

In other words, the normalized weighted sum of overlapping attributes between content $\mathcal{M}(C)$ and interest $\mathcal{I}(N)$ determines the degree of matching or satisfaction.

ICEMAN considers the matching threshold t and a bound on the number of matches as part of the interest. Data objects are retrieved, ranked, and prioritized at each node using a lexicographical ordering based on the degree of matching and the creation time stamp (freshest first). Since an application can issue multiple concurrent threshold queries/subscriptions, it is straightforward to represent arbitrary combinations of conjunctions and disjunctions by transforming them into disjunctive normal form. Unlike Haggle, which uses a different semantics for local vs. remote queries, in ICEMAN matching is uniformly defined as stated above. In our generalization, interests are represented by application node descriptions and are disseminated separately from device node descriptions.

ICEMAN periodically disseminates cache summaries to avoid redundant transmissions and further refine the set of matched data objects. Each node maintains a counting Bloom filter representing the content in its local cache. When a node description is generated and sent to a neighbor, a compact non-counting abstraction of the local Bloom filter is included. Prior to sending a data object to a neighbor, the sender will first check to see if there is a Bloom filter hit for the data object in its local view of the neighbor's Bloom filter. Additionally, the Bloom filter reduces the data base query results to only those data objects that the interested party does not already have. In other words, if a node N has interest $\mathcal{I}(N)$ and content approximated by $\mathcal{BF}(N)$, only data objects satisfying $\mathcal{I}(N) \wedge \neg \mathcal{BF}(N)$ are sent towards N. Together, the interest and the Bloom filter define the *effective interest* of a node in a concise fashion. Bloom filter abstractions are generated periodically, so that eventual consistency is maintained between the long-term local Bloom filters and their disseminated short-term abstractions. To suppress immediate retransmissions each node's local perception of the peer's short-term Bloom filter abstraction is updated optimistically, and will be replaced by the actual peer's Bloom filter abstraction when its next node description is received.

3.2   Content Dissemination

ICEMAN transports data objects in a hop-by-hop fashion. It dynamically selects which transport protocol to use based on the content transport policy, which can be content-based, i.e., depending on attributes and payload size. All of the transport protocols support an application-layer atomic transaction protocol, the *control protocol*, which can suppress redundant transmissions at the cost of additional control messages. Currently we support TCP, UDP unicast, and UDP broadcast. Both UDP unicast and UDP broadcast can optionally disable the control protocol, in which case only Bloom filters are used to ensure delivery.

ICEMAN supports both proactive (push-based) as well as reactive (pull-based) dissemination algorithms, consistent with the observation in [33] that both paradigms are needed in content-based MANETs. ICEMAN dynamically selects which dissemination algorithm to use based on the content dissemination policy (e.g., depending on attributes and payload size).

Flooding and Replication. With *proactive flooding*, the data objects will be flooded to all nodes within the connected component of the publisher. This mechanism has been extended to *proactive replication* (epidemic propagation [39]) to push contact across newly discovered connected components.

A typical dissemination policy is to proactively flood important critical situation awareness information relevant within a squad, thus avoiding the cost of a round trip with the destination in a pull-based policy. If a message ferry is needed, then proactive replication may be a better choice to avoid additional round trip delays. It is also necessary to support one-way message ferrying.

Reactive counterparts of these algorithms are also supported, which means that content is *reactively flooded* or *reactively replicated* as soon as a matching interest is detected. Reactive replication provides an alternative method of dissemination that can deliver requested content with high probability if proactive dissemination is not feasible due to the large amount of available content.

Interest-Driven Routing. DIsruption REsilient Content Transport (DIRECT) is an interest-driven content dissemination protocol for DTNs developed in the DARPA DTN program. With some important changes described below, we have adopted DIRECT's interest propagation, and added DIRECT's reverse path method of content dissemination to ICEMAN. Specifically, interests are periodically epidemically disseminated with a creation timestamp and periodically purged. Upon a data object match with an interest, the data object is forwarded to the neighbor from which the interest was first received. We do not adopt DIRECT's use of CCN-style hierarchical naming and matching, nor its method of marking queries inactive upon satisfaction to provide flow-balance between interest and content.

Unlike DIRECT, ICEMAN decouples interest dissemination from query satisfaction: ICEMAN can support both search and immediate routing of newly published information, even after the subscription has been issued. Another difference relative to DIRECT is the use of knowledge about cached content to minimize the probability of routing content that the subscriber has already obtained from other sources. This knowledge is explicitly disseminated in [37] and approximated through Bloom filters in the ICEMAN architecture. Through randomized propagation of node descriptions and hence interest, ICEMAN achieves multi-path diversity, which is especially useful together with network coding or fragmentation. Last but not least, interest-driven routing in ICEMAN can be combined with data object broadcast, which implies that data objects are pushed to and cached at overhearing nodes even if they were never requested.

Mobility-Driven Routing. ICEMAN supports mobility-driven routing using PRoPHET [29]. PRoPHET is a routing protocol designed for disconnected networks with non-random mobility. Experimental results in [29] show that with constrained cache sizes, PRoPHET can obtain higher delivery ratios with a modest increase (and sometimes decrease) in delay in comparison to epidemic routing. It uses a delivery predictability metric to estimate the probability that any particular destination can be reached through a particular neighbor. This delivery predictability metric is based on each node's encounter history: nodes that meet frequently or for long durations have a high delivery predictability metric. Each node calculates its delivery predictability to every encountered node, and nodes exchange their delivery predictability vectors to transitively compute the probability of reaching a particular destination through a particular neighbor. In the context of ICEMAN, PRoPHET selects the neighbor with the highest delivery predictability when forwarding.

To serve as suitable basis for comparison, we have incorporated some of the newer ideas of the latest PRoPHET Internet Draft [14], most notably an improved transitivity rule, the periodic dissemination of routing information, and the periodic sampling of the current neighborhood to take into account contact duration. This enhanced version of PRoPHET works together with the periodic dissemination of node descriptions and matching as discussed previously. These modifications are needed, because PRoPHET is used in a content-based network where each node is a potential source, as opposed to its original use to route between two endpoints in DTNs.

3.3 Content- and Utility-Based Caching

Content-based caching is a feature of ICEMAN that aims to ensure that the amount of content managed by the network does not grow beyond its bounded capacity and that resources are primarily used for content that is relevant to the user. Content caching is a powerful mechanism to reduce latency and bandwidth, and mandatory if content needs to be transferred over multiple hops without a contemporaneous end-to-end path (e.g., using message ferrying or due to intermittent disruptions). Even with an end-to-end path, typical multi-hop loss rates over TCP will trigger end-to-end retransmissions, and render ICEMAN's caching-based store-and-forward solution more economic in terms of bandwidth if the content is sufficiently large. ICEMAN allows the specification of content-based caching strategies that enable fine-grained in-network purging of obsolete content, as opposed to end-to-end purging at the application level.

With *time-based purging* strategies, content can be purged either by an absolute or relative expiration time (relative to reception). The user can specify (1) a *tag* to denote the class of data objects to be purged, and (2) a *metric* to determine the absolute or relative time-to-live.

Another caching strategy inspired by our earlier work [23] is *order-based replacement*. While the concept is very general, the most common use is to keep only the *freshest* piece of content, while *staler* content is discarded from the data store. The user can specify (1) a *tag* to denote the class of data objects to be totally ordered, (2) an *id* to indicate the attribute that needs to match (e.g., content originator), and (3) a *metric* to determine the ordering of the objects (e.g., content creation time). Formally, (1) and (2) define an equivalence relation $\equiv$ on matching data objects and (3) defines a total order $\prec$ over all data objects in each equivalence class. This total-order replacement strategy only keeps the maximal element in each equivalence class that has been received. Multiple total replacement strategies can be composed in a prioritized fashion, for instance to define a lexicographical ordering, which is generally a partial order.

Content-based caching is further generalized to a *utility-based caching* pipeline which builds on the work in [11,38] and frames the cache replacement and decision problem as a utility maximization problem. A caching policy defines a utility function which assigns a real number between 0 and 1 to each data object in the cache. This utility function is a composition of multiple utility functions that are content and context sensitive (they vary in time and space). Data objects that do not meet a minimum threshold (as specified by the policy) are immediately evicted. Once the cache exceeds a certain watermark capacity, the pipeline chooses which data objects to evict in order to bring the cache capacity under the watermark. This eviction selection is posed as a 0–1 knapsack problem where the watermark capacity is the bag size, the data object payload size is the cost, and the computed utility is the benefit. By specifying suitable utility functions various combinations of popularity-based and cooperative caching strategies can be expressed in this framework.

3.4   Network Coding and Fragmentation

In a MANET environment, *network coding* can take advantage of the broadcast nature of transmissions as well as node mobility [26]. To overcome intermittent connectivity and to allow content dissemination in a decentralized setting, ICE-MAN can perform network coding at the level of data objects (depending on content size and other factors) rather than individual packets. ICEMAN specifically exploits the capability of network coding to mitigate the last coupon collector problem. In our targeted applications, groups may merge and split dynamically. When groups merge they can exchange innovative blocks which will expedite the reconstruction of the transmitted content.

In addition to network coding, ICEMAN supports *randomized informed fragmentation* to support scenarios where network coding is not needed or the overhead incurred by network coding is too high. We call it informed, because the sender examines the receiver's Bloom filter and selects a random subset of fragments from the peer's set of missing fragments. Randomizing the selection subset across multiple nodes increases the likelihood that different fragments are received by a node concurrently from different sources. Network coding can be combined with fragmentation, in which case the fragments are also known as generations. Multiple generations are needed when content is too large to be solely network coded due to the overhead of the associated vectors.

Blocks and fragments are cached and disseminated by intermediate nodes. Both coded blocks and uncoded fragments remain unchanged; i.e. different from random- linear network coding, ICEMAN does not perform mixing of blocks at intermediate nodes, but peers can become new seeds of innovative blocks upon reconstruction.

3.5   Security

ICEMAN leverages any underlying link- or network-layer security mechanisms, but our work focuses on providing an independent layer of security that secures the content directly. End-to-end security properties, namely non-repudiation and confidentiality are based on digital signatures and attribute-based encryption [10]. In both cases, we use protocols that support multiple certification authorities. Our current architecture secures the payload, while security for metadata is a challenging research topic left for future work.

Nodes have their signing keys certified by one or more authorities. Each node only accepts content from a neighbor if they share a certification authority. This prevents an attacker (insider or outsider) from polluting the network without exposing his (assumed) identity. Simultaneously, the availability of multiple authorities ensures that trust can be flexibly and robustly bootstrapped.

Publishers can limit access to content by specifying access policies framed over node attributes. Policies are specified with a range of operators, including conjunction and disjunction, allowing expressive authorization, and can combine attributes from multiple authorities [27]. Similarly, nodes can receive their attributes from multiple authorities. During publication, content is encrypted with a policy, ensuring that access control is enforced cryptographically with an end-to-end guarantee of confidentiality despite the flexible access specification. Hybrid encryption is used to optimize performance, with AES [2] used for the content, and multi-authority attribute-based encryption in the Charm framework [8] used to encrypt the AES keys.

As a basic mechanism to reject unwanted traffic, signatures are not only used on exogenous data objects, but on all other endogenous data objects. The overhead of signing is low due to our choice of a relatively large fragment/block size. This approach is reasonable under a closed-system model, where it is difficult for an attacker to generate new valid identities that are accepted by the original nodes of the network. Taking control of existing nodes is the only practical way to do so. The capability to exclude nodes from the network is a stepping stone for an architecture that attempts to maintain network availability by a notion of trust that evolves over time.

4   Summary of Evaluation

In [40], we conducted an extensive evaluation of ICEMAN through emulation with CORE/EMANE [7] to understand the performance characteristics of different policies. We modeled a tactical network consisting of 30 nodes (3 squads of 10) with different classes of situational awareness traffic. We found that different dissemination, transport and caching policies have significantly different performance characteristics (in terms of total data objects delivered and delay). A combination of content-based policies was necessary to achieve the best performance (e.g., epidemic broadcast for node descriptions and interest driven routing and network coding for large data objects and high channel contention). Combinations of hard- and soft-constraint utility-based caching policies that intelligently rank data according to network context achieved higher performance than only using hard-constraint policies such as time-based purging and order-based replacement. Battery life-time results on Nexus S phones demonstrated the feasibility of ICEMAN on current hardware, where CPU intensive policies such as network coding achieved higher performance than alternative policies. Similarly, security performance tests demonstrate that policy caching can achieve significant performance improvements, making efficient attribute-based encryption feasible on mobile devices. A detailed description of SRI's evaluation framework and further performance studies can be found in [25].

5   Conclusion

We have introduced a new ICN architecture where scope and interest are dual concepts associated with publishers and subscribers, respectively, and uniformly expressed in an attribute-based framework. The design of our ICEMAN architecture emphasizes compositionality in the sense that all features seamlessly interoperate with each other. Without architectural changes, our system supports any combination of the discussed caching, transport, and dissemination mechanisms. All features are independently configurable and for backward compatibility and performance comparisons we support the original feature set of Haggle.

The utility-based caching framework is a first step towards a unified utility-based architecture that formulates content dissemination, caching, and resource management policy selection as an online utility maximization problem.

We also plan to add a higher-level distributed monitoring and optimization component to maximize content availability based on an analysis of the tradeoff space of policies and parameters. By quantifying their benefit and cost, ICEMAN can potentially improve the overall system utility, for instance using an approach similar to cross-layer optimization [22]. Distributed monitoring plays another role in the detection of unexpected behaviors such as using an excessive amount of resources. It can also detect violations of properties (expected invariants) and their combinations that could indicate compromised devices or attacks. An adaptive trust management component could utilize this information to exclude misbehaving nodes from the network or require additional confirmation.

Attribute-based naming is a first step towards a logic, but there is much more potential in the declarative approach to content-based networking by further increasing the expressiveness of queries and subscriptions. For instance, predicate-based naming with OWL/RDF [24] has been implemented on top of ICEMAN in the context of the DARPA CBMEN [12] Program by the Drexel university team. ICEMAN has a transport architecture that can support other transport mechanisms, such as NORM [6] which has been integrated with our architecture in the scope of the same program. In this program, we have also developed an interest modeling component [28] to capture and model information needs in order to perform proactive actions (e.g., prefetching content).

With ICEMAN we are exploring a new area of the networking space that is quite different from existing research on MANETs and peer-to-peer networks. The need for a higher level of abstraction and increased expressiveness means that data objects have a much higher constant overhead than packets in IP; ICEMAN operates at a higher time scale and a level of content-granularity to amortize the cost. On the other hand, the transition to a higher level of abstractions seems essential to solve the problems that face traditional approaches by being too distant from the actual needs of applications. More interestingly, it opens opportunities for new mobile applications of the future, where the network architecture can provide services and optimize resources based on what the content represents and how it is used.

ICEMAN is the core of the ENCODERS architecture, which is available under the Apache Open Source License 2.0 [1]. Please refer to the ENCODERS design documents [3,4] for the details beyond the scope of this overview.

Acknowledgements. Apart from building on results from earlier projects, such those funded in the context of the DARPA Disruption Tolerant Networking program and our NSF- and ONR- funded projects on Networked Cyber Physical Systems at SRI, the ENCODERS project has also leveraged some results and ideas developed by International Fellows in their own research while visiting SRI. Here we would like to use the opportunity to thank Hasnain Lakhani, Jong-Seok Choi, Dawood Tariq, Rizwan Asghar, Je-Min Kim, Francoise Sailhan, Sathiya Kumar, and Sylvain Lefebvre for their valuable ideas and contributions. We would like to thank our team member SET Corp. (David Anhalt, Ralph Costantini, Dr. Hua Li, and Dr. Rafael Alonso) for their support and collaboration on interest modeling. At the beginning of the program we selected SAIC as our Mobile Systems Integrator. Hence we would like to thank the entire SAIC team (led by Dr. William Merrill and George Weston) for providing the development platform and successfully demonstrating an integrated CBMEN system based on the SRI ENCODERS architecture at Ft. AP Hill, VA, and MIT Lincoln Labs (the team led by Dr. Andrew Worthen) for their independent evaluation of the performance in testbed experiments and in the field.

References

1. ENCODERS. http://encoders.csl.sri.com/
2. Federal Information Processing Standards, Publication 197: Advanced encryption standard. http://csrc.nist.gov/publications/fips/fips197/fips-197.pdf
3. ENCODERS software design description v. 1.0 (2013). http://encoders.csl.sri.com/wp-content/uploads/2014/08/CBMEN-SRI-Design-Description-V1.0-Dist-A.pdf
4. ENCODERS software design description v. 2.0 (2014). http://encoders.csl.sri.com/wp-content/uploads/2014/08/CBMEN-SRI-Design-Description-V2.0-Dist-A.pdf
5. Shikfa, A., Oenen, M. (eds.): Haggle 027918: an innovative paradigm for autonomic opportunistic communication prototype of trust and security mechanisms. Technical report (2009). http://www.haggleproject.org/deliverables/D4.3_final.pdf
6. Adamson, B., Bormann, C., Handley, M., Macker, J.: Negative-acknowledgment (NACK)-oriented reliable multicast (NORM) protocol. Internet Society Request for Comments RFC 3940 (2004)
7. Ahrenholz, J., Danilov, C., Henderson, T.R., Kim, J.H.: Core: a real-time network emulator. In: 2008 IEEE Military Communications Conference, MILCOM 2008, pp, 1–7. IEEE (2008)
8. Akinyele, J., Green, M., Rubin, A.: Charm: a framework for rapidly prototyping cryptosystems. Technical report, Johns Hopkins University (2011)
9. Basu, P., Krishnan, R., Brown, D.W.: Persistent delivery with deferred binding to descriptively named destinations. In: Proceedings of the IEEE Military Communications Conference (2008)
10. Bethencourt, J., Sahai, A., Waters, B.: Ciphertext-policy attribute-based encryption. In: 28th IEEE Symposium on Security and Privacy (2007)
11. Chand, N., Joshi, R.C., Misra, M.: Cooperative caching in mobile ad hoc networks based on data utility. Mobile Inf. Syst. **3**, 19–37 (2007)
12. Defense Advanced Research Projects Agency (DARPA). Content-based mobile edge networking (2012). http://www.darpa.mil/Our_Work/STO/Programs/Content-Based_Mobile_Edge_Networking_(CBMEN).aspx
13. Djamaa, B., Richardson, M., Aouf, N., Walters, B.: Towards efficient distributed service discovery in low-power and lossy networks. Wireless Netw. **20**(8), 2437–2453 (2014)

14. Lindgren, A. et al.: Probabilistic routing protocol for intermittently connected networks. Internet Draft draft-irtf-dtnrg-prophet-10 (2012)
15. Finn, W.: Improving battlefield connectivity for dismounted forces, pp. 6–10. Defense Tech Briefs (2012)
16. Frink, S.: Secure cell phone technology gets ready for deployment. Military and Aerospace Electronics (2012)
17. Ghadimi, E., Landsiedel, O., Soldati, P., Duquennoy, S., Johansson, M.: Opportunistic routing in low duty-cycle wireless sensor networks. ACM Trans. Sen. Netw. **10**(4), 67:1–67:39 (2014)
18. Ghodsi, A., Shenker, S., Koponen, T., Singla, A., Raghavan, B., Wilcox, J.: Information-centric networking: seeing the forest for the trees. In: Proceedings of the 10th ACM Workshop on Hot Topics in Networks, p. 1. ACM (2011)
19. Hui, P., Chaintreau, A., Scott, J., Gass, R., Crowcroft, J., Diot, C.: Pocket switched networks and human mobility in conference environments. In: Proceedings of the 2005 ACM SIGCOMM Workshop on Delay-tolerant Networking, WDTN 2005, pp. 244–251, ACM, New York (2005)
20. Ioannidis, S., Chaintreau, A., Massoulié, L.: Distributing content updates over a mobile social network. SIGMOBILE Mob. Comput. Commun. Rev. **13**, 44–47 (2009)
21. Jacobson, V., Smetters, D.K., Thornton, J.D., Plass, M.F., Briggs, N.H., Braynard, R.L.: Networking named content. In: Proceedings of the 5th International Conference on Emerging Networking Experiments And Technologies, pp. 1–12. ACM (2009)
22. Kim, M., Kim, J.-M., Stehr, M.-O., Gehani, A., Tariq, D., Kim, J.S.: Maximizing availability of content in disruptive environments by cross-layer optimization. In: 28th ACM Symposium on Applied Computing (SAC) (2013)
23. Kim, M., Stehr, M.-O., Kim, J., Ha, S.: An application framework for loosely coupled networked cyber-physical systems. In: 8th IEEE International Conference on Embedded and Ubiquitous Computing (EUC-10), Hong Kong, December 2010
24. Kopena, J.B., Loo, B.T.: Ontonet: scalable knowledge-based networking. In: Proceedings of the 2008 IEEE 24th International Conference on Data Engineering Workshop, ICDEW 2008, pp. 170–175. IEEE Computer Society, Washington, DC, USA (2008)
25. Lakhani, H., McCarthy, T., Kim, M., Wilkins, D., Wood, S.: Evaluation of a delay-tolerant ICN architecture. In: International Conference on Ubiquitous and Future Networks, July 2015
26. Lee, U., Park, J.-S., Yeh, J., Pau, G., Gerla, M.: Code torrent: content distribution using network coding in vanet. In: Proceedings of the 1st International Workshop on Decentralized Resource Sharing in Mobile Computing And Networking, MobiShare 2006, pp. 1–5. ACM, New York (2006)
27. Lewko, A., Waters, B.: Decentralizing attribute-based encryption. In: Paterson, K.G. (ed.) EUROCRYPT 2011. LNCS, vol. 6632, pp. 568–588. Springer, Heidelberg (2011)
28. Li, H., Costantini, R., Anhalt, D., Alonso, R., Stehr, M.-O., Talcot, C., Kim, M., McCarthy, T., Wood, S.: Adaptive interest modeling improves content services at the network edge. In: 2014 IEEE Military Communications Conference (MILCOM), pp. 1027–1033, October 2014
29. Lindgren, A., Doria, A., Schelén, O.: Probabilistic routing in intermittently connected networks. ACM SIGMOBILE Mobile Comput. Commun. Rev. **7**(3), 19–20 (2003)

30. Mtibaa, A., May, M., Diot, C., Ammar, M.: Peoplerank: social opportunistic forwarding. In: Proceedings of the 29th Conference on Information Communications, INFOCOM 2010, pp. 111–115. IEEE Press, Piscataway, NJ, USA (2010)

31. Musolesi, M., Hui, P., Mascolo, C., Crowcroft, C.: Writing on the clean slate: Implementing a socially-aware protocol in Haggle. In: Proceedings of the 2008 International Symposium on a World of Wireless, Mobile and Multimedia Networks, pp. 1–6. IEEE Computer Society, Washington, DC, USA (2008)

32. Nordstrom, E., Gunningberg, P., Rohner, C.; A search-based network architecture for mobile devices. Technical report, Uppsala University (2009)

33. Oh, S.-Y., Lau, D., Gerla, M.: Content centric networking in tactical and emergency MANETs. In: Wireless Days, pp. 1–5. IEEE (2010)

34. Shikfa, A., Önen, M., Molva, R.: Bootstrapping security associations in opportunistic networks. In: PerCom Workshops, pp. 147–152. IEEE (2010)

35. Shikfa, A., Önen, M., Molva, R.: Privacy and confidentiality in context-based and epidemic forwarding. Comput. Commun. **33**, 1493–1504 (2010)

36. Solis, I., Garcia-Luna-Aceves, J.J.: Robust content dissemination in disrupted environments. In: Proceedings of the Third ACM Workshop on Challenged Networks, CHANTS 2008, pp. 3–10. ACM, New York (2008)

37. Stehr, M.-O., Talcott, C.: Planning and learning algorithms for routing in disruption-tolerant networks. In: IEEE Military Communications Conference (2008)

38. Obraczka, K., Spyropoulos, T., Turletti, T.: Routing in delay-tolerant networks comprising heterogeneous node populations. IEEE Trans. Mobile Comput. **8**, 1132–1147 (2009)

39. Vahdat, A., Becker, D. et al.: Epidemic routing for partially connected ad hoc networks. Technical report, Technical Report CS-200006, Duke University (2000)

40. Wood, S., Mathewson, J., Joy, J., Stehr, M.O., Kim, M., Gehani, A., Gerla, M., Sadjadpour, H., Garcia-Luna-Aceves, J.J.: ICEMAN: a system for efficient, robust and secure situational awareness at the network edge. In: Proceedings of IEEE Military Communications Conference (2013)

Author Index

Printed in the United States
By Bookmasters